Smart Innovation, Systems and Technologies

Volume 247

Series Editors

Robert J. Howlett, Bournemouth University and KES International, Shoreham-by-Sea, UK

Lakhmi C. Jain, KES International, Shoreham-by-Sea, UK

The Smart Innovation, Systems and Technologies book series encompasses the topics of knowledge, intelligence, innovation and sustainability. The aim of the series is to make available a platform for the publication of books on all aspects of single and multi-disciplinary research on these themes in order to make the latest results available in a readily-accessible form. Volumes on interdisciplinary research combining two or more of these areas is particularly sought.

The series covers systems and paradigms that employ knowledge and intelligence in a broad sense. Its scope is systems having embedded knowledge and intelligence, which may be applied to the solution of world problems in industry, the environment and the community. It also focusses on the knowledge-transfer methodologies and innovation strategies employed to make this happen effectively. The combination of intelligent systems tools and a broad range of applications introduces a need for a synergy of disciplines from science, technology, business and the humanities. The series will include conference proceedings, edited collections, monographs, handbooks, reference books, and other relevant types of book in areas of science and technology where smart systems and technologies can offer innovative solutions.

High quality content is an essential feature for all book proposals accepted for the series. It is expected that editors of all accepted volumes will ensure that contributions are subjected to an appropriate level of reviewing process and adhere to KES quality principles.

Indexed by SCOPUS, EI Compendex, INSPEC, WTI Frankfurt eG, zbMATH, Japanese Science and Technology Agency (JST), SCImago, DBLP.

All books published in the series are submitted for consideration in Web of Science.

More information about this series at http://www.springer.com/series/8767

Mark Shamtsyan · Marco Pasetti ·
Alexey Beskopylny
Editors

Robotics, Machinery and Engineering Technology for Precision Agriculture

Proceedings of XIV International Scientific Conference "INTERAGROMASH 2021"

Springer

Editors
Mark Shamtsyan
Saint Petersburg State Institute of Technology
Saint Petersburg, Russia

Marco Pasetti
Department of Information Engineering
Università degli Studi di Brescia
Brescia, Italy

Alexey Beskopylny
Don State Technical University
Rostov-on-Don, Russia

ISSN 2190-3018 ISSN 2190-3026 (electronic)
Smart Innovation, Systems and Technologies
ISBN 978-981-16-3843-5 ISBN 978-981-16-3844-2 (eBook)
https://doi.org/10.1007/978-981-16-3844-2

This Springer imprint is published by the registered company Springer Nature Singapore Pte Ltd.
The registered company address is: 152 Beach Road, #21-01/04 Gateway East, Singapore 189721, Singapore

Preface

The book presents proceedings of the international conference INTERAGROMASH 2021 dedicated to the issues of precision agriculture, robotics, machines, and agriculture biotechnologies. It contains a collection of original and fundamental studies by international researchers, academics, and experts in the following areas: software and mobile apps for precision agriculture, robotics systems and machinery for agriculture, simulation models and decision support systems, agricultural engineering, crop production improvement, and others. Modern technologies are developing rapidly and are widely used in the field of agriculture. The materials presented in this book are of significant scientific value and can be used in practice, since they give an idea of existing high-tech solutions for agriculture. The book can be of interest for professionals and practitioners, for researchers, scholars, and producers. It can be used in the educational process at specific agricultural universities or during vocational training at enterprises.

Saint Petersburg, Russia
Brescia, Italy
Rostov-on-Don, Russia

Mark Shamtsyan
Marco Pasetti
Alexey Beskopylny

Contents

About the Editors

Dr. Mark Shamtsyan (1967) is Professor of St. Petersburg State Institute of Technology (Technical University), Russia. His scientific works are focused in the field of food biotechnology, food regulations and safety. He is Global Harmonization Initiative (GHI) Ambassador to Russia and Co-Chair of GHI working group on mycotoxins, Chairman of Russian Regional Section of European Hygienic Engineering and Design Group (EHEDG) and EHEDG Authorized Trainer. He is one of the founders of Black Sea regional association of Food Science and Technology (B-FoST) and its ExCo Member. He is Fellow of International Academy of Food Science and Technology and Member of International Academy of Refrigeration.

Dr. Marco Pasetti (1972) is Assistant Professor of Electrical Energy Systems at the Department of Information Engineering of the University of Brescia (Università degli Studi di Brescia), Italy. Main research interests include energy systems, distributed generation, renewable energy sources, solar engineering, photovoltaics, energy storage, energy management systems, supervisory control and data acquisition, electric vehicle charging systems, smart grids and micro-grids.

Dr. Alexey Beskopylny (1958) is Doctor of Science, Professor and Vice-Rector for Academic Affairs and Advanced Degrees of the Don State Technical University, Russia. His scientific interests are dynamic of structures, non-destructive methods, impact indentation and reliability of machines and structures.

Strengthening of Welded Joints of Load-Bearing Structures of Robotic Systems with Ball-Rod Hardening

Alexey Beskopylny, Besarion Meskhi, Nikita Beskopylny, Irina Chukarina, Alexandr Isaev, and Andrey Veremeenko

Abstract The article considers the problem of assessing the hardening of a butt-welded joint made of structural steel 09G2S during ball-rod hardening of a steel plate. The problem of ensuring the strength of welded joints is critical for heavy-loaded frame structures of robotic systems used in agriculture. The numerical analysis of the problem of ball-rod hardening was carried out in the ANSYS software in a static plane formulation. The inhomogeneity of the material of the welded joint was specified by various elastoplastic elements with bilinear isotropic hardening. The analysis of the deformation fields in the contact zone of the ball-rod indenter with the inhomogeneous plate made it possible to estimate the degree of hardening of the heat-affected zone and select the modes of the ball-rod hardening.

1 Introduction

The reliable performance of welded joints in heavy-duty frame structures of robotic systems is critical to ensure the entire structure's reliability. It is known that welded joints have high heterogeneity due to the influence of thermal effects and significantly complicate preliminary strength calculations since the medium becomes significantly anisotropic, and the properties of the material are not known.

In works [1–3], to determine the material's mechanical properties, indentation methods are used as the most effective methods for assessing the strength characteristics. At the same time, the problem of determining the strength of the entire structure, taking into account the heterogeneity of the welded joint, remains.

The welded joint's structure and properties have characteristic features due to the influence of thermal action, which causes heterogeneity, which reduces the strength of the welded joint. It is shown in [4] that when welding DP800 and DP1000 steels, the microhardness tendency for all welded joints was characterized by the highest value in the weld zone and the lowest in the heat-affected zone (Fig. 1). The degree

A. Beskopylny (✉) · B. Meskhi · N. Beskopylny · I. Chukarina · A. Isaev · A. Veremeenko
Don State Technical University, sq. Gagarin, 1, Rostov-on-Don, Russia 344003

M. Shamtsyan et al. (eds.), *Robotics, Machinery and Engineering Technology for Precision Agriculture*, Smart Innovation, Systems and Technologies 247,
https://doi.org/10.1007/978-981-16-3844-2_1

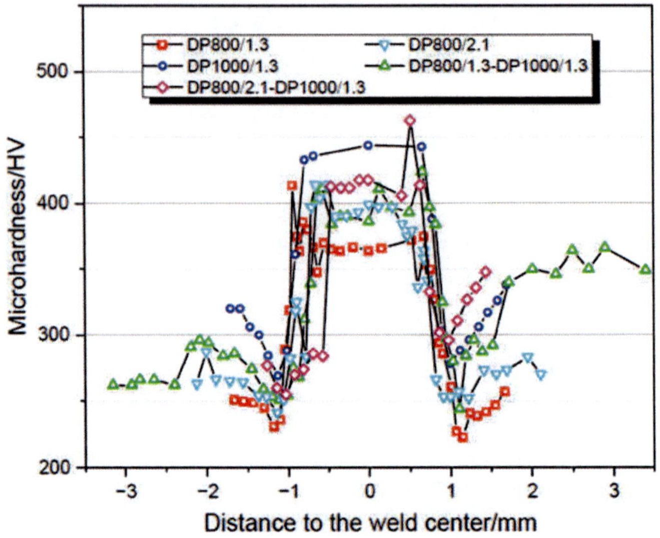

Fig.1 Scattering of microhardness along the length of the welded joint according to [4]

of softening reached 20% and, as can be seen from Fig. 1 has a large statistical dispersion.

In [5], a new mathematical model is proposed that describes the physical phenomena in submerged arc welding of fillet welds. Numerical modeling of the temperature distribution during welding is carried out and the distribution of temperature fields influencing the processes of hardening and softening is obtained.

The softening zone is a weak link in the strength of the welded joint, and many technological operations are aimed at increasing the strength in this particular zone. In work [6], a method and a device for strengthening the welded joints of pipes under the combined action of shock-vibration and thermal effects are proposed. The results showed that this method can increase the strength of the weakened zone by about 20% and reduce the dissipation of the strength characteristics of the welded joint. The method of shock-vibration hardening is also effectively used in the hardening of crankshafts [7] while maintaining the requirements for processing accuracy. Shock-vibration methods are widely used in the problems of diagnostics of inhomogeneous media [8].

Similar processes occur in friction welding technology. In [9], the microstructure of the weld and heat-affected zone, as well as the heat-affected zone located between them, formed during friction stir welding, was investigated. With this welding technology, the core of the welded seams is not formed in the superplasticity mode, and its structure has a shear-strip character. Due to the peculiarities of friction welding technology, the dependence of the microhardness on the coordinate along the length of the welded joint has a slightly different character. However, a softening zone is present.

In [10], the grain structure and texture were investigated using electron backscattering diffraction. Strength characteristics in the area of softening and fusion have been investigated. True stress–true strain curves were determined experimentally.

The study of the evolution of the microstructure of softening zones after cyclic plastic deformation and its effect on the mechanical properties of welded joints of a wound pipeline X60 was carried out in [11]. The authors showed that the presence of a weakened zone led to the fact that the deformation was mainly concentrated in this area during the cyclic loading process. However, a fracture was observed in the base metal zone due to the effect of strain hardening of the material in the HAZ.

In [12–15], the authors experimentally investigated the multiple impacts of a ball-rod hardener with an energy of 2.5 J of structural steels St3, 45 and aluminum alloys D16, B95, AVT. The dependences of the depth of surface hardening and surface roughness are obtained for various processing modes and the characteristics of the acoustic process are analyzed to ensure safe working conditions.

Thus, it can be seen that shock-vibration methods of surface plastic deformation are effective methods for locally increasing the strength of a structure. However, the issues related to the choice of hardening modes, impact energy or indentation force, technological parameters of the hardening process and the conditions for fixing the part remain unclear. The purpose of this article is to simulate the processes of strengthening welded joints of frame structures of robotic agricultural machines, to determine their strength characteristics, taking into account the inhomogeneous state of the welded joint, and to recommend the choice of the mode of strengthening with a ball-bar hardener.

2 Materials and Methods

In this work, we used structural steels of the 09G2S type. The chemical composition of the steel is shown in Table 1.

Mechanical characteristics of steel are given in Table 2. The elastic modulus of steel is 2.05E+5 MPa and Poisson's ratio is 0.3. The modulus of elasticity of the indenter is 2E+6 MPa.

The weld joint is a butt-welded plate with manual arc welding. Fig. 2 shows a diagram of a welded joint in a flat setting (flat deformation). The X-axis is directed horizontally and coincides with the bottom edge of the plate. The Y-axis is directed vertically and coincides with the middle of the weld. Due to the symmetry of the problem, we consider only the ½ part.

Table1 Mass fraction of main chemical elements, %

C	Si	Mn	P	S	Cr	Ni	V	Cu
<12	0.5–0.8	1.3–1.7	<0.03	<0.035	0.3	0.3	0.12	0.3

Table 2 Mechanical characteristics of welded joint zones

No	Zone name	Yield point σ_Y, MPa	Tensile strength σ_u, MPa	Elongation δ	Hardening module E_t, MPa
1	Base metal	325	460	0.22	500
2	HAZ 2	250	360	0.4	275
3	HAZ 3	300	450	0.25	600
4	Weld seam	350	470	0.2	636

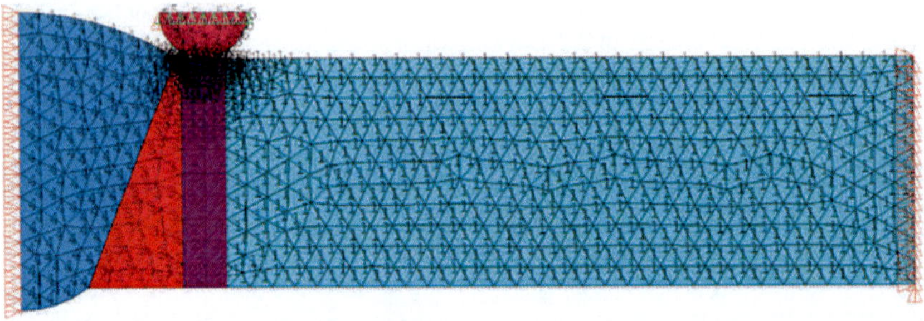

Fig. 2 Model of a welded joint by the finite element method: (1) base material; (2) heat-affected zone; (3) heat-affected zone and fusion zone; (4) welded seam; (5) ball-rod hardener

The geometry of the welded joint is as follows. Length of ½ plate $L = 40$ mm, plate thickness $H = 10$ mm. The heat-affected zone's width is $B2 = 2$ mm, the distance from the weld axis to the HAZ axis is $B1 = 7.3$ mm.

The load is applied step by step to the spherical tip in the form of a displacement of the upper face, maximum displacement $h_d = 0.2$ mm. In this problem, the case of applying a contact load at a point in the center of the HAZ at a distance from the weld axis $B2 = 8.3$ mm is considered. In the first step of active indentation, plastic deformation occurs. In the second step, the load is removed and the residual stresses and deformations are investigated.

The boundary conditions of the problem are shown in Fig. 2. The lower edge of the plate is free from loads and is not fixed. The right side of the plate is rigidly clamped. The nodes lying on the seam axis can be displaced along the Y-axis and are fixed along the X-axis.

3 Results and Discussion

Analysis of the calculation results by FEM made it possible to reveal the features of this hardening scheme. Figure 3 shows the equivalent von Mises stresses at the stage of active loading. It is seen that the effect of inhomogeneity near the weld makes the stress field asymmetric. It is noteworthy that the maximum von Mises

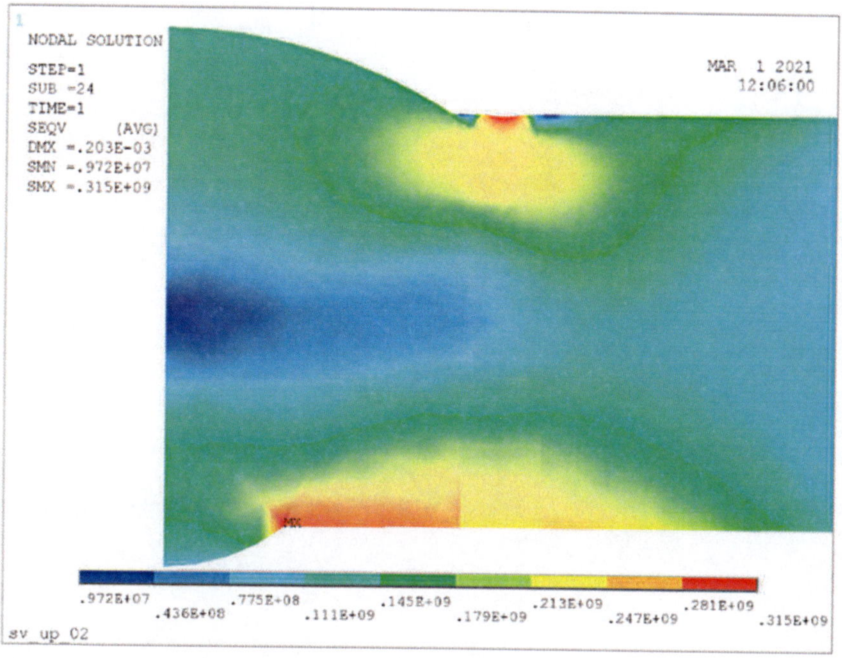

Fig. 3 Fields of equivalent von Mises stress at the loading stage

stresses, which determine the plastic flow of the material, arise not only directly in the contact zone, but at the weld root. These stresses have their maximum values in the zone of stress concentration and can lead to material embrittlement (reduction in the plasticity margin) along the lower boundary of the plate.

Figure 4 shows the fields of elastic deformations ε_Y^{el} at the stage of unloading. Despite the fact that elastic deformations are fully restored after unloading, they demonstrate the nature of the entire structure's deformed state.

It is seen that the entire plate is deformed with a characteristic maximum of compressive deformations in the contact zone. Outside the contact zone, tensile deformation zones appear in the weld and the base metal.

Figure 5 shows the plastic strain fields ε_Y^{pl} at the loading stage. It can be seen that the maximum plastic deformations are localized in the contact zone and reach values of 0.269.

Figure 6 shows the stress fields σ_X that tend to grow along the bottom surface of the plate to a depth of 3 mm. In the upper zone under the indentation, a zone of compressive stresses σ_X is formed to a depth of about 2–3.5 mm. The combination of tensile stresses along the lower edge of the plate and compressive stresses under the indentation creates a complex stress state that prevents uniform work hardening.

This combination of stresses along the upper chord of the plate with the zone of stress concentration and along the lower chord at the root of the weld can be dangerous

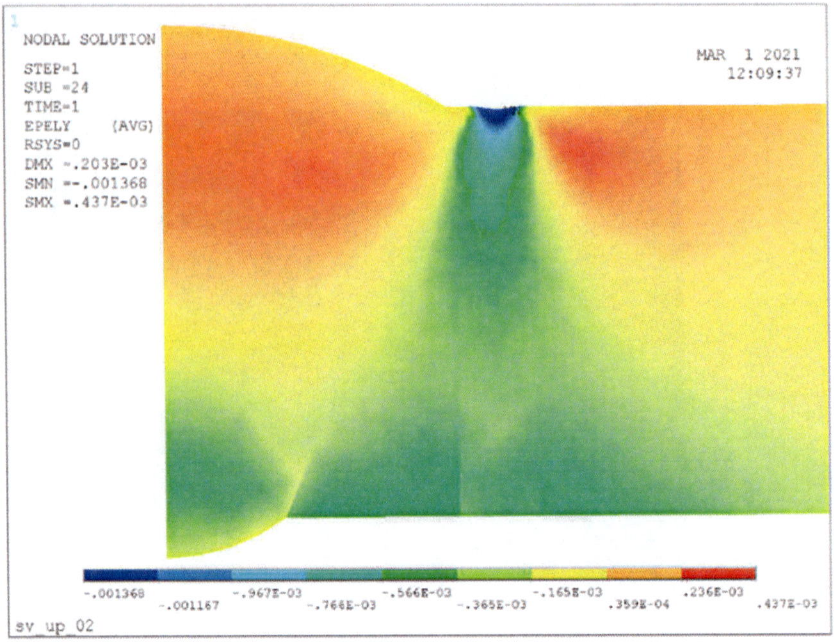

Fig. 4 Elastic strain fields ε_Y^{el} at the loading stage

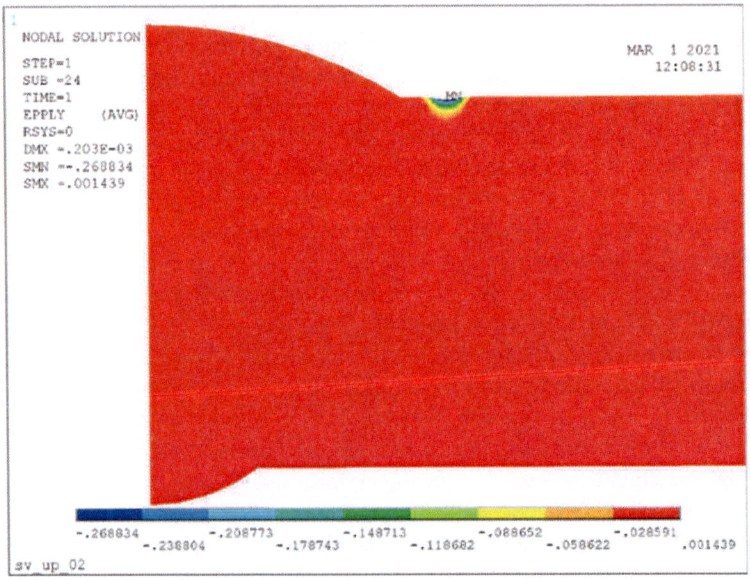

Fig. 5 Elastic strain fields ε_Y^{el} at the loading stage

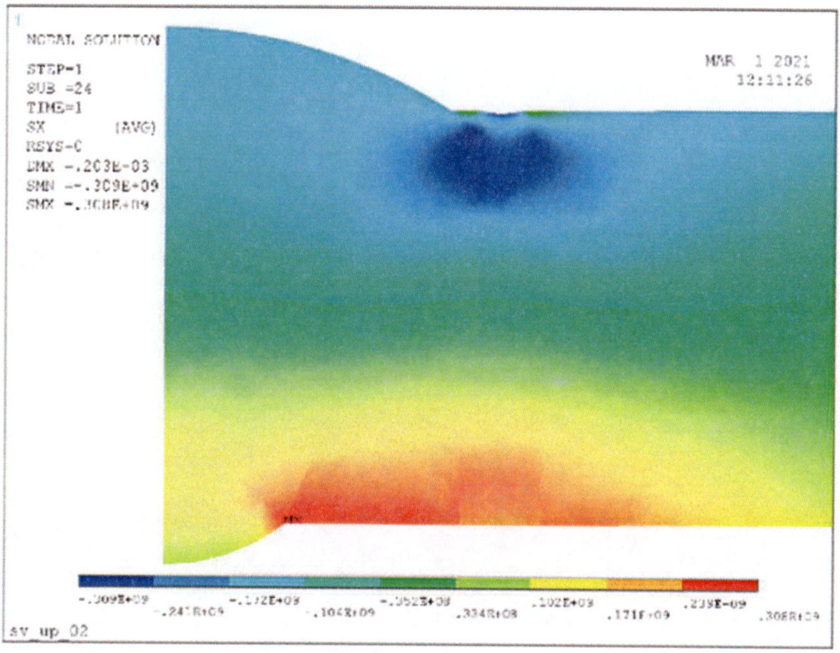

Fig. 6 Stress fields σ_X at the stage of loading

under cyclic loads arising from the operation of frame structures of robotic systems during their movement along irregularities.

Figure 7 shows stresses σ_Y at the stage of unloading, which, due to the inhomogeneity of the welded joint, have a pronounced asymmetry relative to the indentation axis. It is also seen that a zone of tensile stresses σ_Y appears at the weld root.

Figure 8 shows the shear stress fields. In accordance with the Tresca criterion, shear stresses indicate the zones in which plastic flow begins. It can be seen that in this scheme of fixing the plate at the root of the weld, a zone of plastic deformation arises.

Figure 9 shows the von Mises stresses after unloading. The picture of residual stresses shows that the zone of maximum residual von Mises stresses is not in the zone of application of contact stresses, but in the fusion zone, at the point defining the stress concentrator (Fig. 9, label MX). It is also seen that the hardening zone is uneven. At a depth of up to 1 mm, residual plastic deformations form a bowl (yellow zone), then at a depth of up to 2 mm, an unloading zone occurs as a result of hydrostatic stress. In this zone, the metal does not undergo plastic deformation. Further, at a depth of 2.5–3 mm, a hardening zone is again observed.

It is also seen that hardening occurs along the lower edge of the plate due to bending deformations. Figure 9 shows that after removing the load along the lower chord, stresses remain localized under the surface at a depth of about 2–3 mm and

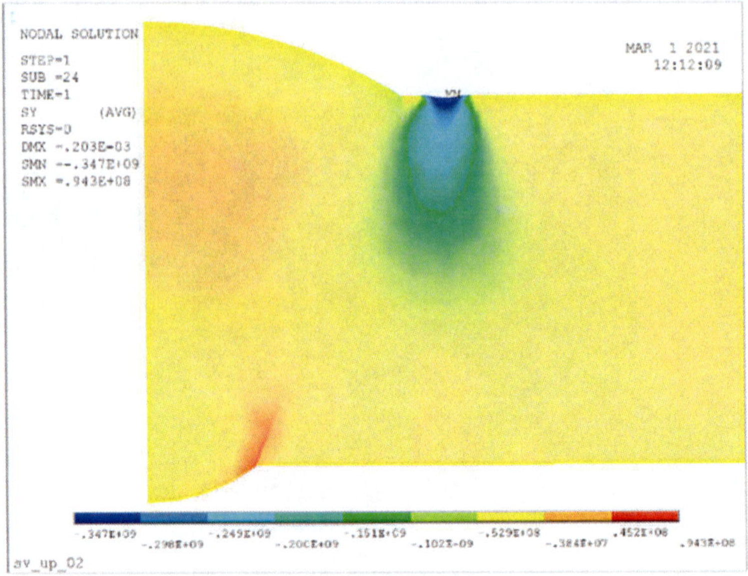

Fig. 7 Stress fields σ_Y at the stage of loading stage

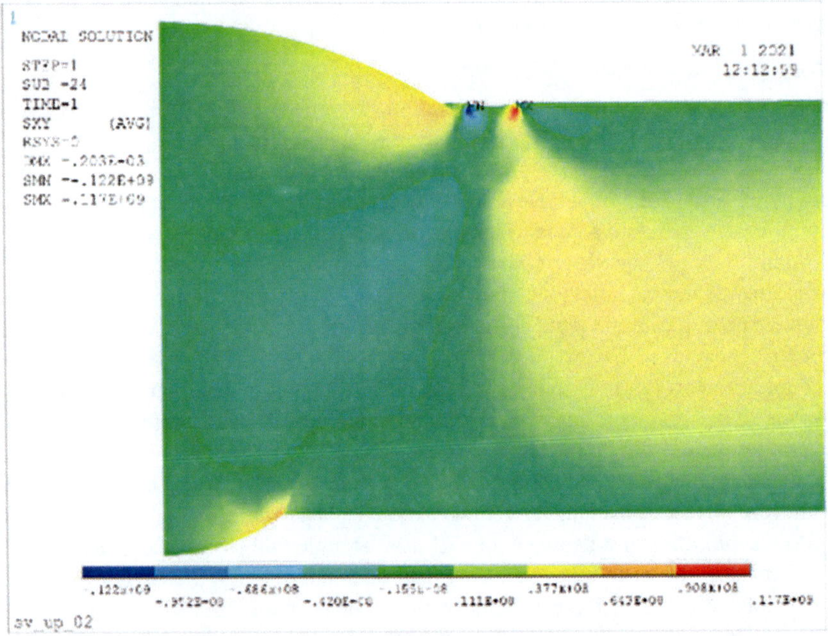

Fig. 8 Shear stress fields σ_{XY} at the stage of unloading

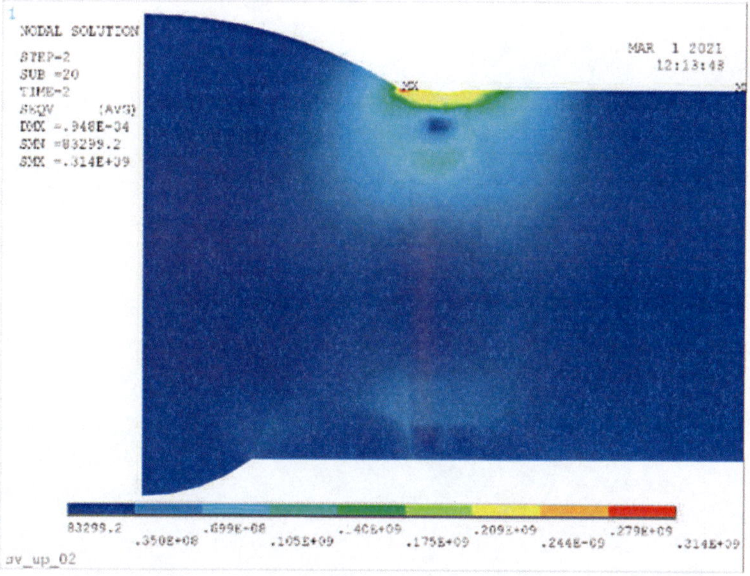

Fig. 9 Fields of equivalent von Mises stress at the unloading stage

close to the surface in the zone of the weld root. The effect of residual stresses upon repeated exposure to a structure can cause fatigue cracks and fracture.

Figure 10 shows the residual strain fields after unloading. It can be seen that the structure is in a stressed state due to plastic deformations in the upper zone and in the lower belt at the weld root.

4 Conclusion

Surface plastic hardening of parts and frame structures is an effective way to locally increase strength and performance, which is very important for heavy-duty frame structures of robotic systems. To increase the reliability of frame structures, a method of ball-rod hardening of a welded joint is proposed, which is characterized by a significant inhomogeneity of the structure and a scatter of mechanical characteristics.

Numerical analysis by FEM showed that the nature of hardening is sensitive to the method of fixing the part. The calculated stresses at different points of the welded joint showed that when the plate is clamped by cantilever, due to bending deformations, the plastic flow zone arises not only at the point of application of the contact force, but also at the weld root, which can reduce the plasticity margin of the frame and lead to fatigue cracks in exploitation. Therefore, it is advisable to first determine the mechanical properties in local areas of the material, and then, knowing their values, use numerical methods to calculate the stresses in these sections.

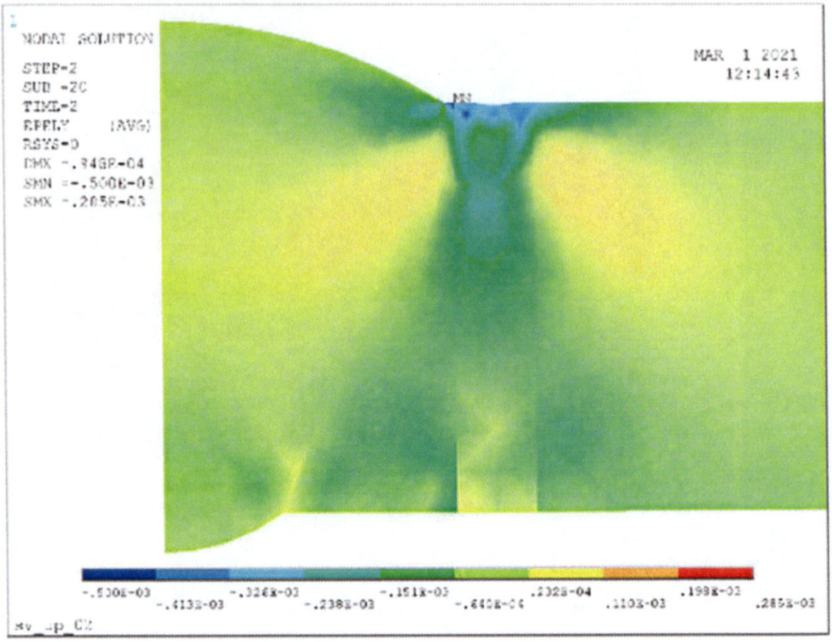

Fig. 10 Fields of elastic deformations ε_Y^{el} at the stage of unloading

The mechanical characteristics of the welded joint have unique properties due to the fact that an increase in the values of a group of properties associated with strength (yield strength and ultimate tensile strength, hardness) inevitably leads to a decrease in the values of another group of properties associated with ductility and impact toughness (elongation, impact strength). For the case of a welded joint, this group of properties includes the cracking resistance characteristics of the material associated with the material's resistance to the onset of rapid propagation of an existing crack.

Analysis of the stress–strain state according to the FEM showed that during loading, stresses of 315 MPa are created, which exceed the yield point in the HAZ by 26%, and propagate to a depth of 1 mm. The results obtained make it possible to assess the nature of residual deformations and select the optimal modes of ball-rod hardening.

References

1. Xue, H., He, J., Jia, W., Zhang, J., Wang, S., Zhang, S., Yang, H., Wang, Z.: An approach for obtaining mechanical property of austenitic stainless steel by using continuous indentation test analysis. Structures **28**, 2752–2759 (2020). https://doi.org/10.1016/j.istruc.2020.11.001

2. Pham, T.-H., Nguyen, N.-V.: Mechanical properties of constituent phases in structural steels and heat-affected zones investigated by statistical nanoindentation analysis. Constr. Build. Mater. **268**, 121211 (2021). https://doi.org/10.1016/j.conbuildmat.2020.121211
3. Litvinov, S., Beskopylny, A., Trush, L., Yazyev, S.: Optimization of thick-walled spherical shells at thermal and power influences. MATEC Web Conf. **106**, 04013 (2017). https://doi.org/10.1051/matecconf/201710604013
4. He, H., Forouzan, F., Volpp, J., Robertson, S.M., Vuorinen, E.: Microstructure and mechanical properties of laser-welded DP steels used in the automotive industry. Materials **14**(2), 456 (2021). https://doi.org/10.3390/ma14020456
5. Mosin, A., Erofeev, V., Sholokhov, M.: Physicomathematical modeling of the formation features of fillet welds of bridge metal structures under submerged-arc welding. Adv. Eng. Res. **20**(3), 259–268 (2020). https://doi.org/10.23947/2687-1653-2020-20-3-259-268
6. Butenko, V., Isaev, A.N., Bagdasaryan, T., Cherepenjko, A.: Combined strengthening treatment of welded joints of metal pipes of large diameter. MATEC Web Conf. **226**, 01030 (2018). https://doi.org/10.1051/matecconf/201822601030
7. Lebedev, V.A., Pastukhov, F.A., Chaava, M.M., Sergaб G.V.: Technological features of crankshaft hardening by vibration shock method. Adv. Eng. Res. **20**(4), 390–396 (2020). https://doi.org/10.23947/2687-1653-2020-20-4-390-396
8. Lyapin, A., Beskopylny, A., Meskhi, B.: Structural monitoring of underground structures in multi-layer media by dynamic methods. Sensors **20**(18), 5241 (2020). https://doi.org/10.3390/s20185241
9. Chularis, A., Rzaev, R., Valisheva, A.: Formation and structure features of the weld joints made by friction stir welding. Inorg. Mater. Appl. Res. **10**(3), 673–681 (2019). https://doi.org/10.1134/S2075113319030067
10. Ahmed, M.M.Z., Ataya, S., Seleman, M.M.E.-S., Allam, T., Alsaleh, N.A., Ahmed, E.: Grain structure, crystallographic texture, and hardening behavior of dissimilar friction stir welded AA5083-O and AA5754-H14. Metals **11**, 181 (2021). https://doi.org/10.3390/met11020181
11. Jiang, Y., Li, C., Di, X., Wang, D., Liu, J.: EBSD analysis of microstructures and mechanical properties of softened zones in X60 reeled-pipeline welded joint after cyclic plastic deformation. Welding World **64**, 1213–1225 (2020). https://doi.org/10.1007/s40194-020-00916-0
12. Tamarkin, M., Tishchenko, E., Murugova, E., Melnikov, A.: Surface quality assurance and process reliability in the processing with a ball-rod hardener. E3S Web Conf., **175**, 05008 (2020). https://doi.org/10.1051/e3sconf/202017505008
13. Tamarkin, M., Tishchenko, E., Chukarina, I., Sosnitskaya, T.: Parts Processing Technology for Transport Engineering. In: Popovic Z., Manakov A., Breskich V. (eds) VIII International Scientific Siberian Transport Forum. TransSiberia 2019. Advances in Intelligent Systems and Computing, 1115, pp. 913–922 (2020). https://doi.org/10.1007/978-3-030-37916-2_90
14. Tamarkin, M.A., Tishchenko, E.E., Shvedova, A.S.: Optimization of dynamic surface plastic deformation in machining. Russ. Eng. Res. **38**, 726–727 (2018). https://doi.org/10.3103/S1068798X18090277
15. Beskopylny, A., Lyapin, A., Anysz, H., Meskhi, B., Veremeenko, A., Mozgovoy, A.: Artificial neural networks in classification of steel grades based on non-destructive tests. Materials **13**, 2445 (2020). https://doi.org/10.3390/ma13112445

Quality Control of Frame Structures of Robotic Systems by Express Nondestructive Methods

Alexey Beskopylny, Besarion Meskhi, Nikita Beskopylny, and Margarita Bezuglova

Abstract The problem of quality control of parts of robotic systems is an essential task for modern technology's effective functioning. The article discusses the fundamental principles of ensuring the quality of parts using express nondestructive methods of impact indentation of conical indenters using artificial intelligence algorithms. The quality of parts is considered from the standpoint of a set of mechanical properties that determine strength, hardness and deformability. In practice, the strength characteristics must be known during production and must be quickly and accurately controlled during operation. The characteristics of the strength and deformability of steels are of a stochastic nature. To implement the proposed approach, a device for impact indentation of an indenter has been developed, and a method based on the probabilistic nature of strength properties has been proposed.

1 Introduction

In Russia, in the absence of competition, product quality management has developed in the form of an improved form of certification, regulated by particular regulatory documents. Often, our manufacturers use outdated technological maps [1–3], which does not meet the requirements of Industry 4.0 and the requirements of lean manufacturing. At the same time, abroad in a market economy, the basis for product quality management is certification, that is, actions aimed at determining products or services' conformity to certain standards or technical conditions. In addition, certification also provides for obtaining complete and reliable information about the quality of a product at every stage of production. The main incentives for certification are economic factors, safety and environmental requirements, and free market conditions.

Modern methods of quality management in robotic systems are based on a testing system and an in-depth analysis of product failures with the introduction of appropriate corrective actions and the use of mathematical statistics [4, 5]. Product quality

A. Beskopylny (✉) · B. Meskhi · N. Beskopylny · M. Bezuglova
Don State Technical University, sq. Gagarin, 1, Rostov-on-Don, Russia 344010

M. Shamtsyan et al. (eds.), *Robotics, Machinery and Engineering Technology for Precision Agriculture*, Smart Innovation, Systems and Technologies 247,
https://doi.org/10.1007/978-981-16-3844-2_2

depends on raw materials and supplies. In the conditions of additive manufacturing, quality control of raw materials plays a leading role. In [6], it is proposed to use real-time machine learning algorithms to analyze data and perform other functions of quality control of the melt material.

In [7], to improve the quality, the problem of express control of the process monitoring system on metalworking equipment is solved. The article presents algorithms for accounting for the number and type of manufactured parts on metalworking machines and diagnostics of the quality of a technological operation according to the time diagrams of the electric current consumed by the machine.

The dependence of the hydraulic resistance of the hydro-rotating antennas on the quality of the surface treatment of parts is given in [8]. It is shown that for critical structures, the quality and efficiency management include a specially manufactured hydromechanical device for modeling and researching the processes of a hydraulic system with energy recovery [9].

At the present stage, to improve the quality control of robotic systems, artificial intelligence methods are used, for example machine learning [10, 11], blockchain technologies [12, 13], digital twins [14] and others. The proposed methods make it possible to increase the accuracy of assembly, the accuracy of control of an individual robot and a system of robots, as well as to increase their efficiency.

The primary structural material is metals, the properties of which must be known at every stage of the life cycle [15]. In [16], a method is proposed based on a conical indenter's shock indentation and processing the results using neural network algorithms. The method makes it possible to evaluate structural elements' mechanical characteristics and optimize structural elements [17]. Thus, this article's main goal is to improve the method of quality control of frame structures of robotic systems by expressing nondestructive tests based on the impact indentation of conical indenter.

2 Materials and Methods

In our opinion, the most significant part of the quality of metals used in robotic systems of agricultural machinery is the complex of its mechanical properties because both the machinability of rolled metal to the finished part and the conditions for their subsequent reliable operation will depend on this. Therefore, the complex of properties that characterize the quality of materials in interrelation must be certified, and their changes should be monitored at each stage of production (for example, after welding or heat treatment). Obviously, the quality of the material of rolled metal should ensure the possibility of processing it to a finished part without destruction and the appearance of cracks and subsequent "work" of the part or element under external factors for a given time without reaching the limit state (in terms of yield stress, fatigue, wear).

On the basis of a large amount of experimental material [18, 19], it was shown that the three-parameter Weibull law well describes the distribution of mechanical properties in most cases with a location parameter

$$F(x) = 1 - \exp\left[-\left(\frac{x-\mu}{\beta}\right)^k\right] \tag{1}$$

where μ, β and k are location, scale and shape parameters.

The location parameter in Eq. (1) has a specific meaning—it is the minimum value of the general population's mechanical property. From this point of view, for quality control, the location parameter is of the most significant interest since it characterizes the quality of the rolled metal material from the position of minimum strength.

Under conditions of a stable technological process for a particular steel grade's mechanical property, the distributions' shape and scale parameters do not change much from one batch to another. Therefore, these parameters' specific values can be estimated in advance using the data available at the plant and used as a priori information.

Consider a small sample of mechanical properties $x_1, \ldots, x_n$ from the general population distributed according to the Weibull law *F(x)* (1). Usually, under conditions of a factory laboratory, $n = 1.0.3$, and the minimum value from the sample is taken as an estimate of the shift.

$$y = \min(x_1, \ldots, x_n), \tag{2}$$

For a new random variable y, the distribution law is

$$G(y) = 1 - [1 - F(y)]^n = 1 - \exp\left[-n\left(\frac{y-\mu}{\beta}\right)^k\right] \tag{3}$$

with expectation

$$E(y) = \mu + \beta n^{-1/k}\Gamma\left(1 + \frac{1}{k}\right) \tag{4}$$

Dubey in [20] showed that the minimum member of the sample (2) is a biased estimate of the shift parameter, but the magnitude of the shift depends on the sample size and the parameters of the shape and scale of the initial distribution F(x). If these parameters are known a priori, then the unbiased estimate of the shift parameter will have the form

$$\mu = y - \beta n^{-\frac{1}{k}}\Gamma\left(1 + \frac{1}{k}\right) \tag{5}$$

where *y* is the minimum value of the mechanical property, determined from a sample of volume *n*; Γ (*)—gamma function.

To show the advantages of the proposed method, let us compare the errors arising in estimating the location parameter. Based on the theory of extremal sample members, it can be shown that for the Weibull distribution, the average relative error in estimating the minimum sample member (2) under the control strategy that currently exists has the form

$$\varepsilon_y = \frac{|y - \mu|}{\mu} = \frac{\beta n^{-1/k}}{\mu}\Gamma\left(1 + \frac{1}{k}\right) \tag{6}$$

For the experimental study, tensile tests were carried out, and data from continuous inspection of critical grades of 09G2S, 17GSU, 40X, D32 and D40 steels of the main mechanical properties were used.

For nondestructive express determination of mechanical characteristics, an original device has been developed that allows one to determine the mechanical characteristics of steels (Fig. 1). The device consists of a rod 1, a spring 2, a striker 3, a body 4, an indenter holder 5, a sensor 6 and 9, a body part 7 and a damper 8.

During measurements, striker 3 hits the indenter holder 5, and sensor 6 registers the velocity signal *V(t)*, which is subsequently processed to obtain the dependences of the indenter displacement *S(t)* and the indenter acceleration *W(t)*.

In practice, the mechanical characteristics' determination is carried out in 1–2 s and leaves an impression of 0.2 mm. This method is nondestructive and fast.

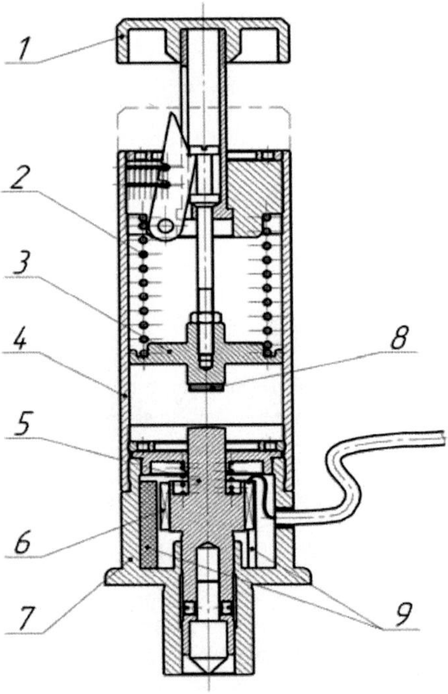

Fig. 1 Device for mechanical characteristics assessment

3 Results and Discussion

The results show that the method for estimating the shear parameter by the method of extreme members of the sample quickly converges and allows one to determine the minimum strength of the material for a small sample.

Figure 2 shows the change in the distribution density from the original to the distribution of the sample's minimum members at $m = 5$. It can be seen that the distribution density of the minimum member of the sample is rapidly narrowing and tends to the parameter of the location of the general population. This approach requires preliminary research on the other two parameters: shape and scale.

To determine the estimation error of the location parameter for each mechanical property of each steel grade, 20 samples of random numbers were generated, with a volume of $n = 5, 10, 15$, having the given distribution parameters, and the average error of the minimum mechanical property ε_y was estimated by the existing method and taking into account a priori information ε_μ.

For the practical application of formula (6), it is necessary at the stage of preliminary studies to establish the parameters β and k of the distribution (1) for various grades of steels. Experimental studies show that steel production technology has little effect on the magnitude of the scale and shape parameters and can be effectively used to determine the minimum strength of materials for robotic systems' supporting structures.

Table 1 shows the results of calculating the error in estimating the shift parameter when approximated by Weibull's law and applying the minimum sample member's strategy. Table 1 shows that the error decreases with an increase in the sample size and already 5 measurements are enough to have a sufficiently stable value of the minimum value of the strength of the supporting structure and not to carry out cumbersome and laborious tests.

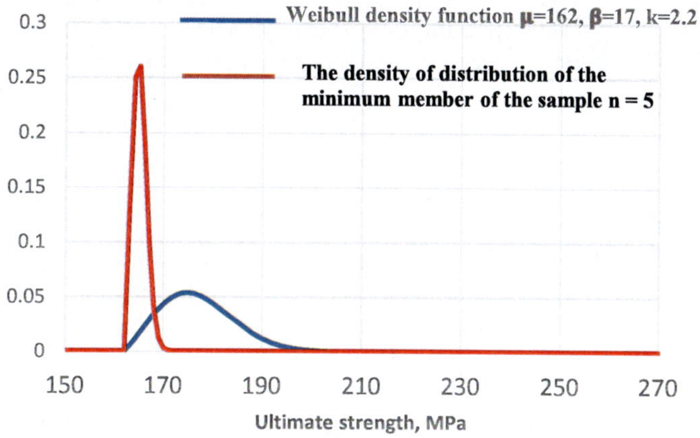

Fig. 2 Weibull distribution densities and minimum members of the sample at $n = 5$

Table 1 Errors in estimating the shear parameter when approximating a mechanical property by Weibull's law

Steel grade	Theoretical average error ε_y at sample size *n*			Selected average posterior error ε_μ location parameter at sample size *n*		
	5	10	15	5	10	15
Tensile strength						
17GSU	2.30	1.77	1.51	0.74	0.57	0.51
D32	2.25	1.60	1.31	0.89	0.63	0.54
09G2S	4.19	2.81	2.23	1.83	1.26	1.01
D40	5.36	4.20	3.64	1.62	1.28	1.06
Yield point						
17GSU	5.46	4.41	3.89	1.47	1.20	1.04
D32	3.33	2.30	1.96	1.23	0.92	0.80
09G2S	3.15	1.82	1.32	1.79	1.04	0.93
D40	4.72	3.49	2.93	1.69	1.29	1.04
Elongation						
17GSU	4.13	2.85	2.30	1.75	1.20	1.15
40X	47.7	34.8	28.7	18.0	13.0	9.40
D40	16.8	15.4	13.3	5.15	4.78	4.38
Impact strength						
17GSU	7.40	4.90	3.80	3.45	2.68	1.68
D32	129	85.9	73.8	59.1	39.9	35.5
09G2S	49.3	33.6	26.9	21.4	14.6	14.3
D40	23.6	33.6	26.9	21.4	14.6	14.3

The use of the device for express assessment of mechanical characteristics will allow to quickly determine the mechanical properties and their probabilistic characteristics, the most important of which is the placement parameter.

4 Conclusion

The problem of quality control of robotic systems' supporting structures is an essential task for modern agricultural machinery's effective functioning. The primary material of agricultural robotic systems is steel, and the mechanical and strength properties of which must be assessed at all life cycle stages. The strength properties of steel are described by a probability distribution with three parameters—shape, scale and location. As shown, the most critical parameter is the location parameter, which shows the minimum strength of the population. The quality control strategy is to a statistical analysis of the placement parameter based on the theory of extreme sample members and provided the parameters of scale and shape are known. To implement the proposed approach, a device for impact indentation of an indenter has been developed and a method based on the probabilistic nature of strength properties has been proposed.

References

1. Golubeva, O., Pogorelova, A.: Development of a software module for quality control and statistical data processing in production. In: E3S Web Conferences, vol. 217, 06008 (2020). https://doi.org/10.1051/e3sconf/202021706008
2. Golubeva, O., Pogorelova, A., Mirnyy, V.: Algorithmic component of the new statistical quality control software module. In: E3S Web of Conferences, vol. 210, p. 01008 (2020). https://doi.org/10.1051/e3sconf/202021001008
3. Bogdanova, I., Dymchenko, O.: Production quality and safety management in the corporate-type integrated structures in the agro-industrial complex. In: E3S Web of Conferences vol. 175, p. 13013 (2020). https://doi.org/10.1051/e3sconf/202017513013
4. Li, Y., Mou, W., Li, J., Liu, C., Gao, J.: An automatic and accurate method for tool wear inspection using grayscale image probability algorithm based on bayesian inference. Robot. Comput-Integr. Manuf. **68**, 102079 (2021). https://doi.org/10.1016/j.rcim.2020.102079.
5. Nguyen, V., Melkote, S.: Hybrid statistical modelling of the frequency response function of industrial robots. Robot. Comput-Integr. Manuf. **70**, 102134 (2021). https://doi.org/10.1016/j.rcim.2021.102134.
6. Adnan, M., Lu, Y., Jones, A., Cheng, F.-T., Yeung, H. A.: New architectural approach to monitoring and controlling AM processes. Appl. Sci. **10**, 6616 (2020). https://doi.org/10.3390/app10186616
7. Lukyanov, A., Vernezi, M., Katin, O., Dolgov, V., Zimovnov, I., Studennikova, S.: Development of methods for analyzing patterns of current consumption in a system for wireless monitoring the effectiveness of metalworking production. IOP Conference Series: Materials Science and Engineering, vol. 900, p. 012017 (2020). https://doi.org/10.1088/1757-899X/900/1/012017
8. Rybak, A., Tsybry, I., Shileev, K., Ermolenko, I.: Analysis of hydraulic resistance in a rotating junction of the cooling system of an active phased array antenna of a circular view. In: IOP Conference Series: Materials Science and Engineering, vol. 1001, p. 012080 (2020). https://doi.org/10.1088/1757-899X/1001/1/012080
9. Pelipenko, A., Rybak, A.: Stand for carrying out life tests of plunger hydraulic cylinders with energy recovery. In: IOP Conference Series: Materials Science and Engineering, vol. 1001, p. 012001 (2020). https://doi.org/10.1088/1757-899X/1001/1/012001
10. Nazir, Q., Shao, C.: Online tool condition monitoring for ultrasonic metal welding via sensor fusion and machine learning. J. Manuf. Processes **62**, 806–816 (2021). https://doi.org/10.1016/j.jmapro.2020.12.050
11. Ke, K.-C., Huang, M.-S.: Quality classification of injection-molded components by using quality indices, grading, and machine learning. Polymers **13**, 353 (2021). https://doi.org/10.3390/polym13030353
12. Li, Z., Zhong, R., Tian, Z., Dai, H., Barenji, A., Huang, G.: Industrial blockchain: a state-of-the-art survey. Robot. Comput–Integr. Manuf. **70**, 102124 (2021). https://doi.org/10.1016/j.rcim.2021.102124
13. Leal, F., Chis, A., González–Vélez, H.: Multi-service model for blockchain networks. Inf. Process. Manage. **58**, 102525 (2021). https://doi.org/10.1016/j.ipm.2021.102525
14. Wang, K., Liu, D., Liu, Z., Wang, Q., Tan, J.: An assembly precision analysis method based on a general part digital twin model. Robot. Comput-Integr. Manufact. **68**, 102089 (2021). https://doi.org/10.1016/j.rcim.2020.102089

15. Tsybrii I.K., Vyalikov I.L., Ignatenko V.I.: Data measurement system for non-destructive quality testing of hard alloys. Vestn. Don State Techn. Univ. **18**(4), 421–425 (2018). https://doi.org/10.23947/1992-5980-2018-18-4-421-425
16. Beskopylny, A., Lyapin, A., Anysz, H., Meskhi, B., Veremeenko, A., Mozgovoy, A.: Artificial neural networks in classification of steel grades based on non-destructive tests. Materials **13**, 2445 (2020). https://doi.org/10.3390/ma13112445
17. Litvinov, S., Beskopylny, A., Trush, L., Yazyev, S.: Optimization of thick-walled spherical shells at thermal and power influences. MATEC Web Conf. **106**, 04013 (2017). https://doi.org/10.1051/matecconf/201710604013
18. Li, Z., Pasternak, H.: Statistical size effect and corresponding influence for reliability in steel structures—part 1: model and experiment. Bautechnik **97**(11), 766–772 (2020). https://doi.org/10.1002/bate.202000030
19. Beskopylny, A.N., Rykov, V.B., Zubrilina, E.M., Chistyakov, A.D.: Agricultural machine parts quality control by dynamic non-destructive methods. MATEC Web of Conf. **226**, 04034 (2018). https://doi.org/10.1051/matecconf/201822604034
20. Dubey, S.: Hyper-efficient estimator of the location parameter of the Weibull laws. Naval Res. Logistics Q **13**(3), 253–264 (1966). https://doi.org/10.1002/nav.3800130304

Strategy of Service and Maintainability of Machines

Mihail Buraev, Alena Tronts, Alexey Shisteev, Galina Buraeva, and Anna Anosova

Abstract Nowadays, the main volume of repair and technical work transferred to enterprises which only operate equipment and is aimed at replacing parts and some simple assemblies. Under such conditions, production costs are increasing. In some cases, this is linked with the rules for routine maintenance of machines after the assigned resource has been used up. The article provides comparative analysis of the possibility of maintenance and repair of machines according to the established regulations and technical condition.

1 Introduction

It is known that the adaptability of a machine to repair and maintenance effects are laid down at the stage of its design [1, 2]. The knowledge of the design features of the machine and the organization of its maintenance during operation gives the possibility to control the duration and labor intensity of repair and service work.

2 Materials and Discussion

According to the planned preventive maintenance and repair system, carried out in a planned manner to ensure the operability of the fleet of machines till the end of entire service life, two strategies for servicing and repairing complex machines are possible (Table 1): C_2—replacement and repair of assembly units (aggregates) of the machine, are carried out strictly after the assigned resource has been worked out; C_3—replacement and repair of assembly units (aggregates) are carried out according to their technical condition. We do not take into account the strategy C_1 of service

M. Buraev (✉) · A. Tronts · A. Shisteev · G. Buraeva · A. Anosova
Irkutsk State Agricultural University Named After A.A. Ezhevsky, 1/1, Molodezhny, Irkutsk, Russia 664038
e-mail: buraev@mail.ru

M. Shamtsyan et al. (eds.), *Robotics, Machinery and Engineering Technology for Precision Agriculture*, Smart Innovation, Systems and Technologies 247,
https://doi.org/10.1007/978-981-16-3844-2_3

Table 1 Variants of strategies for repair and maintenance actions for different types of machines

Types of machines	Types of maintenance and repair					
	Shift maintenance	Numbered maintenance	Seasonal maintenance	Maintenance during storage	Routine repair	Machine overhaul
Tractors	$C_2C_3^3$	$C_2C_3^3$	C_3^1	C_3^1	$C_3^2C_3^3$	C_3^1
Cars	$C_2C_3^3$	$C_2C_3^3$	C_3^1	C_3^1	C_3^2	C_3^1
Combine harvester	$C_2C_3^3$	$C_2C_3^3$	-	C_3^1	$C_3^2C_3^3$	C_3^1
Self-propelled sweepers	$C_2C_3^3$	$C_2C_3^3$	–	C_3^1	$C_3^2C_3^3$	C_3^1
Trailed, mounted harvesting machines	$C_2C_3^3$	$C_2C_3^3$	–	C_3^1	$C_3^2C_3^3$	–
Self-propelled fertilizer applicators	$C_2C_3^3$	$C_2C_3^3$	–	C_3^1	$C_3^2C_3^3$	–
Trailed fertilizer applicators	$C_2C_3^3$	$C_2C_3^3$	–	C_3^1	$C_3^2C_3^3$	–
Irrigation machines	$C_2C_3^3$	$C_2C_3^3$	–	C_3^1	$C_3^2C_3^3$	–
Pumping stations	$C_2C_3^3$	$C_2C_3^3$	–	C_3^1	$C_3^2C_3^3$	C_3^1
Earthmoving machines	$C_2C_3^3$	$C_2C_3^3$	–	C_3^1	$C_3^2C_3^3$	C_3^1

C_1—strategy on demand after failure; C_2—routine maintenance and repair strategy; C_3—strategy based on the state of diagnostics (C_3^1—the scope of repairs is strictly regulated; C_3^2—the content of work is not regulated; C_3^3—the content of work is not regulated, the consequences of failures are eliminated strictly as they occur)

on demand after failure due to its attribution to unscheduled (emergency) types of service [3, 4].

In the first case, the repair and maintenance actions are performed according to a certain regulation, and thus, the number of repair and maintenance operations is always unchanged. In this case, the duration and volume of repair and maintenance impacts for any type of maintenance during the operations are random.

In the second case, the scope of actual repair and maintenance work is determined based on the results of preliminary diagnostic monitoring of the technical condition of the machine or unit [5].

For a comparative assessment of these maintenance strategies, let us consider the maintainability function (distribution of labor intensity) during the period of temporary loss of serviceability by the machine due to elimination of failures, maintenance and repairs $F(t)$. Obviously, the effectiveness of these maintenance strategies will be different [6, 7].

The duration of repair and maintenance actions during the first maintenance strategy can be determined by the formula

$$t_{\Sigma,n} = \sum_{i=1}^{n} t_{\Sigma,i} \tag{1}$$

where n—is the number of elements of the machine being serviced;

$t_{\Sigma,i}$—is the total duration of work on the ith element.

If the number of elements being serviced is greater than the length of the simulated variational series, the sum of independent random variables of duration is distributed asymptotically normally.

The total duration of the maintenance and repair actions during the second maintenance strategy is determined by taking into account the preliminary control with probability P_i. Consequently,

$$t_{\Sigma,N} = \sum_{j=1}^{N} t_{\Sigma,j} = \sum_{i=1}^{n} P_i \cdot t_{\Sigma,i} \tag{2}$$

where N—is a random number of maintenance and repair actions.

In this formula, the probability of occurrence of N is determined by the Poisson distribution.

The mathematical expectations of random variables $t_{\Sigma,n}$ and $t_{\Sigma,N}$ are equal:

$$M\left[t_{\Sigma,n}\right] = n \cdot \bar{t}_{\Sigma,i} \tag{3}$$

$$M\left[t_{\Sigma,N}\right] = M[N] \cdot \bar{t}_{\Sigma} \tag{4}$$

where $\bar{t}_{\Sigma,\mathrm{i}}$ —is the average total duration of the repair and maintenance actions that account on the ith element of the machine.

As in the second maintenance strategy, the volume of repair and maintenance actions follows from the results of preliminary monitoring of the technical condition of the machine or unit, $M[N] \leq n$ therefore,

$$M\left[t_{\Sigma,n}\right] \geq M\left[t_{\Sigma,N}\right] \tag{5}$$

Let us compare the value of maintainability for the first and second strategies of maintenance and repair at points. Let us compare the value of maintainability for the first and second strategies of maintenance and repair at points

$$t_1 = M\left[t_{\Sigma,n}\right] \text{ and } t_2 = M\left[t_{\Sigma,N}\right] \tag{6}$$

For the first strategy, because $F[t_\Sigma]$ is a normal distribution, then

$$F_1(t_1) = 0, 5.$$

For the second servicing strategy with a Poisson distribution of the number of serviced elements, the value $t_{\Sigma,N}$ is distributed according to the exponential law and, therefore, $F_2(t_2) = 0, 63$.

Suppose, that the labor costs for both service strategies will be equal, that is, $t_1 = t_2$, then in this case the distribution functions $F_1(t_1) < F_2(t_2)$, and the numerical data $M\left[t_{\Sigma,n}\right]$ and, therefore, the values $t_{\Sigma,n}$ are always greater than the values of $M\left[t_{\Sigma,N}\right]$ and $t_{\Sigma,N}$. So, while using the second strategy of repair and maintenance actions, the higher characteristics of the machine's maintainability $F_\Sigma/t/$ are provided.

The second strategy of repair and maintenance actions gives the possibility to determine two main directions for the practical solution of the problem: a failure is eliminated during maintenance or the occurrence of it is prevented [8]. Repair and maintenance actions to eliminate failures are carried out only when the resource characteristics (technical condition parameters) of the elements reach their critical value, i.e., the limit of the rejection attribute. In order to prevent failures, repair and maintenance actions are carried out without waiting for the state when parameters can reach the rejection limit. As a result of preventive repair and restoration of deteriorated characteristics two problems may be solved: firstly—at what values of the condition characteristics the repair should be carried out and secondly—whether it is advisable to repeat it periodically till the write—off machine recommendation [9, 10].

The usage of preventative and remediation strategies together or in its combinations can achieve the following benefits:

- to reduce or minimize the number and time of machine downtime due to technical reasons, mainly due to sudden failures;
- to load repair enterprises evenly throughout the year and reduce or eliminate overtime work of repair personnel, because the volume of repair work can be performed during one work shift;
- to minimize the cost of spare parts and repair materials, due to the low probability of failures, to improve technological training, to minimize the manpower needs and reduce the cost of maintenance;
- to lower the need for a reserve of complete units, assemblies and machines.
- to improve warehouse logistics, reduce damage from excessive storage of machine elements, improve the work on the reservation of spare parts.
- the quality of agricultural work is improving;
- losses reduce to a minimum due to whole usage of the resource of elements and machines; It means that the machine will perform an additional amount of work during its service life.

Table 2 To the calculation of the multipurpose maintainability index

Indicator	Designation	Calculation	Factor value		Factor weight
			Condition	Regulation	
Average recovery time	t_r	$t_r = \Sigma t_{ri}/N$	10.5	30.5	0.3
Recovery rate	μ	$\mu = 1/t_r$	0.095	0.03	0.3
Coefficient of temporary downtime	K_t	$K_t = t_r/(t_r + t)$ $\approx t_r/t = \mu\ t_r$	16.5	45	0.2
Intensity of scheduled repairs	μ_{pr}	$\mu_{pr} = 1/t_{pr}$	0.13	0.02	0.1
Duration of scheduled repairs	t_{pr}	According to technical specifications of farm	–	46	0.1
			15%	4.4%	

Taking into account the factors given in Table 2, the assessment of multipurpose maintainability indicator can be presented in the form of a functional dependence [1]:

$$Y = f\left(t_r, \mu, K_t, \mu_{\text{pr}}, t_{\text{pr}}\right), \tag{7}$$

where Y—is the output characteristic;

f—is the required function;

t_r, μ, K_t, μ_{pr}, t_{pr}—are generalized factors.

Table 2 shows the values of the maintainability factors based on the results of a study illustrated by the MTZ tractors. The determination of the complex indicator of maintainability was made by using the coefficients of the weighting factors according to the formula

$$B = S_1 t_r + S_2 \mu + S_3 K_t + S_4 \mu_{\text{pr}} + S_5 t_{\text{pr}}. \tag{8}$$

where S_i—are the weighting factors.

From the calculations it can be seen that the maintainability index of the machine increases when using the maintenance strategy by condition [11, 12]. The maintainability index of the machine increases as machine constructions are adapted to the processes of continuous monitoring of their technical condition and, in its turn, the processes of continuous monitoring create the conditions for operational intervention in the process of eliminating failures and malfunctions. Complex combination of "operating time" and "condition-based" methods of maintenance and repair can

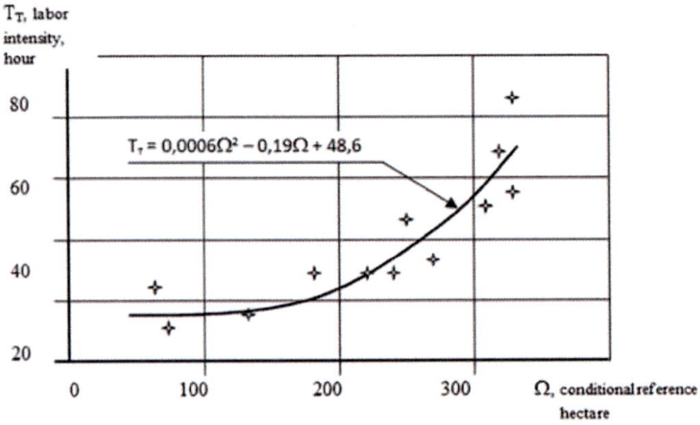

Fig. 1 Change in the laboriousness of eliminating machine downtime due to technical (T_T) with an increase in the annual average loading

also be effective, especially for machines whose overhaul resource is regulated by the manufacturer [13, 14].

At the same time, an increase in the share of complex failures of old equipment (Fig. 1), is observed. The elimination of such failures requires longer and more expensive repair and maintenance actions.

Such a situation creates, so to say, a deadlock situation, especially in remote areas, where the infrastructure of technical service is underdeveloped [15]. The owners of cars find the way out either in self-service, or in the full stop of machine use.

3 Conclusions

Structural elements of modern and future machines strive to do and calculate so, that the limit wear of any surface would mean the limit wear of all surfaces of the structural element, and, therefore, its complete unsuitability for further usage. Under continuous monitoring of the product, its condition characteristics will cease to be random and will have specific values for each product at each moment of time. The failure will cease to be a random event and, consequently, the preventive processes of machine maintenance will be simplified.

References

1. Buraev, M., Shisteev, A., Zhabin, A., Anosova, A., Ilyin, P.: To clarify the standards of spare parts for technical service of autotractors in zone conditions. In: E3S Web of Conferences, vol. 175, p. 05001 (2020)

2. Buraev, M., Ilyin, P., Ilyin, S., Shisteev, A., Anosova, A.: The calculation program of the technical service enterprise of transport-technological machines in agriculture, In: IOP Conference Series: Materials Science and Engineering, vol. 632, no. 1 (2019). https://doi.org/10.1088/1757-899X/632/1/012019
3. Bodyakina, T., Boloev, P., Buraev, M., Shisteev, A.: Diagnostics of hydraulic density of plunger couple of tractor diesel. In: E3S Web of Conferences, vol. 175, p. 05035 (2020)
4. Chubareva, N.V., Khabardin, V.N., Chubareva, M.V.: Technique and results of experimental studies of specific cost determination of tractors maintenance during implementation of different methods. In: IOP Conference Series: Earth and Environmental Science. The Proceedings of the conference AgroCON-2019, p. 012114 (2019)
5. Chubareva, N.V., Khabardin, V.N., Chubareva, M.V.: Resource-saving by choosing methods of tractors maintenance. In: IOP Conference Series: Materials Science and Engineering. 2019 International Conference on Innovations in Automotive and Aerospace Engineering, ICI2AE 2019, p. 012045 (2019).
6. Ding, Y.: Model predictive control and its application in agriculture: A review. Comput. Electron. Agric. **151**, 104–117 (2018)
7. Eltoshkina, E.V., Ilyin, P.I., Ovchinnikova, N.I., Svirbutovich, O.A., Rozhkov, D.M.: Evaluation of the limit properties of a solid body system. In: IOP Conference Series: Materials Science and Engineering. Irkutsk National Research Technical University, p. 012023 (2019).
8. Katarzyna, A.: Classification of spare parts as the element of a proper realization of the machine maintenance process and logistics—case study. IFAC-Papers Line **49–12**, 1389–1393 (2016)
9. Krakovsky, Y.M., Luzgin, A.N., Ivanyo, Y.M.: Cyberattack intensity forecasting on informatization objects of critical infrastructures. In: IOP Conference Series: Materials Science and Engineering, vol. 481, no. 1, p. 012003 (2019)
10. Lengu, D.: Spare parts management: Linking distributional assumptions to demand classification. Eur. J. Oper. Res. **235**, 624–635 (2014)
11. Ovchinnikova, N.I., Bonnet, V.V., Kosareva, A.V.: Simulation diagnostics of power train mechanical drives. In: IOP Conference Series: Earth and Environmental Science, vol. 548, no. 5, p. 052033 (2020)
12. Osmakova, A.: Recent biotechnology developments and trends in the Russian Federation. New Biotechnol. **40**, 76–81 (2018)
13. Shukhanov, S.N., Kuzmin, A.V., Boloev, P.A.: Reliability of the machine-tractor unit Engineering technologies and systems. **30**(1), 8–20 (2020)
14. Hu, Y., Liu, Y., Wang, Z., Wen, J., Li, J., Lu, J.: A two-stage dynamic capacity planning approach for agricultural machinery maintenance service with demand uncertainty. Biosys. Eng. **190**, 201–217 (2020)
15. Zhu, S.: Spare parts inventory control based on maintenance planning. Reliab. Eng. Syst. Safety **193**, 106600 (2020)

Automatic Monitoring of Smart Greenhouse Parameters and Detection of Plant Diseases by Neural Networks

Evgeniy Ivliev, Viktoria Demchenko, and Pavel Obukhov

Abstract This research embodies the idea of greenhouse automation by using the Internet of Things (IoT). The aim of this development is to introduce small agricultural greenhouses and make them "smart". They are intended for monitoring conditions, controlling irrigation, collecting images with cameras installed, and controlling plant disease with leaf data. However, this research focuses on the development of the proposed design and architecture of a greenhouse surveillance system based on the Internet of Things, irrigation management, and an effective method for detecting plant diseases.

1 Introduction

The greenhouse is the main source of not only vitamins due to growing vegetables, fruits, herbs, but also profits. According to experts, the highest EBITDA margin is for greenhouse vegetable growers: It increased by 8% points (pp) to 52% over the year. Additionally, after the introduction of the food embargo and ruble devaluation, the rapid development of the greenhouse business began in 2014.

The high productivity ensures that modern greenhouses are profitable. Out of 1 m^2 you can obtain up to 80 kg of harvest, and the price of vegetables in winter and early spring is too high, especially out of season. Also, investors who have constructed their own mini-CHP plants for greenhouse heating and lighting are even more profitable.

Greenhouses have invested over 200 billion RUB in 2015–2019. The new greenhouses have been laid on over 1100 ha. Around 40% of tomatoes are yet imported into Russia, but cucumbers from their own cultivation are, nevertheless, available to the country. In addition, the most lucrative area of agriculture was the greenhouse business in 2015–2017. Currently, the margin is lower than the Rosselkhozbank estimate for most greenhouses. In 2017–2018, the indicator had been high, approximately 50%. After the Turkish supply ban, a deficit was then formed.

E. Ivliev · V. Demchenko · P. Obukhov (✉)
Don State Technical University, Gagarin square, 1, Rostov-on-Don, Russia 344000
e-mail: 123ivliev123@mail.ru

M. Shamtsyan et al. (eds.), *Robotics, Machinery and Engineering Technology for Precision Agriculture*, Smart Innovation, Systems and Technologies 247,
https://doi.org/10.1007/978-981-16-3844-2_4

Today, there are plenty of cheapest tomatoes and most of the large producers master premium products such as cherry tomatoes, packaged vegetables, and branded products. This is a more costly product, but it is more expensive. Additionally, the greenhouse EBITDA margin is currently 45% and not increasing. The impact of abolishing capital costs compensation will be noticed in this respect, although the margins will be increased to the previous level in future, according to experts. The decline in the profitability of the greenhouse business is due to increasing competition. However, the payback period should also be considered to evaluate the attractiveness of investment by the sector. The cost offset increased from 7.5 to 12 years as a result of the abolition of the capital compensation.

The creation of SMART greenhouses in Russia, which use an automatic climatic control system and preserve the conditions necessary for a healthy harvest, represents a promising direction for all of these things. The production of suitable agricultural products will have a positive effect on the replacement of the importers and the country's development. Also, the cost and maintenance of greenhouses will be reduced with automated climate control systems with lower cost and simplicity of use. The use of Internet Things technology for automated microclimate control is also important as this field is fully and continuously developing in all spheres of society. On the other hand, the IoT technology for this project includes display tools, a single-board microcontroller, a sensor for temperature and humidity, and a test platform. They are designed to control greenhouse temperatures and moisture and to prevent plant diseases. Therefore, both study and development are required for this area. The objective of this study is to validate the proposed method appropriate to detect diseases of plants on plant leaf datasets.

2 Review Literature and Associated Work

The paper [1] proposes the greenhouse monitoring model "IoTOMATO". The main purposes of the systems were to design it so that it could be used in their fields by even a small group of farmers. Analytics is also an important feature that enables the selection and collection of data in a more extensive manner using an Arduino microcontroller in the tomato monitoring system. It has functions for detecting greenhouse diseases using computer vision techniques, and algorithms, and convolutional neural networks AlexNet and SqueezeNet. The system can only be used for half because MATLAB is for image analysis, which requires licensing.

In [2] convolutional neural network models have been designed with simple images of healthy and diseased plants using a methodological approach for the detection and diagnosis of plant diseases. Several models such as AlexNet, AlexNetOWTBn, GoogLeNet, Over feat, VGG have been trained. The VGG architectural model demonstrated the highest degree of accuracy, 99.53%. The extremely high level of accuracy makes it a very useful tool to consult or early warning on the basis of VGG architecture. However, this approach is difficult to apply in real-growing conditions to identify plant diseases, because it requires focusing on one or more plant leaves to achieve efficiency.

The paper [3] proposed a greenhouse parameter monitoring system, irrigation management, and effective detection of leaf diseases for plants in a greenhouse. Additionally, the study concerns the Internet of Things in the agriculture field of application. A greenhouse control system such as temperature, humidity, and irrigation automation, built on the Arduino platform is used in this study. Otherwise, the study also provides for precocious plant disease detection. The MATLAB software simulation is used to analyze the leaf disease detection. According to the authors, the proposed model is related to this study, but not to its implementation in a greenhouse environment. Therefore, the experiments also lack the possibility of open-source implementation because the equipment referred to here is not open-source equipment and the implementation of that prototype is also limited to just one equipment production.

The paper [4] presents a strawberry disease identification scheme. In order to determine the health status of the strawberry plant, digital images of the leaf are processed. On the other side, the process includes three steps: improvement in the image, the segmentation of color from RGB to HSV, a regional segmentation in order to determine the area of deformed and undamaged leaves. Hence, the detection accuracy, according to the authors, is 85%.

3 Materials and Methods

3.1 System Design

The system architecture consists of two parts:

- Automated system of control and management;
- System for plant disease detection.

It is indicated that the automated system solves issues relating to automated temperature control and management, humidity, irrigation, and greenhouse parameter monitoring. The leaf and plant disease detection is conducted by obtaining images and application of imagery and computer vision algorithms of the floor from the greenhouse. The system structure is a network of automatic control and administration modules as well as a cameras network, Fig. 1.

Each module consists of a microcontroller, temperature and humidity sensors, sensors of soil moisture, water flow sensors, water pumps, and light equipment. As the microcontroller responsible for module management, the STM32F103C8T6 has a high frequency of 72 MHz and UART hardware interfaces. This project uses a DHT21 sensor to measure temperature and humidity. A soil humidity sensor is composed of two bimetallic samples that easily pass through the ground and record the level of resistance to determine the humidity level. A large volume of water makes

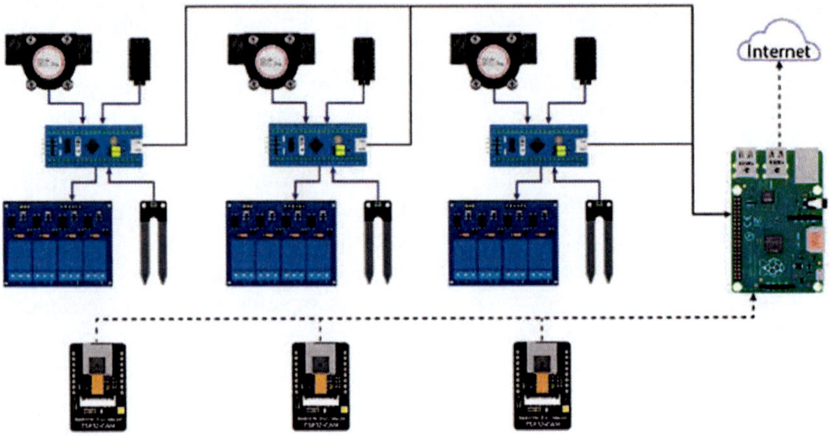

Fig. 1 System structure

electricity with less resistance easier to conduct the soil, while dry soil poorly drives electricity with greater strength. To measure water flow, a Hall sensor is used.

Each module is connected via a wired RS-485 interface to the Network and interacts using the Modbus RTU industrial communication protocol.

The ESP32-CAM module was selected for the disease detection system.

The devices used are based on open-source hardware which can be used with various other hardware components from various manufacturers.

3.2 *Disease Detection System*

The use of neural networks is the most optimal solution to the issue of identifying plant diseases. In the TensorFlow Object Detection API, which is an open-source platform, Google-based TensorFlow [5], we reviewed a number of key neural network architectures, which allow the building, training, or deployment of object detection models. However, in recent years, the most modern object detection algorithms, such as Faster R-CNN [6], R-FCN [7], SSD [8] and YOLO [9], used and are available in mobile devices and consumer products for convolutional neural network systems. In addition to standard metrics such as average accuracy, other factors such as memory consumptions or operating time are also critical in determining what detector is best for detecting plant diseases.

Since many of the leading modern object detection methods have converged on a common methodology consisting of a single CNN which uses a sliding window prediction and is trained for a mixed regression and classification purpose, the authors implement the R-CNN, R-FCN, and SSD Meta Architectures, in combination, with various extractive features: Resnet V1 50, Resnet V1 101 [10], Inception V2 [11], Inception Resnet V2 [12], and Mobilenet V1 [13].

The Faster R-CNN Inception Resket V2 trackings, which are the highest accuracy and SSD mobiles the fastest and can be used with mobile devices, were made based on possible meta-architecturally and feature-extraction combinations based on the researches [14] and [15]. In addition, a training sample consisting of 1200 images has been made for the training of neural networks. The pictures show healthy tomato leaves infected in a true environment. Nine categories were identified among infected leaves shown in Fig. 2. The LabelImg program carried out an annotation of data to identify the boundaries of the object of interest and the class to which that object belongs.

Fig. 2 Tomatoes disease. **a** Canker, **b** Speck, **c** Bacterial wilt, **d** Gray leaf spot, **e** Septoria leaf spot, **f** Gray mold, **g** Curly top, **h** Potato virus Y, **i** Tomato yellow leaf curl

4 Results and Discussion

The result is analyzed in two ways: a smart greenhouse with all relevant sensors and modules based on the Internet of Things, and image analysis for the detection of plant leaves.

4.1 Smart Greenhouse Analysis on the Internet of Things

The high efficiency and modularity of this system were demonstrated through the analysis of the automated monitoring and controlling system. Likewise, using the RS485 interface, 256 modules with a total 1-km line-length can be connected to one network for automatic control and monitoring. On the one hand, the IoT-based implementation of the DHT21 temperature and humidity sensor results based on current room temperatures and humidity. On the other hand, the environment for testing is approximately 29 °C and the humidity is approximately between 73 and 76°. The module and the water meter of the soil work together to keep the plants irrigated.

4.2 Analysis of Leaf Disease Detection Efficiency

In order to assess the efficiency of a plant disease detector, metrology (Recall, *R*) and precision (Precision, *P*) are used for measuring intersections in the predicted and true binding boxes containing baggage tags (Intersection, I).

The intersection of predicted and genuine bounding cases I (1) indicates that the coordinates of the bounding box were predicted in comparison with the true marking in the fully convolutional neural network.

$$I = \frac{S_I}{S_f + S_{gt} - S_I} \quad (1)$$

where S_I is the intersection area of the predicted and true bounding box, S_f is the area of the predicted bounding box, S_{gt} is the area of the true bounding box.

The Recall *R* (2) indicates the sensitivity of the 2nd type algorithm to errors, that is gaps, and is equal to the ratio in the true markup of the amount of properly predicted items to that number.

$$R = \frac{tp}{tp + fn} \quad (2)$$

Table 1 Results of the accuracy of the selected models in the detection of plant diseases

Models	Intersection (I), %	Precision (P), %	Recall (R), %
Faster R-CNN Inception Resnet V2	91.68	82.36	93.88
SSD Mobilenet V1	80.51	62.71	58.03

Table 2 Characteristics of models for detection

Models	Precision (P), %	FPS, 1/c	Memory, MB	FLOPS, (*10^9)	Number of parameters (*10^6)
Faster R-CNN inception resnet V2	82.36	1.36	18,250.45	1837.54	59.41
SSD Mobilenet V1	62.71	41.34	94.7	2.3	5.57

where *tp*—true-positive—those objects that we expected to see and received at the output, *fn*—false-negative objects that we expected to see, but the algorithm did not determine them.

The Precision *P* (3) shows that the algorithm's sensitiveness to errors of the first type is false positive and equals the proportion of the number of predicted properly in the bounding boxes predicted by the algorithm. The algorithm is also correct.

$$P = \frac{tp}{tp + fp} \tag{2}$$

where *fp*—false-positive—objects that should not be at the output, but the algorithm returned them by mistake at the output.

Based on the data in Tables 1 and 2, the SSD Mobilenet V1 model is the best suited in real-time and mobile-driven systems. However, you should use Faster R-CNN Inception Resnet V2 in order to obtain a better result in the detection of plant diseases.

5 Conclusion

In practice, the use of intelligent greenhouses leads to problems such as management and low maintenance. One of the main issues is temperature maintenance. Furthermore, the temperature control effect is caused by the greenhouse structure, as it is covered by plates from top to bottom and from right to right. Thus, sunlight, rain, and natural air do not affect the interior of the greenhouse. This leads to a lower temperature, lower humidity, and less ventilation turbulence. In order to maintain the greenhouse environment for a period of time, continuous monitoring is necessary. Also, due to the extreme closed design, the ground is not reached by watering plants. For manual seeding, therefore, it is necessary to have a constant water supply on time.

Greenhouses are also affected by insect plagues and flies because they are closed and produce turbulence which can severely harm plant health, slowing growth. With regard to these issues, greenhouse practices need to be revised and technologies integrated to resolve them. This study also includes the idea of plant diseases, conditions, and irrigation systems in the greenhouse are detected, brought to life. The prototype of the experiment and the analysis work as expected. Space will increase in future for more functions with this system, and these improvements for further development will be envisaged.

References

1. Kitpo, N., Kugai, Y., Inoue, M., Yokemura, T., Satomura, S.: In: IEEE International Conference on Consumer Electronics, vol. 4, p. 8661999 (2019). https://doi.org/10.1109/ICCE.2019.8661999
2. Ferentinos, K.P.: Computers and electronics in agriculture, **145**, 311–318 (2018). https://doi.org/10.1016/j.compag.2018.01.009
3. Khan, F.A., Ibrahim, A.A., Zeki, A.M.: J. Phys. Commun. **4**, 055008 (2020). https://doi.org/10.1088/2399-6528/ab90c1
4. Kusumandari, D.E., Adzkia, M., Gultom, S.P., Turnip, M., Turnip, A.: J. Phys. Conf. Series **1230**, 012092 (2019). https://doi.org/10.1088/1742-6596/1230/1/012092
5. Abadi, M., Agarwal, A., Barham, P., Brevdo, E., Chen, Z., Citro, C., Corrado, G.S., Davis, A., Dean, J., Devin, M., et al.: Tensorflow: Large-scale machine learning on heterogeneous systems (2015)
6. Ren, S., He, K., Girshick, R., Sun, J.: Neural Inf. Process. Syst. **39**, 1137–1149 (2015)
7. Dai, J., Li, Y., He, K., Sun, J.: Neural Inf. Proc. Syst. **1**, 379–387 (2016)
8. Liu, W., Erhan, D., Szegedy, C., Reed, S., Fu, C.Y., Berg, A.C.: Eur. Conf. Comput. Vision, **1**, 21–37 (2016). https://doi.org/10.1007/978-3-319-46448-0_2
9. Redmon, J., Farhadi, A.: IEEE Conf. Comput. Vision Pattern Recognit. **1**, 6517–6525 (2017). https://doi.org/10.1109/CVPR.2017.690
10. He, K., Zhang, X., Ren, S., Sun, J.: IEEE Conf. Comput. Vision Pattern Recogn. **1**, 770–778 (2016). https://doi.org/10.1109/CVPR.2016.90
11. Ioffe, S., Szegedy, C.: Proc. Mach. Learn. Res. **37**, 448–456 (2015). cite arXiv: 1502.03167
12. Szegedy, C., Ioffe, S., Vanhoucke, V., Alemi, A. A.: Inception-v4, inception-resnet and the impact of residual connections on learning. AAAI Conference on Artificial Intelligence (2017). cite arXiv:1602.07261
13. Howard, A. G., Zhu, M., Chen, B., Kalenichenko, D., Wang, W., Weyand, T., Andreetto, M., Adam, H.: Mobilenets: Efficient convolutional neural networks for mobile vision applications (2017). cite arXiv:1704.04861
14. Huang, J., Rathod, V., Sun, C., Zhu, M., Korattikara, A., Fathi, A., Fischer, I., Wojna, Z., Song, Y., Guadarrama, S.: IEEE Conf. Comput. Vision Pattern Recogn. **1**, 7310–7319 (2017). https://doi.org/10.1109/CVPR.2017.351
15. Arcos-García, Á., Álvarez-García, J.A., Soria-Morillo, L.M.: Neurocomputing **316**, 332–344 (2018). https://doi.org/10.1016/j.neucom.2018.08.009

Increasing Biocompatibility of Endoprostheses by Modeling the Impregnation Process

Vladimir Taran, Irina Grishina, Andrey Nikolaev, Maksim Dodonov, and Alexandra Lyasnikova

Abstract The biocompatibility of endoprostheses used in modern implantology can be increased by impregnating their surface by using liquids with antimicrobial, anti-inflammatory or osteostimulating effect. It is obvious that increasing the amount of impregnated substance will proportionally increase biocompatibility. However, an increase in the total number of pores in the coating is a prerequisite, which will negatively affect the mechanical properties of the coatings. An algorithm for statistical numerical modeling of the porous structure of biocompatible coatings was demonstrated to find the ideal balance between the porosity of the coating, the proportion of impregnated substance, and mechanical characteristics. Experimental studies of porous coatings based on hydroxyapatite on the surface of endoprostheses have proven the applicability of the developed technique. It was possible to achieve a significant increase in the size of pore channels and the proportion of the impregnated substance and, consequently, in biocompatibility without significant loss of the mechanical properties of the coatings.

1 Introduction

Endoprostheses are made mainly of titanium-based alloys. These materials have good stability and are safe for the manufacture of implants. In addition to positive qualities, there are also negative ones, for example, the probability of rejection of the endoprosthesis by bone tissue. To solve this problem, a thin biologically active layer with certain adhesive properties is formed on the surface of the endoprosthesis [1–4].

V. Taran · I. Grishina
Saratov State Technical University, Politechnicheskaya str., 77, Saratov, Russian Federation 410054

A. Nikolaev (✉) · M. Dodonov · A. Lyasnikova
Don State Technical University, Gagarin square, 1, Rostov-on-Don, Russian Federation 344003
e-mail: andreynicolaev@eurosites.ru

M. Shamtsyan et al. (eds.), *Robotics, Machinery and Engineering Technology for Precision Agriculture*, Smart Innovation, Systems and Technologies 247,
https://doi.org/10.1007/978-981-16-3844-2_5

The main requirements for materials for the manufacture of endoprostheses are strength and biocompatibility. Biocompatibility is a complex parameter that depends on number physical and chemical properties of the implant surface. Among them, porosity, roughness and hardness are important [5–7]. The porosity of the implant surface coating can be controlled within a wide range by changing the parameters of the biologically active layer synthesis [8]. Also, one of the important directions for increasing the biocompatibility of endoprostheses is to create a porous coating on the implanted surface filled with a special liquid medicinal substance, which provides favorable compatibility of living tissue with the material of the endoprosthesis [9].

The main structural formation of such coatings is a solid framework of powder particles and pores (voids) formed inside the framework due to loose packing of powder particles. The framework of a porous coating can be imagined as a complex hydraulic system, where the pores are a container for long-term storage of a liquid substance and are interconnected by capillary channels. Medicinal substances with antimicrobial, anti-inflammatory or osteostimulating action can serve as liquid substances introduced into the porous biocomposite coatings of endoprostheses.

Note that the technology for the formation of porous coatings imposes a number of restrictions on their performance properties. For example, to increase the volume of liquid stored within the coating, it is desirable to have pores with a large diameter and sufficient coating thickness. However, with an increase in pore diameter or coating thickness above certain limiting values, the strength characteristics of the coating sharply decrease, which excludes their practical use. In this regard, the problem arises of finding the optimal relationship between the capacitive properties of porous formations inside the coating and the strength of the solid frame.

To solve the problem, it is necessary to consider a whole range of interrelated problems, including operational, design and technological factors that affect the properties of coatings.

2 Algorithm for Evaluating the Porosity of the Coating

An algorithm for providing mathematical modeling of the predicted porosity of the coating is developed. Since the size and shape of the pores in the solid framework of the coating are very diverse, model concepts of the porous structure are of great importance for studying their properties and role in various technical applications [10].

Discrete methods for modeling complex structures and processes, focused on computer modeling, are actively used [11, 12]. The main idea of the methods is to represent physical quantities or structural formations by a set of interconnected discrete elements, which in technical applications are most often called discrete cells, and in discrete mathematics—sets, graphs, etc. Breaking the whole into parts, or discretization, allows you to describe in detail geometric structure and interrelated physical quantities by graph methods. Using these methods, one can go from a visual geometric image to a numerical solution of problems on a PC.

In the process of dividing an area of coating into discrete elements in the form of geometric shapes, it is advisable to use the methods of splitting complex geometric shapes into relatively simple shapes used in the mathematical method of finite elements [13–15]. In our case, the discretization of the studied volume of the porous coating was carried out by cubic figures with flat lateral surfaces, which follows will be called discrete cells [16].

The following is an algorithm for estimating the porosity of the coating, based on the method of statistical numerical modeling of the properties of porous coatings developed by the authors.

The algorithm for assessing the porosity of the coating includes the following provisions:

- for modeling, it is sufficient to consider a relatively small volume of coating with statistical properties characteristic of the entire volume of coating;
- a parallelepiped figure divided into discrete elementary cubic cells with edge size r was selected as the modeled volume of coating;
- the coordinate system of the simulated volume is a grid (lattice) with length L, width B and height H, divided into layers along the planes (xy), (zx), (yz) (Fig. 1a);
- in the process of modeling, each layer of the coating is represented by a two-coordinate grid ij (Fig. 1b);
- Each layer is divided into discrete cells, which are designated by the symbol λ_{ij}^{xyz}. The "xyz" indices define the layer number in the coating volume, and the "ij" indices —the location (position) of the discrete cell within the layer;
- the position and state of each discrete cell are randomly generated.
- sequential movement along the coordinate grids xyz and ij is performed (Fig. 2);

The following designations are also introduced:

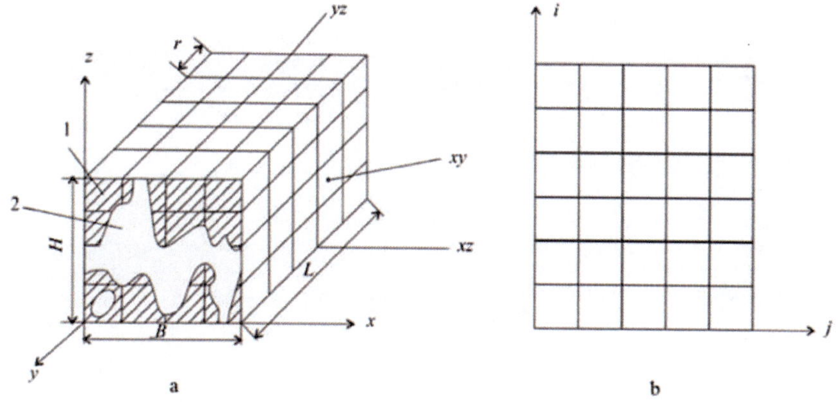

Fig. 1 Scheme of dividing the volume of a porous coating into cubic discrete cells and layers **a**, a discrete coordinate grid ij **b**: 1—coating frame; 2-pores; H, B, L—height, width and length of the cover, respectively; xz, xy, yz—discrete cell layers

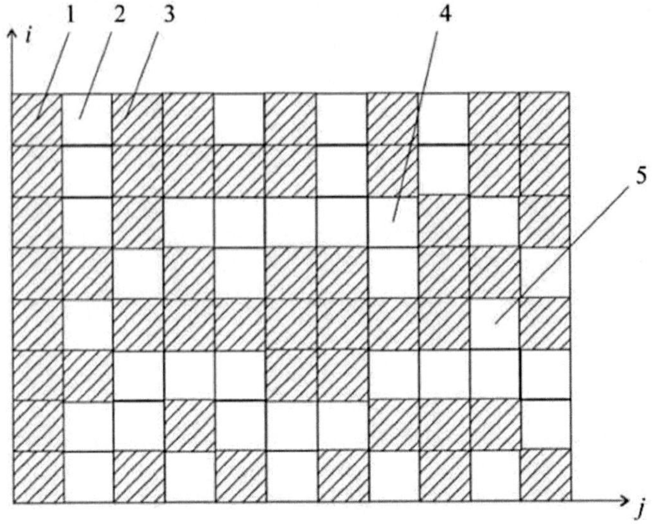

Fig. 2 Model representation of the porous coating formation scheme in the layer: 1—coating surface; 2—discrete cell in the "pore" state; 3—discrete cell in a "solid frame" state; 4—pore channel with access to the coating surface; 5—discrete cell with closed porosity

- pore channel (the concept is determined by the number of interconnected cells in the "pore" state);
- the number of interconnected cells in the pore state (l);
- the number of pore channels with an outlet to the coating surface (g);
- the number of discrete cells with open porosity (a);
- the number of discrete cells with closed porosity (b);
- density of pore channels on the coating surface

$$P = \frac{g}{LB}$$

- total number of pore channels

$$K = \sum l = a + b$$

- total volume of pore channels

$$V = Kr^3$$

- coating porosity

$$\Pi = \frac{V}{lBH} 100\%$$

Table 1 Designations and names of the simulated quantities

Designation	Explanation
L, B, H	Length, width and height of the simulated coating volume
Z	Edge length of the modeled discrete cube
Δ	Divider of the probability P (0,1) into the states "pore", "solid frame"
X, Y, Z	Coordinates of the modeled object
α, β, λ	The number of discrete cells at coordinates x, y, z, respectively
i, j	Coordinates of the modeled layer
l	The number of interconnected discrete cells in the "pore" state, forming pore channels
g	The number of interconnected pore channels exposed to the surface of the coating
a, b	The total number of discrete cells in pore channels that emerged on the surface of the modeled object (open porosity) and did not emerge on its surface (closed porosity), respectively
c	Number of "*i*" coordinates
n, m	The number of discrete cells at the coordinates of the layer i, j, respectively
P	The density of pore channels on the surface of the simulated coating volume
K	The total number of cells in the "pore" state of the simulated coating volume
V	The total volume of discrete cells in the "pore" state
Π	Coating porosity

The coatings were obtained on special automatic stands with computer control. Table 1 shows the designations and names of the simulated quantities. The block diagram of the algorithm for modeling the porosity of the coating is shown in Fig. 3.

To test the applicability of the developed model, experimental studies of the porosity and strength of biocomposite coatings were carried out. Samples with a biocomposite coating of hydroxyapatite were selected as objects of research. Similar coatings are widely used in the technology of manufacturing medical endoprostheses.

The quality of the coating was assessed according to three indicators: strength, porosity of the coating and the degree of its filling with a liquid substance. The last indicator was assessed on a special automatic stand using the equation:

$$Z = \frac{P_0}{P_z} 100\%,$$

where *Z*—the degree of filling a porous coating with a liquid substance;

P_0, P_z—the mass of the test sample before and after filling with liquid, respectively.

The strength of the coating was determined by the shear method on an IR 5082–100 universal testing machine. The determination of the porosity of the coating was carried out by the method of low-temperature adsorption of nitrogen by the

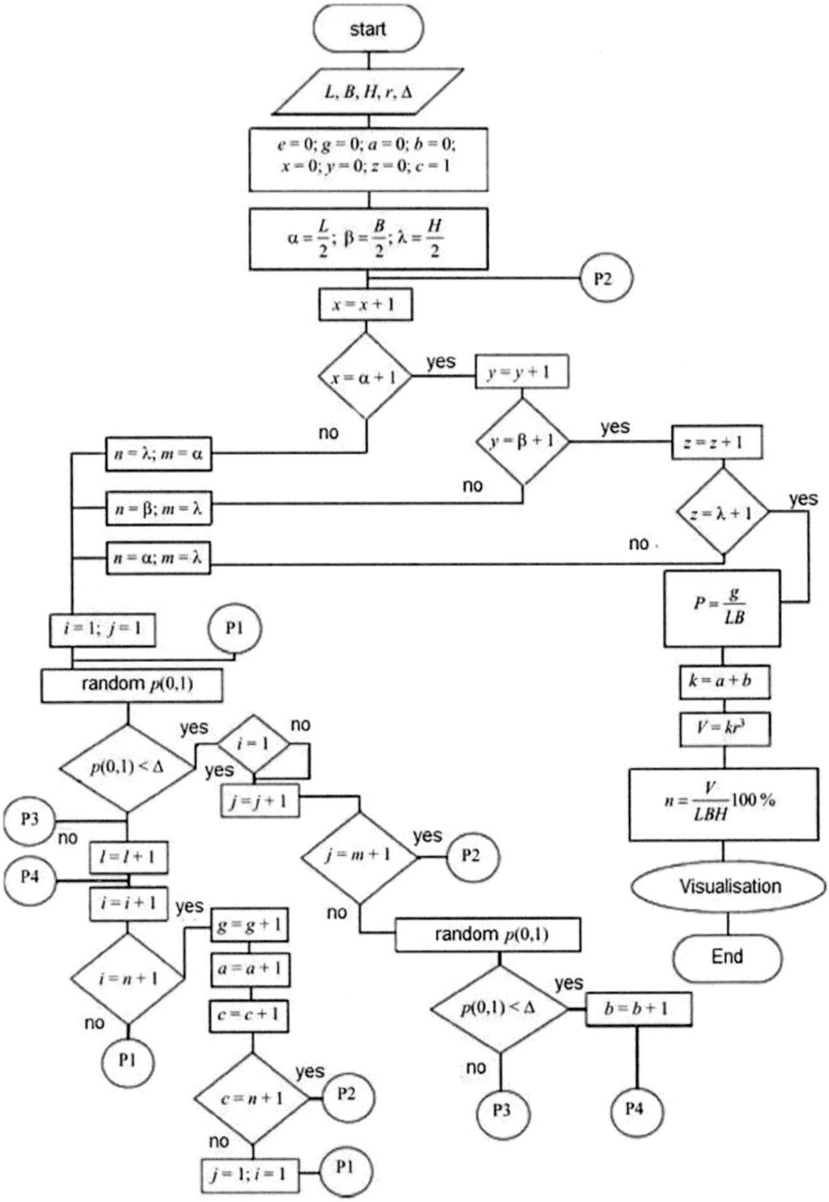

Fig. 3 Block diagram of the algorithm for modeling the porosity of the coating

Table 2 Qualitative indicators of the tested biocompatible coatings based on hydroxyapatite

Coating obtaining Method	Qualitative indicators of the coatings		
	Strength, MPa	Porosity, %	The degree of filling of coatings with liquid substances
Technology using the results of mathematical modeling	14–20	30–40	0.1–0.8
Technology without using the results of mathematical modeling	8–16	10–25	0.01–0.08

Brunauer–Emmett–Teller method on a NOVA 2000e specific surface area analyzer (Table 2).

3 Conclusion

The main characteristics of porous coatings are modeled from the hydraulic system point of view. A model for assessing the porosity of a biocomposite coating has been developed. The data of experimental studies of samples with coatings showed a significant improvement in the quality of coatings during their obtaining using the results of computer modeling. The use of the developed technique makes it possible to find an ideal balance between porosity, mechanical properties and parameters of impregnation of biocompatible coatings used in the manufacture of modern endoprostheses.

Acknowledgements This research was performed into framework of financing RFBR grant number 20-07-00637 A.

References

1. Lyasnikova, A.V., Markelova, O.A., Dudareva, O.A., Lyasnikov, V.N., Barabash, A.P., Shpinyak, S.P.: Powder Metall. Met. Ceram. **55**(5), 328–333 (2016). https://doi.org/10.1007/s11106-016-9809-9
2. Mel'nikova, I. P., Lyasnikova, A. V., Lyasnikov, V. N.: Russian physics journal, vol. 56(10), pp. 1190–1197 (2014). doi :https://doi.org/10.1007/s11182-014-0161-6
3. Koshuro, et al.: Technical physics, vol. 59(10), pp. 1570–1572 (2014). doi:https://doi.org/10.1134/S106378421410020X
4. Sharifianjazi, F., Pakseresht, A.H., Asl, M.S., Esmaeilkhanian, A., Jang, H.W., Shokouhimehr, M.: Journal of Composites and Compounds **2**(2), 26–34 (2020). https://doi.org/10.29252/jcc.2.1.4
5. Arcos, D., Vallet-Regí, M.: Journal of Materials Chemistry B **8**(9), 1781–1800 (2020). https://doi.org/10.1039/C9TB02710F

6. Ferraris, S., Yamaguchi, S., Barbani, N., Cazzola, M., Cristallini, C., Miola, M., Spriano, S.: Acta Biomater. **102**, 468–480 (2020). https://doi.org/10.1016/j.actbio.2019.11.024
7. Alvar, F. S., Heydari, M., Kazemzadeh, A., Vaezi, M. R., Nikzad, L.: Ceramics International, vol. 46(4), pp. 4215–4221(2020). doi:https://doi.org/10.1016/j.ceramint.2019.10.140
8. Ahmadi, R., Afshar, A.: Surface and Coatings Technology, 405, 126594 (2021). doi:https://doi.org/10.1016/j.surfcoat.2020.126594
9. Asimakopoulou, A., Gkekas, I., Kastrinaki, G., Prigione, A., Zaspalis, V.T., Petrakis, S.: Journal of functional biomaterials **11**(3), 65 (2020). https://doi.org/10.3390/jfb11030065
10. Aslannejad, H., Hassanizadeh, S.M., Raoof, A., de Winter, D.A.M., Tomozeiu, N., Van Genuchten, M.T.: Chem. Eng. Sci. **160**, 275–280 (2017). https://doi.org/10.1016/j.ces.2016.11.021
11. Nikolaev, A.L., Mitrin, B.I., Sadyrin, E.V., Zelentsov, V.B., Aguiar, A.R., Aizikovich, S.M.: Adv. Struct. Mater. **136**, 137–146 (2020). https://doi.org/10.1007/978-3-030-48161-2_9
12. Sadyrin, E.V., Mitrin, B.I., Krenev, L.I., Nikolaev, A.L., Aizikovich, S.M.: Springer Proc. Phys. **207**, 495–502 (2018). https://doi.org/10.1007/978-3-319-78919-4_39
13. Wang, L., Li, D.C., Yang, J.S., Shao, F., Zhong, X.H., Zhao, H.Y., Wang, Y.: J. Eur. Ceram. Soc. **36**(6), 1313–1331 (2016). https://doi.org/10.1016/j.jeurceramsoc.2015.12.038
14. Pushkariov, V. I., Nikolaev, A. L., Kaidashev, E. M.: J. Phys. : Conf. Ser. **541**, 012031 (2014). https://doi.org/10.1088/1742-6596/541/1/012031
15. Vestrum, O., Langseth, M., Børvik, T.: Compos. B Eng. **172**, 406–415 (2019). https://doi.org/10.1016/j.compositesb.2019.04.028
16. Wainer, G. A., Mosterman, P. J.: (Eds.). Discrete-event modeling and simulation: theory and applications. CRC press (2018)

Use of Neural Networks and Autopilot for Quick and Accurate Grain Discharge on the Elevator

Sergey Popov, Galina Galchenko, Julianna Marchenko, and Dmitry Drozdov

Abstract The software complex "Auto-Elevator" aimed at automating the unloading and loading of grain at the elevator is considered. The software package is written in C# and allows solving two problems: automatic detection of a free port for unloading grain and the use of an autopilot to supply a vehicle to a loading or unloading terminal. To determine the boundaries of the parking space or to determine a free port for unloading grain, image recognition using neural networks consisting of basic blocks is used. The basic blocks are assembled in the same way as the constructor, by adding layer by layer. The main blocks of convolutional neural networks are convolutional layers, subsampling (pooling) layers, activation layers, and fully connected layers. A convolutional layer extracts features in the input image and generates a feature map.

1 Introduction

Agriculture is one of the most important sectors of the economy. This industry has a particular social significance—the livelihood of the population, it provides basic food products and provides raw materials for the production of consumer goods. Agriculture occupies a central place in the structure of the country's agro-industrial complex (AIC): it produces more than 48% of agricultural products, has 68% of the fixed assets of the complex, and employs almost 67% of workers in the agricultural production sectors.

For a long time, agriculture has not been an attractive business for investors, due to a long production cycle, subject to natural risks and large crop losses during cultivation, collection and storage, the inability to automate biological processes and the lack of progress in increasing productivity and innovation. The use of information technology in agriculture was limited to the use of computers and software, mainly for

S. Popov (✉) · G. Galchenko · J. Marchenko
Don State Technical University, 1, Gagarin sq., Rostov-on-Don, Russia 344003

D. Drozdov
Southern Federal University, Milchakova st. 8-a, Rostov-on-Don 344090, Russia

M. Shamtsyan et al. (eds.), *Robotics, Machinery and Engineering Technology for Precision Agriculture*, Smart Innovation, Systems and Technologies 247,
https://doi.org/10.1007/978-981-16-3844-2_6

financial management and tracking commercial transactions. More recently, farmers have started using digital technology to monitor crops, livestock and various elements of the agricultural process.

Currently, more and more various modern equipment is used in agriculture. In parallel with the use of new technology, modern information technologies are being introduced [1–8]. One of the most promising ways to improve the efficiency of agricultural production management is the use of information systems at elevators. The presence of the Internet, computer equipment and the ability to use software products [9–13] in conjunction with modern technical equipment allows to improve the work processes at the elevators. Much attention is paid to the development of the acceptance and release front. Grain delivery to elevators is a short summer period. Therefore, the process of early automation, reducing the influence of the human factor in the work of the receiving front is very relevant [14–21].

This paper proposes a software package "Auto-Elevator", aimed at automating the unloading and loading of grain at the elevator. The software package is written in C# and allows you to solve two problems: automatic detection of a free port for unloading grain and the use of an autopilot to feed a vehicle to a loading or unloading terminal. Image recognition using neural networks is used to determine the boundaries of a parking space or to determine a free port for unloading grain. If a free port of the bunker for unloading grain is found on the elevator, the autopilot starts working and the vehicle automatically parks and unloads the grain.

2 Recognition of a Free Port for Unloading Grain

Pattern recognition allows you to determine if a port is free and highlight its boundaries. This problem is solved using convolutional neural networks. Such networks consist of basic blocks that are assembled like a constructor by adding layer by layer. The main blocks of convolutional neural networks are convolutional layers, subsampling (pooling) layers, activation layers and fully connected layers. A convolutional layer extracts features in the input image and generates a feature map. Figure 1 shows filters that allow you to highlight features such as borders or fragments of a car.

As a subsampling layer, maxpooling is used, which selects one maximum number in the pixel representation of the image in each square of a given size. In this case, the size of the image is reduced and the selection of features is enhanced. Figure 2 shows an example of how maxpooling works.

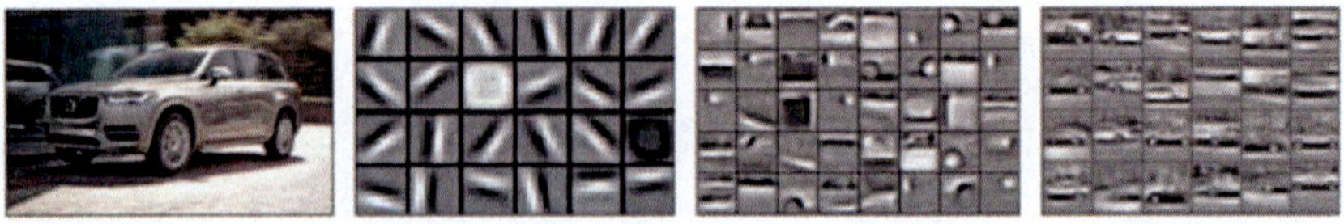

Fig. 1 Filters for object features recognition

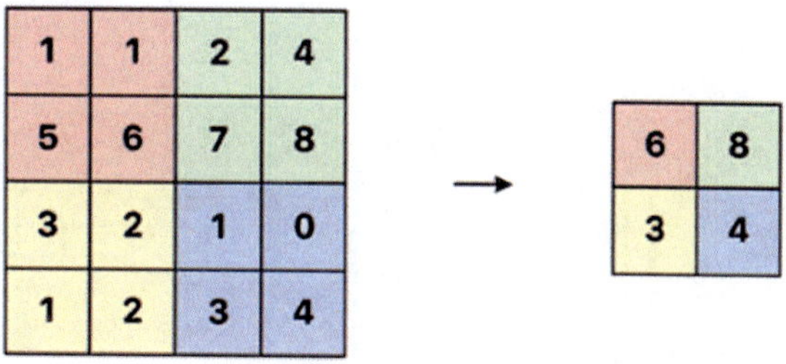

Fig. 2 An example of the maxpooling subsampling.

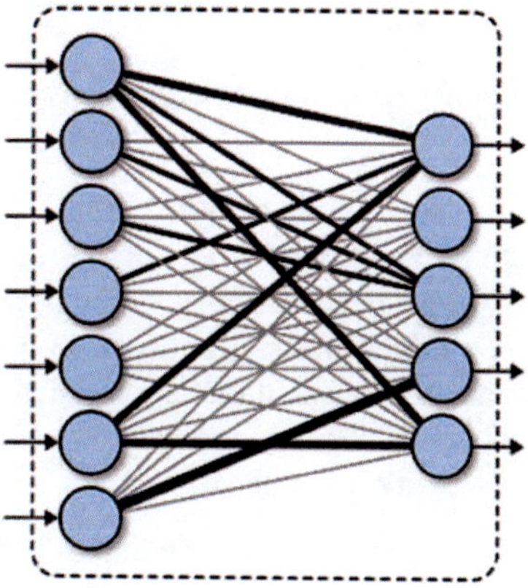

Fig. 3 Fully connected layer

The activation layer is a function that is applied to each element of the intermediate matrix. (At the first stage, the matrix is the pixels of the original image.) The program uses a sigmoid as such a function: $\sigma(x) = \frac{1}{1+e^{-x}}$.

The fully connected layer is the perceptron that usually completes the architecture of the convolutional network. Figure 3 shows a fully connected layer.

The network layers consist of neurons, and each neuron is an object of a class with the following fields: input weights, polarization signal weight, and activation function.

The neuron works according to the following principle:

1. the signals transmitted to the input are multiplied by the corresponding weights and added;
2. the value of the polarization signal is added to the obtained value, multiplied by the corresponding weight;
3. an activation function is applied to the result, for example, the sigmoid described above. Signals 2 and 1 are applied to the input, and the input weights are 0.5 and 0.9, respectively. We set the weight on the polarization signal equal to 0.4.

Then, $x = 2 * 0.5 + 1 * 0.9 - 1 * 0.4 = 1 + 0.9 - 0.4 = 1.5$

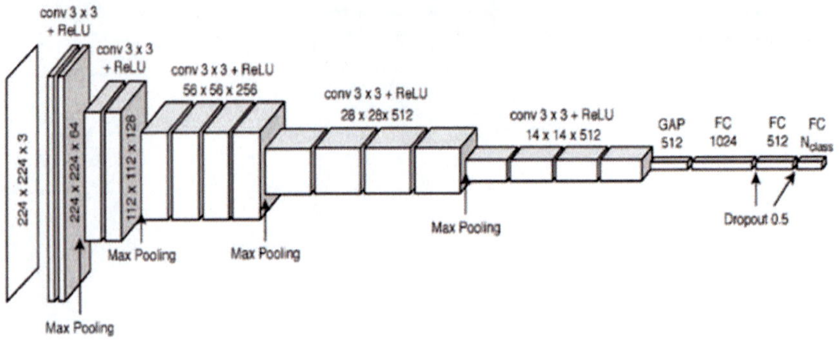

Fig. 4 Architecture of the used neural network

Let's apply the activation function: $\sigma(x) = \frac{1}{1+e^{-x}} = \frac{1}{1+e^{-1.5}} \approx 0.8175$—neuron output. The combination of the described layers will be a convolutional neural network. The architecture of the network used in the program is shown in Fig. 4.

To the compression network described in this work, an inverse expanding network is added. (This architecture is called U-Net.) The resulting network is trained using the backpropagation method.

Network learning stages:

1. Weights are randomly initialized;
2. The values of the features (for example, image pixels) are fed to the network inputs;
3. Each neuron calculates its output and transmits data to the neurons of the next layer. This operation is performed with each layer of the network;
4. The values of the outputs of the neural network are calculated and the error is calculated using the formula $\delta_j = y_j(1 - y_j)(d_j - y_j)$;
5. The error is forwarded to the previous layers by the formula $\delta_j = y_j(1 - y_j)\sum_{k \in OUT_j} \delta_k w_{jk}$;
6. Then, the weights are adjusted according to the formula $w_{ij} = w_{ij} + \Delta w_{ij} = w_{ij} + \eta\delta_j x_i$, where η—learning rate (selected empirically);
7. Items 2–6 are repeated for each image from the training set. (A training set is a set of marked images, that is, with areas already selected.)

Figure 5 shows a network with three layers (S—input, A— associative, R—output).

We put the inputs of the neural network: x = 1, y = 0, z = 1. Let's count the outputs of the neural network R_1, R_2, R_3.

We take the sigmoid as the activation function.

Let's calculate the outputs of neurons on the layer A:

1. 1*0.2 + 0*0.2 – 0.3 = –0.1, f(–0.1) = 0.48
2. 1*0.3 + 0*0.4 + 1*0.1 – 0.2 = 0.2, f(0.2) = 0.55
3. 0*0.2 + 1*0.1 – 0.2 = –0.1, f(–0.1) = 0.48

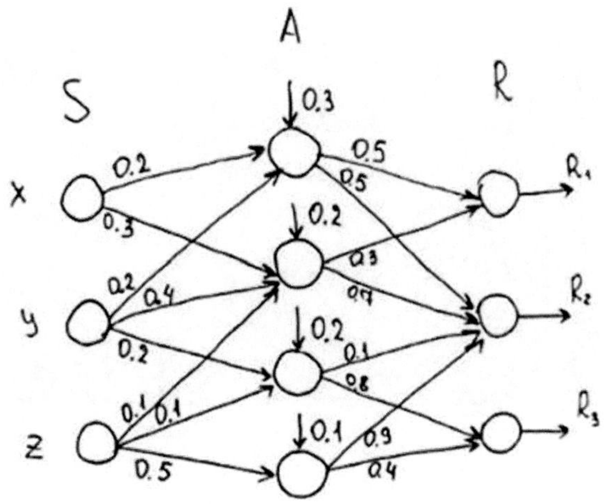

Fig. 5 Neural network with three layers

4. 1*0.5 – 0.1 = 0.4, f(0.4) = 0.6

The weights on the polarization signals on the output neurons are equal to 0.4. Let's calculate the outputs of the neural network:

1. 0.48*0.5 + 0.55*0.3 – 0.4 = 0.005, R_1 = f(0.005) = 0.5
2. 0.48*0.5+0.55*0.7+0.48*0.1+0.6*0.9–0.4 = 0.813, R_2 = f(0.813) = 0.69
3. 0.48*0.8 + 0.6*0.4 – 0.4 = 0.224, R_3 = f(0.224) = 0.56

It is necessary to get the values $R_1 = 0, R_2 = 1, R_3 = 0$.
Calculate the error on the last layer:

1. $\delta_1 = y_1 * (1 - y_1)(d_1 - y_1) = 0.5*(1 - 0.5)*(0 - 0.5) = -0.125$
2. $\delta_2 = y_2 * (1 - y_2)(d_2 - y_2) = 0.69*(1 - 0.69)*(1 - 0.69) = 0.066$
3. $\delta_3 = y_3 * (1 - y_3)(d_3 - y_3) = 0.56*(1 - 0.56)*(0 - 0.56) = -0.138$

Let's forward the error to the previous layers. Transferring to layer A:

1. (0.5*(–0.125) + 0.5*0.066)*0.48*(1 – 0.48) = –0.007
2. (0.3*(–0.125) + 0.7*0.066)*0.55*(1 – 0.55) = 0.002
3. (0.1*0.066 + 0.8*(–0.138))*0.48*(1 – 0.48) = –0.026
4. (0.9*0.066 + 0.4*(–0.138))*0.6*(1 – 0.6) = 0.001

Throwing on a layer S:

1. (0.2*(–0.007) + 0.3*0.002)*1*(1 – 1) = 0
2. (0.2*(–0.007) + 0.4*0.002 + 0.2*(–0.026))*0*(1 – 0) = 0
3. (0.1*0.002 + 0.1*(–0.026) + 0.5*0.001)*1*(1 – 1) = 0

Correcting the weights. Let be $\eta = 1$.
The weights on the *S* layer will not change, since the error there is everywhere equal to 0, and, therefore, $\Delta w_{ij} = 0$. Adjust the weights on layer A:

1. 0.5 + 1*(–0.125)*0.48 = 0.45

Fig. 6 **a** Segmented image, **b** grain unloading

2. $0.3 + 1*(-0.125)*0.55 = 0.23$
3. $0.1 + 1*0.066*0.48 = 0.13$
4. $0.9 + 1*0.066*0.6 = 0.94$

Thus, the weights were corrected, that is, the network was trained.

After training, the network can perform image segmentation. An example of a segmented image is shown in Fig. 6a. The picture is divided into areas, each of which is recognized and assigned to a certain class (car, road, traffic light, etc.). As a result, the program receives an image, performs its segmentation and determines whether the port is free. If the port is open, the "Autopilot-E" program starts working and the grain is unloaded (Fig. 6b).

3 Autopilot for Automatic Vehicle Parking

Many of the world's leading car manufacturers are developing dozens of promising projects that wean from the usual driving [22–24]. No pedal or steering wheel manipulation. The autopilot system will operate much better than a human, and this time is not far off. The machine is controlled using an autopilot based on an expert system. The expert system used is based on fuzzy logic, that is, it operates with concepts such as "far" or "close" instead of specific values. The program receives data from sensor sensors about the distance to obstacles, fuzzifies (converts to a fuzzy set), applies inference rules and, by defuzzing, obtains the values of the output parameters.

Rear Sensor Inference Rules:

1. If l = far & c = far & n = close, then turn left.
2. If l = close & c = far & n = far, then turn right.
3. If l = c = n = far, then v = fast and turn toward the parking space.
4. If one of the sensors is close, then v = slow.

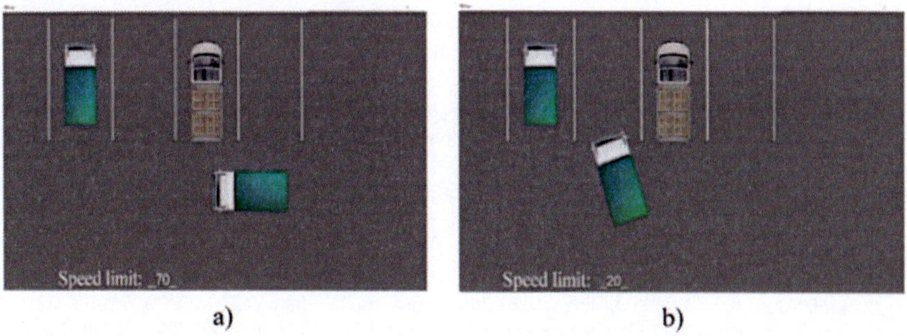

Fig. 7 Image on the monitor screen when the "Autopilot-E" program is running **a** choice of a free port, **b** parking

5. If l = c = close & n! = Close, then turn right.
6. If n = c = close & l! = Close, then turn left.
7. If l = n = far & s! = Far, then turn toward the parking space.
8. If all rear sensors show obstacles (close), then we go forward on the front sensors.
9. Front sensors, as soon as they see an obstacle, transfer control to the rear.

As a result, the program reports the values of rotation (in degrees) and speed (in km/h), which should be set in order to correctly enter the parking space. In Fig. 7 shows the image on the monitor screen when the "Autopilot-E" program is running: (a) selection of a free port, (c) parking [25–27].

4 Conclusions

The use of modern technology and software in agriculture allows the use of new software products that increase the efficiency of various agricultural processes. The use of information systems at elevators will optimize work in a short summer period of time, when grain is delivered and unloaded from the fields. In this work, using modern neural network technology, the problem of determining a free port for unloading grain at an elevator and the operation of an autopilot for automatic parking of a vehicle that approaches a free port is solved.

At present, computer programs are widely used in the world as educational laboratories. The proposed software product can be used, just like a computer laboratory for scientific and educational purposes. [28–31]

References

1. Ivanov, V.V., Popov, S.I., Selemeneva, E.M., Babazhanov, N.T.: In: AIP Conference Proceedings, vol. 2188, pp. 020015-1–020015-5 (2019). https://doi.org/10.1063/1.5138389

2. Ivanov, V.V., Popov, S.I., Marchenko, Ju.V., Marchenko, E.V., Dontsov, N.S., Timofeev, S.A.: IOP Conference Series: Earth and Environmental Science, vol. 403, p. 012115 (2019). https://doi.org/10.1088/1755-1315/403/1/012115
3. Korotky, A.A., Marchenko, E.V., Ivanov, V.V., Popov, S.I., Marchenko, Ju.V., Dontsov, N.S.: In: IOP Conference Series: Earth and Environmental Science, vol. 403, p. 012116 (2019). https://doi.org/10.1088/1755-1315/403/1/012116
4. Kotesova, A.A., Teplyakova, S.V., Popov, S.I., Kopylov, F.C.: In: IOP Conference Series: Materials Science and Engineering, vol. 698, p. 066029 (2019). https://doi.org/10.1088/1757899X/698/6/066029
5. Ivanov, V.V., Popov, S.I., Kirichek, A.V.: In: IOP Conference Series: Materials Science and Engineering, vol. 327, p. 032026 (2018). https://doi.org/10.1088/1757-899X/327/3/03202
6. Lerer, A., Makeeva, G., Galchenko, G.: RSEMW, vol. 8792726, pp. 261-264 (2019). https://doi.org/10.1109/RSEMW.2019.8792726
7. Gubsky, D., Kleschenkov, A., Galchenko, G.: RSEMW, pp. 344–347 (2019). https://doi.org/10.1109/RSEMW.2019.8792726
8. Nikolaev, N.N., Marchenko, Yu.V., Filatov, S.K.: In: IOP Conference Series: Materials Science and Engineering,vol. 698, p. 066027 (2019). https://doi.org/10.1088/1757-899X/698/6/066027
9. Ivanov, V.V., Popov, S.I., Dontsov, N.S., Marchenko, Ju.V., Ryzhkin, N.V., Oleynikova Ju.A.: In: IOP Conference Series: Materials Science and Engineering, vol. 1029, 012052 (2021). https://doi.org/10.1088/1757-899X/1029/1/012052
10. Ivanov, V.V., Popov, S.I., Dontsov, N.S., Ryzhkin, N.V., Oleynikova, Ju.A., Denisenko, Ju.N.: In: E3S Web of Conferences, vol. 210, p. 01006 (2020). https://doi.org/10.1051/e3sconf/202021001006
11. Kuren, S.G., Galchenko, G.A., Popov, S.I., Marchenko, Ju.V., Dontsov, N.S., Drozdov, D.S.: In: E3S Web of Conferences, vol. 210, p. 09005 (2020). https://doi.org/10.1051/e3sconf/202021009005
12. Teplyakova, S.V., Kotesova, A.A., Popov, S.I., Kotesov, A.A.: In: IOP Conference Series: Materials Science and Engineering, vol. 913, 042054 (2020). https://doi.org/10.1088/1757-899X/913/4/042054
13. Ivanov, V.V., Popov, S.I., Dontsov, N.S., Kotesova, A.A.: In: IOP Conference Series: Materials Science and Engineering, vol. 913, p. 042056 (2020). https://doi.org/10.1088/1757-899X/913/4/042056
14. Korotky, A.A., Marchenko, E.V., Popov, S.I., Marchenko, Ju.V., Dontsov, N.S.: In: E3S Web of Conferences, vol. 175, p. 05018 (2020). https://doi.org/10.1051/e3sconf/202017505018
15. Ivanov, V.V., Popov, S.I., Dontsov, N.S., Ekinil, G.E., Oleynikova, Ju.A., Denisenko, Ju.N.: In: E3S Web of Conferences, vol. 175, p. 05023 (2020). https://doi.org/10.1051/e3sconf/202017505023
16. Korotky, A.A., Popov, S.I., Galchenko, G.A., Marchenko, Ju.V., Drozdov, D.S.: In: E3S Web of Conferences, vol. 175, p. 13019 (2020). https://doi.org/10.1051/e3sconf/202017513019
17. Ivanov, V.V., Popov, S.I., Kirichek, A.V.: Key Eng. Mater. **736**, 8–22 (2017). 10.4028/www.scientific.net/KEM.736.18
18. Kostoglotov, A.A., Andrashitov, D.S., Kornev, A.S., Lazarenko, S.V.: Meas. Tech. **62**, 497–502 (2019). https://doi.org/10.1007/s11018-019-01652-8
19. Lazarenko, S.V., Kostoglotov, A.A.: J. Commun. Technol. Electron. **62**(2), 123–127 (2017). https://doi.org/10.1134/S1064226917020061
20. Kostoglotov, A.A., Pugachev, I.V., Lazarenko, S.V.: AIP Conf. Proc. (2019). https://doi.org/10.1063/1.5138398
21. Lazarenko, S.V., Kostoglotov, A.A., Kostoglotov, A.I.: J. Commun. Technol. Electron. **54**(4), 431–438 (2019). https://doi.org/10.1134/S1064226909040081
22. Kalmykov, BYu., Stradanchenko, S.G., Sirotkin, A.Y., Garmider, A.S., Kalmykova, Y.B.: ARPN Journal of Engineering and Applied Sciences **11**(17), 10205–10208 (2016)
23. Visotski, I.Y., Ovchinnikov, N.A., Petriashvili, I.M., Kalmykova, Y.B., Kalmykov, BYu.: ARPN Journal of Engineering and Applied Sciences **10**(12), 5150–5156 (2015)

24. Kalmykov, BYu., Ovchnnikov, N.A., Kalmikova, O.M., Guguyev, I.K., Kushnariva, I.V.: ARPN Journal of Engineering and Applied Sciences **10**(10), 4366–4371 (2015)
25. Lachuga, Y., Akhalaya, B., Shogenov, Y., Meskhi, B., Rudoy, D., Olshevskaya A.: IOP Conferences Series: Materials Science and Engineering, vol. 1001, p. 012121 (2020). https://doi.org/10.1088/1757-899X/1001/1/012121
26. Pakhomov, V., Maltseva, T., Kulikova, N., Ugrekhelidze, N., Enalyeva, L., Babajanyan, A.: Analysis of the micronization process effect on the amino acid composition in compound feed. In: E3S Web of Conferences, vol. 210, p. 07003 (2020). ITESE-2020 https://doi.org/10.1051/e3sconf/202021006012
27. Matrosov, M., Nizhnik, D., Panfilov, I., Pakhomov, V., Serebryanaya, I., Soloviev, A., Rudoy, D.: Calculation of the movement trajectory of the grain mass in the field stripper. In: E3S Web of Conferences ITESE-2020, vol. 210, p. 08015 (2020). https://doi.org/10.1051/e3sconf/202021008015
28. Kholodov, O., Kholodova, M., Kolycheva, Z., Barbashova, E., Mozgovoy, A., Beskopilniy, A., Polushkin, O., Kolodkin, V., Magomedov, M., Rudoy, D., Olshevskaya, A., Khozyaev, I.: Strategic planning system for agricultural production and agro logistic in Russia. In: IOP Conferences Series: Materials Science and Engineering, vol. 918, p. 012144 (2020). https://doi.org/10.1088/1757-899X/918/1/012144
29. Lachuga, Y., Akhalaya, B., Shogenov, Y., Meskhi, B., Rudoy, D., Olshevskaya, A.: Innovative pneumatic device for combined seeding of various tilled crops. In: E3S Web of Conferences ITSE-2020, vol. 210, p. 08016 (2020). https://doi.org/10.1051/e3sconf/202021008016
30. Molochnikov, D., Khalimov, R., Gayaziev, I., Rudoy, D., Olshevskaya, A.: The improvement of the technique for determining technical condition of repair and maintenance equipment. In: E3S Web of Conferences, ITSE-2020,vol. 210, p. 08006 (2020). https://doi.org/10.1051/e3sconf/202021008006
31. Soloviev, A., Matrosov, A., Panfilov, I., Meskhi, B., Polushkin, O., Rudoy, D., Chebanenko, V.: Mathematical and laboratory modeling of resonant impact on the spike for the purpose of grain selection. In: E3S Web of Conferences, ITSE-2020, vol. 210, p.05017 (2020). https://doi.org/10.1051/e3sconf/202021005017

Operation of Membrane Apparatus with Polarization Layer Removal Using Artificial Neural Networks

Anton Shafray, Ekaterina Sazonova, Dmitry Borodulin, Boris Lobasenko, and Igor Bakin

Abstract The paper focuses on the problem of modeling the operation process of membrane apparatus with polarization layer removal using artificial neural networks. The neural network was trained using the results of experiments, where the technological parameters of the apparatus were changed, and the dry matter concentration rate per meter of the filter area was measured. The resulting neural network showed high accuracy, topping the other modeling methods.

1 Introduction

Artificial intelligence (AI) is developing rapidly, with AI-based technologies penetrating all spheres of the national economy and finding application in all sectors of the economy. One of the most popular areas of artificial intelligence is neural networks—an abstract data structure built on the principle of neurons in the nervous system in the human brain. The neural network, like brain neurons, learns from some data and experience to conduct some kind of productive activity in the future.

All the variety of problems that neural networks solve can be divided into two main types: regression and classification. They are the basis for the variety of ways to use networks, including:

- systems for recognition and classification of objects in images, machine vision;
- voice interfaces for the Internet of Things;
- intelligent security and monitoring systems used in operational monitoring and analysis centers;
- systems for automatic interaction with human beings;
- systems for video emotion analytics;
- systems for optimizing the functioning and management of material flows in the economies of regions, industries, objects of enterprises;

A. Shafray · E. Sazonova (✉) · D. Borodulin · B. Lobasenko · I. Bakin
Kemerovo State University, 6 Krasnaya Street, Kemerovo, Russia 650000

M. Shamtsyan et al. (eds.), *Robotics, Machinery and Engineering Technology for Precision Agriculture*, Smart Innovation, Systems and Technologies 247,
https://doi.org/10.1007/978-981-16-3844-2_7

- intelligent, self-learning production and equipment management systems and fault location systems (including maintenance time prediction) [1].

The use of neural networks in technological processes is becoming more extensive. In most cases, they are used for automation control and quality control of technological processes. Most scientific works that reveal the possibilities of using neural networks discuss the same tasks. Along with this, more recent articles discuss the use of artificial neural networks for modeling the experimental data of various food engineering processes [2–12]. In this study, neural networks were used to model the process of operation of the apparatus with polarization layer removal installed in the membrane unit.

Such membrane apparatuses have a device for polarization layer removal, by which the feed flow in the membrane after permeate removal is divided into a concentration polarization layer and a depleted central flow. This approach is based on the phenomenon of polarization concentration, as a consequence of which a movable layer with an increased retentate content is formed close to the membrane surface after the removal of the permeate. The removal of this layer leads to an increase in the performance of the membrane apparatus by reducing the time for obtaining a product with a given concentration of solids.

To prove the efficiency of the developed design, experimental studies of a membrane unit equipped with a polarization layer removal apparatus were performed. The laboratory equipment design is shown in Fig. 1. The equipment was washed and sterilized prior to use.

For unit operation, the solution is fed to tank (1). After switching on, the circulation pump (2) creates a circulation of the medium through the membrane module (3). The rotation speed of the circulation pump and the opening of the pipeline are regulated by the valve (5) to ensure the necessary flow velocity and pressure (the latter is determined by the manometer (4)). During the operation process, the filtrate (permeate) is removed into the tank 6. The polarization layer is removed into the tank (7). The variety of membranes used in dairy production includes: ultra-filtration acetate cellulose membranes (UAM); ceramic ultra-filtration tubular membranes produced by OOO NPO "Ceramicfilter"; aromatic polysulfonamide polymer membranes produced by ZAO STC "Vladipor," and others.

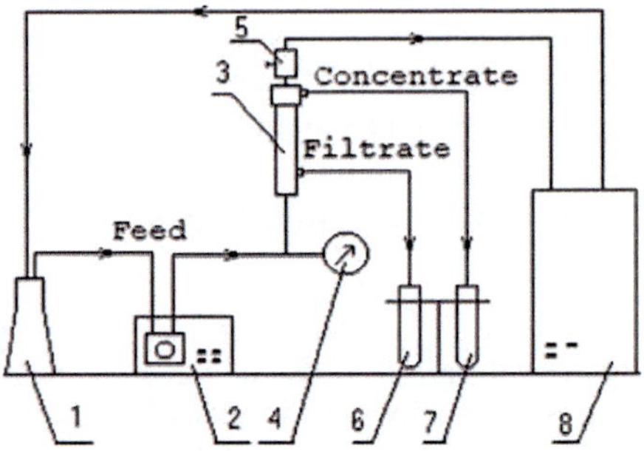

Fig. 1 Laboratory installation scheme

Ceramic membranes have a long working life, and allow a wide range of regeneration and cleaning methods, which ensures a high degree of their recovery, making their application rational.

Whey with dry matter concentration of 5% was used as the feed. The content of dry matter in dairy products is determined by drying. Acid or alkaline washing was used for membrane regeneration.

The purpose of membrane processing of fluid foods is to obtain the product with a necessary concentration and quality. The dry matter content in dairy products is determined by the established method of drying.

Due to the peculiarities of membrane apparatus and the physical and chemical properties of dairy raw materials, a prerequisite for processing dairy products is determining the permissible ranging of technological parameters.

Quality treatment requires operation at a temperature of 50 ± 20 °C to prevent the denaturation of whey proteins, while ensuring high performance indicators of the membrane (to avoid fouling).

The permissible flow velocity range in the apparatus with polarization layer removal is selected with the respect of maintaining the laminar flow regime.

The permissible the working pressure range in the system is 0...0.2 MPa. A higher pressure would lead to a higher concentration of the formed polarization layer, which would adversely affect the performance of the apparatus and make it difficult to clean the inner surface of the membrane.

The economical processing period should not exceed 3 h, since the total working time of the equipment (including washing and disinfection) should not exceed the duration of the technological shift.

The studies were performed with the support of an individual research grant for young scientists of Kemerovo State University in the "FOOD ENGINEERING" Centre of the Department of Technological Design of Food Production, Kemerovo State University.

During the study, it was necessary to design and train a neural network capable of predicting the dry matter concentration rate per meter of filter surface area DMCR (%mass/ (%mass*m2)) in the removed polarization layer, at the specified input parameters: the operation time t (s), the transmembrane pressure P (MPa), the flux per meter of filter surface area V (m3/m2). The neural network was expected to achieve a higher accuracy in predicting the dry matter concentration rate compared to the previously developed models published in the works [13–16]. The neural network was trained using the same experimental data as for the construction of the previous models.

The neural network was designed and trained in Python on Google Colab Laboratory platform. In the course of the work, the freely distributed PyTorch library was used.

2 Identification of the Neural Network Architecture

The identification of a neural network (NN) as a mathematical model consists of several stages in which its main elements are defined. The identification method is chosen based on the problem which requires designing the network. Depending on the problem, a particular type of network architecture will be used. The current study required solving a regression problem, so fully connected layers were used as the basis of the network. For such architecture, the following components must be selected:

- the number of hidden layers;
- the number of neurons in the hidden layers;
- the activation functions;
- the loss functions;
- the step;
- the optimizer;
- regularization;
- the size and number of batches;
- the number of epochs;
- the accuracy criteria.

The selection of all components is based on the researcher's experience. It is performed by trial and error, but the iteration of parameter values is determined by some a priori knowledge about them. In the course of this study, various variants of the network architecture were considered; the most accurate one is described further.

The dataset for network training consisted of numerical values obtained in experiments, where four parameters were measured: dry matter concentration rate per meter of filter surface area DMCR (%mass/ (%mass*m^2)) in the removed polarization layer, operation time t (s), transmembrane pressure P (MPa), flux per meter of filter surface area V (m^3/m^2). Dry matter concentration rate was taken as the output parameter, while the other three were taken as input parameters. The NN task was to model (predict) the output parameter based on a combination of input values.

The data were divided into the training set (80% of total) and the test set (20% of total). The training dataset is used for training the network. The test dataset is used to check the validity of the network and to determine its accuracy. Also after loading the entire dataset, it was randomized.

Before training began, the dataset was normalized: all the values were reduced by a minimum value and divided by the maximum value. Normalization allows increasing the convergence rate of the neural network learning algorithm significantly, i.e., it allows reaching the optimal accuracy of the network quicker. By default, data are values of different orders and, as a result, have different effects on neurons, which may lead to incorrect decisions. Normalized data allow avoiding this effect.

The input layer contained three neurons, each taking the input value of one of the parameters: time, pressure, and flux. After the input layer was followed by four hidden fully connected layers, each of which had seventy neurons. A fully connected

Fig. 2 ReLU function graph

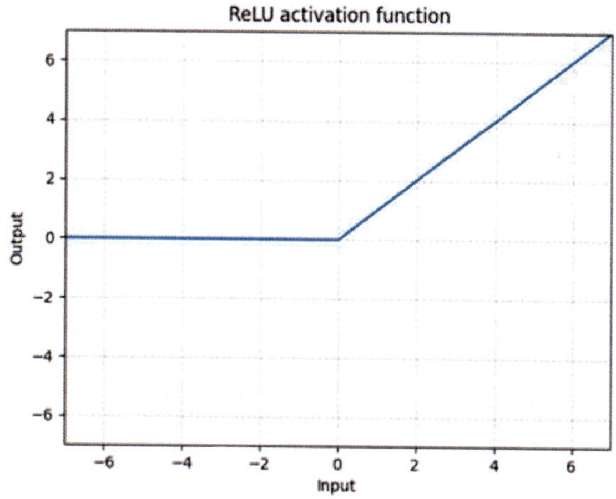

layer is a layer where all neurons are connected to all neurons of the previous layer and to all neurons of the following layer. The output layer contained one neuron, to which the value of dry matter concentration rate modeled by the neural network was transferred.

RRelu was selected as the activation function on all layers. This function has the following formula:

$$PReLU(x) = \{^{x, x \geq 0}_{ax, x < 0}, \tag{1}$$

where x is the argument and α is a random number from the normal distribution. The RReLU (random ReLU) function is a subtype of the ReLU function, and its graph is usually not given due to the random parameter. Therefore, we provide the ReLU function graph below (Fig. 2).

The argument x of the activation function is the sum of input values from the neurons of the previous layer (for the input layer—from the dataset), multiplied by the weights of the layer, with a bias added to neurons. After calculating the function from the argument, its value is transferred to all the neurons of the next layer.

In the output layer, there is one neuron which should model the value of dry matter concentration rate. The correctness of its response can be checked by comparing it to the value from the dataset for the corresponding set of input parameters. The comparison can be performed in various ways, for this purpose the loss function is selected. In this study, MSELoss (root-mean-square error) was chosen as the loss function. It has the following form:

$$\text{MSELoss}(y, \hat{y}) = \frac{\sum_{i=1}^{N} (\hat{y}_i - y_i)^2}{N} \tag{2}$$

where y is the vector of output values from the dataset, $\hat{y}$ is the vector of model output values, and N is the number of data.

The value of the loss function is used in the backpropagation method, where the gradient is calculated based on the difference between model and experimental values, and its calculation happens on all layers. I.e., the error from the output neuron is transferred to the neurons of the previous layer, for each of which a gradient is calculated. Knowing the gradient allows understanding how the weights of the NN should be adjusted to decrease the error rate. This action is called a step. The step is adjusted to decrease the gradient. The purpose of training is to reach the smallest value of the gradient, i.e., its global extremum (minimum).

The most important parameter of the step is its size, or the learning rate, which is selected experimentally, like all other parameters. As a result of the study, the learning rate was taken for 0.001.

In addition to the step size, it is important to know where and how to step, because the minimum can be reached by different ways. It is the optimizer that answers the question of how exactly to take steps. It defines the algorithm by which the minimum of the loss function is searched, and the rule for adjusting weights. It was Adam (adaptive moment estimation) that suited this NN best. It combines the advantages of RMSProp and stochastic gradient descent with momentum. It has the following rule for adjusting weights:

$$W_{t+1} = W_t - \alpha \frac{\mathrm{EMA}_{\beta_1}(\nabla f)^t}{\sqrt{\mathrm{EMA}_{\beta_2}(\nabla f^2)^t + \varepsilon}} \tag{3}$$

where W_{t+1} is new network parameters, W_t is current network parameters, α is the learning rate, $\mathrm{EMA}_{\beta_1}(\nabla f)^t$ is the exponential moving average of the gradient, β_1 and β_2 are the parameters of the exponential moving average, and ε is the smoothing parameter excluding division by 0.

Adam combines the advantages of stochastic gradient descent with momentum and the RMSProp method. Its major advantage is that this method does not require selecting parameters. In most cases, the problem is solved with a standard set of parameters: $\alpha = 0.0003$, $\beta_1 = 0.9$, $\beta_2 = 0.999$.

When training NN, there is a common phenomenon where the designed model explains well the examples from the training dataset, but it works poorly on the examples from the test dataset. This is due to the fact that when constructing a model some random patterns are found in the training dataset that are missing in universal set. This phenomenon is called over-training.

There are several ways to avoid over-training, including: early stop, dataset extension, drop-out, regularization, and others. In the designed NN, we applied the regularization method called L2 (Tikhonov regularization, or weight decay), in which the summand λ is added to the loss function, which contributes an additional penalty to the loss function for too big weights in the neural network. The optimal value of the parameter L2 was 0.000001.

Model training consists in a repetitive passage through the training dataset. Each row of data, which contains the parameters of one experiment, including both input and output data, is fed to the NN input. At the NN output, a model value is generated and compared with the experimental output parameter from the dataset. One passage through the whole dataset is called an epoch. The model showed the best parameters after 5901 epochs of training.

NN training throughout the epoch does not occur immediately throughout the entire dataset. At the beginning of the epoch, the entire training dataset is divided into several equal parts called batches (sometimes the last batch is smaller than the other ones). In each epoch, the division of the dataset into batches is performed randomly; thus, in each epoch batches are formed different datasets. During the training process, the batch is passed through the network, the value of the loss function is calculated, the gradient is calculated, and the weights of the NN are adjusted. This happens with each batch, and after using all batches, a new epoch begins. The number of batches depends on their size; for the designed NN, the batch size was 44.

3 Defining the Neural Network Accuracy

The accuracy of mathematical models is evaluated in different ways. In mathematical models of technological processes, it is often evaluated by calculating relative error. First, relative error is calculated basing on the experimental data used for creating the model, and after that additional experiments can be carried out to confirm the accuracy of the model. In the works [13, 14], the model accuracy was evaluated using this method, and amounted 10% [13] and 6% [14], which is a rather good value for such models.

Evaluating the accuracy of a neural network depends on the problem it solves (regression or classification). In this study, the NN was designed to determine the rate of dry matter concentration per meter of the filter surface area, i.e., for solving the problem of regression. The dry matter concentration rate is expressed in derived units of measurement (%mass/(%mass*m^2)) and varies from 0 to 270 to an accuracy of 1. Comparison of the experimental and the prediction values showed that the mean relative error of the neural network was 4.4026% (Fig. 3) and the average difference between the values amounted 6.7561 (Fig. 4).

An important aspect of comparing neural network accuracy with that of the mathematical models proposed in works [13–16] is that the mathematical models were created using the whole set of the experimental data and their accuracy was assessed using the same set of experimental data. As for the NN, it was trained using one set of the experimental data (the training dataset), while its accuracy was assessed using the other set (the test dataset), which implies an a priori higher accuracy of the output value obtained using the designed NN.

A visual representation of the results obtained using the NN is shown in Fig. 5, where the ordinal numbers of the test values are measured on the abscissa axis, and

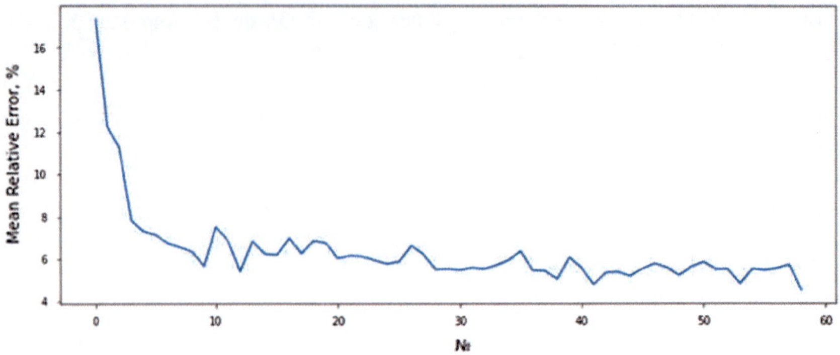

Fig. 3 Neural network mean relative value graph

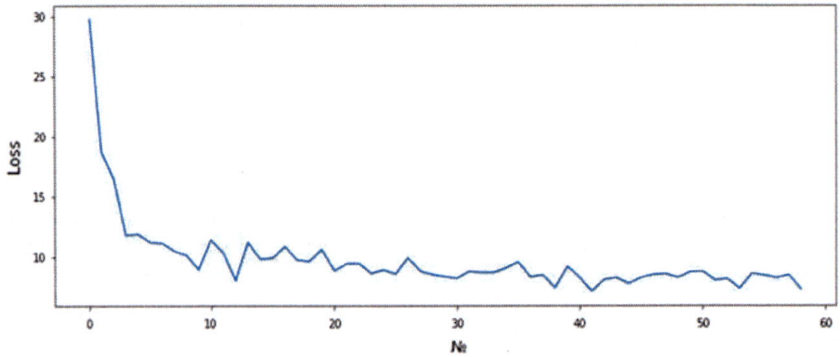

Fig. 4 Mean error graph

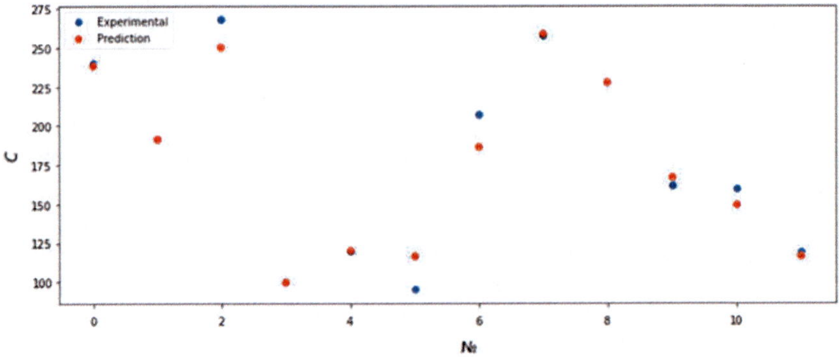

Fig. 5 Comparison of the experimental and prediction values

the test values are measured on the ordinate axis. Blue circles indicate experimental data, and the predicted data are marked with red circles.

4 Technological Parameters Optimization

In the study, the optimal solution of the modeling problem is understood as combination of values of the input parameters (within the permissible values), at which the value of the output parameter will be maximum. Optimization problems have many solution algorithms. In formal mathematical models, the most common solution consists in finding the model extremes and determining the corresponding parameter values. The formal solution for the NN was not provided, because such task had not been set, and the combination of all possible values of input parameters was not too large. Thus, a sample was created, including all possible combinations of independent factors, to determine the optimal values of the membrane apparatus, in other words, the optimization problem was solved with the help of exhaustive method. The operation time varied from 0 to 6000 s, the transmembrane pressure varied from 0.15 to 0.17 MPa, and the flux per meter of the filter surface area varied in the range from 0.4 to 0.5 m^3/m^2.

The optimization sample was fed to the input of the neural network, resulting in the determination of optimal values of the membrane apparatus. According to the NN, the optimal mode of operation will be at a flux of 0.5 m^3/m^2, a pressure of 0.17 MPa, and a time of 2700 s. The laboratory tests, carried out at such rational values, allowed to obtain the concentrate with the maximum dry matter content in the removed polarization layer (5.9423% of the mass), which also proves high efficiency and predictive ability of designed NN.

5 Conclusion

In the study of the operation of membrane apparatuses with a polarization layer removal device, which involved different research methods, one of the best results was achieved by the model created with the help of artificial neural networks. The neural network showed the highest accuracy, and allowed predicting the rational parameters of the device, which were confirmed in practical tests. Successful use of neural networks in membrane process modeling will lead to their further application in modeling other processes.

References

1. Zagorodnikov, A.V., Shafray, A.V.: Neural networks application in technological systems//food innovations and biotechnologies. In: Proceedings of the 8th International Conference of Students, Post-graduates and Young Researchers, Engineering Technologies, vol. 2, pp. 17–18 (2020)
2. Ali, A., Qadri, S., Mashwani, W.K., Belhaouari, S.B., Naeem, S., Rafique, S., Anam, S.: Machine learning approach for the classification of corn seed using hybrid features. Int. J. Food Prop. **23**(1), 1097–1111 (2020). https://doi.org/10.1080/10942912.2020.1778724
3. An, T., Yu, H., Yang, C.S., Liang, G. Z., Chen, J.Y., Hu, Z.H., Dong, C. W.: Black tea withering moisture detection method based on convolution neural network confidence. J. Food Process Eng. **43**(7), (2020). https://doi.org/10.1111/jfpe.13428.
4. Bhargava, A., Barisal, A.: Automatic detection and grading of multiple fruits by machine learning. Food Anal. Methods **13**(3), 751–761 (2020). https://doi.org/10.1007/s12161-019-01690-6
5. Chen, J.D., Zhang, D.F., Nanehkaran, Y.A., Li, D.L.: Detection of rice plant diseases based on deep transfer learning. J. Sci. Food Agric. **100**(7), 3246–3256 (2020). https://doi.org/10.1002/jsfa.10365
6. Chen, X.Y., Chai, Q.Q., Lin, N., Li, X.H., Wang, W.: 1D convolutional neural network for the discrimination of aristolochic acids and their analogues based on near-infrared spectroscopy. Anal. Methods **11**(40), 5118–5125 (2019). https://doi.org/10.1039/c9ay01531k
7. Codina, G.G., Dabija, A., Oroian, M.: Prediction of pasting properties of dough from mixolab measurements using artificial neuronal networks. Foods **8**(10), (2019). https://doi.org/10.3390/foods8100447
8. Ekiz, B., Baygul, O., Yalcintan, H., Ozcan, M.: Comparison of the decision tree, artificial neural network and multiple regression methods for prediction of carcass tissues composition of goat kids. Meat Sci. **161** (2020). https://doi.org/10.1016/j.meatsci.2019.108011.
9. Ikonic, B., Bera, O., Pavlicevic, J., Kojic, P., Jokic, A., Ikonic, P., Saranovic, Z.: Artificial neural network modeling and optimization of wheat starch suspension microfiltration using twisted tape as a turbulence promoter. J. Food Process. Preserv. **43**(11), (2019). https://doi.org/10.1111/jfpp.14219
10. Lu, A.N., Wei, X.X., Cai, R.K., Xiao, S.J., Yuan, H.N., Gong, J.Y., Xiao, G.N.: Modeling the effect of vibration on the quality of stirred yogurt during transportation. Food Science and Biotechnology **29**(7), 889–896 (2020). https://doi.org/10.1007/s10068-020-00741-7
11. Sadeghi, E., Asl, A.H., Movagharnejad, K.: Mathematical modeling of infrared-dried kiwifruit slices under natural and forced convection. Food Sci. Nutr. **7**(11), 3589–3606 (2019). https://doi.org/10.1002/fsn3.1212
12. Sadeghi, E., Movagharnejad, K., Asl, A.H.: Mathematical modeling of infrared radiation thin-layer drying of pumpkin samples under natural and forced convection. J. Food Process. Preserv. **43**(12), (2019). https://doi.org/10.1111/jfpp.14229
13. Lobasenko, B.A., Sazonova, E.K., Musaev, P.A., Pachkin, S.G.: Mathematical modeling and experimental studies of membrane concentration in apparatuses with the withdrawal of the polarization layer. Food Process. Tech. Technol. **46**(3), 152–159 (2017)
14. Lobasenko, B., Kotlyarov, R., Sazonova, E.: Mathematical modeling of membrane concentration based on cybernetic approach. Fundam. Res. **2–1**, 70–75 (2016)
15. Lobasenko, B.A., Ivanets, V.N., Sazonova, E.K., Stefankin, A.E., Kotlyarov, R.V.: Modeling of continuous membrane plant completed with apparatus for diffusion layer removing. Food Process. Tech. Technol **42**(3), 112–117 (2016)
16. Lobasenko, B., Kotlyarov, R., Sazonova, E.: Automation of the production of cottage cheese using the ultrafiltration method. In: 2019 International Science and Technology Conference, pp. 1–5 (2019). https://doi.org/10.1109/EastConf.2019.8725395.

Methodology for Neural Networks Training at Analyzing the Context of Event at Emotions Recognizing

Yulia Kovtun, Larisa Cherckesova, Elena Revyakina, Olga Safaryan, Evgeniya Roshchina, and Vitaly Porksheyan

Abstract This article discusses the problems of improving intelligent information systems for recognizing emotions. The relevance of these studies is determined on the basis of modern trends in science and business in the field of machine learning. A technique is proposed that should improve the existing algorithms for teaching neural networks to recognize emotions based on audio information. Separately, it was noted that the grounds for the proposed methodology were taken from the psychology of human emotions.

1 Introduction

At the moment, the implementation of emotion recognition systems is being carried out in a wide range of areas. In fact, any information system that directly interacts with the user needs not only to analyze the meaning of a person's commands or actions, but also to understand his emotions as non-verbal communication.

Leadership in the use of neural networks that recognize emotions in Russia is taken by banks. Sberbank, Alfa-Bank, VTB, Rosbank and others use empathic artificial intelligence mainly to analyze customer satisfaction from communication with operators. There are also case studies that use emotion analysis to make loan approval decisions. Often, these assessment services take into account pauses in speech, its volume and communication time.

Among the interesting applications of neural networks that recognize emotions, voice assistants are also worth noting. For example, the Massachusetts Institute of Technology has been developing an application that can determine the emotional state of a driver. This application monitors the stress level and strives to reduce it when it exceeds the normal value with the help of soothing music, changes in temperature in the cabin [1]. Amazon has a similar empathy systems application in its Halo fitness bracelet, which provides training advice based on the user's emotional state.

Y. Kovtun · L. Cherckesova · E. Revyakina · O. Safaryan (✉) · E. Roshchina · V. Porksheyan
Don State Technical University, Gagarin Square, 1, Rostov-on-Don, Russia 344003

M. Shamtsyan et al. (eds.), *Robotics, Machinery and Engineering Technology for Precision Agriculture*, Smart Innovation, Systems and Technologies 247,
https://doi.org/10.1007/978-981-16-3844-2_8

One of the largest applications of emotion-sensing neural networks is advertising. Nowadays, information about user requests, banner viewing time, etc. is mainly used to customize advertising content. This approach is approximate, because it is based on a complex system of assumptions. Right now emotion recognition is being introduced among large services that provide advertising platforms, for example, in its streaming service Disney, the user's facial expression determines his satisfaction with the content. By 2024, Gartner IT research firm predicts that more than half of all Internet advertising will be based on emotion recognition. The main catalyst for the rapid growth of this area is voice assistants, which, according to statistics, are used by about 45% of US residents for shopping. So many large stores have been partnering with smart assistant developers to reduce the shopping process to a simple voice command. The market of analytics and emotion recognition systems is estimated at more than $20 billion and is projected to double by 2024 [2–4].

Purpose of the study. The development of the accuracy of empathy in intellectual assistants and consultants is necessary because this industry already has a lot of competition today. Bots are able to prove themselves better than ordinary consultants of call centers, and due to the fact that a bot costs a company 3–5 times cheaper than an employee, more and more companies delegate some of the functions of communicating with clients to chat bots (about 57% of companies on this moment). These include large companies, such as MTS, Tinkoff and others [3]. In this regard, it becomes impossible to control the activities of the artificial intelligence of consultants and assistants manually, it is necessary to develop their systems of self-assessment of their activities. Therefore, it would be natural to try to recreate the empathy function in systems whose functions include communication with a person.

Now there are many systems that determine a person's emotions by voice, they are quite accurate, especially if the model range of emotions is narrowed down to positive and negative, in this case the algorithms work with an accuracy of 90% [1, 6]. But given the scale of the application, it is necessary to strive to improve accuracy.

In psychology, emotions are usually divided according to their duration, as well as the strength of their impact. If the system needs to evaluate a person's attitude to a certain object, for example, to an advertising banner or cashier service, it must be taken into account that the reaction to the object can be increased, decreased, distorted or absent under the influence of long-term emotions, for example, bad mood, chronic stress, inspiration, while maintaining another stable long-term emotional process. In this case, it would be useful not to rely on statistical methods of error, but to train the emotion recognition system, analyze whether a person is in a prolonged state of depression or excitement caused by extraneous causes, and, as a consequence, take this fact into account in the final assessment of satisfaction with a product or service.

The theoretical basis of the research, communication between people is a complex process, it includes a verbal level—words and non-verbal—actions caused by emotions. Emotion recognition—empathy, is an ancient evolutionary mechanism that is common to all mammals, it is intuitive, i.e., does not require meaningful analysis [6]. As mentioned above, empathy helps to understand the feelings of the interlocutor, and, based on this, regulate your own behavior.

When an applied task is to assess the person attitude to a certain product or service, machine learning often uses a continuous approach, separating emotions into positive and negative. The aim of the study is to improve the effectiveness of this approach by adding mood assessments at the same time, as this parameter affects how vividly a person expresses an emotion [7].

The main autonomic nervous system changes, which are reflected in speech, we can track by changes in loudness, pitch, speech speed, the range of these changes during the dialogue [8–12].

Despite the fact that these parameters most accurately indicate the affective state (the least prolonged emotional state), thanks to the capabilities of neural networks, we can effectively solve the problems of prediction and pattern recognition, having correctly marked the data for training, based on these parameters, we can accurately determine more complex emotional states.

2 Mathematical Model and Solution Method

The proposed training model is suitable for a neural network that analyzes the overall assessment of a person's mood based on recording his voice.

Model training requires data collection, since there are no ready-made libraries with the required data labeled as yet.

The data collection technique will be carried out in several stages.

A room with a low noise level of less than −30 dB at a speech volume of 0 dB, a condenser microphone is required so that the researcher's speech does not distort the data. People should be interviewed one at a time.

First of all, it is necessary to carry out the SAN testing (methodology and diagnostics of well-being, activity and mood) developed by the staff of the Sechenov First Moscow State Medical University. SAN testing is widely used in assessing the mental state of sick and healthy individuals, psychoemotional response to stress, to identify individual characteristics and biological rhythms of psychophysiological functions [12–15].

The test consists of 30 questions and takes approximately three minutes. The results that we will get: the scale of well-being (SHS), the scale of activity (SA), the scale of mood (SN) in the range from 10 to 70, where 10–30 is low, 31–50 is average, 51–70 is high. From them, we calculate the total value of mood (M) in the range from 0 to 100.

$$M = 5/3(\mathrm{SHS} + \mathrm{SA} + \mathrm{SN}), \tag{1}$$

Then it is necessary to start recording a dialogue with the subject, the main goal of this dialogue is to get a voice recording about five minutes long, therefore this paragraph does not contain exact requirements, but as a recommendation it was proposed to use a set of cards with photographs of some goods, for example, cars of different brands. The person being tested needs to be asked to describe each photographs, and then, dividing them into groups of five cards, offer to choose one most liked and one least liked and explain the reasons for choosing. It is preferably to give the subject the impression that his opinion and choice are important, so as not to cause rejection. To do this, you can write or mark something in the form.

At least, 100 data instances are required for training accuracy.

3 Results of Research

The proposed model is suitable for a recurrent neural network learning using the Neat algorithm.

The Neat algorithm peculiarity is that it is largely based on the evolutionary theory. A separate configuration of a neural network is called an agent, weights are genes, a agents set of one training round is a population. Those the agents who will score the most points in terms of fitness function (in our case, by indicating the mood assessment that is closest to the real one) are allowed to cross. During crossing, crossover and mutation process is recreated, which allows to train the neural network faster and more efficiently without setting the model initial parameters, because when crossing, hidden nodes can be added or removed (new connections are assigned random weights). The entry and exit nodes are not affected during the crossing (Fig. 1).

At the input of the neural network, we will transmit data on loudness and pitch—values that determine affect/emotion; volume range and tone range are mood-defining values. If the input feature vector for the classifier is a real vector $\vec{x}$, then the output score is as follows:

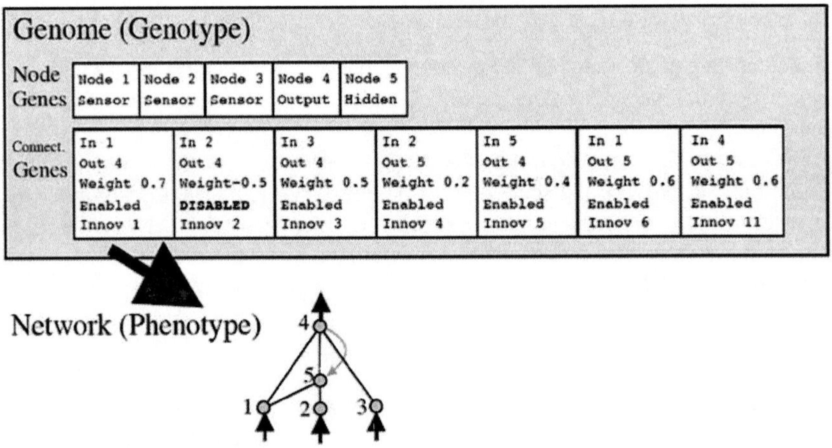

Fig. 1 Neat algorithm mutation scheme

$$y = f(\vec{w} \cdot \vec{x}) = f(\sum_i w_i x_i), \tag{2}$$

$\vec{w}$—a real vector of weights, and an f-function that transforms the dot product of two vectors into the desired result.

$$f(\mathbf{x}) = \begin{cases} 1 \text{ if } \mathbf{w} \cdot \mathbf{x} > T, \\ 0 \text{ otherwise} \end{cases} \tag{3}$$

The key parameters that determine the agent are the generation number and scores in the fitness function. We can also track the change in genes, because each new added node is marked in order to find out in which generation it was added (Fig. 2).

At the output, the neural network will transmit a value corresponding to the mood.

Activation function is tanH.

The population size is about 50, it is worth determining when testing the model.

The maximum of generations is 30.

Fitness Function is approaching Correct Value (step 1; initial value 50).The termination condition is the end of the audio file, or the agent distance from the correct value by more than 51 units [16, 17] (Fig. 3).

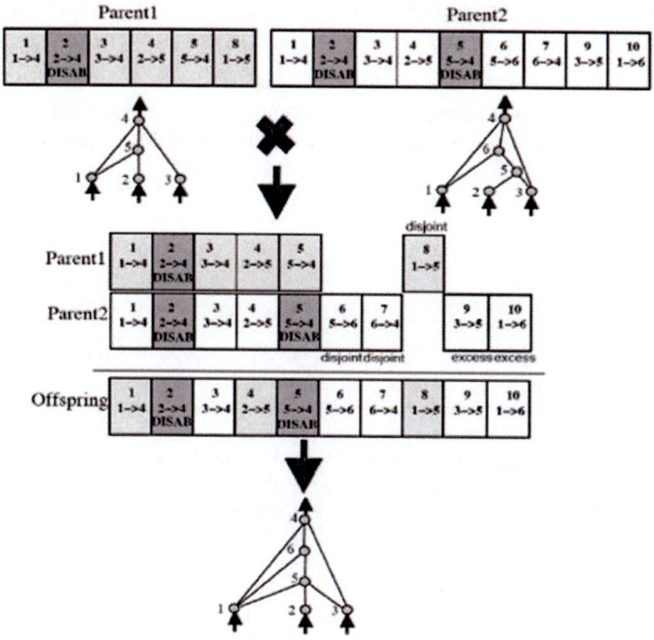

Fig. 2 Generation scheme

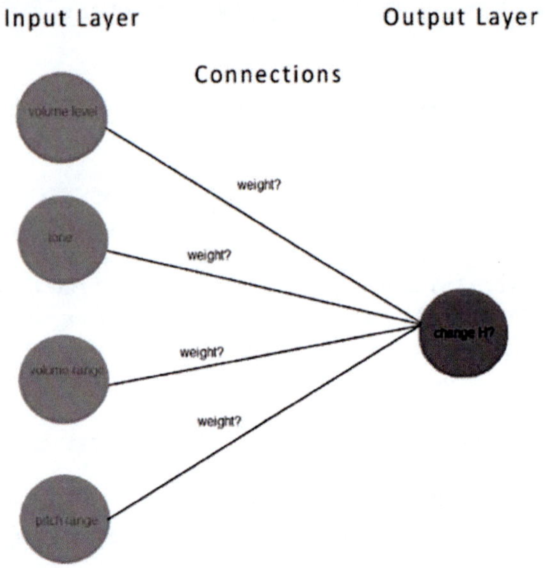

Fig. 3 Input data schema

4 Discussion and Conclusions

In the field of emotion recognition, an increase in efficiency, even by a few percent, correlates with an increase in profits. And, summing up, we can say that the approach to training neural networks proposed in this article needs practical verification, but its theoretical justification is quite reliable.

References

1. Lebedev, G., Zhovnerchuk, E., Zhovnerchuk, I., Moskovenko, A.: Remote recognition of human emotions using deep machine learning of artificial neural networks. Procedia Compu. Sci. **176**, 1517–1522 (2020). https://doi.org/10.1016/j.procs.2020.09.162
2. Uddin, Md. Z., Erik, G.: Nilsson Emotion recognition using speech and neural structured learning to facilitate edge intelligence. Eng. Appl. Artif. Intell. **94**(September), 103775 (2020). https://doi.org/10.1016/j.engappai.2020.103775
3. Griol, D., Callejas, Z.: Combining speech-based and linguistic classifiers to recognize emotion in user spoken utterances. Neurocomputing **326–327**(31 January), 132–140 (2019). https://doi.org/10.1016/j.neucom.2017.01.120
4. Razumov, P., Boldyrikhin, N., Cherckesova, L., Safaryan, O., Reshetnikova, I., Beryoza, A.: Specific features of the practical implementation of observation planning in systems for monitoring networks with dynamic topology. In: E3S Web of Conferences, vol. 224, p. 01033 (2020).
5. Zhilin, V.V., Drozdova, I.I., Sakharov, I.A., Safaryan, O.A., Cherckesova, L.V., Porksheyan, V.M., Lobodenko, A.G., Morozov, S.A.: IEEE east-west design and test symposium. Ewdts

Institute of Electrical and Electronics Engineers Inc., p. 8884375 (2019). https://doi.org/10.1109/EWDTS.2019.8884375
6. Qiang, Q.Z., Wang, Z.G.: The uncertainty of probabilistic rough sets in multi-granulation spaces. Int. J. Approximate Reasoning **77**(October), 38–54 (2016). https://doi.org/10.1016/j.ijar.2016.06.001
7. Zhi, R., Jin, Y.: Action unit analysis enhanced facial expression recognition by deep neural network evolution. Neurocomputing **425**(15 February), 135–148 (2021) https://doi.org/10.1016/j.neucom.2020.03.036
8. Tan, C., Љarlija, M., Kasabov, N.: NeuroSense: Short-term emotion recognition and understanding based on spiking neural network modelling of spatio-temporal EEG patterns. Neurocomputing **434**(28 April), 137–148 (2021). https://doi.org/10.1016/j.neucom.2020.12.098
9. Savchenkoa, A.V.: Fast inference in convolutional neural networks based on sequential three-way decisions. Inf. Sci. **560**(June), 370–385 (2021). https://doi.org/10.1016/j.ins.2021.01.068
10. Selvapriya, M., Maria, G., Materialstoday, P.: Proceedings, Available online 21 Jan (2021).
11. Dolcos, F., Katsumi, Y., Moore, M., Berggren, N., de Gelder, B., Derakshan, N., Hamm, A.O., Koster, E.H.W., Ladouceur, C.D., Okon-Singer, H., Pegna, A.J., Richter, T., Schweizer, S., den Stock, J.V., Ventura-Bort, C., Weymar, M., Dolcos, S.: Neurosci. Biobehav. Rev. **108**(January), 559–601 (2020)
12. Wu, P., Li, X., Shen, S., He, D.: Int. J. Inf. Manag. **51**(April 101978) (2020).
13. Zhu, L., Su, C., Zhang, J. Ianhai, Cui, G., Cichocki, A., Zhou, C., Li, J.: Adv. Eng. Inf. **46**(October), 101191 (2020).
14. Li, W., Zhang, Z., Song, A., Measurement. **172**(February), 108747 (2021).
15. Zhiab, R., Zhouab, C., Liab, T., Liuab, S., Jin, Y.: Neurocomputing. **425**(15 February), 135–148 (2021)
16. Fei, Z., Yang, E., Day-UeiLi, D., Butler, S., Ijomah, W., Li, X., Zhou, H.: Neurocomputing. **388**(7 May), 212–227 (2020)
17. Mannepalli, K., Narahari Sastry, P., Suman, M.: Alexandria Eng. J. **56**(4, December), 485–497 (2017)

Tensometry of Interfaces of the Working Body of Technology Machines for Reclamation Works

Aleksey Apatenko and Nadezhda Sevryugina

Abstract The characteristic of the change in technical condition on the parameters of trimetric loads is given. The simulations made it possible to identify the reparative points of the installation of tensometric sensors. It is substantiated that the boron of load data allows establishing the degree of deviation of the pairing parameters from limited, which allows carrying out corrective service impacts to ensure the health of the machine. Experimental studies were carried out on the working body of the excavator and the model of the quik-droper. Comparisons of the received deformation relationships of the two measurements carried out on different parts of the elements of the working body, in the complex of automated design, allowed to develop an algorithm for the digitization of technical systems of machines. Obtaining information about the locations of the reparation points will additionally create a data bank on the amount of external and internal loads in all connections of the machine's systems and monitor its reliability throughout the life of the machine. When recycling machines, these data allowed getting information about the residual resources of the elements for their reuse or the purposefulness of recovery, which, in turn, will ensure environmental-friendliness and economy for machines owners.

1 Introduction

Exploitation of technological machines, the functional purpose of which provides for the performance of environmental development works have a certain specificity, the factors of which reduce their efficiency.

"The reliability of machines is formed from the durability of its structural and non-constructive elements. The design elements are all the parts that are part of the design of the machine, and their communication or normal functioning is provided by non-constructive elements" [1]. The main non-constructive elements of the machines are the types of compounds of machine parts and the lubricant of friction nodes.

A. Apatenko · N. Sevryugina (✉)
Russian State Agrarian University, Moscow Timiryazev Agricultural Academy, Timiryazevskaya St., 49, 127550 Moscow, Russia

M. Shamtsyan et al. (eds.), *Robotics, Machinery and Engineering Technology for Precision Agriculture*, Smart Innovation, Systems and Technologies 247,
https://doi.org/10.1007/978-981-16-3844-2_9

"The development of machine reliability measures is based on research into the processes taking place in machine elements during operation. The machine's resource indicators are made up of the resources of its constructive and non-constructive elements and can be represented by the expression as follows:

$$R_M = \sum R_{Ki} + \sum R_{HKi} \tag{1}$$

where $R_{Ki} u R_{Hki}$ is resources, *i*-unconstructive и *i*-non-constructive elements in the car" [1].

"Modern repair technologies, used to restore the efficiency of machines, do not ensure the full restoration of the resource of constructive and non-constructive elements, so the values of these resources are reduced. The following processes contribute to the reduction of the resource: changing the parameters of parts due to wear; aging, fatigue of parts materials; changing parts parameters due to plastic deformations; changes in the physical and mechanical properties of parts materials under the influence of the environment during operation. All these processes are the result of physical and chemical processes: heat exchange, interaction of contacting surfaces of parts during the operation of the machine, wear, expiration of lubricants, oxidation and corrosion of materials and lead to a decrease in the health of machine parts" [2].

The resource of the machine is considered exhausted when reaching its limit state, i.e., based on the expression (1) of reaching the limit state of the structural and non-constructive elements forming its structure.

"The limit state of each constructive element and unconstructive element of the machine depends on its purpose, nature and magnitude of loads, working conditions and other factors and must, therefore, be considered for each of these elements" [3].

The purpose of the research is the development of a mechanism for managing the resource potency of technological machines of environmental engineering by taking into account the tribometry factor by means of digitizing the element base of technical systems.

"In the task of researching the nature of the processes of interaction of assembly units in work sites, first of all, the laws of conjugated surfaces, the establishment of degrees of freedom and their physical and chemical effects on changes in the structure of the structure are defined. Structurally, the machines appear to be a set of technical systems, the dynamics of which are provided through tribo-connections. The challenge of addressing resource management issues is to improve the quality of friction elements and reduce loads by applying appropriate materials such as oils or polymaterials viscosity characteristics which compensate for loads" [1–4].

2 Digital Technology Triboanalysis of Machine Systems

"The reliability of the machine's technical systems in places where the matched elements are frictiond is investigated from the point of view of the main provisions of the tribology, laying the metrological characteristics according to the conditions of normal functioning" [4–6].

A.P. Kuznetsov and others in the work [8] conducted research in the field of resource conservation and developed an energy information model with a basic criterion—the efficiency of the mechanical system. In this vein, the results of the development of a mechanism for managing technological changes, conducted Hsin-HuiChou and JudyZolkiewski [7].

Niko Siltala, Eeva Järvenpää, MinnaLanz justify that an integral part of resource evaluation is the concept of executable capabilities, which describes the resource management interface in a vendor-neutral form [10].

S.I. Chalaganidze, J.B. Katsitadze, G.G. Kutelia showed the importance of reliability for agrotechnicians that operate in tough soil—climatic and dynamic conditions unlike some other industry machinery and equipment [11].

Valerie C.Y. Zhu, LinyanSun, LinhuiSun, Xiaohong Li and other scientists presented a model of changing the quality of the system as a function, proportional linear function of time through a polynomial solution to minimize the total consumption of the resource within the specified restrictive limits [12–15].

Modern tensometer sensors located in the reparation points of the pairing provide the collection of information and analytical data on the indicator of tribyometry (from Greek tribos—friction and metreo—measure) [1, 2].

"The internal field will be characterized by the design, i.e., the basic condition of achieving the target function, with the laid extreme of existence: maximum and minimum and formed in the design—production period of the life cycle of the machine. The outer field is formed during the period of operation and is probabilistic, which allows the physical essence of the processes to present in accordance with the variation principle, the interpretation of which is that of all possible movements nature chooses the one at which the purpose of movement is achieved at the lowest cost of action. As you know, the speed attainment of the limit state of pairing, in particular in wear, depends on the friction that occurs when touching the matched surfaces, and friction is caused by movement, which in turn arises from the condition of having one or more degrees of freedom in the pairing" [1].

Installation of tensometry sensors and the use of a software machine collecting information ensure the digitization of technical systems at the level of elemental pairing of parts [3].

The basic provisions of tribology give a mathematical description of the pairing on the criterion of metrological consistency:

$$u_{\lim} = \alpha + \sum_{i=1}^{n-1} \alpha_i u_i \tag{2}$$

where u_i is rationed geometric pairing characteristic as a reactive technical status database information parameter; α_i is regression factor, is measured personalized and includes environmental influence factors and load characteristics; n-1 is the number of agreed-upon reparer points that provide the resource potential of connectivity.

The estimated value of resource potential is formed by the value of the period of intensity of the change in the parameter $\xi(t)$ to the level of reaching its limit value $\xi(t_{\text{lim}})$:

$$t_{\text{bat}} = t\left[\left(\frac{\xi(t_{\text{lim}})}{\xi(t)}\right)^{\frac{1}{\alpha}} - 1\right] \tag{3}$$

where α is the degree to which the nature of the change in the parameter.

3 Simulation Model of Loading Matching Elements of the Machine's Working Body

The task of the designer is to create a working equipment or working body that would meet the requirements of strength and rigidity, were reliable in operation and durable, had less metal intensity and better technical parameters.

In the first stage of the design, determine the basic configuration of the element and the ways of its attachment to the nodes of the machine and assign in the first approximation preliminary sizes and shape of the cross section, depending on the working conditions of the element in the machine, external loading, technology and labor-intensive manufacture. In the second stage of design, the strained state of the element is assessed, its strength indicators are determined, and the shape and size are adjusted.

The simulation model is presented by the example of changes in the dynamics of the attachment axes of the kvik-capler at loads caused by working equipment, in particular the installation of buckets of different volumes (Fig. 1).

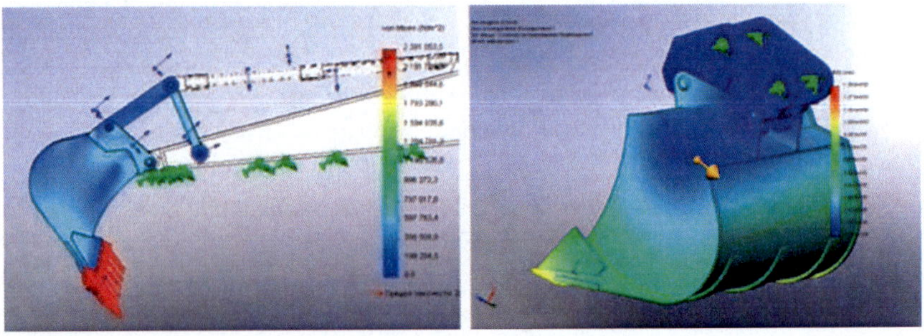

Fig. 1 Simulation of the loading mode when the excavator performs the digging process

Simulation enables, with minimal material and manpower, to conduct research to assess the effectiveness of the developed design. For the operation of the machines, it becomes important to extend the resource of the entire machine, which is provided by service impact. In simulation, places to install sensors, reparation points were identified. The collection of load data allows you to determine the degree of deviation of pairing parameters from limited, which allows corrective service impacts. Further studies have been carried out to establish the consistency of the proposed theory and experimental results. Obtaining information about the locations of the reparation points will additionally create a data bank on the amount of external and internal loads in all connections of the machine's systems and monitor its reliability throughout the life of the machine. When disposed of, this information will provide information about the residual resources of the elements for their reuse or the purposefulness of recovery, which in turn will ensure environmental-friendliness and cost-effectiveness for machines owners [15, 16].

4 Results and Discussions

The definition of a tensely deformed state in the "the hilt-kvik-capler—bucket" system used a measuring system, which includes sensitive element—tensoresiist; measuring module for dynamic measurements; PC data transmission module; smart sensor cable connection to the measuring network; software.

The structurally measuring system includes an intelligent sensor, which consists of a primary converter (tenzomost) and a measuring module ZET 7111 Tensometer-CAN [2, 3].

The tenzomost is installed at the measurement site, and the measuring module is in close proximity. At the exit of the tensomost, a voltage proportional to the impact of the measured value is formed. The measuring module converts this voltage into the values of the measured value and transmits the results digitally. Software was used to display and process measurement results ZETLab. Measurements do not require setting up a sensor and processing the results—all the necessary settings are set by the manufacturer when the smart sensor is released and stored in the memory of the module ZET 7111 and all measurements are made by a signal processor that is part of the measuring module ZET 7111 Tensometer-CAN. Intelligent tensaters ZET 7111 Tensometer-CAN with basic settings measure the values of relative deformation. Distributed measuring networks are built on the basis of intelligent sensors. To connect the measuring network to systems ZETLAB и ZETVIEW is used interface converter ZET 7174 [3].

A model with elements modeling the handle and a working organ has been created to carry out tenzometric measurements. The connectors of the structure used a two-way solid seam of the brand's joint without a slant and edges (for each wall) rolled joint 6–8 mm.

The next step was to prepare the places of fixing tensorestors: stripping areas with sandpaper and skimming.

The tensoressers used were installed on the surface next to the front pairs of the lower and upper sidewall brackets.

The resulting design is installed in a hydraulic press and is created to increase the voltage in the structure to establish the moment of deformation with ever-increasing effort.

With the help of the collected measuring network, tests of excavator working equipment were also carried out to determine the tense-deformed condition in the metal structures of the real machine at the joints of the handle and the working body.

Experimental studies have found that the technological use of the meter system discussed above is not possible due to the low sensitivity of tensorestors, and therefore, it was decided to use force as a measuring sensor. Tensometry was carried out on the example of the study of the tense-deformed state of the working organ of the technological machine, in the case of the excavator bucket using a kvik-a droper (Fig. 2).

The discrepancy of the data is 8.9%, which is within the limits of the acceptable 10%. Comparisons of the received deformation relationships of the two measurements taken on different sections of the stand allow us to conclude the validity of the studies conducted in the complex of automated design. It is also worth noting the errors in the work with tenso-measuring systems, due to possible interference with the frequency range of the current of 2.5 Hz, and take into account the possible looseness of contact of the tensoresictor with the study surface.

The results of the research made it possible to create a universal apparatus of management of the resource potential of the technical systems of the technological machine for the entire life of the environmental management operations in conditions of high humidity and dusty external environment [16–18].

5 Conclusions

- The characteristic of the change in the technical condition on the parameters of tribyometric loads, taking into account the factors manifested in the performance of reclamation work;
- The method of tribyometric control of the health of machines with the introduction of digital state accounting systems has been proposed;
- Presented algorithm of digitization of technical systems of machines, which provides management of resource potential for the entire life of their service;
- The results of the research made it possible to create a universal apparatus of management of the resource potential of the technical systems of the technological machine for the entire life of the environmental management operations in conditions of high humidity and dusty external environment [9].

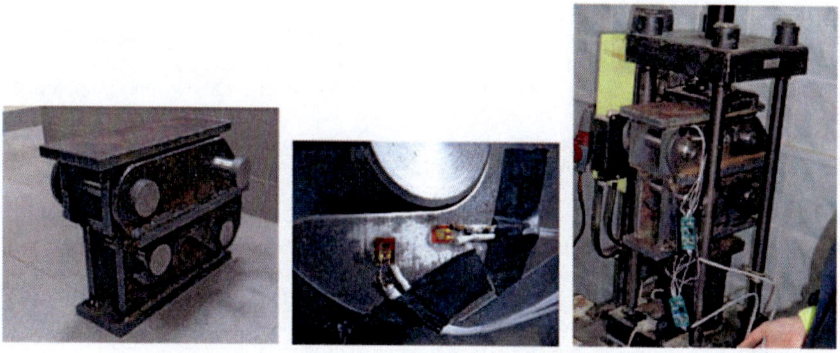

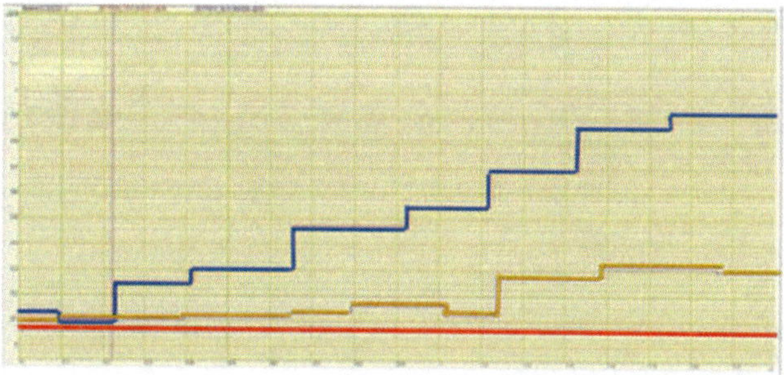

Fig. 2 Example of research into the resource potential of the working body elements of the technological machine by tendometrication by the measuring system ZETLAB sensor

References

1. Zorin, V.A., Baurova, N. I., Pegachkov, A.A.: Assessment of products risks of mechanical engineering by results of diagnosing. J Period. Eng. Nat. Sci. **7**(1), 287–293 (2019). https://doi.org/10.21533/pen.v7i1.391
2. Duganova, E.V., Sevryugina, N.S., Kapyrin, P.D.: A method constructing standard-size series of quick couplers. IOP Conf. Ser. Mater. Sci. Eng. **786**(1), 012052 (2020). https://doi.org/10.1088/1757-899X/786/1/012052
3. Sevryugina, N.: Modified method for calculation of vehicles residual lifetime with regard of the impact factors variability: EMMFT 2017. Adv. Intell. Syst. Comput. **692**, 273–281 (2017). https://doi.org/10.1007/978-3-319-70987-1_29
4. Bruzzone, A.G., Costa, H.L.: Functional characterization of structured surfaces for tribological applications?. Procedia CIRP **12**, 456–461 (2013). https://doi.org/10.1016/j.procir.2013.09.078
5. da Silva, W.M., Costa, H.L., de Mello, J.D.B.: Transitions in abrasive wear mechanisms: Effect of the superimposition of interactions. Wear, **271**(5–6), 977–986 (2011) https://doi.org/10.1016/j.wear.2011.04.010
6. Lazić, V., Arsić, D., Nikolić, R., Aleksandrović, S., Djordjević, M., Hadzima, B., Bujnak, J.: Experimental determination of deformations of the hard faced samples made of steel for operating at elevated temperatures. Procedia Eng. **111**, 495–501 (2015). https://doi.org/10.1016/j.proeng.2015.07.122
7. Chou, H.-H., Zolkiewski, J.: Managing resource interaction as a means to cope with technological change. J. Bus. Res. **65**(2), 188–195 (2012). https://doi.org/10.1016/j.jbusres.2011.05.021
8. Kuznetsov, A.P., Blau, P., Koriath, H.-J., Richter, M.: Criteria for energy-efficiency of technological processes. J. Technol. Mach. Prod. Eng. , Proced. CIRP, **46**, 340–343 (2016). https://doi.org/10.1016/j.procir.2016.04.002
9. Wang, D., Tong, X., Wang, Y.: An early risk warning system for out ward foreign direct investment in mineral resource-based enter processing multi-classifiers fusion. J. Res. Pol. **66**, 101593 (2020). https://doi.org/10.1016/j.resourpol.2020.101593
10. Siltala, N., Järvenpää, E., Lanz, M.: An executable capability concept in formal resource descriptions. IFAC-Papers On Line **51**(11), 102–107 (2018). https://doi.org/10.1016/j.ifacol.2018.08.242
11. Chalaganidze, S.I., Katsitadze, J.B., Kutelia, G.G.: The theoretical and experimental study of the ploughs' ploughshares in order to increase the reliability of resource-saving technology using similarity and dimensions theory. J. Ann. Agrar. Sci. **15**(3), 329–331 (2017). https://doi.org/10.1016/j.aasci.2017.05.024
12. Berg, S., Wustmans, M., Bröring, S.: Identifying first signals of emerging dominance in a technological innovation system: a novel approach based on patents. J. Technol. Forecast. Soc. Change **146**, 706–722 (2019). https://doi.org/10.1016/j.techfore.2018.07.046
13. Avdeenko, T.V., Mezentsev, Y.A., Estraykh, I.V.: Heuristic approach to unrelated parallel machines scheduling under availability and resource constraints. IFAC-Papers On Line **1**(50), 13096–13101 (2017). https://doi.org/10.1016/j.ifacol.2017.08.1998
14. Callaghan, C.W.: Surviving a technological future: technological proliferation and modes of discovery. Futures **104**, 100–116 (2018). https://doi.org/10.1016/j.futures.2018.08.001
15. Wang, D., Tong, X., Wang, Y.: An early risk warning system for outward foreign direct investment in mineral resource-based enterprises using multi-classifiers fusion. Resour. Policy **66**, 101593 (2020). https://doi.org/10.1016/j.resourpol.2020.101593
16. Savitskiy, M.E.: Algorithms for constructing resource-saving cutting machines. Proced. Eng. **129**, 781–786 (2015). https://doi.org/10.1016/j.proeng.2015.12.104

17. Levitin, G., Finkelstein, M., Xiang Y.: Optimal mission abort policies for repairable multistate systems performing multi-attempt mission. Reliab. Eng. Syst. Saf. **209**, 107497 (2021). https://doi.org/10.1016/j.ress.2021.107497
18. Sevryugina, N.S., Apatenko, A.S., Revyako, S.I.: Theory of the technological machine state recognition by the criterion of resource uncertainty. IOP Conf Ser.: Mater. Sci. Eng. **1001**, 012983 (2020). https://doi.org/10.1088/1757-899X/1001/1/012083

Model of Sealing Joint with Incising Ring

Sergey Kireev, Alexey Lebedev, Haydyar Kaderov, and Daniil Mihaylin

Abstract Perspective design of sealing connection with cutting ring is considered. It is noted that to date, the mutual influence of the structural parameters of the connection has not yet been studied. Formalized description of structural relationships of parts of sealing connection with cutting ring is carried out, on the basis of which parametric solid model of connection design is designed. Method of analysis of stress–strain state of parts during joint assembly has been developed. Studies of reliability, strength and weight of the sealing joint were carried out depending on a number of design parameters. Method of analysis of stress–strain state of parts during joint assembly has been developed. Description of fasteners, loads, contacts, method of obtaining finite element mesh, setting properties of materials of joint parts is given. The method of conducting studies of reliability, strength and mass of the joint has been developed depending on a number of design parameters of the joint. Multifactorial regression analysis revealed the relationships, direction and significance of the effect of structural parameters of joint parts on the parameters of reliability, strength and mass of joint parts. Recommendations have been developed for the design of sealing joint parts to achieve higher performance.

1 Introduction

Sealing joints with incising rings are successfully used in hydraulic systems of machines and equipment operating at pressures up to 120 MPa [1–3]. The strength and reliability of the sealing joint are due to the combination of the appropriate heat treatment (up to a hardness of 38 … 42 HRC) of the ring and the peculiarity of its stress–strain state during the assembly of the joint. In this case, the cutting edge of the ring contacts the less durable surface of the pipeline with the formation of a deformation contact wave (collar) along the entire circumference [4, 5].

The GOST 23354-78 (RU) standard regulates the basic requirements and dimensions of sealing joints based on rings obtained by machining by cutting. In the

S. Kireev (✉) · A. Lebedev · H. Kaderov · D. Mihaylin
Don State Technical University, Rostov-on-Don, Russia

M. Shamtsyan et al. (eds.), *Robotics, Machinery and Engineering Technology for Precision Agriculture*, Smart Innovation, Systems and Technologies 247,
https://doi.org/10.1007/978-981-16-3844-2_10

patent 2180067 RU proposed an alternative design of the incising rings, obtained by stamping from rolled pipes. The advantages of joints with incising rings, obtained by stamping, set forth in [6], sufficiently convincingly prove that it is precisely this design of the sealing joint that has prospects in hydraulic systems of machines. This is, first of all, the low-material consumption and the cost of the technology for producing the incising ring. The forming lines of the surfaces of the stamped rings smoothly merge into each other, which has a positive effect on the fatigue strength of the ring. Rip-ring couplings offer improved performance due to simplified assembly technology and improved reliability.

However, in [6] it was indicated that the use of sealing joints based on stamped rings is limited by the absence of a methodology for designing joint details. The creation of such a technique is impossible without a preliminary study of the interconnections of the parameters of the joint parts and their influence on the technical and economic indicators. Therefore, the purpose of this work is to study the relationships between the parameters of the parts of the sealing joint on the basis of computer simulation. A similar approach is successfully used by researchers, for example, in [7, 8].

2 Materials and Methods

Figure 1 shows the connection design in the initial state of the assembly. A pipeline 4 with an outer diameter d is installed with a small gap in the hole of the fitting 1 and abuts against the end of this hole so as to ensure a sufficiently reliable centering of the pipe relative to the surfaces of rotation of the fitting. The incising ring 3 along the inner diameter d mates with the gap with the outer surface of the pipeline 4 and touches the surface of the nozzle cone.

The contact of the tapered and cutting part of the ring provides the conditions for the complete positioning of the ring relative to the fitting and predetermines the

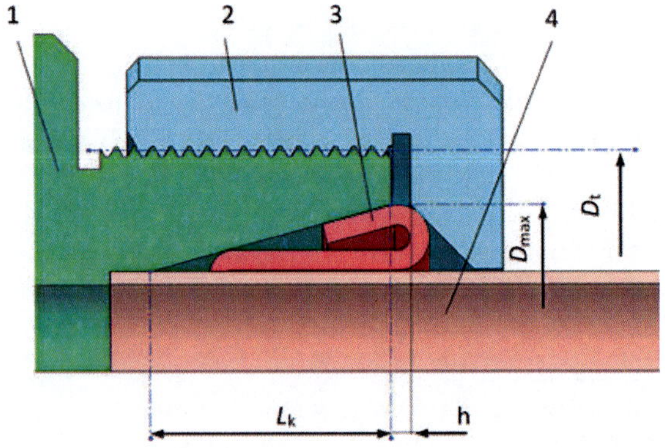

Fig. 1 The design of the sealing joint with the incising ring: 1—fitting; 2—nut; 3—incising ring; 4—pipe

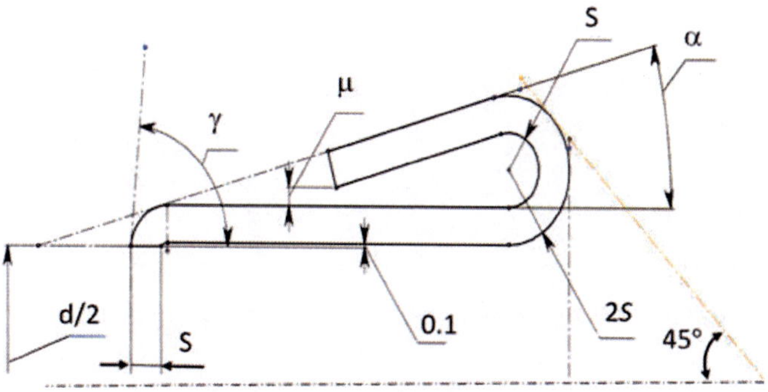

Fig. 2 Sketch of the longitudinal section of the incising ring

creation of a sealing joint as a result of the force movement of the ring when screwing nut 2.

Figure 2 shows a sketch of a longitudinal section of a ring constructed using five parameters: S—wall thickness of a tubular blank, from which a ring is obtained by stamping; α-angle of inclination of the conical surface of the ring; d is the diameter of the ring hole, equal to the outer diameter of the pipe; γ—cutting angle of the ring; μ—ring loop clearance. Size 0.1 mm—technological to ensure the conditions for free movement of the ring along the pipe before starting the assembly of the joint.

The geometric dimensions of the ring directly affect the dimensions of other connection parts. From Fig. 1, it is easy to see that the maximum diameter of the tapered hole of the fitting is taken equal to the maximum diameter of the tapered hole of the nut, as well as the maximum diameter of the ring, and is determined from the condition:

$$D_{\max} = d + 8S \tag{1}$$

and the length of the tapered hole of the nozzle with sufficient accuracy for practice is determined from the expression:

$$L_k = 2S \times (2/tg\alpha + 1) \tag{2}$$

The diameter of the thread of the union and the nut according to the condition of the structural rigidity of the connection is

$$D_t = D_{\max} + 4S \tag{3}$$

and is taken equal to the nearest standard value.

In the initial state of the assembly, when the tapered surface of the ring touches the tapered surface of the fitting and the cone of the nut touches the surface of the ring with a radius of $2S$, a distance $h > S$ must be maintained between the ends of the fitting and the nut to enable the force assembly of the connection.

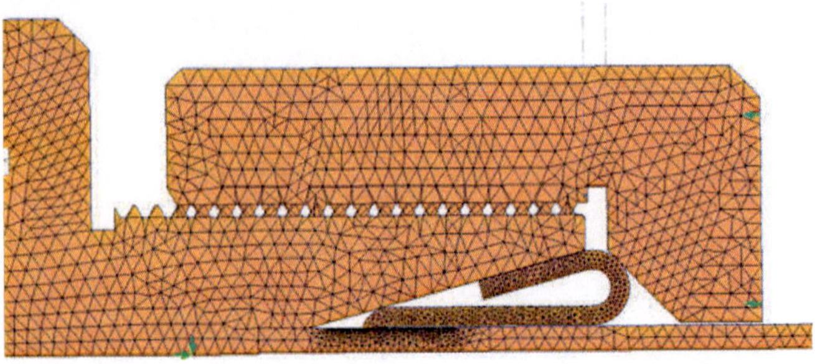

Fig. 3 Finite element mesh in the model of sealing joint with incising ring

The above ratios formed the basis for a parametric solid model of a sealing joint with incising ring.

The analysis of the stress–strain state of parts during assembly of the joint was carried out in the SOLIDWORKS system with the SIMULATION module connected. Study type—static nonlinear. A 2D simplification of the model is applied on the assumption of a plane stress–strain state of the joint parts under conditions of axisymmetric loading. The parameters of the loading scheme were set based on the study of the problems of mechanics presented in [9–14].

Figure 3 shows the conditions for fastening the elements of the sealing joint parts. The points of the generatrix of the central hole of the nozzle are fixed in the model. Load—longitudinal movement of the right end of the nut, simulating its movement when assembling the sealing joint parts. The amount of movement is equal to the distance between the ends of the union and the nut in the initial state of the assembly (see parameter h in Fig. 1).

The contact interaction between the nut and union components allows mutual penetration. Contact interactions between the surfaces of the ring, pipe and fitting are prescribed in the absence of mutual penetration under conditions of friction when surfaces are in contact. In the area of contact interaction between the pipe and the ring in the place of possible cutting, the finite element mesh control is applied. For the ring, as the most loaded connection element, the grid control is also applied.

The material of the fitting and the nut is alloy steel with the standard parameters of the linear elastic isotropic model. The pipe material is carbon steel with the parameters of the von Mises plasticity model.

The authors had to pay special attention to the choice of the parameters of the model of the material of the incising ring, which is in a state of volumetric compression at the moment of completion of the assembly.

In examples of practical application of this type of joints [6], it is known that the material of the ring is a heat-treated alloy steel of the 40X type according to GOST 4543-2016 with the parameters of the yield point of the order of 750–800 MPa. Therefore, initially, the ring material was described using a linear elastic isotropic model. In this case, even with the "Pipe" component excluded from consideration, stresses in the ring walls exceeding several thousand MPa arose (Fig. 4a).

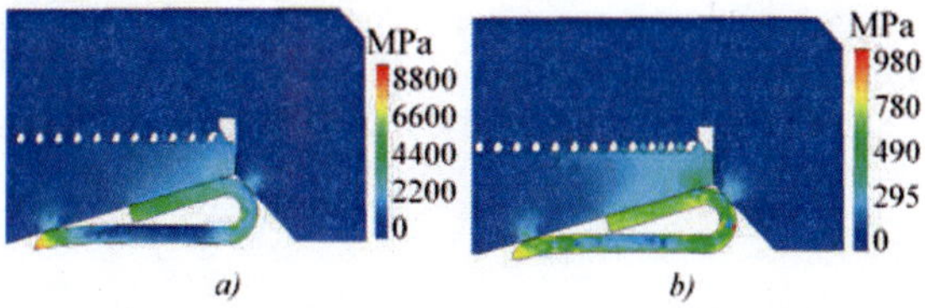

Fig. 4 Diagrams of von Mises stresses during radial compression of the ring in the absence of contact with the pipeline. Ring material model: **a** linear elastic; **b** nonlinear elastic

In the case of using the model of nonlinear elastic behavior of the ring material (Fig. 4b), the calculated stresses in the longitudinal section fully satisfy the conditions for the strength of the joint parts—the stresses do not exceed the yield stress of the material of the parts. It should be noted that the material "Alloy Steel" and the nonlinear elasticity model were taken from the standard library of the CAD system developer, that is, and the "Stress–Strain" curve was not created or corrected by us. The need to use nonlinear elastic models of materials in such problems is also indicated in [15, 16].

Figure 5 shows the stress diagram obtained during the assembly of the sealing joint with the incising ring for $d = 16$ mm; $S = 1.2$ mm; $\alpha = 15°$; $\gamma = 75°$; $\mu = 0.5$ mm. The nut travel was $h = 1.5$ mm.

To measure the target parameters of the study, four sensors are introduced into the model: $D1$ and $D2$—maximum and minimum radial displacement of the outer surface of the pipe, mm; $D3$ is the average stress of Von Mises in the ring, MPa; $D4$—assembly mass, kg.

A parametric model of an incising ring sealing joint was investigated in order to determine the influence of design parameters on the reliability and strength of the joint.

The following parameters were selected as target research parameters:

$Y1 = |D1| + |D2|$—height of the collar formed by cutting the ring into the pipe body when assembling the joint, mm;

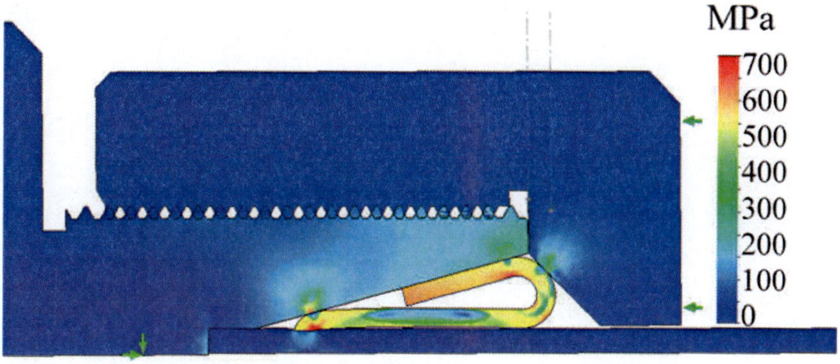

Fig. 5 Von Mises stress plot during assembly of a sealing joint with incising ring

$Y2 = D3$—average value of stress intensity in the longitudinal section of the ring, MPa;

$Y3 = D4$—total mass of a sealing joint parts, kg.

3 Results

The experimental design and the values of the target parameters obtained during the experiments are shown in Table 1.

Table 1 Research plan and data of the model

№	X_1	X_2	X_3	Y_1	Y_2	Y_3	h	D
1	1	10	60	0.081	302	1.61	1.29	32
2	1.4	10	60	0.102	254	2.55	1.69	36
3	1.8	10	60	0.116	303	4.78	2.09	44
4	1	20	60	0.117	463	0.88	1.33	32
5	1.4	20	60	0.146	410	1.47	1.8	36
6	1.8	20	60	0.157	432	2.51	2.27	42
7	1	30	60	0.09	473	0.68	1.46	32
8	1.4	30	60	0.115	474	1.135	2.01	36
9	1.8	30	60	0.12	403	1.94	2.56	42
10	1	10	75	0.099	281	1.54	1.29	32
11	1.4	10	75	0.133	226	2.55	1.69	36
12	1.8	10	75	0.156	291	4.78	2.09	44
13	1	20	75	0.141	521	0.88	1.33	32
14	1.4	20	75	0.168	485	1.47	1.8	36
15	1.8	20	75	0.168	413	2.51	2.27	42
16	1	30	75	0.118	581	0.675	1.6	32
17	1.4	30	75	0.116	501	1.13	2.01	36
18	1.8	30	75	0.117	587	1.94	2.6	42
19	1	10	90	0.103	140	1.54	1.29	32
20	1.4	10	90	0.162	206	2.55	1.69	36
21	1.8	10	90	0.165	285	4.78	2.09	42
22	1	20	90	0.157	362	3.76	1.33	32
23	1.4	20	90	0.178	432	1.47	1.8	36
24	1.8	20	90	0.191	383	2.51	2.27	42
25	1	30	90	0.123	555	0.675	1.46	32
26	1.4	30	90	0.109	501	1.13	2.01	36
27	1.8	30	90	0.099	516	1.94	2.8	42

The height of the collar $Y1$ is an important indicator that determines the reliability of the joint functioning, since an underestimated value of this parameter can lead to depressurization and a decrease in the strength of the cut-in joint between the ring and the pipe.

The average stress intensity $Y2$ characterizes the strength of the ring material subjected to significant compressive loads during assembly of the joint. The choice of the mean value is due to the presence of several "singular points" at which, due to insurmountable errors in dividing the section of the part into finite elements, peak stress values are observed, distorting the picture of the overall uniform stress–strain state of the ring material.

The parameter of the total mass of the parts of the joint $Y3$ characterizes the material consumption of the joint structure, the analysis of which will make it possible to make informed decisions in the technical and economic analysis of the joint structure.

The main target parameter when evaluating the design of the sealing joint in our studies was chosen the parameter of the sealing joint reliability or the height of the collar $Y1$.

Variable research parameters are: $X1$—incising ring thickness from 1 mm to 1.8 mm; $X2$—the angle of inclination of the ring cone within the range from 10° to 30°; $X3$—cutting angle of the ring in the range from 60° to 90°;

The indicated limits of variation of the factors X were obtained on the basis of preliminary studies of the connection, in which the possibility of the existence of part models without violating the relationships between the assembly elements was established.

A number of parameters of the model in the studies were deliberately not subjected to change and were considered as experimental conditions: Z_1—outer diameter of the pipe (16 mm); Z_2—ring loop gap (0.2 mm).

Data processing was carried out using multivariate regression analysis. Factors X_1–$X3$ were subjected to coding operations in order to correctly determine their significance. The results of data processing are shown in Table 2.

Student's calculation criterion for checking the significance of the model coefficients by condition $t_r > t0.95 = 1.7$. Checking the adequacy of the conditional regression model $F_r > F0.95$ (Table 2).

The study of the multifactorial regression dependence of the collar height allows us to draw the following conclusions. The model with a probability of 0.95 adequately describes the obtained data of the numerical experiment, since the value of the calculated Fisher Criterion $F_r = 17.9$ turned out to be higher than the tabular $F\ 0.95 = 2.21$.

4 Discussion

All the coefficients of the regression model with a probability of 0.95 according to the Student's test turned out to be significant; therefore, the collar height during

Table 2 Results of the multivariate regression analysis

Regression equation coefficients	Target parameters of object studies					
	Y_1 (mm)		Y_2 (MPa)		Y_3(kg)	
	Values	t_p	Values	t_p	Values	t_p
B_0	0.167	119.8	454.37	57.25	1.49	11.28
$B_1 \cdot X_1$	0.0144	48.38	−3.61	2.12	0.858	30.2
$B_2 \cdot X_2$	−0.0061	20.5	127.9	75.2	−0.857	30.17
$B_3 \cdot X_3$	0.0135	45.2	−7.44	4.37	0.155	5.47
$B_4 \cdot X_1X_2$	−0.0124	27.7	−21.58	8.46	−0.488	11.45
$B_5 \cdot X_1X_3$	−0.0028	6.1	18.92	7.41	−0.23	5.48
$B_6 \cdot X_2X_3$	−0.0104	23.26	37.5	14.7	0.005	0.12
$B_7 \cdot X_1X_2X_3$	−0.0101	15.07	−14.12	3.69	−0.0081	0.127
$B_8 \cdot X_1^2$	−0.0078	8.68	17.38	3.41	0.501	5.87
$B_9 \cdot X_2^2$	−0.0401	44.78	−51.2	10.05	0.167	1.958
$B_{10} \cdot X_3^2$	−0.0056	6.26	−48.77	9.56	0.164	1.928
$F_{0,95} = 2.21$	$F_r = 17.9$		$F_r = 9.27$		$F_r = 3.42$	

assembly of the joint is significantly influenced not only by the factors $X1$–$X3$, but also by their interactions and squares. With an increase in ring thickness $X1$ and cutting angle $X3$, as well as with a decrease in the angle of the tapered part of the ring, the collar height $Y1$ increases. The most significant direct impact on this target is the wall thickness and the cutting angle. However, the effect of the square of the cone angle of the $X2$ ring is also extremely significant.

The study of the regression model of the mean stress in the ring is carried out. The model with a probability of 0.95 adequately describes the obtained data of the numerical experiment ($F_r = 9.27 > F0.95 = 2.2$). All the coefficients of the regression model with a probability of 0.95 turned out to be significant, therefore, the strength of the ring is significantly influenced not only by the factors $X1$–$X3$, but also by their interactions and squares. All the coefficients of the regression model with a probability of 0.95 turned out to be significant, therefore, the strength of the ring is significantly influenced not only by the factors $X1$–$X3$, but also by their interactions and squares. With an increase in the ring thickness $X1$ and the cutting angle $X3$, as well as with a decrease in the angle of the tapered part of the ring, the average stress in the ring wall $Y2$ decreases. The most significant direct impact on this target is the taper angle. The joint effects of the $X2X3$ interaction and the square of the cone angle of the $X22$ ring are recognized as strong enough.

The study of the third target parameter $Y3$—the masses of the joint details showed that the regression model is adequate with a probability of 0.95 ($F_r = 3.42 > F\ 0.95 = 2.21$).

Factors $X1$–$X3$ and interactions of the form $X1X2$, $X1X3$. $X22$ are recognized as significantly influencing the target parameter. It is interesting to note that the direct effect of the wall thickness factor practically coincides in significance with the inverse effect of the ring cone angle factor.

5 Conclusions

1. Perspective design of sealing connection with incising ring is considered. It is noted that to date, the mutual influence of the structural parameters of the connection on the strength, reliability and material capacity of the connection has not yet been studied.
2. Formalized description of structural relationships of parts of sealing joint with incising ring is carried out, on the basis of which parametric solid model of connection design is designed.
3. Method of analysis of stress–strain state of parts during joint assembly has been developed. Description of fasteners, loads, contacts, method of obtaining finite element mesh, setting properties of materials of joint parts is given.
4. The method of conducting studies of reliability, strength and mass of the joint has been developed depending on a number of design parameters of the joint.
5. Multifactorial regression analysis revealed the relationships, direction and significance of the effect of structural parameters of joint parts on the parameters of reliability, strength and mass of joint parts.
6. In order to improve the reliability of the sealing joint with incising ring, it is necessary to increase the wall thickness and the cutting angle of the ring. The angle of the contact surface of the ring with the nozzle should be taken with the lowest possible value.
7. The strength of the incising ring is determined by its angle of the cone part, the increase in which significantly reduces the level of stress in the ring wall.
8. To reduce the weight of the joint parts, reduce the wall thickness of the ring and increase the angle of its conical part.

References

1. Efimov, N.N., Papin, V.V., Bezuglov, R.V.: Micro energy complex based on wet-steam turbine. Procedia Eng. **150**, 324–329 (2016). Impact-factor Scopus 0238. https://doi.org/10.1016/j.proeng.2016.07.022.
2. Kireev, S.O., Nikishenko, S.L., Vasilev, B.N., Vasilev, M.A.: Analysis of operating conditions of sliding friction units of driving system of high-pressure oil and gas well service plunger pumps. Chem. Pet. Eng. **52**(5–6), 332–338 (2016)
3. Lukyanov, V.F., Lukyanov, A.A., Osadchiy, V.G.: Unions for repair of damaged sections of a main pipeline. ARPN J. Eng. Appl. Sci. (2016). https://www.scopus.com/inward/record.url?eid=2-s2.0-85006293449&partnerID=40&md5=12499a8a9144606da1913cf0e8a0020c

4. Kaderov, K.K., Kireev, S.O., Korchagina, M.V., Osipuk, A.Y.: Improved seals in pipelines and high-pressure hoses. Russ. Eng. Res. **38**(3), 241–243 (2018). https://doi.org/10.3103/S1068798X18030061
5. Kaderov, K., Kireev, S., Korchagina, M., Lebedev, A.: Determination of technological parameters and tool for flanging holes from sheet blanks. In: E3S Web of Conferences, 164, paper № 14019 (2020)
6. Kaderov, K.K., Korchagina, M.V., Kireev, S.O., Nikishenko, S.L.: Solution of problems of gasketing of joints of units of machines and equipment of oil and gas enterprises when in use in high- and low-temperature regimes. Chem. Pet. Eng. **54**(3–4), 188–192 (2018)
7. Efimov, N.N., Papin, V.V., Bezuglov, R.V.: Determination of rotor surfacing time for the vertical microturbine with axial gas-dynamic bearings. **150**, 294–299 (2016). Impact-factor Scopus 0238. https://doi.org/10.1016/j.proeng.2016.07.006
8. Ovchinnikov, N.A., Zhigulsky, V.I., Kozyreva, E.A., Chefranova, O.V.: Modeling and estimate of the strain-stress state of a bus body pillar with account for geometrical nonlinearity. ARPN J. Eng. Appl. Sci. (2017) https://www.scopus.com/inward/record.url?eid=2-s2.0-85015224140&partnerID=40&md5=8d5793552517df70f8f493a4689ec682.
9. Kolesnikov, A.M.: Inflation of a cylindrical membrane partially stretched over a rigid cylinder. Adv. Struct. Mater. (2019) https://www.scopus.com/inward/record.url?eid=2-s2.0-85073154642&partnerID=40&md5=64775f45acd8f677dbdb7b94ee62e52a.
10. Sadyrin, E.V., Vasiliev, A.S., Volkov, S.S., Mitrin, B.I., Aizikovich, S.M.; Simplified analytical solution of the contact problem on indentation of a coated half-space by a spherical punch. WIT Trans. Eng. Sci. (2019). https://www.scopus.com/inward/record.url?eid=2-s2.0-85063547769&partnerID=40&md5=f392b984cd99af2e74896cb5ce29985a
11. Volkov, S.S., Vasiliev, A.S., Sadyrin, E.V.: Contact problem on indentation of an elastic half-plane with functionally-graded coating in presence of tangential stresses on the surface, MATEC Web of Conferences (2018) https://www.scopus.com/inward/record.url?eid=2-s2.0-85056533508&partnerID=40&md5=5d8ce512c4ecb91671bdcb956d007e6d.
12. Nosachev, S.V., Chuveiko, M.V., Zhukova, M.Y., Kuzin, G.A., Khozyaev, I.A.: Effects of deformation dynamics in problems of heavy transport. In: MATEC Web of Conferences (2018). https://www.scopus.com/inward/record.url?eid=2-s2.0-85056529427&partnerID=40&md5=ea90ce74fbdf0f53ebdfeedaf9b80d59
13. Tyurikov, E.V.: One case of extended boundary value problem of the membrane theory of convex shells by I. N. Vekua, Problemy Analiza (2018) https://www.scopus.com/inward/record.url?eid=2-s2.0-85056131172&partnerID=40&md5=a25351cd28707fedf5e2177bf0028152
14. Beskopylny, A., Kadomtseva, E., Strelnikov, G.: Bending of a wavy plate of a periodic profile on an elastic foundation. In: IOP Conference Series: Materials Science and Engineering (2018) https://www.scopus.com/inward/record.url?eid=2-s2.0-85049393856&partnerID=40&md5=b5aaa5cce660e9427021366607060b10
15. Ipatov, A.A., Belov, A.A., Amenitskiy, A.V.: Study of viscoelastic parameter influence on dynamic response in poroviscoelastic prismatic solid. Mater. Physics. Mech. (2016). https://www.scopus.com/inward/record.url?eid=2-s2.0-84998590093&partnerID=40&md5=61081eebad47821a46d24a51b26df10f.
16. Ohrimenko, O.I., Maltsev, I.M., Rokotyanskaya, V.V., Vilisova, M.L.: The theory of nonlinear systems as an instrument for solving engineering problems. In: MATEC Web of Conferences (2018). https://www.scopus.com/inward/record.url?eid=2-s2.0-85056543077&partnerID=40&md5=04774d6461ea13e071bf09ce0e97cd68

Optimization of Hopper System of Pump Unit on Automotive Chassis

Alexey Lebedev, Sergey Kireev, Marina Korchagina, and Valentin Stepanov

Abstract A preliminary parametric model of the bunker system was developed, on the basis of which a pumping unit was arranged on the chassis of the KAMAZ car, used for the transportation and supply of powdered material during plugging operations at oil and gas wells. The target parameters of the volume, weight and dimensions of the hopper were examined when the conditions for its placement on the truck chassis were met. The model is built on nine basic parameters in the top-level assembly, whose values are passed to the lower-level part and subassembly models. Examples of operation of parametric model of bunker system are given. The methodology was developed and studies were carried out on the influence of five variable parameters of the bunker system on the target parameters of the structure, among which the hopper volume parameter was mainly recognized, and others play the role of constraints. Constructed regression models adequately describe the relationships of target and variable model parameters. The task of multidimensional optimization is set and solved, as a result of which the composition of optimal parameters of the bunker system with maximum useful volume is determined.

1 Introduction

In the construction or overhaul of oil and gas wells in accordance with RD 39-1-149-79 (RU), drilling, plugging, washing, forcing solutions, buffer fluids, hydraulic fracturing fluids and bottom-hole treatment fluids are widely used. The preparation of solutions and liquids requires the presence of bunker, pump, mixing equipment for the storage and transportation of powdered materials as part of mobile units.

In the process of designing mobile units, the problems of optimal arrangement of equipment on the vehicle chassis are solved taking into account restrictions on overall and weight characteristics of a specific model of a motor vehicle [1–3]. The use of computer solid-state models of the plant significantly expands the possibilities

A. Lebedev (✉) · S. Kireev · M. Korchagina · V. Stepanov
Don State Technical University, Rostov-on-Don, Russia

M. Shamtsyan et al. (eds.), *Robotics, Machinery and Engineering Technology for Precision Agriculture*, Smart Innovation, Systems and Technologies 247,
https://doi.org/10.1007/978-981-16-3844-2_11

for obtaining an optimal result on the overall and weight characteristics of the entire unit [4, 5].

Design issues of process machines, including hopper systems for storage and transportation of liquid and powdered materials, discussed in works [6–8], necessarily include problems of increase of efficiency of these units. It is noted that the search for the optimal design option is always based on the principle of multivariable design, which allows using computer modeling to determine the most effective options at an early stage of design, significantly reducing the time for field tests [9–13]. Examples of computer modeling of technical objects are considered in works [14–16].

The purpose of the study is to optimize the useful volume of the hopper for storage and transportation of loose mixtures based on the analysis of the relationships of the design parameters of the hopper system as part of the KAMAZ chassis pump plant.

The tasks of the study are to develop solid-state models of the pump unit equipment, to arrange the units on the chassis of the KAMAZ car. The model of the hopper for transportation and supply of powdered materials should be carried out taking into account the target and changing parameters, the system of which is to be developed. Develop a methodology for multivariable design of bunker systems of pump units.

2 Materials and Methods

The pump unit (Fig. 1) is arranged on the chassis of the KAMAZ 43118-0003079-46 (1), which is equipped with a 220 kW engine, a ZF 9S1310 gearbox and a 10-speed manual transfer box 65,112. Power take-off gearbox of PTO MP29-4201918

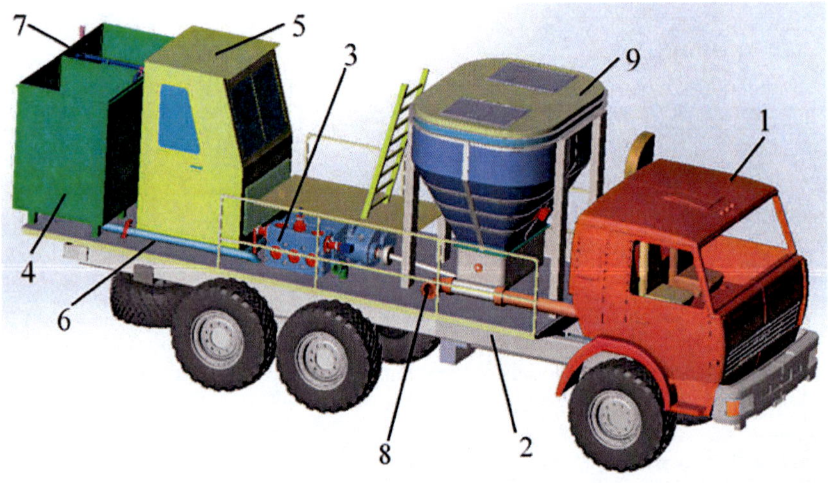

Fig. 1 Draft model of pump unit: 1—*KAMAZ* 43118-0003078-46; 2—subframe; 3—three-plunger pump *TWS*-600; 4—measuring capacity 4 m^3; 5—operator's cabin; 6—dispensing manifold; 7—loading manifold; 8—high-pressure air duct; 9—hopper system

is installed on gearbox for connection of plunger pump drive (3). Subframe (2) is installed on chassis frame through 10 mm gasket from conveyor belt. Power element of subframe is channel bracket 20P GOST 8240-97(RU). The subframe is attached to the chassis frame using clamps.

All process equipment of the unit is installed on the subframe: plunger pump TWS-600 with drive through the cardan shaft from KOM MP29; measuring (4) capacity 4 m^3 divided into two compartments with dispensing manifold (6), loading manifold (7) with a system of bottom valves and gate valves; operator's cabin (5) with protective screen on the windows on the side of plunger pump; operator bridges for servicing and monitoring the liquid level of the measuring tank and the operating parameters of the plunger pump; hopper (9) equipped with screw discharge device with hydraulic drive. Hopper device is unloaded by means of high-pressure air duct (8) connected to automobile compressor.

Consider the task of determining the maximum allowable weight and position of the center of gravity of the bunker plant. To do this, we will arrange previously designed equipment with known mass characteristics on the chassis of the car, as shown in Fig. 2. The unknown place allocated for the bunker installation is depicted by a rectangle with a center of gravity at the intersection of its diagonals.

Enter the coordinate system of the car, the origin of which is located at the point of contact of the front wheels with the ground. Axis X is directed along the car, axis Y is vertically up.

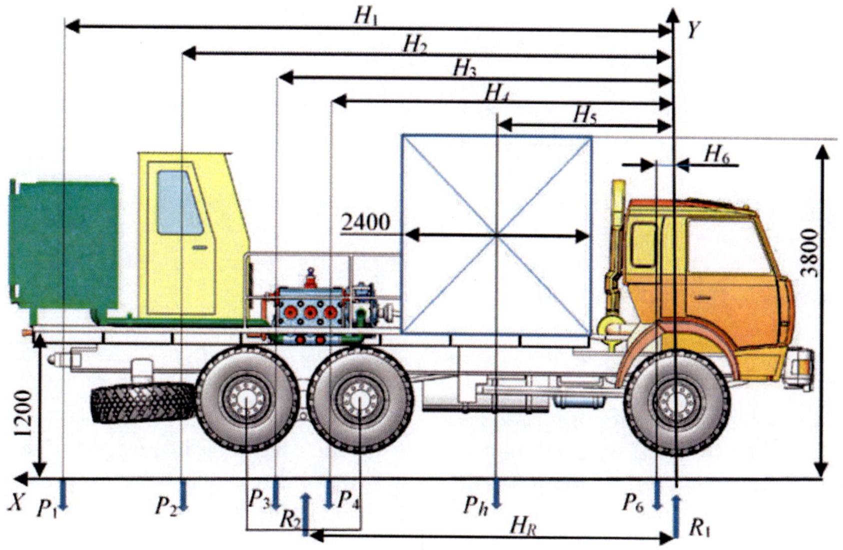

Fig. 2 Load diagram to pump unit weight calculation

Table 1 Weight distribution of installed equipment on chassis

Equipment name	Symbol in Fig. 2	Weight P (kg)	Position of center of gravity H (mm)
Measuring capacity and manifolds	P_1	893	7746
Operator's cabin	P_2	368	6328
Subframe	P_3	1692	5095
Pump *TWS*600	P_4	1948	4685
Hopper system	P_h	Define!	2800
Crew	P_6	150	315

Weights and coordinates of gravity centers of equipment of pump unit based on chassis KAMAZ-43118 are given in Table 1.

Analysis of technical characteristics of KAMAZ 43118-0003078-46 vehicle shows that maximum value of support reactions from impact of only installed pump station equipment on front wheels is R1max = 1005 kg, for rear wheel trolley R2max = 11490 kg.

We present in the form of functions the full weight of the P_0 add-in, the reactions of the supports R_1, and R_2, the argument of which is the weight of the hopper system RB.

The P_0 value is defined from the expression:

$$P_0(P_h) = P_1 + P_2 + P_3 + P_4 + P_h + P_6. \quad (1)$$

From the equation of the moments concerning a point of action of reaction of force of R_1, we obtained dependence for reaction of R_2 (Ph)

$$R_2(P_h) = (P_1H_1 + P_2H_2 + P_3H_3 + P_4H_4 + P_hH_5 + P_6H_6)/H_R \quad (2)$$

From the equation of balance of projections of all forces to axis Y, we obtained dependence for reaction of R_1 (Ph)

$$R_1(P_h) = P_1 - P_2 - P_3 - P_4 - P_5 - P_h - R_2 \quad (3)$$

Consider the behavior of functions (1), (2), (3) in graphs (Fig. 3).

It can be seen that the limiting parameter affecting the allowable weight of the hopper Ph is the load on the front wheels R_1. At point A at Ph about 6 tons, the calculated reaction R_1 becomes greater than the allowable R1max. At point C, the total weight of the mounted equipment begins to exceed the permissible value of P0max. Finally, at point B, with a bin weight of about 8 tons, the load on the rear wheel trolley becomes unacceptable.

To accurately determine the allowable weight of the hopper, we solve the system of Eqs. (2) and (3) in which R_1 is substituted instead of R1max(Ph). Omitting the

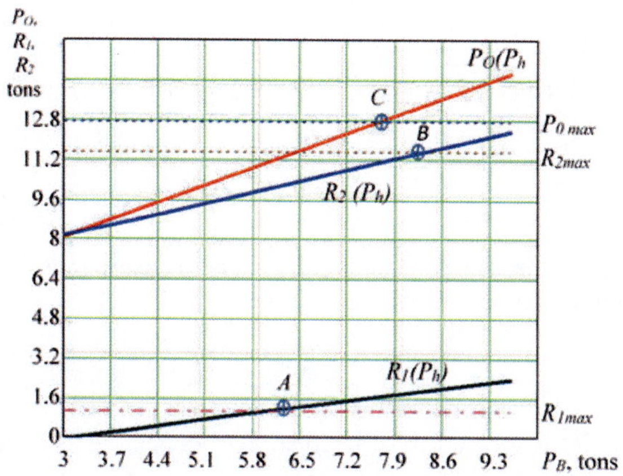

Fig. 3 Plots of influence of PB hopper weight on KAMAZ vehicle weight characteristics

intermediate calculations, we will present the final result for the maximum allowable hopper weight, at which all the necessary requirements for the weight parameters of the pump unit Ph = 5.929 tons are met.

Figure 4 is a diagram for determining the volume of the hopper. We accept the following restrictions on the numerical values of the parameters of the hopper system:

Parameter C is the width of the top of the bin. According to the condition of overall dimensions of the car

$$C = A + 2H_1 tg(a) \leq 2500 \text{ mm} \tag{4}$$

Parameter D is the length of the top of the bin. From condition of KAMAZ vehicle chassis location

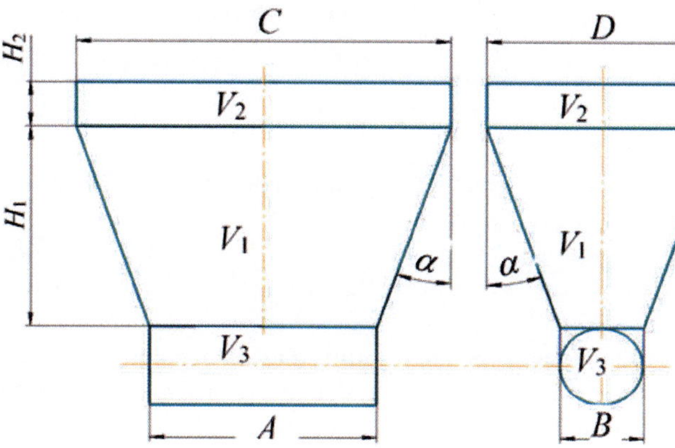

Fig. 4 Diagram for bunker volume determination

$$D = B + 2H_1 tg(a) \leq 2400 \text{ mm} \tag{5}$$

Parameter $H_1 + H_2 + B$—height of bunker system limited by vehicle height dimension

$$H_1 + H_2 + B < 2600 \text{ mm} \tag{6}$$

On the recommendation from [7] to ensure loose cement powder, we accept the limitation on the ratio of heights:

$$H_2/H_1 < 1 \tag{7}$$

Rational angle α of inclination of hopper walls depends mainly on coefficient of external friction of loose material and can be found by formula [7]:

$$\alpha = \arccos(\sqrt{4 \cdot f^4 + 1} - 2 \cdot f^2) \tag{8}$$

where $f = 0.47 \ldots 0.53$ external friction coefficient for dry cement mixture.

When designing hopper system, the angle of inclination of the walls should be taken slightly higher than the design value to take into account the effect of the coefficient of friction due to corrosion of the walls of the hopper and sticking the residues of bulk material to them. The angle of inclination to the vertical that allows the material to move depends on the shape of the outlet. Based on these recommendations, we accept the following limitation:

$$25^\circ < \alpha < 35^\circ \tag{9}$$

When assigning the parameters of the hopper, we also take into account the recommendation from [7] on the condition of the ratio of the length and width of the hopper funnel:

$$A/B \geq 3 \tag{10}$$

To limit the diameter of the screw, the dimensions of the day off of branch pipe of the hopper are accepted

$$B < 0.3 \text{ m} \tag{11}$$

Limitations (4)–(11) are applicable in setting the task of optimizing the parameters of the hopper in order to find its maximum volume. Define the objective function of the hopper system volume as the sum of the three terms (Fig. 4):

$$V_h = V_1 + V_2 + V_3 \tag{12}$$

where V_1, V_2, V_3 are volumes of hopper part conical, cylindrical and funnel part, m_3, respectively.

The volume of V_1 is determined by the formula of the volume of the truncated pyramid in the form:

$$V_1 = \frac{1}{3}H_1(S_1 + \sqrt{S_1 \cdot S_2} + S_2) \tag{13}$$

where S_1 and S_2 are areas of small and large sections of the pyramid, m_2, respectively

$$S_1 = A \times B; \qquad S_2 = C \times D = (A + 2H_1 tg(a)) \times (B + 2H_1 tg(a)) \tag{14}$$

The volume of V_2 is determined by the formula

$$V_2 = S_2 \times H_2. \tag{15}$$

Volume V_3 is defined as the volume of the cylinder with diameter B and length A by the formula:

$$V_3 = p \times A \times B^2/4 \tag{16}$$

To control the weight limitation, we introduce another restriction into the task:

$$\rho V_h + P_{\text{ust}} < [V_h] = 5929 \text{ kg} \tag{17}$$

where ρ is dry cement density (1100 kg/m^3); Pust-weight of cement-unloaded bunker system kg.

Thus, we obtain a multivariate optimization problem with an objective function (12) and constraints (4)–(11), (17).

These relationships formed the basis of the solid-state parametric model of the bunker system created in the CAD system (Fig. 5).

In total, 9 basic parameters are defined in the hopper system assembly model, on which all other dimensions of parts and assemblies depend. The list of parameters and their purpose are given below (the current values for the model in Fig. 5 are indicated in parentheses): t_list—thickness of sheet metal of hopper system skin (5 mm); Ugol—angle of cone part of hopper body (28°); SHIRINA—the width of the funnel part of the hopper and the diameter of the screw (300 mm); DLINA—hopper funnel length (1200 mm); VISOTA—height of the cone part of the hopper body (1320 mm); LUKA—silo cover manhole width (800 mm); LUKB—length of hopper cover hatch (600 mm); t_obod—thickness of ribs and beams of hopper system housings (10 mm); Visota_Cil—height of the cylindrical part of the hopper body (440 mm).

To demonstrate the capabilities of the parametric model of the hopper system, Fig. 6 shows several design options with changed parameters in various ways.

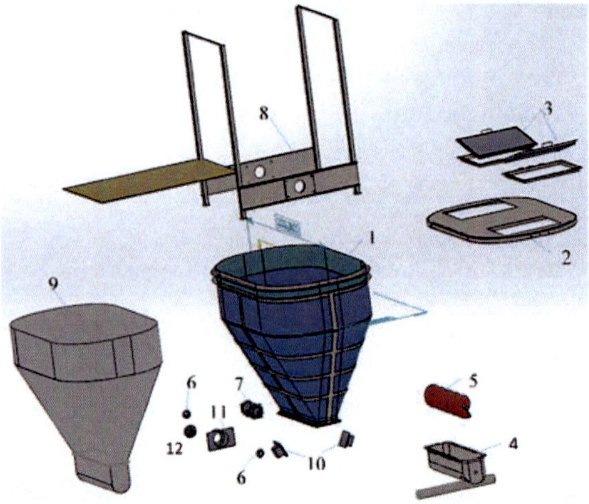

Fig. 5 Sketch model of hopper system. View with spaced parts: 1—hopper body; 2 bin cover; 3—cover panel; 4—screw housing; 5—screw; 6—screw supports; 7—hydraulic motor; 8—silo rack; 9 is a body of cement powder. 10—hopper vibrator; 11—motor bracket; 12—coupling

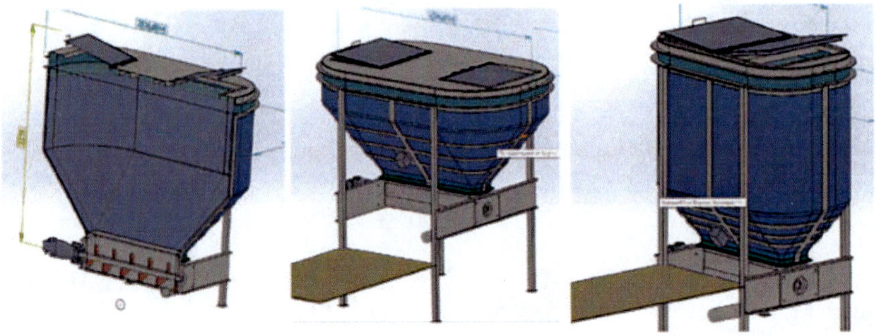

Fig. 6 Design variants of the bunker system obtained by changing the model parameters

The solid 3D model of the bunker allows you to ensure the process of multivariable design, while improving the weight and volume characteristics of the study object, as well as apply strength analysis of the metal structures of the bunker.

A priori ranges of variation of each factor X are established, at which the parametric model of the hopper remains operable. Table 2 shows the composition and limits of change of each factor. The number of factor levels, assuming the linear form of the influence of factors on the target function, is proposed to choose two—the minimum and maximum level. This means that the number of calculated experiments for five factors will be $N = 25 = 32$.

The plan for conducting experiments implies a complete review of all levels of all varying factors. The plan is developed and presented in Table 3.

The values of the variable factors X_1–X_5 are set directly in the table of parameters of the upper assembly of the hopper system model.

Table 2 Composition and ranges of varying factors

No	Name of factor	Limits variations	Units of measure
1	X_1—funnel part length	900–1100	mm
2	X_2—width of funnel part	250–300	mm
3	X_3—angle of cone part	25–30	degree
4	X_4—height of conical part	1000–1400	mm
5	X_5—height of cylindrical part	300–600	mm

Table 3 Plan and data of experimental studies (N1–13)

No	X_1	X_2	X_3	X_4	X_5	Y_1	Y_2	Y_3	Y_4	Y_5
1	900	250	25	1000	300	1.48	2512	1276	2000	1870
2	1100	250	25	1000	300	1.69	2825	1276	2000	2070
3	900	300	25	1000	300	1.57	2645	1276	2000	1870
4	1100	300	25	1000	300	1.8	2968	1326	2000	2070
5	900	250	30	1000	300	1.87	3031	1482	2000	2076
6	1100	250	30	1000	300	2.12	3378	1482	2000	2276
7	900	300	30	1000	300	1.97	3169	1532	2000	2076
8	1100	300	30	1000	300	2.23	3531	1532	2000	2276
9	900	250	25	1400	300	2.72	4114	1643	2560	2243
10	1100	250	25	1400	300	3.07	4573	1649	2560	2443
11	900	300	25	1400	300	2.85	4294	1699	2560	2243
12	1100	300	25	1400	300	3.22	4772	1699	2560	2443
13	900	250	30	1400	300	3.56	5170	1944	2560	2538

The target hopper volume Y_1 function is fixed by calculating the mass characteristics of the cement body upper assembly component. To control the weight and overall characteristics of the model, the target also includes: Y_2—the weight of the hopper system, kg; Y_3—length dimension, mm; Y_4—height dimension, mm; Y_5—width dimension, mm.

The weight of the hopper system is fixed by calling the mass characteristics of the solid model. The dimensions of the hopper system are determined from the dimensions of the upper assembly of the model.

3 Results

The experimental plan and target values obtained during the tests are shown in Tables 3 and 4.

Processing of experimental data was carried out using the method of multifactorial regression analysis. Data on the obtained coefficients of regression dependencies, their significance and adequacy of models are given in Table 5.

The resulting Fisher test regression dependencies adequately describe the relationships of target and variable model parameters. Therefore, it is possible to attempt to solve the problem of multivariate optimization of the main target parameter of the hopper volume Y_1 by applying the regression dependencies of the remaining four target parameters Y_2–Y_5 as constraints. The solution to this problem is carried out in MATHCAD.

As a result of optimization, the following set of optimal parameters of the bunker system was obtained: $X_1 = 1100$ mm; $X_2 = 300$ mm; $X_3 = 30°$; $X_4 = 1193$ mm; $X_5 = 580$ mm. Volume of bunker system $Y_1 = 4.09$ m^3, weight of bunker fully loaded with cement $Y_2 = 5.77$ tons. Overall dimensions of the system: length $Y_3 = 1758$, mm; height $Y_4 = 2600$ mm; width $Y_5 = 2500$ mm.

Table 4 Plan and data of experimental studies (N14–32)

No	X_1	X_2	X_3	X_4	X_5	Y_1	Y_2	Y_3	Y_4	Y_5
14	1100	250	30	1400	300	3.96	5697	1944	2560	2738
15	900	300	30	1400	300	3.71	5367	1994	2560	2538
16	1100	300	30	1400	300	4544	5914	1994	2560	2738
17	900	250	25	1000	600	1.99	3152	1276	4240	1870
18	1100	250	25	1000	600	2.27	3536	1276	2420	2070
19	900	300	25	1000	600	2.1	3310	1326	2420	1870
20	1100	300	25	1000	600	2.4	3712	1326	2420	2070
21	900	250	30	1000	600	2.54	3854	1482	2420	2076
22	1100	250	30	1000	600	2.87	4291	1482	2420	2276
23	900	300	30	1000	600	2.67	4025	1532	2420	2076
24	1100	300	30	1000	600	3.01	4481	1532	2420	2276
25	900	250	25	1400	600	3.53	5100	1649	2980	2243
26	1100	250	25	1400	600	3.97	5660	1649	2980	2443
27	900	300	25	1400	600	3.7	5315	1699	2980	2243
28	1100	300	25	1400	600	4.15	5897	1699	2980	2443
29	900	250	30	1400	600	4.66	6488	1944	2980	2538
30	1100	250	30	1400	600	5.17	7136	1944	2980	2738
31	900	300	30	1400	600	4.84	6725	1994	2980	2538
32	1100	300	30	1400	600	5.38	7369	1994	2980	2738

Table 5 Regression dependencies for target study parameters

Model coefficients	Object research target parameters									
	Y_1 (m^3)		Y_2 (kg)		Y_3 (mm)		Y_4 (mm)		Y_5 (mm)	
	Val	t_p*	Val	t_p*	Val	t_p*	Val	t_p*	Val	t_p*
B_0	3.05	89.5	4500	107.5	1611	208	2400	227	2281	276
$B_1 \cdot X_1$	0.19	5.50	233	5.57	1.75	0.22	0	0	100	12.1
$B_2 \cdot X_2$	0.08	2.45	93	2.22	23.63	3.05	0	0	0	0
$B_3 \cdot X_3$	0.39	11.5	476.1	11.37	127	16.4	0	0	125.3	15.2
$B_4 \cdot X_4$	0.88	26.1	1099	24.2	210.1	27.2	280	25.6	208.8	25.3
$B_5 \cdot X_5$	0.40	11.8	502.8	12.0	1.75	0.23	210	19.2	0	0
$F_{0,95} = 1.89$	$F_r = 26.76$ adequate**		$F_r = 33.8$ adequate**		$F_r = 103.5$ adequate**		$F_r =$ 1e+5 adequate**		$F_r = 118.2$ adequate**	

*Student calculation test to test the significance of model coefficients by condition $t_p >$ t 0.95 = 1.14; ** Checking the adequacy of the regression model by condition $F_r > F$ 0.95

4 Discussion of Results

It should be noted that the resulting set of optimum hopper parameters cannot be considered a final solution. After carrying out the strength analysis check calculations, when the weight of the bunker system design is finally determined, it is necessary to clarify the parameters of the bunker system using computer modeling.

Obviously, if the strength criterion of the parts of the bunker system (for example, the minimum safety factor of the most loaded part) was developed and included in the above-described optimization methodology, the design result would be more adequate and complete for practical application.

5 Conclusions

1. Draft model of pump unit based on KAMAZ chassis has been developed. 2. The weight and position of the center of gravity of the designed hopper system were calculated based on the weight calculation of the chassis of the KAMAZ car with the installed additional equipment. It was established that the weight of the hopper system should not exceed 5.93 tons, and its center of gravity should be located from the front axle of the wheels no closer than 2800 mm. The maximum overall dimensions of the hopper installation for placement on the chassis of the car were also determined.
2. The objective is to optimize the volume of the hopper. Five parameters were identified, the change of which allows to obtain the maximum volume of the hopper within the limits specified, the system of which was developed.
3. Parametric solid-state model of hopper system of pump unit has been developed. The model is built on nine basic parameters in the top-level assembly, whose values are passed to the lower-level part and subassembly models. Examples of operation of parametric model of bunker system are given.
4. The methodology was developed and studies were carried out on the influence of five variable parameters of the hopper system on the target parameters of the structure, among which the hopper volume parameter was mainly recognized, and others play the role of constraints. Constructed regression models adequately describe the relationships of target and variable model parameters.
5. The task of multidimensional optimization is set, as a result of which the composition of optimal parameters of the hopper system with the maximum useful volume is determined.

References

1. Nosachev, S.V., Chuveiko, M.V., Zhukova, M.Y., Kuzin, G.A., Khozyaev, I.A.: Effects of deformation dynamics in problems of heavy transport. MATEC Web Conf. **226**, 02006 (2018)
2. Kasyanov, V., Deryushev, V., Kosenko, E., Kosenko, V., Golubeva, A.Y.: Synthesis of methods and principles of ensuring the reliability of one-off and serial production machines. MATEC Web Conf. **224**, 02106 (2018)
3. Badalyan, L.K., Kurdjukov, V.N., Ovcharenko, A.M.: Assessment of car emission masses at the change of kinetic energy of the engine—transport vehicle system. MATEC Web Conf. **226**, 02014 (2018)
4. Efimov, A.V., Kireev, S.O., Korchagina, M.V.: The efficiency is a single criterion for choosing the type of interaxle drive. MATEC Web Conf. **226**, 04002 (2018)
5. Lemeshko, M.A., Molev, M.D., Iliev, A.G.: Adaptive drive of the technological machine with two working movements of the executive body. MATEC Web Conf. **226**, 02014 (2018)
6. Kaderov, K.K., Kireev, S.O., Korchagina, M.V., Osipuk, A.Y.: Improved seals in pipelines and high-pressure hoses. Russ. Eng. Res. **38**(3), 241–243 (2018). https://doi.org/10.3103/S1068798X18030061
7. Mailyan, D.R., Trufanova, E.V.: Planning of multilayer cylindrical wall reservoirs. Procedia Eng. **150**, 1926–1935 (2016)
8. Kaderov, K., Kireev, S., Korchagina, M., Lebedev, A.: Determination of technological parameters and tool for flanging holes from sheet blanks. E3S Web Conf. **164**, 14019 (2020)
9. Lemeshko, M., Molev, M., Golovin, I.: Hydraulic technological machines with adaptive drive structure. MATEC Web Conf. **224**, 02087 (2018)
10. Ivanov, Y., Pakhomov, V., Kambulov, S., Rudoi, D.: Determination of the parameters of the hydrodynamic mixer. MATEC Web Conf. **224**, 05023 (2018)
11. Rybak, A.T., Temirkanov, A.R., Lyakhnitskaya, O.V.: Dynamics of synchronous hydromechanical drive in mobile machine. Russ. Eng. Res. **38**(9), 702–704 (2018)
12. Kaderov, K.K., Korchagina, M.V., Kireev, S.O., Nikishenko, S.L.: Solution of problems of gasketing of joints of units of machines and equipment of oil and gas enterprises when in use in high- and low-temperature regimes. Chem. Pet. Eng. **54**(3–4), 188–192 (2018)
13. Rybak, A.T., Temirkanov, A.R., Lyakhnitskaya, O.V.: Synchronous: hydromechanical drive of a mobile machine. Russ. Eng. Res. **38**(3), 212–217 (2018)
14. Lapshin, V.P., Turkin, I.A.: Modeling tractive effort torque of wheel in deformation movements of pneumatic tire wheel. Procedia Eng. **206**, 594–599 (2017)
15. Ovchinnikov, N.A., Zhigulsky, V.I., Kozyreva, E.A., Chefranova, O.V.: Modeling and estimate of the strain-stress state of a bus body pillar with account for geometrical nonlinearity. ARPN J. Eng. Appl. Sci. **12**(5), 1567–1572 (2017)
16. Panasyuk, L.N., Kravchenko, G.M., Matua, V.P.: Numerical method for solving the problems of the building structures dynamics with a mobile massive load. Mater. Sci. Forum **931**, 72–77 (2018)

Optimization of Parameters of Parts of Brake Device of Piston of Circular

Haydyar Kaderov, Sergey Kireev, Alexey Lebedev, and Joerg Sperling

Abstract Disclosed is a new design of a preventer with a brake assembly, in which the pusher interacting with the sealing ring is made with teeth on the end surface. The size of the O-ring and the parameters of the gear pusher were optimized. Mathematical model of stress–strain state of sealing ring at interaction with support and piston is built, with the help of which more advanced operational parameters of preventer are achieved. The proposed mathematical model of the stressed-strain state allows optimizing the power and geometric parameters of the braking device: (a) values of the deformation of the ring depending on its dimensions; (b) axial compression force of the ring required to achieve the value of the specified braking moment of the piston; (c) select ring dimensions; adjust its deformation. Claimed design of braking device in combination with created mathematical model of sealing ring can be used in constructions of machines and equipment using sealing parts.

1 Introduction

One of the conditions for an efficient oil and gas production process is the presence of BOP equipment installed at the wellhead. BOP equipment includes prevents used during drilling and operation of oil, gas and gas condensate wells (GOST 13862-90 (RU)). Sealing of well head is achieved by sealing element of complex spatial shape embracing casing string. Problems of design and calculation of sealing elements were also discussed by authors of works [1–4].

There are various designs of ring rotating preventers. A disadvantage of known designs (patents 2,398,954 (RU), 2,398,954 (RU), 2,527,054 (RU)) [4–6] is the lack of reliable brake devices that prevent the rotation of the piston. Works [5–7] also discuss reliability problems of sealing systems. In Patent No 2398954 (RU), the task

H. Kaderov · S. Kireev (✉) · A. Lebedev
Don State Technical University, Rostov-on-Don, Russia

J. Sperling
ACT Well Services GmbH, Diesdorf, Germany

M. Shamtsyan et al. (eds.), *Robotics, Machinery and Engineering Technology for Precision Agriculture*, Smart Innovation, Systems and Technologies 247,
https://doi.org/10.1007/978-981-16-3844-2_12

of braking the piston is accomplished by creating radial forces on the piston by the O-ring. Rubber of various grades is used as material for sealing rings. However, in the process of drilling and operation of wells, ring wear occurs and, accordingly, the effect of braking of the piston is reduced.

In addition, most preventer designs do not have the ability to adjust radial forces and correctly calculate them. The preventer according to (Patent no. 181499 (RU)) is equipped with a special braking unit to prevent rotation of the piston, which allows sufficiently accurate determination of the radial force necessary for braking the piston and the spring stiffness necessary for this [8]. However, it is possible to rotate the piston and the O-ring together in a certain amount.

Disclosed is a preventer design (Patent No. 191857 (RU)) with a brake assembly, in which a pusher interacting with an O-ring is made with teeth on the end surface, which, pressed into the O-ring, fix it in the circumferential direction and simultaneously create radial forces necessary to create a braking moment, both for the piston and the braking device itself.

The purpose of the operation is to optimize the dimensions of the O-ring and the parameters of the gear pusher-l of the preventer with the brake assembly.

Research objectives:

- Build a mathematical model of the stress–strain state (SF) of the O-ring when interacting with the support and piston.
- Perform seal ring VAT analysis using the created mathematical model.

2 Materials and Methods

The design of the proposed preventer in the operating position is shown in Fig. 1. For simplicity, you do not specify a number of entries that explain the design of the preventer. Novelty consists in provision of piston with special braking device containing spring-loaded toothed pusher. In addition, the pusher can be fixed, if necessary, by the limiting threaded bosses 7. The parameters of the toothed pusher include the width of the teeth (depressions), the number of teeth, the depth of the depressions, the compressive force of the sealing ring in the axial direction, which creates the radial force necessary to ensure braking of the piston during drilling and operation of the well.

Initial prerequisites:

1. O-ring material - isotropic incompressible rubber
2. Mechanical properties of rubber given in reference literature
   (a) Allowable relative strain (allowable stress) 20–25%.
   (b) Elastic modulus of rubber $E = 7$ MPa.
   (c) Poisson's coefficient (ratio of transverse deformation to linear) $\mu = 0.5$.
   (d) The relationship between strain and stress is linear.

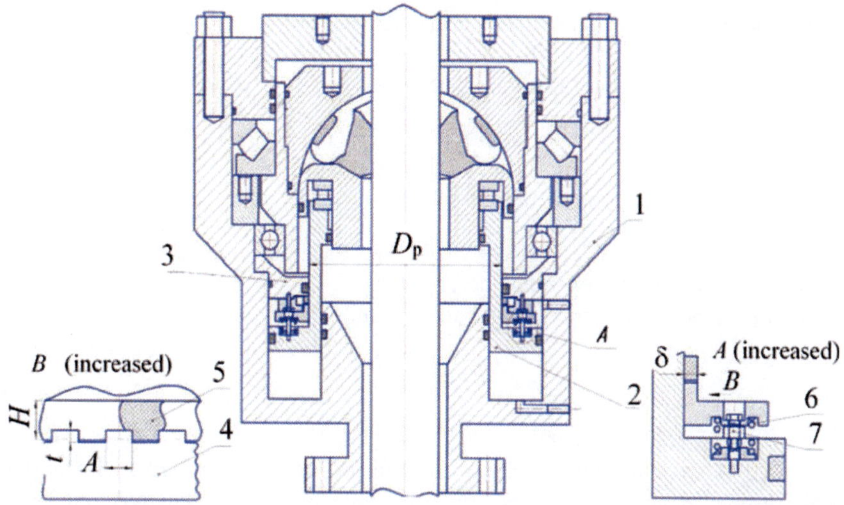

Fig. 1 High-reliability ring preventer: 1—housing; 2—piston; 3—support; 4—gear pusher; 5—ring; 6—spring; 7—limiter

3. Stresses (respectively deformation) are distributed evenly over the volume of the sections of the ring under consideration
4. Boundary conditions:
   (a) Rigid contact of end surfaces of ring and support - limitation in axial direction.
   (b) Rigid contact of inner and outer surfaces of the ring with surfaces: outer piston and inner support - restriction in radial direction.
5. Designations:
   (a) The main relative deformations: (a) ε_1—axial, (b) ε_2—circumferential, (c) ε_3—radial.
   (b) Ring dimensions: internal diameter D_{in} is equal to piston diameter D_p, δ—thickness, external diameter $D_e = D_{in} + 2\,\delta$, H—height.
   (c) Dimensions of the working section of the toothed annular pusher: A—tooth length, t—depth of the depression, and the number of teeth—n.
   (d) Original volume of deformed section V_0 and modified volume of deformed section V_1.

Optimization of shape and dimensions of sealing ring and teeth of toothed pusher is carried out on the basis of mathematical model of their power interaction, which is determined by axial force of ring compression and its friction forces with pusher and parts contacting with ring. These forces are associated with major strains and corresponding normal stresses.

Normal stresses σ are generally determined by the product of relative deformation ε per modulus of elasticity E [9].

Accordingly, the radial force F_r is determined by the relation

$$F_r = \sum_{i=1}^{n} (S_i \cdot \varepsilon_{3i} \cdot E) \quad (1)$$

where n is the number of deformed sections, S_i is the area of the inner surface of the i-th section, ε_{3i} is its relative radial deformation.

The given moment of braking M_t can be found

$$M_t = F_r \cdot K_f \quad (2)$$

where K_f is friction coefficient.

The total area of the inner surface of the deformed sections of the ring will consist of the formed depressions and protrusions (respectively under the tooth and the pusher cavity). Since the length of the tooth is A and the height H, the area of the inner cylindrical surface of the i-th portion of the ring S_i under the tooth is determined by the relative axial deformation ε_1.

Accordingly, the area under all n teeth ΣS_i will be equal to

$$\sum S_1 = \sum_{i=1}^{n} S_1 \cdot n. \quad (3)$$

To determine the area of the inner cylindrical surface of the deformed ring under the recesses of the teeth, it is necessary to construct a mathematical model of the deformed state of the ring sections, both under the projection and under the depression.

As was customary in the initial prerequisites, the restrictions of ring deformation (boundary conditions) are—rigid contact of the ring with the outer surface of the piston and the end support. Then displacement of rubber volume under pusher tooth during compression occurs outside in radial and circumferential directions, and in cavities between teeth in axial and radial directions.

Based on the accepted conditions—isotropy and incompressibility of the material, Poisson's coefficient $\mu = 0.5$, linearity of the deformation coupling (up to 25%) and stresses [6, 10–12], it is easy to construct a mathematical model of the ring stress–strain state in the operating position, which allows determining the axial compression force of the ring to create the necessary braking moment. Let's show by example.

The following dimensions were taken for the ring in the initial state: the inner diameter is equal to the outer diameter of the piston D_p, the thickness $\delta = 20$ mm, and its height $H = 40$ mm—was chosen based on the condition of compliance with the requirements for the size of the sample in standard compression tests [13, 14]. The length of the pusher tooth A was equated with the height of the ring $H = 40$ mm, based on the same considerations. The number of teeth $n = 50$ was chosen taking into account the proximity of tooth lengths and troughs at $D_p = 1270$ mm. In this case, the depth of the depressions t can be assigned based on the results of the analysis of the deformed state.

The compression deformation of the ring α_1 was taken equal to 20%, hence its height under the tooth will decrease to 32 mm.

Due to the small ratio of the ring thickness δ to the inner diameter (less than 0.016) for the analysis of its deformed state with sufficient accuracy, it is proposed in the initial state to limit itself to a section in the form of two adjacent parallelepipeds with dimensions: length $A = 40$, height $H = 40$, thickness $\delta = 20$ mm (Fig. 2). We denote them with the numbers 1 (under the tooth) and 2 (between the teeth). For dimensions of deformed Sect. 1 (refer to Fig. 2) the following designations are adopted: $A1$, $H1$, $\delta 1$, and for pos. 2—$A2$, $H2$, $\delta 2$. At the same time, it should be borne in mind that the symbols $A1$ and $A2$ can appear only conditionally when calculating the displacement of volumes.

The change in the shape and dimensions of the parallelepiped (under the tooth) pos.1 (see Fig. 2) and then the corresponding stresses can be determined based on the hypothesis of flat sections and simultaneously using the constancy of the univariate. Since Poisson's coefficient is $\mu = 0{,}5$ and, accordingly, the uniformity of deformations over a given volume $\varepsilon_2 = \varepsilon_3$, they can be found

$$\varepsilon_2 = \varepsilon_3 = \sqrt{1/(1 - \varepsilon_1)} \tag{4}$$

If, as is customary $\varepsilon_1 = 0.2$, then $\varepsilon_2 = \varepsilon_3 = 0.118034$. Then the virtual deformed parallelepiped pos. 1 will have dimensions: $H1 = 32$, $A1 = 44.72$, $\delta 1 = 22.36$ mm.

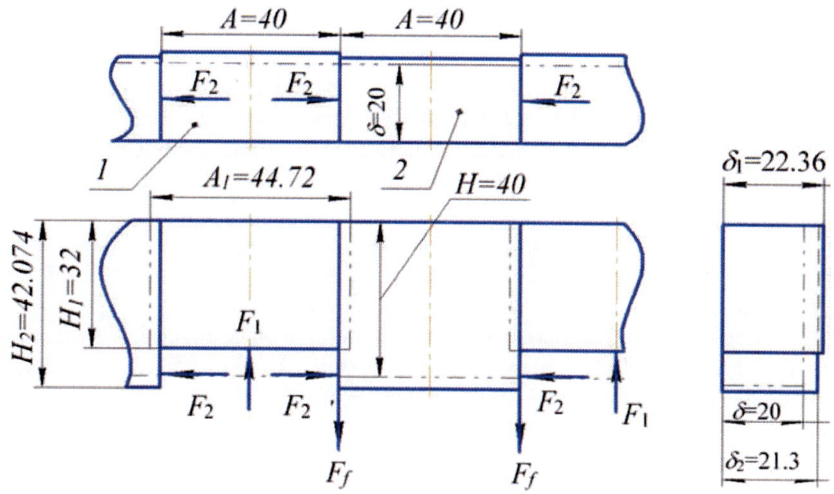

Fig. 2 Calculation scheme for building a model of a deformed state: F_1—vertical compression force; F_2—horizontal compression force; F_f—friction force

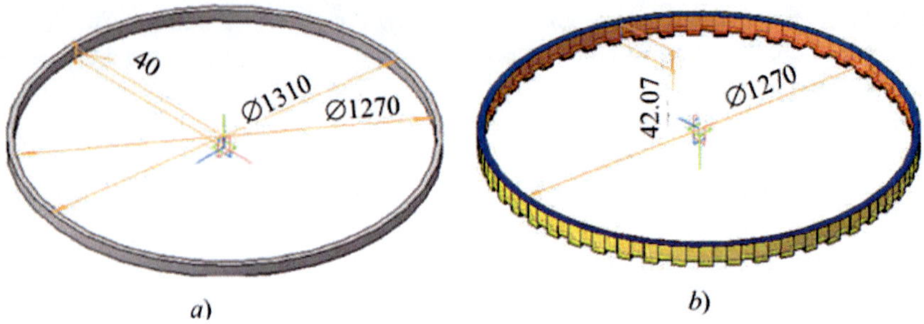

Fig. 3 O-ring models in initial (**a**) and deformed (**b**) states

The deformed volume in the circumferential direction ε_2·and A is displaced into the cavity between the teeth and distributed into an increase in the height of the parallelepiped pos. 2 (see Fig. 2) up to $H2$ and thickness $\delta 2$. The dimensions of the deformed parallelepiped pos.2 can be found by analogy, as for pos.1. Here, the deformation in the circumferential direction (denote its ε_1') is linear. Then the height and thickness deformations (as transverse ones are denoted by ε_2' and ε_3', respectively) are determined by analogy with (4)

$$\varepsilon_2' = \varepsilon_3' = \sqrt{1/(1-\varepsilon_1')} \tag{5}$$

Numeric ε_2' and ε_3' will be 0.0648. Accordingly, $\delta 2 = 21.296$ mm, and height $H2$ is determined as the sum of the initial height H and the deformation of height $H1 =$ 32 mm. As a result, we have $H2 = 42.074$ mm.

Based on the results of calculations, three-dimensional models of the ring in initial and deformed states are built, shown in Fig. 3. When constructing the model of the deformed ring, the shape and dimensions of the parallelepipeds were extrapolated as follows: the length of the pusher teeth with parallel side faces was taken to be 40 mm, and their number was 50. Accordingly, the circumferential pitch of the teeth measured on the inner surface of the pusher was 79.8 mm—this difference is 0.25% of the length of the section used in the calculations previously carried out.

The mass of the O-ring according to Fig. 3 in the initial (upper part) and deformed states is almost equal, which confirms the correctness of the calculations and the usefulness of three-dimensional modeling.

It should be noted that there is no restriction on the outer cylindrical surface of the ring in the calculation scheme. However, it is necessary to achieve radial force providing braking of piston [15]. At the same time, the efficiency of the piston sealing will be improved, which is also one of the functions of the proposed structure. The design diagram for determining power parameters is shown in Fig. 4.

By analogy with Fig. 2 designation pos. 1 in Fig. 4—this is the section (parallelepiped) under the tooth, and pos. 2 is a section between the teeth.

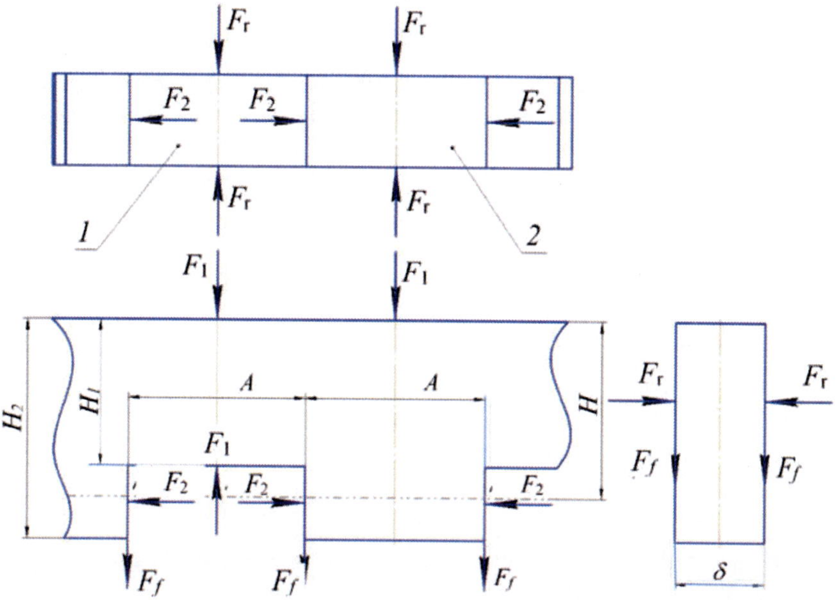

Fig. 4 Calculation diagram for determination of brake torque M_b. A is the length of the pusher tooth, H is the height of the O-ring in the initial state, H_1—height of deformed ring under tooth H_2—honeycomb of deformed ring between teeth, δ—thickness of sealing ring, F_1—vertical force of pusher compression, assigned to one tooth, F_2 mountains horizontal force of compression of one protrusion of the defective-world ring, F_r is a radial force on the cylindrical surfaces of the deformed ring, F_f is a friction force during deformation of the ring

The previously constructed model of the deformed state of the O-ring allows you to determine all the power parameters necessary to solve the tasks:

1. Calculate the reduced radial force $F_{r,i}$, which creates the required braking torque M_b, kN m

$$M_b = \sum_{i=1}^{n} F_{ri} \cdot R \cdot K_f \tag{6}$$

where n is the number of teeth, F_{ri} is the sum of radial forces assigned to one tooth, R is the reduced radius of braking forces, K_f is the friction coefficient, which has wide limits [9, 10], is taken equal to $K_f = 0.125$. At the same time the condition must be met: the braking moment must be greater than the torque required for drilling [12–14].
2. The total compression force $F_{\text{sum p}}$ of the pusher is equal to the sum of the vertical compression force of the ring F_1 and the friction force F_f.
3. Vertical compressive force F_1 is determined by formula

$$F_1 = (S \cdot \sigma \cdot n + S_1 \cdot \sigma_1) \tag{7}$$

where σ—compression tension under tooth, and S—the area of a ring under tooth, equal to the work of thickness of a ring (pusher tooth) at tooth length, respectively, σ_1—compression tension at all square S_1 between teeth's.

4. The friction force F_f, taking into account all its components (see Fig. 4), is calculated as follows:

$$F_f = K_f \cdot H \cdot E \cdot \left(\pi \cdot D_m \cdot \left(\sqrt{1/(1-\varepsilon_1)} - 1\right) + 2 \cdot n \cdot \delta \cdot \varepsilon_1^2\right) \tag{7}$$

where D_m is the average ring diameter.

Based on the built model of the deformed state and the rigid restriction of the three surfaces (including both cylindrical) rings, you can write

$$F_r = \pi \cdot D_m \cdot H \cdot E \cdot \left(\sqrt{1/(1-\varepsilon_1)} - 1\right) \tag{9}$$

5. The main task is to ensure the necessary braking torque of the M_b piston should be solved by the radial force F_r

$$M_b = K_f \cdot F_r \cdot 0.5 \cdot D_m \tag{10}$$

3 Results

The necessary force parameters were calculated: brake torque, axial compression force for a particular preventer with a piston diameter of 1270 mm, an O-ring made of isotropic incompressible rubber with an internal diameter of 1270 mm (elastic modulus and permissible stress is given above), 20 mm thick and 40 mm high with a number of pusher teeth of 50 and a circumferential length of 40 mm. Using the calculation scheme in Fig. 4, additional braking moments of the pusher Mb from the pusher teeth, kN m, were calculated. The calculation data are given in Table 1 of the

Table 1 Power parameters of the braking device

ε_1	F_1 (kN)	F_r (kN)	F_f(kN)	F_r(kN)	M_t(kN m)	M_b(kN m)
0.05	17.79	29.46	3.86	21.65	2.34	0.45
0.08	28.69	48.28	6.4	35.09	3.83	1.15
0.1	36.09	61.35	8.36	44.45	4.86	1.80
0.15	54.98	96.01	13.57	68.55	7.62	4.06
0.2	74.62	133.87	19.5	94.12	10.62	7.22

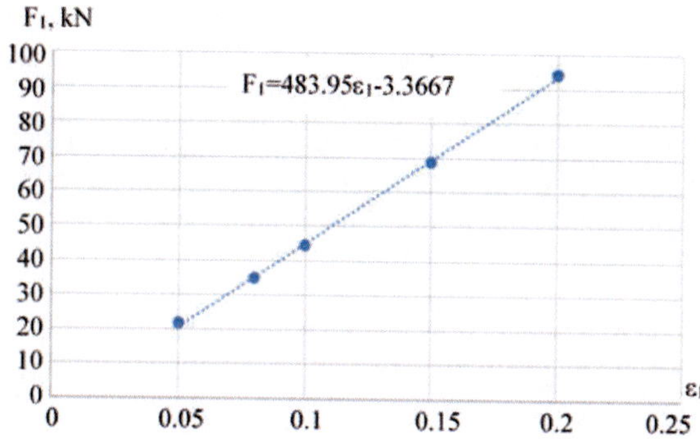

Fig. 5 Diagram of compression force of gear pusher F_1

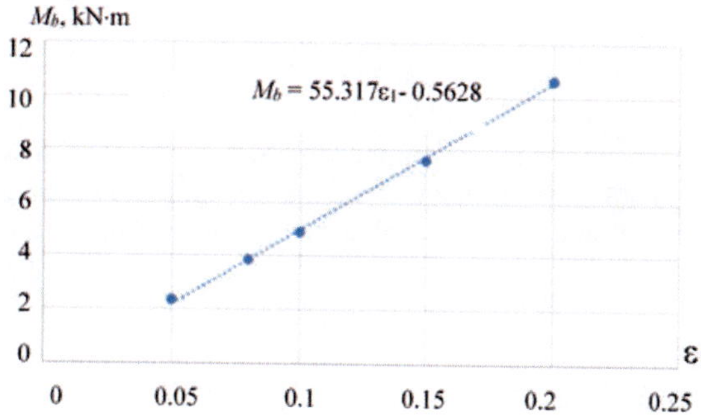

Fig. 6 Brake moment diagram of piston M_b

power parameters depending on the relative axial deformation of the compression ε_1.

According to the table, diagrams are drawn to determine the values depending on the deformation ε_1: (a) the compression force of the O-ring with a toothed pusher F_1, kN, (Fig. 5), (b) the moment of braking of the piston M_b, kN m, (Fig. 6).

Calculated in operation [14] required for drilling torque M_r for this preventer is 4.51 kN m. Taking into account the above safety factor for the required moment of braking of the piston at the given dimensions and material of the ring, the compression deformation should be within the range of $\varepsilon_1 = 0.15\ldots 0.2$, which according to the table is permissible.

4 Discussion of Results

Using the table of power parameters and diagrams in Figs. 5 and 6, you can select linear (axial) compression deformation and select compression springs of the brake gear pusher.

The proposed mathematical model of the stressed-deformed state allows optimizing the power and geometric parameters of the braking device: (a) values of the deformation of the ring depending on its dimensions; (b) axial compression force of the ring required to achieve the specified torus-brain moment of the piston; (c) select ring dimensions; adjust its deformation.

Claimed design of braking device in combination with created mathematical model of sealing ring stressed-deformed state can be used in constructions of machines and equipment using sealing parts.

5 Conclusions

1. A mathematical model describing a stress–strain state of an O-ring when interacting with a toothed pusher of a brake device of a preventer piston is proposed.
2. A three-dimensional model of the sealing ring is built, interpreting its deformed state as a result of shaping.
3. Using the proposed mathematical model, calculations of power and geometric parameters of the brake device of the piston for a particular preventer are made with presentation of their results in the form of a table and diagrams.

References

1. Kaderov, K.K., Kireev, S.O., Korchagina, M.V., Osipuk, A.Y.: Improved seals in pipelines and high-pressure hoses. Russian Eng. Res. (2018). https://www.scopus.com/inward/record.url?eid=2-s2.0-85045990662&partnerID=40&md5=3d37a0cc81149f9c407650a5ede73374.
2. Efimov, A.V., Kireev, S.O., Korchagina, M.V.: Mathematical model for assessing selection of the interaxle drive type. J. Mach. Manuf. Reliab. (2018). https://www.scopus.com/inward/record.url?eid=2-s2.0-85043463727&partnerID=40&md5=268c4032454b1c1408d065e05a7ec86e.
3. Demyanov, A., Demyanov, A., Pavlova, I., Pavlitskiy, B.: Improvement of friction units limiting the durability of mechanical systems. https://doi.org/10.1088/1757-899X/1029/1/012004. Conference Paper, Scopus (2021) https://www.scopus.com/inward/record.uri?eid=2-s2.0-85101615324&doi=10.1088%2f1757-899X%2f1029%2f1%2f012004&partnerID=40&md5=d85ad5e9d0ddd2592cbfc329b2f875a3.
4. Manshin Y.P., Manshina E.Y.: Reliability aspects in machine designs. https://doi.org/10.3103/S1052618820120109. Article, Scopus (2020). https://www.scopus.com/inward/record.uri?eid=2-s2.0-85102071650&doi=10.3103%2fS1052618820120109&partnerID=40&md5=dd728bdbace35bd065f121c468506164.
5. Lemeshko, M.A., Zibrov, V.A., Bashnyak, S.E.: Simulation of adaptive drill machines with hydraulic drive. In: Conference Paper, Scopus 2020, https://doi.org/10.1088/1757-899X/971/4/042092. https://www.scopus.com/inward/record.uri?eid=2-s2.0-85097775053&doi=10.1088%2f1757-899X%2f971%2f4%2f042092&partnerID=40&md5=6765f4118708fd52a0a01eb0573d75ef.
6. Ovchinnikov, N.A., Zhigulsky, V.I., Kozyreva, E.A., Chefranova, O.V.: Modeling and estimate of the strain-stress state of a bus body pillar with account for geometrical nonlinearity. ARPN J. Eng. Appl. Sci. (2017). https://www.scopus.com/inward/record.url?eid=2-s2.0-85015224140&partnerID=40&md5=8d5793552517df70f8f493a4689ec682

7. Kasyanov, V., Deryushev, V., Kosenko, E., Kosenko, V., Golubeva, A.Y.: Synthesis of methods and principles of ensuring the reliability of one-off and serial production machines. MATEC Web of Conferences (2018). https://www.scopus.com/inward/record.url?eid=2-s2.0-85056657184&partnerID=40&md5=d2960acd147256ac3b7cd28e5fb27f65.
8. Kaderov, K.K., Korchagina, M.V., Kireev, S.O., Nikishenko, S.L.: Solution of problems of gasketing of joints of units of machines and equipment of oil and gas enterprises when in use in high- and low-temperature regimes. Chem. Petrol. Eng. (2018). https://www.scopus.com/inward/record.url?eid=2-s2.0-85052809302&partnerID=40&md5=76b3907958433f2392a862dd55f6988d.
9. Zelentsov, V.B., Sadyrin, E.V., Sukiyazov, A.G., Shubchinskaya, N.Y.: On a method for determination of Poisson's ratio and Young modulus of a material. MATEC Web of Conferences (2018). https://www.scopus.com/inward/record.url?eid=2-s2.0-85056511919&partnerID=40&md5=72e64853a8f771fdd278e4bafba1e39f.
10. Skachkov, S.V., Shchutsky, S.V.: Analysis of the stresses of the wall of a steel water tower. Mater. Sci. Forum (2018). https://www.scopus.com/inward/record.url?eid=2-s2.0-85055948391&partnerID=40&md5=6847a4c7d76d8ac156019c6bd02238ad.
11. Skachkov, S.V., Golotaystrova, E.Y.: Reliability comparison of trusses made of thin-walled steel sections. Mater. Sci. Forum (2018). https://www.scopus.com/inward/record.url?eid=2-s2.0-85055944821&partnerID=40&md5=689df0f71740d1873b04192379a14db5.
12. Kolesnikov, A.M., Shubchinskaya, N.Y.: Cylindrical membrane partially dressed on a rigid body of revolution. https://doi.org/10.1007/s00161-021-00974-9. Article, Scopus (2021). https://www.scopus.com/inward/record.uri?eid=2-s2.0-85100517281&doi=10.1007%2fs00161-021-00974-9&partnerID=40&md5=c3aef22433ec6a40d972385255c4e4d2
13. Ivanovskaya, A.V., Bogatyreva, E.V., Rybak, A.T.: Simulation of the cable part of the load-lifting device drive, represented as a rod system. https://doi.org/10.1088/1757-899X/1001/1/012059. Conference Paper, Scopus (2020). https://www.scopus.com/inward/record.uri?eid=2-s2.0-85100020085&doi=10.1088%2f1757-899X%2f1001%2f1%2f012059&partnerID=40&md5=0459514b9584e1db4d5cc2e280a320a5
14. Panasiuk, L., Turina, V.: The stress-strain state simulation of slopes with dynamic loads. https://doi.org/10.1088/1757-899X/698/2/022021. Conference Paper, Scopus (2019). https://www.scopus.com/inward/record.uri?eid=2-s2.0-85078193712&doi=10.1088%2f1757-899X%2f698%2f2%2f022021&partnerID=40&md5=7960d7edf80624f1c6693ea9413a13c1.
15. Efimov, A., Kireev, S., Korchagina, M., Lebedev, A.: Mathematical model of the vehicle taking into account the turn and reduction of tire friction, **1**, https://doi.org/10.1088/1757-899X/918/1/012161. Conference Paper, Scopus (2020). https://www.scopus.com/inward/record.uri?eid=2-s2.0-85094169088&doi=10.1088%2f1757-899X%2f918%2f1%2f012161&partnerID=40&md5=f98a2afee58fcb64f99f1ff38d40586e.

Improving the Processing of Thin Wall Sleeves

Maxim Bogomolov, Andrey Malikov, Alexander Yamnikov, and Aleksandr Brailov

Abstract It is noted that the strength of the workpiece being processed is not taken into account when assigning locating and clamping scheme for most parts when machining by cutting. It is assumed that the limiting factors are the combination of the physical and mechanical properties of the materials of the workpiece and the cutting part of the tool, which, in combination with a given allowance, determine the required clamping forces for the selected locating and clamping scheme. However, for non-rigid workpieces such as thin-walled sleeves, errors caused by the clamping forces of the workpiece in the device can become dominant. An additional factor complicating the cutting process is the occurrence of significant vibrations that deteriorate the quality of the machined surface. Of particular difficulty is the study and description of the milling process, this is due to the presence of variables in modulus and direction of cutting forces, leading to additional forced vibrations of the process system. It is shown that to reduce vibrations in the system, it is possible to reduce cutting conditions and, accordingly, performance, increase the strength and damping capacity of the system. When installing thin-walled sleeves on multi-purpose machines, clamping devices are of special importance. Devices with metal expanding elements act on a limited section of the cylindrical surface length, and therefore, they cannot increase the vibration resistance over the entire surface. The design of a mandrel with an elastic sleeves, located between the body of the workpiece and the main base mandrel, has been developed. By means of axial compression of the elastic sleeve, its thickness is increased, as well as the strength and damping capacity of the process system is increased, which ultimately leads to a decrease in the roughness of the processed surface by 1.5 times.

M. Bogomolov · A. Malikov · A. Yamnikov (✉)
Tula State University, 92, Lenina Avenue, 300012 Tula, Russia

A. Brailov
Odessa Professional College of Computer Technologies, 5/2, Akademika Koroleva Str., 65101 Odessa, Ukraine

M. Shamtsyan et al. (eds.), *Robotics, Machinery and Engineering Technology for Precision Agriculture*, Smart Innovation, Systems and Technologies 247,
https://doi.org/10.1007/978-981-16-3844-2_13

1 Introduction

When choosing a scheme for locating and applying clamping forces, as well as when assigning cutting conditions, the rigidity of the workpiece to be processed is not taken into account [1]. It is assumed that the limiting factors are the combination of the physical and mechanical properties of the materials of the workpiece and the cutting part of the tool, which, in combination with a given allowance, determine the required clamping forces for the selected locating and clamping scheme. However, the paper [2] notes that errors caused by the forces of the workpiece clamping in the device can become dominant. There are also examples of analytical calculation and experimental determination of locating errors for various devices when processing thin-walled rings.

Later studies [2–6] describe the use of modern methods for mathematical modeling of clamped workpieces elastic deformations. Papers [6, 7] studied the capabilities of hydroplastic devices and showed that these devices are suitable only for clamping workpieces that are accurate in shape and size. For less precise workpieces of solids of revolution of 14 $\div$ 10 quality standards, long wide jaws bored to the base diameter of the workpiece are usually used. In [2, 8], the authors theoretically substantiate and experimentally prove that in almost all cases the original shape of the workpiece is violated, except for the ideal match of the base surface diameter and the clamping surface of the jaws. However, in this case, in the presence of an initial error in the workpiece shape, for example, in terms of ovality, the ovality is straightened at the time of clamping, and is restored upon unclamping, but not completely, since during processing the stressed layers of the material are removed from the workpiece and a redistribution of residual stresses occurs [9, 10].

The paper [11] shows that in addition to hereditary errors in the shape of a thin-walled workpiece, at the time of clamping, its diameter changes, so when locating a pipe with a diameter of 122 mm, the latter decreased by 0.05 mm. This condition must also be taken into account when adjusting to size.

2 Materials and Methods

The processing of a part (Fig. 1), which is a thin-walled cylinder with a wall thickness of 2.5 mm, is considered. On the upper surface of the cylinder, there are two stages 1, in which it is necessary to make 8 flats 2 with a radius of Rmax 20 -3. In this case, a requirement for roughness Ra 6.3 and a deviation from cylindricity of no more than 0.1 mm is imposed on surface 2 [11].

A pipe made of steel 30 GOST 1050-2013 is used as a workpiece. Before milling, the workpiece undergoes sawing operation, rough turning, and pressing operation (Table 1).

The first operations are preparatory. The original workpiece is obtained for them for subsequent machining. During turning operation, the ends are trimmed for reliable

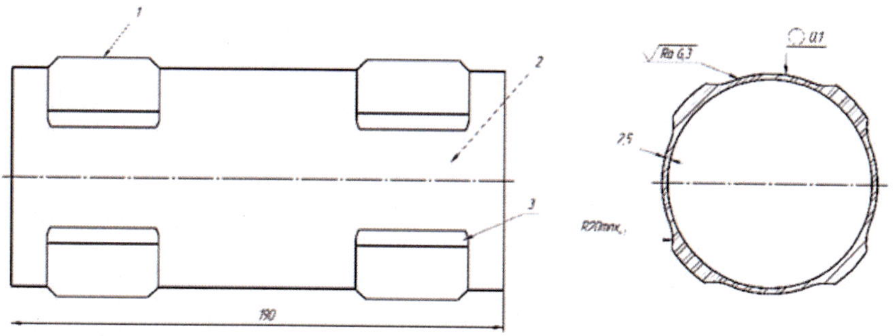

Fig. 1 Analyzed part

basing on the pressing operation, during which the diametrical dimensions of the workpiece are reduced due to the lack of the required grade of hot-rolled pipes. Then, the shaping of the surfaces of the part, specified by the drawing, is carried out (see Fig. 1).

During turning operation, the ends are trimmed and boring when clamped in a three-jaw self-centering chuck with elongated wide jaws, bored along the average diameter of the base surface of the workpiece (Fig. 2). A similar chuck with mirrored jaws is used for trimming the end on the other side and turning the outer surface of the workpiece (Fig. 3).

The turning operation did not cause process difficulties: the process base was a long cylindrical surface implemented by long jaws, the clamping surface of which was made along the average diameter of the base surface, and the end surface abutting against the steps of the chuck jaws. The turning process did not cause significant vibrations and the roughness of the machined surface corresponded to the specified value.

3 Problem Statement

In serial production, the option of shaping the shaped surface of the sleeves by milling with a spherical body cutter RBH BCF3028S32-160 with replaceable inserts ZCET 150CE on a four-axis CNC machine " Akira seiki performa v4 xp" is adopted. A section of pipe 121 × 7 GOST 8732–78 made of steel 30 GOST 1050–88 was used as a workpiece. The workpiece was installed with a gap on a rigid mandrel fixed in the dividing head of the fourth rotary axis of the machine. The clamp was made using a washer, a stud screwed into a rigid mandrel, and a nut.

A full radius cutter was buried into the workpiece to a depth of 4.21 mm and a width of 13.72 mm (Fig. 4). In this case, the cutting mode was as follows: cutting speed $V = 282.6$ m / min, cutter speed $n = 3000$ min $^{-1}$, feed $S_{\min} = 450$ mm / min, cutting depth $t = 2.5$ mm, milling width $B = 3$ mm.

Table 1 Technological process of preliminary machining of the shaped sleeve

No	Ha type of operation	Content of operation	Rigging
005	Sawing	1. Cut the workpiece, keeping the dimension 195_{-2}	Automatic band saw BMSO 320
010	Turning	1. Install, clamp, unclamp, remove the part 2. Turn the surface, holding to the dimensions 194_{-1} 3. Turn the surface, holding to the dimensions 193_{-1}	TAKISAWA CNC turning lathe PCLNR2020K12HP holder; CNMX1204 A1-SM roughing insert
015	Pressing	Direct extrusion of the workpiece, holding to the dimensions Ø116 and Ø102	Hydraulic press P6328 RI 01.00.000 Draw die
020	Annealing	Typical technological process OGMet	
025	Turning	1. Install, clamp, unclamp, remove the part 2. Turn the surface, holding to the dimensions $191_{-0.5}$ 3. Bore the surface, holding to the dimensions Ø $104^{+0.15}$ 4. Turn the surface, holding to the dimensions $190_{-0.5}$ 5. Turn the surface, holding to the dimensions Ø $109_{-0.2}$; Ø $119_{-0.2}$	TAKISAWA CNC turning lathe OP-210 3-jaw chuck P 01.00.000 jaws with linings; P 02.00.000 jaws with linings; 51.2–2020-21 M tool holder; PCLNR2020K12HP holder; CNMX1204 A1-SM roughing insert; CNMG1204 04WF finishing insert; C4-R C2080-59110A tool holder; C4-PCLNR-13080-09M1 holder; CNMM 120,412-WR roughing insert CNMG 120,404-WL finishing insert
030	Milling	1. Mill eight flats, holding to the dimension 2.5 with a radius of R_{max} 15_{-1} and the angle of 90°. °	Akira Seiki Performa V4 CNC milling machine P 03.00.000 special device BT40-KCH32-135 milling chuck BT; BCF3028S32-160 Ball nose cutter with indexable inserts BCF; ZCET150CE/SE insert; SI 01.00.000 caliber for flats

After plunge-in to the entire depth of the machined section, a line-by-line movement of the cutter was used for processing the cylindrical section of the surface with the cutting mode: $V = 282.6$ m/min, $n = 3000$ min^{-1}, $S_{min} = 3000$ mm/min, $t = 2.5$ mm, $B = 0.4$ mm (Fig. 5). To develop the CNC code, MasterCAM software from CNC Software was used. The software was used to create a 3D model of the part, select the appropriate machining strategy, determine the cutting speed, advance

Fig. 2 Three-jaw self-centering chuck with extended wide jaws for internal boring

Fig. 3 Three-jaw self-centering chuck with extended wide jaws for surface machining

to insert, and milling width as recommended by the tool manufacturer. When all settings were completed, the CAM system generated a G code for the NC tool. The resulting part showed clear signs of vibration (Fig. 6).

The surface quality was measured using a portable profilometer "Surftest SJ-210P" 178–930-2D manufactured by "Mitutoyo" (Japan) at each site at five points (Tables 2).

Table 2 shows that milling when installing a thin-walled sleeve into an existing device cannot provide the required surface quality.

The paper [12] used "... the finite element method to simulate the dynamics of complex shaped parts. In accordance with the principles of the finite element method,

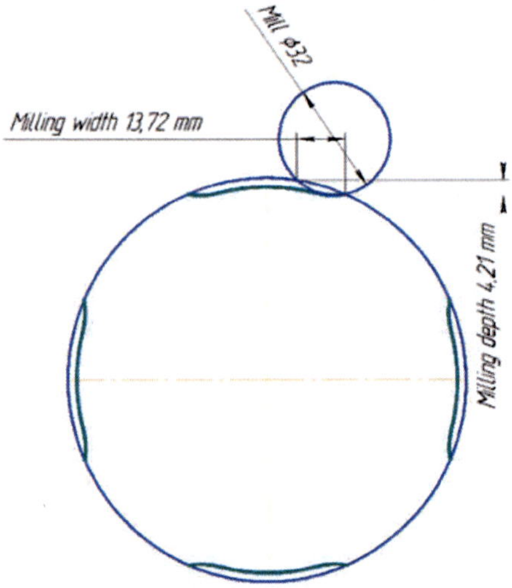

Fig. 4 Shaped surface milling conditions

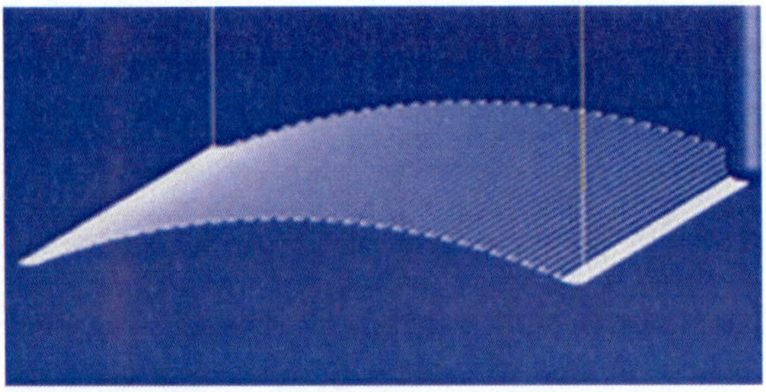

Fig. 5 Multipass milling toolpath

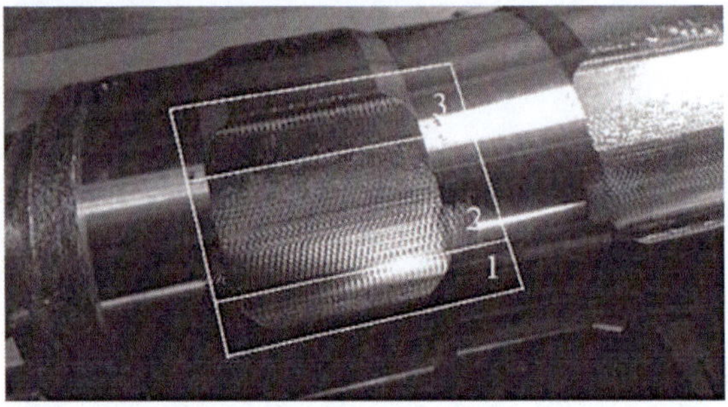

Fig. 6 Machined sleeve

Table 2 Average and max surface roughness values for the sleeve gap positioned on the rigid mandrel

Ra surface roughness (μm)		
Surface 1	Surface 2	Surface 3
9.76 /12.4	6.66/7.6	6.62/7.5

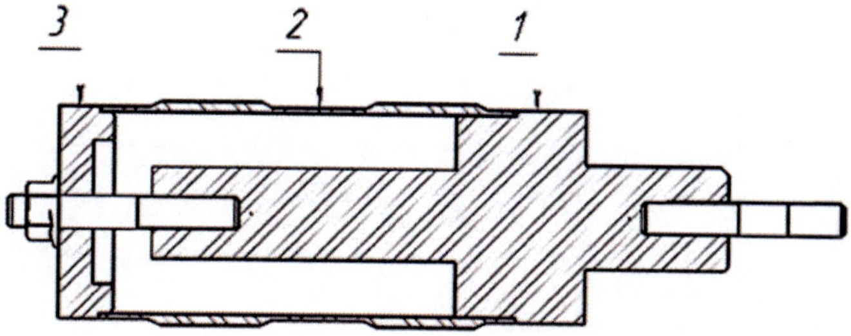

Fig. 7 Working device

the workpiece is broken down into separate elements, which together make up a finite element (FE) model. The finite element model of the workpiece was prepared on the basis of a CAD model using the SolidWorks software." The output data of the simulation of the part under consideration shows that the most dangerous mode in the direction of the Z axis is the mode with a frequency of 1963.4 Hz, where 52% of the part mass is involved. For the *Y*-axis—the mode with a frequency of 1963.6 Hz, where 52% of the part mass is involved. At the same time, the frequency of forced oscillations was 100 Hz, which indicates the absence of the resonance effect.

Reducing the feed rate is the most common way to reduce the cutting speed. In this particular case, this was unacceptable due to its inherent performance loss. In addition, with a sharp decrease in feed, the chip thickness can reach a critical limit when cutting is stopped [13]. Since the cutter was compact and specially designed for CNC machines, it was impossible to modify it to increase the damping ability [10, 11, 14, 15]. The device in the active process is attached to the fourth pivot axis of the CNC milling machine (Fig. 7).

The machined sleeve—2 is installed on the cylindrical surface of the landing shaft and abuts against the end. The other end of the sleeve is fixed with a pressure washer—3. The disadvantage of such a device is that the central part of the sleeve is not fixed, thus, during milling, self-oscillations occur under the action of cutting forces, which leads to a deterioration in the surface quality.

The only way to increase the strength of the clamped workpiece was to optimize the locating and improve the fastening structure.

4 Results

The performed theoretical and experimental studies made it possible to determine the main provisions that must be followed when developing a device for milling thin-walled sleeves [10]. These provisions were taken into account when designing a device with increased vibration resistance (Fig. 8). The operating principle of the device with increased vibration resistance is described in detail in [15].

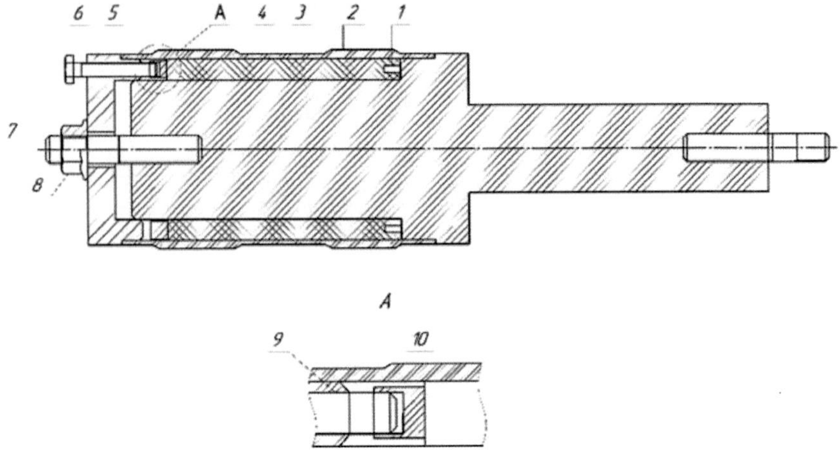

Fig. 8 Designed device: 1-fixing pin, 2-clamped workpiece, 3-base rod, 4-elastic element, 5-cover, 6-screws, 7-central pin, 8-nut, 9-threaded hole, 10-spacer washer

Table 3 Average and max surface roughness values for the part positioned on the vibration-resistant mandrel

Surface 1	Surface 2	Surface 3
Ra surface roughness (μm)		
5.72 /6.9	4.26 /5.2	4.26 /5.2
Average value in the measured area (Ra, μm)		
3.7	3.7	3.7

The workpiece, positioned on the mandrel shown in Fig. 8, was milled with the same milling conditions as before. The roughness of the same surfaces is given in Table 3.

The results show that the roughness exceeds the permissible value only in one case, while on average it is completely below the required value of Ra 6.3.

5 Discussion

The results of the study show that the operation of milling the surfaces of thin-walled sleeves is significantly less vibration-resistant than turning. This is due to the periodically changing value and direction of the cutting force. Since the dimensions of the tool are limited, an increase in the rigidity and damping capacity of the device is required to clamp the workpiece.

In the developed design of the vibration-damping mandrel, both effects are achieved by axial compression of a set of washers made of elastic material, which, when compressed, increase in their outer diameter and are tightly pressed against the inner surface of the sleeve.

6 Conclusion

The proposed device, in all respects, provides the smallest value of the roughness of the machined surface and is used in production, which ensures a decrease in the level of vibrations by 50% and the fulfillment of the given value of the machined surface roughness.

Acknowledgements The reported study was funded by RFBR, project number 20-38-90248.

References

1. Korsakov, V.S.: Precision of Machining, vol. 340. MASHGIZ, Moscow (1961)
2. Astakhov, V.P.: Authentication of FEM in metal cutting. In: Davim J.P. (ed.) Chapter 1 in book Finite Element Method in Manufacturing Processes, pp. 1–43. Wiley, New York (2011)
3. Abushawashi, Y., Xiao, X., Astakhov, V.P.: FEM simulation of metal cutting using a new approach to model chip formation. Int. J. Adv. Mach. Form. Operat. **3**, 71–92 (2011)
4. Abushawashi, Y., Xiao, X., Astakhov, V.P.: A novel approach for determining material constitutive parameters for a wide range of triaxiality under plane strain loading conditions. Int. J. Mech. Sci. **74**, 133–142 (2013)
5. Chuprikov, A.O., Yamnikov, A.S., Troitsky, D. I.: Simulation of deformations in thin shells clamped in a gripper with wrap-around jaws. AIP Conf. Proc. **2188**, 050028 (2019)
6. Glazkov, I.D.: Substantiation of materials for the device used when fixing thin-walled parts of the shaft type. Int. Sci. Res. J. Tech. Sci. I 2 (Yekaterinburg). **1**(67), 76–9 (2018)
7. Pyatykh, A.S., Savilov, A.V., Nikolaev, AYu.: Research of chucks for drilling precise holes in the system. Methods. Technologies (Bratsk) **4**(32), 70–74 (2016)
8. Yamnikov, A.S., Chuprikov, A.O.: Chucks for thin-walled blanks. Russ. Eng. Res. **35**(11), 838–40 (2015)
9. Vasilyev, A.S., Yamnikov, A.S., Matveev, I.A., Yamnikova, O.A.: Influence of hereditary technological errors of production of a basic pipe on parameters of the assembled propulsion section. Chernye Metally. **1**, 67–71 (2019)
10. Yamnikov, A.S., Yamnikova, O.A., Chuprikov, A.O., Matveev, I.A.: Elastic deformations of workpieces of hollow axisymmetric bodies when clamped in three-jaw chucks. Chernye Metally **6**(1038), 25–30 (2018)
11. Yamnikov, A.S., Bogomolov, M.N., Chuprikov, A.O.: Milling-based clamping mandrel. Mech. Eng. Technol. **12**, 3–17 (2019)
12. Yamnikov, A.S., Bogomolov, M.N.: Simulation of oscillation frequency effects on complex shape part milling. IOP Conf. Ser.: Mater. Sci. Eng. **1029**(1), 012006 (2021)
13. Yamnikov, A.S., Myagkov, Yu.V.: University News, vol. 12, pp. 111–115. Mashinostroeniye, Moscow (1979)
14. Leontiev, B.V., Leontieva, A.N.: Cutting Process Management to Remove Vibrations Vologodinsky Conference, pp. 159–162. Vladivostok: Far East Federal University (2012)
15. Yamnikov, A.S., Bogomolov, M.N.: Centering mandrel with increased vibration resistance for milling thin-walled sleeves. Chernye Metally L. **5**, 52–57 (2019)

Study of the Aeration Processes of Filtering Elements for Wastewater Treatment of Industrial Enterprises

Tatyana Novoselova, Larisa Tolmacheva, Alexander Palii, and Jermena Akopdjanyan

Abstract The paper deals with the issues of aeration of filter-carrying elements for wastewater treatment of industrial enterprises and housing and communal services. Conditions for better oxygen utilization in biological wastewater treatment. Aerators of various types are considered. The advantages of permeable cermet materials (PMM) based on alloy steel scale, aluminum oxide and aluminum, capable of performing the functions of wastewater filter elements and their preparation for use in steam generators (SG), are described. A calculation of the parameters the aeration system is given, an estimate of the ejection effect of the flow air bubbles is obtained, and recommendations for the manufacture of elements of aeration tanks and for the placement of filter elements are proposed. Failure to comply with the requirements for the optimal placement of filter elements slightly reduces the efficiency of the system due to the failure to use the dynamic effect of the pressure of air bubbles coming out of the filter elements on the filled volume of the aerotank. If we take as a comparison base aeration systems used at treatment plants based on filter-carrying ceramic tiles and on the basis of perforated asbestos-cement or propylene pipes, then replacing these systems with a new aeration system with SHS-filtering elements leads to a significant increase in reliability due to the strength indicators of filter elements and due to their design.

One of the basic conditions for the best use of oxygen in the biological wastewater treatment is to find it in a dispersed state. Aerators (dispersants) of various types are used to ensure air barbartage. The widespread type of aerator is the filter plate 30×30 cm and a thickness of 25–40 mm (Fig. 1). It is made of a crushed chamotte associated with a mixture of liquid glass with chamotte dust. Sometimes plates are produced with binding on cement. The filtering plates are closed in concrete channels [1].

T. Novoselova · L. Tolmacheva · A. Palii · J. Akopdjanyan (✉)
Polytechnical Institute (Branch) of DSTU in Taganrog, Petrovskaya str., 109a, 347900 Taganrog, Russia

M. Shamtsyan et al. (eds.), *Robotics, Machinery and Engineering Technology for Precision Agriculture*, Smart Innovation, Systems and Technologies 247,
https://doi.org/10.1007/978-981-16-3844-2_14

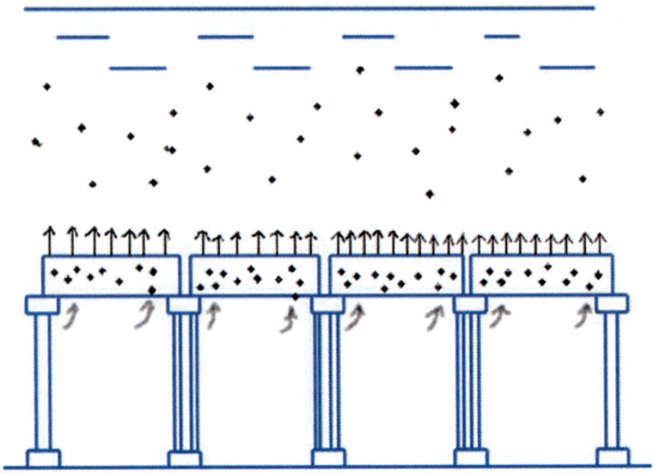

Fig. 1 Ceramic plate installation scheme

The self-propagating high-temperature synthesis (SVS) is the most promising method for the manufacture of porous permeable metal-ceramic materials (PMM) based on the scale of doped steel, alumina and aluminum oxide capable of performing the functions of sewage elements and preparation for use in steam generators (GHG). The advantages of such PMM are their resistance to pressure and corrosion, as well as the uniformity of permeability in volume [2].

To determine the permeability of filters, a calculation method is proposed, which allows you to determine the parameters of the aeration system, substantiate the number of filtering elements, evaluate the ejection effect of air bubbles from the filtered element. At the same time, calculations are given for a standard aerial circuit with detachable transition devices (Fig. 2).

Calculations of the aeration system based on filtering elements, justification of the number of filtering elements [3].

The main calculation of the total area of filtering elements is the Darcy equation

$$\text{grad } P = m * Wk \tag{1}$$

where

grad $P = \mathrm{d}p/\mathrm{d}\delta$—pressure drop across the filter element, MPa;

δ—thickness, m;

m—dynamic air viscosity coefficient, Pa * s;

k—permeability coefficient (value, inverse hydraulic resistance);

W—air dispersion rate by filtering element, $m^3/(m^2 * s)$.

$$W = \frac{Q}{S} \tag{2}$$

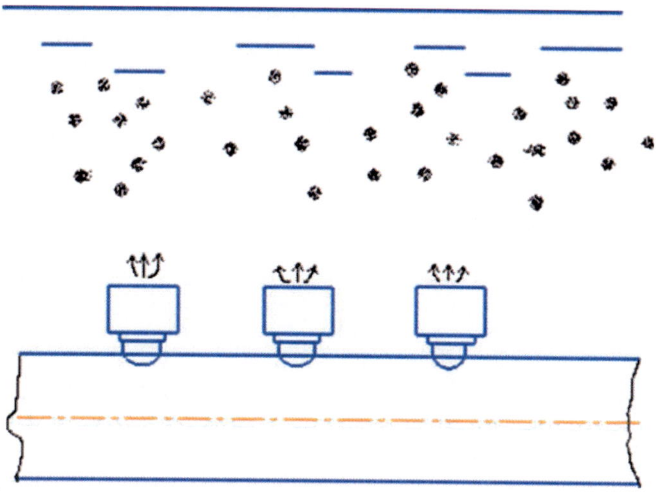

Fig. 2 Aenerative system scheme with deligative transitional filter device

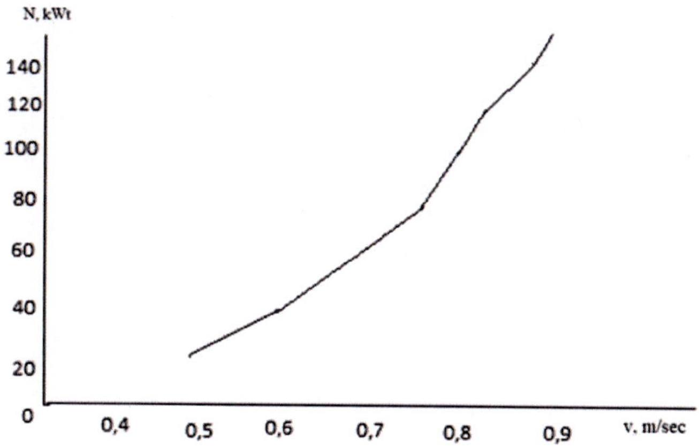

Fig. 3 Dependence of power consumption to ensure the speed of rotational movement in the aerotank

where
Q—air consumption (m^3/s);
S—total aeration area (m^2).
Combining (1) and (2), we have for the total area of filtering elements. Expression:

$$S = \frac{m * \delta * Q}{k * \Delta p} \tag{3}$$

Air consumption on sewage treatment plants:

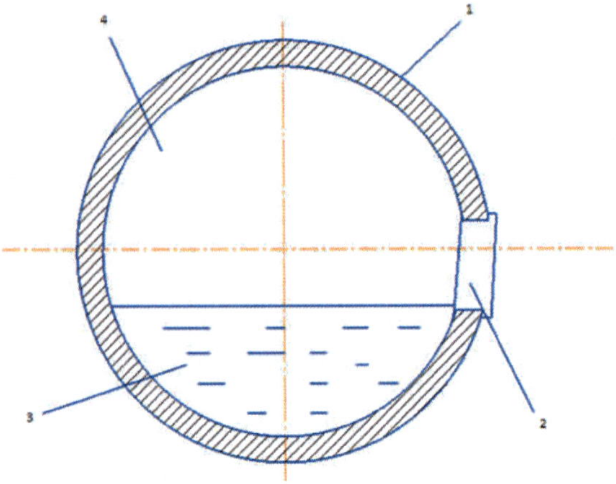

Fig. 4 Diagram of an air supply pipe with a filter element pipe; 2-filtros; 3-water; 4-air

$Q = 181{,}134$ m³/h = 5.04 m/s.
The thickness of the filtering element: $\delta = 0.02$ m.
$m = 0.72*10^{-5}$ kg/(m*s)

$$\Delta P = P_{\text{wat}} - P_{\text{air}} \tag{4}$$

where P_{wat}—Hydraulic pressure at the depth of h installation of filtering elements ($h = 4$ m, $P_{\text{вод}} = h*g*h = 0.04$ MPa).

Using the given data, we get from Eq. (3):

$$S = 4.33 * 10 - 11 * 1/k,\ \text{m}^2 \tag{5}$$

The number of filtering elements is calculated by the formula:

$$n = S/S_{\text{f}} \tag{6}$$

where S_{f}—square of the filtering element ($S_{\text{f}} = 0.00196$ м^2).

The permeability rate was calculated and amounted to $5 * 10^{-11}$. Then, with a security ratio of 1.5 get:

$$P_f = \frac{4.33 * 10^{-11}}{5 * 10^{-11}} * \frac{1}{0.00196} * 1.5 = 660\,\text{pc} \tag{7}$$

The length of the trunk pipeline $L = 80$ m and two air ducts in the Aeroten Corridor filtering elements should be positioned on each air duct in 25 cm [4].

Recommendations for the placement of filtering elements.

Aerotank may pass in the mode of normal operation with air supply through filtering elements; Air removal mode from the air duct tube (after turning on the blower) with the transition to the normal operation mode. Therefore, the optimal placement of filtering elements should satisfy a number of contradictory requirements [5]:

- the direction of movement of air bubbles must correspond to the direction of rotational movement in the aerotank;
- the location of the filter elements must allow the removal of liquid from the pipe;
- the remaining liquid in the pipe should not restrict the air flow through the pipe.

Failure to comply with the first requirement insignificantly reduces the efficiency of the system due to the failure to use the dynamic effect of the impact of the pressure of air bubbles coming out of the filter elements. On the filled volume of the aeration tank.

Estimation of the ejection effect of the flow of air bubbles emanating from the filter element.

To estimate the ejection effect, the following assumptions are used:

- the power developed by the emerging air bubble (Archimedes force) is spent on overcoming local resistances and friction during the movement of waste water in the volume of the aeration tank;
- the movement of fluid in the aerotank is circular;
- liquid is characterized by a kinematic viscosity coefficient $v = 10{*}10^{-6}$ m/s and density $p = 1000$ kg/^{3};
- roughness of the walls of the aeration tank—$\Delta = 0.005$ m;
- cornering resistance coefficient $\varphi_M = 2.5$;
- stream length $l = 7$ m;
- equivalent (hydraulic) diameter $d = 2.96$ m.

At a given air flow rate per aeration tank $Q_B = 5.04$ m^3/s.

The power of the flow of air bubbles (the force of Archimedes):

$$N = Q_B * p * h = 86{,}000 \text{ Wt} \quad (8)$$

Calculation sequence:

1. We set the average speed of rotation of the liquid in the aerotank v, m/s;
2. We calculate the value of the Reynolds criterion $R_c = v\text{d}/V$;
3. We calculate the coefficient of hydraulic friction:

$$\lambda\text{=64/Re, if a Re} < 2300 \text{ – laminar flow,}$$
$$\lambda\text{=0,11*(68/Re+ }\Delta_3\text{/d), if Re} > 2300 \text{ – transient flow} \quad (9)$$

4. Knowing the coefficient of hydraulic friction, we calculate the coefficient of hydraulic resistance:

$$\varphi = \left(\varphi_{\text{M}} + \lambda * \frac{l}{d}\right); \quad (10)$$

Table 1 Estimated data

V (m/s)	Q (m/h)	ƛ	ΔP (Pa)	N_{load} (kW)
0.5	75	0.0400	324	24.3
0.7	105	0.0398	635	66.7
0.8	120	0.0398	930	99.6
0.9	135	0.00398	1050	141.8

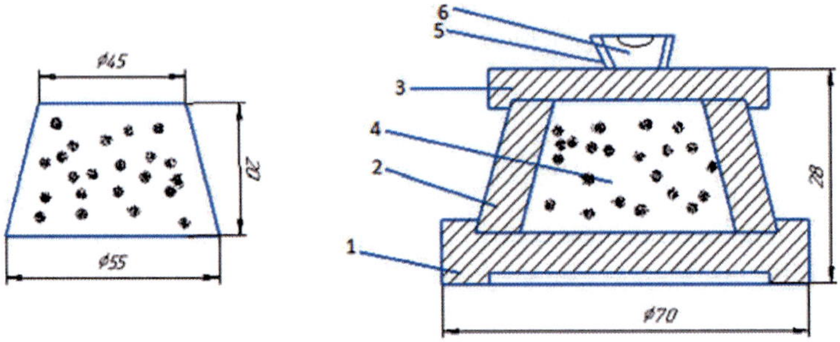

Fig. 5 Filter element and assembled form base; 2-sleeve; 3-cover; 4-composite mixture; 5-funnel; 6-termite mixture

5. We calculate hydraulic losses using the formula:

$$\Delta P = \varphi^* p^* \frac{v^2}{2}; \tag{11}$$

6. We calculate the power that must be spent on the transverse rotational motion of the fluid:

$$N_{\text{spent}} = \Delta P_T * S * V \tag{12}$$

Value N_{spent} as a function of speed are given in Table 1 and Fig. 3.

The desired result is the value of the speed v at which the equality of the value of the flow rate of air bubbles, and the power spent on the transverse rotational movement is fulfilled.

Insofar as $N = 86$ kWt, then the rotational speed $v = 0.76$ m/s. Assuming that the sedimentation rate of activated sludge particles is at the level 0.1 m/s, it can be assumed that the mixing of sludge in the waste water of the aeration tank will be ensured at a given air flow rate [6].

Optimal placement of filter elements—at the bottom of the air supply pipe at an angle to the vertical $0 \pm 15°$ (Fig. 4). In this case, the amount of residual liquid filling is not more than 2–5% of its cross section, which does not significantly reduce the efficiency of air supply[7].

Manufacturing of aerator filter element.

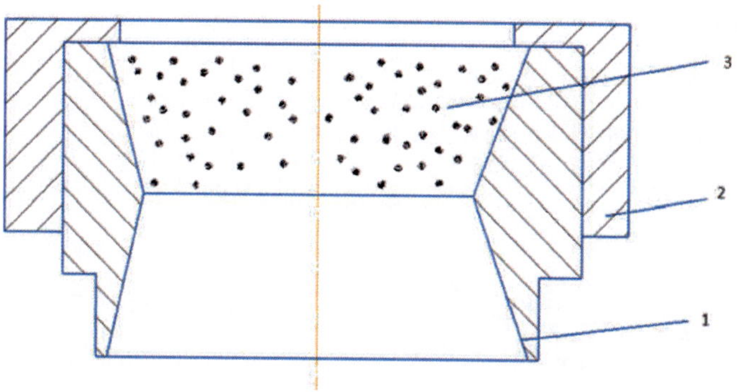

Fig. 6 Detachable adapter device in metal design building; 2-nut; 3-filter element

The design of the filter element and, accordingly, its equipment are shown in Fig. 5.

We have chosen the optimal wall thickness of the technological equipment, equal to 2–4 mm. With a small wall thickness, less than 1 mm, the elements of technological equipment are prone to warping. An increase in the wall thickness by 4–5 mm leads to the formation of an unreacted boundary layer of the composite composition, which distorts the geometric dimensions of the filter-carrying element.

The assembled form (Fig. 5) is installed under the hopper with a slide gate and filled with the composite composition to the brim. The filled mold is placed on a vibrating table and compacted. Compaction is performed within 8010 s, with a frequency of fluctuations of 50 beats per minute. After removing the filling frame, the mixture is cut with a ruler to the level of the upper sleeve. The top cover is installed on the mold [8, 9].

In order to prevent damage to the working surface of the filter element, a funnel 5 is additionally introduced, into which the thermite composition 6 is poured and the SHS reaction is initiated by an electric spiral or open fire. In the funnel, the combustion heat front begins to move, the structure is formed, and the combustion front is aligned. The taper of the funnel makes it easy to separate this part of the mold without damaging the surface of the filter element [7].

After cooling, the product is removed from the mold and cleaned with a metal brush. The produced filter element is installed in a split adapter filtering device (Fig. 6).

f we take as a comparison base aeration systems used at treatment plants in Barnaul, Bataysk, Tomsk and others, based on filter-carrying ceramic tiles and on the basis of perforated asbestos-cement or propylene pipes, but replacing these systems with a new aeration system with SHS-filter-carrying elements leads to a significant increase in reliability due to the strength indicators of the filter elements and due to their design. Re-equipment of the aeration system leads to an increase in the service life between preventive maintenance of aeration tanks [10].

References

1. Chernov, N.N., Palii, A.V., Maevskiy, A.M., Ignatyev, V.V., Spiridonov, O.B.: Proceedings of SPIE—The International Society for Optical Engineering, vol. 11163, 111630K (2019).
2. Gaiser, G., Jorg J.: Oesterle (J. Eberspacher GmbH and Co. KG, Esslinden): Drallelemente zur Gleichmassigen Beladung und Regeneration von Partikelfiltern. MTZ: Motortechn. Z. **66** (5), 362–369 (2005)
3. Moallemi, M., Clifford J., Neighbors J., Pesce J., Towhidnejad M.: AIAA/IEEE Digital Avionics Systems Conference—Proceedings, pp. 7778002 (2016)
4. Aswani, D.J., Van Nieuwstadt, M.J., Cook, J.A., Grizzle, J.W.: Trans. ASME. J. Dyn Syst. Meas. Contr. **127**(1), 1–12
5. Chernov, N.N., Palii, A.V., Sayenko, A.V., Maevskii, A.M.: Tech. Phys. Lett. **44**(4), 328–330 (2018)
6. Schar, C.M., Onder, C.H., Geerig, H.P.: Control of an SCR catalytic converter system for a mobile heavy—duty application. IEE Trans. Contr. Syst. Technol. **14**(4), 641–653 (2006)
7. Tsinoglou, D.N., Koltsakis, G.C.: Influence of pulsating flow on close-coupled catalyst performance. Trans. ASME. J. Eng. Gas Turb. Power. **127**(3), 676–682 (2005)
8. Chernov, N.N., Palii, A.V., Ignatyev, V.V., Maevskiy, A.M., Maksimov, A.V.: Adv. Intell. Syst. Comput. AISC. **1224**, 380–387 (2020)
9. Zueco, S.C., Halbei, J., Neumann, F., Oesterle, J.J., Rudelt, J., Többen, H.: Harnstoff-SCR-Systeme im Fokus Neue Herausforderungen bei der Entwicklung von Abgasanlagen. MTZ: Motortechn. Z. **68** (9), 714–719 (2007)
10. Chernov, N.N., Kovalev, A.V., Palii, A.V., Sayenko, A.V., Maevskiy, A.M.: IOP Conf. Ser.: Mater. Sci. Eng. **862**(3), 032028 (2020)

Layout Features of Drives with Modified Adaptive Friction Clutch

Mikhail Shishkarev and Alexander Rybak

Abstract Urgency of the research is that the task of minimizing the gross mass of the machine drive through determining an efficient mounting location of the adaptive friction clutch in the kinematic chain and establishing the optimal accuracy of its operation has not been solved yet. The research problem is to determine functional dependences of the gearbox weight, adaptive friction clutch (AFC) and rigid movable couplings on the peak torques transmitted by these nodes. The objective of the paper is to develop and analyze a mathematical model of the dependence of the gross mass of the machine drive on the mounting location of the modified first-generation AFC in the machine drive. The research method consists in studying the effect of the peak torques transmitted by gearboxes, a modified adaptive friction clutch and rigid movable couplings on their mass. It is established that the use of a modified first-generation AFC in the machine drive reduces drastically the gross mass of the drive, and a decrease in the gross mass of the drive occurs when the mounting location of the coupling is removed from the drive motor without optimizing its mounting location.

1 Introduction

The property of the first-generation AFC, which consists in increasing its mass while improving the response accuracy through increasing the feedback gain, has led to the existence of a single efficient place to install them in the machine drive [1]. In this case, minimization of the gross mass of the drive elements including the mass of the clutch is obtained. The greatest effect of minimizing the gross mass of the drive is achieved with its extended kinematic chain.

The research objective is as follows: development and analysis of a mathematical model of the dependence of the gross mass of the machine drive on the mounting location of the modified first-generation AFC in the machine drive.

M. Shishkarev (✉) · A. Rybak
Don State Technical University, 1, Gagarin Sq., 344000 Rostov-on-Don, Russia

M. Shamtsyan et al. (eds.), *Robotics, Machinery and Engineering Technology for Precision Agriculture*, Smart Innovation, Systems and Technologies 247,
https://doi.org/10.1007/978-981-16-3844-2_15

2 Research Method

The research method is based on the study of the effect of the peak torques transmitted by the gearboxes, the modified adaptive friction clutch and rigid movable couplings on their mass, the effects of the modified adaptive friction clutch and its mounting location in the kinematic chain of the machine drive on its gross mass.

3 Problem Solution

In contrast to the basic first-generation AFC models based on N.D. Werner's clutch, a modified AFC provides an increase in its own load capacity while increasing the operating accuracy [2].

Consider the circuit diagram of the machine drive shown in Fig. 1. The drive includes an electric motor 1, mechanical gearboxes 2–5, an open gear 6, AFC 7, an executive (working) machine body 8 and flexible double-slider couplings 9. Drive units are mounted on frame 10.

The AFC mounting location divides the drive into two parts:

- a protected part, which is located between the AFC 7 and the motor 1 including the latter;
- a conditionally unprotected part, located between AFC 7 and the working body 8 including the latter.

Recent studies show that the magnitude of the loads acting in the unprotected part of the drive when the AFC is operated, under certain conditions, depends on the accuracy of its operation.

The specified division into parts is made on account of the level of overloads operating in the said parts of the drive.

In the layout, when the AFC is mounted on the motor shaft, it protects only the motor from overloads and is set to the torque equal to the rated torque of the motor

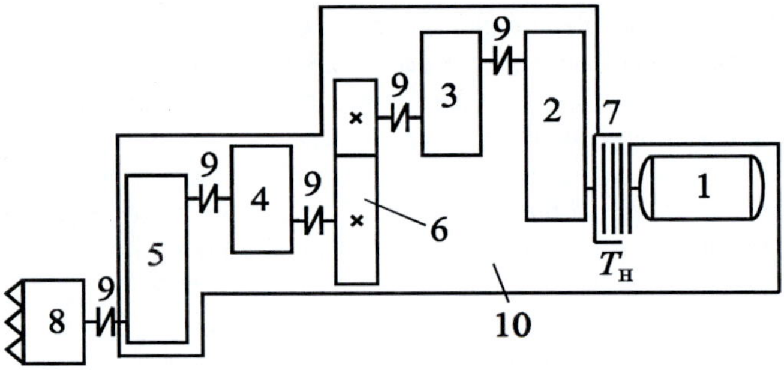

Fig. 1 General basic diagram of machine drive with AFC

T_n. The rest of the drive elements are subjected to a load whose maximum value is determined by the nature of the overloads acting on the working body of the machine 8, by the angular stiffness reduced to the AFC shaft and the reduced moment of inertia of the elements of the unprotected part of the drive.

The written sources analysis of the torque values and the masses of mechanical gearboxes associated with them [3, 4] shows that there is a functional relationship between these parameters, which is approximately described by the following formula:

$$M_i = pT_{\text{vyh.i}}^{2/3}, \tag{1}$$

where M_i is the mass of the i-th gear; p is the dimensional coefficient, kg/(Nm)2/3; $T_{\text{vyh.i}}$ is torque on the output (slow-speed) gearbox shaft.

The coefficient p is thought to be:

- $p = 0.92$ for single-stage spur gearboxes;
- $p = 1.23$ for two-stage spur gearboxes;
- $p = 1.25$ for three-stage spur gearboxes;
- $p = 0.93$ for coaxial two-stage spur gearboxes;
- $p = 0.89$ for planetary single-stage spur gearboxes;
- $p = 0.54$ for waveform gear reduction;
- $p = 0.98$ for worm single-stage gearboxes;
- $p = 1.02$ for worm two-stage gearboxes

The required torque values on the output shafts of the gearboxes are:

$$T'_{\text{vyh.i}} = T_n \prod_i^1 u_i \eta_i, \tag{2}$$

where u_i is the total ratio of the i-th gearbox; η_i is the i-th gearbox efficiency.

With a total number of gears in the drive equal to s, the torque value on the working body of the machine 8 is, according to the formula (2):

$$T_{n.\text{ro}} = T_n \prod_s^1 u_i \eta_i. \tag{3}$$

The magnitude of the peak overload torque acting on the machine working body considering the formula (3) is:

$$T_{\text{ro.max}} = \gamma_{p.n}(K_d + K_r) T_n \prod_s^1 u_i \eta_i, \tag{4}$$

where K_d is the dynamic coefficient whose value depends on the type of engine; v is the dynamic coefficient whose value depends on the type of working machine [5]; $\gamma_{p.n}$ is the factor of the rated load exceedance in the drive [5].

In the considered version of the AFC installed in the drive, under overloading, the load will affect the working body whose value is determined by the formula (4), and the torque values on the slow-speed (output) shafts of the gearboxes will increase to the values determined by the formula (2), taking into account K_d and K_r coefficients of dynamics.

In this case, to avoid damage to the drive elements, the characteristics of the gearboxes, in particular, peak torques on output shafts, should correspond to the maximum overloads.

For AFC based on the analysis of their mass–torque ratio [6], we can approximately put:

$$M_m = q_1 T_n^{2/3}, \tag{5}$$

where M_m is the AFC mass; q_1 is the dimensional coefficient, kg/(Nm)$^{2/3}$; T_n—see above.

According to the estimates made, the q_1 coefficient value is 0.16 … 0.20.

Similarly to the above, the relationship between the mass and torque of the double-slider couplings can be expressed through the following ratio:

$$\sum M_{kdm} = q_2 n T_{m.n}^{1/2}, \tag{6}$$

where $\sum M_{kdm}$ is the gross mass of double-slider couplings; q_2 is the dimensional coefficient, kg/(Nm)$^{1/2}$; n is the number of double-slider couplings; and $T_{m.n}$ is rated torque of the coupling.

According to the estimates made, the q_2 coefficient value is 0.04 … 0.07.

The mass of the frame 10 (Fig. 1) also depends on the rate of the load transmitted by the drive units since, with a change in the torque value $T_{vyh.i}$, the overall dimensions of the gearboxes change resulting in a corresponding change in the frame dimensions. According to the written sources analysis, the mentioned dependence can be represented in the following form:

$$M_r = q_3 T_{\text{vyh.sr.i}}^{2/3}, \tag{7}$$

where M_r is the frame mass; q_3 is the dimension factor of proportionality, kg/(Nm)$^{2/3}$; $T_{\text{vyh.sr.i}}$ is the average value of peak torques on output shafts of gearboxes.

According to various estimates, the q_3 coefficient is 10 … 12.

On the basis of the above, the gross mass of the drive (without the drive electric motor and the working body of the machine), considering the formulas (1), (5), (6) and (7), will be:

$$M_{\text{obsh}} = T_n^{2/3} \sum_1^s p_i \left(\Pi_i^1 u_i \eta_i \right)^{2/3} + q_1 T_n + q_2 n (K_d + K_r) \gamma_{p.n} \sum_1^n T_{m.n}^{1/2} + q_3 T_{\text{vyh.sr.}i}^{2/3} \tag{8}$$

where p_i is the current p coefficient value.

In the formula (8), a change in the mass of the open gear is considered by the first term since the dependence of the said mass on the load transmitted by the gear is described by the relation (1).

If you move the AFC in the kinematic chain of the drive from the motor 1 in the direction of load transfer installing it on the output shafts of the gearboxes 2–4 (instead of the double-slider coupling 9) (see Fig. 1), then the rated torque of the AFC should be increased according to the torque on the output shaft of the corresponding gearbox.

At that, due to an increase in the accuracy of AFC operation, the torques on output shafts of gearboxes located in the protected part of the drive will be reduced, which, accordingly, will affect the value of the rated AFC torque.

With the mentioned movement of the AFC in the kinematic chain of the drive, we will proceed from the fact that the value of the nominal torque on the shaft of the electric motor 1 is equal to the tuning breakdown torque of the AFC.

This means that the torque margin when adjusting the coupling to increase it through the efficiency of the feedback gain constant (GC) is absent. Therefore, we assume that an increase in the AFC torque is accompanied by an increase in its mass in accordance with the dependence (5).

The growth of the AFC mass as a result of increasing the tuning torque is due to the requirement to expand the area of friction surfaces to maintain the contact pressure margin. As a result, the radial dimensions of the elements of the friction group and the half-couplings increase [7].

The torque of the modified first-generation AFC is determined by the formula [8]:

$$T_p = zF_pR_{sr}C\frac{f}{1+zCf}, \tag{9}$$

where z is the number of friction couples of the friction group; F_p is the total tension force of the connecting spring block; R_{sr} is the average radius of the friction surfaces of the friction group; f is the friction factor; C is the feedback GC:

$$C = \frac{R_{sr}}{r}tg\alpha,$$

r is the radius of the circle on which sensitive elements of the feedback device are located; α is the pressure angle of the sensing elements.

If the AFC mounting location in the drive is replaced with its removal from the engine, then the accompanying increase in the nominal torque of the coupling can be carried out, according to the formula (8), through increasing the GC. In this case, the required value of GC can be determined as follows. When installing the AFC drive half-coupling on the output shaft of the i-th gearbox, the following equality should be performed:

$$T_n^{'}K_t = T_{vyh.i}, \tag{10}$$

where $T_n^{'}$ is the rated (tuning) AFC torque under the specified condition; K_t is the AFC accuracy factor; $T_{\text{vyh}.i}$ is the peak torque on the output shaft of the i-th gearbox.

The formula (9) is written considering that the AFC is tuned to transmit the rated torque with the minimum value of the friction factor [7].

The first-generation AFC accuracy factor is determined through the following formula [9]:

$$K_t = \frac{f_{\max}(1 + zCf_{\min})}{f_{\min}(1 + zCf_{\max})} \tag{11}$$

where $f_{\min}$, $f_{\max}$ are the minimum and maximum friction factors, respectively.

Substituting the formula (10) into the equality (9) and taking into account the relation (8) (at $f = f_{\min}$), we find the required value of GC:

$$C_i = \frac{T_{\text{vyh}.i}}{zf_{\max}\left(F_p R_{\text{sr}} - T_{\text{vyh}.i}\right)}.$$

Considering that $T_{\text{vyh}.i} = K_t T_{\text{vyh}.i}^{'}$, and using the formulas (2) and (10), we find:

$$C_i = \frac{T_n \prod_{i}^{1} u_i \eta_i}{zf\left(F_p R_{\text{sr}} - T_H \prod_{i}^{1} u_i \eta_i\right)_{\min}}. \tag{12}$$

When changing the AFC mounting location (relative to the initial one), overloads determined by the accuracy of AFC operation will act in the protected part of the drive [10]. In the unprotected part of the drive, peak loads are determined by the formula (4) [10].

On this basis, the gross mass of the drive units, excluding the engine and the working body of the machine, is:

$$\begin{aligned} M_{\text{obsh}} = {} & \gamma_{p.n}(K_d + K_r)T_n^{\frac{2}{3}} \sum_{i=1}^{k} p_j \left(\prod_{i}^{1} u_i \eta_i\right)^{\frac{2}{3}} + q_1 T_{\text{vyh}.i} + K_t T_n^{2/3} \sum_{i=1}^{s-k} p_j \left(\prod_{i}^{1} u_i \eta_i\right)^{\frac{2}{3}} \\ & + \gamma_{p.n}(K_d + K_r) q_2 \sum_{i=1}^{n_1} T_{m.n.i} + (n - n_1) K_t \sum_{i=1}^{n-n_1} T_{m.n.i} \end{aligned} \tag{13}$$

where k is the number of gears located in the unprotected part of the drive; n_1 is the number of double-slider couplings located in the unprotected part of the drive; $T_{m.n.i}$ is torque on the output shaft of the i-th gearbox counting from the electric motor.

Other notations used in the formula (12) are given above.

The formulas (7) and (12) do not allow for the numerical analysis of changes in the gross mass of the drive depending on the AFC mounting location. In this regard, we will use the graphical method constructing a dependency graph and using the formula (7) for this as a partial dependence (12).

To build a graph, we use the following initial data:

$\gamma_{p.n} = 1.125$ [8], $K_d = 0.25$ [8], $K_r = 1.6$ [8], $q_1 = 0.18$, $q_2 = 0.05$, $q_3 = 10$, $T_n = 60$ N·m, $u_1 = 2.8$, $\eta_1 = 0.98$, $u_2 = 4$, $\eta_2 = 0.97$, $u_3 = 2.8$, $\eta_3 = 0.97$, $u_4 = 2.8$, $\eta_4 = 0.98$, $n = s = 4$, $p_1 = p_4 = 0.92$, $p_2 = p_3 = 1.23$.

When calculating the torque values on the output shafts of the gearboxes, the formula (2) was used.

Changing the AFC mounting location in the drive results in a corresponding change in the value of its nominal torque. Since the AFC rated torque changes due to the variation of the GC value, the clutch accuracy coefficient in the formula (12) is variable: its value was determined by the relation (10). For that, the following values were taken: $z = 6$, $f_{\min} = 0.1$, $F_p = 3000$ N.

For an open gear transmission, it is accepted: $u_{o.p} = 2.4$, $\eta_{o.p} = 0.97$.

For the given value of the spring tension force F_p and the initial value of the GC C_i, the minimum contact area of the friction couple is:

$$S = \frac{F_p C_i r}{R_{\mathrm{sr}}[q]}, \tag{14}$$

where $[q]$ is the value of permissible contact pressure on the friction surfaces.

When the width coefficient of the ring area of the friction disk is 0.25 [8], the contact area of the friction couple is calculated by the following formula:

$$S = \pi R_{\mathrm{sr}}^2.$$

After substituting this equality into the formula (13) and replacing C_i with C_i, we find the dependence of the average radius of the friction surfaces on the GC:

$$R_{\mathrm{sr}} = \sqrt[3]{\frac{F_p r C_i}{\pi[q]}}. \tag{15}$$

Substituting the right side of the formula (14) into the equality (11), we obtain:

$$C_i\left(F_p C_i^{1/3}\sqrt[3]{\frac{F_p r}{\pi[q]}} - T_n \prod_i^1 u_i\eta_i\right) = \frac{T_n \prod_i^1 u_i\eta_i}{z f_{\min}}. \tag{16}$$

The left side of Eq. (15) cannot be simplified since a power function with an exponent less than unity is not linearized. Therefore, to find the required values C_i corresponding to certain AFC mounting locations in the drive, we will use the graphical method to solve Eq. (15) with respect to the unknown term C_i.

We accept the following initial data: $r = 0.04$ m, $[q] = 0.2$ MPa [7]. The rest of the initial data of the parameters are mentioned above.

Figure 2 shows the dependence graphs $A(C_i)$ (line 1) and $B(C_i)$ (lines 2). The parameter A reflects the left side of Eq. (15), and the parameter B reflects its right side. Lines 2 correspond to the torques on the output shafts of the gearboxes and the open gear transmission. The intersection points of line 1 with lines 2 show the GC values when installing the AFC on the corresponding output gear shaft.

The found values of C_i were used to construct a dependence graph of the gross mass of the drive on the AFC mounting location shown in Fig. 3. The graph is built on the basis of the formula (12) according to the accepted source data.

The graph shows a steady decrease in the gross mass of the drive as the AFC mounting location moves away from the electric motor (the consecutive numbers of the AFC mounting locations are marked on the abscissa axis: the first is on the electric motor shaft, the second is on the output shaft of the first gearbox, etc.).

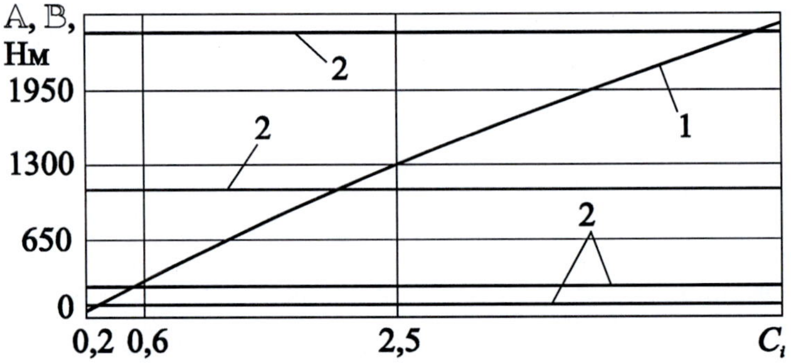

Fig. 2 Graphic solution to Eq. (15)

Fig. 3 Dependence of gross mass of the machine drive on AFC mounting location

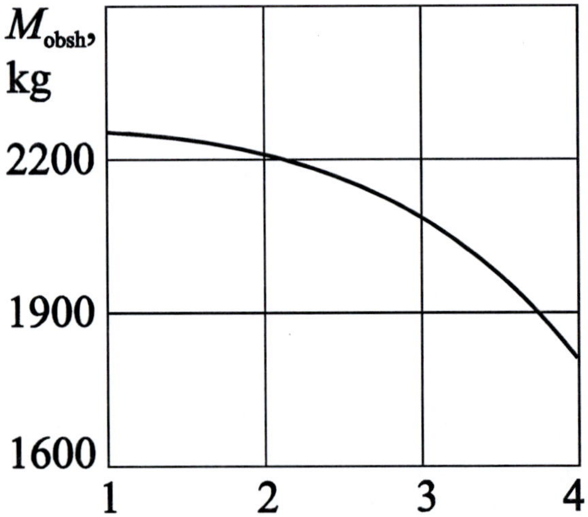

This is a key difference between the modified first-generation AFC and the basic version of the coupling. (When using the basic version, there is only one place to install the coupling corresponding to the minimum gross mass of the drive.)

The graph in Fig. 3 also shows that with an increase in the GC, the gross mass of the drive decreases, which is explained, firstly, by an increase in the protected part of the drive, and secondly, by an increase in the AFC operation accuracy.

When using the basic AFC version, on the contrary, there is a limiting (optimal, from the point of view of the response accuracy) GC value corresponding to the minimum gross mass of the drive [10]. Therefore, when using a modified first-generation AFC, a greater effect of reducing the gross mass of the drive is achieved than when using the basic version of the first-generation AFC.

Finally, from Fig. 3, it follows that the slope of the curve increases with the distance of the AFC mounting location from the electric motor, therefore, from the point of view of reducing the gross mass of the drive, it is more advantageous to install the coupling closer to the working body, thereby increasing the dimensions of the protected part of the drive.

4 Discussion and Results

The found functional dependences of the mass of the machine drive units on the maximum load transmitted by them made it possible to create a mathematical model of the gross mass of the drive in which the AFC mounting location is variable. The operating characteristics of the AFC, in particular, the operation accuracy, together with the mounting location of the coupling directly affect the gross mass of the drive.

The research results can be used in the calculations and design of the modified first-generation AFC considering the operating characteristics of the drive. In addition, the developed methodology of the drive physical data can be used to study the effect of the operating characteristics of other types of AFC on the performance of a machine drive.

5 Conclusions

1. There are functional dependences of the mass of the gearbox, AFC, and rigid movable couplings on the peak torques transmitted by these nodes, which provide the creation of a mathematical model of the dependence of the gross mass of the machine drive on the AFC mounting location in the drive.
2. The use of a modified first-generation AFC in the machine drive reduces drastically the gross mass of the drive; moreover, a decrease in the gross mass of the drive occurs when the AFC mounting location is removed from the drive motor without an optimal coupling mounting location.

3. It is efficient to install the AFC as close as possible to the working body of the machine if it is a source of overload in the drive. In this case, the AFC not only effectively protects drive components from overload, but also minimizes the gross weight of the drive.

References

1. A. Ivanovskaya, E. Bogatyreva, A. Rybak, Simulation of the cable part of the load-lifting device drive, represented as a rod system. In: The Collection: IOP Conference Series: Materials Science and Engineering. Ser. "International Scientific and Practical Conference Environmental Risks and Safety in Mechanical Engineering, ERSME 2020 (2020), pp. 012059
2. A. Rybak, M. Shishkarev, V. Zharov, A. Demyanov, Modeling and calculation of hydromechanical systems dinamics based on the volume rigidity theory. In the Collection: MATEC Web of Conferences (2018), pp. 01001
3. A. Rybak, A. Pelipenko, M. Shishkarev, M. Nevzorova, Projection of the hydraulic pressure drive with application of partial synthesis M. In: The Collection: MATEC Web of Conferences, pp. 01017 (2018)
4. Shishkarev, M.: The choice of optimal actuation accuracy of adaptive friction clutches. Assembly in Mech Eng. Instr. Making **8**, 342–348 (2019)
5. M. Shishkarev, A. Rybak, Dynamic processes in the machine drive with safety clutch. In the collection: AIP conference proceedings. In: XV International Scientific-Technical Conference "Dynamics of Technical Systems", DTS 2019, pp. 060008 (2019)
6. Shishkarev, M.: Synthesis and analysis of adaptive friction coupling with negative-zero feedback. Assem. Mech. Eng. Instr. Making **11**, 499–505 (2020)
7. Shishkarev, M.: Elements of the theory of adaptive friction clutches of the second generation. Assem. Mech. Eng. Instr. Making **1**, 20–29 (2021)
8. Shishkarev, M.: Influence of the method of setting up an adaptive friction clutch (basic version) on its maximum load. Assem. Mech. Eng. Instrument-making **12**, 552–557 (2020)
9. Shishkarev, M.: Synthesis and analysis of adaptive friction coupling with separate friction units. Assem. Mech. Eng. Instr. Making **1**, 18–24 (2019)
10. Zelikov, V., Raghuram Kadali, B., Stasyuk, V., Kazachek, M.: Use of fuses of increased accuracy in the drives of rotary forestry machines. In: The Collection: IOP Conference Series: Earth and Environmental Science. International Forestry Forum "Forest Ecosystems as Global Resource of the Biosphere: Calls, Threats, Solutions", pp. 012065 (2020)

Analyzing the Methods Identification Shaft Position in Active Robotic Assembly of "Shaft-Sleeve" Joints with Chamfer Contact

Mikhail Vladimirovich Vartanov, Vladimir Kirillovich Petrov, Van Dung Nguyen, and Dinh Van Tran

Abstract The article answers the questions about robotization of the cylindrical joint assembly. A method identifying the shaft position in a robotic assembly at the stage of contact on the face is proposed. A mathematical model has been developed for the chamfer contact based on identifying the output signals from the force-torque sensor. The model is constructed according to a quasi-static formulation. Analytical data have been received determining the jamming conditions and the maximum displacements of parts during robotic assembly. An experimental installation equipped with a force-torque sensor and a gripping device has been created. Chose a rigid fastening scheme for the force-torque sensor and the part in the gripper to organize feedback with the robot control system. The experimental values of the friction coefficient obtained by the authors earlier were used. The regularities of changes in the forces, moments, and angles of the shaft skew that characterize the coupling stage hase been obtained and calculated, and experimental values of forces and moments have been compared. Identifying the contact point position will allow you to adjust the movement of the output link of the robot during the assembly operation. The assembly based on positional force control will ensure the technological reliability of the coupling processes of cylindrical joints with a small gap. The given studies provide intermediate results on developing contact states of the assembly at the stage of contact on the chamfer. Further work contains a solution to the issue of identifying the part at the three-point contact stage.

1 Introduction

The process of mating cylindrical joints is a common task in assembly processes. There are several methods for the robotic assembly of shaft-sleeve joints. In [1], a method for assembling cylindrical joints with a gap is considered based on passive adaptation and low-frequency oscillations. In work [2], the influence of rotation and low-frequency vibration in assembling cylindrical joints with a gap is investigated.

M. V. Vartanov · V. K. Petrov · V. D. Nguyen (✉) · D. V. Tran
Moscow Polytechnic University, Moscow, Russia

M. Shamtsyan et al. (eds.), *Robotics, Machinery and Engineering Technology for Precision Agriculture*, Smart Innovation, Systems and Technologies 247,
https://doi.org/10.1007/978-981-16-3844-2_16

Currently, robotic assembly requires high positioning accuracy, so it is practical to apply feedback based on a force-torque sensor. For example, Ghalyan [3] uses an actively adaptive device to provide direct control of the parts' position to be joined, and methods for modeling contact states for rigid and flexible robotic assembly have also been developed. Simakov [4] proposes ways of identifying the position of the contact point of parts at the first step of assembly, when the shaft contacts the end face of the sleeve, and a model for determining the contact angles of a part during assembly is developed based on signals from a force-torque sensor.

Sensor signals can be used to recognize contact states for robotic assembly tasks [5–8]. This work will determine the positions of the points of contact between the shaft and the bushing chamfer using signals from the force-torque sensor. Information about the contact point position will allow you to adjust the movement of the working body of the robot during the assembly operation [9]. The development of an assembly method using active adaptation and applying a force-torque sensor will ensure the technological reliability of the processes of mating cylindrical joints with a small gap, including for parts with chamfers. This article solves the problem of determining the party's position at the mating stage when contacting along the chamfers (Fig. 1).

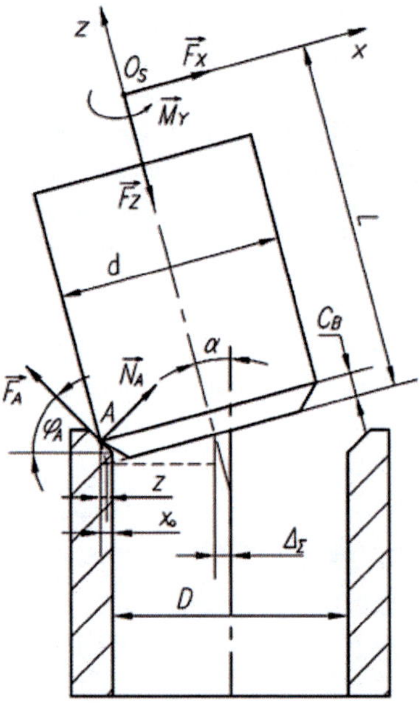

Fig. 1 Schematic of the action of forces on the shaft when contacting along the chamfers

2 Analysis of Forces in Contact with Chamfers

As shown in Fig. 1, the coordinate system is associated with its central point O_S, the $O_S z$ axis coinciding with the gripper axis, and $O_S x$ is chosen so that the shaft and sleeve are symmetrical about the $O_S xz$ plane. The contact is denoted at a point on the cylindrical surface of the shaft (point A). At the point of contact A, the normal force $\vec{\mathbf{N}}_A$ and the friction force $\vec{\mathbf{F}}_A$ are applied [10–14].

The friction force at point A is determined by a relationship:

$$F_A = f N_A \tag{1}$$

where f is the coefficient of friction.

The forces in the single-point contact form the assembly forces F_X, F_Z and the torque M_Y at the center point O_S. For the boundary conditions that determine the jamming conditions, the equilibrium equations can be written:

$$\begin{aligned} F_X &= -N_A \sin(\alpha + \varphi_A) + F_A \cos(\alpha + \varphi_A) \\ F_Z &= N_A \cos(\alpha + \varphi_A) + F_A \sin(\alpha + \varphi_A) \\ M_Y &= -(L - C_B) N_A \sin(\alpha + \varphi_A) + \frac{d}{2} N_A \cos(\alpha + \varphi_A) \\ &\quad + (L - C_B) F_A \cos(\alpha + \varphi_A) + \frac{d}{2} F_A \sin(\alpha + \varphi_A) \end{aligned} \tag{2}$$

where

L distance from the center of the force-torque sensor to the end of the shaft;
φ_A chamfer angle in the bushing hole;
α angle of misalignment of the axis of the shaft and the hole;
C_B chamfer height on mating surfaces.

Substituting the values (1) into (2), we obtain the equilibrium equations:

$$\begin{aligned} F_X &= -N_A \sin(\alpha + \varphi_A) + f N_A \cos(\alpha + \varphi_A) \\ F_Z &= N_A \cos(\alpha + \varphi_A) + f N_A \sin(\alpha + \varphi_A) \\ M_Y &= -(L - C_B) N_A \sin(\alpha + \varphi_A) + \frac{d}{2} N_A \cos(\alpha + \varphi_A) \\ &\quad + (L - C_B) f N_A \cos(\alpha + \varphi_A) + \frac{d}{2} f N_A \sin(\alpha + \varphi_A) \end{aligned} \tag{3}$$

The values of the reactions N_A to F_X and F_Z are determined from the first two equations (3). The reaction values depend on the value of the assembly torque:

$$N_A = \frac{F_X \cos(\alpha + \varphi_A) + F_Z \sin(\alpha + \varphi_A)}{f} \tag{4}$$

Substituting values (4) into (3):

$$M_Y = -\frac{F_X \cos(\alpha + \varphi_A) + F_Z \sin(\alpha + \varphi_A)}{f} \cdot \left[\left(L - C_B - \frac{d}{2} f \right) \sin(\alpha + \varphi_A) - \left(\frac{d}{2} + (L - C_B) f \right) \cos(\alpha + \varphi_A) \right] \quad (5)$$

Transforming Eq. (5):

$$\left(L - C_B - \frac{d}{2} f \right) F_Z \sin^2(\alpha + \varphi_A) + \left[\left(L - C_B - \frac{d}{2} f \right) F_X - \left(\frac{d}{2} + (L - C_B) f \right) F_Z \right] \sin(\alpha + \varphi_A) \cos(\alpha + \varphi_A) - \left(\frac{d}{2} + (L - C_B) f \right) F_X \cos^2(\alpha + \varphi_A) + f M_Y = 0 \quad (6)$$

$$\left[\left(L - C_B - \frac{d}{2} f \right) F_Z + f M_Y \right] \tan^2(\alpha + \varphi_A) + \left[\left(L - C_B - \frac{d}{2} f \right) F_X - \left(\frac{d}{2} + (L - C_B) f \right) F_Z \right] \tan(\alpha + \varphi_A) + \left[-\left(\frac{d}{2} + (L - C_B) f \right) F_X + f M_Y \right] = 0 \quad (7)$$

Having solved Eq. (7), we obtain the value of the skew angle α:

$$\alpha = \arctan \frac{-b \pm \sqrt{\Delta}}{2a} - \varphi_A \quad (8)$$

where: $a = (L - C_B - \frac{d}{2} f) F_Z + f M_Y$; $b = (L - C_B - \frac{d}{2} f) F_X - [\frac{d}{2} + (L - C_B) f] F_Z$; $c = -[\frac{d}{2} + (L - C_B) f] F_X + f M_Y$; $\Delta = b^2 - 4ac$.

However, when using a force-torque sensor, the readings are taken in the form of forces and moments directed along three coordinate axes: the $O_S z_S$ axis is directed along the shaft axis, and the $O_S x_S$ and $O_S y_S$ axes are perpendicular to it and lie in the plane of the shaft cross-section. The projections of the force vector on the coordinate axis F_{X1}, F_{Y1}, F_{Z1} and the projections of the moment vector M_{X1}, M_{Y1}, M_{Z1} are taken from the force-torque sensor.

This problem will become easier to solve if represent the force in the form of two components: axial—$\vec{\mathbf{F}}_Z$ and radial $\vec{\mathbf{F}}_X$ lying in the contact plane [15]:

$$\vec{\mathbf{F}}_Z = F_{Z1} \vec{\mathbf{k}}_1, \quad (9)$$

where $\vec{\mathbf{k}}_1$ is the unit vector of the $O_S z_S$ axis.

The force modulus $\vec{\mathbf{F}}_{\mathrm{X}}$ can be defined as:

$$F_{\mathrm{X}} = \sqrt{F_{\mathrm{X1}}^2 + F_{\mathrm{Y1}}^2} \tag{10}$$

The direction of the vector $\vec{\mathbf{F}}_{\mathrm{X}}$ is set by the direction cosines of the angles between the force vector and the positive directions of the coordinate axes, which determine the position of the contact points A and B relative to the $O_{\mathrm{S}}x_{\mathrm{S}}$ and $O_{\mathrm{S}}y_{\mathrm{S}}$ axes. Similarly, determine the torque vector. Having received the vector $\vec{\mathbf{M}}_{\mathrm{Y}}$ from the projection sensor M_{X1}, M_{Y1}, we can calculate its modulus [15]:

$$M_{\mathrm{Y}} = \sqrt{M_{\mathrm{X1}}^2 + M_{\mathrm{Y1}}^2} \tag{11}$$

By solving Eq. (8) in MATLAB, we obtain the limit value of the skew angle α, which determine by the displacement between the shaft and the bushing during single-point contact. Besides, when programming, also needed the relative limiting displacement of the axes. Then the errors in the position of the axes of the mating surfaces in the assembly device are as follows:

$$\begin{aligned}\Delta_{\Sigma} &= 0.5D - 0.5d\cos\alpha + x_o - z \\ &= 0.5D - 0.5d\cos\alpha + x_o - C_B \cdot \sin\alpha \\ &\approx 0.5D - 0.5d\cos\alpha\end{aligned} \tag{12}$$

3 Experimental Setup

An experimental setup equipped with a force-torque sensor (Fig. 2) [15] was created for experimentation. The installation is mounted on the basis of an IRB-140 industrial robot with an IRC5 controller. The cylindrical mating parts of the shaft and sleeve are rigidly based on the gripper 2 and the base device 4. Information about the forces and moments arising in the assembly area enters the control system of the industrial robot and is also recorded by a personal computer 5.

The shaft and sleeve were made of the same materials CT45. The coefficient of friction was determined at different depths of conjugation between the shaft and sleeve [15]. The conjugation was performed by fit $\varnothing 16\frac{H7}{f7}$ and chamfer of the mating surface $C_B = 0.6$ mm. A gripping device was used to meet the requirements of rigid fixation of the shaft and also allow to make a forming feedback in the control system of an industrial robot (Fig. 3) [16]. The calculated distance from the center of the force-torque sensor to the end of the shaft is $L = 314$ mm. The main components of the adaptive gripper are: a force-torque sensor model Gamma SI-65-5 by ATI Industrial Automation (1) and a pneumatic gripper Schunk GSM-P 64-E-180 (3), on which original prismatic jaws are installed (4). To combine the components into a

Fig. 2 Experimental setup for researching robotic assembly: 1—industrial robot ABB IRB 140; 2—gripper; 3—experimental samples; 4—basic device; 5—personal computer

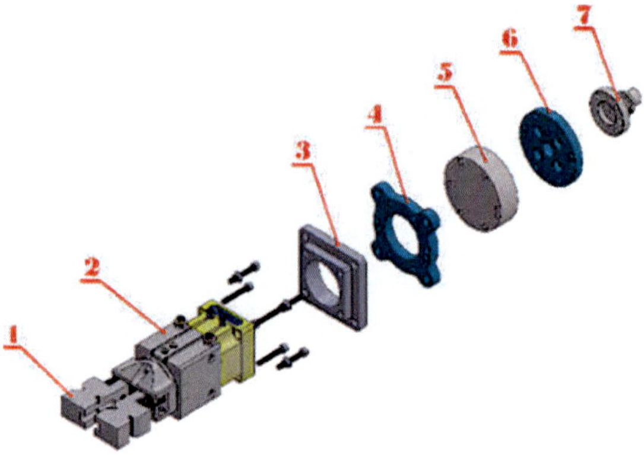

Fig. 3 Adaptive gripper: 1—prismatic jaws; 2—pneumatic gripper model Schunk GSM-P 64 E-180; 3—adapter; 4—Flange 2; 5—force-torque sensor model FTN-AXIA80 SI-200-8/SI-500-20 Ethernet, art. 1324513; 6—flange 1; 7—output link of the manipulator

single device, a mounting device (2) was designed, with the help of which the device is also installed on the manipulator's output link.

Table 1 shows the values of the components of the forces and moments F_{X1}, F_{Y1}, F_{Z1},M_{X1}, M_{Y1}, M_{Z1} obtained from the force-torque sensor and calculated in the MATLAB environment from Eqs. (8–12).

Figure 4 shows how the forces arising at the points of contact changed during the mating process. Theoretical values of moments and forces M_Y and F_X are calculated from formulas (10) and (11). On the horizontal axis of coordinates, the points of displacement of the shaft relative to the bushing were plotted, and on the vertical axis, the coordinate showed the components of the primary vector of force and moment.

Figure 5 shows the changes in the angles of deviations and displacements of parts. Besides, the distance from the center of the hole to the center of the sensor decreases as the inclination angle decreases.

Table 1 Experimental and calculated parameters

No	Experimental data						Calculated data				
	F_{X1}	F_{Y1}	F_{Z1}	M_{X1}	M_{Y1}	M_{Z1}	F_X	F_Z	M_Y	α	$\Delta_{\sum}$
1	0.839	0.017	−0.536	0.044	0.252	0.006	0.839	−0.536	0.256	24.76	0.75
2	0.840	−0.087	−0.581	0.051	0.251	0.004	0.845	−0.581	0.256	22.75	0.64
3	0.880	−0.090	−0.620	0.063	0.247	0.000	0.885	−0.620	0.255	21.72	0.59
4	0.898	−0.166	−0.641	0.066	0.241	−0.002	0.913	−0.641	0.250	21.19	0.56
5	1.918	−0.797	−1.567	0.251	0.561	0.006	2.077	−1.567	0.614	20.00	0.50
6	1.855	−0.864	−1.596	0.254	0.564	0.006	2.046	−1.596	0.619	19.33	0.47
7	1.821	−0.900	−1.627	0.256	0.567	0.006	2.031	−1.627	0.622	18.73	0.44
8	1.821	−0.876	−1.643	0.253	0.567	0.006	2.021	−1.643	0.621	18.34	0.43
9	1.829	−0.774	−1.647	0.249	0.568	0.006	1.986	−1.647	0.621	17.96	0.41
10	1.920	−0.678	−1.692	0.230	0.561	0.005	2.036	−1.692	0.606	17.39	0.39
11	2.419	−1.057	−2.190	0.292	0.665	0.006	2.640	−2.190	0.726	16.65	0.35
12	2.357	−1.074	−2.162	0.297	0.668	0.007	2.590	−2.162	0.731	16.73	0.36
13	2.157	−1.125	−2.118	0.307	0.668	0.007	2.433	−2.118	0.735	16.25	0.34
14	2.153	−1.092	−2.121	0.315	0.662	0.008	2.414	−2.121	0.733	16.05	0.33
15	2.190	−1.046	−2.140	0.318	0.659	0.006	2.427	−2.140	0.732	15.86	0.32
16	3.258	−1.338	−3.199	0.421	0.939	0.010	3.522	−3.199	1.029	14.71	0.28
17	4.216	−1.814	−4.186	0.559	1.224	0.010	4.589	−4.186	1.346	14.63	0.28
18	5.133	−2.264	−5.211	0.694	1.482	0.016	5.611	−5.211	1.637	14.07	0.26
19	5.294	−2.637	−6.046	0.831	1.520	0.013	5.914	−6.046	1.732	11.39	0.18
20	5.837	−3.154	−7.160	0.977	1.683	0.012	6.635	−7.160	1.946	9.87	0.14
21	6.137	−3.515	−8.090	1.017	1.741	0.008	7.072	−8.090	2.016	7.96	0.10

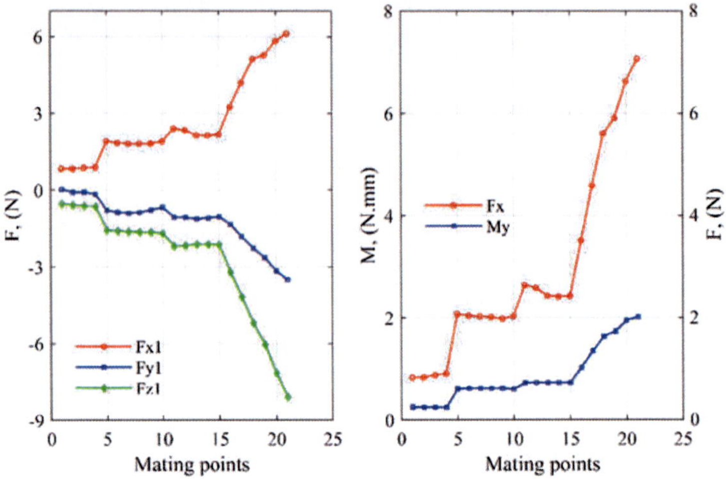

Fig. 4 Change of the main vector of forces and moment at different mating points

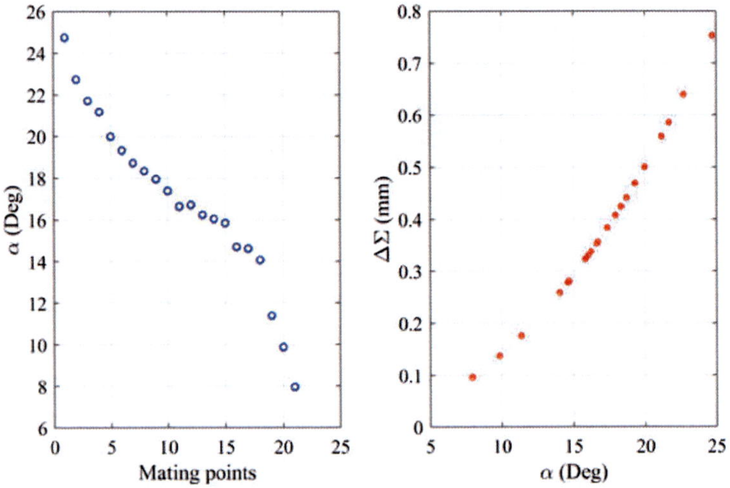

Fig. 5 Changing the angle of deflection and offset of the axes of parts at different points of mating

4 Conclusion

In this work, equations are obtained that connect the parameters of orientation and the position of the connected part with the information issued by the sensor of forces and moments of the assembled robot. These relations can be used as the base for the part's adaptive motion control system's identifier of contact situations. According to the experiment results, determine the angle of deviation of the shaft axis relative to the bushing axis was. The deflection angles decreased from 1 point to 21 contact points (Fig. 5) because the process of coupling "shaft-sleeve" goes into the next phase of contact: the state of three-point contact. The studies carried out provide intermediate results on the development of contact states of the assembly, which help stimulate the contact states of the shaft-sleeve interface's assembly. An algorithm for controlling the assembly robot will be built in further work, which has feedback based on the transitions between contact states during assembly. The method has broad prospects for application in mechanical engineering.

Acknowledgements This research work was carried out in the laboratory "Automation tools and industrial robots" of the Faculty of Mechanical Engineering of the Moscow Poly. The authors express their gratitude to the Department of Technologies and Equipment for Mechanical Engineering, Candidate of Technical Sciences, Associate Professor Alexander Nikolaevich Vasiliev, Moscow Polytechnic University for the opportunity to work in the laboratory, significant support and valuable suggestions for improving research.

References

1. Vinokurova, M. E.: Assembly of Adjustable Cylindrical Adhesive Joints. Ph. D Thesis. Bauman Moscow State Technical University, Moscow (2017)
2. Vartanov, M.V., Trung Ta Tran, Van Dung Nguyen: Research of the effect of rotation and low-frequency vibration on the robotic assembly process. Adv. Sci. Technol. Eng. Syst. J. (ASTESJ) **5**(6), 160–168 (2020). https://doi.org/10.25046/aj050619
3. Ghalyan, I.F.J.: Force-Controlled Robotic Assembly Processes of Rigid and Flexible Objects. Springer, Verlag (2016)
4. Kuznetsova, S.V., Simakov, A.L.: The recognition of component part's the contact point position in robotic assembly with forces and moments sensing. Autom. Industry **6**, 9–13 (2020). https://doi.org/10.25728/avtprom.2020.06.02
5. Song, J., Chen, Q., Li, Z.: A peg-in-hole robot assembly system based on Gauss mixture model. Robot. Comput.-Integr. Manuf. **67**, 101996 (2021). https://doi.org/10.1016/j.rcim.2020.101996
6. Jasim, I.F., Plapper, P.W.: Contact-state monitoring of force-guided robotic assembly tasks using expectation maximization-based Gaussian mixtures models. Int. J. Adv. Manuf. Technol. **73**, 623–633 (2014). https://doi.org/10.1007/s00170-014-5803-x
7. Jasim, I.F., Plapper, P.W., Voos, H.: Contact-state modelling in force-controlled robotic peg-in-hole assembly processes of flexible objects using optimised Gaussian mixtures. Proc. Inst. Mech. Eng. Part B: J. Eng. Manuf. **231**(8), 1448–1463 (2015). https://doi.org/10.1177/0954405415598945
8. Xu, J., Hou, Z., Liu, Z., Qiao, H.: Compare Contact Model-based Control and Contact Model-free Learning: A Survey of Robotic Peg-in-hole Assembly Strategies. ArXiv, abs/1904.05240 (2019)
9. Bozhkova, L.V., Vartanov, M.V., Shandrov, B.V.: Stages of creating algorithmic support for intelligent robotic assembly. In: Proceedings of the Volgograd State Technical University, vol. 21, issue, 148, pp. 59–64 (2014)
10. Zenkevich, S.L.: Robotics Course. Mir, Moscow (1990)
11. Gusev, A.A., Pavlov, V.V., Andreev, A.G., et al.: Mechanical engineering. In: Frolov, K. V., et al (Eds.) Encyclopedia. Assembly Technology in Mechanical Engineering. Mechanical Engineering, Moscow (2001)
12. Wang, Y., et al.: Contact force/torque prediction and analysis model for large length-diameter ratio Peg-in-hole assembly. In: 2018 IEEE International Conference on Robotics

and Biomimetics (ROBIO), Kuala Lumpur, Malaysia, pp. 2285–2290 (2018). https://doi.org/10.1109/ROBIO.2018.8665115
13. Kholodkova, A.G., Kristal, M.G., Shtrikov, B.L., Zenkin, A.S., Arpentyev, B.M., Andreev, A.G.: Automatic Assembly Technology. Mechanical Engineering, Moscow (2010)
14. Chernyakhovskaya, L.B.: Kinematic and Dynamic Analysis of Automatic Assembly of Cylindrical Parts. Samara State Technical University, Samara (2011)
15. Vartanov, M.V., Nguyen Van Dung, Tran Dinh Van: Determination of changing friction coefficient using force torque sensor during robotic assembly of cylindrical connection with clearance. In: 2020 International Russian Automation Conference (RusAutoCon), Sochi, Russia, pp. 977–981 (2020). https://doi.org/10.1109/RusAutoCon49822.2020.9208047
16. Vartanov, M.V., Arkhipov, M.V., Petrov, V.K., Mishchenko, R.S.: Experimental studies of the conditions of assemblability in active robotic assembly. STIN **4**, 14–16 (2017)

Design of High Precision Machining Part Processes in Free Abrasives

Mikhail Tamarkin, Elina Tishchenko, and Elena Murugova

Abstract This paper presents the research results on vibration abrasive processing of particularly accurate parts. The peculiarities of processing technologies and theoretical results are identified. The research shows that intensity of vibration processing depends on modes and duration of treatment, characteristics and sizes of working medium particles, working chamber volume and its filling degree, volume ratio of processed parts and abrasive granules, mechanical properties of the processed part material and their mass. Single trace parameters are defined. The formula for calculation of maximum depth of medium particle penetration into the surface of the machined part is presented. Trace parameters of medium particle interaction with the part surface are set. Metal removal, surface roughness and processing time are determined to solve various technological problems. The experimental results of vibration processing of particularly accurate details confirm the adequacy of theoretical dependencies. The samples from various materials are used for the experimental studies, the accuracy of which corresponds to 6–8 degree. The comparative results of theoretical and experimental studies show that the discrepancy does not exceed 20%, which indicates the adequacy of the obtained theoretical models. The methodology of designing the vibration processing technology of particularly precise parts is applied in the engineering enterprises.

1 Introduction

Vibration processing in an abrasive medium is one of the methods of processing with free abrasives. The widespread use of the method in the operations of cleaning workpieces, deburring, rounding sharp edges, coating, polishing, finishing the parts surface in order to improve its quality has been subjected to conscious scrutiny of this method, the possibility of purchasing equipment and working environments. The main advantage of processing with a free abrasive is a more complete use of the cutting capacity of the grains than when processing with a fixed abrasive, since there

M. Tamarkin (✉) · E. Tishchenko · E. Murugova
Don State Technical University, Rostov-on-Don, Russia

M. Shamtsyan et al. (eds.), *Robotics, Machinery and Engineering Technology for Precision Agriculture*, Smart Innovation, Systems and Technologies 247,
https://doi.org/10.1007/978-981-16-3844-2_17

is a uniform distribution of their cutting edges relative to the treated surfaces, as well as reorientation and movement during processing. Vibration processing in an abrasive medium has quite broad technological capabilities and allows, with relatively simple kinematic schemes of equipment, to perform high-efficiency processing of parts, even of complex configurations [1–12].

Particularly, interest is the processing of precise machine parts. The purpose of any type of processing is to produce parts with a given accuracy. The accuracy and quality of the processed parts surfaces are highly demanded, as the machines' durability depends mainly on the accuracy of the mating surfaces and their roughness. At the same time, it is the finishing operation that determine the dimensions accuracy and the quality of the parts surface, so research in the field of finishing methods of processing, including the free abrasive environment, is especially relevant [1–3].

2 Materials and Methods

The intensity of vibration processing depends on the modes and duration, characteristics and particle sizes of the working medium, the volume of the working chamber and the degree of its filling, the volume ratio of the processed parts and abrasive granules, the mechanical properties of the processed parts materials and their mass, etc. The most significant factors are the grain size of the abrasive granules and the contact interaction forces of the processed surfaces with the working medium.

When conducting theoretical studies of the processing, the maximum penetration depth of the medium particle into the surface of the processed part is determined:

$$h_{\max} = 2V_{\text{ef}}R\sin\beta\sqrt{\frac{\rho_r}{3k_R c\sigma_s}} \tag{1}$$

where V_{ef}—effective abrasive particle velocity; R—particle radius; β—the impact angle of the abrasive particle; k_R—the coefficient that influences the grain size of the abrasive particle on the actual contact area; ρ_r—the density of the particle material; c—the coefficient that evaluates the bearing capacity of the contact surface;σ_s—the yield strength of the part material.

The ellipsoid parameters of the medium particle contact with the surface of the part are determined. The major and minor semi-axis of the ellipsoid can be determined by the following formulas [1–3]:

$$\begin{aligned} b &= \sqrt{R^2 - (R - h_{\max})^2} \\ a &= \frac{\pi}{2}(ctg\beta - f)h_{\max} + b \end{aligned} \tag{2}$$

where f—sproportionality factor.

The dependence for determining the metal removal in a single blow of an abrasive particle is obtained:

$$q = 15.5k_c k_z R\left(V_{\text{ef}}\sin\beta\sqrt{\frac{\rho_r}{3\sigma_s}}\right)^{\frac{5}{2}}(ctg\beta - f)\rho_d \tag{3}$$

where k_c—chip formation coefficient; k_z—coefficient that influences the grain size of the abrasive particle on the removal of the metal; ρ_d—the density of the part material.

Based on the conducted research, the metal removal from the entire surface of the part is determined by the following formulas:

$$Q\prime = P_1 P_2 \omega t q \frac{S_{\text{det}}}{4R^2} \quad \text{where } \mathrm{S}_{\text{det}} \succ 4R^2 \tag{4}$$

$$Q\prime = P_1' P_2 \omega t q \quad \text{where } \mathrm{S}_{\text{det}} \prec 4R^2 \tag{5}$$

where S_{det}—surface area of the part; $P_1\left(P_1'\right)$—the geometric probability of an event that any point in the square of the package is covered with a contact spot in the process of a single cycle of exposure to a mass of abrasive particles; P_2—the probability of an event that the interaction of an abrasive particle with the part surface will lead to micro-cutting; t—processing time, ω—working chamber oscillation frequency.

When processing precise parts, the size of the layer removed from the part is of great importance. It can be defined by the following dependency:

$$h = \frac{Q}{4R^2\rho_d} = \frac{P_1 P_2 \omega t q}{4R^2\rho_d} \tag{6}$$

While studying the mechanism of the roughness formation of the processed surface, the dependence is obtained for calculating the arithmetic mean deviation of the roughness profile in the case of vibration processing:

$$Ra = 0.09\sqrt{\frac{h_{\max} L_{\text{ed}}}{z_0}} \tag{7}$$

where z_0—the number of grains per unit of the bundle surface, L_{ed}—unit length.

The formation of surface roughness during processing in a free abrasive medium has the following features: after the initial roughness, profile is completely removed and the processing continues, the so-called steady-state roughness is formed on the part surface, which is reproduced over time.

Determining the processing time is one of the most important tasks of mechanical engineering technology, without which it is impossible to design technological processes. During the research, dependences were obtained for determining the time of reaching the steady-state roughness, removing burrs and chips and rounding sharp edges.

3 Results

Experimental studies of the vibration processing of precise parts are carried out. For the experiments, samples made of various materials (steel, aluminum alloys, brass), made according to the 6, 7 and 8 accuracy quality, were used. To maintain the part accuracy specified by the designer, it is necessary that the actual size of the sample obtained after vibration processing has a value that is in the specified tolerance zone. In this regard, it is necessary to know the difference between the actual size and the limit size of the sample before performing the vibration treatment. Moreover, it really matters how much the actual size of the sample is close to the smallest limit size. In this regard, the input control is required. For samples that have a valid size close to the smallest limit size, it is necessary to adjust the processing time in the direction of reduction or leave the allowance for vibration processing within the metal removal during processing.

When conducting experiments on the study of removal from the sample surface, 4 samples from each material under study were processed. The following abrasive media were used (shown by numbers in Figs. 1, 2, 3 and 4): 1—an abrasive cone on a polymer bundle with a size of 10 mm with a grain size of 8, 2—an abrasive cone with a size of 12 mm with a grain size of 15, 3—triangular prisms PT 15 × 15 with a grain size of 25. Processing modes: camera oscillation frequency-26 Hz, oscillation amplitude-1.5–2 mm. Every 10 min, the samples were removed from the chamber and weighed. Experimentally, the height of the removed layer was determined by the formula:

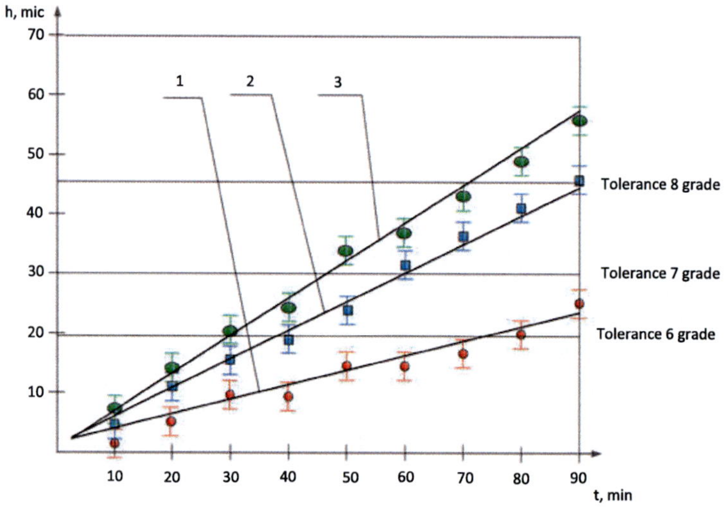

Fig. 1 The dependence of the height of the removed layer on the processing time. Part material-steel 45

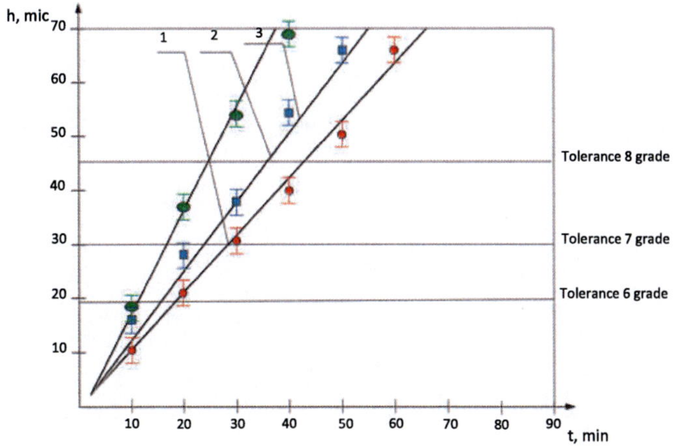

Fig. 2 The dependence of the height of the removed layer on the processing time

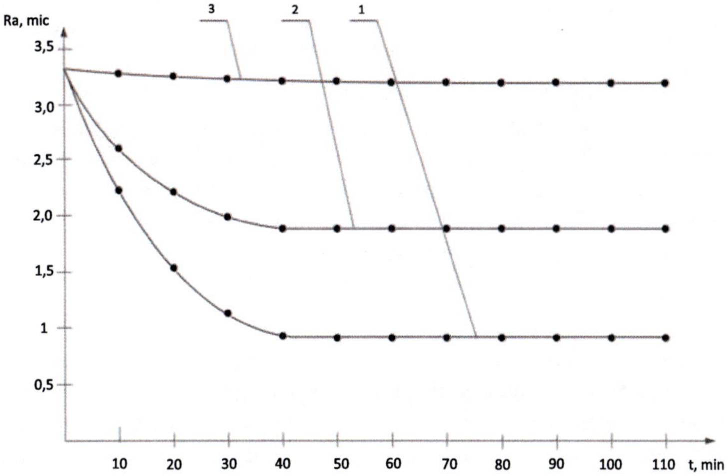

Fig. 3 Dependence of the surface roughness on the processing time

$$h = \frac{m}{S_{\det} \rho_d} \tag{8}$$

where *m*—weight of the removed metal.

Figures 1 and 2 compare the results of theoretical and experimental studies for the samples made of steel 45 and D16. Similarly, dependencies were constructed for the other materials under study. Solid lines show the straight lines obtained by calculation according to the theoretical dependencies. Red circles show the results of

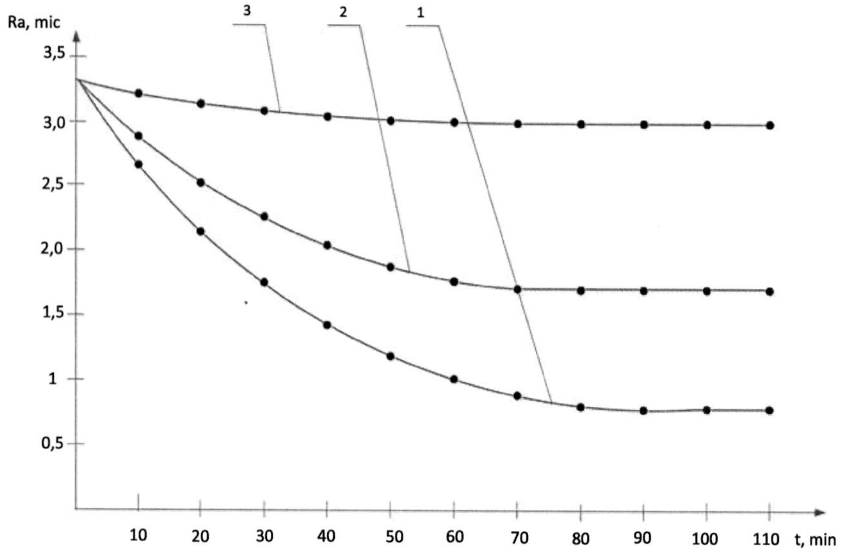

Fig. 4 The dependence of the surface roughness on the processing time

experimental studies in an abrasive medium with a grain size of eight, blue squares—the results of experimental studies in a medium with a grain size of 15, green ovals-the results of experimental studies in a medium with a grain size of 25. Confidence intervals with a 95% confidence probability are constructed.

The material of the part is aluminum alloy D16.

The conducted research shows the following results:

1. When conducting research, there is a high convergence of the theoretical results and experimental data. The discrepancy does not exceed 20%, which indicates the adequacy of the obtained theoretical dependencies.
2. Vibration processing in abrasive granular media allows processing the high-precision parts. At the same time, for samples made of steel with a high yield strength (600 Mpa or more), after 40–70 min of processing, accuracy is provided within the established tolerance, for samples made of non-ferrous alloys with a lower yield strength, this time is reduced to 15–30 min.
3. With the increase of the grain size of the abrasive medium, the processing time decreases. At the same time, for high-precision samples made of non-ferrous alloys, an abrasive medium with a low grain size is recommended.
4. As the camera oscillation amplitude increases, the processing time decreases. It is recommended to process high-precision samples of six and seven grade with an amplitude not exceeding 2 mm.
5. To maintain the accuracy of the part specified by the designer, it is necessary to leave the allowance for vibration processing within the metal removal during processing in accordance with the assigned modes.

While studying the surface roughness of samples made of various materials, the same processing modes, abrasive media and material grades were used as in the removal study. The aim of the experiments is to establish the influence of technological modes and characteristics of working media on the value of the steady-state roughness of high-precision parts made according to the 6–8 accuracy grade quality. Each grade of accuracy, according to the reference data, corresponds to a certain value of the surface roughness. The initial roughness of the samples 8 and 7 of the grade quality was Ra 3.2 microns, 6 of the quality grade -Ra 2.5 microns. Figures 3 and 4 show the curves describing the formation of the sample surface roughness.

The part material is B95 alloy. The precision quality of the parts is −8.

Part material-steel 45. The precision quality of the parts is −7.

Based on the results of the conducted research, the following conclusions can be drawn:

1. The curves obtained in the graphs confirm the above-mentioned mechanism for the formation of surface roughness during vibration treatment in a free abrasive medium and have an exponential character.
2. The formation of a steady surface roughness is observed.
3. The values of the steady-state surface roughness for the samples from all the studied materials in an abrasive medium of different grain size are established. With an increase in the grain size of the abrasive medium, the established surface roughness of non-ferrous alloy samples exceeds the initial one, which indicates that it is unacceptable to use it for processing high-precision parts.

4 Conclusion

To ensure the best results of processing high-precision machine parts at the lowest cost, it is necessary to develop a method for selecting rational technological parameters. As rationality criteria, such indicators of the technological process as high productivity or low cost can be used. In this case, the part must not lose the specified accuracy. The following method of calculating rational modes is proposed:

1. Depending on the size and part configuration, it is necessary to determine the size of the abrasive particles used in the processing and select the abrasive material.
2. Depending on the part material and the required value of accuracy, roughness or other processing parameters (for example, the radius of the sharp edge rounding), we select several combinations of processing modes when using an abrasive of different grain sizes.
3. Using formulas, we determine the theoretical values of metal removal from the part surface, the steady-state roughness or the radius of the edge rounding, depending on the required tasks, and select the processing modes and the abrasive medium. We leave only those modes that correspond to the set tasks and determine the processing time to get the specified processing parameters. From

the previously selected modes, it is selected the one in which the processing is performed in less time. Vibration processing with the selected modes is the most efficient.

The developed methodology of designing vibration technological processes in the free abrasive medium of particularly precise parts has been applied in machine-building enterprises.

References

1. Tamarkin, M., Tishchenko,E.: Fundamentals of Optimization of Processes of Parts Processing with Free Abrasive/Saarboniken/Jermany: LAP LAMBERT Academic Publishing (2015)
2. Tamarkin, M., Tishchenko, E., Chukarina, I., Sosnitskay, T.: Part Processing Technology in Transport Engineering/Proceedings of the VIII International Scientific Siberian Transport Forum - TransSiberia **2019**, 1 (2019)
3. Tamarkin, M., Tishchenko, E., Murugova, E., Melnikov, A.: Surface Quality Assurance and Process Reliability in the Processing with a Ball-rod Hardener /E3S Web of Conferences, 175, 05008 (2020)
4. Hamouda, K., Bournine, H., Babichev, A., Saidi, D., Amrou, H.: Effect of the velocity of rotation in the process of vibration grinding on the surface state/materials. Science **52**(2), 216–221 (2016)
5. Solovev A., Van T.N., Panfilov I., Wang J.: Modeling contact abrasive interactions in centrifugally rotary processing by finite element method. In: Physics and Mechanics of New Materials and Their Applications (PHENMA 2018). Abstracts & Schedule, pp. 334–336 (2018)
6. Lebedev, V., Sokolov, V., Davydova, F.: Forecasting of physical and mechanical characteristics of quality of surface layer, modified PPD/IV Strengthening technologies and coatings. 14 (2-158), 54–58 (2018)
7. Ankudimov, Y., Sadovaya, I., Dieudonne Essola, Bagdasaryan, T.: The effect of vibro-finishing treatment on the porosity of the surface layer of powder metallurgy products. In: MATEC Web of Conferences, 297, 04006 (2019)
8. Prokopetz, G., Azarova, A., Prokopetz, A.: Analysis of ways to control the process of vibration machining in order to increase its efficiency and reduce the technological cost of processed parts. In: E3S Web of Conferences, 175, 05039 (2020)
9. Prokopetz, G., Azarova, A., Prokopetz, A., Murugova, E.:Standardization of the technological parameters of vibratory abrasion as a tool for process reliability enhancement. In: E3S Web of Conferences, 175, 05006 (2020)
10. Lebedev, V., Ivanov, V., Vesnin, V., Pastuchov, F.A., Sanamyan, G.: Technological provision for vibration cleaning of products in preparing them for disposal. In: IOP Conference Series: Materials Science and Engineering 1029(1), 012009 (2021)
11. Lebedev, V., Kirichek, A. , Chunakhova, L.: Effectiveness of Application of Additional Strengthening Processing of Surface Plastic Deformation on Increase in Fatigue Life of Parts. Lecture Notes in Mechanical Engineering, pp. 17–25 (2020)
12. Lebedev, V., Krupenya, E., Shishkina, A., Basenko, G.: Finishing characteristics of machinery parts with granular organic substance made from natural materials in vibration technological systems. Mater. Today: Proc. **38**, 1693–1696 (2020)

Exploring Cutting Forces While Radial Shaped Turning

Denis Moiseev

Abstract An analysis of the features of radial shaped turning with tools with a convex radius profile, as opposed to straight-blade cutting, is given in order to develop analytical dependencies for calculating force characteristics. During radial shaped (form) turning, parameters such as thickness of cut, specific pressures, and plan approach angle are variable along the cutting edge both during plunging and when deepening to the full profile height. Thus, taking this into account for different cutting modes, the possibility of implementing the task using the calculated values of cutting areas and average specific pressures is considered. In this regard, this work presents data of extensive experimental studies of force parameters in accordance with variable cutting areas. The values of average specific pressures were also obtained. The obtained databases of the relationship of geometric and force characteristics are the basis for solving the problem of developing reliable design formulae of forces for radial shaped turning.

1 Introduction

The article presents the results of experimental studies of cutting forces while radial turning with tools with convex radius shaped edges. The aim of the research is to develop analytical dependencies for calculating cutting forces. Convex radial profiles are widely spread when machining tracks of inner and outer rings of ball bearings, pulley blocks for lifting mechanisms, etc. Radial shaped turning has a number of significant features, which complicate the development of analytical force dependencies in contrast to straight-blade turning.

Thus, all geometric and force parameters of cutting change during plunging:

- the length of the active part of the cutting blade progressively increases from zero to the length of the arc of a semicircle of radius *r*. The width of the cut also increases accordingly;

D. Moiseev (✉)
Don State Technical University, Rostov-on-Don, Russia

M. Shamtsyan et al. (eds.), *Robotics, Machinery and Engineering Technology for Precision Agriculture*, Smart Innovation, Systems and Technologies 247,
https://doi.org/10.1007/978-981-16-3844-2_18

- the cutting blade of a shaped tool can be considered as a locus of points with smoothly decreasing plan approach angle φ from $\varphi = 90°$ to $\varphi = 0°$;
- thickness of cut in the normal section is not constant. The maximum thickness of cut *a*, equal to the feed value *S*, takes place at the top of the tool. With a decrease in the angle φ, the thickness of cut a decreases smoothly, obeying the equation $a \approx S \cdot \sin\ \varphi$. When the tool submerges to a depth of $ra = 0$;
- the cutting speed decreases with a decrease in the machining diameter *D*;
- the cut has a crescent shape with a variable thickness along the cutting blade. The cutting area increases due to the increment of the active part of the cutting blade during plunging. In this case special formulae to calculate the cutting areas depending on cutting depth are required;
- as a result, for radial turning, the deformation (and the temperature) conditions [4] change along the cutting blade, and the cutting forces at constant modes increase during plunging of a tool due to an increase of cutting area, width and active part of the cutting blade.

2 Analysis of the Problem and Statement of the Research Task

The features above make it difficult to obtain reliable design formulae of cutting forces. In this regard, there are currently no analytical dependencies for determining the force characteristics that take into account the features of radial shaped turning. Repeated unsuccessful attempts to develop analytical force dependencies for turning shaped profiles were based on formal criteria: the length of the active part of the cutting blade, the depth and the complexity of the profile, the constancy of specific forces, etc. However, these criteria do not take into account the peculiarities of shaped turning.

Traditional methods of developing analytical dependencies for determining the cutting forces $P_z = a \cdot b \cdot q$ using the cutting area $a \cdot b$ and the specific cutting forces q assume the constancy of the geometric parameters of the cut and the specific pressures [2, 9, 10].

Taking into account that the conditions of deformation in the cutting zone depend on the thickness of cut a, then, when a is constant, the specific pressure is also constant $q = f(a) = \text{const}$. The source [7] gives specific pressures, depending on the cut thickness, obtained experimentally for various steels under the conditions of straight-edged cutting. But the area and the thickness of cut smoothly change along the cutting blade while radial shaped turning, and, as a result, the specific pressures change, the value of which cannot be determined experimentally. Attempts to theoretically determine the functional dependences $q = f(a)$ for curved profiles were unsuccessful [5, 6]. In this regard, determination of force dependencies while shaped radial turning should be carried out in 3 stages:

1. Obtaining instantaneous cutting areas while radial plunging *H* and depending on feed *S*.

2. Experimental studies of cutting forces and average pressures in relation to the cutting areas.
3. Development of force dependencies based on the obtained cutting areas and the corresponding experimental force characteristics.

The first stage of solving this problem was completed [3]. An analytical dependence for determining the instantaneous cutting areas for tools with radius profiles ($r = 4.5$ mm) was obtained. Thus, the cutting area can be determined for any i-th section (for the *i*-th revolution after the start of cutting) according to the dependence:

$$F_i = \frac{1}{2} r^2 \left[(\varepsilon_i - \sin \varepsilon_i) - (\varepsilon_{i-1} - \sin \varepsilon_{i-1}) \right] \tag{1}$$

where:

$$\varepsilon_i = 2 \cdot \arccos\left(1 - \frac{S \cdot n'_i}{r}\right)$$
$$\varepsilon_{i-1} = 2 \cdot \arccos\left(1 - \frac{S \cdot n'_{i-1}}{r}\right) \tag{2}$$

$n'_i = \frac{H}{S}$—number of spindle revolutions at the moment of plunge H, at feed S and profile radius r.

The instantaneous cutting areas were calculated for each spindle revolution when plunging to the full depth of the profile for different cutting modes.

The second stage, associated with the determination of the specific cutting forces, is the goal of these experimental studies.

3 Organization of the Experiment

Figure 1 shows a scheme for conducting experimental studies with a symmetrical radius profile. The forces during the radial plunging act mirrored about the axis of symmetry OC. In this case, the components P'_y and P'_z are equally directional and pairwise equal, and the forces P'_x are equal in magnitude, but opposite in direction. In this case, the total values of the main forces determined experimentally are equal: $P_y = 2P'_y$, $P_z = 2P'_z$, the remaining forces can be determined from the main ones: $P'_x = \frac{P'_y}{tg\varphi}$, $R' = \frac{P'_y}{\sin \varphi}$, $R = \frac{R'}{\sin \theta}$, where φ—plan approach angle at the current cutting depth H, $\theta = \operatorname{arctg}\frac{R'}{P'_z}$.

The proposed scheme without force component P_x simplifies the experiments and reduces to the definition of two main components P_y and P_z.

The research was carried out on C 45 (rolled steel bars ø120 mm, normalized HB198).

Specially sharpened radius turning tools made of high-speed steel (HSS) R6M5 and hard alloy T15K6 were used as the investigated tools (Fig. 2). The face of the

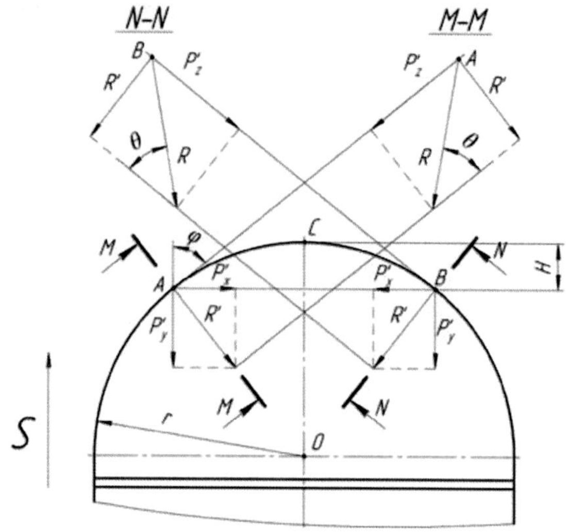

Fig. 1 Scheme of acting forces while radial turning

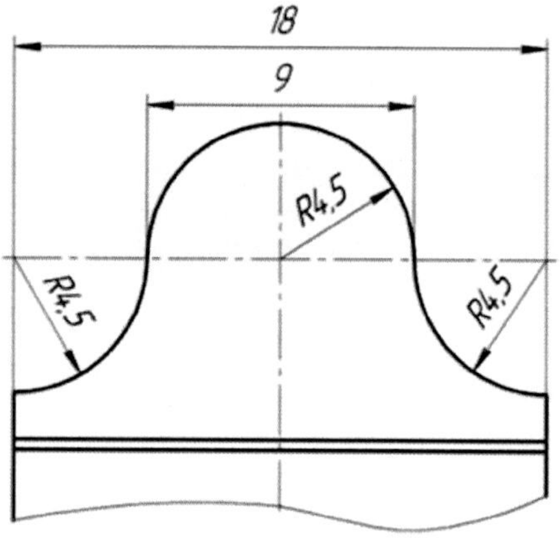

Fig. 2 Test turning tools

tool is flat, rake angle = 0°, clearance angle at the top of the tool $\alpha = 6°$. Surface roughness $R_a = 2.5\ \mu m$.

The research was carried out on a lathe 1K625 equipped with a Mitsubishi frequency converter, which allows to smoothly adjust electric motor speed and obtain accurate cutting speed values during experiments. The spindle rotation frequency was controlled by a laser tachometer with an accuracy of ±0.5 rpm.

Lathe is equipped with a research stand STD.201-1, which allows to measure force components P_z, P_y, P_x while turning. STD.201-1 stand consists of a tool holder head, which is installed on the machine support and an interface unit that processes and sends the measurement output signal to a personal computer (Fig. 3). The data are recorded using special software that comes with the measuring stand.

Fig. 3 General view of the measuring stand

Table 1 Cutting modes while radial shaped (form) turning

Tool material	Cutting speed V (m/min)	Radial feed rate S (mm/rev)
HSS R6M5	16	0.035 0.07 0.105
	24.8	
	28.9	
Hard alloy T15K6	45	
	70	
	100	

Thus, during the machining process, the cutting force data are simultaneously recorded for further processing. The research cutting modes are presented below (Table 1).

During each experiment, the shaped (form) turning tool plunges into the workpiece by an amount not exceeding $r = 4.5$ mm at the modes presented above. The increasing values of cutting forces P_z, P_y, ($P_x = 0$) are recorded in time in the form of oscillograms (Fig. 4).

4 Experimental Data Processing

The obtained oscillograms are close in the form of square root function extending from the origin ($y = a\sqrt{x}$). To eliminate the error in the amplitude oscillations of the force signal P_z, P_y, it was decided to approximate the oscillograms using the

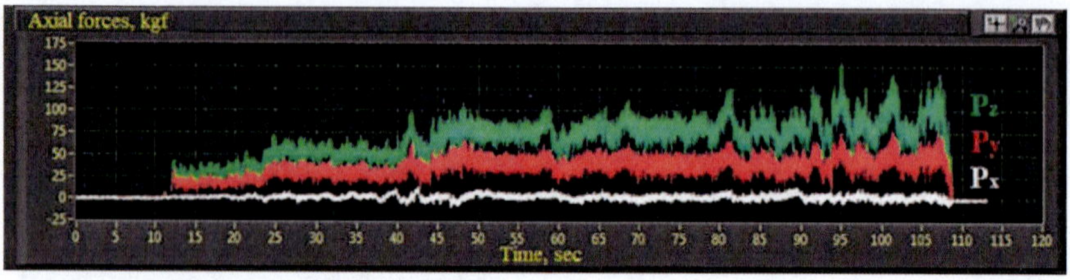

Fig. 4 The interface of the measuring stand STD.201-1

above function in the form:

$$P_z = c_z\sqrt{\tau},\ P_y = c_y\sqrt{\tau} \tag{3}$$

where τ, sec.—time from the start of cutting, c_z, c_y—approximation coefficients for P_z, P_y, respectively, which were averaged depending on the number of experiments [1, 8].

Obtained oscillograms were approximated and presented in the form of calculated dependencies to determine the force components while radial cutting (Table 2).

Table 2 Approximation parameters of functions (3) for various cutting modes

Material of a pair "tool–workpiece"	V (m/min)	S (mm/rev)	The results of data processing		$K_{pr.} = \frac{P_y}{P_z}$
			$P_z = c_z\sqrt{\tau}$	$P_y = c_y\sqrt{\tau}$	
R6M5–C 45	16.06	0.035	$P_z = 10.23\sqrt{\tau}$	$P_y = 6.49\sqrt{\tau}$	0.6344
		0.07	$P_z = 24.48\sqrt{\tau}$	$P_y = 16.26\sqrt{\tau}$	0.6642
		0.105	$P_z = 37.6\sqrt{\tau}$	$P_y = 29.46\sqrt{\tau}$	0.784
	24.82	0.035		$P_y = 8.19\sqrt{\tau}$	0.6958
		0.07	$P_z = 29.18\sqrt{\tau}$	$P_y = 20.3\sqrt{\tau}$	0.696
		0.105	$P_z = 44.95\sqrt{\tau}$	$P_y = 37.5\sqrt{\tau}$	0.8342
	28.9	0.035	$P_z = 12.49\sqrt{\tau}$	$P_y = 9.09\sqrt{\tau}$	0.7276
		0.07	$P_z = 30.2\sqrt{\tau}$	$P_y = 22.5\sqrt{\tau}$	0.745
		0.105	$P_z = 44\sqrt{\tau}$	$P_y = 39\sqrt{\tau}$	0.887
T15K6–C 45	45	0.035	$P_z = 15.1\sqrt{\tau}$	$P_y = 9.6\sqrt{\tau}$	0.6357
		0.07	$P_z = 37.1\sqrt{\tau}$	$P_y = 25\sqrt{\tau}$	0.6738
		0.105	$P_z = 56.7\sqrt{\tau}$	$P_y = 41.6\sqrt{\tau}$	0.7337
	70	0.035	$P_z = 18.2\sqrt{\tau}$	$P_y = 12.2\sqrt{\tau}$	0.6703
		0.07	$P_z = 44.7\sqrt{\tau}$	$P_y = 31.6\sqrt{\tau}$	0.707
		0.105	$P_z = 67.9\sqrt{\tau}$	$P_y = 52.4\sqrt{\tau}$	0.7717
	100	0.035			0.7143
		0.07			0.7563
		0.105			0.8217

The obtained dependencies were used to calculate the cutting forces P_z, P_y, specific average pressures q during radial plunging of a tool H in accordance with the variable cutting area and depending on the cutting modes.

As an example, Table 3 shows (in abbreviated form) the interdependent geometric and dynamic parameters: τ—plunge time, n', φ—respectively, the number of spindle revolutions and the plan approach angle for various feeds S and cutting speeds V, where: $\tau = \frac{n'}{n} \cdot 60$; $n' = \frac{H}{S}$; $\varphi = \arcsin \frac{r-H}{r}$; $n = \frac{1000 \cdot V}{\pi \cdot D}$.

Current studies were carried out for convex profiles with a radius $r = 4.5$ mm without taking into account its changes. In our opinion, the effect of r on the force characteristics should be taken into account through a change of the cutting area without additional experimental studies due to the complete similarity of the profiles, the identity of the change of the cut thickness and average specific pressures along the cutting blade.

5 The Discussion of the Results

The obtained functional dependencies for calculating the instantaneous cutting areas (1), (2) and cutting forces can be reduced to a common argument (time τ, plunging depth H, etc.). It makes possible to determine the correspondence between the cutting area and the effective force for any moment of cutting (Table 3). The resulting database is the basis to develop analytical force dependencies, taking into account the peculiarities of radius turning.

Current studies are relevant not only for turning symmetrical radius profiles, but also for asymmetric or offset ones, for example, machining tracks of angular ball bearings, as well as concave, involute, elliptical and other surfaces, including arbitrary, if they can be replaced by circular arcs.

The obtained dependencies for determining the cutting areas and the corresponding experimental force characteristics are a sufficient database for developing formulae for calculating the components of cutting forces, power, energy costs for shaped radial turning, which is the purpose of this next study.

Table 3 Dynamic characteristics for radius turning of steel C 45 with a shaped radius tool from HSS R6M5

V, m/min (n, rev/min)	S (mm/rev)	Φ (°)	τ (s)	F (mm²)	P_y (kgf)	P_z (kgf)	$q\left(\frac{\text{kgf}}{\text{mm}^2}\right)$	Notice
16.06 (55)	0.035	75	4.7939	0.076729	14.21	22.4	291.9	$P_z =$
		60	18.798	0.15535	28.14	44.35	285.51	$10.23\sqrt{\tau} P_y =$
		45	40.419	0.2215	41.25	65.04	293.62	$6.49\sqrt{\tau}$
		30	70.13	0.27208	54.35	85.67	314.87	
	0.07	75	2.407	0.1427	25.23	38	166.1	$P_z =$
		60	9.402	0.30625	49.86	75.06	145.1	$24.48\sqrt{\tau} P_y =$
		45	20.544	0.44048	73.7	110.96	251.91	$16.3\sqrt{\tau}$
		30	35.066	0.5427	96.94	145	267.1	
	0.105	75	1.43	0.19148	35.23	44.96	234.8	$P_z =$
		60	6.272	0.4525	73.76	94.17	208.09	$37.6\sqrt{\tau} P_y =$
		45	13.697	0.6568	109.03	139.16	211.85	$29.5\sqrt{\tau}$
		30	23.379	0.81182	142.44	181.8	223.94	
24.82 (55)	0.035	75	3.102	0.0767	14.43	20.73	270.16	$P_z =$
		60	12.161	0.1553	28.57	41.05	264.22	$11.77\sqrt{\tau} P_y =$
		45	26.581	0.2215	42.24	60.68	273.96	$8.19\sqrt{\tau}$
		30	45.38	0.2721	55.2	79.29	291.4	
	0.07	75	1.558	0.1427	25.34	36.42	255.21	$P_z =$
		60	6.084	0.3063	50.07	71.98	235.02	$29.18\sqrt{\tau} P_y =$
		45	13.29	0.4405	74	106.38	241.51	$20.3\sqrt{\tau}$
		30	22.69	0.5427	96.7	139	256.11	
	0.105	75	1.061	0.192	37.12	46.3	241.12	$P_z =$
		60	4.058	0.4525	72.63	90.54	200.1	$44.95\sqrt{\tau} P_y =$
		45	8.863	0.65685	107.33	133.76	203.63	$37.5\sqrt{\tau}$
		30	15.125	0.81182	140.2	174.8	215.32	
28.9 (99)	0.035	75	2.457	0.0767	14.25	19.57	266.3	$P_z =$
		60	10.436	0.15534	29.36	40.35	259.73	$12.49\sqrt{\tau} P_y =$
		45	22.795	0.2215	43.4	59.63	269.22	$9.09\sqrt{\tau}$
		30	38.97	0.2721	56.74	77.97	286.57	
	0.07	75	1.336	0.14271	26	34.91	244.62	$P_z =$
		60	5.221	0.30516	51.41	69.02	226.17	$30.2\sqrt{\tau} P_y =$
		45	11.398	0.44049	75.96	101.98	231.5	$22.5\sqrt{\tau}$
		30	19.486	0.5427	99.32	133.3	245.68	
	0.105	75	0.911	0.19139	36.8	42	221.2	$P_z =$
		60	3.487	0.45253	72	82.2	181	$44\sqrt{\tau} P_y =$
		45	7.61	0.65685	106.4	121.37	184.7	$39\sqrt{\tau}$
		30	12.958	0.81182	138.9	158.4	195.1	

References

1. Katsev, P.G.: Statistical Research Methods for Cutting Tools. Mashinostroenie, Moscow (in Russ) (1974)
2. Lapshin, V.P., Moiseev, D.V., Minakov, V.S.: Diagnosing cutting tool wear after change of cutting forces during turning. In: XV International Scientific-Technical Conference "Dynamics of Technical Systems", DTS 2019. AIP Conference Proceedings. Rostov-on-Don, Russia (2019). https://doi.org/10.1063/1.5138394
3. Moiseev, D.V., Ryzhkin, A.A.: Features of constrained radial form turning. Vestnik of Don State Technical University (In Russ.) **18**(1), 22–30 (2018). https://doi.org/10.23947/1992-5980-2018-18-1-22-30
4. Moiseev, D.V., Ryzhkin, A.A., Volokitin, G.I.: Evaluation of cutting temperature of a tool with a curved cutting edge. In: 13th International Scientific-Technical Conference on Dynamic of Technical Systems, DTS 2017, Matec Web of Conferences. Rostov-on-Don, Russia (2017). https://doi.org/10.1051/matecconf/201713201005
5. Petrushin, S.I., Proskokov, A.V.: Theory of constrained cutting: calculating the stress-strain state in the cutting zone. Russ. Eng. Res. **30**(2), 133–137 (2010)
6. Petrushin, S.I., Proskokov, A.V.: Theory of constrained cutting: chip formation with a developed plastic-deformation zone. Russ. Eng. Res. **30**(1), 45–50 (2010)
7. Ryzhkin, A.A., Shuchev, K.G., Klimov: Processing of Materials by Cutting. Feniks, Rostov-on-Don (in Russ.) (2008)
8. Ryzhkin, A.A., Zotov, V.V., Moiseev, V.G.: Processing of Experimental Data of Tool Life. Izd. tsentr DGTU, Rostov-on-Don (in Russ.) (2013)
9. Sadílek, M., Dubsky, J., Sadílková, Z., Poruba, Z.: Cutting forces during turning with variable depth of cut. Perspect. Sci. **7**, 357–363 (2016)
10. Vereshyaka, A.S., Kushner, V.S.: Cutting Materials. Vysshaya shkola, Moscow (in Russ.) (2009)

Technological Support of Machine Element Corrosion Resistance

Oleg Fedonin, Sergey Syanov, and Nina Lakalina

Abstract The operational properties of machine parts can be provided by two-stage and single-stage approaches. The single-stage approach is currently the most relevant way to ensure the operational properties of machine parts. To implement a single-stage approach, it is necessary to establish the interrelation between the operational property and the machining modes and conditions. On the basis of the studies conducted, theoretical dependences are proposed for determining the corrosion rate depending on the modes and conditions for various methods of machining parts. These theoretical dependences allow to define optimal cutting modes and machining conditions to ensure the required corrosion rate for various methods of machining parts. The obtained theoretical dependences help us to conclude that the corrosion rate during edge cutting machining is mostly influenced by the feed, cutting amount and speed, cutting tool geometry, and when grinding, the most influential are average value of the rounding radius of the abrasive grain points, heating temperature of the surface layer and the physical and mechanical properties of the machined material, during finishing and strengthening machining—the feed, operating force, radius and profile radius of the tool, tool roughness, during EDM machining—technological stress, current strength and pulse duration.

1 Introduction

The service life of many mechanical engineering products that are operated in aggressive environment is determined by their corrosion resistance. In order to avoid unnecessary costs for manufacturing or restoring these products, it is necessary to ensure their required corrosion resistance technologically, which guarantees the specified durability of the product.

Technological support of corrosion resistance, as well as other operational properties of machine elements, can be carried out by a traditional two-stage approach (design and technological) or in a new direction—by a one-stage approach (direct

O. Fedonin · S. Syanov (✉) · N. Lakalina
Bryansk State Technical University, 7, 50-let Oktyabrya Boulevard, Bryansk 241035, Russia

M. Shamtsyan et al. (eds.), *Robotics, Machinery and Engineering Technology for Precision Agriculture*, Smart Innovation, Systems and Technologies 247,
https://doi.org/10.1007/978-981-16-3844-2_19

technological support of corrosion resistance). Therefore, the issue of one-stage technological support of corrosion resistance in various processing methods is relevant.

2 Materials and Methods

The theoretical dependence for determining the corrosion rate on the quality parameters of machine element surface layer is given in [2, 10]:

$$v_c = v_{c0} b_0 U_h^{b1} \frac{28 \times 10^6 Rv}{(100 - tm)^2 Sm^2} \mathcal{K}_t \mathcal{K}_w \mathcal{K}_\theta \mathcal{K}_{\text{pH}}, \tag{1}$$

where b0 and b1 are coefficients depending on the material grade and condition (for non-thermostrengthened steel $b0 = 1.0$; $b1 = 5.2$; for thermostrengthened low-alloy steel $b0 = 0.713$; $b1 = 5.72$; and for thermostrengthened medium-alloy steel $b0 = 0.74$; $b1 = 5.57$) [2, 10]; vc0 is the corrosion rate of a reference sample made of the same material as the product in question, but without mechanical processing, in certain standard (reference) conditions (in distilled water); Uh is the degree of hardening of the surface layer material; Rv = 0.55 Rz; tm is the relative reference length of the profile at the mean line; Sm is the average pitch of irregularities; *Kt* is the coefficient that regards the influence of the moisture film thickness (*h*) on the element surface [2, 10]; *Kw* is the coefficient regarding the aggressiveness of the corrosive environment [2, 10]; *K*Θ is the coefficient regarding the influence of the corrosive environment temperature on the corrosion rate of the element [2, 10]; and *K*pH is the coefficient that takes into account the influence of H+ ions concentration on the corrosion rate of the element [2, 10].

$$\mathcal{K}_t = f(h) \tag{2}$$

where
Kt = 20*h*; $h \le 10^{-2}$; micron
Kt = 5*h* 0.695; $10^{-2} \le h \le 10^{-2}$; micron
Kt = 5e-0.0016*h*; $1.0 \le h \le 1000$; micron
Kt = 1.0; 1000 < h; micron
h is moisture film thickness, micron

Moisture film thickness on the metal surface can be defined according to the following dependences:

$$h = 2.37 \times 10^{-8}\, \varphi^{3.8074};\ \varphi \le 98\% \tag{3}$$

$$h = \text{Rz};\ \ 98 < \varphi \le 100\%,\ \ \text{fine dew} \tag{4}$$

$$h = 100 \text{ micron; large dew} \quad (5)$$

$$h = 100 \div 500 \text{ micron; precipitation} \quad (6)$$

$$h = 500 \div 1000 \text{ micron; periodic dipping in liquid} \quad (7)$$

where φ is relative air humidity, %.

The environment aggressiveness can be taken into account using Kw coefficient (Table 1), depending on the chemical composition of the environment, its pollution, etc.

To take into account the influence of the corrosion environment temperature on the metal corrosion rate is possible using $K\Theta$ coefficient:

- in case of atmospheric corrosion

$$K_{\Theta} = 0.674 + 0.019\Theta; \ 20 \le \Theta < 100^{\circ} \quad (8)$$

- in case of liquid corrosion (with full dipping).

$$K_{\Theta} = -1.25 + 0.106\Theta; \ 20 \le \Theta \le 70^{\circ} \quad (9)$$

$$K_{\Theta} = 21 - 0.213\Theta; \ 70 \le \Theta \le 90^{\circ} \quad (10)$$

The influence of pH environment on the corrosion rate can be regarded by using KpH coefficient, which is defined as follows:

$$K_{pH} = 26 - 5_{pH}; \ \le 5 \quad (11)$$

$$K_{pH} = 1.0; 5 \le \ \le 12 \quad (12)$$

$$K_{pH} = -23 - 2_{pH}; 12 \le \quad (13)$$

Theoretical dependences of the surface layer quality parameters on the conditions for various methods of element surface processing are given in [3–5].

Table 1 Influence of corrosive environment aggressiveness on the corrosion rate

Environment name	Steel corrosion resistance, mm/year		Steel corrosion resistance, amount	Definition	Coefficient K_w	Environment composition
Not aggressive	Up to 0.01	Up to 0.01	1–3	H1	0.25 0.1	Deaerated fresh and seawater; Very hard freshwater without impurities *Cl*, SO_2; rooms with humidity $\varphi < 60\%$, and $SO_2 <$ 0.02 mg/m^3 and $Cl^- <$ 0.03 mg/m^2 a day
Weak	0.01 ÷ 0.05	0.01	4–5	C1	1.0	Distilled water; clean air with $SO_2 \leq 0.02$ mg/m^3, $Cl^- <$ 0.03 mg/m^2 a day, $\varphi \approx 70\%$
		0.02		C2	2.0	Freshwater with $Mg^{2+} \cdot 10^{-4}$ M, rooms with humidity $60 < \varphi <$ 80%
		0.03		C3	3.0	Freshwater with Mg^{2+} 5×10^{-5} M, rooms with humidity $\varphi < 85\%$
		0.04		C4	4.0	Freshwater; rural area with $SO_2 \leq 0.02$ mg/m^3 and $Cl^- <$ 0.03 mg/m^2 a day
		0.05		C5	5.0	Fresh mineral water; urban area with $SO_2 = 0.1$ mg/m^3, $Cl^- \approx 0.03$ mg/m^2 a day, $\varphi \approx$ 98–100%
Average	0.05 ÷ 0.5	0.06	6	C6	6.0	Fresh mineral water, intense CO_2; urban industrial area with $SO_2 = 0.1$ mg/m^3, Cl^- = 2 mg/m^2 a day

(continued)

Table 1 (continued)

Environment name	Steel corrosion resistance, mm/year		Steel corrosion resistance, amount	Definition	Coefficient K_w	Environment composition
		0.1		C10	10	Seawater; industrial sea area with $SO_2 = 0.2$ mg/m^3, Cl^- = 2 mg/m^2 a day
		0.2		C20	20	Sea sprays; industrial area with $SO_2 \approx 0.3$ mg/m^3, Cl^- = 20 mg/m^2 a day
		0.3		C30	30	Sea sprays; industrial area with $SO_2 \approx 0.5$ mg/m^3, Cl^- = 30 mg/m^2 a day
		0.4		C40	40	Galvanizing etching shops
		0.5		C50	50	Industrial area and shops with $SO_2 \approx 0.75$ mg/m^3
Strong	More than 0.5	0.6	7	C60	60	Polluted industrial area and shops with $SO_2 \approx 0.85$ mg/m^3
		0.7		C70	70	Polluted industrial area and shops with $SO_2 \approx 1.0$ mg/m^3
		0.8		C80	80	Polluted industrial area and shops with $SO_2 > 1.0$ mg/m^3 and strong dustiness
		0.9		C90	90	Electrolytes with $H_2SO_4 \approx$ 1.0%
		1.0		C100	100	Electrolytes with $H_2SO_4 \approx$ 5.0%

3 Results

Putting in (1) dependences for the quality parameters and physical and mechanical properties of the surface layer on the cutting modes of edge cutting processing from [3–5], we obtain the final theoretical equations for defining the corrosion rate depending on the cutting modes of the edge cutting processing:

if $\varphi \leq \arcsin(s/(2r))$, $\varphi 1 < \arcsin(s/(2r))$

$$
\begin{aligned}
v_c = {} & 22 \times 10^{-5} b_0 v_{c0} \mathcal{K}_t \mathcal{K}_w \mathcal{K}_\Theta \mathcal{K}_{\text{pH}} s^{-2} \left(\frac{7.3 \times 10^{-5} \varphi^{0.35} \rho^{0.17} \delta^{0.36} t^{0.06} \tau_r^{0.75}}{s^{1.1} v^{0.28} \sigma^{0.09} \left(\frac{\sigma_{ul.s.}}{\sigma_{ul.s.el.}} \right)^{0.8}} \right)^{b_1} \\
& \times \Bigg[1.54 \times 10^7 Rz_c + 1.83 \times 10^7 r (1 - \cos(\varphi)) \\
& + \frac{1.83 \times 10^7 (2tg(\varphi)(\cos(\varphi) - \cos(\varphi_1)) + tg(\varphi) tg(\varphi_1)(s - r(\sin(\varphi) - \sin(\varphi_1))))}{tg(\varphi) + tg(\varphi_1)} \\
& - 7.7 \times 10^6 \rho \left(\frac{1000 \sigma_{\text{yiel}}^{-1}}{\left(9.81c + 9.81 \exp\left(\frac{a\left(\frac{p v^l \sigma_{ul.s.}(st)^m}{1000 F \alpha_1} + 293 \right)}{b + 100} \right) \right)} - 1 \right) \left(\frac{1}{tg(\varphi)} + \frac{1}{tg(\varphi_1)} \right)^{-1} \\
& + \frac{1.54 \times 10^7 c_y s^y v^z \left(\text{HB}_{\max}^n t^x - \text{HB}_{\min}^n (t - Rz_{in})^x \right)}{j_{\text{rig}} \text{HB}_{\text{av}}^n} \Bigg]
\end{aligned}
\tag{14}
$$

if $\varphi > \arcsin(s/(2r))$, $\varphi_1 \geq \arcsin(s/(2r))$

$$
\begin{aligned}
v_c = {} & 22 \times 10^{-5} b_0 v_{c0} \mathcal{K}_t \mathcal{K}_w \mathcal{K}_\Theta \mathcal{K}_{\text{pH}} s^{-2} \left(\frac{7.3 \times 10^{-5} \varphi^{0.35} \rho^{0.17} \delta^{0.36} t^{0.06} \tau_r^{0.75}}{s^{1.1} v^{0.28} \alpha^{0.09} \left(\frac{\sigma_{ul.s.}}{\sigma_{ul.s.el.}} \right)^{0.8}} \right)^{b_1} \\
& \times \Bigg[1.54 \times 10^7 Rz_c + 1.83 \times 10^7 r - 9.19 \times 10^6 \sqrt{4r^2 - s^2} \\
& - 144375 \rho r^{-1} \left(\frac{1000}{\sigma_{\text{yiel}} \left(9.81c + 9.81 \exp\left(\frac{a\left(\frac{p v^l \sigma_{ul.s.}(st)^m}{1000 F \alpha_1} + 293 \right)}{b + 100} \right) \right)} - 1 \right) \\
& \left(2s - \frac{1000}{\sigma_{\text{yiel}} \left(9.81c + 9.81 \exp\left(\frac{a\left(\frac{p v^l \sigma_{ul.s.}(st)^m}{1000 F \alpha_1} + 293 \right)}{b + 100} \right) \right)} + 1 \right)
\end{aligned}
$$

$$+\frac{1,54\times10^{7}c_{y}s^{y}v^{z}\left(\mathrm{HB}_{\max}^{n}t^{x}-\mathrm{HB}_{\min}^{n}(t-Rz_{in})^{x}\right)}{j_{\mathrm{rig}}\mathrm{HB}_{\mathrm{av}}^{n}}\Bigg] \tag{15}$$

if $\varphi \geq \arcsin(s/(2r))$, $\varphi_1 < \arcsin(s/(2r))$

$$v_c = 22\times10^{-5}b_0v_{c0}\mathcal{K}_t\mathcal{K}_w\mathcal{K}_{\Theta}\mathcal{K}_{\mathrm{pH}}s^{-2}\left(\frac{7.3\times10^{-5}\varphi^{0.35}\rho^{0.17}\delta^{0.36}t^{0.06}\tau_r^{0.75}}{s^{1.1}\gamma^{0.28}\alpha^{0.09}\left(\frac{\sigma_{ul.s.}}{\sigma_{ul.s.el.}}\right)^{0,8}}\right)^{b_1}$$

$$\times\Big[1.54\times10^{7}Rz_c+1.83\times10^{7}r(1-\cos(\varphi))+1.83\times10^{7}\sin(\varphi_1)(s\cos(\varphi_1)$$

$$+\sqrt{s\sin(\varphi_1)(2r-s\sin(\varphi_1))}\Big)$$

$$-7.7\times10^{6}\rho\left(\frac{1000}{\sigma_{\mathrm{yiel}}\left(9.81c+9.81\exp\left(\frac{a\left(\frac{pv^l\sigma_{ul.s.}(st)^m}{1000F\alpha_1}+293\right)}{b+100}\right)\right)}-1\right)$$

$$\left(\frac{1}{tg(\varphi)}+\frac{2r}{s}\right)^{-1}$$

$$+\frac{1.54\times10^{7}c_{y}s^{y}v^{z}\left(\mathrm{HB}_{\max}^{n}t^{x}-\mathrm{HB}_{\min}^{n}(t-Rz_{\mathrm{in}})^{x}\right)}{j_{\mathrm{rig}}\mathrm{HB}_{\mathrm{av}}^{n}}\Bigg] \tag{16}$$

where s is feeding; r is the apex radius; φ, φ_1 are main and auxiliary angles in the plan; γ is the front angle; $\mathrm{HB}_{\mathrm{av}}$ is the hardness of the working surface; $\mathrm{HB}_{\min}$, $\mathrm{HB}_{\max}$ are minimum and maximum values of working surface hardness; t is depth of cut; ρ is the rounding radius of the cutting edge, $\mathrm{Rz}_{\mathrm{in}}$ is the height of the initial surface roughness; jrig is technological system stiffness; c_y, n, x_p, y_p, z_p are experiment coefficients [4]; σ_{yiel} is the yield strength of the working part material; Rz_c is the roughness height of the cutter apex; a, b, c are empirical coefficients [3, 4]; τ_r is the resistance of the working part material to the plastic shear; α is the rear angle of the cutter; δ is the wear height on the rear edge; $\sigma_{\mathrm{ul.s.}}$ is the stress limit of the working material; $\sigma_{\mathrm{ul.s.el.}}$ is the stress limit of the electrical steel; α_1 is the linear expansion coefficient of the tool material; F is the tool cross section; and p, m, l are coefficients [4].

Similarly, putting the dependences of the quality parameters of the element surface layer on their grinding modes [3–5] in the corrosion rate Eq. (1), we obtain the following dependences of the corrosion rate on the grinding conditions:

$$v_c = \frac{4.13\times10^{8}b_0v_{c0}\mathcal{K}_t\mathcal{K}_w\mathcal{K}\mathcal{K}_{\mathrm{pH}}}{\rho_{\mathrm{av}}}$$

$$\times\left(\frac{0.01(\mathrm{EXP}(-0.0015T_{\mathrm{th}}))100\sigma_{ul.s.}+\sigma_{ul.s.}\psi_{\mathrm{n}}-100\sigma_{\mathrm{yiel}}}{\sigma_{\mathrm{yiel}}}\right)^{b_1} \tag{17}$$

where ρ_{av} is the mean of the rounding radius of abrasive grain apexes [3, 4]; T_{th} is the heat temperature of the surface layer when grinding.

Putting in (1) the dependences of quality parameters and physical–mechanical properties of the surface layer from the conditions of finishing–strengthening treatment of [3–5], we will receive final theoretical equations for determining corrosion rates depending on the conditions of finishing–strengthening treatment:

for conditions when the tool is in contact with the element surface forming a circular or elliptical contact area:

$$v_c = \frac{0.0013 b_0 v_{c0} \mathcal{K}\mathcal{K}_w \mathcal{K}\mathcal{K}_{pH}}{(Sm_{in})^2} \left(\frac{F^{0.15} m}{k_{in} \sigma_{yiel}^{1.15}} - 1 \right)^{b_1} \Bigg(1.54 \times 10^7 Rz_{in} + 1.54 \times 10^7 Rz_t + 1.54 \times 10^7 \sqrt[3]{\frac{0.32 F Rp_{in}^2}{\sqrt{rR}(\mathrm{HB}_{max} - \mathrm{HB}_{min})}} + \frac{1.925 \times 10^6 s^2}{r} - 1.54 \times 10^7 Rp_{in} \times \sqrt{-47.75 F \sqrt{1+f^2} \left(tm_{in} \left(99 H\mu_{in} - \frac{100 F^{0.15} m H\mu_{in}}{k_{in} \sigma_{yiel}^{1.15}} \right) \right)^{-1} (rR)^{-0.5}} \times \left(h_{kin} + h_{el} - \frac{\arccos\left(s \left(0.32 F \left(\frac{100 F^{0.15} m H\mu_{in}}{k_{in} \sigma_{yiel}^{1.15}} - 99 H\mu_{in} \right)^{-1} \right)^{-0.5} - 1 \right)}{180 (h_{kin} + h_{el})^{-1}} \right)^{-0.5} \Bigg) \quad (18)$$

for condition when the tool is in contact with the element surface forming a tear-drop contact area:

$$v_c = \frac{0.0013 b_0 v_{c0} \mathcal{K}_t \mathcal{K}_w \mathcal{K}\mathcal{K}_{pH}}{(Sm_{in})^2} \left(\frac{F^{0.15} m}{k_{in} \sigma_{yiel}^{1.15}} - 1 \right)^{b_1} (1.54 \times 10^7 Rz_{in} + 1.54.10^7 Rz_t + 1.54 \times 10^7 \sqrt[3]{\frac{0.32 F Rp_{in}^2}{\sqrt{rR}(\mathrm{HB}_{max} - \mathrm{HB}_{min})}} + 1.54 \times 10^7 r (1 - \cos\varphi_a) - 1.54 \times 10^7 Rp_{in} \times \sqrt{-47.75 F \sqrt{1+f^2} \left(tm_{in} \left(99 H\mu_{in} - \frac{100 F^{0.15} m H\mu_{in}}{k_{in} \sigma_{yiel}^{1.15}} \right) \right)^{-1} (rR)^{-0.5}}$$

$$\times\left(h_{\text{kin}}+h_{\text{el}}-\frac{\arccos\left(s\left(0.32F\left(\frac{100F^{0.15}mH\mu_{\text{in}}}{k_{in}\sigma_{\text{yiel}}^{1.15}}-99H\mu_{\text{in}}\right)^{-1}\right)^{-0.5}-1\right)}{180(h_{\text{kin}}+h_{\text{el}})^{-1}}\right)^{-0.5}, \tag{19}$$

where s is feeding; r is a profile tool radius; R is a tool radius; φ_{a} is rear indentation angle; HB_{min}, HB_{max} are minimum and maximum values of working surface hardness; Rz_{in}; Rp_{in}; tmin; Smin are initial roughness values; f is the friction ratio between the tool and the workpiece; σ_{yiel} is the stress limit of the working material; m is a ratio depending on the working material [3, 4]; F is the operating force; k_{in} is the initial level of the workpiece surface hardening; hkin is the depth of tool indentation into the workpiece material; hel is the elastic restitution of the working material; Rz_t is the tool working material; and Hμin is the initial microhardness of the material.

Similarly, putting the dependences of the quality parameters of the element surface layer on the modes of their electroerosive treatment [1, 3–9] in the equation of the corrosion rate (1), we obtain the following dependence of the corrosion rate on the conditions of electroerosive treatment:

$$v_c=\frac{494b_0v_{c0}\left(4.04\frac{P_t^{0.171}}{H\mu_{\text{in}}P_d^{0.017}}-1\right)^{b_1}}{\sqrt[3]{\frac{UI\eta\tau}{T\rho c}}}\mathcal{K}_t\mathcal{K}_w\mathcal{K}\mathcal{K}_{\text{pH}}, \tag{20}$$

where I is the process current; U is the process voltage; τ is the duration of an current impulse; η is the efficiency ratio of impulse energy; c is the specific heat capacity of the product material; ρ is the product material density; T is the melting-point temperature of the working material; Pt is Palatnik ratio of the tool material; Pd is Palatnik ratio of the element material; and $H\mu_{\text{in}}$ is the initial microhardness of the material.

4 Conclusion

Analyzing theoretical dependences obtained (14–16), we can conclude that feeding, depth and cutting speed as well as the geometry of the cutting tool greatly influence the corrosion rate after edge cutting processing.

The analysis of theoretical dependence (17) shows that the corrosion rate during grinding is mainly influenced by the mean of the rounding radius of abrasive grain apexes and is the heat temperature of the surface layer, as well as physical and mechanical properties of the processed material.

Analyzing theoretical dependences (18, 19), we come to the conclusion that during finishing–strengthening treatment, the corrosion rate is influenced by the modes (feeding and operating force), as well as the parameters and quality of the tool (radius and profile radius of the tool, its roughness).

The analysis of dependence (20) shows that during electroerosive treatment, the corrosion rate mainly depends on the process voltage and current strength, as well as the duration of current impulses.

References

1. Fedonin, O.N., Syanov, S.Yu., Papikyan, A.M.: Provision of wear resistance and fatigue strength of surfaces during electroerosive processing. In: IOP Conference Series: Materials Science and Engineering. Processing Equipment, Mechanical Engineering Processes and Metals Treatment, Tomsk, Russia, 1–6 (2018)
2. Sergeev, N.N., Sergeev, A.N., Kutepov, S.N., et al.: Physical-Mechanical and Corrosion Properties of Metal Materials Operated in Aggressive Environments: Monograph. Moscow. Infra-Engineering, Vologda (2020)
3. Suslov, A.G., Petreshin, D.I.: Forming and developing engineering technology as a science. In: MATEC Web of Conferences. 2018 International Conference on Modern Trends in Manufacturing Technologies and Equipment, ICMTMTE 2018, Sevastopol, Russia, 1–5 (2018)
4. Suslov, A.G., et al.: Engineering of Machine Parts Surface. Mashinostroeniye, Moscow (2019)
5. Suslov, A.G., et al.: Industrial Engineer Guide. Innovatsionnoe Mashinostroeniye, Moscow (2019)
6. Suslov, A.G., Fedorov, V.P., et al.: Technological Support and Improvement of Operational Properties of Parts and Their Connections. Mashinostroeniye, Moscow (2006)
7. Suslov, A.G.: Surface Layer Quality of Machine Parts. Mashinostroeniye, Moscow (2000)
8. Syanov, SYu., Papikyan, A.M.: Functional-oriented technological processes of electroerosive processing. In: International Conference «Actual Issues of Mechanical Engineering» 2018 (AIME 2018), pp. 480–484. Russia, Novosibirsk (2018)
9. Syanov, S.Yu.: Theoretical determination of surface layer quality characteristics of workpieces, tool electrode wear and efficiency of spark eroding. In: Proceedings of 2015 International Conference on Mechanical Engineering, Automation and Control Systems, Tomsk, Russia, 1–4 (2015)
10. Vereshchagin, A.S., Voznyakovsky, A.P., et al.: Advanced Engineering Technologies, Equipment and Tools. Publishing House Spektr, Moscow (2015)

Manufacturing a Combined Electrode-Instrument by the Method of Fast Prototyping

Evgeniy Smolentsev, Vadim Kuts, Mikhail Razumov, and Dmitriy Krokhin

Abstract The technology is considered electrochemical machining (ECM) of metals. The problem lies in the comparative complexity of manufacturing an electrode-tool (ET), which limits the scope of application of electrochemical machining (ECM). This is largely due to the fact that the design of the electrode-tool (ET) should provide for the electrochemical machining of metals a constant equilibrium supply of the working conductive liquid in the area of the object being processed. The solution lies in using the capabilities of computer-aided design (CAD) and additive technologies. It allows you to implement fundamentally new elements of the design of tools. It eliminates the need for a number of resource-intensive operations and the use of expensive equipment. The place has a solution, where the body of the electrode-tool is made of plastic with dielectric properties, inside which and on separate sections of its surface are placed current-conducting channels and layers of current-conducting plastic. Electroplated copper layers are applied to their outer surface. A special design was developed by the 3D printer for the manufacture of this electrode-tool with an additional rotary horizontal axis and three positions for installing two extruders (the first extruder is designed for ABS plastic, the second extruder is designed for conductive plastic) and a special nozzle designed for copper electroplating. The 3D printer and the software developed for it made it possible to carry out the additive shaping of the combined electrode-tool. The authors propose their own approach to the design of combined ET, including a complex shape.

1 Introduction

Electrochemical machining of metals is based on the use of the phenomenon of anodic dissolution. The metal dissolves on the surface of the anode (processed blank) without

E. Smolentsev · D. Krokhin
Voronezh State Technical University, Voronezh, Russian Federation

V. Kuts (✉) · M. Razumov
Southwest State University, Kursk, Russian Federation

M. Shamtsyan et al. (eds.), *Robotics, Machinery and Engineering Technology for Precision Agriculture*, Smart Innovation, Systems and Technologies 247,
https://doi.org/10.1007/978-981-16-3844-2_20

direct contact between the part and the tool. This occurs in a conductive environment under the influence of electromagnetic fields. The technology has a number of advantages over traditional blade processing methods. Examples of advantages are as follows: high processing performance regardless of the hardness of the workpiece material; the possibility of obtaining various geometries with high purity; treatment of surfaces and cavities in parts, access to which is impossible with a tool for machining. Electrochemical machining also eliminates the use of manual labor (locksmith operations to remove burrs, rounding the edges). This helps to improve product quality and reduce labor costs.

2 Problem Statement

The comparative complexity of the manufacture of the electrode-tool (ET) is one of the factors limiting the area of rational use of electrochemical machining (ECM). This circumstance is complicated when it is used to shape parts with a complex profile, deep holes with variable diameters and a curved axis. The design of the electrode-tool (ET) is non-technological for such products often enough, and it is impossible to create it by traditional methods of blade processing.

This is largely due to the fact that the design of the electrode-tool (ET) for electrochemical machining (ECM) must ensure a constant uniform supply of the working conductive fluid to the areas of the object being processed. The design of the electrodes in this case provides for the supply of the electrolyte grooves, slots, holes. An example is shown schematics for the implementation of electrochemical machining in Fig. 1.

The physical and chemical properties of the electrolyte and its flow properties in the processing zone exist. They have a significant impact on indicators such as productivity, roughness, precision and surface quality. A large number of requirements are imposed in this regard when designing electrodes to slots and holes in them [1]. This applies, for example, the length of electrolyte channel, the location of the electrolyte supply and their mutual position. Fulfillment of all these requirements makes it possible to achieve a reduction in the processing error to 1–2 microns [2]; however, the shaping of such elements either requires too high resource consumption, or, in some cases, is impossible when using traditional subtractive technologies.

As is known [3], such technological parameters of processing as the quality of processing, accuracy and productivity are in practice mutually exclusive in the case of an attempt to simultaneously increase them within the framework of one operation. Nevertheless, a tangible increase (or keeping values at an acceptable level) can be achieved in the case of discrete control of the parameters and the method of electrolyte supply for each of the above indicators during ECM. So, for example, the negative effect can be eliminated from surface hydrogenation, the process of anodic dissolution of metals can be stabilized and the processing error can be reduced several times when channels are formed in the ET to remove gaseous products [1]. Varying

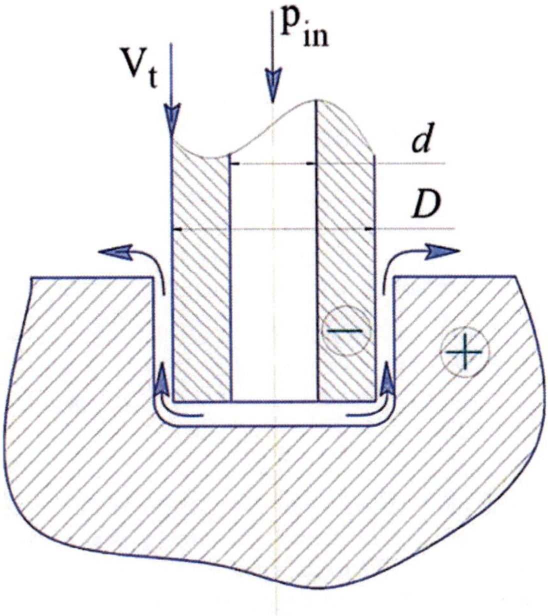

Fig. 1 ECM scheme with direct electrolyte supply: Vt—feed rate of the tool electrode; P_{in}—electrolyte supply under pressure; *D*—diameter of the working part of the electrode-tool; d– diameter of the channel to supply the electrolyte to the treatment zone

the shape of the channels can be realized alternately laminar and turbulent characters of the flow of the working environment. In the first case, stability and quality of processing will be achieved, in the second—accelerated removal of products from the interelectrode gap. The concentration, composition and temperature of the electrolyte can change the processing performance by an order of magnitude. In addition, they affect the accuracy of the formed surface. Therefore, new opportunities may open up to improve technological performance by supplying working fluids differing in the above parameters to different parts of the part. This is especially important in the case of bimetals processing. They are used quite widely, for example, in the production of aerospace technology. It is also possible, due to the formation of the geometry of the channels for the electrolyte supply, to achieve the speed of the electrolyte flow at all processing sites with values that ensure the removal of metal-containing and gas-containing processing products. This provides an increase in technological performance.

3 Research Methods, Results

Electrodes-tools are designed in the following order. First, the shape of the working surface and the position of the channels to supplying the working medium to the processed sections of the workpiece are calculated. When determining the shape of the working part of the tool, one proceeds from a surface that is equidistant to the contour being machined, correcting it according to various methods to the actual one. The most common correction methods are analytical, electric field simulation and production. The first two methods make it difficult to take into account such factors as the hydrodynamics of the working medium, the variation of its temperature and concentration during processing, which reduces the quality of parts manufacturing. Therefore, the third production method is more common [4].

The manufacture of electrodes-tools of complex shape is carried out according to the established technology in the following sequence:

According to well-known methods [4], master templates are calculated and made with a profile inversely equidistant to the working part of the electrode-tool

1. The casting mold is made according to master templates. The workpiece casts on a casting mold, if necessary, to make channels for supplying electrolyte in it.
2. The final geometry of the electrode-tool for electrochemical machining is formed using subtractive technologies.

.

The above makes it obvious that the cost of an electrode-tool (ET) of a complex shape can exceed the cost of electrochemical machining (ECM) of a part with its help, since the creation of a tool requires a wide range of expensive equipment. The scale of costs grows by an order of magnitude in relation to unit production. This is obvious, but in any case does not guarantee the implementation of discrete control of the parameters and the method of supplying the electrolyte to the treatment area, which, as mentioned above, can achieve a significant increase in the technological parameters of treatment.

The solution is to use the capabilities of computer-aided design and additive technologies. It allows you to implement fundamentally new elements of the design of tools. It eliminates the need for a number of resource-intensive operations performed on expensive equipment.

Figure 2 shows a sketch and 3D model of the electrode-tool to remove burrs in the holes. The tool has a traditional design. The base is a solid object. It is made from copper. A layer of dielectric-capron is applied to it for formation to non-conductive areas.

The generalized technology for the manufacture of electrodes-tools for ECM is shown in Fig. 3 [5].

The new design was proposed to enable the manufacture of an electrode-tool in an additive way and to reduce its cost. It is shown in Fig. 4.

The distinctive feature lies in the case of the electrode-tool, made of plastic, which has dielectric properties. Inside the plastic and in some areas of its surface, there are conductive channels, layers of conductive plastic, on the outer surface of which a layer of galvanic copper coating is applied (Fig. 4). Layers of conductive plastic are used as a base for electroplating copper plating. This will increase the electrical conductivity in the areas of the electrode-tool directly involved in the process of electrochemical dissolution of burrs. Thus, for the manufacture of ET by the additive method, it is necessary to construct separate models of two parts of the electrode, made of ABS

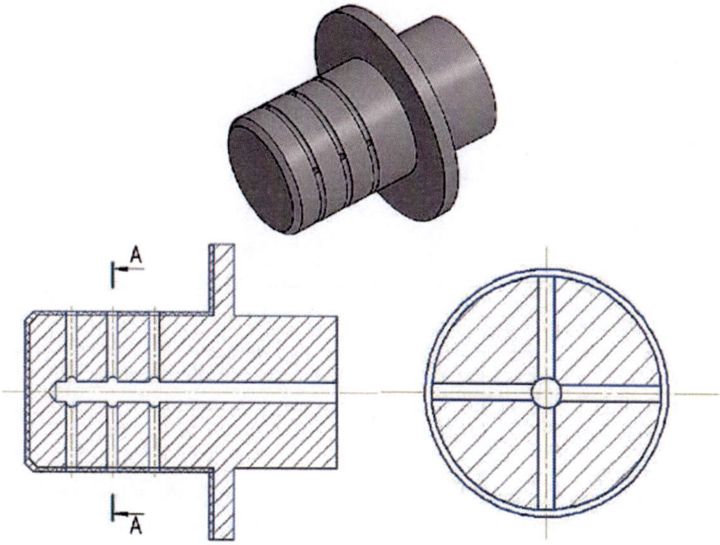

Fig. 2 Traditional electrode-tool structure

plastic and current-conducting plastic. The outer surface will be covered with a layer of copper (Fig. 5).

The special nozzle was designed to apply local galvanic copper-covered on additive equipment. It is installed in the position of an additional extruder in a 3D printer (Fig. 6).

The nozzle is a body consisting of a copper tube 4, which acts as an anode, since a plus from a power source 2 is connected to it, and has a nut 5 and a nozzle 3, in the cavity of which the electrolyte is located. Nut 5 has a thread and is screwed onto a copper tube 4, which makes it possible to add electrolyte if necessary. The nozzle 3 is screwed onto the lower part of the copper tube and has a porous rod 7, which is impregnated with electrolyte and acts as an electrolyte carrier, allowing interaction with part 1, which is a cathode, since it is connected to the negative terminal of power supply 2. Moreover, the size and porosity are selected for the rod 7 taking into account the thickness of the applied single layer. Thus, the deposition of copper ions on the workpiece from the electrolyte is ensured, and the replenishment of copper ions into the electrolyte from the surface of the copper tube. Ring 6 is made of dielectric and it serves to be fixed on the printer.

A special design of a 3D printer was developed for the manufacture of this tool electrode with an additional rotary horizontal axis [6–12] and three positions for installing 2 extruders (one with installed ABS plastic, the second with conductive) and a special nozzle for applying copper electroplating. The proposed 3D printer and the software developed for it made it possible to carry out additive shaping of the combined electrode-tool (Fig. 7).

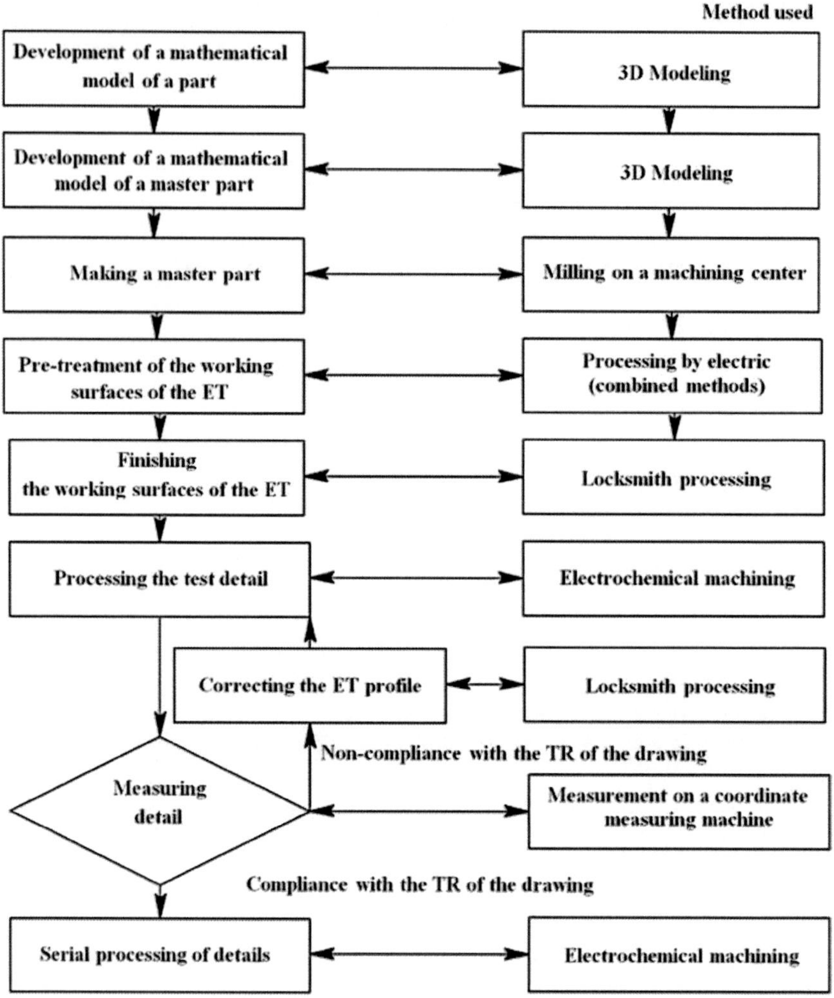

Fig. 3 Generalized technology for the manufacture of electrodes-tools for ECM

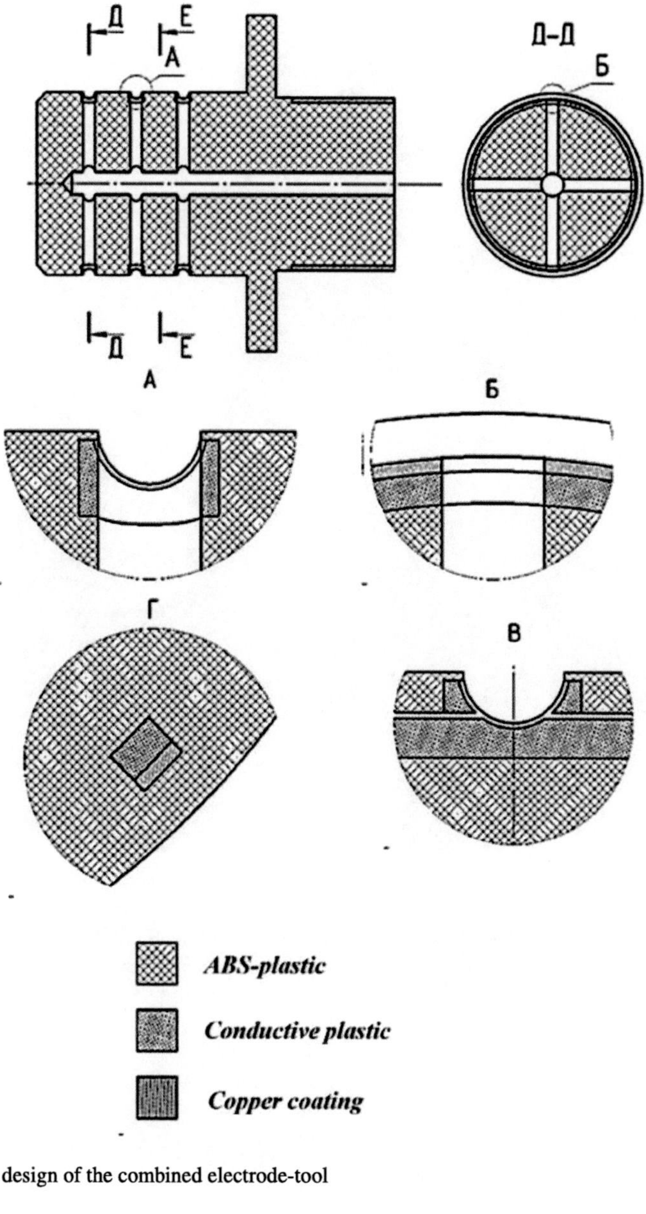

Fig. 4 The design of the combined electrode-tool

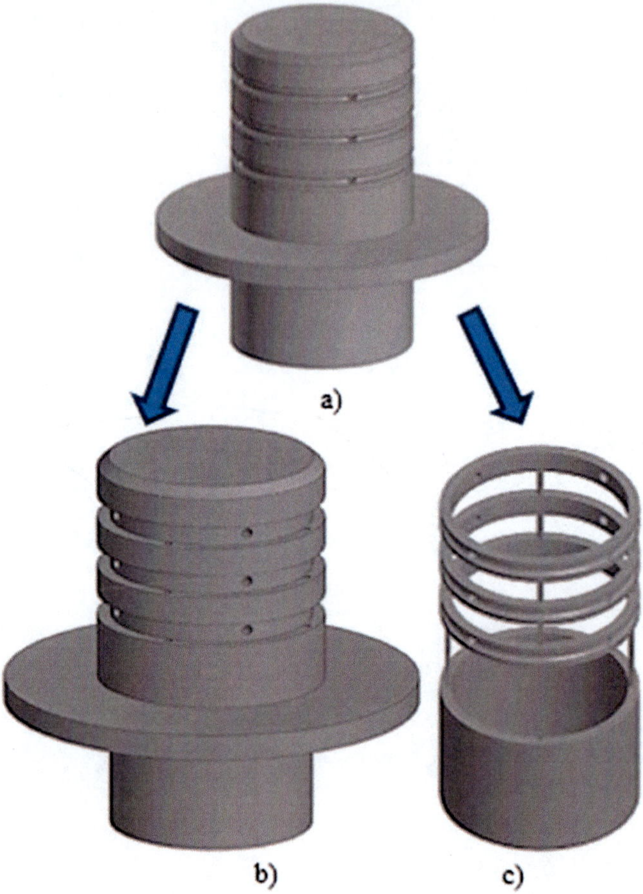

Fig. 5 Structure of the combined electrode-tool: **a** prefabricated combined ET model; **b** part of the model made of ABS plastic; **c** part of the model made of conductive plastic

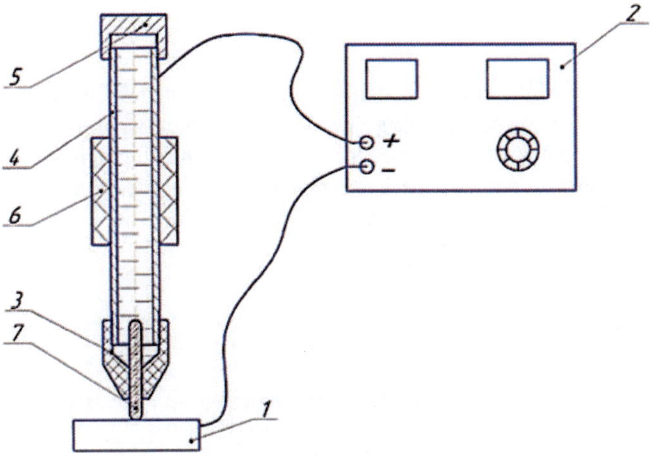

Fig. 6 Attachment for applying local electroplating using a 3D printer

Fig. 7 Model of the combined EI obtained by additive shaping

4 Conclusions

The authors propose their approach to the design of combined ET, including complex forms. The development allows to significantly reduce the cost of materials used for its manufacture, and it makes it possible to subsequently implement the process of its manufacture by additive methods without using a wide range of expensive equipment. This will reduce the cost of manufacturing.

References

1. Amitan, I.A., Baysupov, Yu.M., Baron and others; under the general editorship. Volosatov, V.A.: Reference book on electrochemical and electrophysical methods of processing. Mechanical engineering. Leningrad, Department, 719 (1988)
2. Grigoriev, S.N., Smolentsev, E.V., Voloskova, M.A.: Processing technology with concentrated energy flows: a textbook for students of technical universities. Art. Oskol: TNT. 280 (2009
3. Smolentsev, E.V.: Design of electrical and combined processing methods: monograph. Mechanical Engineering 511 (2205)
4. Smolentsev, V.P., Boldyrev, A.I., Smolentsev, E.V., Smolentsev, G.P., Koptev, I.T.: Theory of electrical and physical and chemical processing methods. In: Part I. Processing of Materials with the Use of Tools: Textbook. GOUV-PO Voronezh State Technical University, Voronezh, 248 (2008)
5. Orlov, A.A.: Increase in productivity and quality of manufacturing of GTE compressor blades by improving the technology of circular electrochemical machining: dis. ... Cand. Tech. Sci. 05.02.08, Rybinsk, 159 (2014)
6. Grechukhin, A.N., Kuts, V.V., Oleshitskiy, A.V., Simonova, Yu.E.: Design of technological equipment for additive shaping with a hybrid layout. Bull. Voronezh State Tech. Univ. **15**(4), 111–118 (2019)
7. Grechukhin, A.N., Kuts, V.V., Razumov, M.S.: Solving problem of curved surface approximation by layers with constant and variable sections during forming by additive methods. Lecture Notes in Mechanical Engineering, pp. 239–248 (2020)
8. Grechukhin, A.N., Kuts, V.V., Razumov, M.S.: Management of space orientation of the end effector of generation of geometry system five axis manufacturing machinery for additive generation of geometry. In: MATEC Web Conference, Vol. 226, 2018, XIV International Scientific-Technical Conference "Dynamic of Technical Systems" (DTS-2018), 7 (2018)
9. Grechukhin, A., Kuts, V., Oleshitskiy, A.: Development and research of technological equipment that implements dynamic control of process of additive fabrication of parts of complex spatial shapes based on mechanisms with a hybrid layout, in the collection: In: IOP Conference Series: Materials Science and Engineering. International Conference on Modern Trends in Manufacturing Technologies and Equipment 2019, ICMTME 2019, 033112 (2020)
10. Grechukhin, A.N., Kuts, V.V., Razumov, M.S.: Calculation of the controlled parameters of the 6-coordinate robot in the process of additive forming of products. In the collection: Journal of Physics: Conference Series, 012053 (2019)
11. Grechukhin, A.N., Kuts, V.V., Oleshitsky, A.V.: Control additive error of morphogenesis with the use of hybrid layout AIP Conference Proceedings. In: XV International Scientific-Technical Conference Dynamics of Technical Systems, DTS 2019, 020003 (2019)

12. Grechukhin, A.N., Kuts, V.V., Oleshitsky, A.V.: Control of geometrical parameters of a single layer at additive forming of products by fdm technology. In the collection: IOP Conference Series: Materials Science and Engineering. In: XV International Scientific-Technical Conference Dynamics of Technical Systems, 012004 (2019)

Technology of Hardening Plowshares by Electrocontact Welding Using Waste from Tool Production

Rashit Latypov, Anton Serov, Nikita Serov, and Olga Chekha

Abstract The article deals with the issues of strengthening the working bodies of agricultural machines, since today the agro-industrial complex is one of the priority areas of development in the Russian Federation. This is especially relevant against the background of a significant increase in agricultural production, while a significant part of agricultural machinery presented on the Russian market does not meet the requirements for reliability. To increase the reliability of agricultural machines, innovative technologies are needed to increase their resource. Such technologies include technologies for obtaining functional coatings that provide higher technical and economic indicators for the production and operation of machine parts. At the same time, the issues of improving the environmental friendliness of production are becoming more and more relevant every year. It is possible to increase the environmental friendliness of production by using secondary resources (recycling). The article proposes a technology for obtaining wear-resistant functional coatings with simultaneous disposal of waste from machine-building and tool production on the parts of the working bodies of agricultural machines on the example of a plowshare. The use of waste as an additive material in electric contact welding allows to increase the efficiency and environmental friendliness of the process of obtaining functional coatings by this method. The effectiveness of the proposed technology is confirmed by studies of the structure and properties of the obtained coatings, during which it was found that the relative wear resistance of coatings obtained by this method from spent hacksaw blades made of steel grade 11R3AM3F2 is 7.3 times higher than that of the reference steel grade 45 with a hardness of 48 HRN15, and the resource of hardened parts increases up to two times.

R. Latypov
Moscow Polytechnic University, 16, Avtozavodskaya st., 111280 Moscow, Russia

A. Serov
Bauman Moscow State Technical University, 5, 2nd Baumanskaya St., 105505 Moscow, Russia

N. Serov (✉) · O. Chekha
Russian State Agrarian University—MTAA, 49, Timiryazevskaya St., 127550 Moscow, Russia
e-mail: n.serov@rgau-msha.ru

M. Shamtsyan et al. (eds.), *Robotics, Machinery and Engineering Technology for Precision Agriculture*, Smart Innovation, Systems and Technologies 247,
https://doi.org/10.1007/978-981-16-3844-2_21

1 Introduction

Today, the development of agriculture in the Russian Federation is one of the priority areas. In recent years, the production of agricultural products has increased significantly, which was made possible, largely due to the increase in the number and quality of agricultural machines in the industry.

In the Russian market of agricultural machinery (tillage, harvesting, transport, processing, etc.), there are many domestic and foreign manufacturers.

According to available data, up to 25% of all tillage agricultural machines presented on the Russian market do not meet the requirements for reliability. The main reasons for the failure of parts of tillage machines are their wear out [1–3], due to insufficient wear resistance (strength)—about 50%—and low-quality manufacturing (marriage)—about 38%.

Increasing the reliability of agricultural machinery can be achieved by strengthening the parts of working bodies, drives, and power plants [4]. In the hardening of machine parts, there are two main directions—this is the strengthening treatment of the materials from which they are made or the production of functional coatings on their working surfaces [5–7]. At the same time, the increase in reliability should not lead to a decrease in technical and economic indicators, as well as harm the environment. What is possible when using mechanical engineering waste in the processes of hardening and obtaining functional coatings. There are many methods for obtaining functional coatings, the most promising of which is electric contact welding (ECW) of compact materials and powders [8–11], since it allows you to obtain coatings from a wide range of materials based on steels and cast irons without melting them.

In this regard, the aim of the study is to develop a technology that combines the process of recycling wastes from machine-building and tool production with the process of obtaining wear-resistant coatings on the details of the working bodies of the agricultural machines, as well as the study of the structure and properties of the resulting coatings.

2 Materials and Methods

The microstructure of the coating, the joint zone, and the base material were determined using an Axiovert 40 MAT metallographic microscope (CarlZeiss). Segments for macro- and microsections were cut out on an LC-300 metallographic cutting machine and hot-pressed on an LHM-3000 automatic hydraulic press into bakelite. Finished macro- and microsections were ground and polished on a double-disk grinding and polishing machine "LAP-2X." The hardness of the welded layer of steel 11R3AM3F2 (GOST 19265–73), on the base of steel 65G (GOST 14959–2016) of the obtained ECW of the proposed technology, was measured by the Rockwell method (GOST 9013-59) and Super-Rockwell (GOST 22975-78), and microhardness according to GOST R ISO 6507-1-2007. Rockwell and Super-Rockwell hardness was measured on a TH320. Measurements of the microhardness of the joint zone of the coating and the base of the samples under consideration were carried out on a METOLAB 502 device. The wear resistance study was carried out on IM-01 units in accordance with GOST 23.208-79. The tests were carried out on specimens made of 65G steel with a hardness of 52 HRN15, 65G steel after electrical contact hardening (ECH) of the surface to 88 HRN15, 65G steel coated with 50HFA steel (GOST 14959-2016) with a hardness of 89 HRN15 and a U12A steel coating (GOST 1435-99) with a hardness of 92 HRN15, obtained by the ECW, as well as a sample from steel 65G with a coating obtained by the ECW of a used hacksaw blade made of steel 11R3AM3F2 with a hardness of 91.5 HRN15. Samples of steel 45 with a hardness of 48 HRN15 were taken as a standard. ECW and ECH were carried out on the 011-1-02 "Remdetal" unit, and the coatings from worn-out hacksaw blades were obtained on the AKS-1 manual contact welding machine.

3 Investigation of the Properties of Coatings Obtained by Electric Contact Welding of Used Hacksaw Blades Made of Steel 11R3AM3F2 and Development of Technology for Strengthening the Plowshare

To combine the process of waste disposal of machine-building and tool production with the process of obtaining wear-resistant coatings on the parts of the working bodies of agricultural machines, as well as the subsequent study of the structure and properties of the resulting coatings, the method of waste disposal of machine-building and tool production was used, Patent No. 2663234 of the Russian Federation. When implementing this method, steel waste in the form of hacksaw blades, files, broach files, and drills is used as an additive material for electric contact welding to obtain functional coatings on the surface of the part. The method does not require remelting of waste in metallurgical units, while it provides an increase in the service life of machine parts operating under conditions of intense wear.

This method was used to obtain samples according to the above method. Metallographic studies of samples obtained by electrocontact welding of used hacksaw blades made of 11R3AM3F2 steel onto a base made of 65G steel showed that carbides are observed in non-etched martensite in the structure of a coating made of 11R3AM3F2 steel (Fig. 1a), in turn, a base made of 65G steel in the heat-affected zone martensitic–trostitic (Fig. 1b), and the base outside the heat-affected zone has a ferrite–pearlite structure (Fig. 1c).

The hardness of the coatings made of steel 11R3AM3F2 obtained by electric contact welding reached 64 HRC (with an average value of 60.6 HRC), and the microhardness—954 MPa.

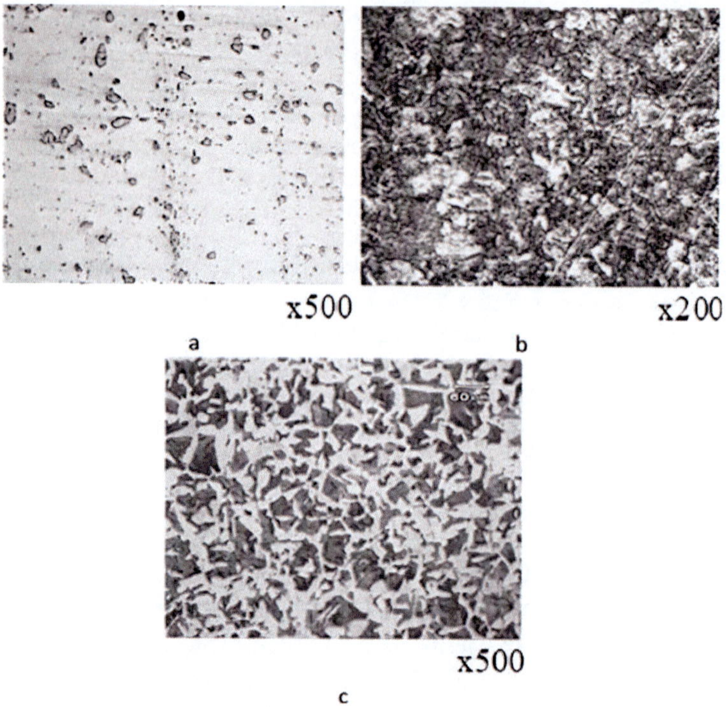

Fig. 1 Microstructure of materials after ECW

The relative wear resistance of 65G steel after ECH is 2 times higher than that of the reference sample made of 45 steel heat-treated to a hardness of 48 HRN15 and 1.5 times higher than that of 65G steel in the initial state. The coatings obtained by the ECW of a steel strip 0.5 mm thick from steel 50HFA and steel U12A on a base of steel 65G have a relative wear resistance of 2.7 and 4.1 times higher than the standard, respectively, and the coatings obtained by the ECW of hacksaw blades from steel 11R3AM3F2—in 7.3 times (Fig. 2).

Based on the results of the conducted research, a technology was developed for strengthening the flat surfaces of the working bodies of agricultural machinery by electric contact welding of tool steel waste, the scheme of which is shown in Fig. 3.

This technology was used to strengthen the plowshare "PLZH RZZ 31-702" [12, 13]. As an additive material, used hacksaw blades made of steel 11R3AM3F2 were used, which were welded on the plowshare in places prone to the greatest wear according to (Fig. 3b).

To intensify the process, before welding, gas dynamic (sandblasting) treatment of the filler material and the plowshare was performed in the places to be welded, which allows creating an optimal microrelief, cleaning it from mechanical contamination and oxide films, as well as creating an energy reserve in the near-surface layers due to their plastic deformation, while an intermediate layer of tape or powder solder

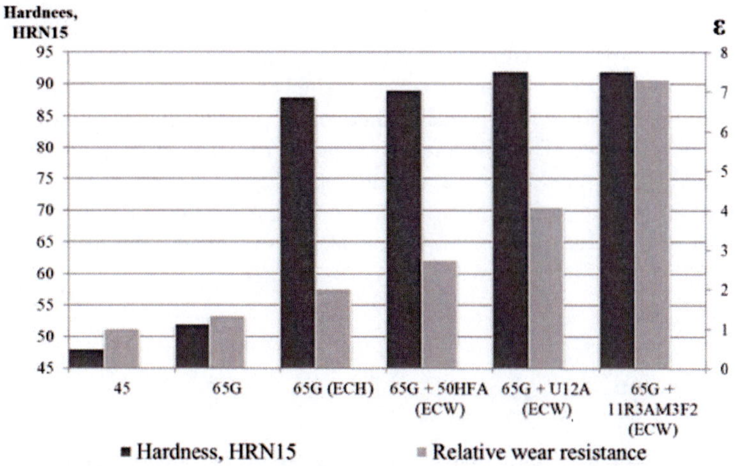

Fig. 2 Hardness and relative wear resistance of the test samples

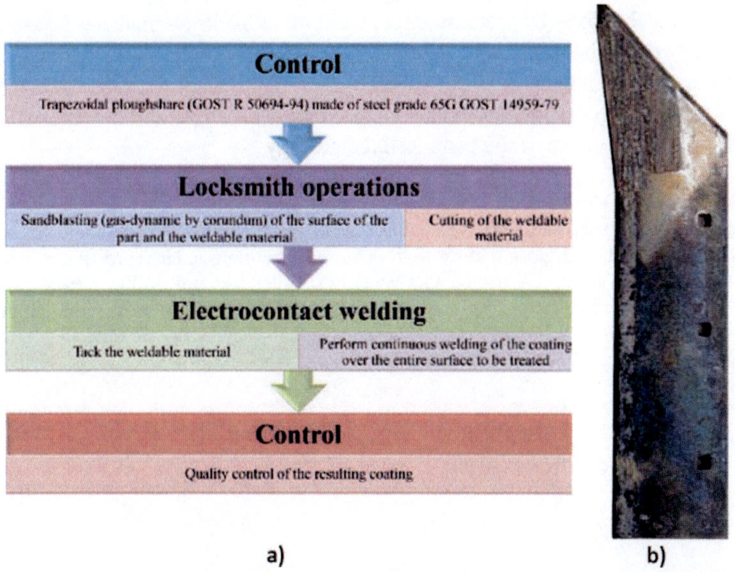

Fig. 3 Scheme of the technological process of hardening, **a** and the plowshare "PLZH RZZ 31-702," **b** with a wear-resistant coating obtained by this technology

is placed between the coating and the hardened surface of the plowshare (Russian Federation Patent No. 2605259) [14, 15].

Hacksaw blades (fragments) were cut to size, their edges were leveled and then laid on the surface of the part, with fixing welding points, after covering the required area of the part, continuous welding was performed over the entire surface due to step-seam heating below the melting temperature, but above the quenching temperature when passing through the connection zone welding current pulses of 6–12 kA at a voltage of 3–6 V with simultaneous application of pressure and supply of coolant to the connection zone. Electrocontact welding of the spent cloths was carried out on the installation intended for parts of the "shaft" type, using a device for suture electrocontact welding of the filler material on the surface of a flat part Patent of the Russian Federation No. 2622074. This technology allows you to increase the operating time of hardened plowshares by 1.8–2.0 times.

4 Conclusion

A promising method of hardening the working bodies of agricultural machines is the electrocontact welding of tool production waste, which allows you to combine the process of recycling waste from machine-building and tool production with the processes of strengthening and restoring parts. The hardness of coatings obtained by the proposed technology by electrocontact welding of hacksaw blades made of steel 11R3AM3F2 onto a base of steel 65G reached 64 HRC, microhardness 954 MPa, and the relative wear resistance was 7.3 times higher than that of the reference sample made of steel 45. Experimental studies of properties coatings and operational tests confirm the effectiveness of the proposed technology.

Acknowledgements The technical novelty is confirmed by the patents of the Russian Federation No. 2605259 "Method for restoring and strengthening the working bodies of agricultural machines," No. 2663234 "Method for disposing of waste from machine-building and tool production" and No. 2622074 "Device for suture electric contact welding of filler material on the surface of a flat part." The study was conducted using the equipment of the center for collective use "High-tech Technologies in Mechanical Engineering." Efficiency is confirmed by the act of testing hardened plowshares on the basis of the Kaluga Research Institute of Agriculture of the Kaluga Agricultural Station.

References

1. Martin, K., Tomáš, P., Tomáš, H.: Resistance of coated electrodes suitable for renovation of tillage tools. J. Central Eur. Agric. **14**(4), 1312–1319 (2013)
2. Hoormazdi, G., Küpferle, J., Röttger, A., Theisen, W., Hack, K.: A concept for the estimation of soil tool abrasive wear using ASTM G65 test data. Int. J. Civil Eng. 103–111 (2019)
3. Napiórkowski, J., Lemecha, M., Konat, L.: Forecasting the wear of operating parts in an abrasive soil mass using the Holm-Archard model. Materials **12**(13), 2180 (2019)
4. Saifullin, R.N.: Improvement of performance of machine parts by electrocontact welding of composite materials. J. Friction Wear **28**(2), 206–211 (2007)
5. Berezshnaya, O.V., Gribkov, E.P., Kuznestov, V.D.: Investigation of thermostressed state coating formation at electric contact surfacing of shaft type parts. Adv. Mater. Sci. Eng. **2016**, 6597317 (2016)

6. Berezshnaya, O.V., Gribkov, E.P.: Matematicheskoe modelirovanie formoobrazovaniya sloya pri electrocontaktnoi naplavke provolokoi detalei typa val. Visnik Volodymyr Dahl East Ukrainian Natl. Univ. **6**(213), 93–96 (2014)
7. Nafikov, M.Z., Saifullin, R.N.: Properties of coatings obtained by electrocontact welding of filler material from steel wires. Strengthening Technol. Coatings **2**, 38–43 (2014)
8. Nafikov, M.Z.: Formalized description of the process of formation of the welded joint in resistance welding of wires. Weld. Int. **29**(6), 466–470 (2015)
9. Sayfullin, R.N.: Electro-contact welding of powder materials during the restoration of parts and the production of protective coatings. In: M-in the Agriculture of the Russian Federation, Bashkir State University. Ufa, Russia, pp. 167–174 (2009)
10. Sayfullin, R.N., Farkhshatov, M.N., Levin, E.L.: Properties of coatings obtained by electrocontact welding of filler materials from steel tapes. Strengthening Technol. Coatings **3**, 13–16 (2014)
11. Saifullin, R.N., Pavlov, A.P.: Prospects for the use of additives in the Write debug-ging mesh materials for the restoration of parts of electric-at-cooking. Repair Restoration Modernization **9**, 11–13 (2011)
12. Serov, A.V., Serov, N.V.: Appointment and evaluation of kinematic parameters of electric contact welding. Mech. Eng. Technol. **2**, 32–37 (2019)
13. Serov, A.V., Serov, N.V., Burak, P.I.: Determination of technological parameters of electric contact welding during restoration and strengthening of flat surfaces. In: Bulletin of the Federal State Educational Institution of Higher Professional Education Moscow State Agroengineering University named after V. P. Goryachkin. 2017. Moscow, Russia, vol. 1, issue 77, pp. 35–40 (2017)
14. Burak, P.I., Serov, A.V., Latypov, R.A.: Optimization of the process of electric resistance welding of metallic strips through an amorphous solde. Weld. Int. **26**(10), 814–818 (2012)
15. Burak P.I.: Methods of applying an intermediate layer for electric contact welding. In: Bulletin of the Federal State Educational Institution of Higher Professional Education Moscow State Agroengineering University named after V. P. Goryachkin. Moscow, Russia, vol. 3, issue 34, pp. 69–72 (2009)

Impact of Fe_3O_4 Nanoparticles on Viscous and Tribotechnical Properties of Oleic Acid

Mikhail Shishkarev, Elena Zadoshenko, Maksim Gavrilenko, and Alla Molchanova Sokolenko

Abstract The paper considers the impact of Fe_3O_4 nanoparticles on the viscosity and tribotechnical properties of oleic acid. Fe_3O_4 nanoparticles were obtained by precipitation of Fe^{+3}/Fe^{+2} iron salts mixture with an excess of ammonium hydroxide from aqueous solutions. Based on the studies of sedimentation analysis and atomic force microscopy of the synthesized particles, it was established that their size is polydisperse, and the major amount of the particles is in the range of 50 nm. Based on oleic acid and synthesized particles with a content of 0.1–0.5% of the mass, lubricating compositions were created, the viscosity of which was evaluated with a HAAKE Viscotester C viscometer at a rotation speed of 100 rpm in a temperature range from 25 to 60 °C. The tribotechnical performances of the lubricating compositions were evaluated according to the results of a study on a four-ball friction machine and in a steel-steel friction pair with an UMT-200 machine. Based on the viscosity studies results, the activation energies of lubricant compositions were determined, which affect the surface layers formation in the friction zone. The tribological studies results have shown that the introduction of an additive of Fe_3O_4 nanoparticles leads to an increase in the antiwear performances of oleic acid and an increase in its carrying capacity, and also improves its antifriction characteristics. Thus, it has been established that Fe_3O_4 nanoparticles lead to a decrease in the viscosity of oleic acid with a temperature increase and to an improvement in its tribotechnical performances, which is related to the protective surface layers formation in the friction zone.

1 Introduction

The development of new high-quality lubricants is an important area of tribology. Improvement in properties and development of new types of lubricants is being realized in the following areas: creation of high-temperature and environment friendly lubricants based on synthetic and vegetable oils, selection and creation of new types

M. Shishkarev · E. Zadoshenko · M. Gavrilenko (✉) · A. M. Sokolenko
Don State Technical University, Rostov-on-Don, Russia
e-mail: aaakot@yandex.ru

M. Shamtsyan et al. (eds.), *Robotics, Machinery and Engineering Technology for Precision Agriculture*, Smart Innovation, Systems and Technologies 247,
https://doi.org/10.1007/978-981-16-3844-2_22

of thickeners, improvement in functional characteristics of lubricants due to the use of new additives [1]. Among the promising additives are nanosized additives of metals, their oxides and sulfides, as well as carbon materials [2–4]. The preferred use of nanoparticles as additives to lubricants stems from a number of their unique properties. So, the small size of particles from 1 to 100 nm allows them to easily penetrate into the friction zone and fill microroughnesses. The high surface energy of nanoparticles, on the one hand, enables them to easily settle or form protective films on the friction surface, and, on the other hand, can lead to their aggregation in lubricants. The effect of nanoparticles on surface processes in the friction zone depends on their size, hardness, shape, and friction conditions.

As studies show [5–7], appending of nanoparticles of copper, titanium, zinc oxides in base oils results in the wear reduction, a decrease in the coefficient of friction and an increase of the load-speed range of friction conditions. In a number of works [8–10], the use of pure Fe_3O_4 nanoparticles and their modifications as additives, which effectively influence the tribotechnical properties of lubricants, has been noted. However, under certain conditions of friction, pure Fe_3O_4 nanoparticles tend to aggregate and cause abrasive wear of friction pairs.

2 Relevance and States of the Problem

Improvements in the tribotechnical characteristics of lubricants in the presence of nanoadditives are explained by the rolling-sliding effect, the formation of a third body between the friction members, the formation of a tribofilm in the friction zone, and a change in the rheological properties of lubricants [1]. Processes of friction surface polishing, smoothing and abrasive wear are also possible [5–7].

Today, despite significant research into the addition of nanoparticles of metal oxides to lubricants, the mechanism of their lubricating action has not been unambiguously determined and requires clarification.

3 Statement of a Problem

Therefore, the aim of this work is to obtain Fe_3O_4 nanoparticles, to study their properties, and to assess their effect on the viscosity and tribotechnical characteristics of oleic acid.

4 Experimental Part

Tribological studies were carried out using a four-ball machine in accordance with National Industry-specific Standard. Four steel balls with a diameter of 10 mm made of ShKh-15 steel were used as rubbing surfaces, the upper ball was fixed in a rotating spindle (the rotational speed is 1460 ± 70 min^{-1}), and the three lower ones were motionlessly located in the machine cage with the tested lubricant. The wear spot diameter D_I was determined using an optical microscope 60 min after the testing at a constant load of 196 N (20 kg-force). When determining the critical load P_c and the welding load P_w, 10s tests of a series of loads were carried out. Evolutionary changes in the coefficient of friction were determined using an UMT-200 machine, where a sample was used as a fixed disk of 50 mm diameter, and three rollers of 10 mm diameter each were used as a counterbody, and a friction area of 1.5 cm^2 at the end of each roller. The disc and fingers were made of steel 45. The samples preparation for the study was carried out as follows: the surface of the rotating disc and rollers was treated with sanding paper, small scratches were smoothed out with a grinding powder, after which they were degreased with hexane and dried in air at room temperature.

The test lubricant was placed on the disk surface. The tests were carried out at a rotation speed of 200–800 rpm and a load of 10 N, the test duration was 1 h. Pure oleic acid was used as a core of the base, and synthesized Fe_3O_4 nanoparticles with a concentration of 0.1–0.5-weight percentage were used as an additive. The viscosity of the lubricating compositions was studied using the HAAKE Viscotester C viscometer at a rotation speed of 100 rpm in a temperature range from 25 to 60 °C.

For the synthesis of Fe_3O_4 nanoparticles, a mixture of iron chloride (II) salts and iron chloride (III) of chemically pure grade in a ratio of 1:2 was used, to which a 1.5-fold excess of 25% aqueous ammonia of analytically pure grade was added. The resulting black suspension was kept for 30 min on a permanent magnet, then the formed particles were separated from the aqueous solution of salts. The resulting precipitate was washed and stabilized with isopropyl alcohol, and then dried at a temperature of 100 °C.

The composition of the synthesized particles was evaluated by X-ray phase analysis (XPA) using the automated device ARL X'TRA with radiation of Cu K_α lines of the anode with a wavelength of 1.54056 Å with a Ni filter in a reflected beam. The exposure was carried out in the range of 2θ from 5 to 70 °C at a scanning speed of 5 ^{0}/min and the temperature of 25 °C. Sedimentation analysis of an aqueous suspension of Fe_3O_4 nanoparticles on a CPS Disk Centrifuge Model DC24000 disk centrifuge was used to study the particle dimensions. The morphology of the obtained Fe_3O_4 particles was studied using atomic force microscopy (AFM) on a PHYWE Compact device.

5 Discussion and Results

The synthesis of highly dispersed magnetite Fe_3O_4 was carried out by the method of chemical condensation (CC), proposed by Elmore, according to the reaction equation:

$$2\,FeCl_3 + FeCl_2 + 8\,NH_4OH = Fe_3O_4 + 8\,NH_4Cl + 4H_2O$$

A feature of the synthesis is strict adherence to the ratio of Fe^{+3}/Fe^{+2} salts 2:1, and the presence of a 1.5-fold excess of the precipitator—ammonium hydroxide.

The composition of the synthesized particles was evaluated according to the results of XPA. The analysis showed that intensity peaks appeared on the X-ray diffraction pattern, which correspond to magnetite of the composition Fe_3O_4. Magnetite is a mixed oxide of iron (II) and (III) and it belongs to the class of inverted spinels with a crystal lattice of the noble spinel and general formula $MeFe_2O_4$.

The Fe_3O_4 particle size was determined by sedimentation analysis. As can be seen from the histogram (Fig. 1), the obtained Fe_3O_4 particles have polydisperse properties and 95% of the particles are in the range from 50 to 100 nm, with a maximum at 50 nm −29%.

The primary clusters of Fe_3O_4 are particles up to 50 nm in size, which are able to enlarge then.

The AFM results (Fig. 2) confirm the polydisperse nature of the synthesized Fe_3O_4 particles. A group of elliptical objects, which is an agglomerate of Fe_3O_4 particles, are highlighted on the 3D image.

One of the main physical and chemical characteristics that determines application of lubricants for equipment is viscosity. At ordinary temperatures, the viscosity values

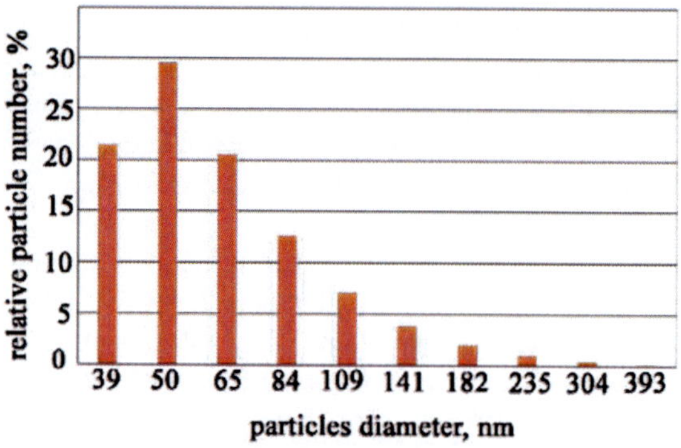

Fig. 1 Histogram of the dispersion composition of Fe_3O_4 nanoparticles

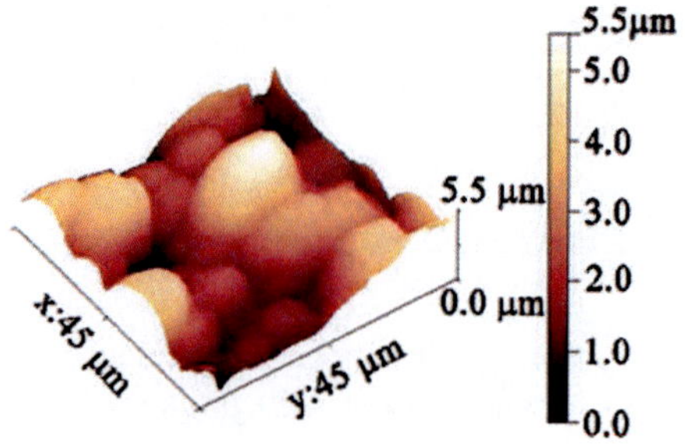

Fig. 2 3D – picture of Fe_3O_4 nanoparticles

Table 1 Viscosity values of the liquids under testing against temperature

Temperature, °C	Oleic acid	Oleic acid and 0.1% Fe_3O_4	Oleic acid and 0.5% Fe_3O_4
	Viscosity, mPa•sec		
25	23.9	23.7	23.1
30	22.8	20.6	21.8
40	15.8	12.3	13.9
50	10.9	8.3	9.2
60	8.9	6.1	7.4

of pure oleic acid and oleic acid with Fe_3O_4 particles are practically the same. When the temperature increases, the viscosity of the liquids under testing decreases due to the mobility of oleic acid molecules (Table 1).

The presence of Fe3O4 nanoparticles in oleic acid leads to a decrease in viscosity to a greater extent, which is linked with the destruction of associative structures between oleic acid molecules and nanoparticles. This agrees with the work results [11], where kerosene and undecane were the carrier liquids.

The associative structures destruction is directly related to an increase in the activation energy of the liquids under study.

The activation energy was calculated on the basis of viscosity studies depending on temperature using the Frenkel-Andrade equation:

$$\eta = C \cdot e^{\frac{W}{kt}} \tag{1}$$

where η is the viscosity of the liquid, C is a coefficient depending on the jump distance of molecules, the frequency of its oscillations and temperature, T is the temperature, W is the activation energy, that is, the energy required to make the jump of the molecule, k is the Boltzmann constant.

For small temperature ranges, the activation energy can be calculated using the Formula 2.

$$W = \frac{\ln \frac{\eta_1}{\eta_2} k T_1 T_2}{T_1 - T_2} \tag{2}$$

Using the calculated values of the activation energies (Table 2), we can see that the oleic acid with a 0.1% content of Fe3O4 nanoparticles has the highest activation energy; if the mass fraction of nanoparticles increases then the value of the activation energy decreases.

Table 2 Calculated values of activation energy

Liquid under testing	Oleic acid	Oleic acid and 0.1% Fe_3O_4	Oleic acid and 0,5% Fe_3O_4
W•10^{-20}, J/K	3.867	5.314	4.457

According to modern provisions of tribology, the coefficient of friction is a nonlinear function of the physical and mechanical properties of the materials of the friction pair, operating conditions (lubricant type, properties and ambient temperature) and friction modes (relative motion speed, contact pressure) [12].

As can be seen from Fig. 3, for all studied liquids, the friction coefficient decreases with time and its values have an oscillatory specifity, which is due to the process of breaking-in of the contacting friction surfaces. The Fe_3O_4 nanoparticles appending into oleic acid results in decrease in the friction coefficient. This fact is consistent with the research results [9]. The improvement in the antifriction properties of oleic acid in the presence of Fe_3O_4 is because their increased activation energy enables them to be easily adsorbed on the metal surface, thereby forming a protective structural layer that prevents the contact of two metal surfaces.

Further tribological studies made it possible to establish that with an increase in the sliding speed, a decrease in the friction coefficient is observed for all studied lubricants. This is due to the fact that the mixed friction regime is replaced by a hydrodynamic one, where the friction surfaces separated by a lubricating layer and the frictional force is determined by the shear resistance within the lubricant (Fig. 4).

The results of tribological studies on a four-ball friction machine (Figs. 5 and 6) demonstrated that the addition of Fe_3O_4 nanoparticles improves the antiwear properties of oleic acid and increases its carrying capacity.

These results confirm the fact that the addition of Fe_3O_4 nanoparticles improves the tribotechnical properties of oleic acid.

Based on the studies carried out, the following mechanism of the influence of Fe_3O_4 nanoparticles on the lubricating effect of oleic acid can be assumed. Fe_3O_4 nanoparticles and oleic acid form associative structures with high activation energies. In the friction zone, they are adsorbed on the surface of metals and form a protective surface layer that does not enable metal surfaces to contact with each other. The consequence of this process is improvement of the tribotechnical properties of oleic acid.

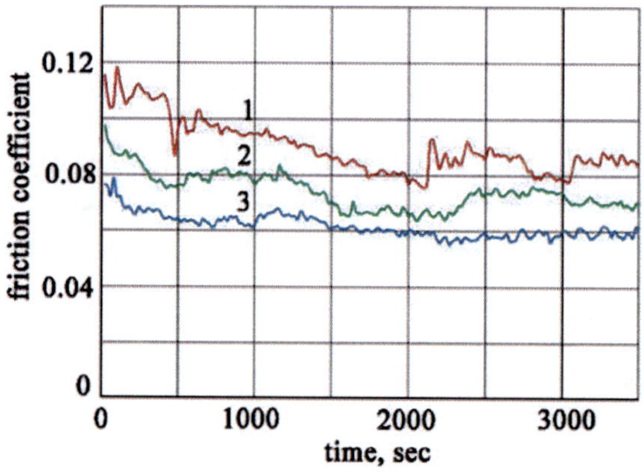

Fig. 3 Evolutionary dependence of the coefficient of friction at a sliding speed of 200 rpm: 1—oleic acid, 2—oleic acid and 0.1% Fe_3O_4, 3—oleic acid and 0.5% Fe_3O_4

Fig. 4 Evolutionary dependence of the friction coefficient at a sliding speed of 500 rpm: 1—oleic acid, 2—oleic acid and 0.1% Fe_3O_4, 3—oleic acid and 0.5% Fe_3O_4

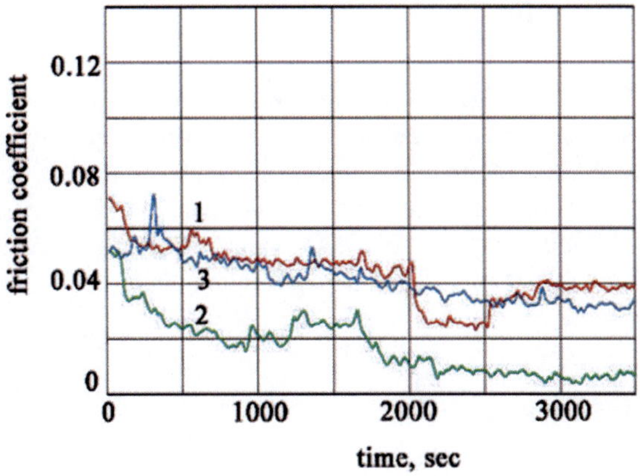

Fig. 5 Wear spot diameter of the studied lubricants

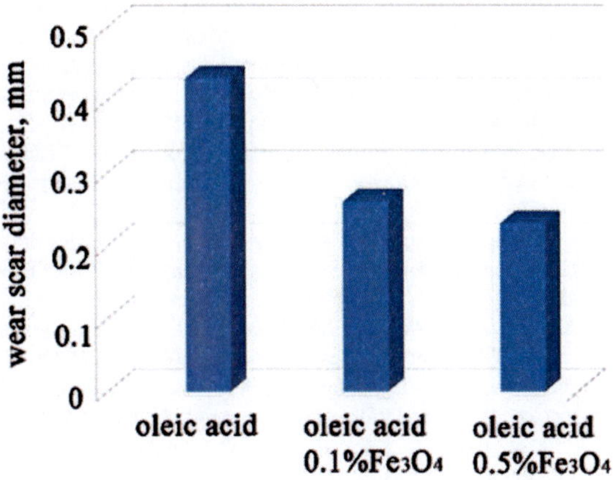

Fig. 6 Critical load (1) and welding load (2) of the studied lubricants

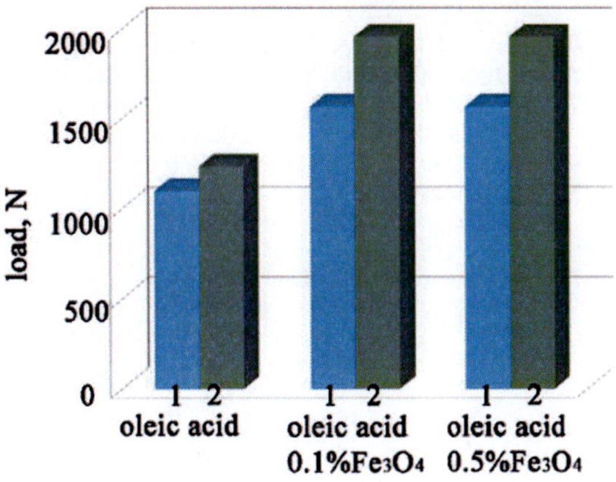

6 Findings

Fe_3O_4 nanoparticles were obtained by chemical condensation. The Fe_3O_4 nanoparticles appending leads to a decrease in the viscosity properties of oleic acid along with temperature rising, and it contributes to the antiwear and antifriction properties improvement of oleic acid. An improvement in the tribotechnical properties of oleic acid in the presence of Fe_3O_4 nanoparticles is in connection with the formation of protective surface layers in the friction zone.

References

1. Wani, K.S., et al.: Friction and wear characteristics of vegetable oils using nanoparticles for sustainable lubrication. Tribol.-Mater. Surf. Interf. **12**, 27–43 (2018)
2. Borda, F.L.G., et al.: Experimental investigation of the tribological behavior of lubricants with additive containing copper nanoparticles. Tribol. Int. **117**, 52–58 (2018)
3. Wang, J., et al.: Tribological characteristics of graphene as grease additive under different contact forms. Tribol. Int. **127**, 457–469 (2018)
4. Zadoshenko, E.G., Burlakova, V.E., Novikova, A.A.: Effect of nickel nanopowder on lubrication behaviour of low-temperature grease in steel-steel tribosystem. Tribol.-Mater. Surf. Interf. **14**, 51–58 (2020)
5. Bhaumik, S., Pathak.: Analysis of anti-wear properties of CuO nanoparticles as friction modifiers in mineral oil (460cSt Viscosity) using pin-on-disk tribometer. Tribol. Indus. **37**, 196–203 (2015)
6. Xia, W., et al.: Analysis of oil-in-water based nanolubricants with varying mass fractions of oil and TiO_2 nanoparticles. Wear **396–397**, 162–171 (2018)
7. Bhaumik, S., et al.: Analyses of anti-wear and extreme pressure properties of castor oil with zinc oxide nano friction modifiers. Appl. Surf. Sci. **449**, 277–286 (2018)
8. Zadoshenko, E.G., Burlakova, V.E.: Analysis of ferromagnetic nanoparticles impact on tribitechnical properties of lubricants. Bullet. Don State Tech. Univer. **15**(1), 85–92 (2015)
9. Huang, J., et al.: Preparation and tribological properties of core-shell Fe_3O_4@C microspheres. Tribol. Int. **129**, 427–435 (2019)
10. del Río, J.M.L. et al.: Tribological behavior of nanolubricants based on coated magnetic nanoparticles and trimethylolpropane trioleate base oil. Nanomaterials **10**, 683–705 (2020)
11. Kolchanov, N.V., Kolesnichenko, E.V.: Viscosity magnetic colloid fluids at different temperatures and volume fractions of solid particles. Bullet. Perm Univer. Phys. **4**, 37–45 (2017)
12. Garkunov, D.N.: Triboengineering (wear and wearlessness). Russian State Agrarian University Publishing, Moscow (2001)

Study of the Influence of the Grinding Wheel Composition Components on Its Performance During ID Grinding

Andrey Romanenko, Dmitry Shatko, Andrey Nepogozhev, and Pavel Strelnikov

Abstract The influence of the composition components and fillers of grinding wheels on their performance are considered in the article. In the course of the literature review, the issue of assessing the performance of grinding tools was studied. The relevance of the problem under consideration is indicated. For this purpose, the features of the ID grinding process and its difference from other grinding types are considered. The main trends in the field of cutting and, in particular, grinding, aimed at intensifying the machining processes, improving the performance characteristics of the tools used and improving the quality of the machined surfaces, have been determined. The main purpose of the tests is to study the performance of experimental grinding wheels during high-speed ID grinding operation and to determine the most optimal machining conditions. Based on the purpose in view, a technique for conducting experiments has been developed; the equipment used has been presented. Experimental data on changes in roughness, average grinding temperature, waviness, out-of-roundness and size of the part, depending on the cutting conditions and wheel characteristics, are presented in sequence. The scientific substantiation of the empirically obtained results is proposed. The probability of occurrence of burns depending on the grinding tool composition and grinding conditions is estimated. Based on the results obtained, conclusions about the degree of influence of grinding conditions, fillers and wheel characteristics were drawn and presented in the form of practical recommendations. The proposed recommendations are aimed at improving ID grinding of bearing steels with wheels with various fillers.

A. Romanenko · D. Shatko (✉) · P. Strelnikov
T.F. Gorbachev Kuzbass, State Technical University, Kemerovo, Russia
e-mail: shdb.tm@kuzstu.ru

A. Nepogozhev
Moscow State Technological University "STANKIN", Moscow, Russia

M. Shamtsyan et al. (eds.), *Robotics, Machinery and Engineering Technology for Precision Agriculture*, Smart Innovation, Systems and Technologies 247,
https://doi.org/10.1007/978-981-16-3844-2_23

1 Introduction

The grinding wheel performance is understood as a quantitative assessment (by one or several indicators) of the degree of conformity of the wheel to its purpose when performing a particular operation.

The ID grinding process, unlike other grinding types, has a number of features:

1. The outer diameter of the grinding wheel is always less than the diameter of the machined surface.
2. The supply of cooling lubricant to the grinding interface and the rejection of heat generated during grinding are impeded.
3. ID grinding requires such structures as spindles, which must provide high rotational speed, rigidity with limited design parameters and overall dimensions.
4. The angle and length of the grinding interface curve is much greater than with other types of grinding, as a result of which the load on abrasive grains and the amount of metal removed by a single grain at the same grinding depth increase significantly compared to other grinding types.

The listed features determine the specifics of machining and cause certain difficulties in ID grinding operations. The proposed work is aimed at studying the main parameters of machining with experimental ID grinding wheels.

2 Relevance and Analysis of the State of the Problem

Current trends in the machining of materials and, in particular, grinding, are aimed at intensifying manufacturing processes, improving the performance of the tools used and improving the quality of the machined surfaces.

It is known that one of the main ways of increasing the ID grinding efficiency is to increase the grinding speed. This is due to the fact that with an increase in grinding speed, all other things being equal, the thickness of the chips removed by one grain decreases. Simultaneously, the cutting forces, the load on the individual grain and the time of its contact with the machined surface reduce. At the same time, the number of abrasive grains involved in cutting per unit of time increases. The high speed grinding performance can be increased the coarse cross feed. However, it should be taken into consideration that with an increase in the cutting speed, the temperature in the grinding interface increases, which can cause a decrease in the physical and mechanical properties of the machined part.

Based on this, the work aimed at a comprehensive study of the of ID grinding wheel performance is quite relevant and in demand.

A large number of works are devoted to the study of the issues of increasing the performance of grinding tools [1–12]. In some works, the authors focus on the tool design features—its composition, structure, used abrasive grains, their shape and orientation [1, 5–7, 9].

Other researchers are studying extensively the effect of heat flux on machined surfaces as it affects the performance of finished parts. Based on this, the study of the temperature pattern during grinding is an important task [2, 10, 11]. The problem of abrasive processing of bearing, high-alloy corrosion-resistant steels and heat-resistant alloys is especially acute.

The characteristics of the grinding wheel and grinding methods significantly affect the quality of the machined surface, therefore, increased attention is also paid to the study of the quantitative parameters of the machined surface roughness [3, 8, 12].

3 Problem Statement and Research Methods

The goal of the tests is to study the performance of experimental grinding wheels in the high-speed ID grinding operation and to determine the most optimal machining conditions. To achieve this goal, the following tasks were formulated:

1. To study the grinding process parameters.
2. To choose the optimal machining conditions.
3. To give recommendations on the practical use of grinding wheels.

Actual material removal, root-mean-square deviation of the finished part size, surface roughness, waviness, out-of-roundness, average temperature of the machined part surface, occurrence of burns and microhardness were chosen as indicators characterizing the grinding wheel performance.

The characteristics of the grinding wheels were selected on the basis of the research results and the obtained statistical mathematical models. The characteristics of the experimental straight grinding wheels (SGW) of the standard size 63 × 15 × 20 are shown in Table 1.

The grinding wheel specifications are marked as follows: A—regular aluminium oxide abrasive material, F—grain sizes, L—medium hardness, 12—wheel structure number, V—vitrified bond.

The following designations of fillers in the wheel marking are adopted: GM—glass microsphere, FP—fruit pits, EM—electrocorundum microsphere. The figure shows the percentage of filler in the product by volume.

The bearing 209.01 outer races, previously ground to the uniform size, were used as samples.

The experiments were carried out using the S/W3/1CAC ID grinding machine.

Table 1 Grinding wheel characteristics

S. No	Grinding wheel characteristic
1	A F120 L 8 V-S,
2	A F80 L I2 V EM-20
3	A F80 L 12 V EM-10
4	A F80 L 12 V FP-5/GM-10

The successive grinding tasks: wheel profiling—accelerated entry—roughing—finishing—sparkout.

The average temperature of the machined part surface during grinding was measured with the artificial Chromium-Copel thermocouple, which was attached to the size gage probe with a flat spring, which ensures continuous contact of the thermocouple with the machined part surface.

The resulting size of the finished part after machining was controlled with an accuracy of 0.001 mm by an indicator head mounted on a special device. Roughness Ra of the machined surface was measured by 202 profilometer. Waviness and out-of-roundness of the machined bearing race were determined by BV-2003 roundness gauge.

The presence of burns was determined by a typical cold etching process according to the Volzhsky Research Institute of Abrasives and Grinding standards No. I 101–65.

4 Results and Discussion

The resulted dependence of the average temperature of the part on cutting speed, cross feed amount and wheel characteristics is shown in Fig. 1.

As can be seen from the obtained dependences, the highest heat flux of the grinding process is observed when working with the standard wheel, an increase in the grinding wheel structure number and an increase in porosity leads to a decrease in the grinding temperature. Grinding temperature when working with the AF80L12VFP-5/GM-5

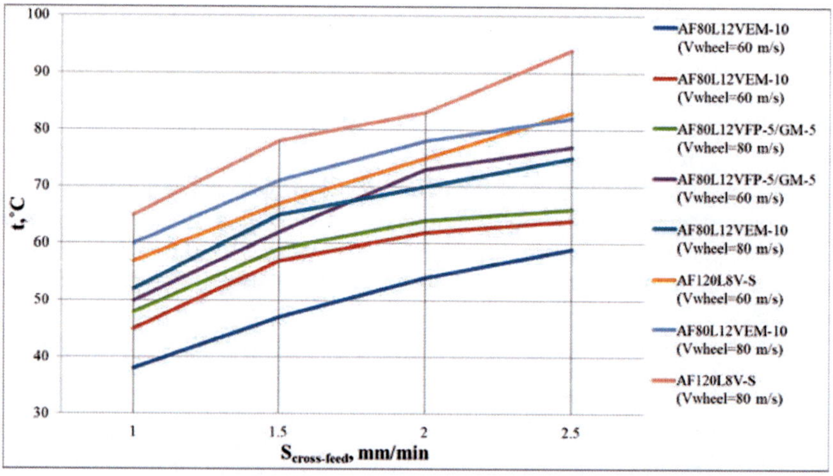

Fig. 1 Change in the average surface temperature of the part depending on cutting speed, cross feed amount and wheel characteristics

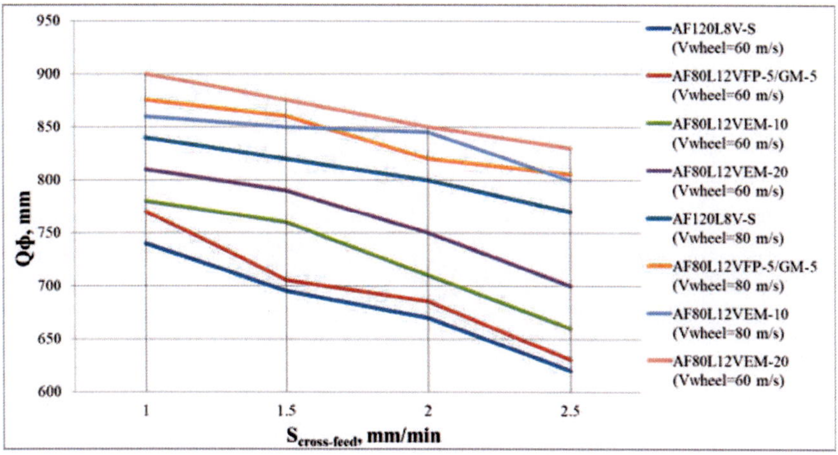

Fig. 2 Change in actual material removal depending on cutting speed, cross feed amount and wheel characteristics

wheel is 1.5 times lower than that of the standard one; two other experimental wheels also provide a decrease in the grinding temperature by 1.2–1.3 times.

Machining performance is determined by removal of material per minute.

The obtained dependence of the actual material removal on the grinding conditions and the wheel characteristics is presented in Fig. 2.

As can be seen from this figure, with an increase in cross feed, the actual removal rate decreases on average by I5-20% for all tested wheels. Maximum actual removal is achieved by the wheel containing 20% microspheres by reducing wheel wear and cutting force. The introduction of fruit pits, as is known, increases the wheel wear, but, nevertheless, it provides better grinding conditions due to the good placement of chips in the pores. Therefore, glass microspheres are added to the composition in order to increase the redress life of a wheel.

The actual removal rate of the wheel containing glass microspheres and fruit pits is 7% lower than that of the wheel containing 20% microspheres and 5% higher than that of the standard wheel. Increasing cutting speed contributes to an average 10% increase in actual material removal for all wheels.

The machining accuracy depends on temperature deformations, delay in the execution of the wheel return command, magnitude of the linear wear of the grinding wheel, locating surface error, error in the operation of the wheel wear compensation mechanism, as well as the sparkout time relay. All of the listed errors, except for the magnitude of the linear wear of the wheel and the temperature deformations, can be taken into account when setting up the machine, while the temperature deformations and the magnitude of linear wear of the wheel depend on the uniformity of the tool used. The dependence of the root-mean-square deviation of the diametrical dimensions of the finished part is shown in Fig. 3.

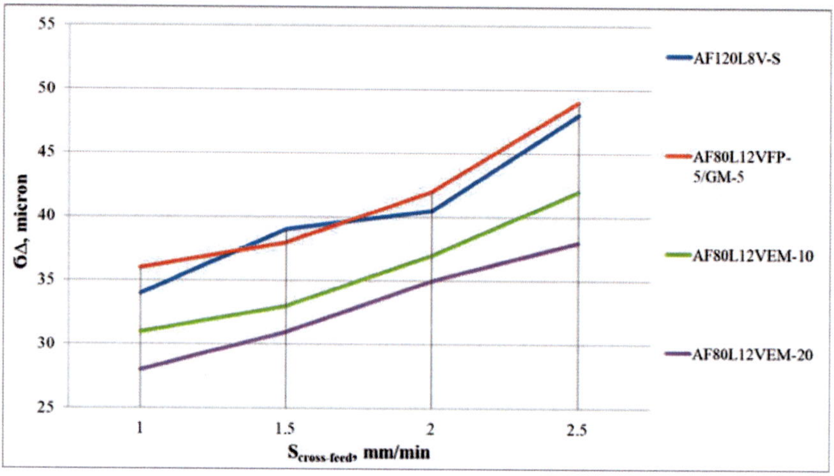

Fig. 3 Change in the finished part accuracy depending on cross feed and wheel characteristics at a cutting speed of up to 60 m/s

As can be seen from the figure, with an increase in the cross feed, the value of deviations increases due to increased wear of the wheel and an increase in the grinding temperature. The value of deviations in the dimensions of the parts machined by the experimental wheels with microspheres is 1.25 times lower than that of the standard one, which is achieved due to the longer redress life of the experimental wheels. The machining accuracy of the standard wheel and the wheel containing glass microspheres and fruit pits is approximately the same.

The out-of-roundness depends primarily on the bearing race surface error and the interstices angles. The influence of grinding conditions is indirect here; this is due to the fact that the deviation from roundness includes not only ovality and lobing, but also waviness. The conducted studies (Fig. 4) showed that the influence of grinding conditions on the characteristics of a tool is insignificant.

The machined surface waviness is associated with the onset of oscillations while cutting. Flexural oscillations can be caused both by external factors (forced oscillations) and by the cutting process.

The dependence of waviness on wheel characteristics, coarse cross feed amount and cutting speed is shown in Fig. 5.

As can be seen from the figure, the minimum waviness is provided by the AF80LI2VEM-20 wheel; this can be explained by the fact that the introduction of microspheres increases the wheel structure uniformity, which reduces the unbalance, and also reduces the wear of the abrasive wheel during operation. The introduction of fruit pits into the composition increases the heterogeneity of the abrasive tool structure and increases the wear of the wheel. Glass microspheres, as already noted, on the contrary, provide a high structure uniformity and a decrease in its wear. Therefore,

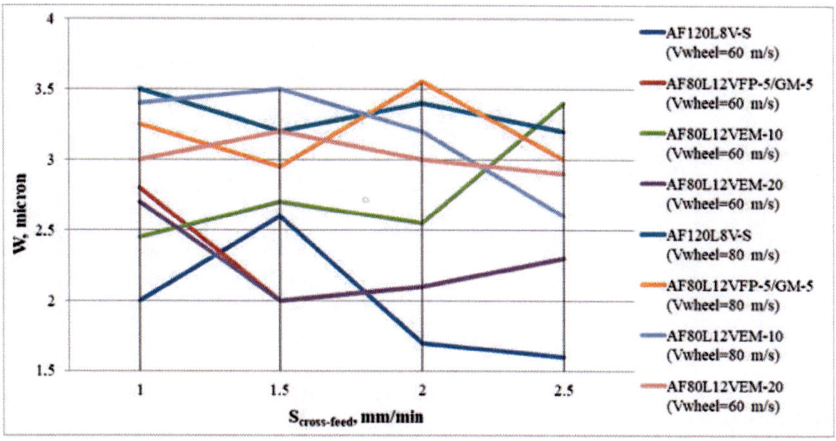

Fig. 4 Dependence of out-of-roundness on coarse cross feed amount, cutting speed and wheel characteristics

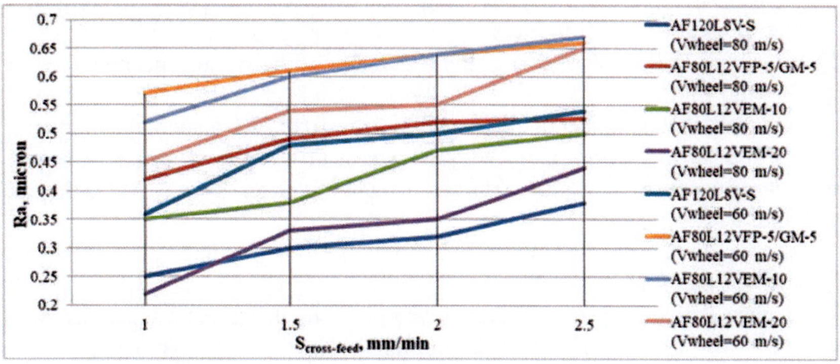

Fig. 5 Dependence of the surface waviness of the finished part on coarse cross feed amount, cutting speed and wheel characteristics

the AF80L12VFP-5/GM-10 wheel provides an increase in waviness. The high waviness obtained when machining with standard tools is due to the fact that the wheels for grinding raceways are pre-impregnated with sulfur in order to reduce wear and eliminate burns. However, with the available technologies, it is impossible to achieve uniform impregnation, which leads to an increase in unbalance and uneven wear of the wheel.

The dependence of the surface roughness on coarse cross feed amount, cutting speed, wheel characteristics is shown in Fig. 6.

As can be seen from the results obtained, the wheel structure has the greatest effect on the surface roughness at a cutting speed of 60 m/s and the minimum cross feed. The roughness ratio between the standard wheel and the AF80L12VFP-5/GM-10

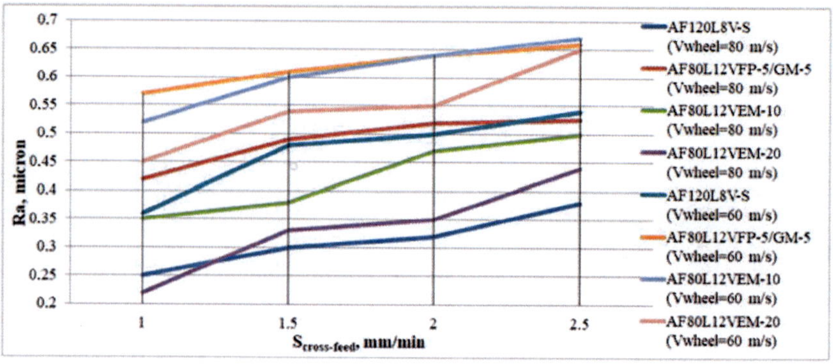

Fig. 6 Change in surface roughness depending on coarse cross feed amount, cutting speed, wheel characteristics

wheel is 1.6 times, and under conditions of maximum feed and cutting speed—1.3 times. This can be explained by the fact that with an increase in the cross feed, the wear of the standard wheel increases to a greater extent than that of the experimental one, and in addition, with an increase in the cutting speed, the number of cutting edges increases, which also reduces the roughness of the surface machined by the experimental wheel.

An increase in the number of microspheres from 10 to 20% reduces the roughness obviously due to the fact that, along with an increase in the wheel structure uniformity and its wear evenness, the spheres also have a cutting effect, thereby increasing the number of cutting edges.

When grinding, the choice of grinding conditions is determined by the possibility of occurrence of burns.

The results of studies on the occurrence of burns in the parts machined with experimental wheels, depending on the cutting speed are shows in Table 2.

Signs in Table 2: (–)—no burns; (+)—burns.

Based on the results obtained, we can say that the experimental wheels, in comparison with the standard tool, allow working in more heavy-duty conditions. A decrease

Table 2 The presence of burns during grinding with experimental wheels

Grinding wheel characteristics	Cross feed, mm/min, cutting speed, m/s							
	1		1.5		2		2.5	
	60	80	60	80	60	80	60	80
A F80 L I2 V EM-20	–	–	–	–	–	–	+	+
A F80 L 12 V EM-10	–	–	–	–	–	+	+	+
A F80 L 12 V FP-5/GM-5	–	–	–	–	–	–	–	–
A F120 L 8 V-S	–	–	–	+	+	+	+	+

in the heat flux of the process is achieved through the use of more open structures, which reduces the number of cutting grains and the wheel loading.

The occurrence of burns is associated with thermal phenomena that occur during the grinding of metals. The grinding process is known to be characterized by increased heat generation, the formation of significant contact temperatures in the grinding interface, instantaneous heating and a high concentration of heat in the thinnest surface layers of the parts being machined. All this leads to various structural changes in the machined metal, which worsen its physical and mechanical properties. All the grinding conditions, as already noted, except for the speed of rotation of the part, give an increase in temperature. Therefore, with more heavy-duty grinding conditions, the probability of occurrence of burns increases.

Reducing the heat flux of the grinding process is facilitated by the use of oil-based cooling lubricant, or with surfactants, an increase in the grinding wheel structure number and its porosity.

5 Conclusions

The obtained experimental data made it possible to draw the following conclusions, which can be used as practical recommendations.

1. With an increase in the grinding wheel structure number and the abrasive tool porosity, a decrease in temperature is observed. The temperature of the surface machined with the AF80L12VFP-5/GM-10 wheel is 1.5 times lower than that of the surface machined with the AF120L8V-S wheel. Wheels with electrocorundum microspheres can reduce the temperature by an average of 1.3 times
2. The greatest actual material removal is achieved when working with grinding wheels with electrocorundum microspheres.
3. With an increase in cross feed, the actual removal rate decreases.
4. The lowest value of the root-mean-square deviation of the size of the finished parts is obtained when working with the wheel with a 20% volumetric content of microspheres.
5. With an increase in the cross feed, the machining accuracy decreases. The minimum waviness is obtained when working with wheels containing microspheres. The maximum waviness was obtained using the AF120L8V-S wheel.
6. With an increase in cutting speed and cross feed, waviness and roughness of the machined surface increase.
7. The AF80L12VFP-5/GM-5 wheel provides the maximum roughness during processing. The lowest roughness is achieved when working with the AF120L8V-S wheel. Wheels with electrocorundum microspheres, when machined, provide a roughness that is less than that of the wheel with fruit pits and more than that of the standard tool.

8. When working at a cutting speed of up to 60 m/s, it is recommended to use wheels with electrocorundum microspheres; and at high cutting speeds, wheels with glass microspheres and fruit pits are preferable.

References

1. Ding, Y., Miao, W., Luo, M., Zhao, Y., Bao, H.: Influence of structure of vitrified bond diamond grinding wheel on its performance. Jingangshi yu Moliao Moju Gongcheng/Diamond and Abrasives Eng. **40**(4), 19–23 (2020)
2. Li, P., Chen, S., Xiao, H., Dai, H., Jin, T.: Effects of local strain rate and temperature on the workpiece subsurface damage in grinding of optical glass. Int. J. Mech. Sci. **182**, 105737 (2020)
3. Lyukshin, V., Shatko, D., Strelnikov, P.: Development of a technique for the study of the coated abrasive surface wear. In: XI International Scientific and Practical Conference "Innovative technologies in mechanical engineering". "Solid State Phenomena". vol 313, pp. 66–71. (2021)
4. Mingazhev, A., Taimasova, L., Gunderov, D., Gafarova, V.: Quality assurance in the process of machining of parts made of heat-resistant alloys with the help of abrasive electrochemical grinding. J. Phys.: Conf. Ser. **1515**(4), 042062 (2020)
5. Romanenko, A., Shatko, D., Strelnikov, P:. The influence of the formulation components and the mixing technology on physical and mechanical properties of abrasive wheel. In: X International Scientific and Practical Conference "Innovations in mechanical engineering", Kemerovo, Russia. vol 297, pp. 09005. (2019)
6. Shatko, D., Lyukshin, V., Strelnikov, P.: The influence of the grinding grains shape and orientation on performance of coated abrasive tools. In: X International Scientific and Practical Conference "Innovations in mechanical engineering", Kemerovo, Russia. vol 297, pp. 09006. (2019)
7. Shatko, D., Lyukshin, V., Shutko, L., Strelnikov, P.: Methods and approaches to improving the design of flexible backing grinding tools. In: International Workshop "Advanced Technologies in Material Science, Mechanical and Automation Engineering", Krasnoyarsk, Russia. vol 537, pp. 032063. (2019)
8. Shatko, D., Lyukshin, V., Shutko, L., Strelnikov, P.: Development of an innovative approach to the diagnosis of a coated abrasive surface. In: VII International Scientific and Practical Conference "Actual Problems in Machine Building", Novosibirsk, Russia. vol. 843(1), pp. 012011. (2020)
9. Tonkonogyi, V., Sidelnykova, T., Dašić, P., Yakimov, A.: Improving the performance properties of abrasive tools at the stage of their operation. Lecture Notes in Netw. Syst. **76**, 136–145 (2020)
10. Tonkonogyi, V., Yakimov, A., Bovnegra, L., Sidelnykova, T., Dašić, P.: The use of intermittent wheels, impregnated by the contact method to reduce the thermal stress of the grinding process. IOP Conf. Ser.: Mater. Sci. Eng. **708**(1), 012034 (2019)
11. Yan, X., Xiao, B., Wu, H., Xiao, B.: Research on grinding temperature field of rail grinding with compound wheel. Jingangshi yu Moliao Moju Gongcheng/Diamond and Abrasives Eng. **40**(2), 11–16 (2020)
12. Zhang, L., Kang, M., Tang, W.: Effect analysis of grinding wheel under different dressing parameters on surface grinding quality. Int. Conf. Comput. Commun. Eng. ICCCE **2019**(9010888), 113–116 (2019)

Improvement of Design of Device for Production of Cycloidal Pinion Gear

Alexey Lebedev, Sergey Kireev, Marina Korchagina, and Vitaly Lubinec

Abstract Two versions of the design of devices for cutting cycloidal gears of pin engagement described in the patents of the Russian Federation are considered. An analysis of the advantages and disadvantages of each version was carried out, on the basis of which a new design of an accessory intended for mass production of peel wheels is proposed. A kinematic scheme of the new device and a method for determining power and speed parameters of the mechanism have been developed. An example of design is given, on the basis of which a solid-state model of the new device is developed, which allows you to control the operability of the mechanism, its kinematic and power relationships. The proposed design of the new device has a set of reduction gears only, which allows selecting a low-power engine and a small torque. The accessory can be quickly rebuilt to handle the peg wheels of a wide range of output due to the possibility of changing the axial distance of the guitar gears and the eccentricity of the table.

1 Introduction

Planetary gears with extra centroid cycloidal engagement, having unique transmission ratio characteristics (up to 500), ratio of mass to transmitted moment (up to 0.015 kg/Nm), efficiency (up to 0.99), kinematic accuracy (7 degree of accuracy of conventional gear), resistance to impact loads (up to 500% of rated torque) is widely used in drives of devices in which it is necessary to move massive units of a wide variety of mechanisms with high accuracy and significant torque [1].

The cycloidal profile of the pinion gear, which determines the quality of the transmission in modern machine-building production, is usually obtained on equipment with numerical program control, which has built-in capabilities of calculating the coordinates of the reference points of the profile, visualizing the tool movement during processing, automated design of the control program. [2, 3]. CNC equipment has the necessary manufacturing flexibility and, in combination with a modern

A. Lebedev (✉) · S. Kireev · M. Korchagina · V. Lubinec
Don State Technical University, Rostov-on-Don, Russia

M. Shamtsyan et al. (eds.), *Robotics, Machinery and Engineering Technology for Precision Agriculture*, Smart Innovation, Systems and Technologies 247,
https://doi.org/10.1007/978-981-16-3844-2_24

CAM system, can quickly ensure the serial production of pinion gears with various parameters [4, 5].

However, in our opinion, do not forget about the economic feasibility when organizing the production of pinion gears. The total cost of a modern multi-coordinate CNC machine with powerful and high-precision actuators of the main movement and supply, a CAM system, the work of high-paid installers and operators can reach a significantly high value.

The shaping movement along the trajectory of the profile of the pinion gears is complex and includes the result of summing up the movement of the rotation of the point of the profile around the center with eccentric movement of the same point relative to the circumferential line of the center of the pins. It is proposed to create a relatively simple mechanism-device that reproduces such a complex movement using mechanical gears and use it to cut the teeth of a pinion gear, using in addition any simple machine with only one mechanical rotary cutting movement (for example, a table drilling or milling machine). Such a technological complex of mechanisms would have a cost of tens of times less than using a CNC machine and a CAM-system.

Some 25 years ago, technical solutions implementing such ideas were patented. In the conditions of a small machine-building enterprise, which received orders for the manufacture of pinion gear with a wide range of production, the use of such devices can be a fairly effective technical and organizational solution.

An analysis of two patents proposing designs for cycloidal pinion gear cutters has shown that a more efficient technical solution for the implementation of such installations can be found. This solution is achieved only thanks to the capabilities of modern computer solid modeling using a CAD system. Similar approach was widely used by authors of works [6–9]. Simulation of machine mechanisms design allows to significantly reduce volumes of full-scale tests, reduce material consumption, correct design errors by reproducing on the model of movements of all parts of the future device [10, 11].

The purpose of the work is to improve the design of the technological device for cutting the pinion gears.

The tasks of the work are to analyze the existing structures of devices for cutting gear wheels of the pin engagement, to develop a diagram of the new device and its design methodology, to create a sketched solid-state model of the device, to analyze its technical and economic characteristics.

2 Materials and Methods

Diagrams of the gear fittings that can be used for cycloid gear processing are shown in Fig. 1.

Please refer to the installation of the first patent (Fig. 1a). Motor 1 is connected to half-coupling of Oldham coupling 2 made in form of cam 3. Gear 4 is secured on clutch driven shaft 2 and engages with wheel 7 installed movably on frame shaft 5.

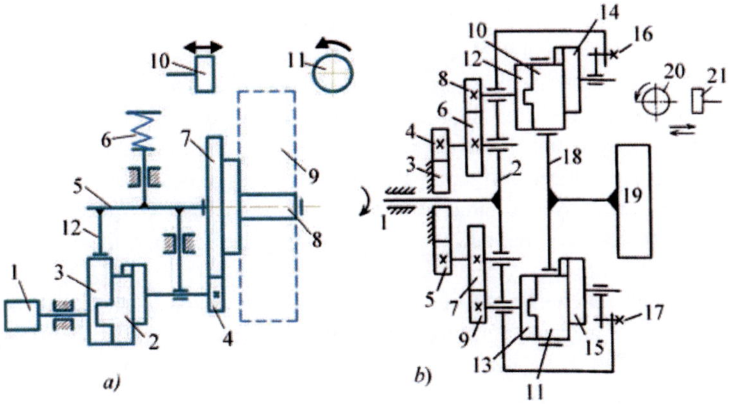

Fig. 1 Circuits for processing cycloidal pinion gears: **a**—patent RU 2,009,796; **b**—patent RU 2,018,422

Shaft 2 is installed in support of frame 5 moving under action of half-coupling—cam 3 and spring 6. The spindle of the article 8 is rigidly connected to the wheel 7, on which the billets of cycloidal wheels 9 are fixed, which are processed by a chisel 10 or a mill 11.

The accessory operates as follows. The motor 1 rotates the half-coupling 3 which reciprocates the frame 5 through the pusher 12 by means of guides and a spring 6. At the same time, rotation from the engine is transmitted through the half-coupling 2 to the spindle 8 through the gear train 4–7. Billet 9 performs complex motion summing reciprocal motion of frame 5 and rotary spindle. The trajectory of the point making such movement reproduces the contour of the pin wheel reproduced by the cutting tool 10 or 11.

The device according to the second patent (Fig. 1b) comprises a drive shaft 1, on the end of which a carrier 2 is fixed, which drives a planetary gear mechanism consisting of a fixed sun wheel 3 and satellites 4 and 5, on the axes of which wheels 6 and 7 are installed in the carrier 2, which are engaged with gears 8 and 9. Two Oldham clutch 10 and 11 are also located in the carrier, some half-couplings (12 and 13) of which are connected to gears 8 and 9, and other half-couplings (14 and 15) are installed in the carrier body 2 with the possibility of adjusting displacement of axes due to clamping screws 16 and 17. On the inserts 10 and 11 of the Oldham couplings there is a movable frame 18 with which a table 19 is rigidly connected to place the bead wheel blank.

Rotation of shaft 1 drives carrier 2. The satellites 4 and 5 roll around the sun wheel 3 and rotate the wheels 6 and 7, and through them the gears 8 and 9, which transmit rotation to the Oldham couplings 10 and 11. At the same time inserts of these couplings perform flat movement along circumference and drive frame 18, which together with table 19 for blanks performs translational movement along circumference. Thus, the blank on the table 19 makes a complex movement: rotary together with the carrier 2 and translational along the circumference of the displacement radius of the axes of the half-couplings, Oldham couplings, together with the frame 18. The tool (disc mill 20 or shoulder 21) makes cutting movements, as a result of which the teeth of the cycloidal wheel are formed on the workpiece. The lobe 21 is a copy of the collet of the wheel mated with the cycloidal wheel to be manufactured, and in the case of the disc cutter 20, its tooth is sharpened along the radius of the collet. Adjustment of the device for the manufacture of wheels of other sizes and numbers of teeth is carried out by changing the satellites 4, 5 and wheels 6, 7, which are made according to the guitar type, and adjusting the displacement in Oldham couplings due to screws 16 and 17.

The proposed scheme of the device for cutting the pinion gear show in Fig. 2. Engine 1 rotates faceplate 3 secured in housing of accessory on rolling supports through gear 2. The faceplate 3 has, on one side, a gear 4 mounted thereon, which transmits rotation to the gear 5, and, on the other side, dovetail guides on which the table 12 can be moved and fixed by screw mechanisms, on which the left half-coupling 11 of the Oldham coupling rotates freely on the ball bearings.

Gear wheel 5 is fixed on shaft and transmits rotation to gear 6 and, further, to gear wheel 7 rigidly connected to drive shaft 8 fixed on rolling supports in housing of accessory and on sliding supports in inner hole of chuck 3. Right half-coupling 9 of Oldham coupling is rigidly fixed on right end of drive shaft 8. Thus, the Oldham coupling insert 10 performs a complex motion summing up the rotation of the drive shaft 8 and the eccentric motion of the half-clutch 11. Fixture table 13 is rigidly connected to insert 10 to secure blank 14 to be machined by end mill 15. The adjustment of the eccentricity *e* of the pinion gear being cut is effected by moving the table 12 along the chuck guides and then securing it by screw mechanisms. The

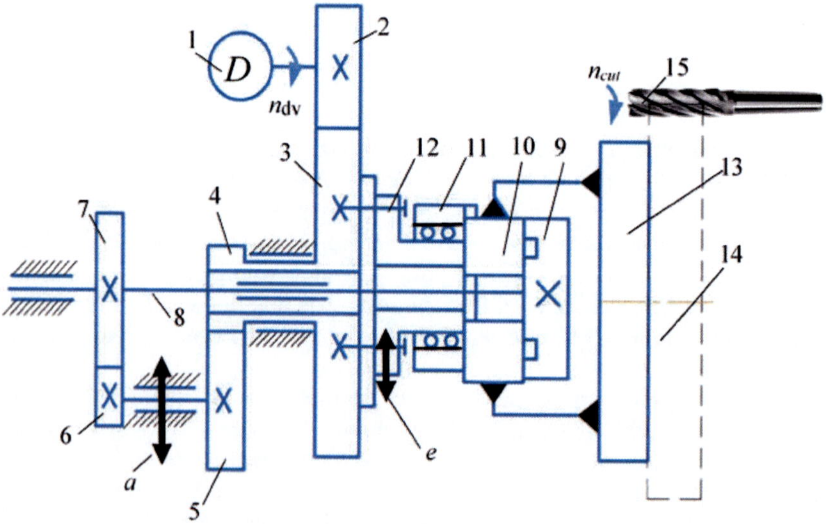

Fig. 2 Diagram of proposed accessory for cutting of pinion gears

number of teeth of the wheel being cut is determined by the gear ratio of the guitar of the gears 4–5–6–7. In order to expand the number of variants of the guitar transmission ratios, the device is provided with the possibility of adjusting the axial distance, and in the housing of the device.

Let's consider the procedure for designing a new device for manufacturing cycloidal gears of a pin engagement. The initial parameters are the structural parameters of the wheel being cut: eccentricity e of the wheel of the pin engagement (mm); the number of teeth of the wheel Z to be cut; the width of the rim of the pin wheel B (mm); radius of the center circle of the pins r_3 (mm); radius of pin r_p (mm).

Functions for calculation of coordinates of points of the theoretical profile of the pin wheel relative to the circle of the center of the pins [12]

$$\begin{aligned} x(\varphi) &= r_3 \cos(\varphi) - \mathrm{e}\,\cos((Z+1)\varphi)\,, \\ y(\varphi) &= r_3 \sin(\varphi) - \mathrm{e}\,\sin((Z+1)\varphi), \end{aligned} \tag{1}$$

where φ is the rotation angle of the radius vector of the tooth profile point.

In order to create the profile of the wheel to be cut, it is necessary to define the trajectory points equidistant to the profile (3) by the distance r_p according to the Formulas:

$$\begin{aligned} X(\varphi) &= x(\varphi) - \frac{r_p \frac{\mathrm{d}y(\varphi)}{\mathrm{d}\varphi}}{\sqrt{\left(\frac{\mathrm{d}x(\varphi)}{\mathrm{d}\varphi}\right)^2 + \left(\frac{\mathrm{d}y(\varphi)}{\mathrm{d}\varphi}\right)^2}} \\ Y(\varphi) &= y(\varphi) + \frac{r_p \frac{\mathrm{d}x(\varphi)}{\mathrm{d}\varphi}}{\sqrt{\left(\frac{\mathrm{d}x(\varphi)}{\mathrm{d}\varphi}\right)^2 + \left(\frac{\mathrm{d}y(\varphi)}{\mathrm{d}\varphi}\right)^2}} \end{aligned} \tag{2}$$

Figure 3 shows an example of a pin wheel profile for $Z = 11$; $\mathrm{e} = 5$ mm; $rp = 16$ mm; $r3 = 110$ mm.

In order to handle the profile of the pin gear with a rotating tool (for example, an end mill), it is necessary to define the minimum radius of the profile $R_{\min}$. Work [12] gives the formula for determining the minimum radius of the profile of the pin gear:

$$R_{\min} = \left| \frac{(r_3 - \mathrm{e}(Z+1))^2}{\frac{r_3}{Z+1} + \mathrm{e}(Z+1)} \right| \tag{3}$$

The selected mill diameter D_m must be equal to the diameter of the roller 2rp. To avoid violations of the profile geometry of the pin gear, the diameter of the mill must not exceed the value of $2R_{\min}$ [13] (see Fig. 1).

Using the above ratios, the parametric solid-state model of the pin gear was designed, which allows to obtain complete geometric information about the created object and apply it when designing the process accessory for cutting the pin gears [14].

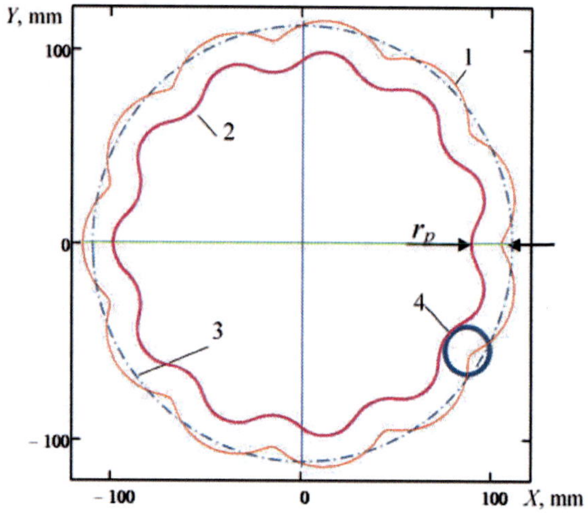

Fig. 3 Construction of the profile of the pin gear: 1—theoretical profile according to formulas (1); 2 is a profile according to formulas (2); 3 is the circumference of the center of the pins; 4—mill diameter for gear profile treatment

The procedure for designing the device includes carrying out technological calculations of the process of processing the profile of the pin gear, determining the power and speed characteristics of the drive [15, 16].

The milling speed at end milling is determined by the known cutting speed Formula, which is located in the reference literature

$$V = \frac{C_v D_m^{q_v}}{T^{m_v} t^{x_v} S_z^{y_v} B^{u_v} Z_m^{p_v}} K_v\,, \tag{4}$$

where T—tool resistance, min; t—cutting depth, mm, S_z—feed to mill tooth mm/tooth; B—milling width, mm; Z_m is the number of milling cutter teeth; C_v, q_v, m_v, x_v, y_v, u_v, p_v are empirical coefficients. K_v is the correction factor for the material to be treated and other processing conditions.

Mill rotation speed, min-1

$$n_{\text{cut}} = \frac{1000V}{\pi D_m} \tag{5}$$

Minute mill feed, mm/min

$$S_{\min} = S_z \times n_{\text{cut}} \times Z_m \tag{6}$$

Without taking into account the different directions of the linear velocity vector of the tool relative to the machined profile of the pin gear, assume that the linear velocity of the tool movement directed tangentially to the circumference of the center of the pins with radius $r3$. Then the rotation speed of the table with the blank is, min-1

$$n_{\mathrm{t}} = \frac{S_{\min}}{2\pi r_3}. \tag{7}$$

The circumferential force P_z in end milling is also determined from the reference literature.

$$P_z = \frac{10 \cdot C_p t^{x_p} S_z^{y_p} B^{u_p} Z_m}{D_m^{q_p} n_{\mathrm{cut}}^{w_p}} K_p, \tag{8}$$

where C_p, q_p, x_p, y_p, u_p, w_p—are empirical coefficients. K_p—is the correction factor for the material to be treated and other processing conditions.

The torque generated in the table with the workpiece attached to it during milling will be

$$M_t = P_z \cdot r_3 \tag{9}$$

By determining the speed and force load factors for the workpiece table, it is possible to carry out a kinematic and power calculation based on the selection of the source of motion of the device.

Kinematic balance equation of the plant (see Fig. 2)

$$n_{\mathrm{dv}} \frac{Z_2}{Z_3 i_{\mathrm{g}}} = n_{\mathrm{t}} \tag{10}$$

where Z_2 and Z_3 are the number of teeth of wheels 2 and 3; n_{dv} is the rotation speed of the motor shaft to be determined; i_g—gear ratio of the guitar of replacement wheels

$$i_{\mathrm{g}} = \frac{Z_5}{Z_4} \frac{Z_7}{Z_6}, \tag{11}$$

where Z_4–Z_7—gear teeth and wheels 4–7 of the unit.

When selecting the numbers of teeth of the guitar wheels, the following conditions must be fulfilling unconditionally:

- the gear ratio of the guitar of the replacement wheels is equal to the number of teeth of the cut pin wheel plus one:

$$i_g = Z + 1 \tag{12}$$

- the axial distances or the sum of the gear numbers of the guitar must be equal to.

$$Z_4 + Z_5 = Z_6 + Z_7. \tag{13}$$

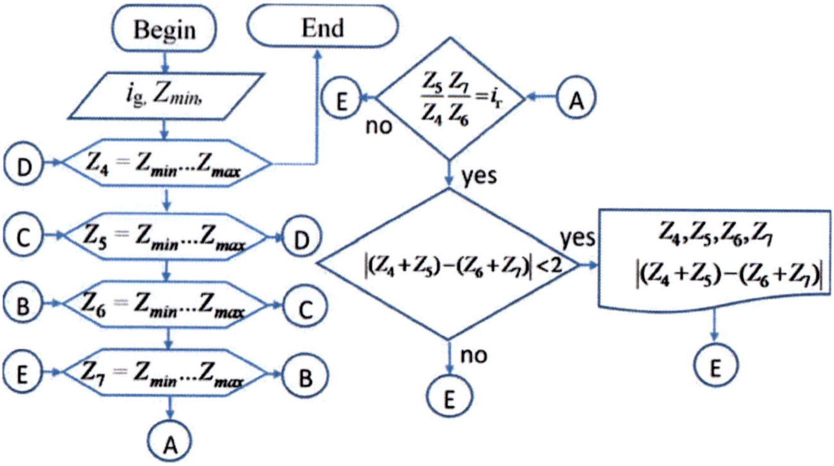

Fig. 4 Diagram of the algorithm for determining the numbers of guitar teeth of replaceable wheels

To solve the problem, an algorithm has been developed (Fig. 4), the implementation of which using computer technology allows you to determine an acceptable solution.

The algorithm consists of four nested cycles, in each of which the number of wheel teeth varies from Z_{min} to Z_{max}. Thus, all possible combinations of the numbers of the teeth of the wheels and gears of the guitar are checked. If conditions (13) and (14) are met simultaneously, the option is selected for consideration.

Condition 14 is sufficiently rigid and may not always be satisfied. Therefore, in the algorithm, the second condition requires that the difference in the sum of the numbers of the guitar gear teeth be zero or one. If the difference is zero, it means that transmissions with the same inter-axis distances are found. If the difference is one and there is no better zero option, then an inter-axial distance correction will have to be applied by introducing a displacement of the initial contour of the teeth.

The required torque (Nm) on the engine shaft is

$$M'_{\text{dv}} = M_t \frac{Z_2}{Z_3} \frac{Z_4}{Z_5} \frac{Z_6}{Z_7} \tag{14}$$

The required rotation speed of the shaft (min^{-1}) of the engine is

$$n'_{\text{dv}} = n_t \frac{Z_7}{Z_6} \frac{Z_5}{Z_4} \frac{Z_3}{Z_2} \tag{15}$$

The required power (kW) on the engine shaft is

$$P'_{\text{dv}} = \frac{M'_{\text{dv}} n'_{\text{dv}}}{9550} \tag{16}$$

The obtained preliminary values of power, torque and shaft speed make it possible to reasonably select the engine for driving the device under development, which should provide the possibility of adjusting the rotation speed of the engine shaft $n_{min} < n_{dv'} < n_{max}$. The power P_{dv} and torque M_{dv} of the selected engine shall be greater than P'_{dv} and M'_{dv}, respectively.

3 Results

Let us give an example of the design of the main parameters of the device for cutting the pin gear, having the following parameters: Z = 11; e = 5 mm; *rp* = 16 mm; mm *r*3 = 110. The diameter of the mill for processing the pin gear D_f = 2rp = 32 mm. Check if condition (3) is met:

$$R_{\min} = \left| \frac{(110 - 5(11+1))^2}{\frac{110}{11+1} + 5(11+1)} \right| = 36.1\text{ mm}$$

The selected mill diameter does not exceed 2Rmin.

The minute feed for end milling of structural steel by milling with diameter over 30 mm is Sz = 0.15 mm/tooth. The milling speed of the end mill with diameter D_m = 32 mm with hard alloy plates with number of teeth Z_m = 4 during processing of structural steel 45 with milling width B = 20 mm, milling depth $t = Dm/2$ = 16 mm (semi-open contour milling mode) according to the reference tables is according to Formula (5):

$$V = \frac{234 \cdot 32^{0.44}}{60^{0.37}16^{0.24}0.15^{0.26}20^{0.1}4^{0.13}}1 = 123\text{ m/min}$$

Mill rotation speed as per Formula (6)

$$n_{cut} = \frac{1000 \cdot 123}{\pi \cdot 32} = 1224\ \overset{-1}{\text{min}}.$$

Minute tool feed by Formula (7)

$$S_{\min} = 0.15 \times 1224 = 735\ \text{mm/min}.$$

Rotation speed of the table with the blank as per Formula (8)

$$n_{\mathrm{t}} = \frac{735}{2\pi \cdot 110} = 1.06\ \overset{-1}{\text{min}}$$

Circumferential force P_z according to Formula (9)

$$P_z = \frac{10 \cdot 12.5 \cdot 16^{0.75} \cdot 0.15^1 \cdot 20^{0.73} \cdot 4}{32^{0.85} \cdot 1224^{-0.13}} 1 = 708 \text{ N}$$

The torque on the workpiece from the cutting force according to Formula (10)

$$M_t = 708 \cdot 0.110 = 78 \text{ Nm}$$

The algorithm (Fig. 4) is implemented in Visual Basic Application and is built into the Excel workbook using a macro. Solution received for $i_g = 12$. The best option is that the sums of gear numbers are equal and minimal: $Z_4 = 20$, $Z_5 = 60$, $Z_6 = 16$, $Z_7 = 64$. Numbers of gear teeth from gears 2 to 3 are equal to $Z_2 = 15$, $Z_3 = 120$. Check compliance with conditions (13) and (14)

$$i_g = 60/20 \cdot 64/16 = 12; \quad 20 + 60 = 16 + 64$$

Required torque on the engine shaft as per Formula (15)

$$M'_{dv} = 78 \frac{15}{120} \frac{20}{60} \frac{16}{64} = 0.81 \text{ Nm}.$$

Required engine shaft rotation speed as per Formula (16)

$$n'_{dv} = 1.06 \frac{64}{16} \frac{60}{20} \frac{120}{15} = 102 \ \text{min}^{-1}.$$

Preliminary power (kW) on the engine shaft according to Formula (17).
$P'_{dv} = \frac{0.81 \cdot 102}{9550} = 0.0086 \cdot kW$.

An engine is required with the ability to adjust the speed of the engine shaft in the range from 50 to 200 min^{-1}, with a power of more than 8.6 W, with a nominal torque of more than 0.81 Nm.

Such conditions, for example, are satisfied by a 9SDGE 90F2 drive with a reduction gear and a speed adjustment of 15–230 min, having a torque of 1.07–2.98 Nm on the output shaft. Engine power is 90 W.

Figure 5 shows a sketched solid-state model of the machine for cutting the pin gears, adjusted to handle the wheel with the above parameters.

Electric motor 1 is installed on housing 2. Gear 3 engaged with wheel 4 secured on faceplate 5 rotating on bearings 6 mounted in housing 2 of accessory is rigidly fixed on shaft of engine with. On the faceplate 5 there are two guide plates, along which the table7 is moved and fixed by screws to set the eccentricity value of the processed collet wheel.

Lower half-coupling 8 of Oldham coupling is installed on bearing concentrically with table 7 and can rotate freely. This ensures eccentric movement of the workpiece being processed. Rectangular projection of Oldham clutch insert 9 provides its translational movement in slot of lower half-clutch 8. On the other hand, the insert 9 has a rectangular projection entering the slot of the upper half-coupling 10. Insert

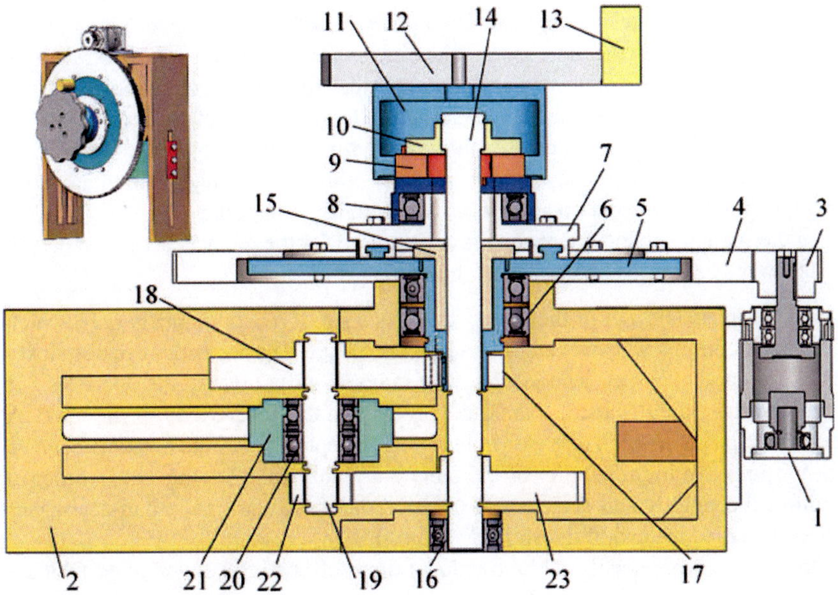

Fig. 5 Sketched model of process accessory for cutting of pin gears

slots are mutually perpendicular. Thus, the insert 9 makes a complex movement by combining the movements of the two half-couplings. Table 11 is secured on insert 9 to accommodate bead wheel blank 12 processed by end mill 13. Upper half-coupling 10 is rigidly fixed on shaft 14 to which rotation of main kinematic chain of accessory is transmitted. Shaft 14 rotates in sliding support 15 and bearing 16.

Gear 17 transmitting rotation to gear wheel 18 fixed on shaft 19 installed on rolling supports 20 in plate 21 is rigidly fixed on faceplate 5. The installation is capable of adjusting the axial distance of the guitar of replaceable gears consisting of gears 17, 18, 22, 23.

This is achieved by moving the bar 21 along the guides of the body 2 and then securing it by means of screw mechanisms. The gear 22 transmits rotation to the wheel 23 to provide rotation of the shaft 14.

4 Discussion of Results

Analysis of existing structures of units for cutting cycloidal peg wheels of structures shows the following:

- In the first and second design, the transfer to the manufacture of wheels with different geometrical parameters requires almost complete replacement of the main bearing parts of the structure of the device since when the number of teeth of the cut wheels changes, other axial distances of the gears determined by the frame structures in the first case and the carrier in the second case may be required.

- The second version of the device design provides for duplication of radially divergent gears of the two wheels, which, with a large number of teeth of the wheel being cut, significantly increases the overall dimensions and metal consumption of the structure. There may also be problems with the accuracy of simultaneously adjusting the same parameters of the two Oldgham couplings.
- In the second version, increasing gears are used for the replaceable guitar, which increases the torque and power of the motion source.

The proposed design of the new device has a set of reduction gears only, which allows selecting a low-power engine and a small torque. The accessory can be quickly rebuilt to handle the peg wheels of a wide range of output due to the possibility of changing the axial distance of the guitar gears and the eccentricity of the table. A number of pre-fabricated threaded holes on the faceplate 5 and the housing 2 can be used to improve the reliability of the table 7 and the plate 21 (Fig. 5). In order to improve the rigidity and accuracy of the rotation of the shaft 14, a ball-screw pair can be installed instead of the sliding bearing 15.

In the sketched model, the lubrication system of bearing supports and Oldham couplings is not yet provided. However, this is easily realized by installing the lubricant in the closed space between the table, the faceplate and the Oldham coupling parts. Lubrication of the surfaces of the slow-moving gears of the guitar can be arranged in any available way from periodic lubrication to the installation of a circulating drip lubrication system.

5 Conclusions

1. The complex profile of the cycloidal pinion gear in the modern machine-building industry is obtained using multi-coordinate machines with numerical program control, and the control program is obtained using CAM systems. An alternative and tens of times less expensive method of forming such wheels is the use of a special mechanical device in which the complex flat movement of the wheel of the pin gear is reproduced.
2. An analysis of two versions of devices for cutting the peg wheels was carried out, according to the results of which ways of correcting some shortcomings of such structures were outlined: complexity of re-adjustment, oversized dimensions and power characteristics of the device drives.
3. A scheme of a new design of a device for cutting gear wheels of a pin engagement and a method for calculating its power and speed characteristics has been developed.
4. An example of design of a device for cutting a pinion gear wheel is considered. Kinematic and power calculations showed that a drive 9SDGE 90F2 with a

reduction gear and a speed adjustment of 15–230 min, having a torque of 1.07–2.98 Nm on the output shaft, can be used as a source of motion in the device. Engine power is 90 W. The total cost of such a drive and parts of a new design is tens of times less than the cost of a modern CNC machine and CAM system.

References

1. Borislavov, B., Borisov, I., Panchev, V.: Design of a Planetary-Cyclo-Drive Speed Reducer, p. 90. Linnaeus University, Sweden (2012)
2. Kargin, P.A.: Application, synthesis, and manufacture of involute gear transmissions with asymmetric gear profiles. Russian Eng. Res. (2009). https://www.scopus.com/inward/record.url?eid=2-s2.0-74549177066&partnerID=40&md5=a0215a812e2f82ac45f170d3e9e79ddb
3. Kargin, P.A.: Prospects for involute transmissions with asymmetric tooth profiles. Russian Eng. Res. (2018). https://www.scopus.com/inward/record.url?eid=2-s2.0-84923830704&partnerID=40&md5=9b948d30799d12fcf885037f511f8c72
4. Beskopylny, A., Onishkov, N., Korotkin, V.: The contact mechanics of Novikov's surface-hardened gearing during running-in process, Adv. Tribol. (2018). https://www.scopus.com/inward/record.url?eid=2-s2.0-85053879977&partnerID=40&md5=baf8f035033c445747204081fd1f5dbd
5. Chirkin, A.V., et al.: Profiles of gear teeth. Russ. Eng. Res. **35**(3), 167–170 (2015). https://doi.org/10.3103/C1068798X15030077
6. Polushkin, O.O.: Static balancing of rotating bodies. Russian Eng. Res. (2010). https://www.scopus.com/inward/record.url?eid=2-s2.0-77952063095&partnerID=40&md5=e010e881934b64ffa2a6e98728c91a69
7. Polushkin, O.O.: Rotor drive in a balancing machine. Russian Eng. Res. (2017). https://www.scopus.com/inward/record.url?eid=2-s2.0-84857835391&partnerID=40&md5=4cb181f8714f87d8b2b19467c0379d64
8. Chufistov, E.A., Chufistov, O.E., Gurevich, Y.E.: Materialy i tekhnologii XXII veka: sb. st. XIV MNTK [Materials and technologies of XXII century: proceedings of XIV ISTC]. Penza: PDZ. pp. 82–89. (2017)
9. Gubanova, A.A.: Investigation of stationary trajectories with associated milling by spur gears. In: MATEC Web of Conferences (2018). https://www.scopus.com/inward/record.url?eid=2-s2.0-85056505018&partnerID=40&md5=b4cf09f1f55b10894b92775e1934879d
10. Gubanova, A.A.: Machining by end mills with overlapping cutter–workpiece contacts. Russian Eng. Res. (2018). https://www.scopus.com/inward/record.url?eid=2-s2.0-85055038657&partnerID=40&md5=f5f66a01393da86dfb14e143b403ab4b
11. Efimov, A.V., Kireev, S.O., Korchagina, M.V.: Mathematical model for assessing selection of the interaxle drive type. J. Mach. Manuf. Reliabil. (2018). https://www.scopus.com/inward/record.url?eid=2-s2.0-85043463727&partnerID=40&md5=268c4032454b1c1408d065e05a7ec86e
12. Kireev, S.O.: Teoreticheskie osnovy metodov analiza i sinteza planetarnykh mekhanizmov s vnetsentroidnym vnutrennim tsevochnym zatsepleniem: dis. d-ra tekhn. Nauk [Theoretical foundations of analysis and synthesis of planetary mechanisms with eccentric internal corking: dissertation to apply for the degree of the candidate of engineering sciences]. Novocherkassk: Yuzhno-rus. gos. tekhn. univer. pp. 441. (2002)
13. Chufistov, E.A., Chufistov, O.E., Gurevich, Y.E.: Problemy issledovaniya i proektirovaniya mashin. Novye khimicheskie tekhnologii, zashchitnye i spetsial'nye pokrytiya: pro-izvodstvo i primenenie: sb. st. Kh Mezhdunar. nauch.-tekhn. konf. [Issues of research and design of

mashines. New chemical technologies, protective and special coatings: production and application: proceedings of X International scientific and technical conference]. Penza: PDZ. pp. 95–101. (2016)

14. Trilisskiy, V.O., Golubovskiy, V.V., Istomina, Y.V.: Komp'yuternoe modelirovanie tekhnologicheskogo oborudovaniya: ucheb. posobie [Computer simulation of technological equipment: tutorial]. Penza: Inf.-izd. tsentr PGU.124 (2009)
15. Efimov, A., Kireev, S., Korchagina, M., Lebedev, A., Kaderov, K.: Influence of differential with the limited gear ratio on the efficiency reflecting losses on wheel slip. In: E3S Web of Conferences (2019). https://www.scopus.com/inward/record.url?eid=2-s2.0-85076518405&partnerID=40&md5=26acf0ad520994a2215f127191016ae0
16. Efimov, A.V., Kireev, S.O., Korchagina, M.V.: The efficiency is a single criterion for choosing the type of interaxle drive. In: MATEC Web of Conferences (2018). https://www.scopus.com/inward/record.url?eid=2-s2.0-85056505757&partnerID=40&md5=87323ec409e48c8452dc4ba556272c27

Universal Device for Fixation of Tooling and Articles to Tables of Equipment in Machine Building

Haydyar Kaderov, Sergey Kireev, Alexey Lebedev, and Elena Kutsaeva

Abstract Information on the invention of the fastening device (application 2,020,137,783 of 18.11 is presented. 2020 (RU), which automatically maintains the parallelism of the nut support surfaces and the grip face when securing products. Calculations of device parts for strength by methods of technical mechanics and using CAD system COMPASS were carried out. The proposed design with dimensions corresponding to standard slots as per GOST 1574–91 (RU) showed satisfactory results of strength calculations of parts both using analytical formulas and using the finite element method. The adopted calculation scheme made it possible to find the maximum permissible angles of deviation of the bolt from the nominal (vertical) position relative to the plane of the equipment desktop. An engineering analysis with the division of the fastener assembly model into two assemblies using the APM FEM application in KOMPAS-3D showed mainly coincidental results with calculations using technical mechanics. Claimed device is used for fixation of articles to tables of equipment with standard T-shaped slots taking into account restrictions imposed by them on dimensions of device parts.

1 Introduction

Fasteners are an important element in many manufacturing processes. In metalworking industries, reliable fixation on machine equipment of blanks, devices, as well as dies and molds are an integral part of ensuring both the quality of manufactured products (GOST 12,467–67 (RU), GOST 12,772–67 (RU)—GOST12778-67 (RU) and safety requirements (GOST 12.2.109–89 (RU)). At the same time, in other industries where there is a need to install and attach products, various clamping devices are used to ensure the operational reliability of the production process. The above relates, in particular, to fastening devices of rod-type blue-bin pumps to hydraulic drives. In the listed cases, as a rule, T-shaped slots are provided in tables or stamping

H. Kaderov · S. Kireev (✉) · A. Lebedev · E. Kutsaeva
Don State Technical University, Rostov-on-Don, Russia
e-mail: kireeso@yandex.ru

M. Shamtsyan et al. (eds.), *Robotics, Machinery and Engineering Technology for Precision Agriculture*, Smart Innovation, Systems and Technologies 247,
https://doi.org/10.1007/978-981-16-3844-2_25

plates, regulated by GOST 1574–91 (RU) and GOST 9226–92 (RU), using standard T-shaped bolts (GOST 13,152–67 (RU)—GOST 13,165–67 (RU)).

Accordingly, there is a wide range of fasteners. In addition to these standard ones, special clamping devices are widespread. For example, in the part fixing device (A. S. 1,572,780 (RU)), the clamping element (gripping with a through groove) by means of a cutting assembly interacts with poppet springs mounted in specially made cavities of the machine table, the elastic force of which fastens the part. A disadvantage of this arrangement is the need for special cavities in the machine table. A hydraulic drive was provided to secure the part in the device, which greatly increased the cost of the device and made it often impossible to use it in oil and gas equipment.

In the device "Clamping unit for attachment of replaceable packages of universal dies" (A. p. 1,532,140 (RU)), the clamping element (grip) is made with the possibility of rotation together with the clamping bolt in horizontal and vertical planes. In this invention, however, it is necessary to provide cavities in the sub-die plate to accommodate the attachment units. In addition, the structure provides a replaceable tooling with a certain height of the shelf interacting with the tack, which limits the scope of application of this device.

"Die fixing device" (A. p. 709,388 (RU)) contains a bar located on the press table with a post and a pressure screw placed on it. The rod, in turn, enters the T-shaped slot of the stamped plate with its mating projection with the possibility of movement along the slot. A T-shaped groove is made in the post, along which the grip can move by means of a projection conjugated by its shape and size with the post slot. However, this device also requires a special T-shaped slot on the stamped plate and on its side surface. This limits the use of standardized table designs (stamped plates) in which slots are provided on working surfaces. In addition, in some cases, bolt fastening is used directly through slots made in its shelf for greater reliability. This option is not provided here.

A significant disadvantage of these devices is the need to make special slots or cavities for their installation on tables or stamping plates. At the same time, another important factor needed to be considered—the difficulty of achieving parallelism between the nut and clamp support surfaces. In case of violation of this condition, the effect of eccentric loading of the bolt occurs. This leads to, in addition to tensile stresses in the bolt rod, bending stresses [10], which significantly reduce the strength of the bolt and, accordingly, the reliability of the threaded connection, contrary to the requirements without danger of labor (GOST 12.2.109–89 (RU)). To prevent eccentric loading of the bolt, the parallel-bearing of the nut support surfaces and grip is achieved by selecting supports for it, which is quite difficult, or adjusting the height of the fixed shelf using its additional processing.

In addition to these disadvantages in the listed inventions, it is necessary to note the absence of calculations of deformations and strength of parts.

The main element in fastening devices is threaded connections, widely used in various mechanical devices. The problems of calculating them and their parts are currently being paid special attention [1]. Increased requirements to bolt tightening force leading to deformation of parts are noted in Article [2]. The optimal tightening force (maximum permissible) ensuring the durability of the threaded connection is discussed in works [3, 4]. In work [5], it is noted that one-time determination of friction coefficient in thread helps to increase reliability for rational selection of dynamometer moment.

The purpose of the work is to improve the design of the fastening device for attaching tooling and products to the tables of equipment in machine building.

The objectives of the research are to develop a methodology for the analytical calculation of the geometric parameters of parts of the claimed device, based on the known principles of technical mechanics with observance of the conditions of permissible stresses on bending, crushing, and cutting. Perform engineering analysis of the new device using APM FEM in the COMPAS CAD system.

2 Materials and Methods

New fastening device (Application 2,020,137,783 (RU)), (Fig. 1), contains housing 1, into which L-shaped gripping 2 is hinged with formation of rack and fastening parts and bolt 3 by means of axles 4. At the same time in the fixing part of the grip, there is an open through slot, and the configuration of the rack frame allows it to be placed in the slot of the table. One of the options for fixing the axes is the possibility of using screws 5. Reference numerals 6 and 7 denote pieces of equipment table and fastened article, respectively. Nut 8 and washer 9 perform attachment of the article. The drawing of the device is made in parametric press [6], which made it possible to select the position of the parts depending on the height of the shelf of the attached article.

The proposed device simplifies and simultaneously extends the variability of fixation of products on equipment tables with standard T-shaped slots (Fig. 2).

The hinge connections of the bolt and grip in the claimed fastening device automatically maintain the parallelism of the nut support surface and the corresponding grip face when attaching products to the equipment tables, regardless of the change

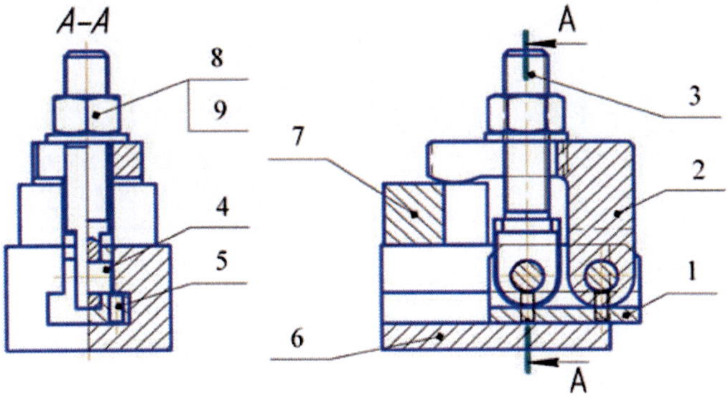

Fig. 1 Universal fastening device

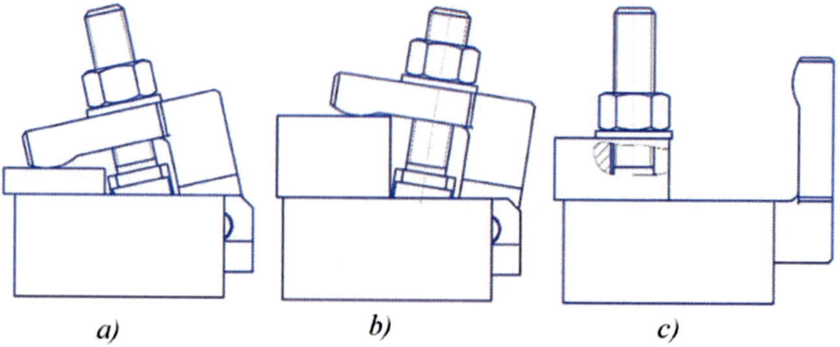

Fig. 2 Types of fastening of articles

in its height or special shelf within certain limits (Fig. 2a, b). In addition, the grip design allows you to attach the product if it has slots (Fig. 2c), and there is no need to use the grip. This type of fixation is often used for die tooling and molds: It is more reliable and allows them to be more accurately positioned relative to the axis of the equipment, which is important for the reliability of its operation.

The diagram for calculating the parts of the new device is given in Fig. 3. It presents the designations that will be used in the calculations: *a*, *b*, *h*, and *c*—the dimensions of

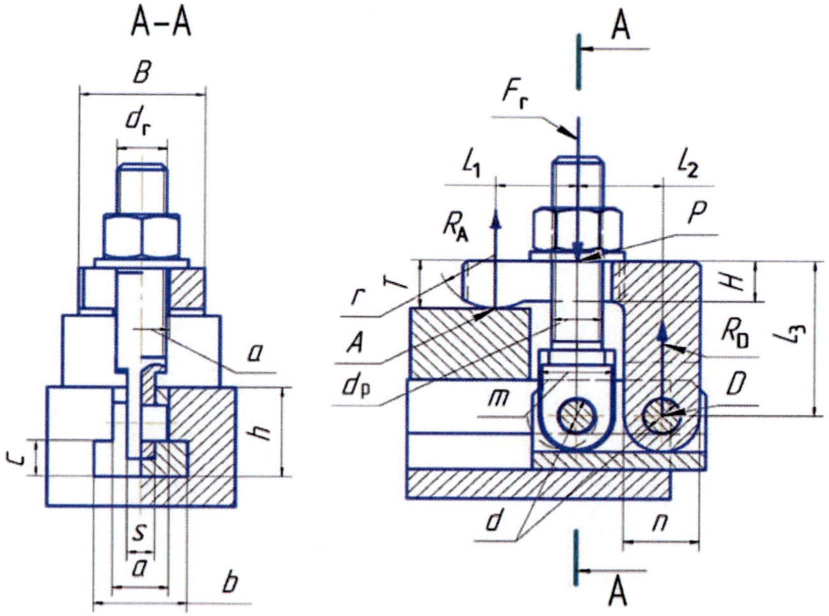

Fig. 3 Design diagram for bolt and grip sizing

the elements of body 1 corresponding to the dimensions of the slot of the equipment table; *Fr*—thread tightening force; RA and RD—reactions of supports at points of contact of article with tack and axis *D*, respectively; L1 and L2—distances to points *A* and *D*; d*r* is the diameter of the bolt thread for the corresponding width of the table slot a; *d*—diameter of bolt and tack axes; *T*, *H* and L3—gripping dimensions, *r*—radius of gripping contact surface; *S*—thickness of the eyes of the hinge part of the bolt and grip; *P* is the point of the bearing surface of the screw washer.

Design scheme with orthogonal arrangement of calculated elements makes it possible with sufficient accuracy determine dimensions of device elements providing its functional properties and operational reliability.

The most vulnerable in terms of the strength of the parts of the device is their elements in the hinge area. These include axles (4) with holes for them and tack tabs (2) and bolt (3) installed in housing (1) (see Fig. 1). It should be noted that the limited space of standard slots into which these elements are mounted determines the need to consider strength factors when assigning dimensions to these elements.

The initial data for calculations are:

1. width of the equipment table slot *a*, depth *h*, and corresponding bolt thread diameter d_r;
2. distances: L_1 between the bolt axis and point A on the contact line of the tack with the fastened article and L_2 between the axes of bolt 3 and tack 2 (see Fig. 3);
3. the width of the open through slot of the grip 2 is equal to the width of the slot of the equipment table a;
4. part material and heat treatment.

The distances L1, L2, and L3 should be chosen structurally taking into account the minimum metal consumption, and the width *B* and the height H, taking into account their strength. The thickness S of the gripping lugs 2 and the bolt 3 can be taken to be equal to half of the width of the slot a. Accordingly, the length of the axes 4 should be equal to the size a.

Complex shape, sharp transitions of gripping stages are a condition of application of various methods of strength analysis [7–11] for evaluation of its reliability.

In general, after selecting the distances L1 and L2, the task of calculating the width *B* and the grip height *H* is reduced to calculating the corresponding resistance moment *W*

$$[\sigma_b] \geq \frac{M_b}{W}, \tag{1}$$

where M_b is the bending moment in the section passing through the point of application of the concentrated load (bolt tightening force) F_r, $[\sigma_b]$ is the permissible bending stress. M_b is determined from known ratios [12] of reactions of supports RA, RD, and distances L1 and L2 (see Fig. 3).

The moment of resistance *W* in this case is

$$W \geq \frac{(B-a)\cdot H^2}{6}, \tag{2}$$

where B and a are the width of the grip and its groove, respectively, and H is the height of the grip at the point of application P of the tightening force of the thread F_r. With the known width of the slot a, it is easy to calculate the value of B.

The radius of the contact surface of the grip r can be found from the formula for calculating the stresses σ_H at the sea contact of the grip elements and the fastened article [10]

$$\sigma_{\mathrm{H}} = 0,418\sqrt{\frac{q \cdot E_{\mathrm{g}}}{r}} \leq [\sigma_{\mathrm{H}}], \tag{3}$$

where q is the distributed force, $[\sigma_H]$ is the allowable stresses on the contact strength, and E_g is the given modulus of elasticity for the case when the materials of the grip and the fastened product are different.

The distributed force q is determined by the ratio of the reaction R_A to the length of the contact line, which is equal to the difference in the values of the grip width B and its slot a—see Formula (2)

$$q = \frac{R_A}{B - a}, \tag{4}$$

where RA is the reaction of the shelf of the article; a is the width of the gripping slot, equal to the width of the table slot.

To define the radius of the grip contact surface r, convert Formula (3) as follows:

$$r \geq 0,418^2 \frac{q \cdot E_{\mathrm{g}}}{[\sigma_H]^2}, \tag{5}$$

Dimensions of bolt and gripping tabs: width c and thickness S, as well as diameter d of holes are determined based on crushing stresses

$$[\sigma_{\mathrm{cr}}] \geq \frac{F_r}{d \cdot S}, \tag{6}$$

where $[\sigma_{\mathrm{cr}}]$—permissible crushing stress; F_r—thread tightening force; d is the diameter of the eye hole equal to the diameter of the axis; S is the thickness of the bolt and grip tabs (see Fig. 3).

The selected diameter of the d-axes should be checked for allowable shear stress $[\tau_{\mathrm{sh}}]$

$$[\tau_{\mathrm{sh}}] \geq \frac{2F_r}{\pi \cdot d^2}. \tag{7}$$

The Formula (7) takes into account that the axis experiences shear stresses simultaneously in two sections.

The results of calculation according to the above scheme of geometric parameters of the proposed device for a standard T-shaped slot (see Fig. 3) with nominal dimensions: $a = 18$ mm, $b = 30$ mm, $c = 12$ mm, $h = 30$ mm, and bolt thread diameter $dr = 16$ mm are shown below. Controlled tightening force for thread M16—Fr = 23,500 N [10]. It is accepted that the material of the fastened article and parts of the fastener are steel. Then the given modulus of elasticity $Eg = 2{\cdot}05$ MPa [12].

It has been established that with the dimensions of the grip $L1 = 45$ mm, $L3 = 80$ mm, the distance between the axes $L2 = 30$ mm, the width and thickness of the bolt and grip tabs, respectively, $m = n = 25$ mm and $S = 8$ mm, respectively (see Fig. 3). The following dimensions were obtained for the material-parts of the device—Steel 45: grip width—$B = 50$ mm, its height $H = 20$ mm and contact surface radius $r = 25$ mm, and axis diameter d = 14 mm.

At the same time, it should be noted that small articles (units), which are in constant use, are recommended to be hardened by analogy with tools. This increases their service life and increases operational reliability, which is one of the conditions for securing products [13, 14].

Figure 4 shows the changes in the geometrical and power parameters of the device at different height of the fastened article. Increase of the fastening force of the article at decrease of its height $k1$ due to decrease of the ratio L'1/L'2 (Fig. 4a) and, conversely, decrease of the fastening force in case of increase of height $k2$ (Fig. 4b).

In this case, a horizontal force F_X appears, creating a factor of mutual ejection of the device and the fastened product. Force F_X is determined by

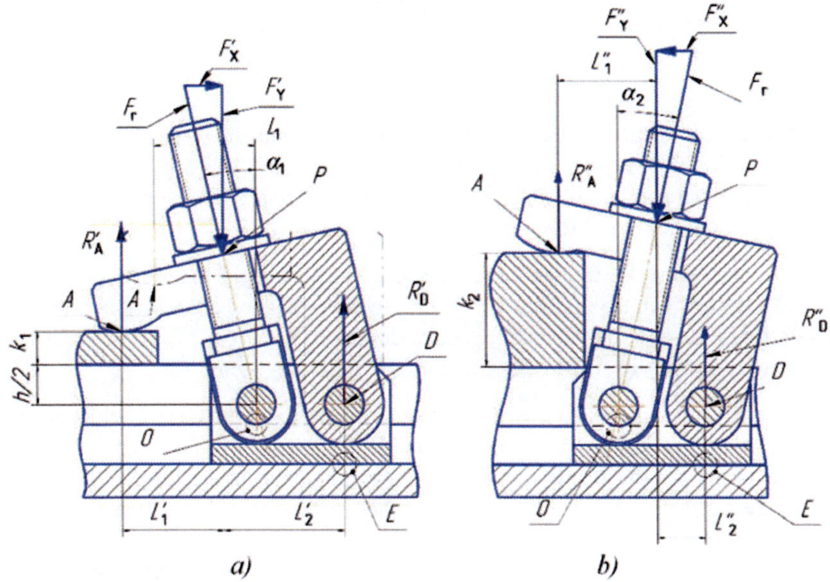

Fig. 4 Changing product docking settings

$$F_X = F_r \cdot \sin\alpha, \tag{8}$$

where α is the rotation angle of the device.

When the bolt deviates from vertical position by angles $\alpha 1 = \alpha 2 = 15°$, horizontal force FX will be equal to 25% of tightening force of thread Fr. The force FX is counteracted by the friction force from the reactions of the supports and the force of interaction with the grip Fr, as well as the shelves and the base of the housing with the corresponding elements of the slot. It is calculated mentally that the total friction force will be 30% from the tightening force of the thread Fr with the coefficient of friction of steel over steel 0.1… 0.12. These deviations of angles $\alpha 1$ and $\alpha 2$ can be considered optimal. In this case, at the specified geometric parameters, the height of the shelf will vary: from $k1 = 20$, to $k2 = 60$ mm (Fig. 4a, b). As a rule, this is quite enough [9, 15]. For example, such shelf height limits are widely used in stamps and molds. At the same time, it should be noted that the article is secured by several bolt connections. At the same time, their horizontal components FX mutually compensate for each other.

3 Results

Strength analysis of the elements of the claimed device was carried out using APM FEM application in the COMPASS CAD system. The justification for the application of engineering analysis was contradictions in the interaction of the parts of the device during tightening of the bolt connection. The nut together with the washer create a horizontal load on the grip surface, which transmits it to its elements. At the same time:

1. the contact surface of the fixing part of the grip forms the force of pressing the article shelf to the equipment table
2. the post portion is hinged to press the bottom support surface of the housing against the bottom of the T-groove of the table.

At the same time, the tightening force is reproduced through the hinge joint of the bolt with the body in the opposite direction. Then there is force interaction between the housing flanges and the slot in opposite directions.

Strength analysis in APM FEM includes the following steps:

1. Preparing the model. At this stage, it is necessary to fix the faces of elements in which no stresses can arise that lead to plastic deformations—by analogy with the journals of the gear box shafts. Such faces, shown in Fig. 5, are: 1—the face of the product shelf, 2—the surface of the equipment table, 3—the side faces of the bottom of the table slot, 4—the bottom of the slot, and 5—the contact surfaces of the grip. In addition, the mating faces of the fastener body 1 (see Fig. 1) were fixed with the corresponding faces of the T-groove—with the side 6 and its bottom (not shown).

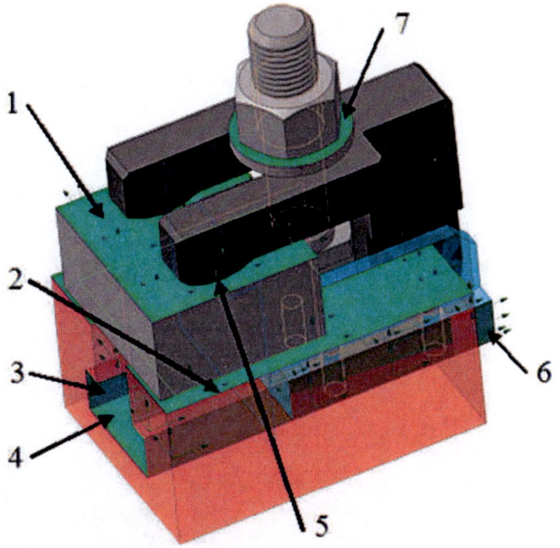

Fig. 5 Docking faces in the model

2. In the next step, the elements of the parts of the device to be loaded, such as pressure or distributed force, are indicated. In this case, the upper abutment 7 of the bolt washer (see Fig. 5) is selected to apply a pressure *P*, the value of which is assigned taking into account the force of controlled tightening of the bolt $F_r = 23{,}500$ N on the support surface area of the washer with an outer diameter of 30 mm and a hole diameter of 18 mm. The design pressure was $P = 77$ MPa.
3. The result of finite element mesh (FE) generation is shown in Fig. 6. Partition parameters: (a) element type −10-node tetrahedral; (b) maximum length of element side −5 mm; (c) quantity of final elements −17,836; (d) number of nodes −33,943.
4. After splitting the device model, the application performs strength calculation of the elements with presentation of the results map in accordance with the specified properties of the element material—steel 45 (Fig. 7):

- Equivalent Mises stress (type—SVM): minimum value 0, max. −464 MPa (Fig. 8). In this case, the query for presenting the results is selected on the undeformable model;
- Figures 9 and 10 reflect the yield and strength safety factors, respectively. Here, when the results are displayed, a deformable workpiece is selected

The results of the strength analysis showed accurate correspondence with the calculated ones made according to Formulae (1—7). In addition, the most vulnerable places are reflected—the angles of the steps when moving from one face to another. They are catalysts for stress concentration. It should be noted that the choice of mechanical properties of steel 45 corresponded to heat treatment—normalization.

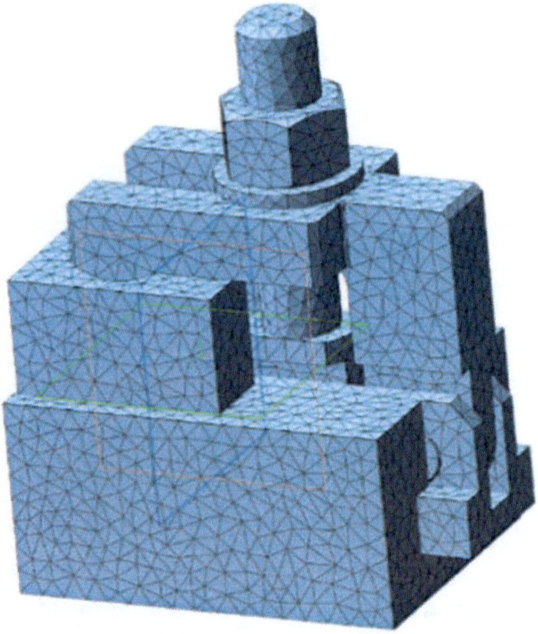

Fig. 6 Generation of finite element mesh

Fig. 7 Part material properties

Young module, [MPa]	20000
Poisson ratio, [-]	0.3
Density, [kg/m^3]	7825
Thermal expansion factor, [1/°C]	1.2e-05
Thermal conductivity, [W/(°C·m)]	0.055
Compressive strength, [MPa]	530
Ultimate strength, [MPa]	530
Yield strength, [MPa]	275
Fatigue strength limit (n). [MPa]	305
Fatigue strength limit (k). [MPa]	180

The earlier conclusion that the use of quenching for similar units was preferable was confirmed. It is recommended to use alloyed steels 40H or 40HN with subsequent hardening for critical parts. 2) and bolt (pos. 3) (see Fig. 1).

Engineering analysis of the interaction of the faces of parts was carried out: Housing 1, Gripping pos. 2 and Bolt pos. 3, and T-shaped slot of equipment table 7 (refer to Fig. 1). To do this, the calculation scheme in Fig. 11 was used. Takeover 2 is presented as a fragment to facilitate visualization of the analysis results. Working surface 1 of equipment plate and side faces 2 of its T-shaped slot are fixed here. To

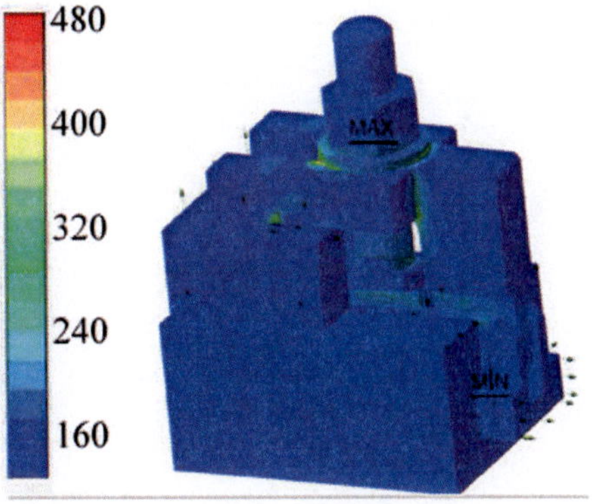

Fig. 8 Stress calculation results (MPa)

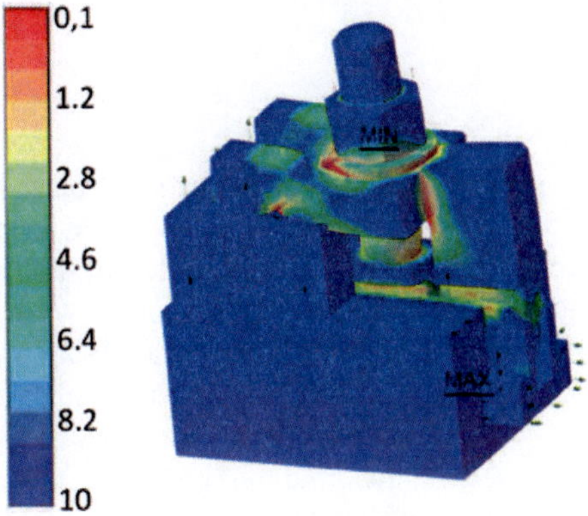

Fig. 9 Results of the yield stock factor

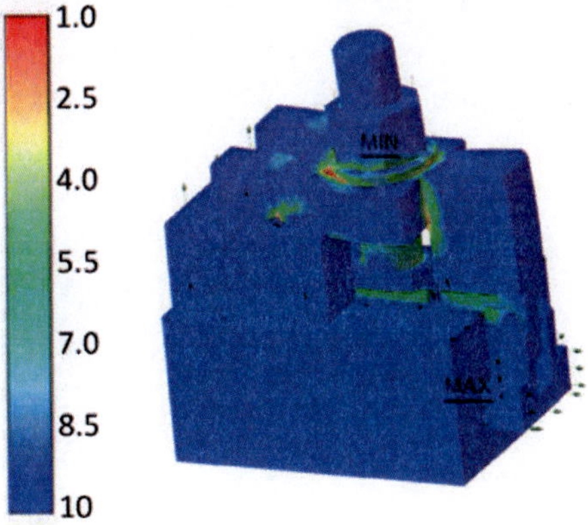

Fig. 10 Safety factor results

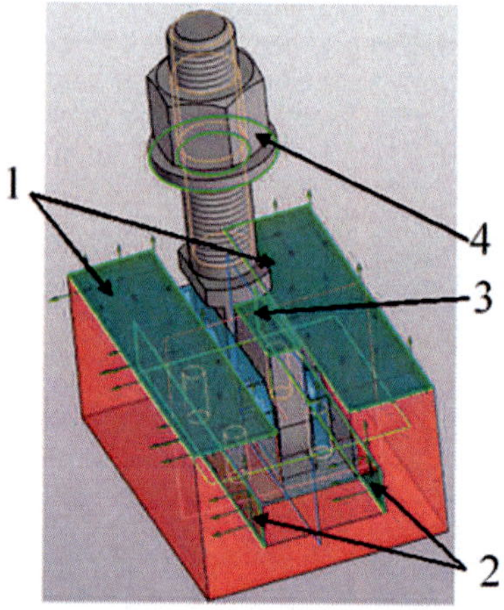

Fig. 11 Securing the faces

apply the load (P), face 3 of the grip fragment and bottom surface 4 of the washer are provided. Considering that the force placed on the face 3 of the gripping fragment is modulo 9400 H for the ratio $L1 = 45$ mm and $L2 = 30$ mm (see Fig. 3), the dimensions of the face 3 were chosen so that the pressure value $P = 77$ MPa on this face is single, as on the surface of the washer 4 (Fig. 12).

After securing the faces and applying pressure, create a split and generate mesh (Fig. 13). The material of the parts and their properties is the same as those of the previous calculation. Figure 14 shows the results of the stress analysis—minimum value 0.0657, maximum −786 MPa in the area of the washer-nut interface.

The parameters of FE generation are accepted, unlike item 3 (see Fig. 6), with a maximum length of the element side of 3 mm 10-node tetrahedral.

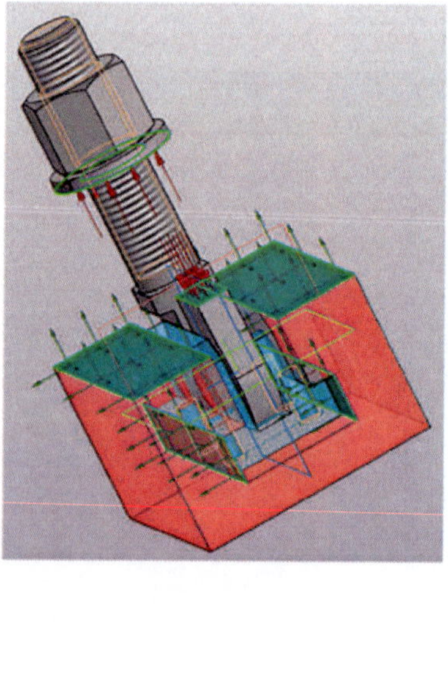

Fig. 12 Application of pressure

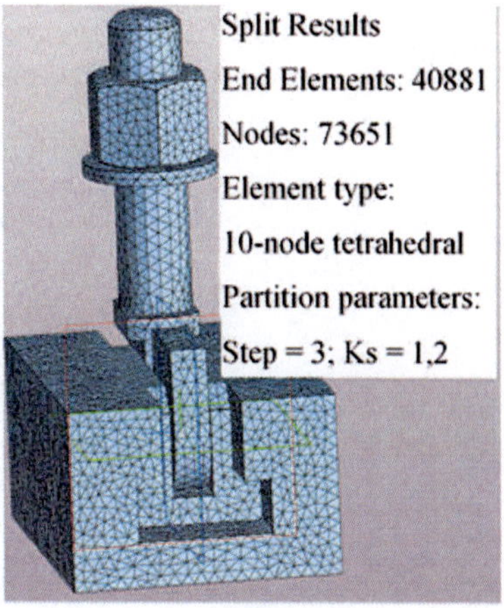

Fig. 13 Generation of finite element mesh

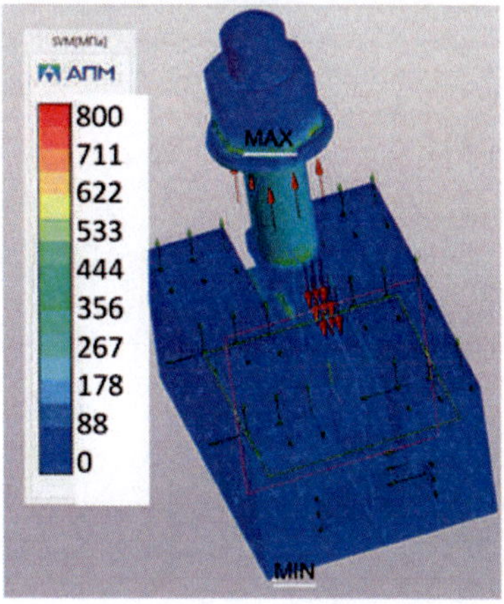

Fig. 14 Stress calculation results (MPa)

For comparison 464 MPa in the previous case. The difference is explained both by the difference in the areas of the mating support surfaces of the washer with tack and the washer with nut, and by the difference in the stressed-deformed states.

Figures 15 and 16 show the results of the strength analysis for yield and strength safety factors, respectively.

At the same time, a discrepancy in the results of the analysis can be noted, which suggests that it is advisable to separately consider the interaction of the body, the axis with the bolt and the washer, where the main stresses are tension and shear, and grip with the body and the axis experiencing mainly compression and shear stresses.

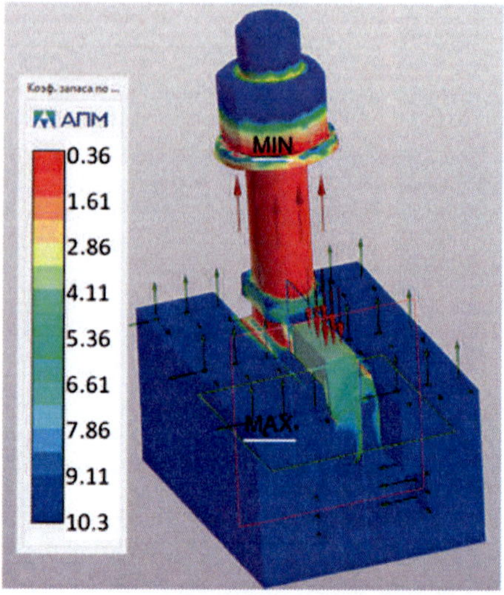

Fig. 15 Results of yield stock ratios

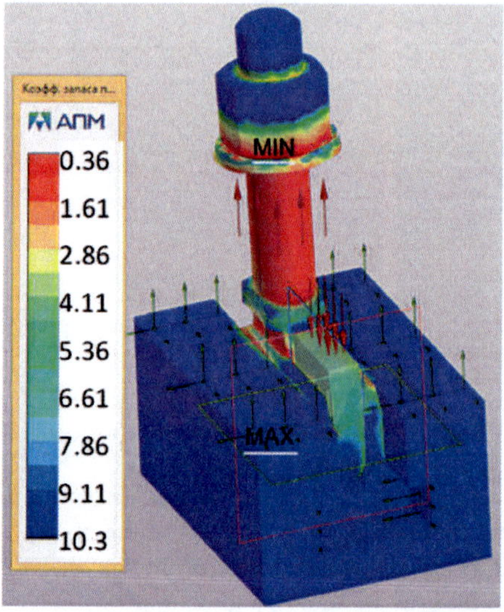

Fig. 16 Safety factor analysis

4 Discussion of Results

Strength calculations using APM FEM for cases where the bolt axis deviates from the vertical position by 15° (Figs. 2a and b) showed similar results as in the vertical position.

1. The presented scheme of calculation of geometric parameters of the device parts taking into account the types of their stress state at controlled bolt tightening

using means of technical mechanics allows to determine the main geometric parameters of the parts.
2. The adopted calculation scheme made it possible to find limit angles of the bolt deviation from the nominal (vertical) position relative to the equipment desktop.
3. The engineering analysis conducted with the division of the fastener assembly model into two assemblies using the APM FEM application in KOMPAS-3D showed mainly the same results as calculations using technical mechanics.
4. Safety factors are lowest in the areas of threaded connection and in the angles of transition from one stage to another.

The claimed fastening device based on the strength analysis of the parts using the technical mechanics positions and the APM FEM application in the COMPASS CAD system can be effective. This allows you to assign it the status "Node" and be accepted into production as an application to equipment equipped with tables with standard slots for attaching products and tooling.

To increase the strength reserve factors, it is recommended to use hardening for parts of the device up to 45 HRC. Simultaneously, it is advantageous to use alloyed steels such as 40H or 40HN as materials.

5 Conclusions

Claimed device is used for fixation of articles on tables of equipment with standard T-shaped slots taking into account restrictions imposed by them on dimensions of device parts. The results of the analysis of the design and strength properties of the device showed its acceptability for wide use in various industries, including oil and gas industries.

References

1. Ryahovskiy, O.A., Syromyatnikov, V.S.: A calculation of the main load coefficient in threaded connections (2019). https://doi.org/10.18698/0536-1044-2019-8-10-17
2. Ryahovskiy, O.A., Syromyatnikov, V.S.: Determining deformations of parts when tightening bolted joints (2018). https://doi.org/10.18698/0536-1044-2018-10-18-23
3. Syromyatnikov, V.S., Garsia Martines, X.M., Samora Kantana, L.A., Ortega Rosales, M.G.: Calculation of a bolted connection under variable load (2017). https://doi.org/10.18698/0536-1044-2017-11-28-35
4. Garsia Martines, X.M., Samora Kantana, L.A., Ortega Rosales, M.G.: The optimum preload of a threaded joint against (2016) .https://doi.org/10.18698/0536-1044-2016-12-43-50,izvuzmash.ru›articles/1371/eng/1371.pdf
5. Ryahovski, O.A., Syromyatnikov, V.S. The experimental determination of the coefficients of friction in bolted joints (2016).https://doi.org/10.18698/0536-1044-2016-10-18-25
6. Kaderov, K., Kireev, S., Korchagina, M., Lebedev, A.: Determination of technological parameters and tool for flanging holes from sheet blanks. In: E3S Web of Conferences, vol 164, (2020)

7. Kaderov, K.K., Kireev, S.O., Korchagina, M.V., Osipuk, A.Y.: Improved seals in pipelines and high-pressure hoses. Russian Eng. Res. (2018). https://www.scopus.com/inward/record.url?eid=2-s2.0-85045990662&partnerID=40&md5=3d37a0cc81149f9c407650a5ede73374
8. Efimov, A.V., Kireev, S.O., Korchagina, M.V.: Mathematical model for assessing selection of the interaxle drive type. J. Mach. Manuf. Reliabil. (2018). https://www.scopus.com/inward/record.url?eid=2-s2.0-85043463727&partnerID=40&md5=268c4032454b1c1408d065e05a7ec86e
9. Ovchinnikov, N.A., Zhigulsky, V.I., Kozyreva, E.A., Chefranova, O.V.: Modeling and estimate of the strain-stress state of a bus body pillar with account for geometrical nonlinearity. ARPN J. Eng. Appl. Sci. (2017). https://www.scopus.com/inward/record.url?eid=2-s2.0-85015224140&partnerID=40&md5=8d5793552517df70f8f493a4689ec682
10. Kasyanov, V., Deryushev, V., Kosenko, E., Kosenko, V., Golubeva, A.Y.: (2018) Synthesis of methods and principles of ensuring the reliability of one-off and serial production machines. In: MATEC Web of Conferences (2018). https://www.scopus.com/inward/record.url?eid=2-s2.0-85056657184&partnerID=40&md5=d2960acd147256ac3b7cd28e5fb27f65
11. Kaderov, K.K., Korchagina, M.V., Kireev, S.O., Nikishenko, S.L.: Solution of problems of gasketing of joints of units of machines and equipment of oil and gas enterprises when in use in high- and low-temperature regimes. Chem. Petrol. Eng. (2018). https://www.scopus.com/inward/record.url?eid=2-s2.0-85052809302&partnerID=40&md5=76b3907958433f2392a862dd55f6988d
12. Zelentsov, V.B., Sadyrin, E.V., Sukiyazov, A.G., Shubchinskaya, N.Y.: On a method for determination of poisson's ratio and Young modulus of a material. In: MATEC Web of Conferences (2018). https://www.scopus.com/inward/record.url?eid=2-s2.0-85056511919&partnerID=40&md5=72e64853a8f771fdd278e4bafba1e39f
13. Skachkov, S.V., Shchutsky, S.V.: Analysis of the stresses of the wall of a steel water tower. Mater. Sci. Forum (2018). https://www.scopus.com/inward/record.url?eid=2-s2.0-85055948391&partnerID=40&md5=6847a4c7d76d8ac156019c6bd02238ad
14. Skachkov, S.V., Golotaystrova, E.Y.: Reliability comparison of trusses made of thin-walled steel sections. Mater. Sci. Forum (2018). https://www.scopus.com/inward/record.url?eid=2-s2.0-85055944821&partnerID=40&md5=689df0f71740d1873b04192379a14db5
15. Kravchenko, P.D., Fedorenko, D.N.: Heuristic method of design of the load gripping and manipulating devices for work in special conditions. Int. J. Appl. Eng. Res. (2015). https://www.scopus.com/inward/record.url?eid=2-s2.0-84928911600&partnerID=40&md5=d008cee0db42cd029588bd4012e2f96e.

Study of the Preliminary Elastic Deformation State of Workpieces Made of Non-ferrous Alloys During Drilling

Vadim Kuts, Mikhail Razumov, and Aleksandr Byshkin

Abstract Drilling is one of the most common operations of fabrication processes. That's why increasing the efficiency of drilling is always an urgent task. To increase the efficiency of the drilling process, a method of drilling in a pre-stress-strained material was developed using the technological facilities of Southwest State University. The workpiece is subjected to elastic deformation by applying the load not exceeding the proportional limit of the workpiece material. After loading is finished, the workpiece dimensions remain the same. It has been shown that this drilling method decreases the axial component of cutting force in the range from 20 to 30%. This happens because the workpiece material enters the cutting area (namely, the primary deformation area) in a pre-deformed state, which contributes to the saturation of the crystal lattice of the processed metal with energy. This paper describes the modeling of the drilling process using workpieces made of non-ferrous alloys in a preliminary elastic deformation state. Speaking of the mechanism of this phenomenon, the study and description of changes in the preliminary elastic deformation region of the processed material at different drilling depths are the most relevant. This study was performed using the DEFORM-3D FEM environment, which showed that the elastic deformation state of the processed material is ensured in the cutting area throughout the whole drilling process.

1 Introduction

Drilling is a widespread operation in me-chanical engineering. Drilling is a traditional method of producing holes. However, it is worth noting that drilling small diameter holes (below 5 mm) imposes certain technological limitations associated with the low rigidity of the tool, low strength of the small diameter axi-al tool, complications with the extraction of borings from the cutting area. This means that while these problems are still unsolved, the at-tempts to improve this process are relevant in the

V. Kuts (✉) · M. Razumov · A. Byshkin
Southwest State University, Kursk, Russian Federation
e-mail: kuc-vadim@yandex.ru

M. Shamtsyan et al. (eds.), *Robotics, Machinery and Engineering Technology for Precision Agriculture*, Smart Innovation, Systems and Technologies 247,
https://doi.org/10.1007/978-981-16-3844-2_26

engineering framework to create highly ef-ficient competitive production lines, both in the domestic and external markets.

To achieve this goal, various engineering solutions are proposed allowing to solve the described problems associated with drilling small diameter holes. For example, the following can be applied to ensure favorable conditions during drilling of non-ferrous alloys: vibration drilling methods with additional tool oscillations, workpiece preheating in the processed area in order to change physical and mechanical properties of the processed material, ensure coolant supply to the processed area, etc. [1–7].

However, using these approaches when drilling small diameter holes requires rather complex special-purpose devices to be developed and created.

Also, the cutting processes (mainly, turning operations) can be improved by a well-established method of preliminary plastic de-formations, which has the same disadvantage when applied to drilling. In engineering terms, it is quite difficult to implement preliminary plastic deformations in the closed region.

As a result of this analysis, it was proposed to load the workpiece before the drilling process, thereby forming a preliminary elastic de-formation region in the workpiece material in the hole drilling area. In [8, 9], the concept of a conductor is discussed, which elastically deforms the workpiece with the load not exceeding the workpiece's proportional limit, and the workpiece dimensions remain the same af-ter the load is stopped. Also, these papers pro-vide the results of experimental studies of drill-ing of workpieces made of aluminum and cop-per alloys, which had an elastic deformation region created in workpiece material using this type of conductor. It was shown that this method of drilling of small diameter holes (3 mm, in particular) allows decreasing the axial component of the cutting force by 20–30% (Figs. 1, 2).

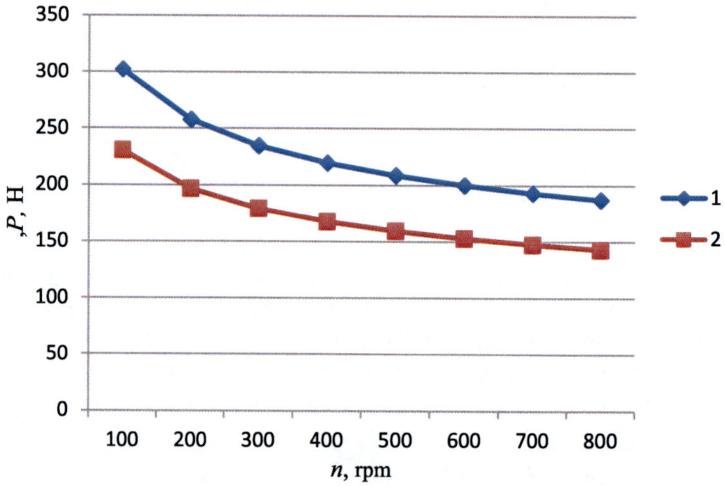

Fig. 1 Dependence of the axial force on the rotation rate during conventional drilling of workpieces made of copper alloy (1) and drilling in a preliminary elastic deformation state (2)

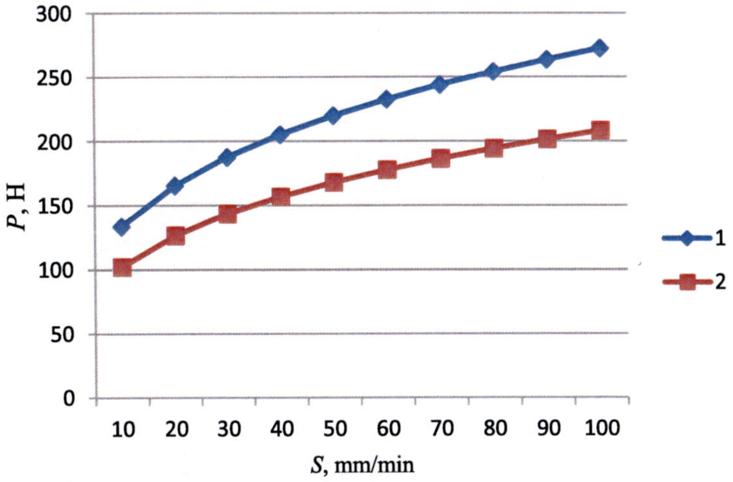

Fig. 2 Dependence of the axial force on the feed per minute during conventional drilling of workpieces made of copper alloy (1) and drilling in a preliminary elastic deformation state (2)

Drilling was performed using a 3 mm twist drill with standard sharpening fed at $S = 100$ mm/min (see Fig. 2) and the spindle rotation rate of $n = 800$ rpm (see Fig. 3).

The reason for this is that the workpiece material will arrive into the cutting zone (namely, in the primary deformation zone) in a pre-deformed state, which contributes to the saturation of the crystal lattice of the processed metal with energy [10]. This energy will bring the lattice close to the equilibrium state. This leads to less work required to bring metal to fracture and, accordingly, leads to a decrease in the reaction force on the tool's front surface and the drill web surface.

2 Problem Definition

Speaking of the mechanism of this phenomenon, the study and description of changes in the preliminary elastic deformation region of the processed material at different drilling depths are the most relevant without taking into account the stresses created by the drill during operation. In this study, we modeled the process of drilling 3 mm through holes using the DEFORM-3D FEM environment. Al-3003 aluminum alloy (GOST 4784–97) was used as workpiece material. The outer workpiece diameter and its height was taken equal to 10 mm. The workpiece was deformed by moving down the top conductor plate for 0.2 mm.

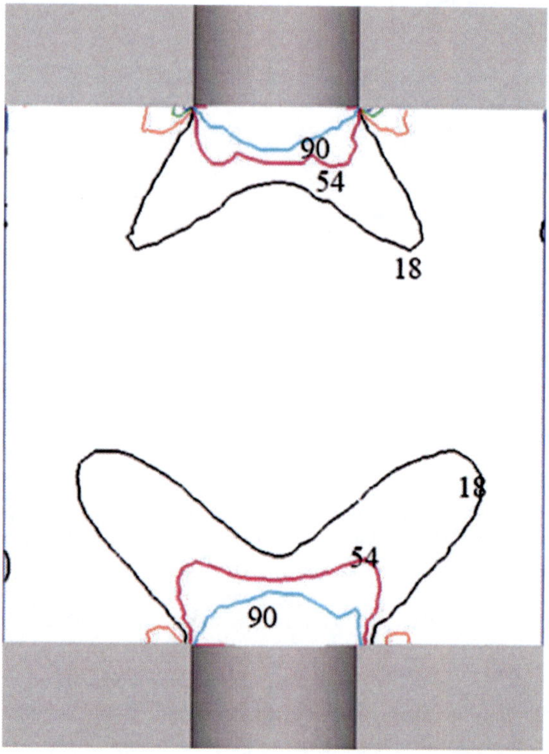

Fig. 3 Distribution of the maximum primary stress (MPa) lines in the initial moment of time

3 Research Methods, Results

Figure 3 shows the distribution of maxi-mum principal stress (MPa) lines after the pre-liminary elastic deformation state is ensured in the material.

This shows that the greatest values of the maximum principal stress (>90 MPa) are ob-served in the workpiece in the conductor open-ing area at a depth of 0.5 mm where the drill contacts the workpiece when it cuts in and dur-ing the subsequent drilling, as well as at the op-posite workpiece side where the drill exits. Thus, it can be argued that workpiece defor-mation using a conductor with a hole allows creating a preliminary elastic deformation re-gion in the processed material at the initial step of processing.

Figure 4 shows the distribution of the max-imum principal stress lines (MPa) in the work-piece at the drilling depths of 1.5; 5; 8 and 9.5 mm.

The figures show that as the drill travels deeper into the workpiece, the region of the largest maximum principal stress values with a depth of 0.5 mm also moves deeper into the workpiece and is located behind the drilled hole bottom surface. This is primarily due to the fact that the hole bottom having a conical shape acts as a stress concentrator.

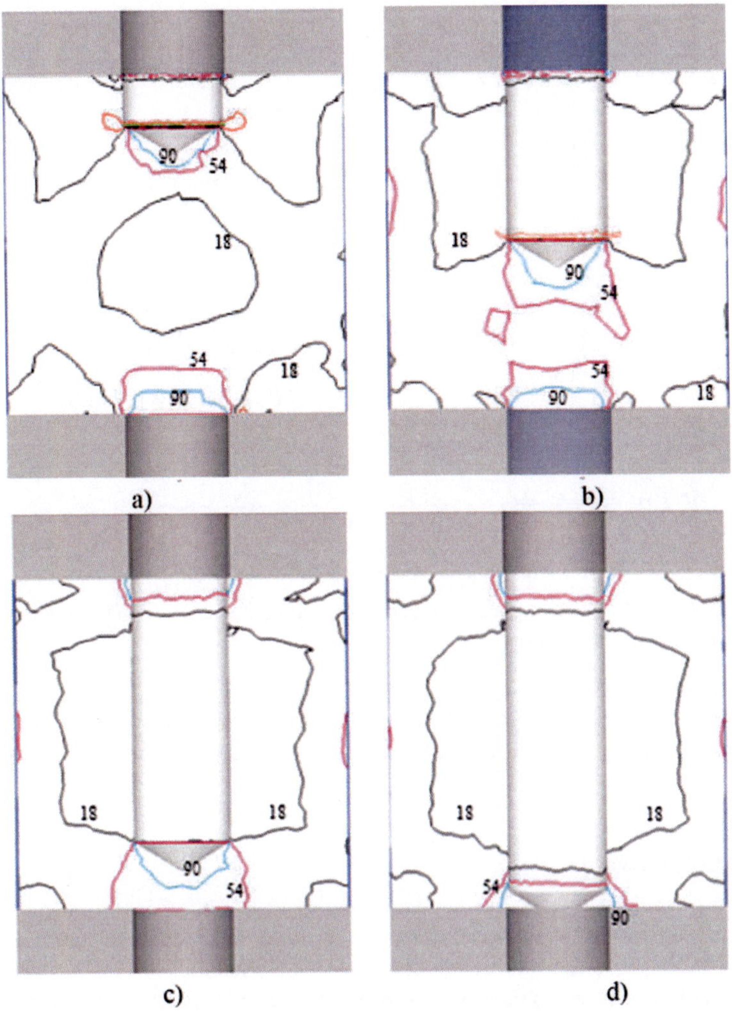

Fig. 4 Distribution of maximum principal stress (MPa) lines during drilling at a depth: **a** 1.5 mm; **b** 5 mm; **c** 8 mm; **d** 9.5 mm

Figure 5 shows the change in the maximum principal stress at the hole bottom surface points located at different distances from the axis during drilling.

Figure 5 shows that, during almost the entire drilling process, the region of the greatest maximum principal stress (≈180 MPa) is formed along the axis of the drilled hole, where the drill web contacts the processed material. The exception is the initial moment when these values do not differ from the values at the hole bottom surface points located at different distances from the axis (≈127 MPa).

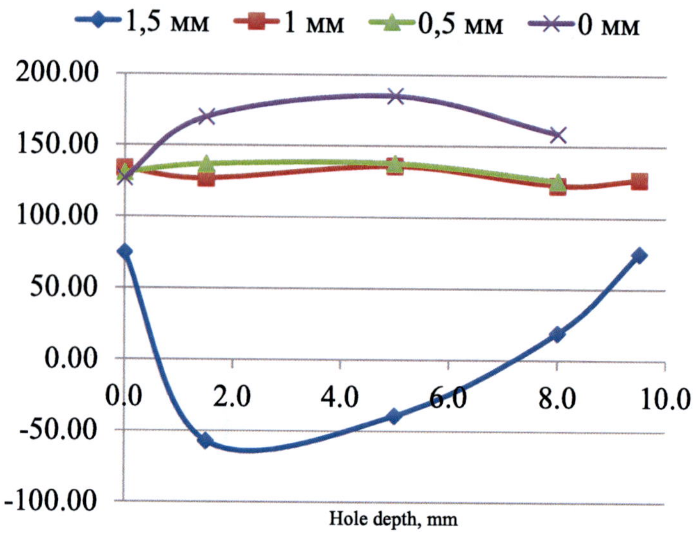

Fig. 5 Changes of maximum principal stress (MPa) during drilling at the hole bottom surface points located at different distances from the axis

The maximum principal stress values are lower at points located near the surface of the drilled hole and range from 75 to −60 MPa down the hole.

4 Conclusions

The obtained results allow concluding that drilling small diameter holes (below 5 mm), the elastic deformation state of the processed material will be located in the cutting area during the whole drilling process, which ensures a decrease in the axial component of the cutting force during drilling.

References

1. Emel'yanov, S.G., Sidorova, V.V., Ponomarev, V.V., Razumov, M.S.: Determining the geometric parameters of a sheared layer in drilling of nonferrous metals and alloys with the use of axial vibrations. Chem. Pet. Eng. **52**(11–12), 796–800 (2017)
2. Sidorova, V.V., Razumov, M.S., Grechukhin, A.N., Emelyanov, S.G.: Analysis of the dependences of the parameters of permanent magnets and their installation on the vibration force during drilling with the imposition of axial vibrations. In: In the Collection: The Future of Mechanical Engineering in Russia. Collection of Reports of the Eighth All-Russian Conference of Young Scientists and Specialists. pp. 34–37. (2015)

3. Sidorova, V.V., Razumov, M.S., Grechukhin, A.N.: Automated means of controlling technological parameters when drilling with the imposition of vibrations. In: In the Collection: Electrical Engineering. Power Engineering. Mechanical Engineering. Collection of scientific papers of the I International Scientific Conference of Young Scientists. pp. 154–156. (2014)
4. Razumov, M.S., Zavorin, M.K., Gatiev, M.S., Byshkin, A.S.: Device for drilling holes with the imposition of vibrations on the workpiece. In: In the Collection: Modern Materials, Technology and Technology Collection of Scientific Articles of the 8th International Scientific and Practical Conference. pp. 347–352. (2018)
5. Vinogradova, N.V.: Theoretical study of the magnitude of residual stresses from structural-phase transformations in the surface layer of parts after mechanical processing. Topical issues of engineering and technology: collection of dokl. mezhdunarodn. nauch. part-time conf.,In: Gorbenko, A.V., Dovzhenko, S.V. (eds.) Lipetsk: Izdat. center De-facto. vol 1, pp. 111–115. (2010)
6. Razumov, M.S., Sidorova, V.V., Grechukhin, A.N.: Automated control of process conditions during drilling with imposition of vibrations Metallurgical and Mining Indus. **6**(5), 19–23 (2014)
7. Ivakhnenko, A.G., Kuts, V.V., Altukhov, AYu., Ivakhnenko, E.O.: Dynamic synthesis of technological equipment for the manufacture of precision articles. Chem. Pet. Eng. **51**(7), 445–451 (2015)
8. Kuts, V.V., Razumov, M.S., Byshkin, A.S.: Experimental studies of hole drilling operations in stress-strain workpiecematerial. In the collection: MATEC Web of Conferences. vol 01011 (2018)
9. Kuts, V., Byshkin, A., Khodyrevskaya, S.: The method and device for automated material elastic and proportionality limits calculation Solid State Phenomena. **284** (SSP), 1361–1366 (2018)
10. Bondarev, A.A.: Investigation of the influence of advanced plastic deformation on the efficiency of the cutting process of structural steels. Abstract of the dissertation of the candidate of technical Sciences. Volgogr. State. Tech. Univ. Volgograd (2016)

Influence of the Geometry of a Similars Specimen on the Parameters of the Dynamics of the Thermal Field in Friction Stir Welding

Radmir Rzaev, Azatulla Dzhalmukhambetov, Irina Krutova, and Marina Fisenko

Abstract The paper considers two mathematical models of temperature distribution in the process of friction stir welding (FSW) of dissimilars thin-sheet and bulk samples. The model makes it possible to take into account the geometric dimensions of the welded samples, the thermophysical properties and the temperature distribution in the sample-disk and the sample-hemisphere during the transition of the welded material into the superplastic state. The paper presents comparisons of the temperature distribution for disks with a thickness of 2 and 3 mm, a hemisphere. Temperature measurements were carried out at various points of the hemispherical sample and the disk during the FSW process. These models are used to assess the linear velocity of the FSW associated with the thermophysical properties and mass and dimensions of the welded samples. The calculation results are in satisfactory agreement with the experimental data. A decrease in the thickness of the disk specimen leads to an acceleration of heating in the region adjacent to the rotating welding tool during the FSW. A sample 3 mm thick reaches the state of superplasticity within 3 s, and a sample 2 mm thick - within 1.5 s. The linear speed of the FSW decreases from 6.32 mm/s ($h = 2$ mm) to 1.85 mm/s ($h = 3$ mm). Comparison of the calculated dynamics of the temperature of a massive hemisphere with thin disks shows that heating to a superplastic state of an element adjacent to the tool occurs almost equally quickly.

1 Introduction

Friction stir welding (FSW) is a type of friction welding. Non-detachable joint of structural parts in the case of FSW occurs due to the formation of bonds between the atoms of the surfaces of the structure, which come closer to a distance close to the crystal lattice parameter. In this process, the heating of parts necessary for welding is carried out by direct conversion of mechanical energy into heat due to the work of friction forces [4].

R. Rzaev (✉) · A. Dzhalmukhambetov · I. Krutova · M. Fisenko
Astrakhan State University, Astrakhan, Russia

M. Shamtsyan et al. (eds.), *Robotics, Machinery and Engineering Technology for Precision Agriculture*, Smart Innovation, Systems and Technologies 247,
https://doi.org/10.1007/978-981-16-3844-2_27

When friction stir welding of heat-hardened materials, it is important to have information about the process temperature and the distribution of temperature fields for working out the optimal parameters in order to obtain the required strength characteristics of the weld. To solve this problem, a mathematical model is proposed that describes the process of friction stir welding. The task of such modeling is to predict the temperature distribution fields arising during welding at various mass and size parameters of the welded products [1–3, 5].

2 Mathematical Model of Temperature Dynamics of FSW

In the FSW process, heat is released due to friction in the area of contact with the workpiece material of a rotating tool, the working part of which consists of a narrow tip (pin) and a wider disc (shoulder). Heat is generated in the sample area adjacent to the pin surface [9, 10].

In the case of FSW of products of various shapes and sizes, one has to resort to models that reflect the geometry of the samples to be welded. When welding sufficiently thin plates, the model in the form of a disk with the introduction of thermal power in its center is adequate. In the case of massive products, the model of heat propagation in the hemisphere is suitable.

Computer calculations of the temperature dynamics at the FSW for a sample-disk and a sample-hemisphere with a supply of thermal power in their center were carried out in [6–8]. Heat losses due to radiation and heat exchange with the environment were also taken into account. Here we will consider the possibility of combining these models, since in the transition from one model to another, the structure of the system of differential equations remains practically unchanged. The dimension of the problem is taken into account in the coefficients of this system.

In the case of a disk, the heat propagates axially-symmetrically, so we distinguish concentric cylindrical surfaces with radii increasing exponentially around the central cylinder with radius r_0 (pin)

$$r_k = r_0 \cdot q^k, \quad (k = 0, 1, \ldots N) \tag{1}$$

where the factor of the geometric progression when dividing the sample with radius R into N narrow elements is $q = (R/r_0)^{1/N}$. This selection of elements is possible due to the decrease in the temperature gradient with the increase in the radial variable. The number of allocated elements $N = 10$ is quite sufficient for the application of the considered models for solving technological problems (Fig. 1).

The temperature of the k-th selected element is assumed to be the temperature T_k on the outer cylindrical surface with radius r_k. The enthalpy of this element is expressed as

$$H_k = m_k c T_k + \text{const}, \tag{2}$$

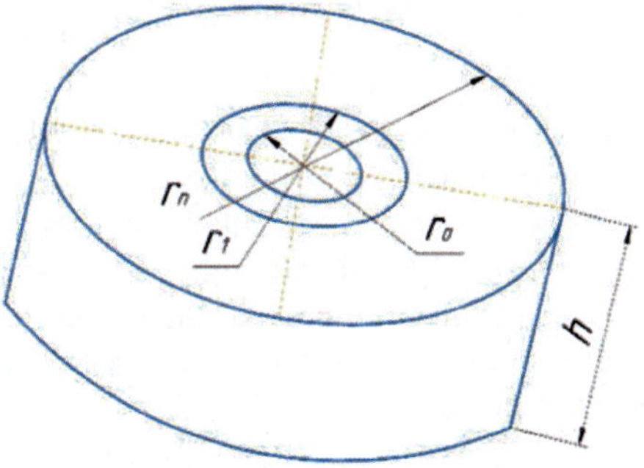

Fig. 1 The scheme of allocation of ring elements of the disk

where c the specific isobaric heat capacity of the metal. The mass of the annular element is related to its thickness h and the density of the metal ρ by the ratio

$$m_k^d = \pi \rho r_0^2 h q^{2k-2}\left(q^2 - 1\right) \cdot (k = 1_{2\ldots 2}N), \quad m_0^d = \pi \rho r_0^2 h \tag{3}$$

In the hemispheric model used to calculate the temperature field of massive samples, heat propagation from the center, in which heat power is released, to the periphery occurs almost radially. The selection of hemispherical elements, the radii of which are determined by a geometric progression (1), is shown in Fig. 2.

The temperature of the k-th hemispherical element is assumed to be the temperature T_k on the outer surface with radius r_k. The enthalpy of an element is expressed by the Formula (2), where the mass of the element is

$$m_k^s = \frac{2}{3}\pi \rho r_0^3 q^{3k-3}\left(q^3 - 1\right), (k = 1, \ldots, N), \quad m_0^s = \frac{2}{3}\pi \rho r_0^3 \tag{4}$$

The rate of change in enthalpy for the selected k-th element of the disk and hemisphere can be represented uniformly in the form of the energy balance equation:

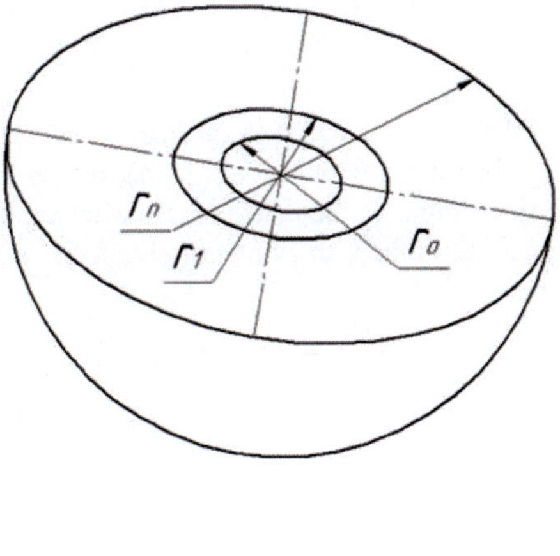

Fig. 2 Scheme of the selection of hemispherical elements

$$m_k c \frac{dT_k}{dt} = wP\delta_{k,0} + (1-w)P\delta_{k,1} + P_k^a + P_k^b + P_k^r + P_k^e. \tag{5}$$

Here w is the fraction of the power released in the area of contact with the pin sample from the value P of the power used to heat the sample. The rest of it stands out in the area of contact with the shoulder. $\delta_{k,k'}$—the Kronecker delta symbol.

The terms P_k^a and P_k^b on the right side of Eq. (5) represent the amount of heat received per unit of time, respectively, from the $(k-1)$—th and $(k+1)$—th element as a result of heat transfer. In accordance with the law of thermal conductivity, we obtain the expressions for them:

$$P_k^{\mathrm{ad}} = \frac{2\pi h\chi q}{(q-1)}(T_{k-1} - T_k)\big(1 - \delta_{k,0}\big), \quad P_k^{as} = \frac{2\pi\chi_0 q^{k-1}}{(q-1)}(T_{k-1} - T_k)\big(1 - \delta_{k,0}\big) \tag{6}$$

$$P_k^{\mathrm{bd}} = \frac{2\pi h\chi q}{(q-1)}(T_{k+1} - T_k)\big(1 - \delta_{k,N}\big), \quad P_k^{\mathrm{bs}} = \frac{2\pi_0\chi q^{k}}{(q-1)}(T_{k+1} - T_k)\big(1 - \delta_{k,N}\big) \tag{7}$$

where χ—the thermal conductivity of the metal. Positive values of these values correspond to the input, and negative values correspond to the loss of energy by the k-th element. The additional superscript d refers to the disk model, and s refers to the hemisphere model.

The power of energy loss of an element for thermal radiation as an absolutely gray body is equal to

$$P_k^r = \pi r_0^2 q^{2k}(q^2 - 1)\varepsilon\sigma(T_c^4 - T_k^4), \tag{8}$$

where ε—the coefficient of radiation absorption by the metal, σ—the Stefan-Boltzmann constant, and T_h—the ambient temperature (air). The power lost by the surface of the element due to convective heat exchange with the surrounding air is represented in Eq. (2) by the last term equal to

$$P_k^e = \pi r_0^2 q^{2k}(q^2 - 1)\omega(T_c - T_k), \tag{9}$$

where ω—the metal-to-air heat transfer coefficient.

Substituting expressions (3–4), (6–9) into Eq. (5), describing energy balance of the sample element, we obtain after transformation the equation of temperature dynamics of the k-th element in the form.

$$\begin{aligned}\frac{dT_k}{dt} &= Q_k^{d(s)} + \kappa_k^{\mathrm{ad}(s)}(T_{k-1} - T_k) + \kappa_k^{bd(s)}(T_{k+1} - T_k) \\ &+ \beta_k^{d(s)}(T_c^4 - T_k^4) + \gamma_k^{d(s)}(T_c - T_k)\end{aligned} \tag{10}$$

For the values $k = 0, 1, \ldots N$, we obtain a system of $N + 1$ Eqs. (10) for the temperature dynamics of the sample elements, the coefficients of which are determined by the following formulas:

$$Q_k^d = \frac{P}{\pi \rho \mathrm{chr}_0^2}\left[w\delta_{k,0} + \frac{(1-w)\delta_{k,1}}{q^2-1}\right],$$
$$Q_k^s = \frac{3P}{2\pi \rho c r_0^3}\left[w\delta_{k,0} + \frac{(1-w)\delta_{k,1}}{q^3-1}\right] \tag{11}$$

$$\kappa_k^{ad} = \frac{2\chi(1-\delta_{k,0})}{\rho c r_0^2 q^{2k-2}(q^2-1)(q-1)},$$
$$\kappa_k^{as} = \frac{3\chi(1-\delta_{k,0})}{\rho c r_0^2 q^{2k-2}(q^3-1)(q-1)} \tag{12}$$

$$\kappa_k^{bd} = \frac{2\chi(1-\delta_{k,N})}{\rho c r_0^2 q^{2k-2}(q^2-1)(q-1)},$$
$$\kappa_k^{bs} = \frac{3\chi q(1-\delta_{k,N})}{\rho c r_0^2 q^{2k-2}(q^3-1)(q-1)} \tag{13}$$

$$\beta_k^d = \frac{\varepsilon\sigma}{\rho ch}\left[1 + \frac{2hq^2\delta_{k,N}}{R(q^2-1)}\right],$$
$$\beta_k^s = \frac{3\varepsilon\sigma}{2\rho c r_0}\left[\delta_{k,0} + \frac{(q^2-1)(1-\delta_{k,0})}{q^{k-1}(q^3-1)} + \frac{2\delta_{k,N}}{q^{N-3}(q^3-1)}\right] \tag{14}$$

$$\gamma_k^d = \frac{\omega}{\rho ch}\left[1 + \frac{2hq^2\delta_{k,N}}{R(q^2-1)}\right]$$
$$\gamma_k^s = \frac{3\omega}{2\rho c r_0}\left[\delta_{k,0} + \frac{(q^2-1)(1-\delta_{k,0})}{q^{k-1}(q^3-1)} + \frac{2\delta_{k,N}}{q^{N-3}(q^3-1)}\right] \tag{15}$$

The system of differential Eqs. (10) is a mathematical model of the dynamics of the temperature field of the samples in the form of a disk and in the form of a hemisphere. The dimensions of the sample and the field are taken into account by the coefficients (11–15) of the equations.

The power generated by the friction of the rotating pin against the shoulder depends on the condition of the sample material. As its temperature in the area of contact with the tool approaches the melting temperature, the metal softens, which leads to a sharp decrease in the released power P. In calculations based on the system of Eq. (10), this was taken into account by the dependence of power on temperature of the form:

$$P(T) = P_0\left[\frac{1}{2} - \frac{1}{\pi}\mathrm{arctg}[b(T - T_m)]\right], \tag{16}$$

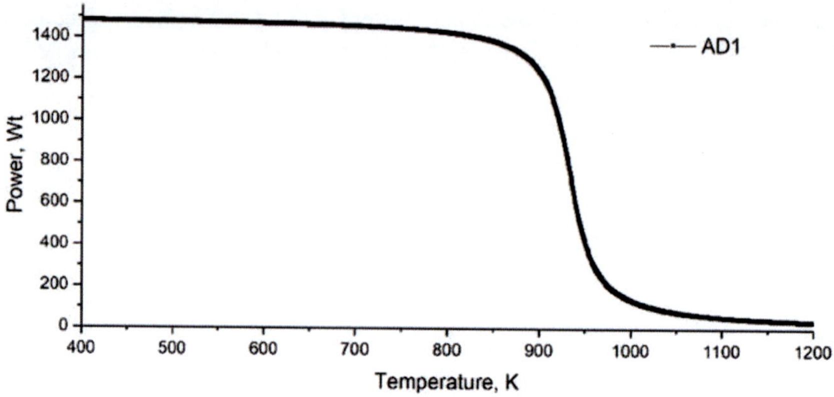

Fig. 3 Dependence of heating power at FSW on temperature

where P_0—the power released in the cold sample (taken equal to 1500 W), T_m—the melting temperature (for aluminum $T_m = 933.2$ K). The smoothing coefficient was taken equal to $b = 0.005$. The corresponding curve is shown in Fig. 3.

3 Comparison of Calculated and Experimental Curves of Temperature Dynamics

For comparison with model calculations, the temperature of aluminum alloy AD1 was measured when a tool with a pin of radius $r_0 = 2.5$ mm and a shoulder of radius 7 mm was moved over the metal surface for 15 s. The temperature was measured by the thermocouple method using a device that displays the data on a screen. The thermocouple was a chromel–alumel wire with a diameter of 0.25 mm. The experiment was repeated three times. Based on this, average temperature values were obtained. The thermocouple was attached to the surface of the welded metal by caulking at a distance of 3 and 5 mm from the heat input axis.

When comparing the calculated and experimental data (Figs. 4 and 5), it was found that the model is accurate (starting from the moment when the maximum temperature is reached at the point, the discrepancy is less than 10%). A slightly lower accuracy is achieved when describing the process on the ascending branch of heating, which is due to the complexity of specifying changes in the heat transfer of the transition layer when the heat source moves.

The curves in Figs. 4 and 5 show that a decrease in the thickness of the sample-disk, as expected, leads to an acceleration of heating in the region of the FSW, accompanied by an increase in the linear velocity of the FSW.

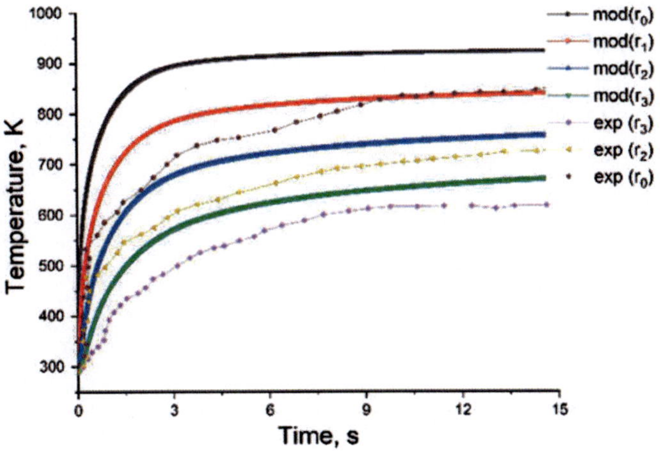

Fig. 4 Graphs of the heating process of a 3 mm thick disc made of AD1 aluminum alloy: mod(r_0) and mod(r_3)—terms simulated for elements 1 through 4; exp(r_0) and exp(r_3)—experimental terms for 1, 3 and 4 elements

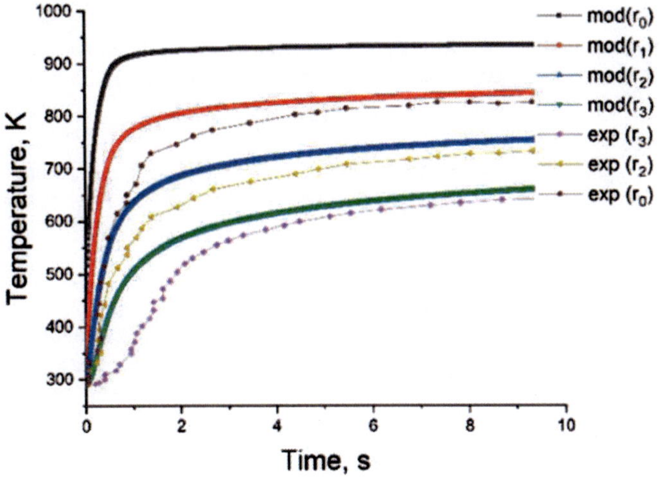

Fig. 5 Graphs of the heating process of a 2 mm thick disc made of AD1 aluminum alloy: mod(r_0) and mod(r_3)—terms simulated for elements 1 to 4; exp(r_0) and exp(r_3)—experimental terms for 1, 3 and 4 elements

The disk model does not allow an adequate description of the temperature dynamics of a massive sample. In this case, the hemisphere model is adequate. The calculated and experimental curves corresponding to this model are shown in Fig. 6.

Calculations for the hemisphere model were also performed at $P_0 = 1500$ W. They show that the temperature of the adjacent element reaches the temperature

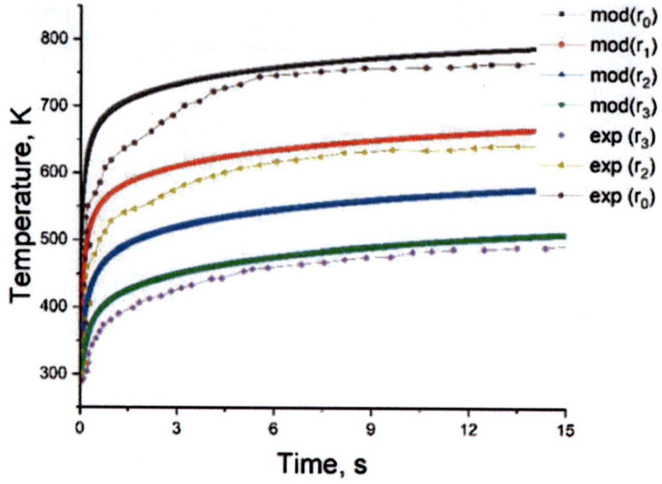

Fig. 6 Graphs of the heating process for a hemisphere made of aluminum alloy AD1: mod(r_0) and mod(r_3)—terms simulated for elements 1 to 4; exp(r_0) and exp(r_3)—experimental terms for 1, 3 and 4 elements

of the superplastic state (≅700 K) with a sufficiently long heating. Therefore, the linear velocity of the FSW is very low (about 0.18 mm/s). In Fig. 7 shows the

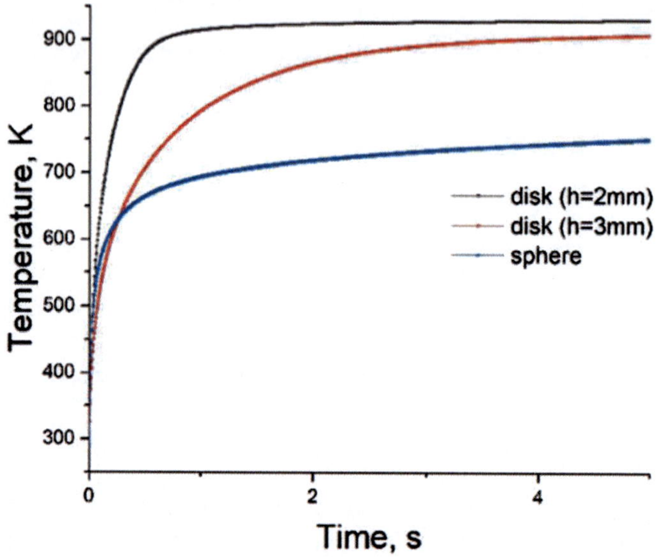

Fig. 7 Comparison of the curves of the temperature dynamics of disk elements 2 and 3 mm thick with the temperature dynamics of the hemisphere

calculated curves of the temperature dynamics of aluminum disks (2 and 3 mm) and a hemisphere.

4 Conclusions

A decrease in the thickness of the sample disk leads to an acceleration of heating in the area adjacent to the rotating tool. A sample 3 mm thick reaches the state of superplasticity within 3 s, and a sample 2 mm thick - within 1.5 s. The linear speed of the FSW decreases from 6.32 mm/s (h = 2 mm) to 1.85 mm/s (h = 3 mm).

Comparison of the calculated dynamics of the temperature of a massive hemisphere with thin disks shows that heating to a superplastic state of an element adjacent to the tool occurs almost equally quickly. The temperature of the elements of the hemisphere remote from the center increases significantly more slowly than in the samples-disks, which is due to the quadratic increase in the area of the elements. The design welding speed is 0.18 mm/s, which is significantly lower than for thin discs.

References

1. Goyal, W., Garg, R.: Modeling and optimization of friction stir welding parameters in joining 5086 H32 aluminium alloy. In: Goyal, W., Garg, R., Iranica, S. (eds.) Transactions B: Mechanical Engineering, vol. 26, pp. 2407–2417
2. Nosal, P., Ganczarski, A.: Modelling of the temperature field that accompanies friction stir welding. ITM Web of Conf. **15**, 03004 (2017)
3. Kumar, M., Khurana, M., Singh, G.: Modeling and optimization of friction stir welding process parameters for dissimilar aluminium alloys. Mater. Today: Proc. **5**, 25440–25449 (2018)
4. Maystrenko, A.L., Nesterenkov, V.M., Dutka, V.A., Lukas, V.A., Zabolotny, S.D., Tkach, V.N.: Modeling of thermal processes to improve the structure of metals and alloys by the method of friction with stirring. Autom. Weld. **1**, 5–14 (2015)
5. Meshram, A., Yarasu, R.: Finite element modeling of friction stir welding—thermal analysis. IJSRD **2**(03), 2321–2327 (2014)
6. Rzaev, R.A., Dzhalmukhambetov, A.U., Dergunova, O.Y., Semenova, L.E.: Mathematical modeling of changes in temperature distribution during friction stir welding in massive samples. In: IOP Conference Series: Materials Science and Engineering. Sevastopol: IOP Conf. Series, vol 971, pp. 032067. (2020)
7. Rzaev, D., Chularis, A., Valisheva, A.: Mathematical modeling of process of the friction stir welding. Mater. Today: Proc. **11**(P1), 3559–3567 (2019)
8. Rzaev, R.A., Chularis, A.A., Dzhalmukhambetov, A.U., Atuev, Sh.M.: Dynamic model of temperature distribution in metal during friction stir welding. Fundamental Res **3–1**, 47–55 (2016)
9. Sankar, B.R., Umamaheswarrao, P.: Modelling and optimisation of friction stir welding on AA6061 alloy. Mater. Today: Proc. **4**, 7448–7456 (2017)
10. Velji, D., Sedmak, A., Rakin, M., Baji, N., Medjo, B., Baji, D., Grabulov, V.: Experimental and numerical thermo—mechanical analysis of friction stir welding of high—strength alluminium alloy. Therm. Sci. **18**(1), 29–38 (2014)

Development and Mathematical Modeling of the AC Sensor for Refinery Automation Systems

Dmitry Fugarov

Abstract Today the current problem is the control of parameters of power equipment within the automation systems of oil refineries, such as the current one. This problem is quite relevant, since the operation of power equipment at oil refineries implies a form of network currents significantly different from sinusoidal thus impacting the quality of the network parameters, and this, in turn, has significant effect on the results of current measurements in power circuits of process equipment within the automation systems. This work is devoted to modeling and study of the proposed magnetodielectric measuring transducer of alternating current (MDT) based on modern materials. The purpose of the study covers theoretical analysis and mathematical modeling of alternating current measuring transducers based on magnetodielectric cores for automation systems in oil refineries.

1 Introduction

The distinctive feature of such transducers is that they have the sufficient range and linearity throughout the entire measurement scale [1]. This ensures the creation of intellectual systems of any complexity on their basis [2]. Another absolute advantage is their minimum weight-size parameters. The main problem here is high technological efficiency of MDT production. This is caused by the fact that the MDT production shall ensure full stability of all magnetic characteristics of cores in time at major temperature change [3].

MDT is based on magnetodielectric cores manufactured from powder carbonyl iron via compaction. This process takes place under high pressure. Cores are made from the mixture of a dielectric filler and iron particles. This material has some disadvantages, including its short lifetime (2 years) [4].

D. Fugarov (✉)
Don State Technical University, Ploshchad Gagarina, 1, 344002 Rostov-on-Don, Russia
e-mail: ddf_1@mail.ru

M. Shamtsyan et al. (eds.), *Robotics, Machinery and Engineering Technology for Precision Agriculture*, Smart Innovation, Systems and Technologies 247,
https://doi.org/10.1007/978-981-16-3844-2_28

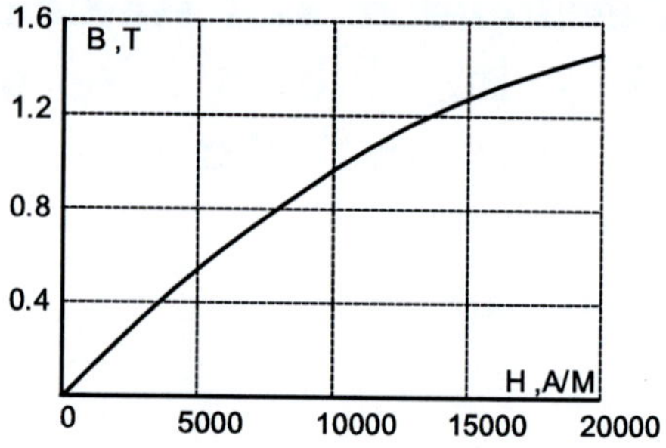

Fig. 1 Normal magnetization curve of magnetics

The alternative could be the use of modern materials—Iron Powder of Magnetics (USA). The use of these materials will allow reducing the overall dimensions of measuring transducers [5].

Figure 1 shows the normal magnetization curve (NMC) for Magnetics material.

Further task of the study covered analytical expressions approximating the normal magnetization curve with the minimum error [6].

2 Modeling of the Normal Magnetization Curve of AC Magnetodielectric Measuring Transducer

The following analytical expression was applied to approximate the normal magnetization curve [7]:

$$B = A\ln(\alpha H + 1), \tag{1}$$

where B and H—magnetization curve coordinates;

A and α—approximation coefficients.

The more precisely the approximating function and the experimental curve coincide, the more precisely be further calculations of the MDT output voltage. To ensure the minimum approximation error, it is necessary to impose additional restrictions on approximated expressions:

$$\lim_{H \to 0} B(H) = 0, \tag{2}$$

$$\lim_{H \to 0} \frac{dB}{dH} = \mu_H, \tag{3}$$

where $\mu_{н}$—initial magnetic permeability of NMC.

From the analytical expression (1) it is possible to receive the following expression:

$$\frac{dB}{dH} = \frac{A\alpha}{\alpha H + 1}, \tag{4}$$

At $H \to 0$ we get:

$$\lim_{H\to 0} \frac{dB}{dH} = A\alpha \tag{5}$$

From the expressions (3), (5) it is possible to receive the following expression:

$$A = \frac{\mu_H}{\alpha} \tag{6}$$

If to substitute the expression (6) into the expression (1), we get:

$$B = \frac{\mu_H}{\alpha} \ln(\alpha H + 1) \tag{7}$$

In case of high intensity (H > 300,000 A/m), the NMC will take a straight form, therefore we can write as follows:

$$B = B_s + \mu_0 H \tag{8}$$

where B_s—ordinate received by continuation of the saturated site of the magnetization curve;
μ_0—magnetic constant.

$$\lim_{H\to \infty} \frac{dB}{dH} = 0 \tag{9}$$

Let us correct the expression (9) and to satisfy the condition:

$$\lim_{H\to \infty} \frac{dB}{dH} = \mu_0 \tag{10}$$

Let us write the expression for the normal magnetization curve as follows:

$$B = \frac{\mu_H - \mu_0}{\alpha} ln(\alpha H + 1) + \mu_0 H \tag{11}$$

The Formula (11) allows satisfying the main NMC requirements, namely:

$$\lim_{H\to 0} B(H) = 0;\ \lim_{H\to 0} \frac{dB}{dH} = \mu_H;\ \lim_{H\to \infty} \frac{dB}{dH} = \mu_0.$$

In the received Formula (11) the approximation coefficient α, which shall be defined, remains unknown. The least-squares method and NMC coordinates received experimentally can be used for this purpose. Let us calculate the absolute error in this point using the following expression:

$$\varepsilon_i(\alpha) = \left| \frac{\mu_H - \mu_0}{\alpha} ln(\alpha H_i + 1) + \mu_0 H_i - B_i \right| \tag{12}$$

where H_i, B_i—NMC coordinates.
The squared error will be equal to:

$$\varepsilon_i^2(\alpha) = \left(\frac{\mu_H - \mu_0}{\alpha} ln(\alpha H_i + 1) + \mu_0 H_i - B_i \right)^2 \tag{13}$$

To consider all NMC points received experimentally, we can make the aggregate squared error:

$$F(\alpha) = \sum_{i=1}^{n} \varepsilon_i^2 \text{ or } F(\alpha) = \sum_{i=1}^{n} \left(\frac{\mu_H - \mu_0}{\alpha} ln(\alpha H_i + 1) + \mu_0 H_i - B_i \right)^2 \tag{14}$$

where n—number of experimentally received points.
Minimizing the expression (14) we find parameter α, and differentiate this expression according to α:

$$\frac{\mathrm{d}F}{\mathrm{d}\alpha}(\alpha) = 2 \sum_{i=1}^{n} \left[\left(\frac{\mu_H - \mu_0}{\alpha} ln(\alpha H_i + 1) + \mu_0 H_i - B_i \right) \times \left(\frac{\mu_H - \mu_0}{-\alpha^2} ln(\alpha H_i + 1) + \frac{\mu_H - \mu_0}{\alpha} \frac{H_i}{\alpha H_i + 1} \right) \right] \tag{15}$$

Minimization $F(\alpha)$ requires the satisfaction of the necessary condition:

$$\frac{\mathrm{d}F}{\mathrm{d}\alpha}(\alpha) = 0 \tag{16}$$

The substitution of (15) into (16) results in the need to solve the following transcendental equation:

$$\sum_{i=1}^{n} \left[\left(\frac{\mu_H - \mu_0}{\alpha} ln(\alpha H_i + 1) + \mu_0 H_i - B_i \right) \times \left(\frac{\mu_H - \mu_0}{-\alpha^2} ln(\alpha H_i + 1) + \frac{\mu_H - \mu_0}{\alpha} \frac{H_i}{\alpha H_i + 1} \right) \right] = 0 \tag{17}$$

The solution of the Eq. (17) by the bisection method at $\mu_H = 1,22 \cdot 10^{-5} \frac{Gn}{M}$ leads to the following result: $\alpha = 1,13 \cdot 10^{-5} m/A$.

3 Modeling of Eclectic Equivalent Circuit of AC Magnetodielectric Measuring Transducer

Let us consider the equivalent circuit for magnetodielectric transformer presented in Fig. 2 [8].

The mathematical model represents the system of equations describing the electromagnetic processes of the transducer:

$$i_1 = i_0 + i_2 \tag{18}$$

$$\frac{d\psi}{dt} = (L_{2S} + L_H)\frac{di_2}{dt} + (r_2 + r_H)i_2 \tag{19}$$

$$\psi = f(i_0)$$

where i_0—magnetization current; i_1—primary current; ψ—flux linkage; L_H—load inductance; L_{2S}—leakage inductance; i_2—secondary current; r_2—secondary resistance; r_H—load.

The output value of a transducer will be its information signal—u_2, while the input signal will represent the measured current i_1.

The system of equations describing electromagnetic processes of the transducer (18–20) can be written as follows:

$$\psi = f(i_0) \tag{20}$$

$$i_1 = i_0 + \frac{u_2}{r_H} \tag{21}$$

$$\psi = f(i_0) \tag{22}$$

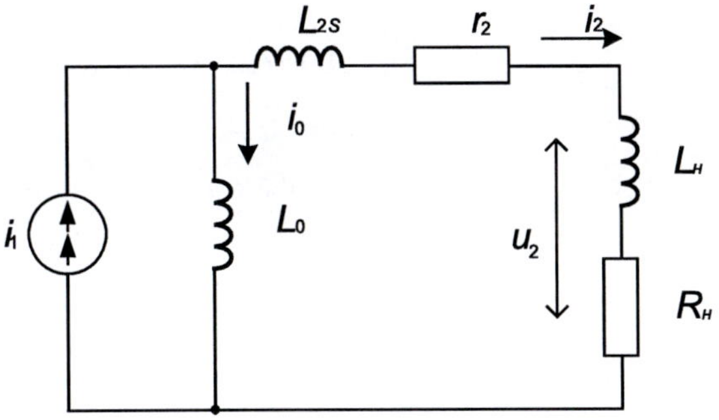

Fig. 2 Electric equivalent circuit of AC magnetodielectric measuring transducer

$$\frac{dU_2}{dt} + \left(1 + \frac{r_2}{r_H}\right)u_2 \tag{23}$$

The magnetodielectric core of the transducer shall be made toroidal, then taking into account the uniformity, L_2S may be neglected. Considering that $r_H \to \infty$, the external fields may be neglected, therefore i1 = i0. Hence, the system of Eqs. (21–23) will be as follows:

$$\frac{d\psi}{dt} = u_2 \tag{24}$$

$$\psi = f(i_1) \tag{25}$$

First, let us define the flux linkage:

$$\psi(t) = F(t)w_2, \tag{26}$$

where *F(t)*—magnetic flux.

$$F(t) = B(t)S. \tag{27}$$

Earlier [9] expressions and parameters for the core magnetization curve were defined. Proceeding from these expressions let us define the induction.

$$B = \frac{\mu_H - \mu_0}{\alpha} ln(\alpha H + 1) + \mu_0 H. \tag{28}$$

where H—voltage; α—approximation coefficient.
The expression for flux linkage will be as follows:

$$\psi(t) = w_2 S\left(\frac{\mu_H - \mu_0}{\alpha} ln(\alpha H + 1) + \mu_0 H\right). \tag{29}$$

Using the Ampere's law we get the following:

$$\oint_{l_0} H d\bar{l} = i_1, \; H l_0 = i_1, \tag{30}$$

where l_0—average length of the magnetic line, hence:

$$H = \frac{i_1}{l_0}. \tag{31}$$

Taking this into account the expression for flux linkage can be as follows:

$$\psi(t) = w_2 S\left(\frac{\mu_H - \mu_0}{\alpha} ln\left(\alpha\frac{i_1}{l_0} + 1\right) + \frac{\mu_0 i_1}{l_0}\right). \quad (32)$$

If to substitute this expression in the Eq. (24), we will get:

$$u_2 = w_2 S\left(\frac{\mu_H - \mu_0}{\alpha} \cdot \frac{\frac{\alpha}{l_0}}{\alpha\frac{i_1}{l_0} + 1} \cdot \frac{di_1}{dt} + \frac{\mu_0}{l_0} \cdot \frac{di_1}{dt}\right). \quad (33)$$

Let us assume that the primary current changes under the law:

$$i_1 = I_{1m}\left(\cos\varphi \cdot e^{\frac{-t}{T_1}} - \cos(wt + \varphi)\right), \quad (34)$$

where ω—angular frequency; T_1—time constant; I_{1m}—amplitude of primary current; φ—initial phase.

If to differentiate this law of changing the primary current over time we get the following:

$$\frac{di_1}{dt} = I_{1m}\left(\omega \sin(\omega t + \varphi) - \cos\varphi \cdot \frac{1}{T_1} \cdot e^{-\frac{t}{T_1}}\right). \quad (35)$$

Having substituted these expressions in the Eq. (33) we get:

$$u_2 = w_2 S I_{1m}\left(\omega \sin(\omega t + \varphi) - \cos\varphi \cdot \frac{1}{T_1} \cdot \mathrm{e}^{-\frac{t}{T_1}}\right) \times \left(\frac{\mu_H - \mu_0}{\alpha I_{1m}\left(\cos\varphi \cdot e^{\frac{-t}{T_1}} - \cos(wt + \varphi)\right) + l_0} + \frac{\mu_0}{l_0}\right). \quad (36)$$

4 Quantitative Assessment of the Proposed Current Transducer

Modeling via standard application program packages was conducted for quantitative assessment of the measurement error of the AC magnetodielectric transducer [10].

The following parameters of the magnetodielectric current transducer were used as a model:

- measured current: 12 kA;
- measured current frequency: fp = 50 Hz;
- form and size of the measuring transducer: preliminary calculation showed that the transducer shall have a toroid form and the following dimensions: D = 45 × 10–3 m; d = 38 × 10–3 m; h = 45 × 10–3 m;

- magneto-field vector H created by primary MDT winding:

$$H = \frac{\sqrt{2} \cdot I_1 \cdot w1}{l_{avr}}, \tag{37}$$

where I_1—test current (A); l_{cp}—length of the average line of a magnetic circuit (m), $w1$—primary winding turns;

$H = 4.041 \cdot 10^5$ A/m.

—quantity of secondary winding turns:

$$w2 = \frac{U_2 \cdot l_{avr}}{2 \cdot \pi \cdot S \cdot f_p \cdot I_1 \mu \cdot \mu_0}. \tag{38}$$

where U_2—secondary voltage (B); S—cross-section of a magnetic circuit (m^2); f_p—operating frequency (Hz); I_1–test current (A); I_1—relative magnetic permeability of a magnetic circuit; μ_0—magnetic constant.

Thus, the quantity of winding turns calculated for current $I = 12\ kA$:

$$w2 = 147, 8.$$

Hence, we can round to $w2 = 148$ turns.

Figure 3 shows flow diagrams illustrating different curve amplitudes of current and the corresponding output voltage of the AC magnetodielectric measuring transducer obtained through modeling. In this case the current values are given in thousands of amperes (kA), while the voltage—in volts (V), time is set in milliseconds.

Modeling of the AC magnetodielectric measuring transducer conducted for the case when the alternating current equals 12 kA [11].

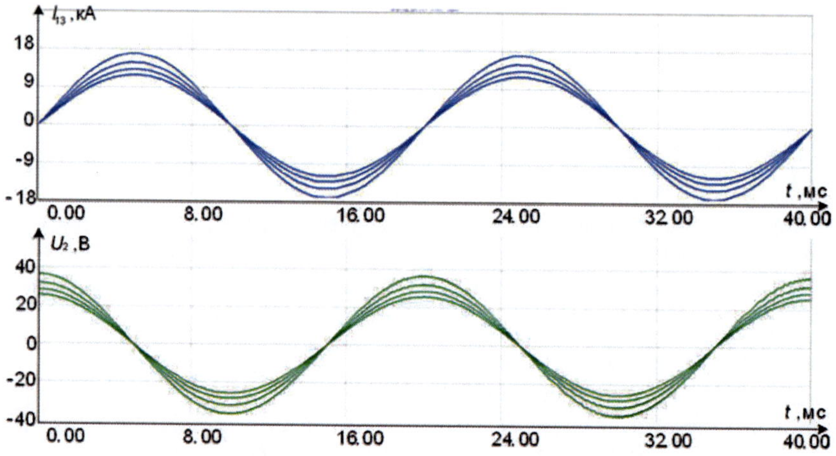

Fig. 3 MDT flow diagrams at alternating current of 12 kA

The key technical indicator for any current transducer is its measurement error [12]. The quantitative assessment of the measurement error of the AC magnetodielectric transducer can be carried out only when the dependence of secondary voltage on the measured current $u_2(i_1)$ is built. For this purpose, it is sufficient to take one half-cycle of the measured current, integrate the secondary voltage of the AC magnetodielectric measuring transducer, and shift the initial phase of the measured current by $\pi/2$ [13]. Figure 4 shows the analyzed current curve, and Fig. 5—the dependence $u_2(i_1)$.

Analyzing Fig. 6 it is possible to conclude that the obtained dependence $u_2(i_1)$ is almost linear. However, to prove this statement it is critical to carry out numerical evaluation of the conversion ratio error of the AC magnetodielectric measuring

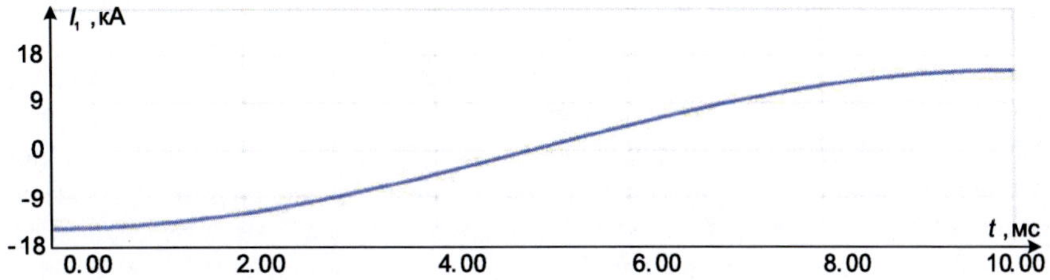

Fig. 4 Analyzed current curve

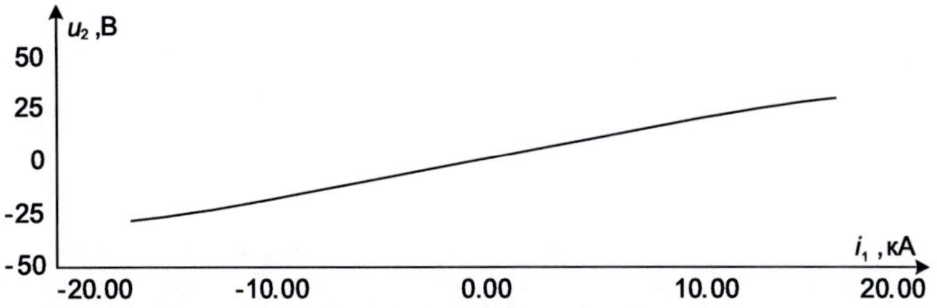

Fig. 5 Dependence of secondary voltage on the measured current $u_2(i_1)$

Fig. 6 Dependence of the fractional error of a conversion ratio on the measured current

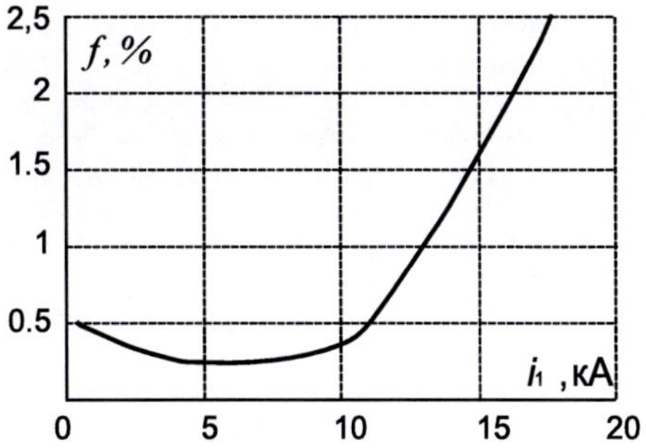

transducer [14]. For this purpose, we need to approximate the dependence $u_2(i_1)$ in Fig. 6 with the function $y = kx + b$, obviously having a linear nature.

The standard functions *slope* and *intercept* of *MathCAD* were used for numerical expression of coefficients k and b [15]. Then, the dependence $u_2(i_1)$ and ideal dependence $y(x)$ were compared with each other [16]. Figure 6 shows the dependence reflecting the quantitative assessment of the fractional error of a conversion ratio.

Analyzing the received dependence it is possible to conclude that the measurement error of the AC magnetodielectric transducer makes not more than 1%, provided the measured current does not exceed 12 kA.

5 Conclusions

The study proposes the magnetodielectric measuring transducer of the alternating current. It also describes the mathematical model of the normal magnetization curve of the AC magnetodielectric measuring transducer, which can be used for engineering and scientific analysis of transition processes within automated measuring electrical circuits of oil refineries [17]. Besides, it refers to the model allowing calculating the secondary voltage of the magnetodielectric current transducer [18]. Such expressions can be used in the design of measuring parts of the automation systems in oil refineries. Moreover, the study shows that the conversion ratio error within the entire range of the measured currents does not exceed 1%.

References

1. Tsygulev, N.I., et al.: Physical modeling of testing currents for diagnosing hidden failures in automatic switches and relay protection (in Russian). Bullet. Higher Educ. Instit. Electromech. **1**, 104–107 (2013)
2. Kuznetsov, A.P.: Introduction of new current measurement devices up to 20 kA used checking the electrical characteristics of the equipment, Moscow, SPO ORGRES, **8** (1998)
3. Onyshko, D. A., et al.: Synchronization system in wireless sensor networks of oil and gas complex. In: E3S Web of Conferences, vol. 164, pp. 03030 (2020)
4. Ganzhur, M., et al.: In: E3S Web of Conferences, vol. 175, pp. 05038. (2020)
5. Solomentsev, K.Y. et al.: Interference elimination in digital controllers of automation systems of oil and gas complex. J. Phys. Conf. Ser. **1015**, 032179 (2018)
6. Agibalov, O.: In: E3S Web of Conferences, vol 224, pp. 01008 (2020)
7. Chernyshev, Y.O.: Swarm-intelligence-based algorithm of connections permutation between pins. J. Theor. Appl. Inf. Technol. **80**(1), 466–473 (2015)
8. Gazizov, A., et al.: In: E3S Web of Conferences,vol 210, pp. 11001. (2020)
9. Fugarov, D.D.: Approximation of the saturation curve of powder materials of current sensor magnetic conductors in the device for diagnosing switching elements of electric installation (in Russian). Bulletin of Higher Educational Institutions. North Caucasus region. – Series: Natural science, vol. 2, pp. 52. (2013)
10. Sukhinov, A.I., et al.: Math. Models Comput. Simul. **12**(2), 232–245 (2020)
11. Fugarov, D. D., Gerasimenko, Y. Y.: Methods for revealing hidden failures of automation system for technological processes in oil and gas sector. J. Phys. Conf. Ser, **1118**, 012055 (2018)

12. Poluyan, A. Y.: Adaptive algorithm of selecting optimal variant of errors detection system for digital means of automation facility of oil and gas complex. J. Phys. Conf. Ser. **1015**, 022013 (2018)
13. Chernyshev, Y.O.: Swarm-intelligence-based algorithm of connections permutation between pins. J. Theor. Appl. Inf. Technol. **80**(1), 13–20 (2015)
14. Gerasimenko, Y., et al: Mathematical modeling and synthesis of an electrical equivalent circuit of an electrochemical device. In: Advances in Intelligent Systems and Computing, vol 1259 AISC, pp. 471–480 (2021)
15. Poluyan, A. Yu., et al.: Application of bionic and immune algorithms for the solution of ambiguous problems of transportation routing. J. Phys. Conf. Ser. **1333**, 032057 (2019)
16. Poluyan, A. Yu., et al.: Solution of task on the minimum cost data flow based on bionic algorithm. J. Phys. Conf. Ser, **1333**, 032056 (2019)
17. Isolated current and voltage transducers. Characteristics-Application-Calculation CH24101 E/US 05.04
18. Vencov, N.N. et al.: In: International Conference on Industrial Design, Application and Production, ICIEAM, vol 8076476 (2017)

Discrete-Event Intersection Operation Model (Yasnyy Proyezd-Dezhnev street, Moscow)

Dmitry Kuzmin and Vera Baginova

Abstract The paper discusses the issue of optimizing the work of a traffic light that regulates the movement of vehicles at an intersection. For this, a discrete-event model of vehicle traffic has been developed. The architecture and content of the model, its interface and graphical parts are described. To test the work of the model, a real intersection of Yasnyy proezd—Dezhneva street in Moscow is used. In order to improve the operation of the intersection, an optimization experiment was carried out using the model. As an objective function during optimization, we used the minimum average time for cars to pass through the intersection. The obtained results of the conducted research allow drawing the following conclusions: a direct increase in traffic intensity several times (+200%) does not have a significant effect on the efficiency of the intersection operation (average time for cars to pass through the intersection, average number of cars at the intersection); the minimum average time for cars to pass through the intersection is 54.66 s. It is achieved with the following parameter values (traffic light phase operating times): *TL1* = 50 s and *TL2* = 20 s; at the moment, the work of the traffic light provides a high efficiency of vehicle passage. However, a slight adjustment of the phase operation time will slightly reduce (~2 s) the time for cars to pass through the intersection. The following tools were used: AnyLogic software simulation environment, Java object-oriented programming language, OptQuest optimizer, Google Maps mapping service.

1 Introduction

The efficient operation of transport infrastructure facilities depends on many factors: season, weather conditions, traffic intensity, vehicle characteristics, etc. For large urban agglomerations around the world, the issue of effective organization of transport services for the population of megalopolises is urgent.

D. Kuzmin (✉) · V. Baginova
Russian University of Transport (MIIT), 127994, GSP-4, Obraztsova str., 9, bldg. 9, Moscow, Russia

M. Shamtsyan et al. (eds.), *Robotics, Machinery and Engineering Technology for Precision Agriculture*, Smart Innovation, Systems and Technologies 247,
https://doi.org/10.1007/978-981-16-3844-2_29

The most acute issues are the organization of road traffic, reducing the number and duration of traffic jams, increasing the throughput of highways, increasing the comfort of using a personal car and urban land transport.

In urban settings, most road junctions are single-level and require regulation. One of the most effective tools for regulating traffic in the city is the traffic light, the efficient operation of which determines the quality of the traffic situation at the intersection. This is why this study is relevant.

Research object: Infrastructure of urban land transport.

Subject of research: improving the process of passing cars at the intersection of Yasnyy proezd—Dezhneva street, Moscow.

Purpose of the study: development of a model for optimizing traffic lights at the intersection of Yasnyy proezd—Dezhneva street, Moscow.

Equipment and equipment used in the study: Simulation environment, Java object-oriented programming language, Microsoft Office package.

2 Object Description

Intersection of Yasnyy proezd—Dezhneva street in the Yuzhnoye Medved kovo district of Moscow is the T-shaped intersection and is formed by the intersection of two streets.

- Dezhneva street. According to the Russian State Standard GOST R 52398–2005 "Classification of motor roads. Basic parameters and requirements" 3, Dezhneva street can be classified as a normal type road (not a freeway) of IB category. The total number of lanes is more than 4. The width of the lanes is 3.75 m. There is a central dividing strip. Intersection with automobile, pedestrian and bicycle roads at the same level with traffic light regulation is allowed;
- Yasnyy proezd. May be classified as a normal type road (not a freeway) of the II category. The total number of lanes is 2, the width of the traffic lane is 3.75 m. There are parking lay-bys. The road provides connection with the inner part of the residential area: there are a large number of pedestrian crossings, residential buildings and social institutions, public transport stops, etc. These factors form intensive pedestrian traffic, which negatively affects the speed of vehicles.

According to GOST, the intersection of these roads and pedestrian crossings is allowed at the same level, but requires mandatory traffic light regulation. Figure 1 shows an aerial photograph of the intersection in Google Maps [5].

The carriageway of Dezhneva street is a two-way motor road (three lanes in both directions, 3.75 m wide each). Dezhneva street provides connection between the large Moscow districts: Otradnoye (185 thousand people), Yuzhnoye Medvedkovo (85 thousand people) and Babushkinsky district (88 thousand people). Quarters No. 1, 2, 3, 17 and 18 of the Yuzhnoye Medvedkovo district adjoin the road directly at the location of the intersection. The permitted speed is 60 km/h.

Fig. 1 Aerial view of the intersection

Yasny proezd is a secondary two-way road (two lanes in each direction). Adjacent to the yards of residential buildings of quarters No. 1, 2, 3 and 18 of the Yuzhnoye Medvedkovo district. The permissible travel speed varies from 20 to 60 km/h. Has places intended for parking cars, both along the road and using special parking lay-bys.

2.1 Characteristics of the Actual Operation of the Intersection

Field studies were carried out to investigate the current operation of the intersection. Date: 21 February 2020. Time: 14:00. During the field study, the general situation of the intersection operation was assessed, traffic intensity and the real time of the operation of the intersection sections were measured. The results of the study are shown in Table 1.

Under the given conditions, there are on average 26–28 cars at the intersection, and the average travel time is 45–50 s.

The intersection operation can be monitored remotely using various online cartographic services (Yandex Maps, Google Maps, etc.). The carried out monitoring allows concluding that the traffic situation at the intersection depends on many factors, among which the most significant are: time of day, day of the week, season, weather conditions and the presence of road accidents. At the same time, the number of cars and the average travel time at the intersection can differ tenfold. The most

Table 1 Intersection performance characteristics

Direction	Traffic intensity, cars/h	Traffic light performance	Traffic routing
East–West Dezhneva street	2640	The duration of the "green phase" is 40 s in the main direction of Dezhneva street and 15 s to turn onto Yasnyy proezd Pedestrian crossing—20 s	80%—Dezhneva str.; 20%—Yasnyy proezd
West–East Dezhneva street	840	The duration of the "green phase" is 15 s in the main direction of Dezhneva street and 25 s to turn onto Yasnyy proezd Pedestrian crossing—20 s	55%—Dezhneva str.; 45%—Yasnyy proezd
North–South Yasnyy proezd	840	The duration of the green phase in both directions is 15 s Pedestrian crossing—20 s	45%—Dezhneva str. West; 45%—Dezhneva str. East

contrasting is the comparison of traffic at night (0 points by Yandex Maps/<300 cars/h) and on a weekday evening (7–9 points by Yandex Maps/>5000 cars/h).

The impact of road traffic accidents should be noted separately. It is also necessary to consider the complex characteristics of the object: the conflict of flows, the many working sections of the traffic light, the presence of pedestrian crossings, etc. The above factors increase the risks of an accident, which leads to various negative consequences, up to a complete stop of traffic.

3 Discrete-Event Model of the Intersection

To study the operation of the intersection in the Anylogic environment, a software-simulation discrete-event model was developed. The architecture of the model and the layout of the space are shown in figures. The model is a set of blocks that simulate various operations (events) occurring with agents (cars). These events include: appearance of cars in the model (CarSource block), routing (selectOutout, carMoveTo), and removal of agents from the model (carDispose block).

The architecture of the model is determined by two factors: the use of the process-event concept of modeling and the logic of the studied object. Accordingly, within the framework of a unified simulation model, the movement of traffic flows from each direction is actually simulated, and the interaction of these flows determines the overall behavior of the system. Thus, the architecture of the simulation model is a combination of three micromodels that simulate the movement of car (agents) from different directions. A detailed description of the model blocks is presented in Table 2 (Figs. 2 and 3).

The model interface provides the ability to track the number of cars at the intersection at a specific time period, the average time of passing through the intersection and traffic density.

The number of agents at the intersection is implemented using a text element displaying the dynamic value of the expression carSource.count()—carDispose.count(). To track traffic density, an element of the roadNetworkDescriptor library was used, which allows graphically displaying traffic density.

A number of simple experiments were carried out using the model to study the behavior of a modeled transport infrastructure object with different traffic intensities. The experimental results are presented in Table 3.

The results of the experiments are shown in Figures 4, 5, 6 and 7.

Based on the obtained results, it can be concluded that a direct increase in traffic intensity several times **does not significantly affect the efficiency of the intersection.** The complete results of the above series of experiments are given in the **appendix**.

3.1 Optimization of the Intersection Operation

In order to improve the operation of the intersection with the model, an optimization experiment was carried out. AnyLogic has a built-in OptQuest optimizer, which automatically finds the best values for the model parameters, taking into account the specified constraints. Optimization consists of several consecutive runs of the model with different values of the parameters. OptQuest allows finding the values of the model parameters corresponding to the maximum or minimum of the objective function, both under uncertainty and in the presence of constraints [1, 2].

In the model, the values of the time of the phases of the traffic light allowing the movement of vehicles are stored by parameters *TL1, TL2*, the variation of which is possible. Parameter *TL3* stores the time of the phase of the pedestrian section, which initiates the prohibition of movement for all directions during the pedestrian crossing. Therefore, during the optimization experiment, this parameter will not change (Fig. 8).

As an objective function for the optimization, the **minimum average travel time for vehicles at the intersection** is used. This characteristic is stored in the model by the *VremyaProezda* block, which collects data to build the corresponding histogram.

The conditions for the optimization experiment are shown in Table 4.

The duration of the simulated period of operation of the intersection is 1 h.

Table 2 Description of model elements

Block name	Functions	Characteristic
CarSource	Formation of the flow of agents in the direction East–West Dezhneva street	The appearance of agents according to the intensity: 2640 cars/h Place of appearance: the eastern part of Dezhneva street
selectOutput	Branching agent flows	Branching is determined by a given probability: 80%—Dezhneva street; 20%—Yasnyy proezd
CarMoveTo	Agent flow routing. Turn to Dezhneva street	Purpose of movement: the end of the current traffic lane of the considered polygon of Dezhneva street
CarMoveTo2	Agent flow routing. Turn to Yasnyy proezd	Purpose of movement: the end of the current traffic lane of the considered polygon of Yasnyy proezd
carDispose	Removing agents from a model	
cars	Population of agents. Models a collection of a given number of agents	When destroying the agents, it accesses the root class of the main model and informs the VremyaProezda element of the difference between the current model time and the agent's appearance time main.VremyaProezda.add (time () - VremyaPoyavleniya)
VremyaProezda	Histogram data. Performs statistical analysis on added values	Analyzed expression (time () - VremyaPoyavleniya)
VremyaPoyavleniya	Parameter. Data storage	Fixes the time when the agent appears in the model The stored data type—double The function used - time ()
trafficLight	Traffic light. Regulates traffic at an intersection	The operating time of the sections is determined by the parameters TL1, TL2, TL3
TL1, TL2, TL3	Parameter. Data storage	The values of the time of operation of the traffic light phases are stored. TL1 is green, TL2 is red, TL3 is yellow Initial value: TL1—45 s, TL2—20 s, TL3—20 s The stored data type—double
Simulation	Optimization experiment	The objective function is the minimum average time of passing through the intersection by cars Objective function expression—root.VremyaProezda.mean ()

Fig. 2 Layout of the simulated space in the AnyLogic environment

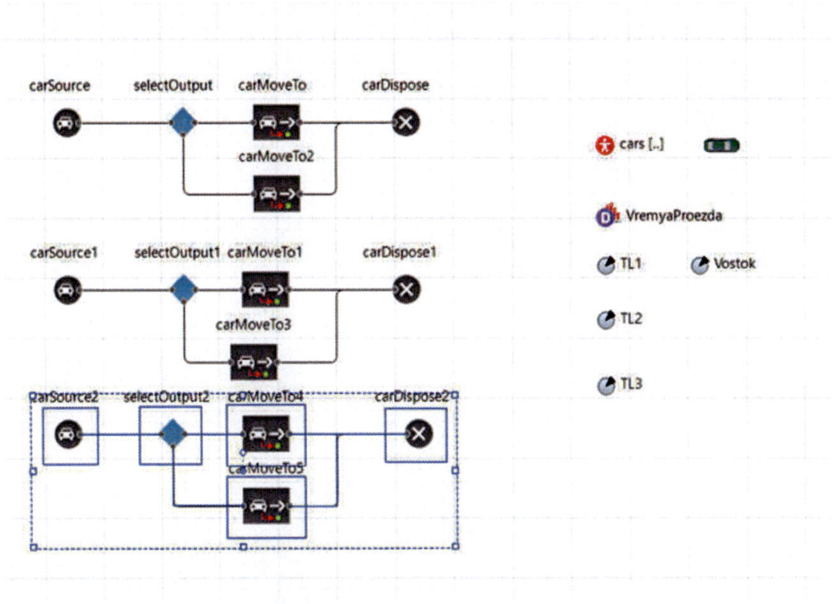

Fig. 3 Architecture of the discrete-event simulation model

Table 3 Experiment log

Direction	Car traffic intensity								
	50%***			100%***			200%***		
	Traf. Int	Traffic*	Time**	Traf. Int	Traffic*	Time**	Traf. Int	Traffic*	Time**
East–West Dezhneva street	3960	23.8	61.2	5280	25.3	63	7920	26.1	63
West–East Dezhneva street	1260			1680			2520		
North–South Yasny proezd	1260			1680			2520		

*average number of cars at the intersection
**average travel time of cars at the intersection, sec
***an increase in the intensity of car traffic from the initial indicators

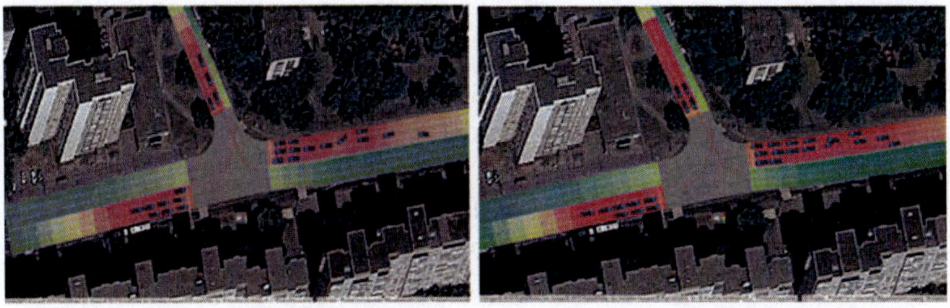

Fig. 4 Map of traffic density (actual traffic and with an increase in intensity by 200%)

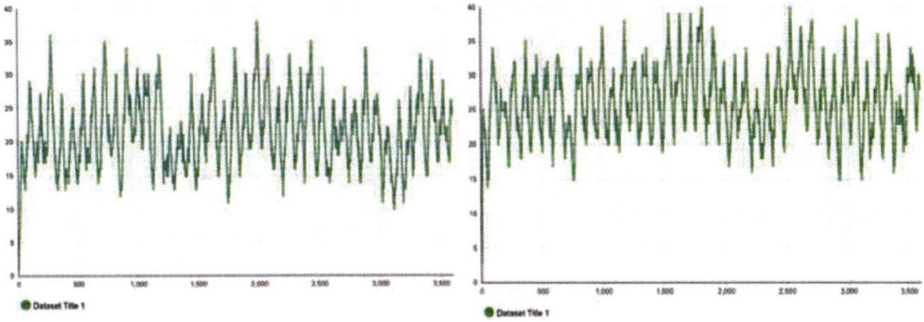

Fig. 5 Actual number of cars at the intersection (actual traffic and with an increase in intensity by 200%)

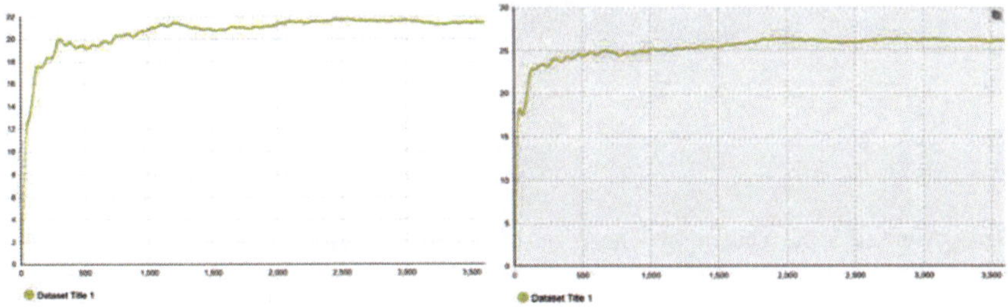

Fig. 6 Average number of cars at the intersection (actual traffic and with an increase in intensity by 200%)

4 Results of the Optimization Experiment

During the experiment, the optimizer carried out 500 iterations (repeated data processing), according to the results of which an iteration was chosen with a combination of parameters *TL1*, *TL2* providing the minimum value of the objective function (minimum average travel time for cars at the intersection).

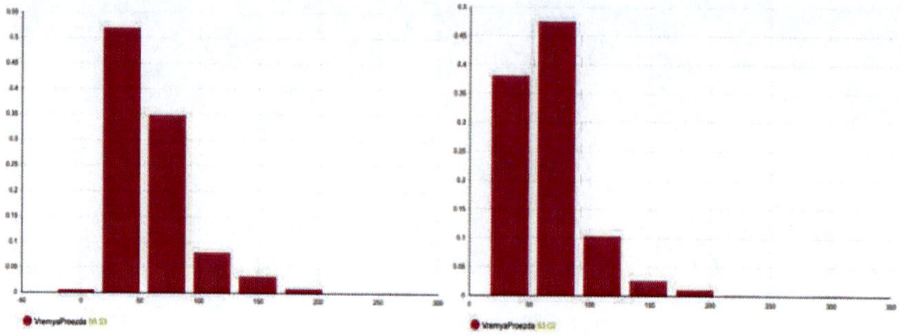

Fig. 7 Distribution of time to pass through the intersection (actual traffic and with an increase in intensity by 200%)

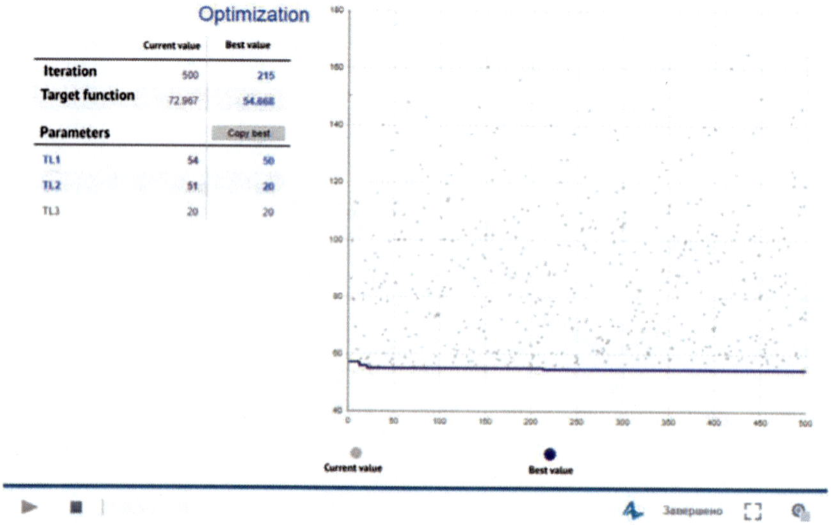

Fig. 8 Interface of the window with the results of the optimization experiment

Table 4 Optimization experiment conditions

Parameter	Type	Values, s			
		min	max	Step	Actual
TL1	discrete	20	60	1	45
TL2	discrete	20	100	1	20
TL3	fixed	-	-	-	20

According to the results of the study, the minimum average travel time for cars at an intersection is 54.66 s. It is achieved with the following parameter values (traffic light phase operating times): *TL1* = 50 s and *TL2* = 20 s.

5 Conclusions

The obtained results of the conducted research allow drawing the following conclusions:

- a direct increase in traffic intensity by several times (+200%) does not have a significant effect on the efficiency of the intersection operation (average travel time for cars at the intersection, average number of cars at the intersection);
- the minimum average travel time for cars at the intersection is 54.66 s. It is achieved with the following parameter values (traffic light phase operating times): *TL1* = 50 s and *TL2* = 20 s;
- at the moment, the work of the traffic light provides a high efficiency of the passage of vehicles. However, a slight adjustment of the time of operation of the phases will slightly reduce (~2 s) the time of passage of cars at the intersection.

References

1. The world's leading simulation optimization engine. OptQuest https://www.opttek.com/products/optquest/. Last accessed 2020/11/21
2. Optimization experiment. https://help.anylogic.ru/index.jsp?topic=%2Fcom.anylogic.help%2Fhtml%2Foptimization%2FCreating+an+Optimization+Experiment.html. Last accessed 2020/11/21.
3. Russian State Standard GOST R 52398-2005. https://allgosts.ru/93/080/gost_r_52398-2005. Last accessed 2020/11/21
4. Some traffic lights were reconfigured in the North-East Administrative District. https://smedvedkovo.mos.ru/presscenter/news/detail/8517458.html. Last accessed 2020/11/21
5. Google Maps mapping service. https://www.google.ru/maps. Last accessed 2020/11/21
6. Rakhmangulov, A.N., Kopylova, O.A.: Assessment of the socio-economic potential of the region for the placement of objects of the logistic infrastructure. Econ. Reg. **2**(38), 254–263 (2014)
7. Rakhmangulov, A.N., Kornilov, S.N., Kolga, A.D.: Ensuring the timeliness of freight traffic in transport and technological systems. Vestnik Magnitogorsk State Tech. Univ. G.I. Nosov **1**(45), 115–121 (2014)
8. Kuzmin, D.V., Baginova, V.V., Rakhmangulov, A.N., Leuven, B.A., Fedorov, L.S., Andreyeva, L.A.: Forming and developing the controller terminal regional network for conditions of the Russia Federation. Int. J. Econ. Persp. **11**, 2 (2017)
9. Muravev, D., Hu, H., Rakhmangulov, A., Mishkurov, P.: Multi-agent optimization of the intermodal terminal main parameters by using anylogic simulation platform: case study on the ningbo-zhoushan port. Int. J. Inform. Manag. **53**, 102133 (2020)
10. Kuzmin, D.V.: Simulation model of regional piggyback transportation. Transp. Urals **1**(44), 114–118 (2015)

11. Kuzmin, D.V.: Modeling the logistics of piggyback transportation. Soiskatel Suppl. J. "World of Transport" **2**(10), 68–73 (2015)

MultiSPAS: Multi-Descriptor Physical Approaches to Computerized Plant Ecology

Feodor Orekhov and Oleg Gradov

Abstract It is proposed to combine "computational and mathematical" and "monitoring" approaches to kinetic physico-chemical systematics based on the correlations between the environmental conditions and the plant response in real-time. Combining a real experiment and a mathematical model provides higher accuracy, heuristic value, and data integration (for data mining and descriptor-based representation based on combination of cheminformatics and/or chemometrics and bioinformatics/biometrics). Development of this method involves using as many variables as required for representing the environmental parameters and the set of factors affecting the biome. Hence, it is not limited to the aquatic or atmospheric environment as in the case of MultiSPAS approaches or their analogs.

1 The Need for Multiparametric Chemometric Reactivity Analysis and Automated Biogeographic Systematics

1.1 The Problem of Objective Multidescriptor Control of the Environment

It is well known that plant species possess specific reactivity to the number of environmental factors, which is evolutionarily determined by the geographical conditions of their growth. Therefore, the reactivity or resistance of the plants specific to the given geo/hydrochemical and geographic area can be used as a criterion of their adaptability to this area during phylogenesis [1–12], and hence, as a criterion for evolutionary species systematics with biogeographical reference. With the reactivity parameters and resistance ranges available for different taxons, one can develop a plant classification database based on the real physico-chemical factors and phylogenesis criteria that do not depend on the specific researcher and his concepts concerning the taxon. A person-independent or, as it is called abroad, an operator-independent approach in the

F. Orekhov (✉) · O. Gradov
N.N. Semenov Institute of Chemical Physics, Kosygina 4, Moscow 119991, Russia

M. Shamtsyan et al. (eds.), *Robotics, Machinery and Engineering Technology for Precision Agriculture*, Smart Innovation, Systems and Technologies 247,
https://doi.org/10.1007/978-981-16-3844-2_30

analysis of biological data (for example, biomorphological, substituted by methods of mathematical morphology [13–16], regardless of the nature of the variables—optical, radiographic, magnetographic and others), their quantitative measurement, identification and clustering (including in soft real time) using fuzzy logic [17–19], at the moment, is becoming extremely widespread in research technology in the field of life sciences in the US and the EU. In many cases, the analysis method and algorithm contain a "correlation context" inherently (i.e. a database descriptor is not a variable, but its correlation with some factor). In the ecology domain, such comprehensive descriptors are correlations of phytophisiological, biochemical and morphological data with the scales of the parameter indication being the natural environmental conditions such as temperature, atmospheric parameters, etc. [20].

Simultaneous consideration of more variables leads to the multidimensional data correlation in ecological models. This fact has been getting obvious in a while, but objective correlation methods exist only for objective hardware-registered datasets (there are examples in optical spectroscopy [21–24], nuclear magnetic resonance spectroscopy [25–29] and NMR spectroscopy [30, 31], their X-ray formal algorithmic equivalents [32], electrophysiology, and bioacoustics [33, 34]). To get a sufficiently clear correlation based on "non-objective" (especially irregular data resulting from the scattered in time samples and bootstrap) is almost unreal, or its interpretation would not be heuristically valuable. Such data are very common in population/ecological studies. The problem of subjective index data correlation is of great deontological importance [35] because decision-making or the object state prediction accuracy in the future depends on the environment quality assessment, thus the action taken can be unsuccessful or harmful because of the wrong assessment.

1.2 The Problem of Separate Analysis of the Additively Acting Factors with Different Weighting Coefficients

It is obvious that "weighting coefficients" assigned to various factors that affect a biological/ecological object of interest differ because sensitivity to various parameters is a taxonomically or biogeographically specific descriptor in some cases. Selective parameter sensitivity is important in experiments and has been used in such models for about 30 years [36]. It is especially important for the multi-parametric response (for example, for the thermo-photo-selective one) during the analysis of plant reactivity to these parameters in different (arid and humid, for example) environmental conditions [37]. The absolute optimum point or the absolute rate range is impossible to be found for any plant species with multiparametric/multidescripyor correlation analysis, but multiplicative seminorms can be identified through analyzing the mutual effects and considering the feedback [38]. Some devices, e.g., the thermometer-hygrometer or psychrometer, work as correlators of two and more descriptors. Another data have to be normalized and processed by mathematical software packages after data acquisition from multiple devices.

The device sensitivity adjustment to match the specific sample is usually complicated because it requires a posteriori data about the object sensitivity expressed as "effect vs. response" curves. The conventional correlation analysis does not work well enough with multi-descriptor classifications [39]. The models of sensitivity to physical factors usually offer lower selectivity than the models of sensitivity to chemical or biochemical ones [40] because the metabolism net/graph topology is more complicated and its dimensionality is higher than for the reactions with easily estimated "dose vs. effect" curves in biophysics (especially for such statistical groups as plant species [41]). Additive ecological response (e.g. the plant age in the community which has been considered as the criterion of the whole community sensitivity towards environmental factors since the 1970s [42]) is not heuristically valuable and handles only the indicative role without considering mechanisms and differentiation between the groups by the plant sensitivity towards every factor measured.

Categorizing the whole effects by factors matches the structure levels affected by the above factors and the actors' applicability to the plants or to the entire community. For each factor, its localization and colocalization (soil, water, air, and other conditions) is identified. When a studied factor (with a certain weighing coefficient of sensitivity for the object studied in the given environmental conditions) is transferred from one environment to another, the weighting coefficients are to be re-evaluated to match the new environmental conditions. For example, radioisotope diffusion in soil differs from that in water, so the weighting factors for soil and water plants are unequal by definition, and the sensitivity kinetics is drastically different, too. UV radiation cannot reach the X depth in water that is why the emission weighting factor is 0 if the depth exceeds X. The heat capacities of air, water, and soil are different which leads to the thermophysical weighting factor recalculation for each environment, etc. Since most plant models by definition operate with sensitive structures in the air that are attached to soil and supply simple chemical agents (considered in the models) from the soil, it is clear that the most complete types of such models are SPAS (Soil-Plant-Atmosphere Scheme) and MSPAS (Modified Soil-Plant-Atmosphere Scheme) [43, 44]. Their additive versions (including the ones that consider the environment chemistry) emerged from the models developed in the late 1970s, particularly at Oak Ridge National Laboratory (e.g., refer to the well-known 1979 preprint by Begovich and Luxmoore: "Some Sensitivity Studies of Chemical Transport Simulated in Models of the Soil-Plant-Litter System" affiliated with the Department of Energy, Oak Ridge National Laboratory). It is obvious that this problem statement requires the experimenter to record big data over many channels [45], and the correlation mismatching does not appear just because of plant immobility that makes the 'stationary' model available to use (unlike in zooecology [46]).

1.3 The Problem of Multi-Channel Measurements and Cross-Correlated Big Data Analysis

Prior to MSPAS, the only available multichannel data acquisition devices were mostly general-purpose crate systems with the CAMAC bus (relevant EUR standards 4100/4600/6100/6500 [ESONE]; IEEE standards 583/596/595/575 [IEEE]; IEC standards 516/552/64p/729 [CEI]) [47]. Some examples of such systems were realized in Soviet Union and in our labs (see Figs. 1, 2 and 3). Many modular CAMACcrate-based constructions were modified and computerized in 2000th for working under Scientific Linux (CERN), Windows XP and NT Workstation (see Fig. 4).

The additive and simplified/abstract probabilistic (stochastic) models seemed more preferable because of data processing complexity. For example, the Monte-Carlo method [48] with the response estimated just as an additive response surface even in dynamical modes with no differentiation by taxonomy and biogeography, dependent and independent descriptors, parametric and nonparametric statistics. Currently, new algorithms for bioinformatics and ecology are under development capable of operating with big data [49, 50], so there is an incentive to introduce new methods for interpreting multi-parametric or multi-taxonomic big data acquired in the course of longitude experiments in physical ecology of plant communities, including those considering geographical location of not only the data, but also the multi-dimensional correlations as a part of the general colocalization [51]. This trend is the key subject of this paper.

The primary issue with the implementation of the correlation and colocalization approaches is reproducibility and the statistical value of the results. The simpler the effect inductor and the base of an organism/a group of organisms that interacts with the inductor, the higher the data replicability is, and the elementary approximation is more reliable. A simple example is the many relations used in radiobiology and radioecology where the inductor reaction base is at a submolecular level. A classic example is radiation carcinogenesis that has a statistical "non-infectious epidemic" nature [52]). In this case RIA-like (radioimmunoassay) systems similar to that shown in Fig. 5 are often required for analysis of the natural radioactivity. For such systems O. Gradov's group in the past decade used to perform digitalization of scintillation analyzers for permanent in situ monitoring of the biological structures cultivated inside the rack positioner of RIA assay analyzer (P. Alexandrov et al.).

The complexity of an organism's metabolic pathway and compensation abilities lead to the toxicokinetics and pharmacokinetics complexity in case of the chemical effects. That is why the most rational approach to the multiparametric correlation-descriptor environment monitoring is not the ecotoxicologist way, but the physical ecology analyst way, since the method development would be simpler and clearer in this area).

Fig. 1 Fragment of the CAMAC crate for multi-channel multi-descriptor signal registration with different analog-to-digital converters (ADCs) and multiplexers with MCU (multiplexer control unit). N.N. Semenov Institute of Chemical Physics RAS, Moscow; 1980th–2000th

Fig. 2 Fragment of the CAMAC crate for multi-channel multi-descriptor registration with multi-channel ADCs, timers, clock generators, counters and linear ratemeters. N.N. Semenov Institute of Chemical Physics RAS, Moscow; 1980th–2000th

2 Key Parameters of SPAS, MSPAS Models, and Their Advanced Differentiated Versions

We will list and review below the basic physical descriptors that enable to build complex correlations in complex, but separate plant analysis as a differentiated pool of physical ecology objects.

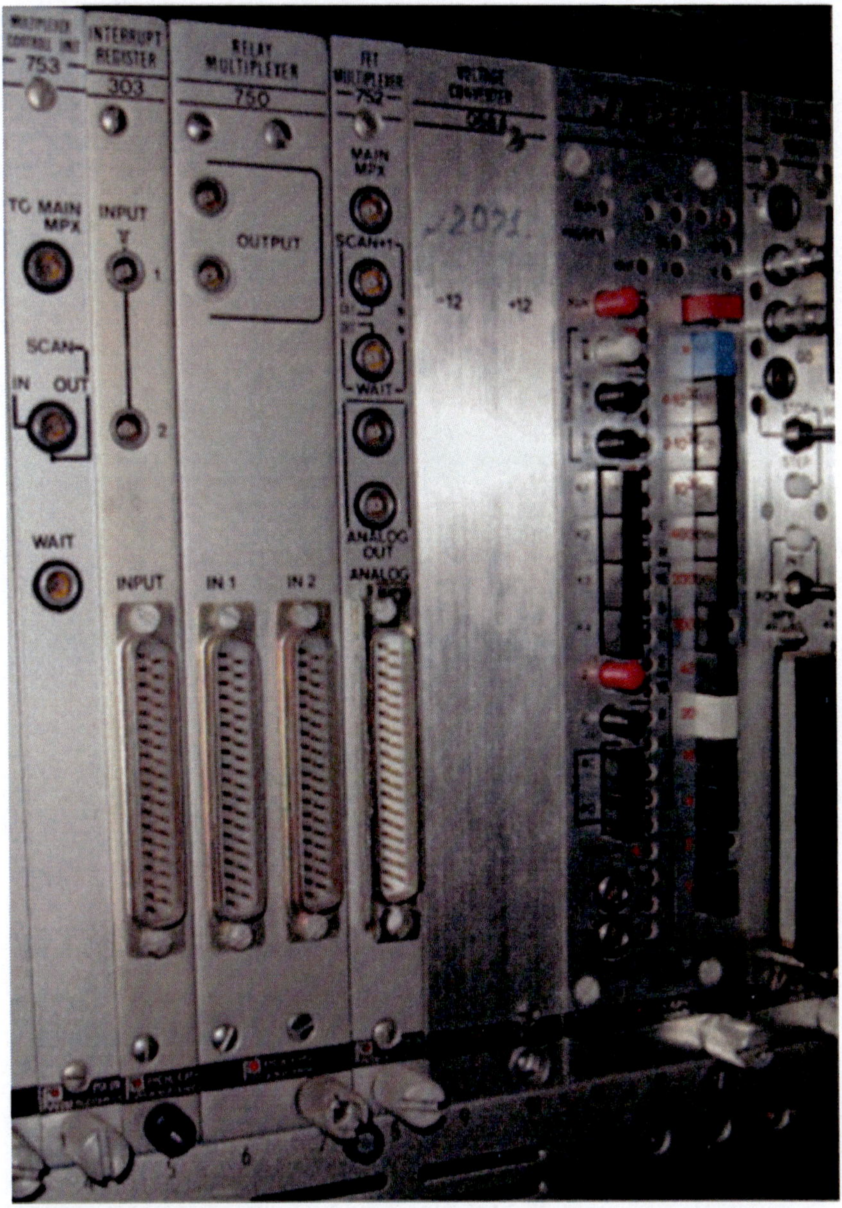

Fig. 3 Fragment of the CAMAC crate for multi-channel multi-descriptor registration with the crate controller, autonomous processor, FET multiplexers and relay multiplexers. N.N. Semenov Institute of Chemical Physics RAS, Moscow; 1980th–2000th

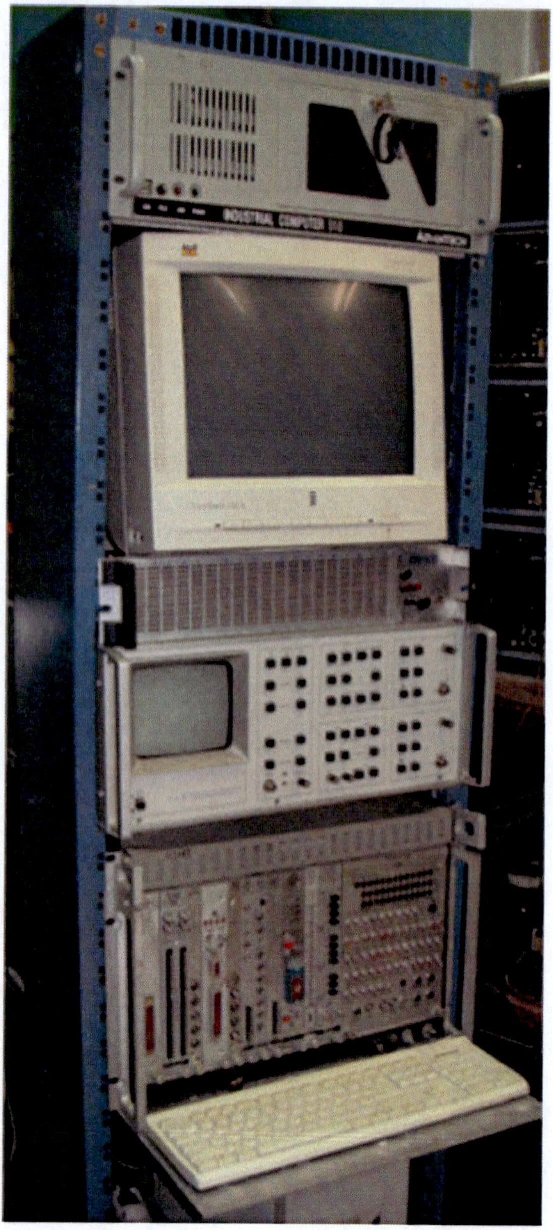

Fig. 4 Starting of the renovation and computerization of the CAMAC crate-based setup with industrial computer module and memorized oscilloscope with GPIB/HPIB-compatible interface (IEEE-488; GOST 26.003-80; IEC-60488-1/IEC-60625-1). N.N. Semenov Institute of Chemical Physics, Moscow; 1980th–2000th (upgrade). This modification has an alternative commutation of the power supply bus in the crate and the corresponding interface place is modified

Fig. 5 Computerization of Beckman Coulter LS liquid scintillation counter using COM-to-USB-converter (RS-232-to-USB-converter) and Terminal-like software (from P. Alexandrov's report, in press)

2.1 S-parameters of the Advanced SPAS-Like Approaches

The vegetation substrate quality analysis [53] covers specific and nonspecific sorption properties in terms of metal particles physically affecting the plant [54]; the formation of mineral/bio-mineral particles with variable electric charges [55]; colloidal and dispersal properties [56]; sensitivity "dose vs. effect" curves [57]; electrochemistry and soil ionics including soil water [58], etc.

At first glance, the "strictly physical" factors (a charge, sorption, degree of dispersion) give chemical descriptors such as sensitivity weighting factors and plant resistance to any environment chemical properties in a real ecosystem with biogeochemical concentrations/clarkes of the certain elements (including the plant feedback). It is especially important for plants because charge effects do not describe the environmental effect of selectivity in general. Even the specific conductivity or the environment resistivity depends on substrate ion composition. In soil as heterogeneous media specific conductivity also depends on the spatial coordinates and is described not by a scalar, but by a symmetric second order tensor (the scalar representation is used only in aeroponics and aero-hydroponics, but in this case, soil S-parameters are substituted by the atmosphere A-parameters and the sorption in the absence of the heterogeneous soil medium is neglected). An example of such devices for analog and semi-analog correlation of S-parameters for hydrobionts (developed by Gradov's group before 2014) is demonstrated in Fig. 6. Another setup for similar aims operating

Fig. 6 Four-channel "GREEN ZONE" correlator for hydrochemical S-parameters, relevant for model experimental aquarium plants (based on complex descriptors of the plant cell/tissue response to hydrochemical factors in the framework of physical chemistry) with imbedded analog-to-digital converter (soundcard-based one)

with the isolated plant tissues and microalgae cells using modern digital registration instruments which can be integrated by means of a SCADA-system with a unified GUI (Gradov's group, 2018) is shown in Fig. 7.

Many examples of correlation and relative weighting factors for such phenomena can be cited. Comparative analysis of the flora sensitivity descriptors to different metals [59] should consider the soil as an ion exchanger and the difference between the flora responses to soil pH and precipitation [60]. The flora also emits several acidification agents and acts as an ion exchanger with various specific curves as applied to "descriptometry" (terms "descriptometry" and "descriptometers" were introduced in [61, 62] according to the device concept demonstrated in Fig. 8); according to metallomics and elementomics [63–67] of the entire biomass involved in the exchange, that is, considering the sensitivity of not only plant rooting and soil but also the accompanying bacteria as well [68].

For example, the model of sorption, transport and deposition of cadmium by the plants takes into account the structural mechanisms determining the element sensitivity [69], but cadmium sensitization metabolic conjugation with iron consumption is typical for some plants (for example, soya [70]). The iron consumption depends on the environment redox state [71, 72] and the atmosphere gas composition [71, 73,

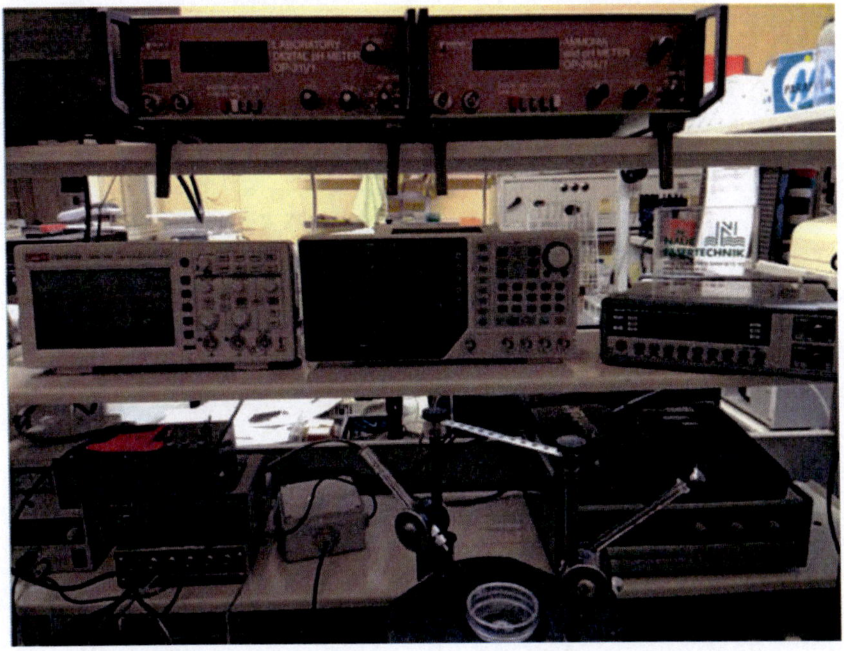

Fig. 7 Digital setup for microinjection-based plant cell reactivity estimation (electrophysiology and morphology studies, including osmotic stress induced dynamic morphological deviation—as in Pfeffer's osmometry) based on the modified S-parameters (pH, Eh, pX, specific conductance of an electrolyte solution or total dissolved solids level). Such devices can be implemented not only with inverted optical microscopes, but also using lens-less microscopes. Gradov's group (RFBR and FASIE Projects), INEPCP RAS, 2018

74], which correlates with the plant activity growing on the soil with different iron content.

The plant photosynthesis in this model depends on the photosensitivity index which correlates with the cadmium impact [75] in Cd-Fe-mediated biological effect models. Moreover, cadmium fixing agent (e.g. phosphates) deprivation reduces cadmium consumption and the level of Cd sensitivity due to the increased iron fixation efficiency accompanied by the photochelatin synthesis inhibition [76]. Photochelatin dysfunction leads to excess metal accumulation in the plants, soil, and microbiome. For instance, it induces the "metallomic stress" due to the root zone sensitivity optimum shift [77]. The result of it is an ion transport trace change in microbiome determined by the element antagonism at the ion transport level (for example, the *E. coli* mutants with excess chromium sensitivity require special iron consumption different from the normal *E. coli*.) The above process is caused by the channelome and membranome mechanisms at the level of cooperative biogeochemical microbiome activity which determines the microelement transfer intensity and weathering as the pedogenesis stage [79].

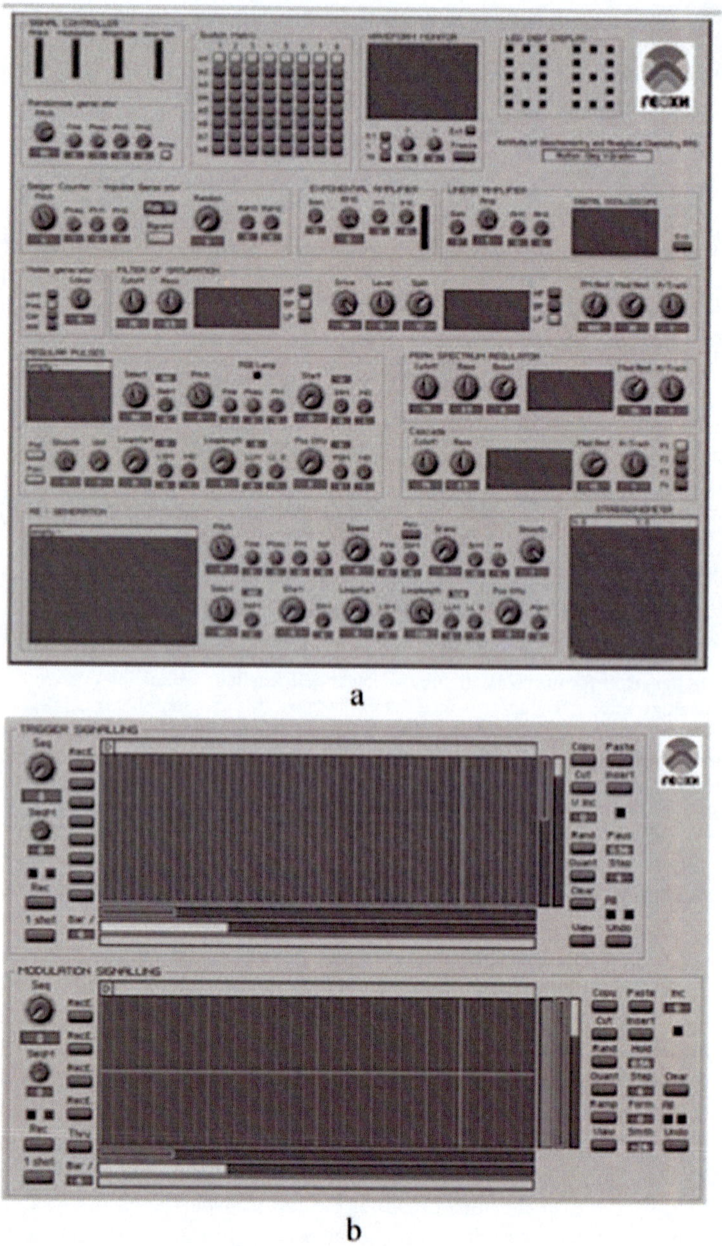

Fig. 8 Virtual devices («descriptometers») created using «Reactor» graphical programming system for data acquisition from different types of analytical devices with optron switching (Institute of Geochemistry and Analytical Chemistry of the Russian Academy of Sciences, GEOKHI RAS, O. Gradov's volunteer students collaboration, 2010–2011) [61, 62]

The above example indicates the problem complexity and the need to consider many factors in the course of the S-parametric analysis in SPAS-like approaches. It is noteworthy that the membrane mechanisms of endocytosis, symport, uniport, and antiport, as well as their dependencies on the potential of the medium and on the cell membrane as an interface, have not been considered despite the fact that they play a crucial role in metallomics of S-parametrizable subsystems used in SPAS-like models [80, 81].

2.2 *A-parameters of Complicated SPAS-Like Approaches*

Some authors associate the atmosphere parameters with the "gas biology" [82–87], an emerging domain of life sciences and biomedical technologies, but this first impression is not full and adequate, since the biochemical bias of this area leads to consolidation of A-parameter analysis to the atmosphere composition and gas chemistry (for example, "ethylene biology" [88].) Although this domain of chemistry is enormous, it does not embrace all possible gas effects (also non-chemical ones) in air ecology [89–91] and air biology (in particular, plant disorders caused by air biological factors [92–94]), mesoscale bioclimatology [95, 96] (for example, in spore spreading [97]) and biometeorology (which also considers pollen and spore spreading, as well as "epiphytotics" of air-borne agents causing plant diseases [97–101]).

Gases emitted by plants are the criterion of their growth and activity as much as the consumed gases—an example of this is the criteria of correlation gas chromatographic auxanometry and GC mass spectrometric auxanometry [102–105]. Biometeorology is a reductionistic science in terms of plant-atmosphere feedback analysis and a sub-field of air physical chemistry and physical geography [106, 107]. However, it is important to consider the reactivity aspects of this physico-chemical media towards the external factors effecting the plant growth and development, such as temperature, insolation, cosmic radiation, and other variables of the "equation of state" if this equation could be formulated. The above method requires both physical and chemical approaches (in the case of the sun radiation: photochemistry, radiochemistry, nuclear chemistry, and isotopy (isotope fractionation by biosystems is a well-known fact [108]). Hence, it is necessary to study "gas biology" in the framework of the atmosphere physics and chemistry, as well as to perform automatization of the corresponding setups (e.g., see Figs. 9 and 10).

Environmental conditions shall be matched to the plant response if they are affected by such global factors as the above-mentioned radiation as a whole system (certainly, with geodetic reference to the specific heights and latitudes at which the target effects are observed). In other words, we need to complement the A-group variables list of the MSPAS models or to introduce a distinct descriptor group for high-energy chemistry effects. The effects of such descriptors are discussed below.

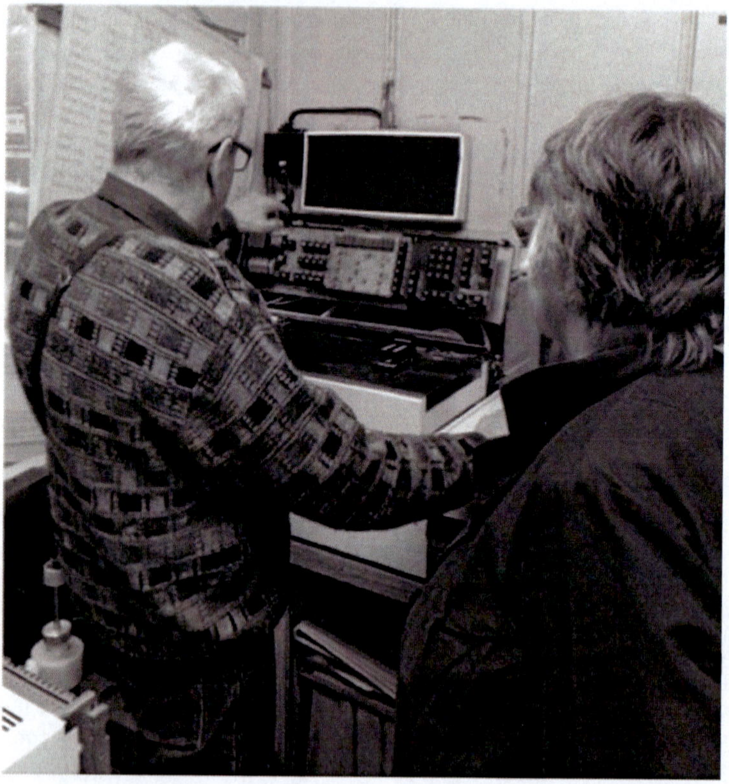

Fig. 9 Computer control testing of the mechatronics of the Specord spectrophotometer for gas and atmosphere chemistry at INEPCP RAS by W. Ryndenkoff (ex-PI at Etalon Research Institute, now—Chemical Laboratory Services)

2.3 *Solar Irradiance Descriptor Group*

The role of irradiance descriptors in the photosynthesis is well studied and does not require explanation, but we also need to take into account periodism that governs any staged development. It is caused by photoperiodism [109] deviating by the spectral selectivity and specific sensitivity to the effects in photomorphogenesis [110] and induction photoperiodism [111] depending on the age, size, and spectrocolorimetric plant characteristics [112], i.e., a weighting coefficient of the effect efficiency.

The "irradiance descriptor" group can both activate and inhibit the growth and development of the plants selectively in accordance with the weighting factors of any area of a variable that form a descriptor (wavelength/frequency, energy, penetrating power, polarization) [113]. That is why we can talk about descriptor inversion or "relative descriptors' with the weighting factors calculated specifically for resistance descriptors of this plant species, or another specific group.

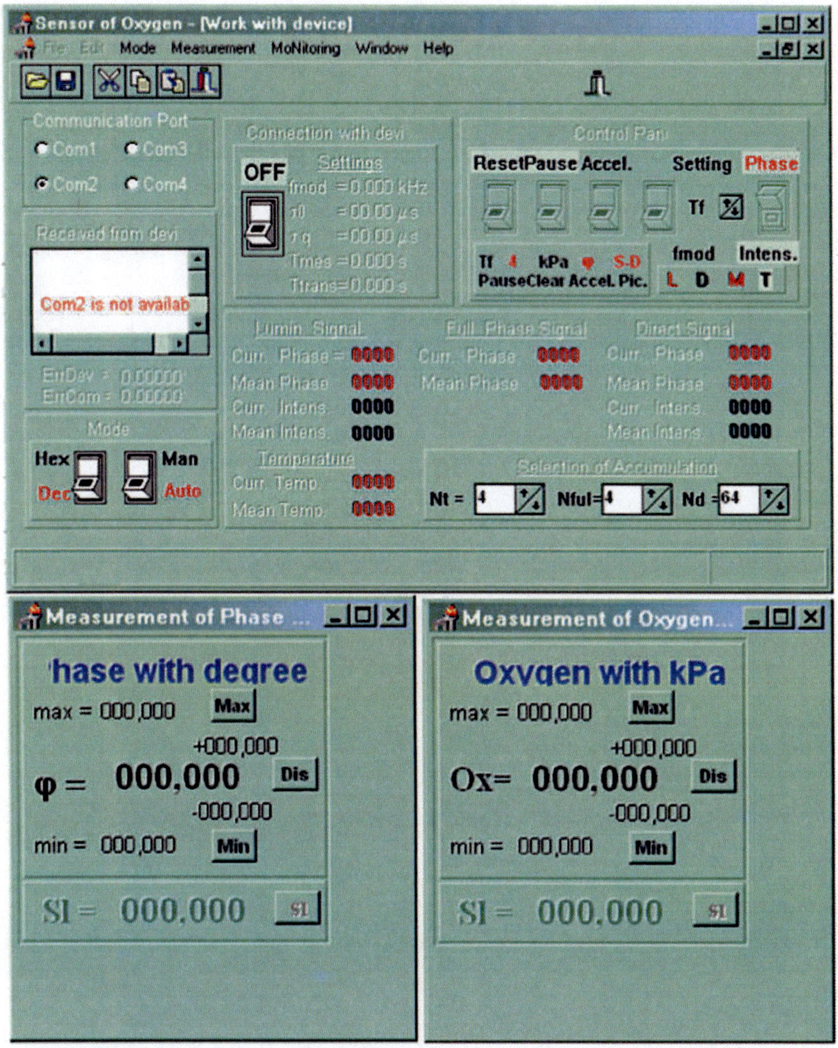

Fig. 10 Software for oxygen monitoring based on phase measurements developed by V.I. Ogurtsov (Papkosky, D.B., Ogurtsob, V.I., Ovtchinnkov, A.N., Kurochkin, I.N., Ponomarev, G.V. Sensor Systems, 1998). The software has been transferred to ICP RAS and INEPCP RAS from Central Research Laboratory (TsNIL) together with the technical equipment by assistant professor A.S. Pavljuk

The irradiance format of the descriptor implies not only a direct radiation impact including the sub-band clustering (for example, UV light is not recorded fully like in the sensitivity studies of the 1970s [114], but is separated into UV-A and UV-B bands [115–118] like in the later studies), but also the effect of the irradiance agents (the UV light from the above example) on the air chemistry. The latter effect produces free radical intermediates (ROS) in the amounts proportional to its intensity. The most classical example of irradiance descriptor effects on the A-parameters of MSPAS-like models is the ozone effects [119, 120] (the setup for ROS monitoring is shown in Figs. 11 and 12).

As opposed to the simplified MSPAS systematics, it is proposed to add extra letters to this abbreviation matching any descriptor. The irradiance descriptor abbreviation would include its emission band abbreviations such as UV, IR. The MSPAS-like model abbreviation shall be modified if there is a new descriptor subsystem or a new range of the spectral effects considered. Besides the atmosphere, (pure A-parameters

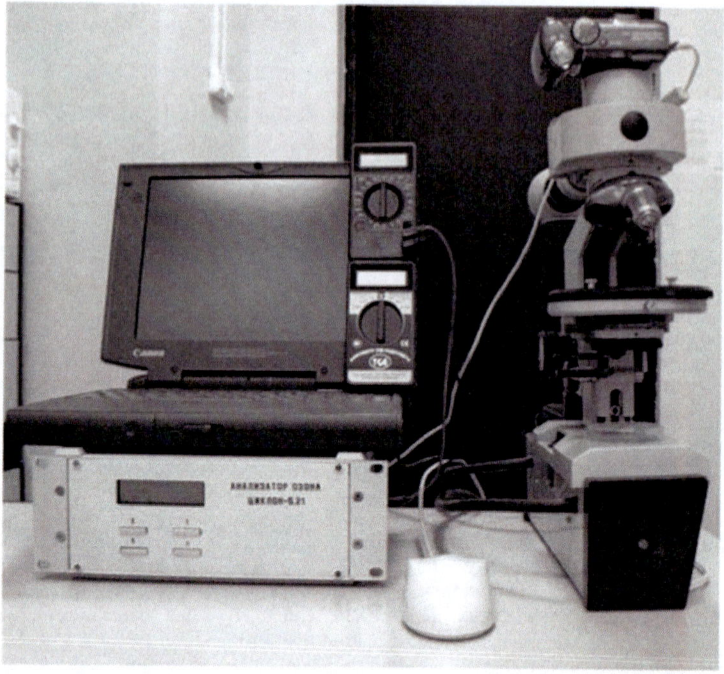

Fig. 11 One of the earliest versions of ozonometric microscopy station (2008) [119]; from the archives of the Research Institutes of Department of Power Engineering, Mechanical Engineering, Mechanics and Control Processes (Russian Academy of Sciences). Such setups can be used not only for ecological measurements, but also for the space material science and high energy chemistry applications

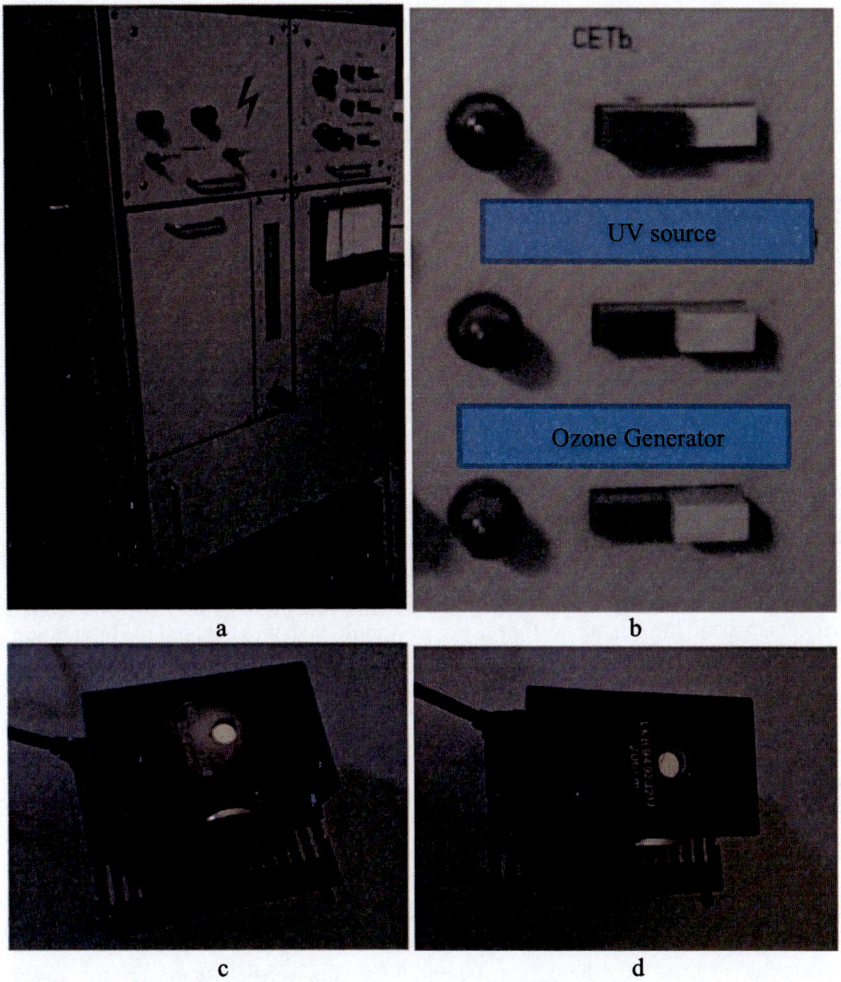

Fig. 12 One of the earliest versions of computerizable UV-beam and ozone generators (a,b) developed at the experimental facilities of the Institute of Chemical Physics in 1980th and computerized/modernized by the Gradov's group in 2018 for the measurements using ozonometer CYCLON 5.21 and UV-radiometers TKA for operating in the setup with digital lens-less microscopes with 254 and 206 nm interference filters (c,d)

in SPAS), a fine stratification should be taken into account: mesosphere, thermosphere, troposphere, exosphere; depending on the zoning of the irradiance descriptor effects with the atmospheric level determining their indirect biogeospheric effect.

2.4 Air-Chemical Descriptor Group (Indirect High Energy Chemistry Descriptors)

Atmospheric chemical stratifications of descriptormetrics analysis in accordance to the databases, the can be reduced to the multi-level kinetic schemes and graphs that describe these N order reactions. A transition from the forms of descriptors of the atmospheric reaction processes to the descriptors characterizing the interaction of the atmospheric chemical reactions (first of all, at the planetary boundary layer level) and their products with biological systems ("living matter"), is of great importance, particularly when it combines multiscale mathematical (computer simulations) and direct analog approaches—from the cellular level (using microscopic setups; see Figs. 11 and 12) to the ecosystem level (using "mesoscopic" and macroscopic setups; see Figs. 17 and 18) for multichannel and multiparametric comparative KDD-assisted model analysis (data mining). Then, for reproducibility, it is necessary to take into account not single arbitrarily chosen descriptors, but their full complex.

There are many challenges that the experimentalist meets during the MSPAS modeling at the "living matter"/"atmospheric chemistry" interface. Various plants have different ozone sensitivity [121] due to the differences in their epidermis, cuticle structure, etc., and even deeper at the level of the redox-active and transpiration mechanisms and ion channels. Therefore, we should take into account the interaction of their target environmental factors and ozone impact effectiveness matched to the exposure in the given environment. It has been shown that a plant water status also correlates with its sensitivity to ozone [122] which is specific to the flux/exposure level [123, 124] that corresponds to the plant lifespan [125]. Since illumination levels and plant survival rates vary in various biotic communities, researchers draw maps (by species and by location) to evaluate or forecast the sensitivity of a certain plant population to ozone [126–128].

Similar sensitivity maps can be based on biophysical taxonomy/biogeography/biogeocenology data correlated to thermohygrometry. The descriptor, in this case, is the precipitation level [129–131] (besides factual description [132, 133], there is a potential for the sensitivity descriptor space and time distribution forecasting [134]. Particularly, it can be used to classify the phenological changes in the flora by their response to these factors [135]). As this excursus shows, to re-evaluate the high-energy chemical properties we need factors that are not studied by atmospheric chemistry, but they indirectly ensure its influence on living organisms. The above factors can be divided into scalar (non-directional isotropic) and vector ones.

2.5 Scalar and Vector Predictors

Sun radiation in atmosphere and water propagates in a known direction (that can be simulated using Monte Carlo methods accounting for the scattering, several quasi-optic effects, and any radiation-to-matter interactions) with some error caused by the immanent environmental properties. In some cases when the radiation produces only heat effects (not selective to λ or ω), some propagation features in MSPAS-like models can be neglected. An example of the direct photon propagation simulation (using Geant4, CERN) with the wavelength of 203.5865 nm (6.09 eV; 49,119.16473 cm^{-1}; 1472.55551 THz) when passing through the hydrated vacuolized algae cell on the surface is given in Fig. 13.

The temperature response is one of the key components of the ecosystem/local plant community response to all the environmental factors [135–137]. Thermal sensitivity is intermediated by the atmosphere and affects the meteoclimatic descriptors. SPA (soil-plant-atmosphere) models are temperature-sensitive, but the temperature effects are colocalized with the solar radiation effects (not only in the visible spectral range). That is why the relationship between the scalar and vector predictors, in this

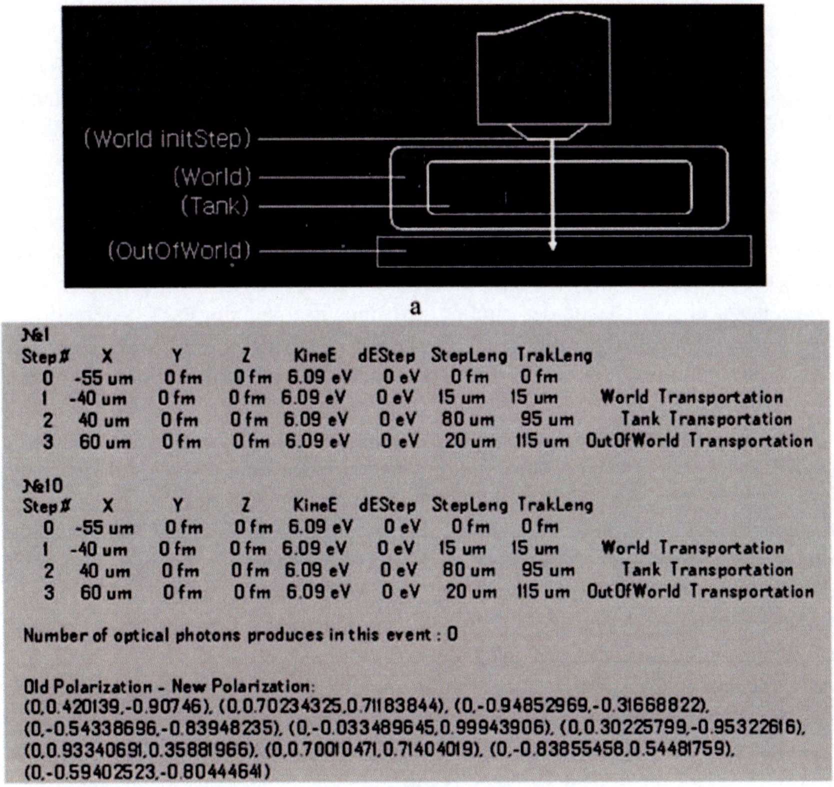

№1

Step#	X	Y	Z	KineE	dEStep	StepLeng	TrakLeng	
0	-55 um	0 fm	0 fm	6.09 eV	0 eV	0 fm	0 fm	
1	-40 um	0 fm	0 fm	6.09 eV	0 eV	15 um	15 um	World Transportation
2	40 um	0 fm	0 fm	6.09 eV	0 eV	80 um	95 um	Tank Transportation
3	60 um	0 fm	0 fm	6.09 eV	0 eV	20 um	115 um	OutOfWorld Transportation

№10

Step#	X	Y	Z	KineE	dEStep	StepLeng	TrakLeng	
0	-55 um	0 fm	0 fm	6.09 eV	0 eV	0 fm	0 fm	
1	-40 um	0 fm	0 fm	6.09 eV	0 eV	15 um	15 um	World Transportation
2	40 um	0 fm	0 fm	6.09 eV	0 eV	80 um	95 um	Tank Transportation
3	60 um	0 fm	0 fm	6.09 eV	0 eV	20 um	115 um	OutOfWorld Transportation

Number of optical photons produces in this event : 0

Old Polarization - New Polarization:
(0,0.420139,-0.90746), (0,0.70234325,0.71183844), (0,-0.94852969,-0.31668822),
(0,-0.54338696,-0.83948235), (0,-0.033489645,0.99943906), (0,0.30225799,-0.95322616),
(0,0.93340691,0.35881966), (0,0.70010471,0.71404019), (0,-0.83855458,0.54481759),
(0,-0.59402523,-0.80444641)

b

Fig. 13 An example of Monte Carlo modeling of the UV-beam propagation through the algae cell (World) with a vacuole (Tank) on the mineral surface (Out of world) under the ozonometric microscope with the UV light source (World InitStep) case, is obvious [138]. Various descriptors and predictors are presented as comprehensive in terms of their agents' direction of propagation (that is, they are scalar within the terminology framework used, [139–142]) due to the methodological simplicity of the unsophisticated methods for measuring these properties used in the past (refer to [142]). Still, the presence of clearly affected points or supramolecular compartments usually means that the agent is a vector (statistically, although there could be many counterexamples found). For instance, metal–organic and inorganic structures cause local overheating at the irradiated surface and form a correlation-interpreted descriptor based on the interrelation of the local cytophysiological agent accumulation (e.g., Hg [143], that perfectly matches the SPA, SPAS, and MSPAS models since it is accumulated from Hg vapors), temperature, and irradiation. Heliotropism itself is a vector phenomenon while its temperature/thermodynamic effects are a scalar "background".

The geoelectromagnetic field is a scalar (background) factor, while bioelectromagnetic fields and localized artificial sources are vector agents. Simulation of the local bioelectric effects with local electrodes placed at various spacing and spatial orientation in various parts of the plant [44] shows that different surface areas in different plant species demonstrate specific sensitivities to such effects. Faraday balances (Fig. 14) providing dynamic measurements and monitoring of magnetic susceptibility can be applied for determination of the targets of orientational magnetobiological effects.

According to Kholodny-Went theory, gravitropism and geotropism are vector phenomena by definition. Sensitivity to external forces is the key element of clinostatting unit operation [145]. Still, the distribution of gases and atmospheric phytohormones is a scalar stochastic descriptor [146] since its agent does not systematically propagate to a specific plant or plant organ from the atmosphere. It can be characterized only with is integral/cumulative content in a certain area not defined by any objective criteria.

The pH sensitivity value is cumulative and scalar [147], but within a specific subcell-levell area [148] it is controlled by the ion channels, i.e. the ion flows in and out of the compartmentalized membraneous environment, which can be measured using FRAP-like techniques (Fig. 15). Such flows are vectors (even Mitchell who proposed the chemiosmotic hypothesis used to talk about vector chemistry as applied to membrane permeability and gradients). Sensitivity to the specific ions is usually analyzed additively "by mass" [149, 150]. We propose to analyze "vector chemistry" using lock-in/phase selective non-FFT patch-clamp spectra and also using "Prony energy spectroscopy" (see Fig. 16). In this context, the analysis should be dynamic and colocalized (similar to calcium imaging in advanced fluorescent cell microscopy) to concurrently obtain information about the shift of the scalar pH/Eh value (just like the FRAP vector methods developed by the authors).

Many ions can contribute to the formation of secondary vector agents, descriptors, predictors, and factors (for example, sensitivity to Fe ions and the related sensitivity to Fe-induced magnetic fields; see above Fig. 14). In most cases, the effect of such field should be considered as a vector descriptor, while the redox potential change

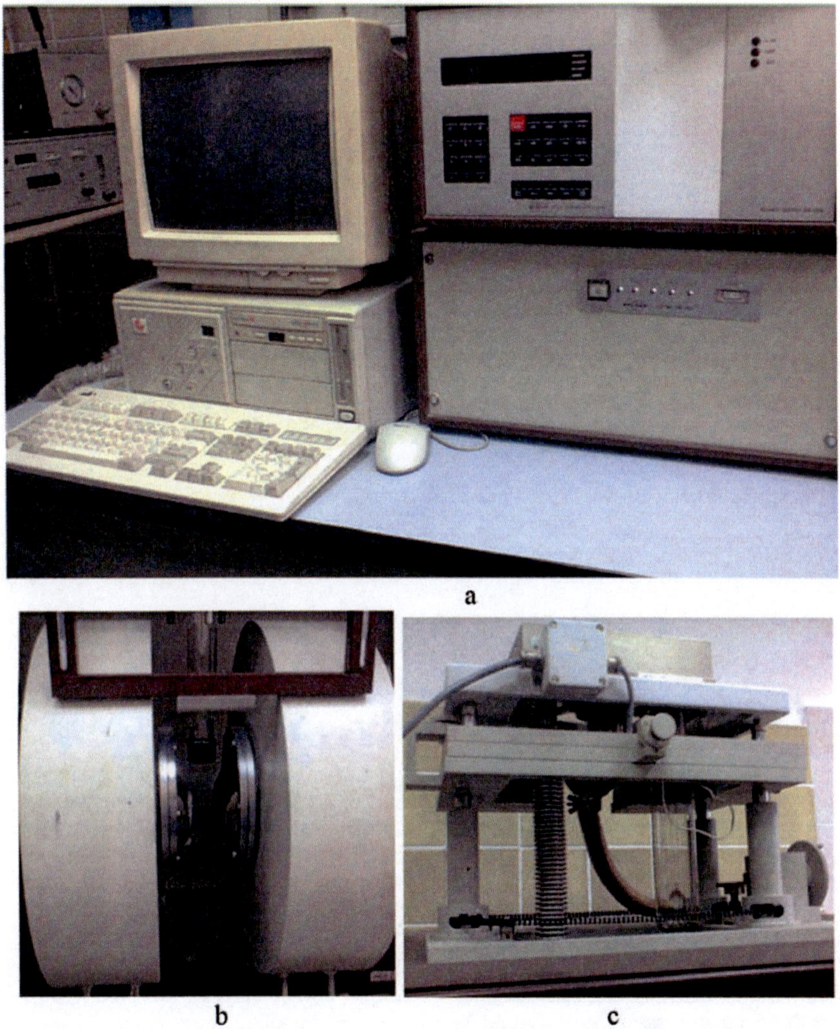

Fig. 14 Faraday balances (re-automatized by P. Alexandrov) for measurements of the magnetic susceptibility (a quantitative measure of the material's response to the applied external magnetic field) of the biological materials (P) and geomineralogical or soil substances (S). Such setups allow to analyze the targets of the magnetic field effects in the natural samples

within a homogeneous fraction—as a scalar one (in terms of sensitivity to Fe in the specified context refer to [151–157]).

Radiation as a cumulative measurable biophysical factor can be considered as a scalar descriptor (undirected gamma and X-rays [158–160]). If there is a preferred direction of propagation (as in the case of gyrosynchrotron radiation) it should be analyzed as a vector agent [161]. Cosmic rays are considered as a vector agent, but

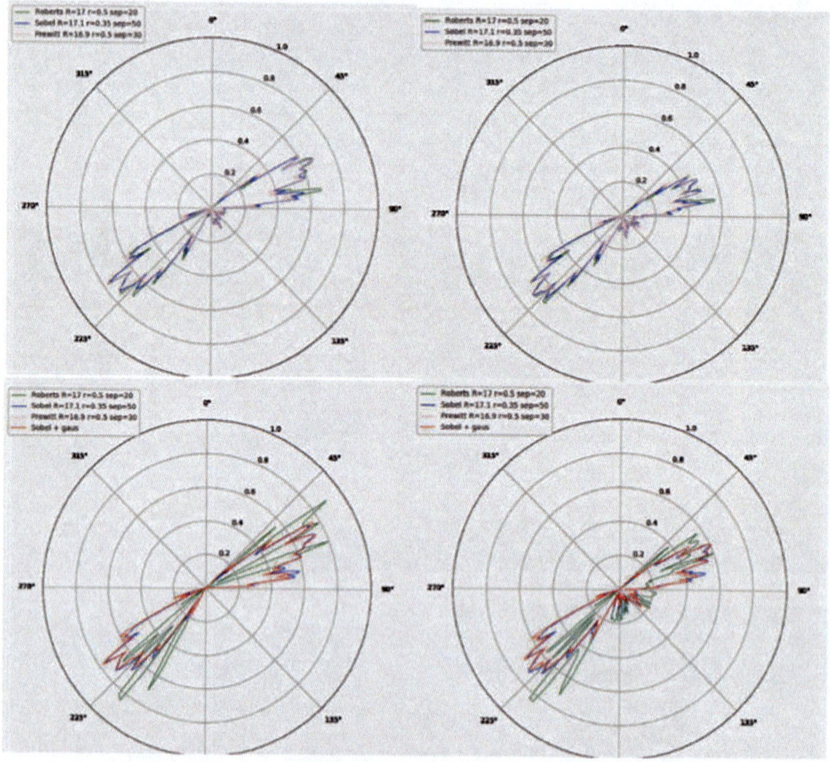

Fig. 15 "Radiation patterns" of different vector fields of diffusion in FRAP (From the IEEE Dataport-deposited Supplement to the preprint: Notchenko A., Endaltsev N., Gradov O.: Novel FRAP techniques…) [https://doi.org/10.21227/kfdk-eq33]. Software for these measurements has been developed by A. Notchenko and N. Endaltsev before 2017–2018

for a large-scale anisotropy only (anisotropy changes are usually registered in real-time with dozens of neutron and muonic telescopes and detectors across the world). Therefore, cosmic ray effects should be mapped in the same scale, and it is not a problem of systematics, but a problem of biogeographical zonation. This is the reason for an incomplete cross0scale compatibility of the effects observed at the singular RIA vials and at the whole environment level, performed by the technically different setups and devices with different spatial scales, resolution and angular characteristics of particle detection (compare the setup from the Figs. 5 and 17a and device from the Fig. 17b).

If the target compartment affected by radiation is more local, e.g., at the molecular level (such as virus cistrons [162] with their expression specifically vulnerable to fatal effects, or pigments highly sensitive to radiation according to the high-energy chemistry principles [163–165]), then we should consider vector effects. In this case, the problem is scaled down to the ultrastructural level, and like for the macro scale,

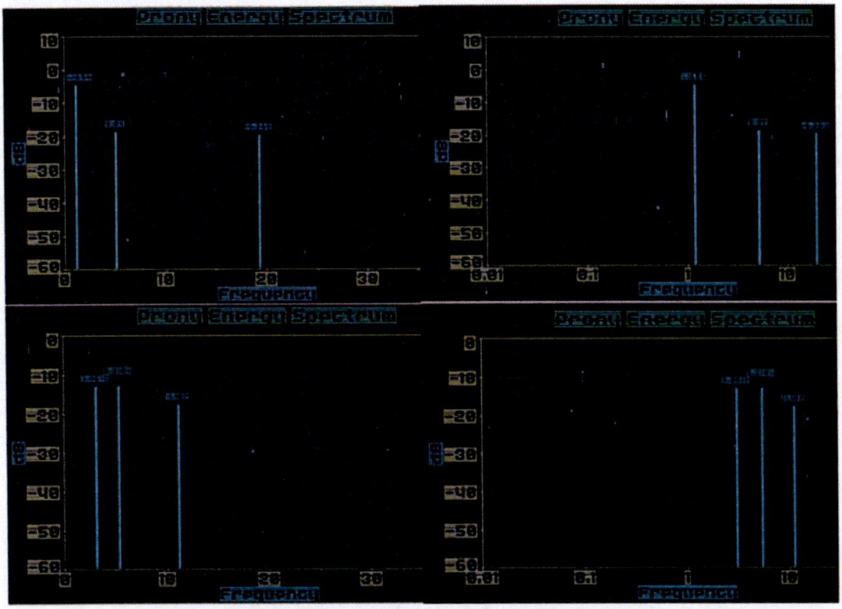

Fig. 16 "Prony energy spectra" of digitalized patch-clamp data in the linear (left) and logarithmic (right) scales. Old graphical representation from O.V. Gradov's presentation at the biogeochemical conference at the Institute of Geochemistry and Analytical Chemistry (GEOKHI RAS), May 2017

it does not more fit into the taxon systematics category (The entire organism is also affected by the background radiation, but after zero calibration the background effect can be removed as applied to the target plant compartment). We can introduce a concept of "resolution" for the types and forms of the vector effect studies applied to specific biological sample forms and organization scales. In some cases, we can propose that the field vector factor effects are related to biogeochemical properties. For example, an ion that causes sensitivity to magnetic fields can also cause sensitivity to radiation [166]. At the same time, the vector or scalar type of reaction depends on the plant development phases (the lowest vector phase is from grains to sprouts; the highest scalar phase is fruits) [167]. The same principle operates at the compartmental level: chromosomes at different stages of the cell cycle with different spiralization/compactization degree demonstrate different reaction to radiation [168]. The same applies to symbionts/consumers and parasites. The efficiency of radiation exposure for parasite control depends on the parasite stage of development [169] and the matching between these stages and the host plant development stages. It also determines the sensitivity in large ecosystems with various cycles and phenospectral periodicity of some species [170].

Quantiative gasobiological descriptors are additive and scalar (obtained by the setups, proposed at the Figs. 18 and 19) by their distribution in the atmosphere or in a shallow soil layer [171–173] (here we do not consider medical "localomics"-based

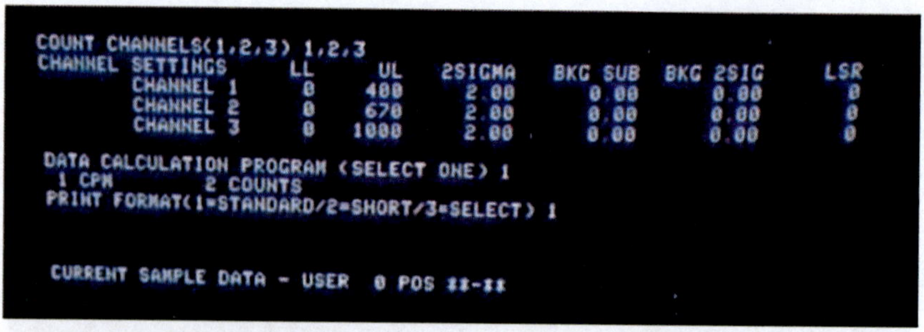

a

b

Fig. 17 Environmental gamma-spectrometry data obtained using the old RIA-station shown in Fig. 5 **a** and using contemporary Russian commercial smartphone-assisted gamma-spectrometer for the field use **b**. Photo by S. Pankratov and T. Orekhov

gasobiological descriptors and their derivatives mapped in the space [174, 175]). The relation of irradiation descriptors interpreted as vectors to the scalar breathing and gas emission descriptors [176] (in this sense, photosynthesis is a CO_2 gasochemical descriptor [177–179]) means that gasochemical analysis should not be limited to scalar values only. Sensitivity to sulfur-containing agents is also a partially vector agent since it depends on the prevailing wind pattern and the acid rain direction [180–183]. Historical wind patterns and climate changes result in phenological changes which subsequently lead to adaptive changes in sensitivity [184–192] both in scalar and vector/tensor sensitivity criteria. Therefore, the reactivity and adaptability of these factors also change. That said, for different morphophysiological types and plant systematics groups different ad hoc reactions to the environmental changes

a

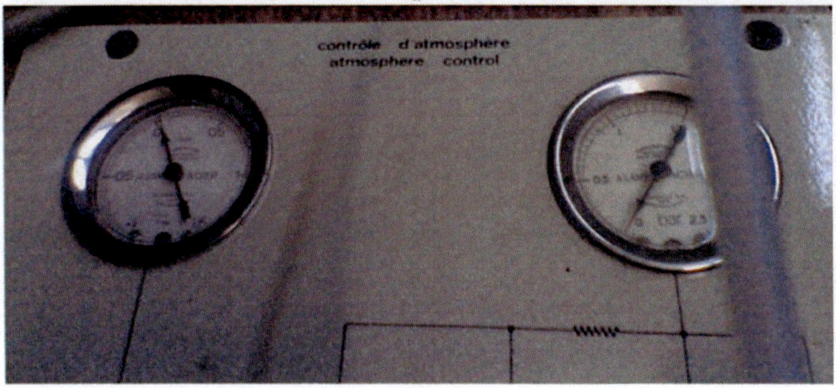

b

Fig. 18 Historical comparison: a modern computerized gasochemical/gasobiological GC setup (N.N. Semenov Institute of Chemical Physics RAS, 2013) and an atmosphere control unit of an old experimental setup (L.Ya. Karpov Institute of Physicsl Chemistry, Rosatom, the same period of time)

[193] occur, which proves the statement. Moreover, there are species-, tissue- (gisto-), cyto-, biogeographic-, age-, reactivity- [194–201], and other descriptor-specific differences in responses to various factors that act as a context for the climate and the natural bioclimatic environment formation. Chemical factor effects should be considered separately as either scalars or vectors (compare qualitatively different references [122, 202–204]). Today, it is a mid-term research objective. It is obvious, but not trivial, that water can be not only a scalar (by its availability and turgor pressure) [205–210], but also a vector effector (in plant phototropism during photosynthesis processes and cell/tissue destruction caused by dehydration [117, 211]) not mentioning its directed transport along the stem (both water and dissolved substances transported that can be traced with isotopic methods [212]) and its participation in the tissue shearing upon growth, which is also a vector factor ([213, 214]).

Fig. 19 Integration of scalar and vector on-chip-descriptometry based on mesofluidic flow microcalorimeter and gas chromatograph under SCADA-like system ZetVIEW. Technical project (proposal) started in 2012 (it was stopped after the reform of RAS in 2013–2014). In 2019 restarted as on-chip-ESEM/GC-MS/TA project (proposal)

3 Conclusion

Thus, the prospects of computer-assisted integration and correlation between the measurements of many parameters within the frame of SPAS/MSPAS schemes (with the corresponding multichannel multidescriptor approach defined as MultiSPAS) are considered. Our scientific group at INEPCP, prior to its collapse in 2018 due to organizational and economic reasons, started to develop the above approach. Despite the incompleteness of the proposed approach implementation in our early works, it is possible to point out several fundamental directions in which this work could be continued.

From the very beginning of development of S-parameters in MSPAS, both hydrochemical [215, 216] and soil [217, 218] factors have been studied. All the registered phenomena could be registered both from the side of the environment and from the side of the biological object through its response to changes in its parameters. Environmental changes were compared with the computer models, whenpossible, as well as were parameterized and included into the databases for further fingerprinting and footprinting of the structures (for example, approximation of macroscale weathering [219] can be supplemented by its binding to the A-parameters of SPAS approach). Measuring of the reactivity of a biological object or an ecosystem component towards MSPAS parameters at the cellular level can be carried out using the methods of the ion channel response measurements with the reference to the chemical composition of the surrounding medium [220], including MS-patch-clamp techniques [221]. We propose to develop for this purpose a special kind of spectroscopic systems providing registration of the electrophysiological parameters and the single ion channel response by means of the local potential fixation techniques (patch-clamp/voltage-clamp) with the advanced spectral processing and the subsequent data mining, with the simultaneous spectral detection of the ion channel state as a coordination and conformationally labile supramolecular structure, with the final data processing results presented not as the single spectra, but as the spectral correlogram [222].

The substance/agent exchange with the medium during the measurement of S-parameters and the integrity of the cellular structures can be measured using MALDI-FRAP and MALDI-FLIP methods with the subsequent correlation processing of the corresponding big data [223, 224], combining the data acquisition conditions, sample preparation and comparison with the medium model for the purpose of qualimetry [225]. When the soil and the cells of the measured sample are in contact, it is possible to use the same methodology, which has not yet been modified for industrial implementation (MALDI-FRAP-like instrumentation [226, 227]). At the same time, P-parameters (plant response) can also be studied at the additive level by electrophysiological methods (monitoring of impedance, potential oscillations), due to which any descriptors of the SPAS model operating in module A, including photobiological ones (the effect of irradiation at a given wavelength, etiolation, photoreactivation, etc.) can be studied by the response of P-parameters [228–230]. At the postmortem stage, dendrochronological data can be correlated with the climatological A-parameters of SPAS, both for the recent and fossilized plant forms [232, 233]. The conformal mapping methods used for this purpose make it possible to take into account teratomorphoses and taphonomic distortions. We planned to continue the above work in the form of a large-scale research program, but a number of factors preceding the reorganization of the scientific institutions and the staff transfer to the more highly paid and popular areas made this impossible. Unfortunately, we could not complete this work by creating fully functional SPAS monitoring devices networked according to the criterion of their attribution to either hydrochemical or soil-geochemical medium. However, this goal has not lost its significance at the moment and hence, can be solved by any reader of this work interested in continuing research in this direction.

Acknowledgements The authors gratefully acknowledge their colleagues from the Department of Metrology and Measuring Instruments of GEOKHI RAS, who worked there in 2010–2014, as well as the colleagues from biogeochemical laboratory of GEOKHI RAS, who expressed their interest in this work, especially in the aspect of biogeochemical applications of the ion channel response analysis.

The authors gratefully acknowledge S.K. Pankratov, E.D. Adamovich and W. Lin for fruitful discussions after the main seminar in the theme of this article.

References

1. Glor, R.E.: Phylogenetic insights on adaptive radiation. Annu. Rev. Ecol. Evol. Syst. **41**, 251–270 (2010)
2. Butler, M.A., King, A.A.: Phylogenetic comparative analysis: a modeling approach for adaptive evolution. Am. Nat. **164**(6), 683–695 (2004)
3. Agrawal, A.A., Fishbein, M., Halitschke, R., Hastings, A.P., Rabosky, D.L., Rasmann, S.: Evidence for adaptive radiation from a phylogenetic study of plant defenses. PNAS **106**(43), 18067–18072 (2009)
4. Sanderson, M.J., Donoghue, M.J.: Reconstructing shifts in diversification rates on phylogenetic trees. Trends Ecol. Evol. **11**(1), 15–20 (1996)
5. Rundell, R.J., Price, T.D.: Adaptive radiation, nonadaptive radiation, ecological speciation and nonecological speciation. Trends Ecol. Evol. **24**(7), 394–399 (2009)
6. Baum, D.A., Larson, A.: Adaptation reviewed: a phylogenetic methodology for studying character macroevolution. Systemat. Biol. **40**(1), 1–18 (1991)
7. Losos, J.B., Glor, R.E.: Phylogenetic comparative methods and the geography of speciation. Trends Ecol. Evol. **18**(5), 220–227 (2003)
8. Losos, J.B.: Adaptive radiation, ecological opportunity, and evolutionary determinism. Am. Nat. **175**(6), 623–639 (2010)
9. Bock, W.J.: Functional-adaptive analysis in evolutionary classification. Am. Zool. **21**(1), 5–20 (1981)
10. Blomberg, S.P., Garland, T.: Tempo and mode in evolution: phylogenetic inertia, adaptation and comparative methods. J. Evol. Biol. **15**(6), 899–910 (2002)
11. Wink, M.: Evolution of secondary metabolites from an ecological and molecular phylogenetic perspective. Phytochemistry **64**(1), 3–19 (2003)
12. Agajanian N.A., Makarova I.I.: Environment and the organism reactivity. Tver, Family, p. 186 (2001)
13. Belghitia, H., Brette, S., Lafitte, S., Reant, P., Picard, F., Serri, K., Lafitte, M., Courregelongue, M., Dos Santos, P., Douard, H., Roudaut, R., DeMaria, A.: Automated function imaging: a new operator-independent strain method for assessing left ventricular function. Arch. Cardiovasc. Dis. **101**(3), 163–169 (2008)
14. Sorensen, T.S., Korperich, H., Greil, G.F., Eichhorn, J., Barth, P., Meyer, H., Pedersen, E.M., Beerbaum, P.: Operator-independent isotropic three-dimensional magnetic resonance imaging for morphology in congenital heart disease: a validation study. Circulation **110**(2), 163–169 (2004)
15. Kahraman, D., Eggers, C., Holstein, A., Schneider, C., Pedrosa, D.J., Dietlein, M., Kobe, C., Timmermann, L., Schmidt, M.: 123I-FP-CIT SPECT imaging of the dopaminergic state. Visual assessment of dopaminergic degeneration patterns reflects quantitative 2D operator-dependent and 3D operator-independent techniques. Nuklearmedizin **51**(6), 244–251 (2012)
16. Ariani, A., Silva, M., Bravi, E., Saracco, M., Parisi, S., De Gennaro, F., Lumetti, F., Idolazzi, L., Seletti, V., Caramaschi, P., Benini, C., Bodini, F.C., Scirè, C.A., Lucchini, G., Santilli, D., Mozzani, F., Imberti, D., Arrigoni, E., Delsante, G., Pellerito, R., Fusaro, E., Sverzellati, N.: Operator-independent quantitative chest computed tomography versus standard assessment of interstitial lung disease related to systemic sclerosis: a multi-centric study. Mod. Rheumatol. **25**(5), 724–730 (2015)

17. Suckling, J., Sigmundsson, T., Greenwood, K., Bullmore, E.T.: A modified fuzzy clustering algorithm for operator independent brain tissue classification of dual echo MR images. Magn. Reson. Imag. **17**(7), 1065–1076 (1999)
18. Wallace, W.E., Kearsley, A.J., Guttman, C.M.: An operator-independent approach to mass spectral peak identification and integration. Anal. Chem. **76**(9), 2446–2452 (2004)
19. Di Stefano, D.A., Arosio, P., Piattelli, A., Perrotti, V., Iezzi, G.: A torque-measuring micromotor provides operator independent measurements marking four different density areas in maxillae. J. Adv. Prosthodont. **7**(1), 51–55 (2015)
20. Stillman, J., Somero, G.: Adaptation to temperature stress and aerial exposure in congeneric species of intertidal porcelain crabs (genus Petrolisthes): correlation of physiology, biochemistry and morphology with vertical distribution. J. Exp. Biol. **199**(8), 1845–1855 (1996)
21. Karadimitriou, M.E., Kavousanaki, E.G., Dani, K.M., Fromer, N.A., Perakis, I.E.: Strong electronic correlation effects in coherent multidimensional nonlinear optical spectroscopy. J. Phys. Chem. B. **115**(18), 5634–5647 (2011)
22. Becker, W., Bergmann, A., Haustein, E., Petrasek, Z., Schwille, P., Biskup, C., Kelbauskas, L., Benndorf, K., Klocker, N., Anhut, T., Riemann, I., Konig, K.: Fluorescence lifetime images and correlation spectra obtained by multidimensional time-correlated single photon counting. Microsc. Res. Tech. **69**(3), 186–195 (2006)
23. Mukamel, S.: Multidimensional femtosecond correlation spectroscopies of electronic and vibrational excitations. Ann. Rev. Phys. Chem. **51**, 691–729 (2000)
24. Wada, A., Tachibana, H., Hayashi, H., Saito, Y.: Multidimensional spectroscopic data correlation in the conformation transition of biological macromolecules. J. Biochem. Biophys. Meth. **2**(5), 257–269 (1980)
25. Gopinath, T., Traaseth, N.J., Mote, K., Veglia, G.: Sensitivity enhanced heteronuclear correlation spectroscopy in multidimensional solid-state NMR of oriented systems via chemical shift coherences. J. Am. Chem. Soc. **132**(15), 5357–5363 (2010)
26. Caldarelli, S., Emsley, L.: Intrinsic asymmetry in multidimensional solid-state NMR correlation spectra. J. Magnet. Reson. **130**(2), 233–237 (1998)
27. Ramamoorthy, A., Wu, C.H., Opella, S.J.L: Experimental aspects of multidimensional solid-state NMR correlation spectroscopy. J. Magnet. Reson. **140**(1), 131–140 (1999)
28. Schmidt-Rohr, K., Spiess, H.W.: Nature of non-exponential loss of correlation above the glass transition investigated by multidimensional NMR. Phys. Rev. Lett. **66**(23), 3020–3023 (1991)
29. Giraud, N., Pitoux, D., Ouvrard, J.M., Merlet, D.: Combining J-edited and correlation spectroscopies within a multi-dimensional spatial frequency encoding: toward fully resolved ^{1}H NMR spectra. Chemistry **19**(37), 12221–12224 (2013)
30. Marigheto, N., Venturi, L., Hibberd, D., Wright, K.M., Ferrante, G., Hills, B.P.: Methods for peak assignment in low-resolution multidimensional NMR cross-correlation relaxometry. J. Magnet. Reson. **187**(2), 327–342 (2007)
31. Venturi, L., Hills, B.: Spatially resolved multidimensional cross-correlation relaxometry. Magn. Reson. Imag. **28**(2), 171–177 (2010)
32. Schweigert, I.V., Mukamel, S.: Coherent ultrafast core-hole correlation spectroscopy: X-ray analogues of multidimensional NMR. Phys. Rev. Lett. **99**(16), 163001-1–163001-4 (2007)
33. Stuart, L., Walter, M., Borisyuk, R.: The correlation grid: analysis of synchronous spiking in multi-dimensional spike train data and identification of feasible connection architectures. Biosystems **79**(1–3), 223–233 (2005)
34. Uloza, V., Vegienė, A., Saferis, V.: Correlation between the basic video laryngostroboscopic parameters and multidimensional voice measurements. J. Voice **27**(6), 744–752 (2013)
35. Wang, W.C., Yao, G., Tsai, Y.J., Wang, J.D., Hsieh, C.L.: Validating, improving reliability, and estimating correlation of the four subscales in the WHOQOL-BREF using multidimensional Rasch analysis. Qual. Life Res. **15**(4), 607–620 (2006)
36. Baker, F.D.: Parameter sensitivity in plant process models. Am. Soc. Agric. Eng. **1417**, 824570 (1982). Winter meeting, Chicago
37. Venkataraman, R.: Studies on thermo-photo-sensitivity of the paddy plant under field conditions. Proc. Plant Sci. **59**(3), 117–136 (1964)

38. Zames, G.: Feedback, optimal sensitivity, and plant uncertainty via multiplicative seminorms. IFAC Proc. Vol. **14**(2), 1171–1175 (1981)
39. Sun, L., Ji, S., Ye, J.: Canonical correlation analysis for multilabel classification: a least-squares formulation, extensions, and analysis. IEEE Trans. Pattern. Anal. Mach. Intel. **33**(1), 194–200 (2011)
40. Hsu, F. C., Kleier, D.A.: Phloem mobility of xenobiotics. III. Sensitivity of unified model to plant parameters and application to patented chemical hybridizing agents. Weed Sci. 315–323 (1990)
41. Deng, S., Gou, S., Sun, B., Lv, W., Li, Y., Peng, H., Xiao, H., Yang, G., Wang, Y.: Modeled dosage–response relationship on the net photosynthetic rate for the sensitivity to acid rain of 21 plant species. Bull. Environ. Contam. Toxicol. **89**(2), 251–256 (2012)
42. Suffling, R., Smith, D.W., Stevens, D., Dai, T.S.: Plant community age as an index of sensitivity to environmental damage. Am. Journ. Bot. **61**(5), 65–65 (1974)
43. Shuhua, L.I.U., Xu, Y.U.E., Huizhi, L.I.U., Fei, H.U.: Using a modified soil-plant-atmosphere scheme (MSPAS) to study the sensitivity of land surface and boundary layer processes to soil and vegetation conditions. Adv. Atmos. Sci. **21**(5), 717–729 (2004)
44. Zeppel, M., Macinnis-Ng, C., Palmer, A., Taylor, D., Whitley, R., Fuentes, S., Yunusa, I., Williams, M., Eamus, D.: An analysis of the sensitivity of sap flux to soil and plant variables assessed for an Australian woodland using a soil-plant-atmosphere model. Func. Plant Biol. **35**(6), 509–520 (2008)
45. Violle, C., Choler, P., Borgy, B., Garnier, E., Amiaud, B., Debarros, G., Diquelou, S., Gachet, S., Jolivet, C., Kattge, J., Lavorel, S., Lemauviel-Lavenant, S., Loranger, J., Mikolajczak, A., Munoz, F., Olivier, J., Viovy, N.: Vegetation ecology meets ecosystem science: permanent grasslands as a functional biogeography case study. Sci. Tot. Environ. **534**, 43–51 (2015)
46. Benson, E.S.: Trackable life: data, sequence, and organism in movement ecology. Stud. Hist. Phil. Biol. Biom. Sci. **57**, 137–147 (2016)
47. Zalud, P., Krekule, I.: Universal preprocessing CAMAC modules. Physiol. Bohem. **29**(2), 181–184 (1980)
48. Lim, J.T., Gold, H.J., Wilkerson, G.G., Raper, C.D.: A Monte Carlo/response surface strategy for sensitivity analysis: application to a dynamic model of vegetative plant growth. Appl. Math. Model. **13**(8), 479–484 (1989)
49. Samecka-Cymerman, A., Stankiewicz, A., Kolon, K., Kempers, A.J., Leuven, R.S.: Market basket analysis: a new tool in ecology to describe chemical relations in the environment—a case study of the fern Athyrium distentifolium in the Tatra National Park in Poland. J. Chem. Ecol. **36**(9), 1029–1034 (2010)
50. Li, C., Riethoven, J.J., Naylor, G.J.: EvolMarkers: a database for mining exon and intron markers for evolution, ecology and conservation studies. Mol. Ecol. Resour. **12**(5), 967–971 (2012)
51. Heinrich, M.P., Papiez, B.W., Schnabel, J.A., Handels, H.: Multispectral image registration based on local canonical correlation analysis. Med. Imag. Comp. Comput. Assist. Interv. **17**(1), 202–209 (2014)
52. Submolecular biology and cancer. CIBA FOUNDATION SYMPOSIUM 67 (new series) in honour of Albert Szent-Györgyi on the occasion of his 85th birthday. CIBA FOUND. SYMP. 67, 1–349 (1978)
53. Trewavas, A.J.: Growth substance sensitivity: the limiting factor in plant development. Physiol. Plant. **55**(1), 60–72 (1982)
54. Piccolo, A.: Reactivity of added humic substances towards plant available heavy metals in soils. Sci. Total Environ. **81**, 607–614 (1989)
55. Violante, A., Gianfreda, L.: Role of biomolecules in the formation of variable-charge minerals and organo-mineral complexes and their reactivity with plant nutrients and organics in soil. Soil Biochem. **10**, 207–270 (2000)
56. Wojtasik, M.: Sensitivity of three cultivated plant species to the density of soils. Zeszyty Problemowe Postepow Nauk Rolniczych **507**, 575–580 (2005)

57. Nissen, P.: Dose responses of plant growth substances: sensitivity and kinetics. Plant Physiol. **72**, 117 (1983)
58. Hyder, S.Z., Greenway, H.: Effects of Ca on plant sensitivity to high NaCl concentrations. Plant Soil 258–260 (1965)
59. Gorlach, E., Gambus, F.: A comparison of sensitivity to the toxic action of heavy metals in various plant species. Polish J. Soil Sci. **25**(2), 207–213 (1992)
60. Ganev, S., Kalichkova, T.: About the problem of cation exchange mechanisms of plant acidic sensitivity and tolerance. Plant Physiol. **18**(4), 21–30 (1992)
61. Orehov, T.C., Gradov, O.V.: Hybridization of COBAC, QSPR/QSAR and SBGN technologies: the unity of theory and practice for biomedical technique design and biochemical diagnostic information analysis. J. Med. Bioeng. **5**(2), 128–132 (2016)
62. Orehov, F.C., Gradov, O.V.: In situ real time analysis in frame of COBAC, QSPR, QSAR and SBGN as a novel tool for the biosimilarity studies and physio-chemical prognostics in the biomedicine-assisted screening and experimental toxicology and allergology. J. Bioanal. Biomed. **7**(5), 95 (2015)
63. Li, Y.F., Chen, C., Qu, Y., Gao, Y., Li, B., Zhao, Y., Chai, Z.: Metallomics, elementomics and analytical techniques. Pure Appl. Chem. **80**(12), 2577–2594 (2008)
64. Haferburg, G., Kothe, E.: Metallomics: lessons for metalliferous soil remediation. Appl. Microbiol. Biotechnol. **87**(4), 1271–1280 (2010)
65. Gall, N.R., Fomina, N.S., Bazhenov, A.N., Masyukevich, S.V., Kretinina, A.V., Gall, L.N.: ERIAD mass spectrometry (electrospray with controlled fragmentation) is the common method for metallomics and biochemistry of elementoorganic molecules. Biophysics **56**(5), 883–891 (2011)
66. Salt, D.E.: Plant metallomics. Metallomics **5**(9), 1088–1089 (2013)
67. Basu, P.: Microbial metallomics. Metallomics. **5**(4), 274–275 (2013)
68. Mondal, R., Mukherjee, N.: Sensitivity of two plant pathogenic bacteria to some inorganic and organic compounds. J. Phytopathol. **83**(1), 87–90 (1975)
69. Palm, V.: A model for sorption, flux and plant uptake of cadmium in a soil profile: model structure and sensitivity analysis. Water Air Soil Pollut. **77**(1–2), 169–190 (1994)
70. Smith, G., Brennan, E.: Cadmium sensitivity related to ability of soybean plant to utilize iron efficiently. Phytopathology **72**(2), 266–266 (1982)
71. Han, J.S., Davey, D.E., Mulcahy, D.E., Yu, F.D.: Effect of the pH value of the precipitation solution on the CO sensitivity of α-Fe_2O_3. Sens. Actuators B Chem. **61**(1), 83–91 (1999)
72. Bruska, M.K., Stiebritz, M.T., Reiher, M.: Analysis of differences in oxygen sensitivity of Fe–S clusters. Dalton Trans. **42**(24), 8729–8735 (2013)
73. Sun, H.T., Cantalini, C., Faccio, M., Pelino, M.: NO_2 gas sensitivity of sol-gel-derived α-Fe_2O_3 thin films. Thin Solid Films **269**(1), 97–101 (1995)
74. Han, J.S., Yu, A.B., He, F.J., Yao, T.: A study of the gas sensitivity of α-Fe_2O_3 sensors to CO and CH_4. J. Mater. Sci. Lett. **15**(5), 434–436 (1996)
75. Solti, Á., Gáspár, L., Vági, P., Záray, G., Fodor, F., Sárvári, E.: Cd, Fe, and light sensitivity: interrelationships in Cd-treated Populus. Omics J. Integr. Biol. **15**(11), 811–818 (2011)
76. Yang, Y., Chen, R., Fu, G., Xiong, J., Tao, L.: Phosphate deprivation decreases cadmium (Cd) uptake but enhances sensitivity to Cd by increasing iron (Fe) uptake and inhibiting phytochelatins synthesis in rice (Oryza sativa). Acta Physiol. Plant. **38**(1), 28–28 (2016)
77. Robin, A., Vansuyt, G., Corberand, T., Briat, J.F., Lemanceau, P.: The soil type affects both the differential accumulation of iron between wild-type and ferritin over-expressor tobacco plants and the sensitivity of their rhizosphere bacterioflora to iron stress. Plant Soil **283**(1–2), 73–81 (2006)
78. Wang, C.C., Newton, A.: Iron transport in Escherichia coli: relationship between chromium sensitivity and high iron requirement in mutants of Escherichia coli. J. Bacteriol. **98**(3), 1135–1141 (1969)
79. Gradov, O.V.: Editorial-Isotope channelomics of chemoautotrophs as the predictor and regulator of the metal deposit formation and the weathering factor. Mod Approaches Oceanogr. Petrochem. Sci. **1**, 3 (2018)

80. Becker, D., Dreyer, I., Hoth, S., Reid, J.D., Bush, H., Lehnen, M., Palme, K., Hedrich, R.: Changes in voltage activation, Cs^+ sensitivity, and ion permeability in H5 mutants of the plant K^+ channel KAT1. PNAS **93**(15), 8123–8128 (1996)
81. Blackford, S., Rea, P.A., Sanders, D.: Voltage sensitivity of H^+/Ca^{2+} antiport in higher plant tonoplast suggests a role in vacuolar calcium accumulation. J. Biol. Chem. **265**(17), 9617–9620 (1990)
82. Kashiba, M., Kajimura, M., Goda, N., Suematsu, M.: From O_2 to H_2S: a landscape view of gas biology. Keio J. Med. **51**(1), 1–10 (2002)
83. Suematsu, M.: Gas biology: how do the gases conduct protein function in vivo? J. Jpn. Biochem. Soc. **74**(11), 1317–1328 (2002) [in Japanese]
84. Suematsu, M.: Quartet signal transducers in gas biology. Antioxid. Redox Sig. **5**(4), 435–437 (2003)
85. Kajimura, M., Fukuda, R., Bateman, R.M., Yamamoto, T., Suematsu, M.: Interactions of multiple gas-transducing systems: hallmarks and uncertainties of CO, NO, and H_2S gas biology. Antioxid. Redox Sig. **13**(2), 157–192 (2010)
86. Nakao, A., Toyoda, Y.: Book review: gas biology research in clinical practice, edited by Toshikazu Yoshikawa and Yuji Naito. Med. Gas Res. **1**, 1 (2011)
87. Kajimura, M., Nakanishi, T., Takenouchi, T., Morikawa, T., Hishiki, T., Yukutake, Y., Suematsu, M.: Gas biology: tiny molecules controlling metabolic systems. Respir. Physiol. Neurobiol. **184**(2), 139–148 (2012)
88. Chang, C., Bleecker, A.B.: Ethylene biology. More than a gas. Plant Physiol. **136**(2), 2895–2899 (2004)
89. Kunz, T.H., Gauthreaux, S.A., Hristov, N.I., Horn, J.W., Jones, G., Kalko, E.K., Larkin, R.P., McCracken, G.F., Swartz, S.M., Srygley, R.B., Dudley, R., Westbrook, J.K., Wikelski, M.: Aeroecology: probing and modeling the aerosphere. Integr. Comput. Biol. **48**(1), 1–11 (2008)
90. Hristov, N.I., Betke, M., Kunz, T.H.: Applications of thermal infrared imaging for research in aeroecology. Integr. Comput. Biol. **48**(1), 50–59 (2008)
91. Swartz, S.M., Breuer, K.S., Willis, D.J.: Aeromechanics in aeroecology: flight biology in the aerosphere. Integr. Comput. Biol. **48**(1), 85–98 (2008)
92. Stakman, E.C., Christensen, C.M.: Aerobiology in relation to plant disease. Bot. Rev. **12**(4), 205–253 (1946)
93. Ponti, I., Cavanni, P.: Aerobiology in plant protection. Aerobiologia **8**(1), 94–101 (1992)
94. Fitt, B.D.L.: Aerobiology and plant pathology. Gewasbescherming **28**, 48–48 (1997)
95. Rundquist, B.C., Harrington, J.J.A., Goodin, D.G.: Mesoscale satellite bioclimatology. Prof. Geogr. **52**(2), 331–344 (2000)
96. Biel, E.R.: Microclimate, bioclimatology, and notes on comparative dynamic climatology. Am. Sci. **49**(3), 326–357 (1961)
97. Valencia-Barrera, R.M., Comtois, P., Fernández-González, D.: Biogeography and bioclimatology in pollen forecasting. Grana **40**(4–5), 223–229 (2001)
98. Mandrioli, P.: Biometeorology and its relation to pollen count. Adv. Aerobiol. (Birkhäuser, Basel) 37–41 (1987)
99. Wellington, W.G.: Biometeorology of dispersal. Bull. Entomol. Soc. Am. **29**(3), 24–30 (1983)
100. Schein, R.D.: Biometeorology and plant disease. AIBS Bull. 34–37 (1963)
101. Schein, R.D.: Biometeorology in plant disease forecasting. Bull. Am. Meteorol. Soc. **43**(11), 628 (1962)
102. Gradov, O.V.: A novel tool for monitoring of forest plant growth and development stages. In: Complexation of Spectroscopy, Gas Chromatography and Direct Mass Spectrometry. GRIN Verlag [GRIN Publishing GmbH], München (2018). ISBN 9783668874909; ISBN 9783668874916
103. Gradov, O.V.: [Chromatographic auxanometry and GC-MS-auxanometry in forest plant species vegetation phenological monitoring based on flavor and gas chemistry principles with automatic pattern recognition (climatic, meteorological, taxonomic and phenospectral)]. Optim. Prot. Ecosyst. **10**(29), 30–45 (2014) [in Ukrainian]

104. Grädow, O.: Novel "phenospectral auxanometry" using complexation of optical spectroscopy and chromatographic auxanometry or GC-MS-auxanometry in forest plant species vegetation phenological monitoring based on gas & flavor chemistry principles. Int. J. Green Herb. Chem. Section A Green Chem. **3**(2), 555–579 (2014)
105. Gradov, O.V.: Multiparametric multivariate patch-clamp-spectroscopy as a method for plant signaling characterization and complementary systematic descriptor for biochemical taxonomy and biogeographic biochemical plant signaling mapping based on correlational phenospectal auxanometry. In: Signal Systems of Plants, pp. 79–81. Saint Petersburg State University, SPbU Press (2016)
106. Tout, D.G.: Biometeorology. Prog. Phys. Geogr. **11**(4), 473–486 (1987)
107. Gosling, S.N., Bryce, E.K., Dixon, P.G., Gabriel, K.M.A., Gosling, E.Y., Hanes, J.M., Hondula, D.M., Liang, L., Mac-Lean, P.A.B., Muthers, S., Nascimento, S.T., Petralli, M., Vanos, J.K., Wanka, E.R.: A glossary for biometeorology. Int. J. Biometeorol. **58**(2), 277–308 (2014)
108. Galimov, E.M. (ed.) The Biological Fractionation of Isotopes. Academic Press (1985). ISBN 9780122739705
109. Liu, H., Tian, Z.: Analysis of plant development stage based on the photoperiod-sensitivity of rice and ramie. Life Sci. Res. **9**(1), 60–62 (2004)
110. Mancinelli, A.L.: Spectral sensitivity of the high irradiance responses of plant photomorphogenesis. Plant Physiol. **63**(5), 155–155 (1979)
111. Borchert, R., Renner, S.S., Calle, Z., Navarrete, D., Tye, A., Gautier, L., Spichiger, R., von Hildebrand, P.: Photoperiodic induction of synchronous flowering near the Equator. Nature **433**(7026), 627–629 (2005)
112. Evans, L.T.: Inflorescence initiation in Lolium temulentum L. I. Effect of plant age and leaf area on sensitivity to photoperiodic induction. Aus. J. Biol. Sci. **13**(2), 123–131 (1960)
113. Quastler, H., Baer, M.: Inhibition of plant growth by irradiation. II. Development and sensitivity. J. Cell Comp. Physiol. **33**, 349–364 (1949)
114. Robberecht, R., Caldwell, M.M.: Leaf epidermal transmittance of ultraviolet radiation and its implications for plant sensitivity to ulraviolet-radiation induced injury. Oecologia **32**(3), 277–287 (1978)
115. Mirecki, R.M., Teramura, A.H.: Effects of ultraviolet-B irradiance on soybean V. The dependence of plant sensitivity on the photosynthetic photon flux density during and after leaf expansion. Plant Physiol. **74**(3), 475–480 (1984)
116. Deckmyn, G., Martens, C., Impens, I.: The importance of the ratio UV-B/PAR during leaf development as determining factor of plant sensitivity to increased UV-B irradiance: effects on growth, gas-exchange and pigmentation of bean plants (Phaseolus vulgaris L). Plant Cell Environ. **17**, 295–301 (1994)
117. Hofmann, R.W., Campbell, B.D., Fountain, D.F.: Sensitivity of white clover to UV-B radiation depends on water availability, plant productivity and duration of stress. Glob. Change Biol. **9**(3), 473–477 (2003)
118. Krizek, D.T.: Influence of PAR and UV-A in determining plant sensitivity and photomorphogenic responses to UV-B radiation. Photochem. Photobiol. **79**(4), 307–315 (2004)
119. Gradov, O.V.: Experimental setups for ozonometric microscopy. Biomed. Eng. **46**(6), 260–264 (2013)
120. Gradov, O.V., Gradova, M.A.: Methods of electron microscopy of biological and abiogenic structures in artificial gas atmospheres. Surf. Eng. Appl. Electrochem. **52**(1), 117–125 (2016)
121. Van Goethem, T.M.W.J., Azevedo, L.B., Van Zelm, R., Hayes, F., Ashmore, M.R., Huijbregts, M.A.J.: Plant species sensitivity distributions for ozone exposure. Environ. Pollut. **178**, 1–6 (2013)
122. Tingey, D.T., Thutt, G.L., Gumpertz, M.L., Hogsett, W.E.: Plant water status influences ozone sensitivity of bean plants. Agric. Environ. **7**(3–4), 243–254 (1982)
123. Grantz, D.A.: Diel trend in plant sensitivity to ozone: implications for exposure-and flux-based ozone metrics. Atmos. Environ. **98**, 571–580 (2014)

124. Grantz, D.A., Vu, H., Heath, R.L., Burkey, K.: Diel trends in plant sensitivity to ozone: toward parameterization of the defense component of effective flux. AGU Fall Meet. Abs. **1**, 0393 (2011)
125. Davis, D.D., Wood, F.A.: Influence of plant age on the sensitivity of Virginia pine to ozone. Phytopathology **63**(3), 381–388 (1973)
126. Timonen, U., Huttunen, S., Manninen, S.: Ozone sensitivity of wild field layer plant species of northern Europe. A review. Plant Ecol. **172**(1), 27–39 (2004)
127. Kline, L.J., Davis, D.D., Skelly, J.M., Savage, J.E., Ferdinand, J.L.: Ozone sensitivity of 28 plant selections exposed to ozone under controlled conditions. Northeast. Naturalist **15**(1), 57–66 (2008)
128. Städtler, S., Ziegler, H.: Illustration of the genetic differences in ozone sensitivity between the varieties Nicotiana tabacum Bel W3 and Bel B using various plant systems. Botanica Acta **106**(3), 265–276 (1993)
129. Fletcher, J.S., Johnson, F.L., McFarlane, J.C.: Influence of greenhouse versus field testing and taxonomic differences on plant sensitivity to chemical treatment. Environ. Toxicol. Chem. **9**(6), 769–776 (1990)
130. Suseela, V., Tharayil, N., Xing, B., Dukes, J.S.: Labile compounds in plant litter reduce the sensitivity of decomposition to warming and altered precipitation. New Phytol. **200**(1), 122–133 (2013)
131. Cleland, E.E., Collins, S.L., Dickson, T.L., Farrer, E.C., Gross, K.L., Gherardi, L.A., Hallett, L.M., Hobbs, R.J., Hsu, J.S., Turnbull, L., Suding, K.N.: Sensitivity of grassland plant community composition to spatial vs. temporal variation in precipitation. Ecology **94**(8), 1687–1696 (2013)
132. Potts, D.L., Huxman, T.E., Scott, R.L., Williams, D.G., Goodrich, D.C.: The sensitivity of ecosystem carbon exchange to seasonal precipitation and woody plant encroachment. Oecologia **150**(3), 453–463 (2006)
133. Milly, P.C.D.: Sensitivity of greenhouse summer dryness to changes in plant rooting characteristics. Geophys. Res. Lett. **24**(3), 269–271 (1997)
134. Feeley, K.J., Malhi, Y., Zelazowski, P., Silman, M.R.: The relative importance of deforestation, precipitation change, and temperature sensitivity in determining the future distributions and diversity of Amazonian plant species. Glob. Change Biol. **18**(8), 2636–2647 (2012)
135. Gradow, O. (ed.): Towards Phenospectral GC-Auxanometry. Scholars' Press, Düsseldorf (2018). ISBN 978-620-2-30586-0
136. Springate, D.A., Kover, P.X.: Plant responses to elevated temperatures: a field study on phenological sensitivity and fitness responses to simulated climate warming. Glob. Change Biol. **20**(2), 456–465 (2014)
137. Schleip, C., Rais, A., Menzel, A.: Bayesian analysis of temperature sensitivity of plant phenology in Germany. Agric. For. Meteorol. **149**(10), 1699–1708 (2009)
138. Luxmoore, R.J., Stolzy, J.L., Holdeman, J.T.: Sensitivity of a soil-plant-atmosphere model to changes in air temperature, dew point temperature, and solar radiation. Agric. Meteorol. **23**, 115–129 (1981)
139. Janssens, I.A., Carrara, A., Ceulemans, R.: Annual Q_{10} of soil respiration reflects plant phenological patterns as well as temperature sensitivity. Glob. Change Biol. **10**(2), 161–169 (2004)
140. McMurchie, E.J.: Temperature sensitivity of ion stimulated ATPases associated with some plant membranes. In: Low Temperature Stress in Crop Plants: The Role of the Membrane, pp. 163–177. Academic Press, New York (1979)
141. Boyd, F.T., Dickson, D.W.: Plant-parasitic nematodes: occurrence in Florida, sensitivity to temperatures, and effect on tropical forage yields. Proc. Soil Crop Sci. Soc. Florida **31**, 267–268 (1971)
142. Rutishauser, T., Schleip, C., Sparks, T.H., Nordli, O., Menzel, A., Wanner, H., Jeanneret, F., Luterbacher, J.: Temperature sensitivity of Swiss and British plant phenology from 1753 to 1958. Clim. Res. **39**, 179–190 (2009)

143. Rutter, A.P., Schauer, J.J., Shafer, M.M., Creswell, J., Olson, M.R., Clary, A., Robinson, M., Parman, A.M., Katzman, T.L.: Climate sensitivity of gaseous elemental mercury dry deposition to plants: impacts of temperature, light intensity, and plant species. Environ. Sci. Technol. **45**(2), 569–575 (2010)
144. Woodcock, A.E.R., Wilkins, M.B.: The geoelectric effect in plant shoots: II. Sensitivity of concentration chain electrodes to reorientation. J. Exp. Botany **20**(4), 687–697 (1969)
145. Laurinavicius, R., Svegzdienė, D., Gaina, V.: Force sensitivity of plant gravisensing. Adv. Space Res. **27**(5), 899–906 (2001)
146. Salisbury, F.B., Gillespie, L., Rorabaugh, P.: Gravitropism in higher plant shoots V. Changing sensitivity to auxin. Plant Physiol. **88**(4), 1186–1194 (1988)
147. Hunter, D.M., McFarlane, W., Sykes, A.G., Dennison, C.: Effect of pH on the self-exchange reactivity of the plant plastocyanin from parsley. Inorg. Chem. **40**(2), 354–360 (2001)
148. Pantoja, O., Smith, J.A.C.: Sensitivity of the plant vacuolar malate channel to pH, Ca^{2+} and anion-channel blockers. J. Membr. Biol. **186**(1), 31–42 (2002)
149. Rutkowska, A.: Sensitivity of plant and soil indices in evaluating the long-term consequences of soil mining from reserves of phosphorus, potassium, and magnesium. Commun. Soil Sci. Plant Anal. **44**(1–4), 377–389 (2013)
150. Silva, S.G., Nóbrega, J.A., Jones, B.T., Donati, G.L.: Magnesium nitrate as a chemical modifier to improve sensitivity in manganese determination in plant materials by tungsten coil atomic emission spectrometry. J. Anal. At. Spectrom. **29**(8), 1499–1503 (2014)
151. Poot, M., Gross, O., Epe, B., Pflaum, M., Hoehn, H.: Cell cycle defect in connection with oxygen and iron sensitivity in Fanconi anemia lymphoblastoid cells. Exp. Cell. Res. **222**(2), 262–268 (1996)
152. Horvath, I., Magyar, K., Gado, I.: The effect of methylene blue on the iron sensitivity of Streptomyces rimosus fermentations. Acta Microbiol. Acad. Sci. Hung. **6**, 47–50 (1959)
153. Cassin, G., Mari, S., Curie, C., Briat, J.F., Czernic, P.: Increased sensitivity to iron deficiency in Arabidopsis thaliana overaccumulating nicotianamine. J. Exp. Bot. **60**(4), 1249–1259 (2009)
154. Flint, D.H., Emptage, M.H.: Dihydroxyacid dehydratase: isolation, characterization as Fe-S proteins, and sensitivity to inactivation by oxygen radicals. In: Biosynthesis of Branched Chain Amino Acids, pp. 285–314. VCH Publishers, New York (1990)
155. Blight, A.: Sensitivity of iron uptake, plant growth and nodulation of narrow-leafed lupin genotypes to bicarbonate. B.Sc. thesis, The University of Western Australia, Perth, Australia (1989)
156. Park, M.J., Jung, H.S., Kim, Y.J., Kwon, Y.J., Lee, J.K., Park, C.M.: High-sensitivity fluorescence imaging of iron in plant tissues. Chem. Commun. **50**(62), 8547–8549 (2014)
157. Gradov, O.V., Jablokov, A.G.: Novel morphometrics-on-a-chip: CCD or CMOS-lab-on-a-chip based on discrete converters of different physical and chemical parameters of histological samples into the optical signals with positional sensitivity for morphometry of non-optical patterns. J. Biomed. Technol. **3**(2), 1–29 (2016)
158. Katayama, T., Nagamatsu, T.: Radiosensitivity in plants: I. Relation between the water content of some crop seeds and their sensitivity to different doses of X-rays and γ-rays. Jpn. J. Breed. **16**(2), 77–82 (1966)
159. Francis, C.M., Gladstones, J.S.: Studies on the use of mutagenic agents in plant breeding. I. The effect of seed moisture content on sensitivity to X-ray in Lupinus angustifolius. Crop Pasture Sci. **14**(1), 12–19 (1963)
160. Shaikh, M.A.Q.: Radiation sensitivity studies of Lathyrus species and Vicia ervilia (L.) Willd. III. Dose-effects on yield components and seed yields per plant. Bangladesh J. Agril. Sci. **3**(2), 130–138 (1976)
161. LebaiJuri, M., Omar, M., Yusof, N.: Sensitivity of the conidia of plant pathogenic fungi to Gamma-rays, electron particles and X-ray (Bremsstrahlung) irradiation. World J. Microbiol. Biotechnol. **11**(6), 610–614 (1995)
162. Gyoergyne Czeck, B: Plant virus sensitivity to gamma irradiation. In: Symposium on the Possibilities of the Application of Irradiation in Agriculture and Food Industry, Budapest, Hungary; 15–16 March 1979, pp. 34–37 (1979)

163. Kumta, U.S., Sawant, P.L., Ramakrishnan, T.V.: Radiation sensitivity studies of plant pigments I. In Vitro lability of β-carotene in lipid solvents. Radiat. Bot. **10**(2), 161–167 (1970)
164. Sawant, P.L., Ramakrishnan, T.V., Kumta, U.S.: Radiation sensitivity studies of plant pigments II. Effects of radiation on carotenoid fractions of orange juice and mango pulp. Radiat. Bot. **10**(2), 169–174 (1970)
165. Ramakrishnan, T.V., Sawant, P.L., Kumta, U.S.: Radiation sensitivity studies of plant pigments. III. In vitro stability of β-carotene in aqueous dispersions. Radiat. Bot. **10**(5), 395–399 (1970)
166. Hazama, Y., Hazama, K., Ehrenberg, L.: The influence of paramagnetic ions on the radiation sensitivity of plant seeds. Radiat. Bot. **3**(1), 7–18 (1963)
167. Mewissen, D.J., Damblon, J., Bacq, Z.M.: Comparative sensitivity to radiation of seeds from a wild plant grown on uraniferous and non-uraniferous soils. Nature **183**, 1449 (1959)
168. Sparrow, A.H.: Relationship between chromosome volume and radiation sensitivity in plant cells [Brookhaven National Lab, Upton, NY, №. BNL-7866; CONF-456–1 (1963)]. In: 18th Annual Symposium on Fundamental Cancer Research and Cellular Radiation Biology, Houston, Tex, (1964)
169. Myers, R.F.: The sensitivity of some plant-parasitic and free-living nematodes to gamma and X-irradiation. Nematologica **5**(1), 56–63 (1960)
170. Woodwell, G.M.: Sensitivity to ionizing radiation: Major ecosystems and dominant plant species. In: Environmental Biology, FASECB, pp. 181–182 (1966)
171. Stolte, K.W.: Sensitivity of plant ecosystems in desert areas to gaseous pollutants. In: Acid rain and air pollution in desert park areas. National Park Service, pp. 42–51. United States Department of the Interior (1991)
172. Fang, J., Zhang, H., Yang, N., Shao, L., He, P.: Gaseous pollutants emitted from a mechanical biological treatment plant for municipal solid waste: Odor assessment and photochemical reactivity. J. Air Waste Manag. Assoc. **63**(11), 1287–1297 (2013)
173. Bushnell, J.: Sensitivity of the potato plant to soil aeration. Agron. J. **27**(4), 251–253 (1935)
174. Yoshikawa, T., Naito, Y. (eds.): Gas Biology Research in Clinical Practice. Karger Medical and Scientific Publishers (2011)
175. Lanza, R.C.: Imaging gas detectors for medicine and biology. Med. Prog. Technol. **15**(3–4), 185–198 (1988)
176. Farkas, G.L., Konrad, E., Kiraly, Z.: The effect of light on the malonate sensitivity of plant respiration. Physiol. Plant. **10**(2), 346–355 (1957)
177. Beerling, D.J., Terry, A.C., Hopwood, C., Osborne, C.P.: Feeling the cold: atmospheric CO_2 enrichment and the frost sensitivity of terrestrial plant foliage. Palaeogeogr. Palaeoclimatol. Palaeoecol. **182**(1), 3–13 (2002)
178. Zhan, X., Wilks, D.S.: CO_2 sensitivity of plant leaf transpiration. In: Proceedings of the 21-st Conference on Agricultural and Forest Meteorology, San Diego, CA, pp. 7–11 (1994)
179. Letendre, J., Poulin, M., Rochefort, L.: Sensitivity of spectral indices to CO_2 fluxes for several plant communities in a Sphagnum-dominated peatland. Can. J. Remote. Sens. **34**(2), S414–S425 (2008)
180. Darrall, N.M.: The sensitivity of net photosynthesis in several plant species to short-term fumigation with sulphur dioxide. J. Exp. Bot. **37**(9), 1313–1322 (1986)
181. Steubing, L., Fangmeier, A.: SO_2-sensitivity of plant communities in a beech forest. Environ. Pollut. **44**(4), 297–306 (1987)
182. Seliskar, D.M., Smart, K.E., Higashikubo, B.T., Gallagher, J.L.: Seedling sulfide sensitivity among plant species colonizing Phragmites-infested wetlands. Wetlands **24**(2), 426–433 (2004)
183. Cox, R.M.: Sensitivity of forest plant reproduction to acid rain. In:: Proceedings of the International Conference "Acid rain and forest resources", Quebec City, Canada (1983)
184. Rutishauser, T.: Historical phenology: plant phenological reconstructions and climate sensitivity in Northern Switzerland. Ph.D. thesis (2007)
185. Lapenis, A., Henry, H., Vuille, M., Mower, J.: Climatic factors controlling plant sensitivity to warming. Clim. Change **122**(4), 723–734 (2014)

186. Jonsdottir, I.S., Magnusson, B., Gudmundsson, J., Elmarsdottir, A. Hjartarson, H.: Variable sensitivity of plant communities in Iceland to experimental warming. Global Change Biol. **11**, 553563 (2005)
187. Goswami, A., Main, D.J., Noto, C.R., Moore, T.L., Scotese, C.: Cretaceous climate sensitivity study using dinosaur and plant paleobiogeography. AGU Fall Meet. Abs. **1**, 1358 (2009)
188. Su, W.-A., Mi, R.-O., Wang, W.-Y., Wang, H.-C.: The influence of cold hardening on plant stress sensitivity. Acta Plant Physiol. **3**, 11 (1990)
189. Kouressy, M., Dingkuhn, M., Vaksmann, M., Heinemann, A.B.: Adaptation to diverse semi-arid environments of sorghum genotypes having different plant type and sensitivity to photoperiod. Agric. For. Meteorol. **148**(3), 357–371 (2008)
190. Beckerson, D.W., Cain, N., Hofstra, G., Ormrod, D.P., Campbell, P.A.: A guide to-plant-sensitivity to environmental-stress. Landscape Archit. **70**(3), 299–303 (1980)
191. Milne, J.A., Hartley, S.E.: Upland plant communities-sensitivity to change. CATENA **42**(2), 333–343 (2001)
192. Joyce, C.: The sensitivity of a species-rich flood-meadow plant community to fertilizer nitrogen: the Lužnice river floodplain, Czech Republic. Plant Ecol. **155**(1), 47–60 (2001)
193. Esther, A., Groeneveld, J., Enright, N.J., Miller, B.P., Lamont, B.B., Perry, G.L.W., Blank, F.B., Jeltsch, F.: Sensitivity of plant functional types to climate change: classification tree analysis of a simulation model. J. Veg. Sci. **21**(3), 447–461 (2010)
194. Kim, D.-H.: SYMPOSIUM 4-Biochemistry and Molecular Biology of Plant-: biochemical basis for tissue-specific difference in auxin sensitivity between rice shoots and roots In: The Biochemical and Molecular Biology Society; formerly Korean Society of Biochemical and Molecular Biology, vol. 1, p. 56 (1999)
195. Otani, H., Nishimura, S., Kohmoto, K.: Nature of specific susceptibility to Alternaria kikuchiana in Nijisseiki Cultivar among Japanese pears: VI. Effect of known plant hormones on sensitivity of pear leaves to host-specific toxin, vol. 11, pp. 1–6. Archives of Tottori University, Faculty of Agriculture (1976)
196. Tingey, D.T., Andersen, C.P.: The physiological basis of differential plant sensitivity to changes in atmospheric quality. In: Ecological Genetics and Air Pollution, pp. 209-235. Springer, New York (1991)
197. Krüger, T.P., Ilioaia, C., Valkunas, L., van Grondelle, R.: Fluorescence intermittency from the main plant light-harvesting complex: sensitivity to the local environment. J. Phys. Chem. B **115**(18), 5083–5095 (2011)
198. Stone, J.M., Liang, X., Nekl, E.R., Stiers, J.J.: Arabidopsis AtSPL14, a plant-specific SBP-domain transcription factor, participates in plant development and sensitivity to fumonisin B1. Plant J. **41**(5), 744–754 (2005)
199. Li, L., Lin, L., Li, X., Lu, Y.: Sensitivity of photosystem II activity in immature and mature leaves of desert sun plant Alhagi sparsifolia Shap. to light stress. Acta Physiologiae Plantarum **37**(8), 1–10 (2015)
200. Evans, L.T.: Inflorescence initiation in Loliurn temulentum L. 1. Effect of plant age and leaf area on sensitivity to photoperiodic induction. Aust. J. Biol. Sei. **13**, 123–131 (1960)
201. Feldman, N.L.: The influence of wound injury on sensitivity of plant cells. In: Problems of Cytology and Protistology, pp. 216–223. Acad. Sci. USSR, Moscow-Leningrad (1960)
202. Mehlhorn, H., Wellburn, A.R.: Stress ethylene formation determines plant sensitivity to ozone. Nature **327**(6121), 417–418 (1987)
203. Perelygin, V.M., Shpirt, M.B., Aripov, O.A., Ershova, V.I.: Effect of some pesticides on immunological reactivity. Gig Sanit **12**, 29–33 (1971). [in Russian]
204. Gradov, O.V.: Is it possible to consider kinetomics and argyromics as novel trends of functional morphology and systems biology of parasitic and nonparasitic Protozoa? A brief analytical review. J. Biomed. Technol. **3**(1), 1–15 (2016)
205. Goldhamer, D.A., Fereres, E., Mata, M., Girona, J., Cohen, M.: Sensitivity of continuous and discrete plant and soil water status monitoring in peach trees subjected to deficit irrigation. J. Am. Soc. Hortic. Sci. **124**(4), 437–444 (1999)

206. Hansen, G.K.: A dynamic continuous simulation model of water state and transportation in the soil-plant-atmosphere system: I. The model and its sensitivity. Acta Agriculturae Scandinavica **25**(2), 129–149 (1975)
207. Burló-Carbonell, F., Carbonell-Barrachina, A., Vidal-Roig, A., Mataix-Beneyto, J.: Effects of irrigation water quality on loquat plant nutrition: sensitivity of loquat plant to salinity. J. Plant Nutr. **20**(1), 119–130 (1997)
208. Aggarwal, P.K., Koundal, K.R.: Relative sensitivity of some physiological characteristics to plant water deficits in wheat. Plant Physiol. Biochem. **15**(1), 161–168 (1988)
209. Granier C., Aguirrezabal, L., Chenu, K., Cookson, S.J., Dauzat, M., Hamard, P., Thioux, J.-J., Rolland, G., Bouchier-Combaud, S., Lebaudy, A., Muller, B., SimonneauT., Tardieu F.: PHENOPSIS, an automated platform for reproducible phenotyping of plant responses to soil water deficit in Arabidopsis thaliana permitted the identification of an accession with low sensitivity to soil water deficit. New Phytol. **169**(3), 623–35 (2006)
210. Till, M.R.: The sensitivity of fruit growth rate to the water balance of the plant. Animal Prod. Sci. **5**(16), 85–86 (1965)
211. Xu, H.L., Yamagishi, T., Kumura, A.: Effects of water deficit on photosynthesis in wheat plants. 2: the physiological basis for the difference in photosynthetic sensitivity to water stress among plant parts. Jpn. J. Crop Sci. **56**(4), 461–466 (1987) [in Japanese]
212. Hafner, C., Erdmenger, E., Jung, K., Gehre, M., Schüürmann, G.: Stable isotope tracer technique as a tool to increase the sensitivity of plant growth bioassays-investigations with Pentachlorophenol and Trichloroacetic acid. Fresenius Environ. Bull. **9**(3/4), 225–231 (2000)
213. Markx, G.H., ten Hoopen, H.J.G., Meijer, J.J., Vinke, K.L.: Dielectric spectroscopy as a novel and convenient tool for the study of the shear sensitivity of plant cells in suspension culture. J. Biotechnol. **19**(2), 145–157 (1991)
214. Rosenberg, M.Z.: The hydrodynamic shear sensitivity of suspension cultured plant cells. Ph. D. thesis, Washington University, St. Louis, USA (1987)
215. Gradov, O.V.: Complex cross-correlated monitoring in phytophysiology of aquarium plants and artificiael ecosystems: from hydrochemistry and biophysics to DIY automatization. Aquaflora **2**, 47–51 (2014). [in Russian]
216. Gradov, O.V.: Toward the biophysical, biochemical, biocenotic and biogeographic systematics of model experimental aquarium plants based on complex descriptors of plant cell/tissue response to hydrochemical factors in the framework of physical chemistry. Int. J. Environ. Prob. **4**, 45–55 (2018)
217. Gradov, O.V.: Monitoring of nonstandard greenhouse substrates using submersible soil lab-on-a-chip approach. Greenhouse World **5**, 43–46 (2012). [in Russian]
218. Gradov, O.V.: [Telemetric lab-on-a-chip systems as contemporary alternatives of submersible soil chambers and Rossi-Cholody slides for soil microbiome investigations]. In: Actual Aspects of Contemporary Microbiology, pp. 62–65 (2012) [in Russian].
219. Bolkhovitinoff, A., Krukowskikh, V., Gradov, O.: Towards the analytical approximation of weathering forms based on fitting of the geomorphological structures by the "Tafeln Höherer Funktionen" profile database. Eur. Geogr. Stud. **5**, 21–31 (2018)
220. Alexandrov, P.L., Gradov, O.V.: The role of ion channels in biogeochemical evolution and phenetic systematics based on key libraries of patch-clamp fingerprints for different reference model geochemical conditions. Proc. Biogeochem. Lab. **26**, 85–101 (2017). [in Russian]
221. Gradov, O., Gradova, M.: MS-patch-clamp or the possibility of mass spectrometry hybridization with patch-clamp setups for single cell metabolomics and channelomics. Adv. Biochem. **3**, 66–71 (2015)
222. Gradov, O., Orehov, F.: Correlational patch-clamp spectrometry of the ion channels. Biomed. Eng. Electron. **13**, 5–28 (2016). [in Russian]
223. Jablokow, A., Gradow, O.: Verifying continuity of membranous organelles and measurements of exchange rate between the nucleus and cytoplasm using FLIP-like MALDI-based imaging [63rd ASMS Conference on Mass Spectrometry and Allied Topics]. J. Am. Soc. Mass Spectrom **26**, 169-ThP662 (2015)

224. Jablokow, A., Gradow, O.: MS-FRAP or MALDI imaging setups with programmable laser sources: a new way to the diffusion, molecular mobility and binding measurements [63rd ASMS Conference on Mass Spectrometry and Allied Topics]. J. Am. Soc. Mass Spectrom **26**, 63-MP175 (2015)
225. Bolkhovitinov, A.: Mathematical qualimetry for mass spectrometry: optimization and harmonization of sample preparation, data processing and data mining. J. Am. Soc. Mass Spectrom **26**, 73-MP425 (2015)
226. Jablokow, A.G., Skrynnik, A.A., Orekhov, F.K., Nasirov, P.A., Gradov, O.V.: "MALDI-FLIP-on-a-chip" and "MALDI-FRAP-on-a-flap": novel techniques for soil microbiology and environmental biogeochemistry. I. Biogeosyst. Tech. **4**(2), 140–188 (2017)
227. Jablokow, A.G., Nasirov, P.A., Orekhov, F.K., Gradov, O.V.: MALDI-FLIP-on-a-chip" and "MALDI-FRAP-on-a-flap": novel techniques for soil microbiology and environmental biogeochemistry. II. Biogeosyst. Tech. **5**, 3–56 (2018)
228. Adamovich, E.D., Gradov, O.V.: Teraohmmeter-assisted measurements of plant leaf resistance. 2. Kinetic registration of plant leaf resistance (10x time-lapse registration). IEEE DataPort. https://doi.org/10.21227/x4nc-w679 (2018)
229. Adamovich E.D., Gradov O.V.: Teraohmmeter-assisted measurements of plant leaf resistance. 3. Oscillations of plant leaf resistance. IEEE DataPort. https://doi.org/10.21227/5j37-ep41 (2018)
230. Adamovich, E.D., Gradov, O.V.: Teraohmmeter-assisted measurements of plant leaf resistance. 4. Accelerated photoinduced kinetics of electric impedance shifting under UV irradiation. IEEE DataPort. https://doi.org/10.21227/4nc9-9c71 (2018)
231. Jablokov, A., Skrynnik, A., Gradov, O.: Applications of tunable diode lasers and the tunable laser diode spectroscopy in Tschachotin's microbeam setups for irradiation of biological tissues, cells and cellular compartments with positional selectivity. J. Phys. Chem. Biophys. **5**(3), 74 (2015)
232. Gradov, O.V., Notchenko, A.V.: [Semiautomatic dendrochronography for research on morphogenesis and teratomorphosis on the saw cuts of higher plants]. Forestry Eng. J. **4**(8), 47–57 (2012) [in Russian]
233. Orehov, F., Gradov, O.: Methods, approaches and tools of multifactorial chemoinformatics and bioinformatics in geoarchaeological and archaeomineral studies on spectroscopic and spectrographic basis. Geoarcheol. Archeol. Mineral. **2**, 45–48 (2015). [in Russian]

Influence of the Problem of Safety Control of Heuristic Algorithms on the Development of Robotics

Oleg Tikhanychev

Abstract The relevance of choosing the use of autonomous robotic systems for various purposes as the subject of research and algorithmic problems arising in the safety control of algorithms as the object of research is determined by the existing contradiction between the need for autonomation of robotic systems and the complexity of the software implementation of this requirement. The study was carried out using the example of the application of robotics in the field of unmanned vehicles. The implementation of autonomy trends in this area gives rise to certain legal and technological problems associated with the safety of algorithmic support for autonomous systems. A manifestation of such problems is critical application errors, which are difficult to identify at the development stage. In robotics, the main potential cause of these errors is the functioning of the software of autonomous robotic systems outside the framework of the deterministic algorithms that were tested during their development. And if for ordinary technical systems the consequences of application errors are predictable, and the responsibility for their results is determined, then for the case of using autonomous robots, the situation requires a solution. In this paper, using general scientific methods of analysis and synthesis, the main aspects of the current state and prospects of the problem of algorithmization of the use of autonomous systems are studied. On the basis of an overview analysis of algorithmic problems of using autonomous robots, possible consequences and causes of the appearance of algorithmic errors that arise during their use, the statement of the scientific problem of solving the problem of developing heuristic algorithms that determine their application is synthesized.

1 Introduction

Nowadays, autonomous robotic systems are being actively developed. The need for their development and implementation is caused by many factors: an increase in the dynamism of all areas of economic activity, an increase in the scale and variety of

O. Tikhanychev (✉)
Company Group «Technoserv», Yunosti street 13, Moscow 111395, Russia

M. Shamtsyan et al. (eds.), *Robotics, Machinery and Engineering Technology for Precision Agriculture*, Smart Innovation, Systems and Technologies 247,
https://doi.org/10.1007/978-981-16-3844-2_31

controlled systems and other factors. Under the prevailing conditions, the human operator becomes an inertial link in the decision-making chain, and the autonomy of robotic systems becomes objectively necessary: both to ensure safety and to optimize the interaction between controlled objects, including robotic ones. And, as already noted, one of the most obvious ways to solve the problem is the use of artificial intelligence, i.e. "a set of technological solutions that makes it possible to simulate human cognitive functions (including self-learning and search for solutions without a predetermined algorithm) and to obtain, when performing specific tasks, results comparable, at least, with the results of human intellectual activity" [1].

But the practical implementation of autonomous robotic systems is hampered by a number of factors: the objective complexity of the development of application software that ensures autonomous operation, a certain distrust of users in a system that is capable of conditionally independent actions, subjective fear of potential responsibility for possible critical errors in the behavior of autonomous systems, aggravated by the unresolved problem of quality control of software developed for robotic systems, and a number of others [2–5]. The emergence of most of these factors is determined by the fact that an autonomous system, by definition, should be not just independent, but adaptive. It must act not according to predetermined deterministic algorithms. The rules of its behavior must be modified during development, training and application, which requires the implementation of stochastic or even heuristic algorithms [5].

Under the influence of the listed factors, a contradiction between the need for the implementation of autonomous systems and the possibility of introducing such functionality into the practice of robotization is formed. Moreover, with the importance of all the components of the assessment of the developed algorithms, the greatest deterrent factors at the moment are the issues of assessing the safety of algorithms for autonomous robots. The formulated contradiction gives rise to a problem at the junction of the legal and technical areas of regulation: all the methods for assessing quality, ensuring the safety of robotic systems, dividing responsibility between the developer and the operator for the use of autonomous systems are based on the principle of immutability of algorithms embedded in each system, rigidly defining areas of responsibility of each participant in the development, production and use of robotics, i.e. deterministic algorithms. But, if we accept that the used algorithm can be adaptive and, therefore, modifiable, then it already belongs to the class of heuristic, and the developer cannot be fully responsible for the application of the robotic system after modifying what he has developed and tested, but the operator, in turn, cannot fully control the system with the principles of behavior that are not fully known to him. The contradiction between the need to use autonomous robots and the problems of their creation is reflected, first of all, in the development of algorithmic basis and software for them.

2 Materials and Methods

A systematic approach was chosen as a methodological basis for the study. The use of a systematic approach, methods of decomposition and comparative analysis made it possible to fully consider the issues of assessing the safety of algorithms for the use of autonomous robotic systems. To assess the state of the subject area and formulate proposals for improving the process of developing algorithms for autonomous robotic systems, the methods of comparative analysis and generalization are used in the paper.

The source base of the study consists of regulatory and guidance documents and scientific papers that are in the open access.

3 Results

Practice shows that the most urgent are the problems of introducing autonomous robots and the accompanying problems of controlling their algorithmic support in those areas where the objective requirement of autonomation of the use conflicts with the need to ensure the safety of using robotic systems. Realizing this, the developers of systems that implement the components of artificial intelligence are trying to take certain measures. Nowadays, due to the lack of other opportunities—mainly organizational ones. For example, the concept of using artificial intelligence in military affairs, developed by the US Department of Defense, which sets the basic principles for using artificial intelligence systems in military and dual-use systems [6]:

- responsible, or limiting the scope of using artificial intelligence to the minimum necessary value for implementing the range of tasks;
- equitable, early prevention of abnormal behavior of artificial intelligence systems;
- traceable, predictability of behavior of objects controlled by artificial intelligence, understanding of the principles of system behavior by operators;
- reliable, safety of objects controlled using artificial intelligence, determined by the fact that the capabilities of military artificial intelligence systems must be unambiguously interpreted, and the permissible limits of their behavior must be checked and periodically confirmed by appropriate tests.
- governable, the ability of operators to detect and prevent undesirable consequences of using artificial intelligence.

Recommendations of the Parliamentary Assembly of the Council of Europe No. 2102 (2017) of 28.04.2017 "Technological convergence, artificial intelligence and human rights" also contain requirements for ensuring the security of algorithms for controlling intelligent systems (clause 9.3), including robotic ones [7]. Similar provisions are contained in the documents of the Commission on Civil Law Rules on Robotics, approved by the European Parliament [8].

It is quite logical that in Russia, documents that regulate the use of artificial intelligence have also been developed, including in terms of the features regulating the safety of its algorithms [1]:

- safety: the inadmissibility of using artificial intelligence for the purpose of deliberately causing harm to citizens and legal entities, as well as preventing and minimizing the risks of negative consequences of using artificial intelligence technologies;
- transparency: explainability of the work of artificial intelligence and the process of achieving results by it.

But, as the analysis of the content of regulatory and guidance documents shows, none of them describes how the formulated safety principles should be implemented. There are no such instructions in any document regulating the use of robotics in the conduct of armed confrontation, nor in the transport, air and water codes of different countries. At the same time, the problem of controlling algorithms for autonomous robotic systems requires formulation and search for a specific solution method.

Nowadays, the development of algorithms implemented as part of software, including for robotic systems, is carried out according to certain rules specified in regulatory documents: standards (for example, Russian State Standards GOST R 60.2.2.1-2016/ISO 13482:2014; R 60.1.2.2-2016/ISO 10218–2:2011), technical regulations and other documents. But, as the comparative analysis shows, all existing regulatory documents are initially focused on the development of systems that implement deterministic algorithms (Fig. 1).

That is, all control methods defined by existing documents are based on the principle of invariability of the basic properties of algorithms:

- discreteness;
- determinism (certainty);

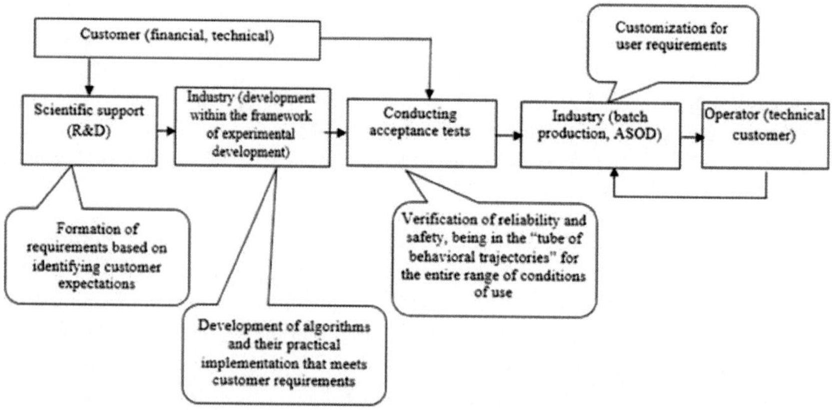

Fig. 1 Principles of control of software algorithms

- comprehensibility;
- termination (finiteness);
- mass character (universality);
- unambiguity;
- effectiveness.

The analysis of algorithms for the presence of possible errors with this approach is carried out by formal methods that provide for the use of formal specifications and, usually, a set of tools for parsing and proving the properties of algorithms.

Proof of the correctness of the developed programs allows us to reveal their properties in relation to the entire range of input data in two ways:

- partial correctness, when the program gives the correct result for those cases when it finishes;
- complete correctness, when the program finishes the work and produces the correct result for all elements from the input data range.

The process of proving the correctness of algorithms is based on comparing the texts of the programs that implement them with the matrix of the predicted ratio of input–output data. The so-called Hoare Triple describes the preconditions and postconditions that guarantee the correct operation of the tested algorithm:

$$\{P\}Q\{S\},$$

where

P—preconditions that must be met before running the program *Q*;
R—postconditions true after completion of the program implementing the algorithm.

As part of the implementation of such methods of analyzing algorithms, the developer's responsibility for the functioning of the robotic system is determined by the boundaries of the acceptance tests, and the operator's responsibility—by knowing the limits of the system's applicability and using it only within the permitted range.

For systems that implement heuristic algorithms, such principles cannot work correctly by definition, primarily due to the fact that it is very problematic to form the output matrix of the Hoare Triplet. The case is similar to stochastic algorithms, but in them, as a rule, the parameters of the distribution laws are known, and it is possible, at least, build a matrix of the output data in the form of a set of ranges. And for heuristic algorithms there is no guarantee of even such a variant, because even with a known set of behavior strategies, a "combinatorial explosion" occurs, and the dimension of the postcondition matrix simply does not allow one to assess the safety of the algorithms.

Taking this into account, the applied approaches nullify all attempts to ensure the safety of the use of autonomous robots by organizational measures, both in the field of armed confrontation [9–12] and in the field of peaceful use of artificial intelligence [13–15]. Analysis of the list and content of guidelines and normative documents shows that within the framework of the existing system of normative and technical documentation, the problems of ensuring the safety of algorithms of autonomous robotic systems cannot be solved.

At the same time, when formulating the requirements for the systems being developed, the customer and the operators of autonomous robotic systems operate with such requirements for algorithms as customizability, modifiability (learnability), and even self-learning. And then, in addition to the standard properties of the algorithms of autonomous systems, there are requirements, for example, to the "honesty" of a commercial system (for example, a forex robot) and other properties inherent in a human rather than a robot. But the determining factor, nevertheless, remains the safety of the behavior of robotic systems, which is ensured, first of all, by the predictability of the algorithms of its behavior.

As the analysis of the subject area shows, the solution to this problem can be found by providing dynamic control of algorithms embedded in autonomous systems, taking into account their possible modification during operation (training).

In essence, the formulated problem is similar to one of the problems of artificial intelligence: the constraint satisfaction problem. But to control the safety of modifiable and self-learning algorithms implemented in autonomous robotic systems, such a task in a direct formulation has not yet been posed.

The problem of controlling algorithms is complicated by the fact that, with a single purpose of using a robotic system, the algorithm of its operation, as a rule, is not something homogeneous, it consists of several parallel algorithms running simultaneously that control the robot's subsystems: an algorithm for processing information about the external environment (for example, computer vision), an algorithm for organizing movement in space, an algorithm for using actuators, and others. When robots are used in groups, these algorithms are supplemented by the requirements for predicting the behavior of interacting and sometimes opposing systems. Each of these private algorithms, in turn, can include embedded algorithms for describing individual operations, and so on. Theoretically, it is possible to control the safety of each particular algorithm with the assumption that the overall result of actions will be also safe, but there is no confirmed guarantee for this, as well as there is no strictly mathematical justification for the safety of the system behavior as a whole.

4 Discussion

Taking into account the lack of reliable methods for controlling the safety of heuristic algorithms, it is proposed to provide a solution to the problem by another method, simpler and more reliable—to form a system of target safety constraints that ensure continuous monitoring of the behavior of robotic systems. Based on the logic of the development and application of such schemes, it is proposed to divide the constraints into the following groups: "internal" and "external" constraints of algorithms, as well as their combination.

The "internal" constraints of algorithms include, for example, the boundary conditions specified when using optimization methods, i.e. constraints on the parameters of the objective function in private control algorithms: boundary conditions, "penalty" functions, etc.

For a typical optimization problem, "internal" constraints can be written as a pair:

- target function $F(x) = \sum_{j=1}^{n} c_j x_{j \to \max(\min)}$ with the calculated indicators of the control vector x_j at constant coefficients c_j;
- a system of constraints of the form $\sum_{j=1}^{n} a_{ij} x_j \leq b_i, i = 1, 2, \ldots, m_i$.

When using this approach, it is necessary to take into account that the constraints are set for a specific type of objective function, for which, in the process of finding a solution (performing a task), only the controlled parameters change. Assuming that the modification of behavior algorithms provided for in autonomous robotic systems can, with a high probability, lead to a change in the parameters of the behavioral function, the use of "internal" constraints cannot unambiguously guarantee the safety of the behavior of the system as a whole. And given the multiplicity of algorithms used in parallel, the probability of a critical security error increases.

The "external" constraints include setting the boundaries of the system's behavior, which describe the permissible area of its application in the form of a set of boundary rules. In other words, the limitation is not of individual parameters of the function $F(x)$ and not of particular algorithms for performing operations, but of options for the final value of the objective function within the space of possible behavioral trajectories (areas).

The most vivid example of the implementation of such an approach is the use of mechanical limiters that reduce the amplitude of movement of the actuators of industrial robots in specified directions. An example in terms of the "external" constraints of software algorithms is the "three rules of robotics" proposed by science fiction writer Isaac Asimov.

The combination of "external" and "internal" constraints can be provided by an integrated approach, the essence of which may be the use of the dynamic formation of "internal" constraints by the system itself when modifying algorithms taking into account some boundary "external" framework of behavior. However, the latter option has not been practically implemented anywhere and is not described in theory until now.

The analysis of the subject area showed that the choice between "internal" and "external" constraints of algorithms, as well as constraints automatically generated by the system itself within the framework of modifying the algorithms of its behavior, is one of the main problems of ensuring the safety of functioning of autonomous systems.

If the method of "internal" constraints is chosen to solve the problem of ensuring the safety of using autonomous robotic systems, the known approaches can be used to limit the algorithms, which are implemented as constraints in optimization methods for finding the optimum of a behavioral function: search methods with pruning, methods of penalty functions, matching methods, genetic algorithms and other methods.

When choosing the method of "external" constraints, as the latter, the principle of setting the general boundaries of the system's behavior can be used: in space, in time, in the form of a set of allowed (forbidden) spatial and logical variables.

The mathematical apparatus for automatic formation of constraints currently exists in an explicit form, but it can be assumed that it will be similar to the method of "internal" constraints with the addition of elements of data aggregation and forecasting the consequences of modification.

All the approaches under consideration have their advantages and disadvantages from the point of view of ease of implementation, the impact of the functionality of the controlled system on the constraint, and ensuring the guaranteed safety of its use.

An analysis of these features, taking into account the role and place of software that implements algorithms for controlling robotic systems, allows drawing conclusions about the effect of constraint methods on the efficiency and safety of the system as a whole. The comparison results are shown in Table 1.

Comparative analysis of the described approaches shows approximately equal preference for control methods of heuristic algorithms. But, given the current state of the methods for controlling the algorithms being developed, one can definitely be sure only in the reliability of the method of "external" constraints, on the basis of which only this approach remains applicable at the current stage.

At the same time, the analysis of the peculiarities of the applied methods for setting constraints shows that "external" constraints can impose certain limits on the use of a robotic system, reducing its functionality. For autonomous systems, this factor can be critical, both from the point of view of the implementation of target functions and in terms of security. And, taking this into account, the methods of "internal" control and self-control can be considered more promising, since they are more reliable in the development of specific algorithms and the formation of general algorithms, providing greater flexibility of behavior.

Taking this into account, we can conclude that the area associated with the use of "internal" constraints and, especially, in terms of self-forming constraints of heuristic algorithms will develop as technologies and mathematical methods improve. Most likely, as it usually happens—according to the principle "from simple to complex", from elementary constraints of individual algorithms to the development of more complex, but more effective methods for limiting interrelated algorithmic groups.

In the future, with the development of artificial intelligence technologies and operational forecasting, the possibilities of using self-forming constraints will expand, and it is likely that this particular approach will become the main one.

5 Conclusions

In this paper, the problem of controlling algorithms for autonomous robotic systems is formulated in the most general form. The problem, for the solution of which the statement of the problem was formulated, is extremely relevant for any autonomous systems that pose a potential danger to humans: transport and industrial ones,

Table 1 Comparison of algorithm security methods

Method name	Method content	Qualitative characteristics of the method			
		Ease of implementation	Reliability of control of system behavior within constraints	Ability to control heuristic algorithms	No negative impact on the functionality of systems
"External" constraints for an autonomous system, forcibly setting the safe boundaries of its functioning	Spatio-temporal and logical constraints on system behavior	+	+	+	–
"Internal" constraints of the system, providing the constraints of each particular algorithm that determines certain aspects of the safety of its behavior	Constraints similar to those used in optimization calculation methods	+	+	–	+
Self-forming algorithm constraints	Constraints that are automatically generated when modifying particular and general algorithms of the system's behavior in the training process and ensuring control of each modification of its components	–	+	+	+

possessing kinetic energy, combat, the very purpose of which is destruction [16–18]. This problem requires a solution: organizational one, in terms of making changes to the regulatory documentation, and technological one, in terms of the development of control methods for modified algorithms [19–21]. The latter can be done, for a start, by any available method. Later, with the development of technologies—by the most rational of the newly developed methods. The statement of the problem formulated in the paper is one of the prerequisites for solving this problem.

References

1. Decree of the President of the Russian Federation of October 10, 2019 No 490 "On the Development of Artificial Intelligence in the Russian Federation". http://www.kremlin.ru/acts/bank/44731. Last accessed 21 Nov 2020
2. Tyutyunnikov, N.: Military thought in terms and definitions. In: Informatization of the Armed Forces. vol. 3, 472p. Pero Publ, Moscow (2018)
3. Simulin, A., et al.: Some aspects of the use of robotics in military affairs. In: Conference Collection Center Sociosphere, vol. 27, pp. 67–71 (2015)
4. Hover and stare: FCS testing UAVS Military.com (2008). https://www.military.com/defensetech/2008/05/30/hover-and-starefcs-testing-uavs . Last accessed 21 Nov 2020
5. Artificial Intelligence: A Legal Perspective: A Scientific Monograph ROO "Institut gosudarstvenno-konfessional'nykh otnoshenii i prava", p. 257. Buki Vedi Publ, Moscow (2017)
6. DOD Adopts Ethical Principles for Artificial Intelligence. US Dept of Defense (2020). https://www.defense.gov/Newsroom/Releases/Release/Article/2091996/dod-adopts-ethical-principles-for-artificial-intelligence. Last accessed 28 Jan 2020
7. Recommendation № 2102 of parliamentary assembly of the council of Europe. Technol. Convergence Artif. Intell. Human rights **28** (2017). http://assembly.coe.int/nw/xml/XRef/Xref-XML2HTML-en.asp?fileid=23726. Last accesed 19 July 2021
8. Recommendation № 2069 of parliamentary assembly of the council of Europe. Technol. Convergence Artif. Intell. Human rights (2015). https://assembly.coe.int/nw/xml/XRef/Xref-XML2HTML-en.asp?fileid=21747&lang=en. Last accesed 19 July 2021
9. Chirov, D., Novak, K.: Promising areas for the development of special-purpose robotic systems. Sec. Issues. **2**, 50–59 (2018). https://doi.org/10.25136/2409-7543.2018.2.22737
10. Khripunov, S., Blagodaryashchev, I., Chirov, D.: Military robotics: modern trends and development vectors. Trends Manag. **4**, 410–422 (2015)
11. Army Equipment Program in support of President's Budget, US Army G-8, p. 56 (2015). https://www.g8.army.mil/pdf/Army_Equipment_Program2015.pdf. Last accessed 19 July 2021
12. Pflimlin, É.: Drones et robots: La guerre des futurs. Estudes Inter. **49**(2), 433–435 (2018)
13. Tikhanychev, O.: On the legal and ethical aspects of the autonomous use of robotic systems in the field of armed confrontation. Sec. Issues. **3**, 33–42 (2019). https://doi.org/10.25136/2409-7543.2019.3.28960
14. Weaver, J.: Abhor a vacuum: the status of artificial intelligence and AI Drones under. Inter. Law New Hampshire Bar J. **1**, 14–21 (2013)
15. Tikhanychev, O.: On improving indicators for assessing the decision support systems' software quality. IOP Conf. Ser. Mater. Sci. Eng. **919**(5), 052009 (2020). https://doi.org/10.1088/1757-899x/919/5/052009
16. Beard, J.: Autonomous weapons and human responsibilities. Georgetown J. Int. Law **45**, 617–681 (2014)
17. Johnson, D.: Computer systems: moral entities but not moral agents. Ethics Inf. Technol. **8**, 195–204 (2006). https://doi.org/10.1007/s10676-006-9111
18. Schuller, A.: At the crossroads of control: the intersection of artificial intelligence in autonomous weapon systems with international humanitarian law. Harvard Nat. Sec. J. **8**, 379–425 (2017)
19. Levkov, A.: Organization of effective polymorphhic algorithms in complex control systems information. J. Appl. Inform. **5**(35), 127–132 (2011)

20. Tikhanychev, O.: Statement of the problem of ensuring the security of intelligent systems algorithms. Nat. Sec. **5**, 45–55 (2020). https://doi.org/10.7256/2454-0668.2020.5.33286
21. Klimenko, A., Stoyanova, O., Dli, M., Boyarinov, Yu.: Neuro-fuzzy construction method models of complex objects. J. Appl. Inform. **3**(9), 119–127 (2007)

New Equipment for Aromatherapy and Related Mobile App: A Tool to Support Small Peasant Farms in Kazakhstan in Crisis

Ibragim Suleimenov, Kaisarali Kadyrzhan, Sherniyaz Kabdushev, Akhat Bakirov, and Eldar Kopishev

Abstract The design of a new type of flavouring enhancer is suggested, which ensures the direct production of aromatic components, including essential oils, from salted plant materials. The work of the flavouring enhancer is based on induction heating of a mixture containing salted raw materials and crushed metal (for example, dissected foil) due to vortical currents. This approach dramatically simplifies the procurement of plant materials, up to the possibility of using agricultural waste for these purposes. The approach is one of the practical implementations of the concept aimed at freeing farms from the dictates of large purchasing and trading networks, providing for the creation of the most simplified schemes for processing and marketing agricultural products. The flavouring enhancer is controlled via a mobile app installed on a smartphone. In the future, it allows ensuring the automation of prophylactic measures carried out using aromatherapy means. This also can provide very visible support to small farms.

1 Introduction

The positive effect of aromatherapy on health has been demonstrated by the results of many experimental studies, in particular, [1–3]. Essential oils, in particular can have a beneficial effect on persons under constant stress [4, 5], they are used in supportive therapy [6], etc.

The use of essential oils for therapeutic, spiritual, wellness and ritual purposes dates back to a number of ancient civilizations, including the Middle East, who used

I. Suleimenov · A. Bakirov (✉)
V.I. Vernadsky Crimean Federal University, Prospekt Vernadskogo 4, Simferopol, Russian Federation

K. Kadyrzhan · S. Kabdushev
Al-Farabi Kazakh National University, 71 al-Farabi Ave., Almaty, Republic of Kazakhstan

E. Kopishev
L. N. Gumilyov, Eurasian National University, 2, Satpayev str., Nur-Sultan, Republic of Kazakhstan

M. Shamtsyan et al. (eds.), *Robotics, Machinery and Engineering Technology for Precision Agriculture*, Smart Innovation, Systems and Technologies 247,
https://doi.org/10.1007/978-981-16-3844-2_32

them in cosmetics, perfumery and medicines. Refined essential oils have been used as medicines since the invention of distillation when Avicenna separated the refined oils through steam distillation [7].

It is obvious that countering stress and Occupational burnout, which the World Health Organization has recognized as a disease [8], is becoming even more relevant in the face of stringent quarantine measures taken in various countries around the world as a result of the Covid-19 pandemic. A sharp change in the usual way of life could not but affect the psychophysiological state of a very large number of people [9–12].

It should also be noted that the market for aromatherapy products is closely related to the market for various types of fragrances used in everyday life to improve the comfort of living quarters. In essence, the only difference is that in one case, the saturation of the room with aromatic substances is carried out for medical reasons, and in the other—for reasons of comfort [13, 14].

Both of these market segments could not be influenced by consumer orientation towards organic products. This tendency in certain respects is aggravated in the post-Soviet states, since a significant number of counterfeits are presented on the market, including through smuggling channels.

As a result, there is a demand for aromatherapy products and household fragrances, the nature of which maximally excludes the likelihood of counterfeiting.

In this paper, we propose a radio-electronic system controlled from a cell phone and designed to saturate the air in rooms with Essential oils, as well as other aromatic substances obtained from natural raw materials obtained with minimal pre-treatment, which, de facto, is reduced to salting. Optionally, the device can be equipped with optical sensors [15].

This approach provides small peasant farms with the opportunity to generate additional income by supplying to the market finished components for aromatherapy products obtained by extremely simple means, by the same token are products of long-term storage [16].

The need for this is felt very acutely, especially since peasant farms lose their traditional income due to the more than significant difference between purchase and retail prices, i.e. due to the de facto dictatorship of large retail chains. Any products of long-term storage, requiring only minimal processing costs will allow damping this dictate by using online trading tools. The use of advertising on Instagram already allows farms to carry out direct deliveries to the end consumer, but so far this form of activity is encountering difficulties with storage and logistics. Obtaining additional profit through the use of the proposed tools can be considered as a very definite contribution to ensuring the economic independence of farms.

2 Materials and Methods

To work out the proposed scheme of a remote-controlled flavour enhancer through a mobile application, plant materials were used that had undergone minimal preliminary processing. Specifically, they used raw materials preserved by pouring a saturated solution of table salt containing its undissolved excess (needles, tangerine peel, lavender, rose petals, etc.).

This method of conservation is not only the simplest, but also provides the possibility of direct visual control of the content by the consumer, i.e. the above task is automatically solved. Moreover, this method of conservation of plant raw materials greatly simplifies its procurement, until it becomes expedient for small peasant farms to harvest raw materials that were previously sent to waste (currant leaves, etc.).

Let us emphasize that one of the varieties of harvested raw materials was forbs collected in the foothills of the Trans-Ili Alatau (environs of Almaty, height above sea level 1250 m). These raw materials were collected for subsequent use in SPA-salons, focused on "alpine saunas", which use the components produced by herbs. Raw materials for this purpose were also procured by the salting method.

The production of aromatic components was carried out by the method of direct sublimation from a container filled with salted plant material rolled into 0.9 mm thick aluminium foil using induction heating with an alternating electromagnetic field with a frequency of 0.2 MHz.

The efficiency of filling the room with vapours of aromatic components was registered organoleptically by comparison with the smell created by aromatizers on the market.

In parallel, the amount of the hydrophobic component released during sublimation was recorded, which is sufficient to assess the effectiveness of the proposed method, since its result is not the production of purified aromatic oils, but the provision of a comfortable environment due to the saturation of the room with vapours of natural plant raw materials sublimation. The amount of the hydrophobic component was measured gravimetrically after the separation of the hydrophilic component.

3 System Construction

The basic scheme of a flavouring enhancer using salted plant materials is shown in Fig. 1. This construction assumes the use of commercially available aroma patrons, i.e. disposable containers containing raw materials of this type. The basic construction assumes that, if necessary, the evaporated substances will pass through an additional filter made of hydrophilic material and designed to block the separation of water vapour. The simplest version of such a filter is a hydrogel layer. However, it should be emphasized that for household use of the flavour enhancer, an additional filter is not necessary since it is intended only to exclude an increase in humidity in the room.

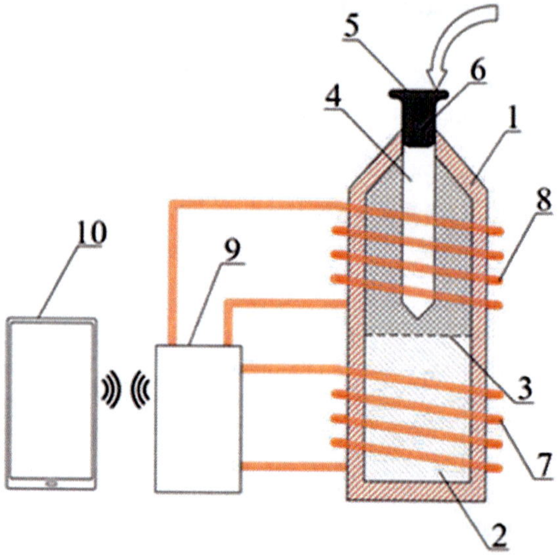

Fig. 1 The basic circuit of induction heating flavour enhancer

The composition of the aroma-patron includes an outer shell (1), an area (2) filled with salted plant material, which is protected by an easily torn septum (3), as well as an area filled with a hydrogel or other substance that traps water vapour. In the upper part of the aroma patron, there is a puncher (4), the lower tip of which in the initial state is adjacent to the septum (3). The upper part (6) of the puncher (4) is permeable to the gas-phase medium. The punch is supplied with a stopper head (5). The patron is inserted into the flavour enhancer's socket, which includes two induction heating coils (7), (8), as well as the electronic control system (9), which includes a Bluetooth-module.

The system is controlled by transmitting data from a smartphone (10), which almost completely eliminates the need to use any moving parts: buttons, toggle switches, etc., in the structure itself. Thanks to this, it can be manufactured in a cheap monolithic corps. In addition, the use of two separate induction coils makes it possible to maintain a given temperature regime of both the raw material itself and the filter that captures the hydrophilic component.

Along with the basic one, a simplified modification can also be used, in which no filter captures the hydrophilic component. In this case, a single induction coil is used, and the aroma-patron is simply a water-permeable shell filled with salted plant material.

The layout of parts inside the housing for the case of a simplified modification is shown in Fig. 2. This diagram is a 3D model of a real prototype, made in the KOMPAS 3D program, which provides subsequent printing on a 3D printer.

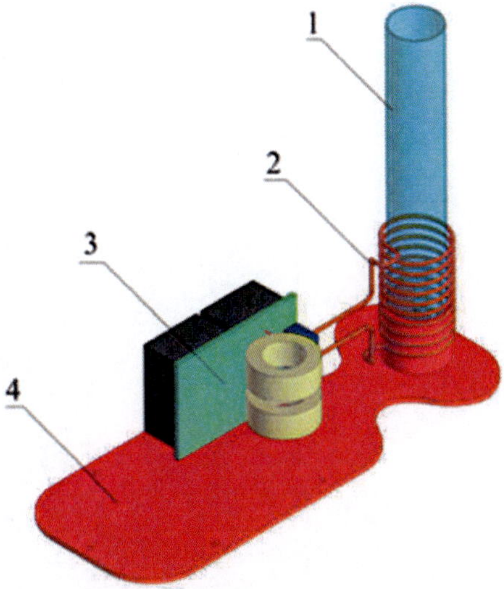

Fig. 2 Placement of parts inside the fragrance body (simplified modification): 1—container for raw materials, 2—inductor, 3—electronic heating control block, 4—the bottom cover of the flavour enhancer body

4 Radio Electronic Scheme and Concomitant Mobile App

The electronic circuit of the device is shown in Fig. 3. The main element of the circuit is the inductor L1. When metal objects are placed inside it, they begin to heat up under the influence of vortical currents induced by an alternating current flowing through this coil. This coil jointly with capacitors C1, C2 form an oscillating circuit. Throttles L2, L3 restrict and smooth the supply current. Oscillations are supported by alternately opening field-effect transistors Q3, Q4.

To control the power of the induction heater system, a field-effect transistor Q1 is used, which is switched on in the key mode, which in turn plays the role of a wide pulse regulator. The Atmega328 microcontroller provides the setting of the operating temperature, its control, etc.

Flavour enhancer control is shown in (Fig. 4). The operating temperature is set by software using a mobile app installed on a smartphone. In particular, this application allows you to set the operating temperature. The data is transmitted to the HC-05 Bluetooth module, which transmits it to the Atmega328 microcontroller, in the form of bytes via the UART.

Further, the received data is converted into an integer type, which provides the transfer of parameters to the PID controller. To set the established temperature, the Atmega328 microcontroller reads the current temperature through the ADC to which the K-type thermocouple is connected, and uses the received data in feedback. The K-type thermocouple is connected to the ADC of the microcontroller through an operational amplifier (op-amp) lm358n. The gain of the signal from the thermocouple is set through the resistors R10 and R11.

The mobile application is written in Android Studio for smartphones with the Android 6 and higher operating system. This application acts as a remote control for

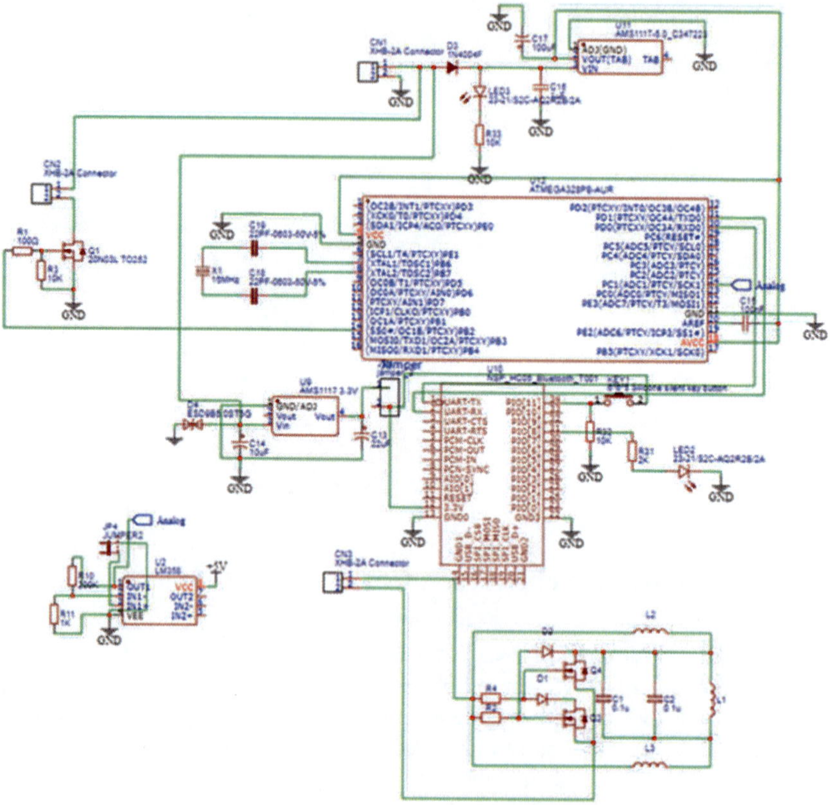

Fig. 3 The schematic electronic circuit of the flavour enhancer

the flavour enhancer. There are functions for setting and measuring the temperature of the medium inside the aroma cartridge. The connection between the device and the smartphone is carried out using a Bluetooth adapter.

Used by the BluetoothServerSocket for the listening server. When a connection is established, the BluetoothSocketServer returns the new state of the BluetoothServer. GetInputStream () and getOutputStream () are used to input and output streams. Setting the temperature value is carried out using Seekbar.progress with conversion to String. The resulting variable is passed to the OutputStream. Next, connected thread.write () is applied to the data. Write converts the data to bytes and sends them to the device.

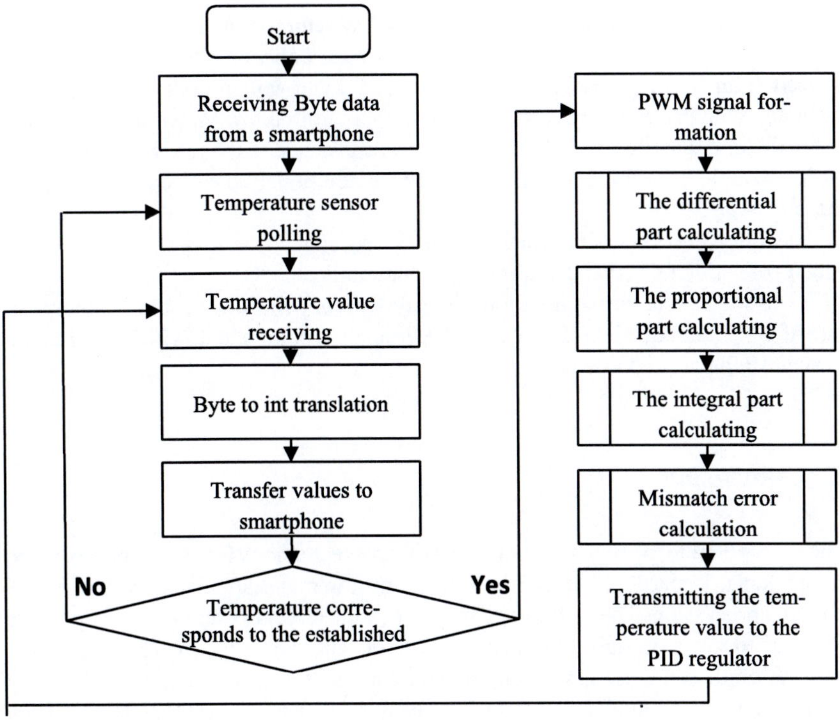

Fig. 4 Bloc-scheme of flavour enhancer's work

5 Results of Experiments

A picture of a prototype of an industrial flavour enhancer of the proposed type, made by printing on a 3D printer, is shown in Fig. 5. The picture emphasizes that the device has a single on/off button. The rest of the control is carried out through the mobile application.

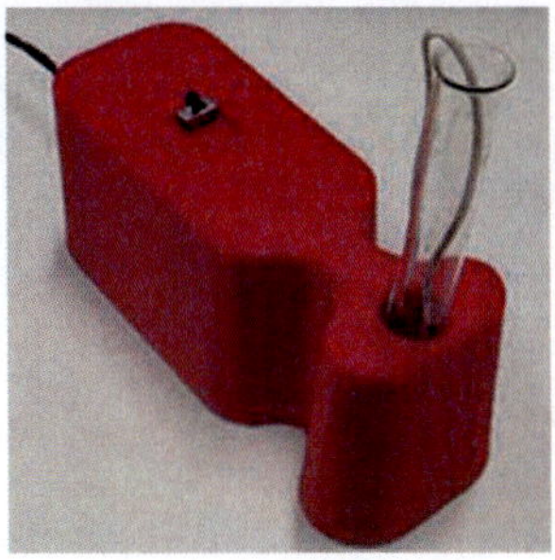

Fig. 5 A photo of the final product's prototype

For testing, a standard type installation was assembled, including a direct-flow refrigerator with a water jacket and a container for collecting condensed vapours. Instead of an aroma patron, a test tube was used, filled with a mixture of salted raw materials with dissected metal foil. In some cases, salted raw materials were rolled into a 0.9 mm thick foil roll. The test tube was connected to the refrigeration system with a flexible hose through a sleeve, in which additional holes were made for the thermocouple.

The following types of raw materials were used: salted orange peel, salted needles, salted collection of herbs, including wild chamomile.

It was shown that the exploitation of the system for 3 min at an operating temperature of 90 °C already yields 2–3 g of the hydrophobic component, which is sufficient to achieve the desired effect.

6 Conclusion

Thus, it is possible to propose a simple and reliable technology for producing aromatic components, in particular Essential oils, from raw materials that have undergone minimal processing. The technology is based on the direct extraction of components from salted raw materials. On the one hand, this approach maximally excludes the penetration of counterfeits into the market. On the other hand, it can provide very visible support to small farms, which in countries such as Kazakhstan have been significantly affected by the 2020 crisis caused by the COVID-19 pandemic. Specifically, such farms receive an additional market for their products or products that are grown additionally.

There must be an opportunity to implement a simple implementation scheme based on the interaction of small businesses that distribute products through online trading.

The proposed flavour enhancer is also one of the options for the realistic implementation of the concept suggested by [17]. This concept is based on the fact that the design of a significant part of industrial products containing radio electronic units can be significantly simplified by transferring a fraction of the functions to software installed on a smartphone. This approach can sufficiently reduce both the cost of the final products and the cost of prototyping. Besides, the suggested construction of the flavour enhancer allows beyond improving mobile apps. Particularly, it is permissible to raise the question of the diagnostic mobile apps' development that will further carry out prophylaxis in an automatic mode.

References

1. Sharare, A., Rezaei, M., Khatony, A.: Comparing effects of aromatherapy with lavender essential oil and orange essential oil on fatigue of hemodialysis patients: a randomized trial.

Complement. Ther. Clin. Pract. **36**, 64–68 (2019)
2. Mikyoung, L., Lim, S., Song, J.A., Kim, M.E., Hur, M.H.: The effects of aromatherapy essential oil inhalation on stress, sleep quality and immunity in healthy adults: randomized controlled trial. Eur. J. Integrative Med. **12**, 79–86 (2017)
3. Van Snick, W., Aibuldinov, Y.K., Dehaen, W.: An efficient synthetic route towards novel thienobenzothiazoles, thienobenzothiazepines, and thienobenzothiazines. Tetrahedron **69**(21), 4176–4184 (2013)
4. Nicolette, P., Perry, R.: Aromatherapy in the management of psychiatric disorders Clinical and neuropharmacological perspectives. CNS Drugs **20**(4), 257–280 (2006)
5. Wheatley, D.: Medicinal plants for insomnia: a review of their pharmacology, efficacy and tolerability. J. Psychopharmacol. **19**(4), 414–421 (2005)
6. Reis, D., Jones, T.: Aromatherapy: using essential oils as a supportive therapy. Clin. J. Oncol. Nurs. **21**(1), 16–19 (2017)
7. Suleimenov, I.E., Moldazhanova, A.A., Kopishev, E.E., Egemberdyeva, Z.M., Niyazova, G.B.: The scientific heritage of the Islamic golden age. Foreign Lang. Teach. **46**(3), 288–298 (2019)
8. Anon. n.d. Burn-out an 'Occupational Phenomenon': International Classification of Diseases. Retrieved 8 March 2021. https://www.who.int/news/item/28-05-2019-burn-out-an-occupational-phenomenon-international-classification-of-diseases
9. Iglesias-Sanchez, P., Vaccaro Witt, G.F., Cabrera, F.E., Jambrino-Maldonado, C.: The contagion of sentiments during the COVID-19 pandemic crisis: the case of isolation in Spain. Int. J. Environ. Res. Public Health **17**(16), 5918 (2020)
10. Pizzoli, S.F.M., Marzorati, Ch., Mazzoni, D., Pravettoni, G.: Web-based relaxation intervention for stress during social isolation: randomized controlled trial. JMIR Mental Health **7**(12), e22757 (2020)
11. Sebri, V., Cincidda, C., Savioni, L., Ongaro, G., Pravettoni, G.: Worry during the initial height of the COVID-19 crisis in an Italian sample. J. Gener. Psychol. 1–33 (2021)
12. Warnecke, T.: Acute crisis states as a presenting issue and some psychophysiological interventions. Body Movement Dance Psychother. **16**(1), 19–31 (2021)
13. Sayin, K.K., Yildirim, Y., Aykar, F.S., Uyar, M., Sagin, F.G., Atay, S.: Effect of inhalation aromatherapy on pain, anxiety, comfort, and cortisol levels during trigger point injection. Holist. Nurs. Pract. **34**(1), 57–64 (2020)
14. Phillips, S.: Comfort and aromatherapy—home fragrance. Soap Cosmet. **77**(11), 25–28 (2001)
15. Suleimenov, I.E., Sedlakova, Z.Z., Kopishev, E.E.: New polymer materials for optical sensor systems. J. Inorg. Organomet. Polym Mater. **29**(3), 758–764 (2019)
16. Safarov, R.Z., Shomanova, Z.K., Nossenko, Y.G., Berdenov, Zh.G., Bexeitova, Zh.B., Shomanov, A.S., Mansurova, M.: Solving of classification problem in spatial analysis applying the technology of gradient boosting catboost. Folia Geographica **62**(1), 112–126 (2020)
17. Suleimenov, I.E., Kabdushev, Sh.B., Kadyrzhan, K., Shaltikova, D.B., Moldakhan, I.: New technologies for measuring viscosity. In: Proceedings of the 2020 6th International Conference on Computer and Technology Applications, pp. 129–133. ACM, New York, NY, USA (2020)

Security Concept of Digital Twins of Industrial Fuel and Energy Complex Objects

Nikolai Korneev and Vyacheslav Merkulov

Abstract The paper examines security problems related to the use of digital twins as part of promising industrial technologies for fuel and energy companies, such as «digital oil field» and «the intelligent refinery». An industrial technology that uses a digital twin in its composition is decomposed into three logical levels. There is examined the practical case of analyzing some security problems of digital twins which are created and operated within the IIoT Predix cloud platform. In the context of each of the three levels, it was concluded that the current security approaches are sufficient. There is identified the mathematical model level which requires a new security methods development. There are described mathematical models as a basic component of digital twin technologies and examined the most important features. By analyzing digital twins mathematical models based on the "black box" principle, there are identified their vulnerabilities. On their basis we formulated the concept of digital twins security that includes 4 key security aspects.

1 Introduction

In the context of digital economy technologies development and the "Industry 4.0" concept, enterprises of the fuel and energy complex (FEC) are interested in digital transformation to maintain competitiveness, which implies the optimization of industrial economic entities activities [1, 2]. Summarizing the existing interpretations, we can say that enterprise digital transformation—is a complicated and multifaceted process aimed at multiple information flows integration into a single platform, including data coming from industrial systems, installations and other equipment [3].

The "digital twin" concept is based on the theory of automatic control. The Russian science significantly contributed in its development, including the domestic concept of the man–machine system as complex information and energy system with artificial

N. Korneev (✉) · V. Merkulov
National University of Oil and Gas "Gubkin University", Leninsky Prospect, 65, Moscow, Russian Federation

M. Shamtsyan et al. (eds.), *Robotics, Machinery and Engineering Technology for Precision Agriculture*, Smart Innovation, Systems and Technologies 247,
https://doi.org/10.1007/978-981-16-3844-2_33

intelligence elements [4, 5]. The "digital twin" approach is a logical extension of the mentioned Russian man–machine system concept [4, 5] and the basis for a number of promising industrial digital technologies for the fuel and energy sector with artificial intelligence elements, such as "digital oil field", "the intelligent refinery", etc. In general, regardless of the specific technology, the "digital twin" implementation consists of creating a digital equivalent of a real industrial object displaying its parameters and characteristics in real time [6–8].

The projects of FEC sphere that rely on the digital twins operation are already being implemented [9–14]. Digital twins can also be used as a tool for solving security problems. For example, it will allow to create a digital copy of an industrial object for its security analysis [15–20].

The use of digital twin technologies creates certain risks, as the digital twin is the result of the displaying a physical industrial object to cyberspace, which characteristics are absolutely different from the features of physical space. Thus, there are conditions for the attacks implementation, which components are distributed across several spaces [21].

At the moment, on the basis of the information search results, there has not been conducted an overall study of threats connected with the digital twins operation as part of industrial automation technologies. Meanwhile, the digital twins cybersecurity safety is one of the key conditions for the introduction of such promising technologies [22]. As a result, there is a need to define the security concept of digital twins, for example, on the basis of the main security aspects description, which should be formulated primarily according to the "digital twin" approach essence and to the industrial technologies analysis that implement this concept in practice.

2 Methods

To analyze the digital twins security, first of all, it is necessary to consider their role, place and purpose within promising digital technologies for the fuel and energy sector.

A significant part of industrial technologies is primarily used in processes related to the industrial objects management [23], which can be generally called a technical system [4, 5]. Considering the situation when digital twins are not used in the process of technical system management, we can say that a person as a management subject interacts with the technical system:

1. Physically, for example, rotates a certain valve.
2. Through certain hardware and software control systems, for example, automatic process control system (APCS) which are widely used nowadays (see Fig. 1).

In turn, the process of technical system management, considering the use of industrial digital technologies that are promising for FEC, which contain a digital twin, can be represented as follows (see Fig. 2).

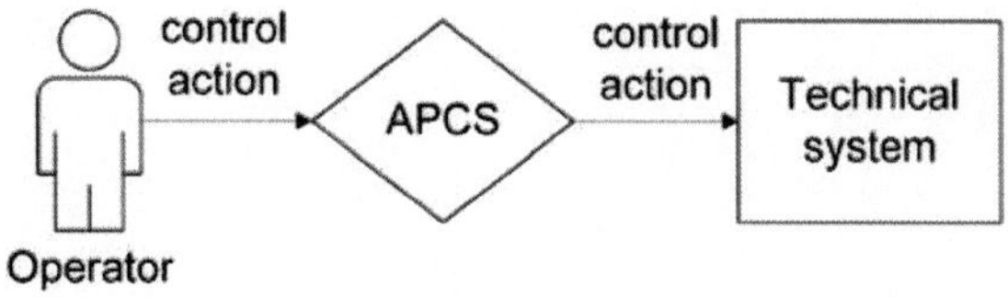

Fig. 1 The process of technical system management using automatic process control system (APCS)

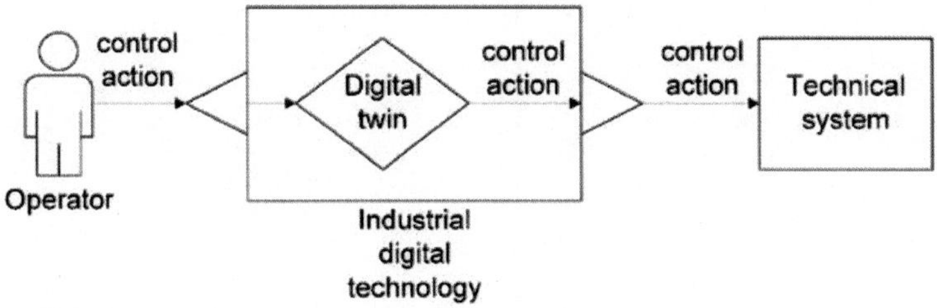

Fig. 2 The process of technical system management in case of a digital twin operation as part of a digital technology

In this case, the control action is also generated by the operator and then goes to the digital technology input, where further processing takes place. The key aspect is the presence of a digital twin, as its internal structure and current state determines the value of the output signal, which is subsequently transmitted to the technical system (physical object) and, accordingly, changes its state. Industrial digital technology is essentially a shell around the digital twin: it does not control it directly, but only transmits the input signal to it and convey the output signal received in the result of processing to the technical system.

3 Results

3.1 Security of Digital Twins at the Level of Software Product and Software Implementation

The security of any industrial technology must be considered at all stages of its life cycle [24], at each stage, the characteristics that are important in the security context will vary.

For further research, we will consider any industrial digital technology based on the principle of a digital twin operation. Regardless of the implementation areas and features of such technology, it can be divided into 3 logical levels corresponding to different stages of the life cycle: a software product, the software implementation and mathematical model (see Fig. 3).

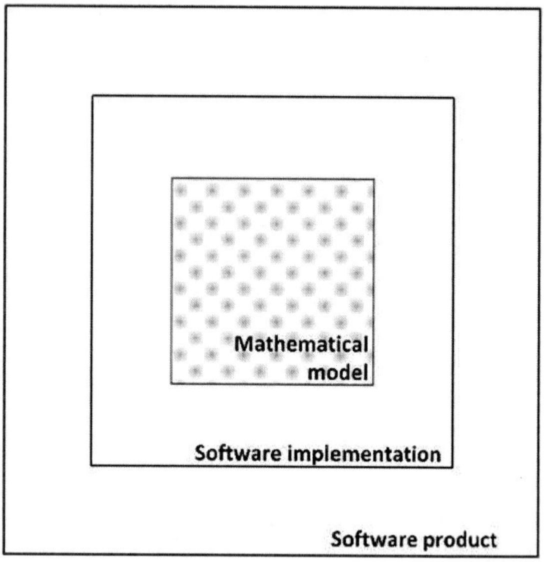

Fig. 3 Decomposition of industrial digital technology on the basis of a digital twin

As the management of a physical object and viewing its characteristics (state) is carried out by a person using the tools included in the technology, which is decomposed into several levels from a software, technical and logical point of view, approaches to the security of digital twin technology are determined through aspects specific to each level. First of all, it is necessary to consider the security aspects of the technology from the point of view of its presentation as a software product and software implementation. If there is no significant differences, security approaches at these levels will correspond to generally accepted practices and techniques.

Software product level When considering the security of a technology as a finished software product, the abstraction level is minimal: in this case, the list of information available for security analysis includes:

1. Information about the vendor and cooperation terms, including, for example, the features of the software product delivery contract.
2. Specifications of communication links and transmitted data that the technology operates with.
3. Requirements for the hardware platform that is required to deploy the technology.
4. Information about the architectural solutions that are the basis of technology, etc.

In fact, the list of available information will be extremely extensive, therefore, to ensure security at this level, it can be concluded that measures based on generally accepted approaches, techniques and standards are sufficient. In general, when implementing the industrial digital twin technology supplied by the vendor, typical tasks related to the security of the software product are solved at the procurement, implementation and operation stages.

Software implementation level Industrial digital technology is implemented at the software level, which includes software developed by the vendor (or by its order). In this context, the software package of the technology is the object of security analysis, during which it is necessary to:

1. Confirm that the development processes meet the security requirements contained in the relevant standards.
2. Verify the developed program code in compliance with the specifications of the mathematical model that is implemented by this code.
3. Implement further continuous monitoring of vulnerabilities, also, ensure timely software updates.
4. Implement other typical activities throughout the entire life cycle.

We can make an intermediate conclusion, that at two "upper" levels (software product and software implementation), the measures related to the security of industrial digital technologies, which include a digital twin, coincide with another ones for any other technologies, if they can be considered as a "product" and "software implementation". Definitely, some peculiarities may occur in practice, depending on specific conditions and factors. For example, if the digital twin operates on the basis of cloud computing. In this case, it is necessary to analyze the architecture of the cloud platform in the part related to the creation and operation of digital twins.

As a practical example we can consider some aspects of the IIoT Predix platform security. According to the publicly available documentation, Predix digital twins can function both in the cloud and on-premises, i.e. in close physical proximity to an industrial oblect. Local execution is characterized by a lower time delay and can be used to comply with legal regulations, as well as in cases when the information transmission via communication channels is unacceptable. However, a locally deployed digital twin, according to the documentation, must still be created in the cloud. Moreover, the Predix documentation states that a locally deployed digital twin "improves over time on the basis of historical analysis in the cloud", i.e. even when a digital twin is deployed locally, it implies transferring the processed data to the cloud to improve the local mathematical model that has already been in use.

First of all, it should be noted, that such a function does make sense. Any model can be improved over time, if it is modified on the basis of the accumulated historical amounts of information, for example, the prediction module [5, 25, 26]. However, if the modification, like a digital twin development, can only be performed in the cloud, it reduces the value of local deployment, in cases when data sending (including historical data) is not allowed even via secure channels. When a digital twin developed in the Predix cloud is deployed locally, the possibility of improving its model is lost, that can significantly affect the efficiency of its operation. Thus, for a number of industrial objects which information flows are undesirable to be send outside the controlled area, this disadvantage of Predix is fundamental.

Considering the decomposition of abstract industrial digital technology into logical levels, as well as the considered example of security analysis of digital twins offered by the Predix platform, we can say that the development of new requirements, methods and tools for security is generally not justified when considering digital twins technologies at the "software product" and "software implementation" levels. The general methodology for security at these levels does not seem to be significantly different from the currently known and applied approaches, as the solved tasks do not differ in principle.

Since no significant security aspects were found at two "upper" levels of the decomposed industrial digital twin technology, it is necessary to analyze the mathematical models as the basis of the "digital twin" concept itself.

3.2 Mathematical Model as a Basic Component of a Digital Twin

To consider the security aspects of mathematical models of digital doubles, first of all, it is necessary to briefly consider their creation purpose and principles.

The "digital twin" approach, as is has already been mentioned above, is a logical continuation, including the domestic man–machine system concept [4, 5]. It is based on the principle of creating and operating a mathematical model that displays the parameters of a real object. This mathematical model is implemented by software and supplemented with appropriate control mechanisms, that allow to use it as a part of a software product to monitor the object state and control its behavior.

Talking about the presented in Fig. 3 decomposition, the mathematical model level, on the one hand, is the most abstract. At this level, the analyzed context includes only the model as a set of logical and mathematical rules and functional dependencies, that together describe a real industrial object. Features of the software implementation, used hardware, architectural solutions, etc. aspects that are traditionally considered as definitely affecting security are determined at higher levels. On the other hand, as it was noted, the "digital twin" is based on mathematical modeling of an industrial object. The modeling method also seems to be universal. Only the objects for which the model is being developed vary.

Considering the information above, to analyze the security of digital twins mathematical models, it is acceptable to abstract from the features of their internal structure and functioning. It is necessary to consider the fundamental issues of mathematical models safe operation as the main components of industrial digital technologies, as well as develop universal approaches to ensure their safety, regardless of the high-level features that are determined by the context of the software implementation and the software product.

For these purposes, the safety analysis of mathematical models of digital twins can be carried out on the principle of "black box" (see Fig. 4), where the input will be the parameters of different data sources (sensors), the control action and the parameters

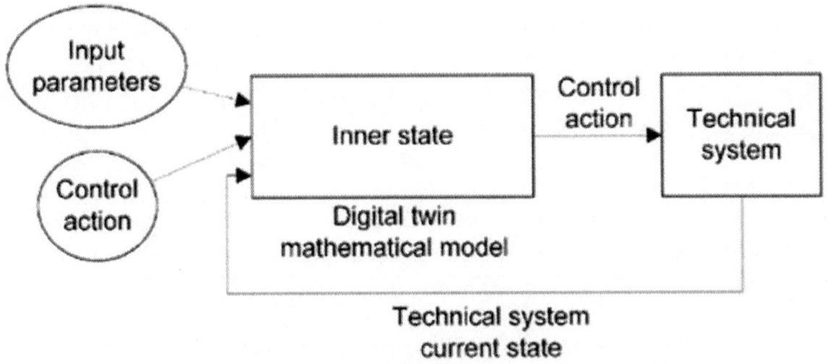

Fig. 4 Representation of a mathematical model of a digital twin according to the "black box" principle

of the current state of the technical system; in the "black box" encapsulated internal state of the mathematical model. The output signal is also a control action directed to the technical system, and is determined on the basis of the input data and the internal model state.

It is should be noted that the internal structure of the digital model, i.e. the set of logical and mathematical rules and functional dependencies included in it, cannot be changed without program code modification that directly implements the digital model. The security aspects of the program code are at the level of "software implementation", and, as mentioned, the methods of the model program code protection from various types of threats do not differ from any other program code.

On the other hand, the internal state of the digital model is constantly changes under external influence [5]. Functional dependencies parameters that are included in the digital model change to meet the requirements set by control actions and match the digital model of a real object as much as possible. At the same time, the ranges of values within which the digital model can change its internal parameters are known and fixed.

Before considering the security aspects, we assume that the mathematical models that are the basis of digital twins can be generally called as "digital models of real objects" or just "digital models". It allows to make a reference to the digital technologies specifics, instead of using the generally accepted understanding of the term "mathematical model" that does not fully describe the object of the research.

3.3 *The Concept of Digital Twins Security*

We define the concept of digital twins security identifying the most important aspects of the digital models security, which can be deduced by considering them in accordance with the "black box" principle.

As mentioned, the internal structure of a digital model cannot be modified without changing the program code that implements this model. It implies the first important aspect: the digital model cannot be directly the object of attack. To distort the digital model, an intruder can perform an attack:

1. To the program code of the digital model.
2. To the source of input parameters that describe the current context of the technical system (industrial object) or the source of the control action.
3. To communication links that transmit input parameters or control actions.

Thus, the negative impact on the digital model can be achieved indirectly: having received distorted data at the input and processing it, the model is at risk of distortion, that will be expressed in the unreliability of the output control action transmitted to the technical system. Additionally, we can note that the internal structure of the model will not change: the output signal will be distorted as a result of compromising the input signal, but the set of logical and mathematical rules and functional dependencies included in the digital model will not change.

From the above, we can identify the second aspect of security, which is to ensure the data sources security, information flows of which form the input signal of the digital model, as data sources and the communication links attract the intruder. As mentioned, distorted input data can unpredictably change the internal state of the digital model, so the attack result on individual data sources can discredit the entire digital model.

If the internal state of a digital model can be discredited only by distorting the input signal, i.e. during the operation of a digital twin, so the internal structure of the digital model may only be modified at the development stage, as the set of logical and mathematical rules and functional dependencies, encapsulated by the model, is determined at this stage and does not change in the future. In this way, the third aspect of security is in the process of a digital model developing, during which there is a possibility that the intruder will make certain modifications, a kind of "bookmarks" which will be implemented programmatically and exploited without notice. Since the model itself is also a complex system, the presence of a "bookmark" may be hidden behind the features of the internal model structure. The "bookmark" presence in the internal device of the digital model potentially allows an intruder to implicitly control a real object by generating certain types of impacts against the digital model, created considering negative modifications made during the model development. Additionally, we can note that the input data coming to the model will not necessarily be suspicious, that reduces the probability of detecting manipulations.

Considering the dynamic nature of a typical industrial object [5], the digital model created for it should be able to flexibly adapt in accordance with the object's behavior criteria in order to ensure the maximum level of compliance of its internal state parameters with the current state of the real object. Also, it has to adapt to requirements set by control actions. From a practical point of view, a similar characteristic has, for example, the IIoT Predix cloud platform, which, as mentioned earlier, can adapt the parameters of digital twins to specific operating conditions, adapting to the changing parameters of the working environment. When changing any characteristic of a "digital oil field" or "the intelligent" refinery, its digital model adapts to new conditions. For this purpose it needs an information and energy system with artificial intelligence elements [4, 5] or artificial intelligence. This self-configuration mechanism can be considered the fourth aspect of security, as it can be used in the development of new security methods for digital models.

4 Discussions

Digital twins that function as part of industrial digital technologies promising for the fuel and energy sector, can be considered as an intermediate link between a person and a technical system controlled by human, that makes it necessary to ensure its security, as the consideration of the threats that arise in this case has not been previously conducted.

In turn, the technologies which are implemented by digital twins of industrial objects can be decomposed into three logical levels for the purpose of security analysis: software product, software implementation and mathematical model.

According to the results of the study, significant differences in the digital twins security approaches can be formulated only when considering the level of the mathematical model. As shown, in the context of the software product and software implementation, it can be concluded that the application of a generally accepted security methodology is sufficient.

5 Conclusion

The mathematical model is the basic component of the digital twin, while approaches to its security have differences. There is analyzed the security of mathematical models of a digital twin in accordance with the "black box" principle, which results formulate the security concept of digital twins including four most important security aspects. These aspects are formulated on the basis of the fundamental vulnerabilities analysis of digital models, which are common for this class of objects in general, without reference to specific implementations and operating conditions. Therefore, they can be used as a basis for creating a counteraction method against new threats [21] that are focused on digital models.

The main feature of such threats is their indirectness: even if a significant part of the attack components is implemented in relation to the data source, communication links or digital technology software. In any case, the ultimate goal is to negatively affect the digital model in order to distort it and, as a result, disrupt the functioning of an industrial object.

Security aspects of digital models formulated in the work can be further used to develop current threats models against digital models and identify security requirements [20] and when modeling complex threats [21] and development of the complex security neurograph theory [19].

References

1. Berge, J.: Digital transformation and IIoT for oil and gas production. In: Proceedings of the Annual Offshore Technology Conference, pp 1204–1213. OnePetro, Houston (2018)
2. Patel, K., McCarthy, M.P.: Digital Transformation: The Essentials of e-Business Leadership. McGraw-Hill, New York (2000)
3. Reis, J., Amorim, M., Melão, N., Matos, P.: Digital transformation: a literature review and guidelines for future research. In: WorldCIST'18 2018: Trends and Advances in Information Systems and Technologies, pp. 411–421. Springer, Cham (2018)
4. Korneev, N.V.: Principles of construction actual technical systems with elements of artificial intellect. Tekhnika mashinostroeniya **2**(66), 2–7 (2008). (in Russian)
5. Korneev, N.V., Kustarev, Yu.S., Morgovsky, Yu.Y.: Theory Automatic Control with Workshop. Academia, Moscow (2008).(in Russian)
6. Uhlemann, T.H.J., Lehmann, C., Steinhilper, R.: The digital twin: realizing the cyber-physical production system for industry 4.0. Procedia Cirp **61**, 335–340 (2017)
7. J. Manuf. Syst. https://doi.org/10.1016/j.jmsy.2020.06.017. Last accessed 2021/01/17
8. Korneev, N.V., Yanitskiy, A.I.: A model of a combined electric drive for a dynamic advertising structure. Russ. Electr. Eng. **90**, 696–701 (2019)
9. Bevilacqua, M., Bottani, E., Ciarapica, F.E., et al.: Digital twin reference model development to prevent operators' risk in process plants. Sustainability (Switzerland) **12**(3), 1088 (2020)
10. Mayani, M.G., Svendsen, M., Oedegaard, S.I.: Drilling digital twin success stories the last 10 years. In: Society of Petroleum Engineers—SPE Norway One Day Seminar, pp. 290–302. OnePetro, Bergen (2018)
11. Shirangi, M.G., Ettehadi, R., Aragall, R., et al.: Digital twins for drilling fluids: advances and opportunities. In: SPE/IADC Drilling Conference, Proceedings, SPE-199681-MS 2020. OnePetro, Galveston (2020)
12. Cameron, D.B., Waaler, A., Komulainen, T.M.: Oil and gas digital twins after twenty years. How can they be made sustainable, maintainable and useful? In: Proceedings of the 59th Conference on Simulation and Modelling, pp 9–16. Linköping University Electronic Press, Linköpings universitet, Oslo (2018)
13. Smolenskaya, N.M., Smolenskii, V.V., Korneev, N.V.: Increase in the thermodynamic efficiency of the working process of spark-ignited engines on natural gas with the addition of hydrogen. IOP Conf. Ser. Earth Environ. Sci. **121**, 052009 (2018)
14. Smolenskaya, N.M., Korneev, N.V.: Modelling of the combustion velocity in UIT-85 on sustainable alternative gas fuel. IOP Conf. Ser. Earth Environ. Sci. **66**, 012016 (2017)
15. Eckhart, M., Ekelhart, A.: Towards security-aware virtual environments for digital twins. In: Proceedings of the 4th ACM Workshop on Cyber-Physical System Security, pp. 61–72. Association for Computing Machinery, Incheon Republic of Korea (2018)
16. Korneev, N.V., Goncharov, V.A.: Mathematical model support management of complex safety objects social sphere with mass stay of people. Estestvennye i tekhnicheskie nauki **4**(118), 277–281 (2018). (in Russian)
17. Wei, L., Chuipin, K., Qiang, N., Jingguo, J., Xionghui, Z.: A method of NC machine tools intelligent monitoring system in smart factories. Rob Comput Integr Manuf **61**, 101842 (2020)
18. Offergeld, T., Knittel, M., Bertram, R.: Cimpyorm—a queryable python CIM-cache for smart grid applications. In: Proceedings of 2019 IEEE PES Innovative Smart Grid Technologies Europe, ISGT-Europe 2019, 8905684. IEEE, Bucharest (2019)

19. Korneev, N. V.: A neurograph as a model to support control over the comprehensive objects safety for BIM technologies. IOP Conf. Ser. Earth Environ. Sci. **224**, 012021 (2019)
20. Korneev, N.V.: Algorithmic Both Program Methods and Tools Estimation of Alternative Projects of the Guard Data Reduction System of Firm on the Basis of the Multicriteria Analysis. Sputnik+, Moscow (2013).(in Russian)
21. Korneev, N., Merkulov, V.: Intellectual analysis and basic modeling of complex threats. CEUR Workshop Proc. **2603**, 23–28 (2019)
22. Wanasinghe, T.R., Wroblewski, L., Petersen, B.K., et al.: Digital twin for the oil and gas industry: overview, research trends, opportunities, and challenges. IEEE Access **8**, 104175–104197 (2020)
23. Poddar, T.: Digital twin bridging intelligence among man, machine and environment. In: Offshore Technology Conference Asia 2018, OTCA 2018, OTC-28480-MS. OnePetro, Kuala Lumpur (2018)
24. Canedo, A.: Industrial IoT lifecycle via digital twins. In: International Conference on Hardware/Software Codesign and System Synthesis, CODES+ISSS 2016, p. 1. Association for Computing Machinery, Pittsburgh Pennsylvania (2016)
25. Siegel, E.: Predictive Analytics: The Power to Predict Who Will Click, Buy, Lie, or Die, Revised and Updated. Wiley, Hoboken (2016)
26. Dakheel, A.H., Dakheel, A.H., Abbas, H.H.: Intrusion detection system in gas-pipeline industry using machine learning. Periodicals Eng. Nat. Sci. **7**(3), 1030–1040 (2019)

Wedge Mechanism Based on the Regulation of Suspension Bracket of Thin Quadrupole Group

Xiaowei Xu, Yaqing Yang, Bin Zhang, Yajun Zhen, Haijiao Lu, Yongxiang Pan, and Jiuwei Yu

Abstract The installation spacing of thin quadrupole group is very small, the traditional base support adjustment support is changed to suspension regulating bracket, which meets the stability and adjustment accuracy of thin quadrupole. The motion and force, transmission efficiency, self-locking conditions of wedge mechanism for small spacing direction adjustment are studied, and the parametric design of wedge mechanism is studied.

1 Introduction

SSC-Linac is a new CW heavy ion linear injector built for SSC by Institute of Modern Physics, Chinese Academy of Sciences, which is composed of ECR, LEBT, RFQ, MEBT, DTL and HEBT, as shown in Fig. 1. The main body of MEBT section is composed of thin quadrupole with small length, large height and width, and the outlet is downward, and the minimum installation spacing. When using the ordinary base support, the base is limited by the space, and can only be designed as a narrow plate and the bottom is too far from the center of the magnet. The magnet is unstable after installation, and it is difficult to align in place due to the excessive swing of the magnet in the adjustment process. In view of this situation, this paper designs a bracket that is suspended and installed with quadrupole iron and can be adjusted in three directions, and focuses on the wedge mechanism for adjusting the direction of small spacing.

X. Xu (✉) · Y. Yang · B. Zhang · Y. Zhen · H. Lu · Y. Pan · J. Yu
Institute of Modern Physics, Chinese Academy of Sciences, Lanzhou 730000, China
e-mail: xuxiaowei2010@impcas.ac.cn

J. Yu
University of Chinese Academy of Sciences, Beijing, China

M. Shamtsyan et al. (eds.), *Robotics, Machinery and Engineering Technology for Precision Agriculture*, Smart Innovation, Systems and Technologies 247,
https://doi.org/10.1007/978-981-16-3844-2_34

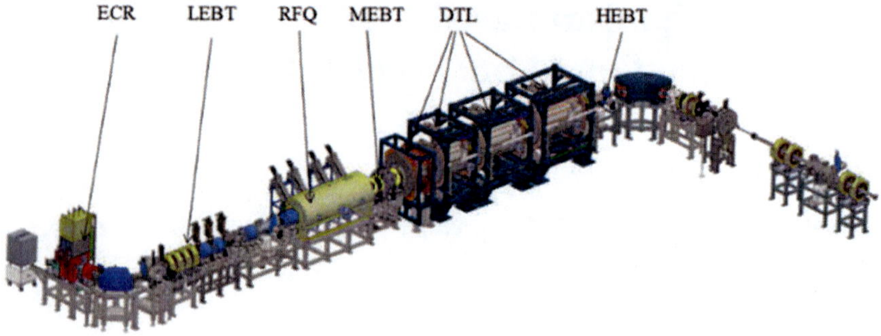

Fig. 1 Layout of the SSC-Linac injector

2 Structural Design of Suspension Regulating Bracket

The structure of the thin quadrupole iron suspension regulating bracket is shown in Fig. 2, which is mainly composed of ① hanging column, ② adjusting bearing, ③ supporting block, ④ changing direction regulating mechanism, ⑤ hanging beam and other main parts.

The support block is fixed on both sides of the quadrupole iron and placed on the adjustment support. The adjustment support is adjusted and fixed by hanging two adjustment bolts and one fixed bolt on the suspension column convex plate.

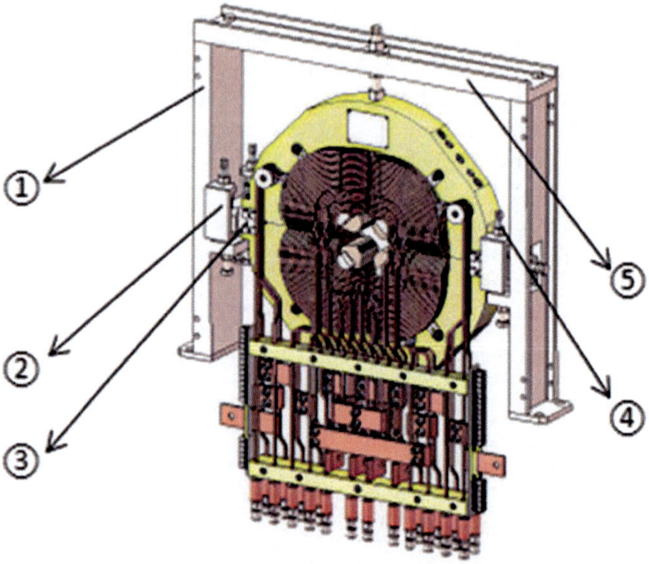

Fig. 2 Thin quadrupole suspension bracket structure

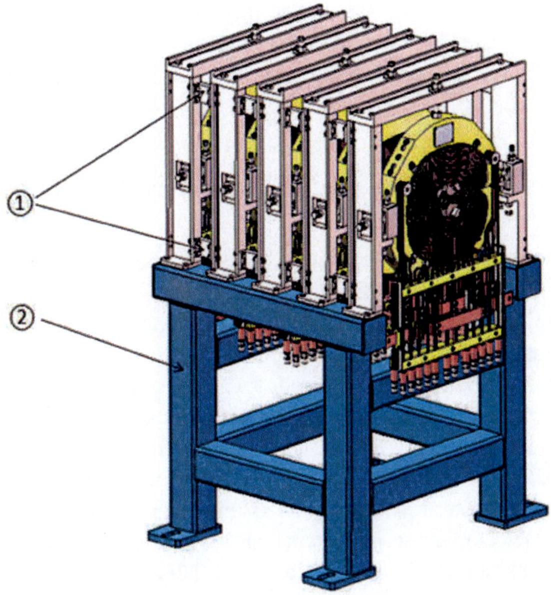

Fig. 3 Suspension installation of quadrupole group. ①Suspension bracket connection plate, ②Support large bracket

The elevation adjustment is mainly adjusted by the screw on the beam to suspend the quadrupole iron, which is installed in the magnetic hook hole. In the horizontal adjustment, the direction of the vertical beam is adjusted by ordinary thread. Due to the small spacing along the beam direction in the MEBT quadrupole group, a special variational regulation structure is adopted to change the vertical motion into the horizontal motion, so as to achieve the adjustment in the minimum space, the assembly shown in Fig. 3. The wedge mechanism studied below is one of the simplest mechanisms that can achieve this kind of directional motion.

3 Research on Regulating Mechanism of Wedge

The wedge mechanism is one of the simplest mechanical mechanism. The standard wedge mechanism is composed of three parts: ① wedge (or active wedge), ② sliding wedge (or driven wedge) and ③ frame. The mechanism is very common, but it is not systematically introduced in mechanical textbooks. Because of its simple structure and fast moving speed, it has been widely used in fixtures and dies. When the mechanism is designed as a clamping device, it is used to locate quickly and clamp reliably. When it is designed as a motion device, it is used to drive accurate, smooth, noiseless, self-locking, force-saving, speed ratio and change the direction or mode of motion of components, which can transform the vertical motion (or horizontal motion) of the inclined wedge into the horizontal motion (or vertical motion or oblique motion) of the sliding wedge. Based on the research on the movement and force of each component when the wedge mechanism is used as a regulating and fixing parts in service, a new method for parametric design of the wedge mechanism is established to solve the design problem of the wedge mechanism under large stroke.

3.1 Design of Wedge Regulation Structure

In the suspension regulating bracket of the first part above, the regulating component in the small space direction is the regulating structure composed of the wedge mechanism shown in Fig. 4. The main components are: ① sliding wedge (connected with quadrupole), ② adjusting support, ③ Adjusting screw, ④ bearing end cover, ⑤ angular contact ball bearing, ⑥ wedge, ⑦ large space horizontal adjusting component.

The adjusting screw 3 is tightly matched with the inner ring of the angular contact ball bearing (ACBB) 5, and the tight fit between the outer ring of angular contact ball bearing 5 and inclined wedge 6. By adjusting the rotation of the screw, the linear motion of the wedge is transformed. Angular contact ball bearing (ACBB) is selected as an important part of the rotor system in this paper, which bears the combined effect of axial load and radial load. ACBB's nonlinear bearing stiffness, dynamic characteristics, the influence of axial force on the dynamic parameters of the bearing, and the variation of bearing stiffness with speed and axial force all have a great influence on its performance. Li Zhen et al. [1] established a quasi-static

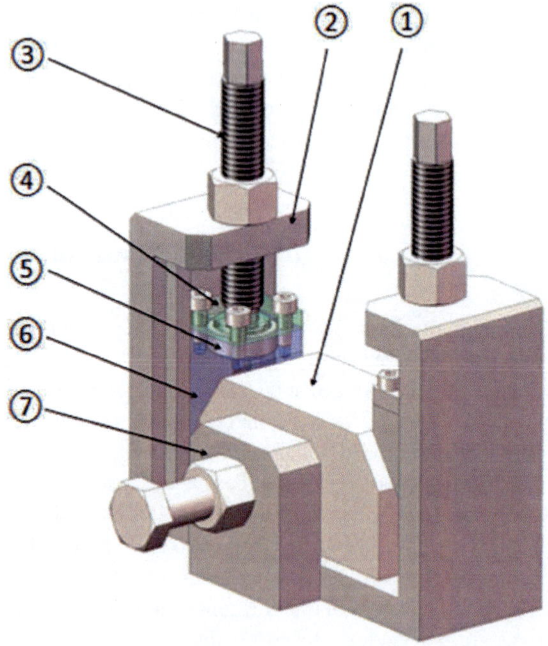

Fig. 4 Wedge adjustment structure

analysis model of angular contact ball bearing with 5-DOF and analyzed the dynamic characteristics of angular contact ball bearing under combined load. In this paper, the use of ACBB is slow in speed, simple working conditions and small force, so it is not considered as the key research content.

3.2 *Study on the Motion of Components in Wedge Regulation Structure*

In the wedge mechanism, the oblique wedge ① and the sliding wedge ② form moving pairs with the frame (relatively fixed surface) respectively, and the low pair mechanism of the moving pair is also formed between the two wedges [2]. There are two types of wedge mechanisms: horizontal motion and oblique motion (including vertical motion) [3]. Considering the fixed regulation direction of quadrupole iron, this paper only discusses the horizontal motion as shown in Fig. 5.

There are three factors in the motion of the wedge adjusting structure:

1. The motion distance of sliding wedge is ±10 mm;
2. Minimize the space occupied by the wedge adjustment structure;
3. Large stroke and high mechanical efficiency.

$$S = L \bullet \tan\theta \tag{1}$$

Substitute $S = 20\,\text{mm}$ into (1) to get the wedge stroke

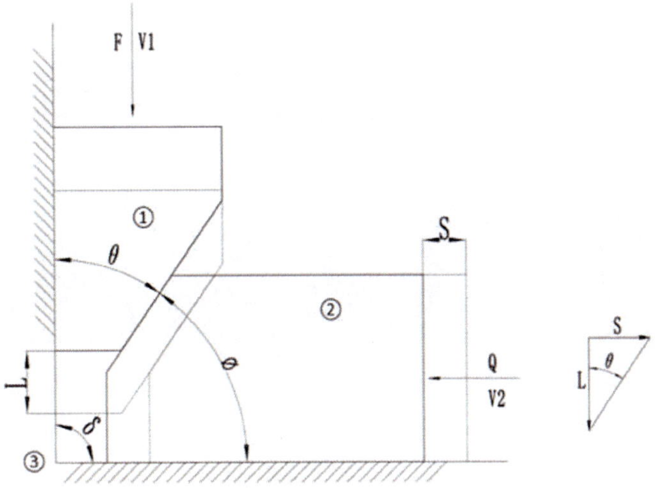

Fig. 5 Horizontal motion diagram of wedge mechanism

$$L = S/\tan\theta = 20/\tan\theta \quad (2)$$

The wedge stroke ratio can be obtained from Eq. (1)

$$i_s = \frac{S}{L} = \tan\theta \quad (3)$$

The theoretical value of $\theta = 0°\sim 90°$.

The stroke ratio of the wedge represents the motion efficiency of the wedge [4]. It can be seen from the curve of stroke ratio of the wedge in Fig. 6, if the motion is only considered, the stroke ratio increases with the increase of the angle θ of the inclined wedge and the stroke ratio reaches the limit at $\theta = 90°$. Within the range $\theta = 0°\sim 30°$, the stroke ratio curve tends to be gentle in the horizontal axis direction and the motion efficiency is low; In the range $\theta = 60°\sim 90°$, the stroke ratio curve has a gentle trend in the vertical axis direction and the motion efficiency is the highest. However, in this range, the combination of inclined wedge and sliding wedge tends to be positive pressure, so it is difficult to achieve directional change motion. Within the range of $\theta = 30° \sim 60°$, the direction changing motion can be realized, and the stroke ratio is relatively high. In this range, to reduce the wedge travel L, the wedge

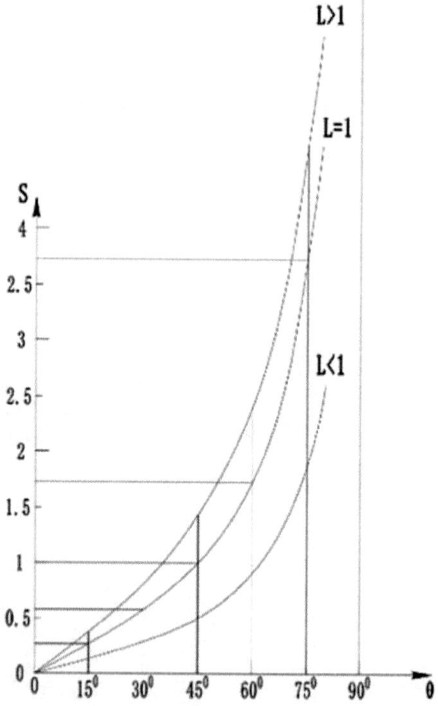

Fig. 6 Curve diagram of wedge stroke ratio

angle can be increased accordingly; In the case of limited space, small stroke but heavy sliding wedge, the wedge angle can be reduced accordingly.

3.3 Study on the Force of Wedge Block in Wedge Regulation Structure

In this paper, the wedge mechanism is used as a regulating element requires not only good movement efficiency, but also good mechanical efficiency, i.e., high force-increasing ratio [5], which can meet the use of force tools in narrow space. The force in this structure is transferred from the rotation force of the screw to the wedge, and then applied to the sliding wedge. The loss and offset of the bearing force during the transfer process are negligible.

Force transmission analysis of regulating screw According to the application form of the regulating screw in this paper, the common screw thread is selected.

Boosting ratio of spiral positive stroke is:

$$i_{p1} = \frac{F}{M} \tag{4}$$

$$i_{p1} = \frac{1}{r\tan\left(\alpha + \grave{\rho}\right)} \tag{5}$$

$$\grave{\rho} = \arctan \dot{\mu} = \arctan \frac{\grave{\mu}}{\cos\beta} \tag{6}$$

According to Eqs. (1), (2) and (3), we can get:

$$F = \frac{M}{r\tan\left(\alpha + \arctan\frac{\mu}{\cos\beta}\right)} = \frac{M}{r} \bullet \frac{\cos\beta - \mu\tan\alpha}{\tan\alpha\cos\beta + \mu} \tag{7}$$

Among the above: M—Driving Troque, F—Bearing Load, r—Pitch Diameter of Thread, α—Helix Angle, β—Half Angle of Threaded Teeth, μ—Friction Factor, $\grave{\rho}$—Equivalent Friction Angle.

Refer to the Handbook of Mechanical Design [6], Substitute $\beta = 30°$, $\alpha = 5°$, $r = 11.026\,\text{mm}(M12\ \text{screw})$, $\mu = 0.05{\sim}0.15$, into (7) to get:

$$F = 0.621\,\text{M}{\sim}0.342\,\text{M} \tag{8}$$

Force transmission analysis of three-component wedge mechanism It can be concluded from Fig. 5 that:

$$\delta = \theta + \varphi = 90°$$

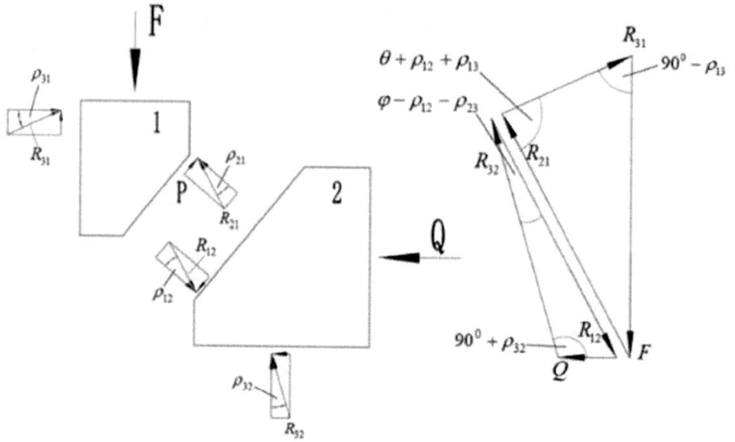

Fig. 7 The wedge adjusting positive stroke force diagram

$$\eta_z = \frac{\tan(\varphi - 2\rho)}{\tan\varphi} = \frac{\tan\theta}{\tan(\theta + 2\rho)} > 0 \tag{9}$$

The positive stroke force of the wedge regulation structure in this article is shown in Fig. 7.

R_{12} and R_{21} are obtained from the wedge force triangle respectively:

$$R_{12} = Q \bullet \frac{\sin(90^\circ + \rho_{32})}{\sin(\varphi - \rho_{12} - \rho_{23})} \tag{10}$$

$$R_{21} = F \bullet \frac{\sin(90^\circ - \rho_{13})}{\sin(\theta + \rho_{12} + \rho_{13})} \tag{11}$$

$$R_{12} = R_{21} \tag{12}$$

The wedge boosting ratio is:

$$i_p = \frac{Q}{F} = \frac{\sin(\varphi - \rho_{12} - \rho_{23})\cos\rho_{13}}{\sin(\theta + \rho_{12} + \rho_{13})\cos\rho_{32}} \tag{13}$$

When the angles of friction are equal, it can be obtained from Eq. (5):

$$i_p = \frac{Q}{F} = \frac{\sin(\varphi - 2\rho)}{\sin(\theta + 2\rho)} \tag{14}$$

Due to $\rho = \arctan\mu = 2.86^\circ \sim 8.53^\circ$.that is inferred:

$$i_p = \frac{Q}{F} = \cot\left[\theta + (5.72^\circ{\sim}17.06^\circ)\right] = \frac{1 - (0.1{\sim}0.3)\tan\theta}{\tan\theta + (0.1{\sim}0.3)} \tag{15}$$

$$\eta_z = \frac{\tan(\varphi - 2\rho)}{\tan\varphi} = \frac{\tan\theta}{\tan(\theta + 2\rho)} > 0 \tag{16}$$

Mechanical efficiency of positive stroke:
Lift angle of maximum efficiency of positive stroke:

$$\theta_{\eta\,\max} = 45^\circ - \rho = 42.14^\circ{\sim}36.47^\circ(\rho \text{ is equal}) \tag{17}$$

Self-locking does not occur during positive stroke. In the working condition shown in this paper, the backstroke input force of this mechanism has to overcome its own characteristics and the backstroke force of the thread, so the backstroke self-locking condition is not considered.

From Eqs. (8) and (17):

$$Q = \frac{(0.621 \sim 0.342) \bullet [1 - (0.1 \sim 0.3) \bullet \tan\theta]}{(0.1 \sim 0.3) + \tan\theta} M \tag{18}$$

Only $G > P$ (P is the friction between the sliding wedge and the frame), the structure can be adjusted. According to [3], the positive pressure of the sliding wedge on the frame is equal to the driving force F of the inclined wedge, and G is the gravity of the sliding wedge (including the magnet).

$$P = \mu N = (0.05{\sim}0.15) \bullet (F + G) \tag{19}$$

From Eqs. (18) and (19):

$$\left[\frac{1 - (0.1{\sim}0.3) \bullet \tan\theta}{(0.1{\sim}0.3) + \tan\theta} - (0.05{\sim}0.15)\right] M > \frac{0.05{\sim}0.15}{0.621{\sim}0.342} G \tag{20}$$

The characteristic curves of the three-component wedge regulation structure are shown in Fig. 8 when the friction angles 3, 5, and 8° are taken as reference (16).

In the wedge regulation structure of this paper, the screw adjustment arm is less than 200 mm, the rotation force is less than 700 N, the mass of magnet and sliding wedge is 50 kg, and take the gravity acceleration $g = 9.8\,\text{m/s}^2$, $\mu = 0.1$. Through Eq. (20), it is calculated that $\theta < 36.96^\circ$, and the angle of the wedge $\theta = 36.5^\circ$ can be selected in combination with Eq. (18). The value is substituted into Eq. (2), and $L = 26.6\,\text{mm}$ is calculated. The stroke distance meets the design size of the wedge.

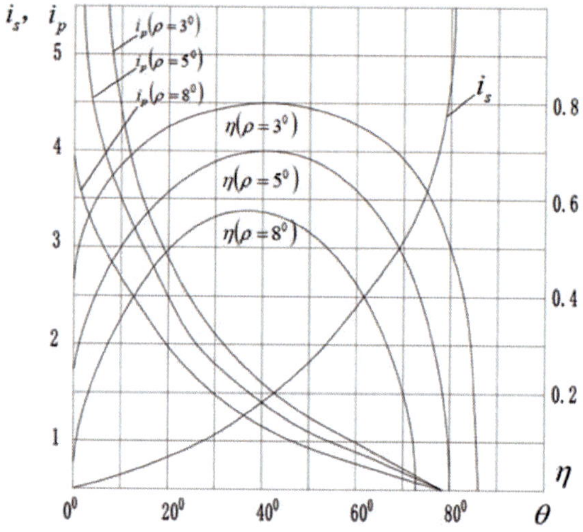

Fig. 8 The wedge regulation structure characteristic curve [6]

4 Advances in Parametric Design of Wedge Mechanism

Due to the change of installation space and magnet specifications, if the wedge mechanism should be adjusted, the traditional design method requires a lot of time and energy, low efficiency, low level of manual calculation and design application, data review and management are difficulties. At present, the research and development of CAD technology is constantly developing. With the application of this technology is more and more widely, the requirements of various industries for this technology are also constantly improving. Therefore, integration, parameterization, intelligence and standardization are the inevitable trend of the development of modern CAD technology [7]. SolidWorks, UG-NX, Pro/E, CATIA and Auto CAD are commonly used 3D CAD modeling software in mechanical design [8]. None of them have a standard part library for wedge mechanism parts, but they have powerful functional configurations, good openness and user interface. They can be connected with many third-party application programming interfaces (API), so that they can be secondary developed to achieve accurate and fast modeling, design and analysis of mechanical products.

Zhao Shiyuan [9] from Southwest Jiaotong University and Ruan Qinchao [10] from Shanghai Jiaotong University both took SolidWorks software as the platform to establish part model. VB was used as the development tool and Access as the database to carry out secondary development to complete the parametric design of the required structure. In the wedge design, the cluster of wedge parts, which is characterized by parameter combination and structure size, is established by SolidWorks. Combined with VB and SolidWorks API functions, the parametric design of wedge mechanism parts is realized by using the two secondary development parametric modeling methods of program-driven and size-driven. The human–computer interaction interface of wedge is developed by VB language, and the automatic generation of wedge model is realized through human–computer communication. The Access data management system is used to automatically update the wedge parameter database, link and share with the developed human–computer interaction interface system in real time, to complete the automatic call of part size parameters and feature creation.

Sun Yadong [11] form Huazhong University of Science and Technology and Hao Zhaoxiang [12] form Chongqing University used UG-NX software as the system development platform for entity modeling, and then used the function library interface provided by UG / OPEN API for secondary development, and used C++ language to compile and influence the data model. In the wedge design, the UG / OpenAPI function is called by VC++, and the STL data structure and algorithm are used to realize the call of the size parameter system. Thereby establishing the oblique wedge design module environment of UG-NX platform. In this environment, the 3D solid modeling design method is adopted, and the parametric 3D design module of the wedge structure is completed by combining the modular design.

The above are two commonly used CAD software based on 3D, and should run the programming language for its secondary development. The combination of key motion parameters, force parameters and structural dimension parameters is established, and the wedge model that meets the parameter requirements is automatically generated through a simple and understandable human–computer interface.

5 Conclusions

The wedge mechanism is used as the regulation structure for the horizontal regulation design of thin quadrupole group in small space, and the motion efficiency and force transmission efficiency are analyzed. The conclusions are as follows:

- The structure is composed of a screw, a bearing and a wedge mechanism, which can transform the rotary motion into a linear motion that drives the translation of the magnet in the minimum space.
- The wedge adjustment structure in this paper, the angle θ of the oblique wedge is between $0°\sim90°$. The larger the wedge angle is, the larger the stroke ratio is and the higher the motion efficiency is. However, only in the range of $\theta = 30°\sim60°$, the directional change motion can be easily realized and the motion efficiency is also relatively high.
- In this paper, the wedge angle in the wedge adjustment structure is in a functional relationship with force boost ratio. When the force transmission efficiency is the largest,$\theta = 42.14°\sim36.47°$ is taken. The output force to promote the translation of the magnet is proportional to the driving torque of the adjustment screw. There will be no jamming during the positive stroke, when the structure is subjected to the input force of the reverse stroke, the wedge and the spiral are combined to form a reverse self-locking double wedge mechanism.

- Based on the research of SolidWorks, UG-NX and other powerful parts modeling platform and open secondary development interface, and the visual programming language was used to provide an effective way for the parametric design and rapid modeling of the wedge mechanism.

The design and research of the above wedge adjustment structure is based on the study of the on-site installation of magnet. In the follow-up work, the transient analysis and prototype test of the magnet adjustment will be carried out to make the research results are more applicable.

References

1. Zhen, L., Xianlei, G., Rui, Z., Qingshan, W.: Analysis of dynamic characteristics of angle contact bearings with combined loads. J. Mech. Eng. **56**(17), 116–125 (2020)
2. Huan, S., Zuomo, C.: Theory of machines and mechanisms, 6rd ed., pp. 9–14. Higher Education Press, Beijing, China (1996)
3. Zhi-LI., W.: Motion and force analysis of component in decling wedge mechanisms. Mech. Res. Appl. **19**(3), 18–19 (2006)
4. Xin, L., Guoran, H.: An analysis of the wedge mechanisms and the design of a wedge mechanism with duplex wedge angles. J. Nan-tong Inst. Technol. **13**(2), 14–18 (1997)
5. Guozhen, J.: Character analysis and design of bi-oblique wedge mechanism. Mach. Des. Manuf. **7**(3), 63–64 (1999)
6. Daxian, C.: Hand book of mechanical design, 5rd ed., vol. 1(4), pp. 165–169. Chemical Industry Press, Beijing, China, 11 (2007)
7. Fenqi, F.: Brief analysis of the development status and trend of CAD technology. Sci. Technol. Innov. **27**, 44–45 (2011)
8. Chenhai, S.: Development and application of CAD technology in mechanical engineering design. Technol. Dev. Enterp. **8**, 54–56 (2015)
9. Shiyuan, Z.: Standard parts parametrzed model library study based on solidworks. Master's thesis, Southwest Jiaotong University, Chendu, China (2006)
10. Qinchao, R.: Research of parametric design based on solidworks for wedge mechanism. Master's thesis, Shanghai Jiaotong University, Shanghai, China (2016)
11. Yadong, S.: A Thesis Submitted in Partial Fulfillment of the Requirements for the Degree of Master of Engineering. Master's thesis, Huazhong University of Science and Technology, Wuhan, China (2011)
12. Zhaoxiang, H.: A Thesis Submitted to Chongqing University in Partial Fulfillment of the Requirement for the Degree of Master of Engineering. Master's thesis, Chongqin University, Chongqin, China (2010)

CO_2: Utilization, Processing and Industrial Use in Shale Deposits

Dmitry Klimov

Abstract The problem of carbon dioxide utilization worries many scientists of humanity, since measures to reduce greenhouse gas emissions are already insufficient to prevent a global increase in temperature on the planet. Most modeling scenarios show that this will require significant deployment of negative emissions technologies. The article discusses current issues that underground carbon dioxide storage projects may face, as well as presents its own laboratory experiment results showing the possibility of storing CO_2 in low-permeable shale formations with a possible economic effect from such storage.

1 Introduction

Today's global problem is global warming on Earth. Scientists of almost all countries are concerned about this problem. Since in the case of uncontrolled warming, mountain glaciers and the ice of the Arctic and Antarctica may melt, which will cause a sharp rise in sea level with many negative consequences.

One of the main reasons for global warming is considered by many researchers to be the increase of carbon dioxide emissions into the atmosphere over time. The world scientific community is trying to simultaneously maintain the level of consumption of hydrocarbon resources and, at the same time, limit the emissions of CO_2 and other greenhouse gases in energy production. Understanding this factor predetermined the emergence of the famous Kyoto Protocol, designed to reduce, and control over time the amount of CO_2 emissions into the atmosphere.

The special report on global warming of 1.5 °C (IPCC SR15) reinforced the important role of CCS in avoiding dangerous climate change. It underlined that reducing emissions alone is no longer enough. To limit global temperature rises to 1.5 °C above pre-industrial levels, the world must reach net zero emissions by around

D. Klimov (✉)
Oil and Gas Research Institute, Russian Academy of Sciences (OGRI RAS), Gubkina st., 3, 119333 Moscow, Russia

M. Shamtsyan et al. (eds.), *Robotics, Machinery and Engineering Technology for Precision Agriculture*, Smart Innovation, Systems and Technologies 247,
https://doi.org/10.1007/978-981-16-3844-2_35

2050. Most modeling scenarios show that this will require significant deployment of negative emissions technologies [1].

2 Materials and Methods

One of the recognized alternatives to combating global warming is the technology of CO_2 capture and storage in the bowels of the Earth (CCS). This alternative is diverse; it provides for the burial of carbon dioxide in salt caverns, in aquifers, in the depths of the oceans, in abandoned, depleted or depleted oil, and gas fields. There are also promising technologies for its further possible use after burial as an oil displacing agent [2, 3]. Such technologies require a serious and thorough scientific approach with an analysis of applicability for specific environmental conditions.

Technologically, the process of utilization of carbon dioxide includes technical measures for the separation of CO_2 from industrial and energy sources, transportation to specialized storage sites, and long-term isolation from the atmosphere.

It has been repeatedly stated that CCS is proven and well understood. Since the 1930s, carbon capture equipment has been used commercially to purify natural gas, hydrogen, and other gas streams in industrial settings. CO_2 was first injected underground in commercial-scale operations in 1972. [4].

Over 260 million tons (Mt) of CO_2 emissions from human activity (anthropogenic sources) has already been captured and stored. The global capture and storage capacity of projects currently operating or under construction stand at around 40 million tons per annum (Mtpa) [4]. Basically, such large numbers are achieved through the use of carbon dioxide in enhanced oil recovery (EOR) projects [5].

Geological carbon storage (CCS) aims to inject CO_2 deep underground to preserve gas for many centuries while maintaining the integrity of the terrestrial ecosystem. Theoretical studies and observational data show [6] that if carbon dioxide was injected into underground traps correctly and in a reasonably chosen location, then, most likely, CO_2 will be stored reliably, as planned. A special report by the International Panel on Climate Change (IPCC) says that "there is a high probability that 99% or more of the CO_2 injected will be stored for 1000 years" [2].

Three types of geological formations that are favorable for storing CO_2 are typically considered: oil and gas reservoirs, deep saline formations, and non-commercial coal seams [3]. Potential candidates for CO_2 storage are those rocks that retain or (as in the case of depleted oil and gas formations) previously retained fluids such as natural gas, oil, or salt solutions. Coal seams can also be used to store CO_2 when there is no likelihood that coal will subsequently be mined if there is sufficient permeability [3]. Under the international projects and programs for the long-term conservation and utilization of carbon dioxide, shale rocks can also fully serve as underground storage traps [7].

Pilot studies by foreign authors on core samples showed good results regarding the suitability of shale deposits for storage of carbon dioxide [7]. Widely known are low-permeability sedimentary rocks with low porosity, including "organic" shale.

It has been shown experimentally that it has the ability to permanently retain a significant amount of gas, due to its ability to retain gas in an adsorbed state in an array of organic matter (i.e., kerogen). Storage in organic-rich shales has an added advantage. It consists in the fact that organic matter acts as a molecular filter, passing CO_2 even into the smallest pores, which is sometimes inaccessible to other natural gases. At the same time, the molecular interactions between the organic substance and the CO_2 molecules are such that they lead to an increase in the absorption capacity of carbon dioxide. Thus, the suitability of shales for storing CO_2 is partly also recognized because the spatial and thermodynamic effects are similar to coals when they are considered during methane extraction from coal seams.

The ideas and pilot projects for organizing mineral sequestration or mineral carbonization with the aim of long-term storage of large amounts of CO_2 can be considered quite similar to the topics of research on the traditional CO_2 disposal; especially where the traditionally considered underground CO_2 sequestration is not possible [8].

The main idea in the methodology of mineral sequestration of carbon dioxide is to simulate the passage of natural processes where calcium (Ca) or magnesium (Mg) from silicate minerals bind to CO_2 (dissolved in water) and form calcium and magnesium carbonates that are stable in this environment [8]. This chemical process can be transformed into an economically sustainable technology that captures large volumes of CO_2.

As a positive example of the implementation of such a technological process, we can safely consider the pilot project "CarbFix"—an industrial training pilot program designed to assess the possibility of mineral sequestration of CO_2 in the basaltic rocks of Iceland [9]. Unique to "CarbFix" is its connection to a geothermal power plant, which makes it possible to capture CO_2 emitted by an industrial facility in addition to mineral CO_2 sequestration. Instead of injecting CO_2 directly into geological formations, in close collaboration between scientists, engineers and technicians, a technology was developed for dissolving CO_2 in reservoir fluids and well water directly during injection. This has opened up great opportunities in the future for the safety of geological storage and carbon sequestration.

The technology becomes even more attractive when you consider that in many territorial areas around the world, there are large deposits of essential minerals, usually magnesium silicates (serpentine, olivine) and sometimes even calcium silicates (wollastonite). Examples of such natural complexes are Finland, the east coast of Australia, Portugal, and the west coast of the USA and Canada. Oman has great potential on the Arabian Peninsula, where some deposits of olivine-containing rocks can contain all the CO_2 that can be obtained from burning all the coal reserves currently represented on Earth.

Many experts agree that both oil and gas production by hydraulic fracturing and geological storage of carbon can be technologically combined in the near future. Such synergies can bring good mutual benefits [10].

Geological storage of CO_2 and the method of hydraulic fracturing combines not only a similar choice of the location of the objects of influence. They also face common challenges such as groundwater pollution risks, water management, seismic risks, public acceptance, or non-recognition [10].

Development and achievements in the field of horizontal drilling, which has been significantly modernized thanks to hydraulic fracturing technologies, can provide additional injection volumes during geological storage. For example, horizontal wells were used at geological storage facilities such as Sleipner and In Salah. Their widespread use in hydraulic fracturing technologies makes them more accessible for underground storage.

The fracturing methodology on an industrial scale is also driven by ongoing innovations, such as the application of stronger cementing technologies in practice. All this, ultimately, will strengthen bilateral communication in the development of effective methods to ensure well integrity and reduce environmental risks.

Some scholars have argued that shale formations can be used for geological underground storage operations after hydraulic fracturing due to the close composition of shale gas and CO_2 [10].

As a result of actual spatial unification, two potential scenarios for the exploitation of underground horizons may arise: an underground carbon storage reservoir will be located above or below the area of hydraulic fracturing (Fig. 1). These scenarios can be combined with other scenarios in which either hydraulic fracturing or geological storage will be a primary operation [10].

If geological storage is of paramount importance, then the concern of industrialists will be to prevent pollution with a CO_2 agent. The injection of fluid into higher or lower aquifers essentially means a loss of well, as happened in some cases at the Barnett field, when vertical wells reached the main aquifer. Currently, everything is not in favor of the scenario of hydraulic fracturing after storing CO_2, and hydraulic fracturing is likely to be the primary mechanism in most cases.

The implementation of the scenario when the formation with hydraulic fracturing (shale or other dense rock) lies below the geological storage area is of greatest concern (Fig. 1a). This area is intersected by several wells, which prompts the consideration of rock integrity by geological storage operators. The foregoing is a realistic assumption of the ability of a CO_2-saturated brine to corrode pipes and create new leak paths, despite progress in cementing technology.

If the formation expected to store CO_2 lies below the area of the hydraulic fracturing (Fig. 1b), the main task for risk assessment will be to assess the integrity and tightness of the sequestration system. However, as mentioned earlier, not all geological storage systems are good candidates for hydraulic fracturing.

One of the subset of generalized scenarios emerges when the fracturing zones and the target area for geological storage are in hydrological integrity. If the fracturing zone is above the target area of the geological trap (b), it becomes part of the storage area. This is the case, for example, in In Salah, where the lower horizons of a solid shutter are fractured and contain liquid. If the fracturing zone is directly below the target underground gas trap (a), lower pressure of the underlying layers can contribute to an overall decrease in pressure.

Despite the well-developed technological aspects of the geological sequestration of CO_2, various projects around the world [4, 11] do not often find information

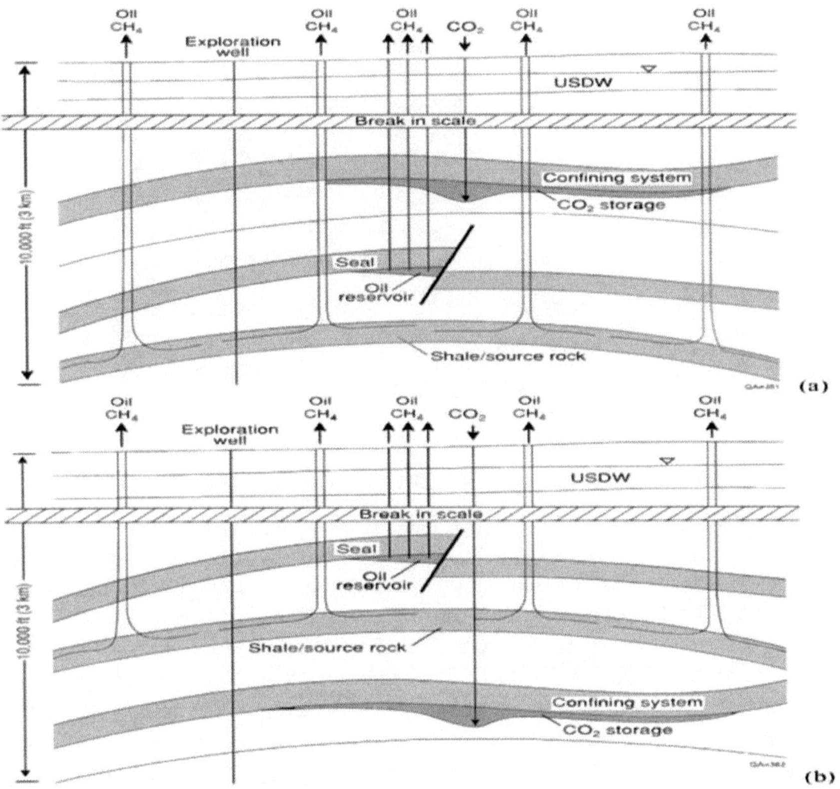

Fig. 1 Possible synergies between hydraulic fracturing and geological sequestration of CO_2: **a**—the formation with hydraulic fracturing lies below the area of organization of geological storage, **b**—CO_2 formation lies below the area of hydraulic fracturing [10]

and studies describing the physicochemical mechanisms of the interaction of carbon dioxide and trap rocks that can occur during this type of underground injection.

3 Results

Based on laboratory studies, we consider the physicochemical effects that can be encountered when organizing a storage facility for CO_2 in low-permeability rocks such as shales. As part of the study and modeling of the physicochemical processes of the interaction of carbon dioxide with models of a porous medium, a series of experimental studies with shale rocks were carried out at OGRI RAS. Shale was a cylindrical rock sample (core) of gray-black color from one of the Russian deposits.

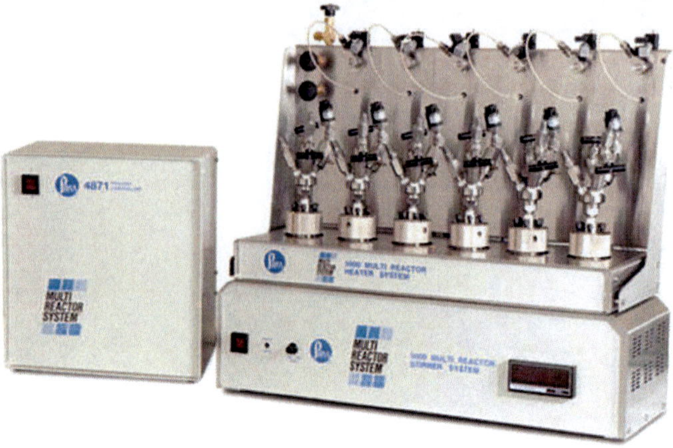

Fig. 2 Multi-reactor multi-system catalytic installation "Parr Multiple Reactor Heater System 5000 Series"

The experiments were carried out on a modern certified multi-reactor catalytic installation "Parr Multiple Reactor Heater System 5000 Series" (Fig. 2), equipped with six independent reactor cells with the ability to set high thermobaric reaction conditions (up to 350 °C and 200 atm.). The catalytic unit was equipped with pressure and temperature sensors that transmitted signals and measurement data to a stationary computer.

One of the goals of the experiments was to study the behavior of shale rock in the presence of distilled water saturated with carbon dioxide (carbonized water), or only with pure carbon dioxide. In this way, the contact of the shale mother rock with in-situ fluids was simulated.

Consider a series of pilot studies related to the effects of exposure to pure carbon dioxide on shale soaked in distilled water. This series of experiments with shale rock was carried out in the following sequence of technological operations (Fig. 3).

The shale rock was crushed to obtain variously rounded rock fragments with a grain size of 0.2–1 cm. Then, the shale samples underwent the procedure of impregnation with distilled water for several days. Next, the soaked ground shale was loaded into several reactor cells 8, which are a cylindrical vessel made of a high-strength alloy of carbon alloy steel with a volume of 45 ml.

Inside the reactor cell 8 was placed up to 35 g. crushed rock, then the cell was twisted with the counterpart 1–7 with the help of a threaded joint nut 11 and fixing screws 12, transforming into a complete sealed system with measuring devices 1. Through a valve with a valve for selection/injection of components 5 and through an immersion pipe for feeding/selection of components 9 into the reactor cell 8 was injected carbon dioxide from a certified carbon dioxide cylinder. After a short (1 min) purge of all connecting lines of the reactor cell 8 under low pressure in order to displace stray air volumes, the supply of carbon dioxide was stopped; valves 2 and 5 were shut off, thereby creating excess pressure in the reactor cell 8, equal to about

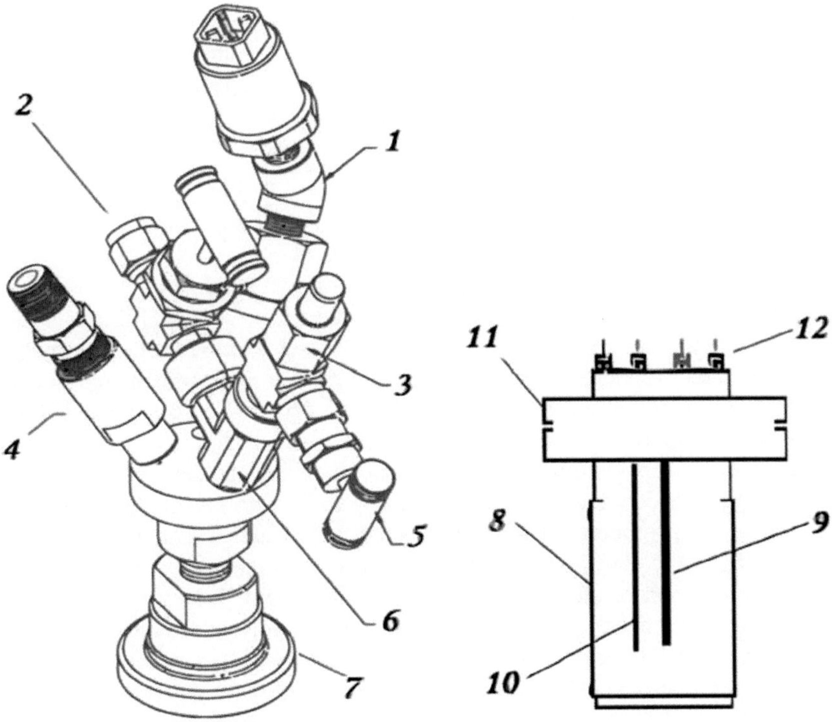

Fig. 3 Diagram of a reactor cell with sensors and valve valves of the Parr Multiple Reactor Heater System 5000 Series: 1—module for connecting sensors, 2—valve and a valve for gas sampling, 3—non-return valve for injecting external gas, 4—protective bursting disc, 5—valve and a valve for sampling / pumping components, 6—system of valves and taps, 7—docking valve a platform with a cell, 8—reactor cell, 9—immersion tube for feeding / selecting components, 10— thermocouple, 11—threaded joint nut, 12—fixing screws

60 atmospheres. The temperature remained within the limits of room conditions (25 °C).

4 Discussion

In the framework of this series of experiments, several comparative parallel studies were carried out with shale rock and CO_2 in several reactor cells under the same initial thermobaric conditions ($P \sim 60$ atm, $T = 25$ °C). The difference was only in the time that carbon dioxide was in contact with the rock after injection until the final sampling of the gas. The purpose of the experiments was a comparative chromatographic analysis of gas samples after final sampling.

The time interval between the sampling of gas samples from different reactors in this series of experiments was 25 days, and between the beginning of the experiment and sampling for each of the reactors was 5 and 30 days, respectively. The results of chromatographic analysis are also presented in Table 1.

From the presented table of the percentage of components of gas samples, several important conclusions can be made.

1. In all experiments, the carbon dioxide content in the gas cap dominated sharply due to its initial injection into the reactor cells.
2. The fact of the presence in all gas analyzes of a wide range of hydrocarbons (methane and its homologues) that are not contained in the initially injected gas, although in small percentage concentrations, is noteworthy. That is, on a small scale, the processes of formation and adsorption of hydrocarbon components from the porous structure of the rock were observed. If it was only about the adsorption of hydrocarbon components, then we would notice a decrease in the analysis of samples from series to series of experiments. On the contrary, we sometimes notice their increase in chromatographic analyzes in comparison with previous experiments (albeit in small quantities). And this cannot be explained only by the effect of greater cumulative adsorption activity with an increase in the duration of the contact of the rock with gas. It is here that the effect of physicochemical generation and geochemical polycondensation synthesis is manifested, the mechanism of which we have widely described in several previous articles [12–14].

On the whole, the longer CO_2 exposure in contact with shale rock had a positive effect on the output of heavier hydrocarbons. The content of some gas components after 30 days of continuous interaction increased by 3–5 times! By the way, experiments with higher initial temperatures (~100° C), which we do not touch on in this article, showed even greater growth in the formation of hydrocarbon components.

Thus, a comparative series of 4–5 experiments on different reactor cells using the described algorithm showed approximately comparable results, both in the nature of the dynamics of pressure changes and in the yield of the final reaction products, which excludes the influence of various random factors (the presence of impurities in the reactor cells, dissolved hydrocarbons in carbon dioxide, short-term single effects of hydrocarbon evolution from shale in connection with their possible oil pollution during field development).

The general mechanism of the physical and chemical processes in shale when they are in carbonated water or pure CO_2 can be explained by their ability to trap gas in an adsorbed state by means of finely dispersed organic matter, which is reflected in experimental studies by authors from the University of Oklahoma [7].

According to studies of shale samples, it was revealed that shale has several mass transfer paths and gas absorption mechanisms, which is associated with the presence of double porosity. After the injected gas collides with the geomechanical porous medium with a double (organic and inorganic) pore system at the "crack/parent rock" interface, it therefore has a choice of the mechanism of penetration into the parent rock: or gas molecules (1) are adsorbed in the organic matter and are spread by

Table 1 Results of chromatographic analysis of the gas phase of the reactor cells after 5 **a** and 30 **b** days of contact of shale rock with carbon dioxide

a Static experiment with shale and CO_2 at 25 ° C and P = 60.5 atm. (gas sampling 5 days after the start of the experiment)

Component	Sample №1	Sample №2	Sample №3	Sample №4
	%	%	%	%
H_2	0.004	0.004	0.006	0.004
O_2	0.518	0.495	0.261	0.383
N_2	1.939	1.916	1.012	1.425
CO	0.000	0.000	0.000	0.000
CH_4	0.000	0.000	0.000	0.000
CO_2	96.637	96.954	98.105	97.316
C_2H6	0.000	0.000	0.000	0.000
C_3H_8	0.001	0.001	0.001	0.001
i-C_4H_{10}	0.0004	0.0009	0.0009	0.0009
C_4H_{10}	0.002	0.002	0.003	0.003
i-C_5H_{12}	0.001	0.001	0.001	0.001
C_5H_{12}	0.001	0.002	0.003	0.003
C_6H_{14}	0.000	0.000	0.000	0.000
C_7H_{16}	0.002	0.000	0.000	0.000
***b** Static experiment with shale and CO_2at 25 ° C and P = 60.5 atm. (selection 30 days after the start of the experiment)*				
H_2	0.002	0.000	0.005	0.009
O_2	0.410	0.410	0.156	0.486
N_2	1.929	1.876	0.940	2.109
CO	0.001	0.001	0.000	0.001
CH_4	0,003	0.001	0.001	0.000
CO_2	97.369	97.560	98.480	96.565
C_2H_6	0.000	0,000	0.000	0.000
C_3H_8	0.006	0.006	0.006	0.006
i-C_4H_{10}	0.005	0.005	0.006	0.006
C_4H_{10}	0.012	0.012	0.013	0.013
i-C_5H_{12}	0.007	0.007	0.008	0.008
C_5H_{12}	0.008	0.008	0.009	0.009
C_6H_{14}	0.005	0.006	0.007	0.005
C_7H_{16}	0.001	0.001	0.001	0.001

diffuse flows through nanopore networks, or (2) the inorganic substance penetrates with the gas stream through a network of irregularly shaped voids. If mass transfer by the second mechanism is due to flow motions according to the Darcy law, then the penetration of gas into organic pores occurs mainly due to transport-molecular mechanisms: pore and surface diffusion. Thus, up to 97% of the gas contained in the rock can be in an adsorbed state inside organic pores, depending on the pressure and temperature inside the shale rock [7].

5 Conclusion

Based on the laboratory studies conducted in the publication, the following conclusions can be drawn.

1. Shale is a fairly reliable medium for storing gas with the possibility of adsorption of associated hydrocarbons. The fact that the spatial and thermodynamic effects in them are similar to those occurring in coal deposits, which are considered during methane extraction operations from coal seams, also speaks in favor of the suitability of shales for CO_2 storage.
2. The organic material of shale has a nanoporous structure. Although it is not entirely suitable for storing large volumes of CO_2 in a free state, it is characterized by a large capacity, which is suitable for capturing significant volumes of gas in an adsorbed state.
3. The prevailing opinion is that the well-known CO_2 storage technologies are so far unprofitable in terms of financial implementation and, while being forced, they only partially pay off by avoiding taxes on CO_2 emissions into the atmosphere. In the light of our experimental results, taking into account additional methane homologues during the implementation of CO_2 storage in low-permeability formations can, to some extent, reduce some of the costly part of such projects.

References

1. Intergovernmental Panel on Climate Change (2018), Global warming of 1.5 °C. An IPCC Special Report on the impacts of global warming of 1.5 °C above pre-industrial levels and related global greenhouse gas emission pathways, in the context of strengthening the global response to the threat of climate change. Accessed from https://www.ipcc.ch/srl5/. Last accessed 2020/11/11
2. Metz, B., Davidson, O., de Coninck, H., Loos, M., Meyer, L.: IPCC special report on carbon dioxide capture and storage. United States (2005)
3. Ha-Duong, M., Chaabane, N.: Captage et stockage du CO_2: enjeux techniques et sociaux en France. Editions Quae (2010)

4. Global CCS Institute: The Global Status of CCS: 2019. Australia. The information in this report is current as of November, p. 46 (2019). https://www.globalccsinstitute.com/resources/global-status-report/. Last accessed 2020/11/11
5. International Energy Agency, Storing CO_2 through Enhanced Oil Recovery: Combining EOR with CO_2 storage (EOR+) for profit. (2015)
6. Price, P.N., Oldenburg, C.M.: The consequences of failure should be considered in siting geologic carbon sequestration projects. Int. J. Greenhouse Gas Control **3**(5), 658–663 (2009)
7. Kang, S.M., Fathi, E., Ambrose, R.J., Akkutlu, I.Y., Sigal, R.F.: CO_2 storage capacity of organic-rich shales. Soc. Petroleum Eng. (2010). https://doi.org/10.2118/134583-MS
8. Zevenhoven, R., Fagerlund, J., Songok, J.K.: CO_2 mineral sequestration: developments toward large-scale application. Greenhouse Gases Sci. Technol. **1**(1), 48–57 (2011)
9. Sigurdardóttir, H., Sigfússon, B., Gunnlaugsson, E.: CarbFix: a CCS pilot project imitating and accelerating natural CO_2 sequestration. Iceland Greenhouse Gases Sci. Technol. **1**(2), 105–118 (2011)
10. Nicot, J.-P., Duncan, I.J.: Common attributes of hydraulically fractured oil and gas production and CO_2 geological sequestration. Greenhouse Gases Sci. Technol. **2**, 352–368 (2012)
11. Global CCS Institute (2019a) CO_2RE Database, Facilities Report. https://CO2RE.co/FacilityData. Last accessed 2020/11/22
12. Barenbaum, A.A., Zakirov, S.N., Zakirov, E.S., Klimov, D.S., Serebryakov, V.A.: Physical and chemical processes during the carbonated water flooding in the oilfields. In: SPE Russian petroleum technology conference, 26–28 October, Moscow, Russia. SPE 176729-MS. (2015)
13. Barenbaum, A.A., Klimov, D.S.: Experimental measurement speed destruction of carbonated water at geosynthesis. Exp. Geosci. **21**(1), 62–65 (2015)
14. Zakirov, S.N., Barenbaum, A.A., Zakirov, E.S., Indrupskiy, I.M., Serebryakov, V.A., Klimov, D.S.: Revisiting the development of oil deposits with low permeability reservoirs. Indian J. Sci. Technol. **9**(42) (2016). https://doi.org/10.17485/ijst/2016/v9i42/104219

Forecasting Agricultural Production as a Tool for Effective Industry Management (On the Example of the Chuvash Republic)

Tatiana Ivanova, Evgeniy Kadyshev, Tatyana Ladykova, Svetlana Brenchagova, Viacheslav Nemtsev, and Anna Ivanova

Abstract The article presents a forecast for the development of agriculture in the Chuvash Republic until 2022 according to three scenarios: optimistic, probabilistic and pessimistic. Agriculture occupies a special position in the region's economy and is a backbone industry. The share of agriculture in the gross regional product is 9%, which is noticeably higher than in other regions of the country, and the share of residents living in rural areas is 36.6%. Structurally, the agriculture of the Chuvash Republic is quite balanced. During the study period from 2010 to 2019 significant positive shifts in the industry can be noted. Thus, the production of basic agricultural products increased 1.93 times, and the gross harvest of basic agricultural crops increased 2.16 times. However, despite the positive trends, problem areas remain in certain areas. The production of basic livestock products during the study period decreased by 135.6 thousand tons; the fleet of basic types of equipment in agricultural organizations decreased by 1.97 times; livestock population decreased by 147.3 thousand heads. In order to predict future changes in the industry, the authors, based on the dynamics of indicators for 2010–2019, characterizing agricultural production in the Chuvash Republic, and using the Excel program, proposed three options for the development of the situation for 2020–2022. As studies have shown, pessimistic and probabilistic scenarios of sustainable functioning are more likely to be implemented, which requires immediate corrective measures for effective management of the industry, in particular, in particular, by creating a single "Chuvash Center for the Development of Rural Areas".

T. Ivanova (✉) · E. Kadyshev · T. Ladykova · V. Nemtsev · A. Ivanova
Chuvash State University Named I.N. Ulyanov, Moskovsky prospect, 15, Cheboksary city, Russia

E. Kadyshev
e-mail: enk808@yandex.ru

A. Ivanova
e-mail: ivanova3011anna@mail.ru

S. Brenchagova
Cheboksary Institute (branch) of the Moscow Humanitarian—Economic University, Civil street, 85, Cheboksary city, Russia
e-mail: brenchagova@list.ru

M. Shamtsyan et al. (eds.), *Robotics, Machinery and Engineering Technology for Precision Agriculture*, Smart Innovation, Systems and Technologies 247,
https://doi.org/10.1007/978-981-16-3844-2_36

1 Introduction

Agriculture has always been and remains the most important resource of any society, influencing the natural, economic, human and ethnocultural potential. For centuries, Russia was an agrarian country, and agricultural products occupied the leading positions in its exports. In the Chuvash Republic, this industry is systemically important: the share of residents in rural areas in 2019 is 36.6% [1], and the share of agriculture in the GRP is 9% [2]. Structurally, the republic's agriculture is fairly balanced and does not have any bias towards crop production (54%) or livestock (46%). In the Russian Federation, the Chuvash Republic ranks 8th in potato production, 14th in production of mustard seeds, and 20th in rye production [1].

The purpose of the article is, on the basis of statistical data on the main indicators of agricultural production in the Chuvash Republic, to calculate their forecast values to substantiate scenarios for the development of the industry and its effective management.

To achieve this goal, the following tasks must be solved: to determine the list of indicators characterizing agricultural production in the Chuvash Republic; analyze them; develop optimistic, probabilistic and pessimistic forecast options; give recommendations on the use of the obtained data. In the course of the study, the following tools were used: analysis and synthesis, correlation and regression analysis, statistical and systematic approaches. To build graphs of the development of the situation in the agriculture of the Chuvash Republic until 2022, the MS Excel application was used.

2 Research Methodology

Analyzing foreign publications over the past five years [3–13], and, taking into account the absence of similar works on these issues in the Chuvash Republic (except for the author's [14–17]), we came to the conclusion that the research technologies used in them are based only on analysis, which does not allow forecasting. In contrast to the presented publications, this work proposes to use the mathematical apparatus to substantiate the sustainable functioning of agriculture and, based on forecasts, to exercise effective leadership.

Consider the main indicators characterizing agricultural production in the Chuvash Republic for the period from 2010 to 2019.

Analyzing the data Table 1, it is seen that agricultural production in the Chuvash region increased from 2010 to 2019 by 19,811.9 million rubles, or 1.93 times; the gross main harvest of agricultural crops increased by 626.9 thousand tons, or 2.16 times; production of basic livestock products decreased by 135.6 thousand tons; the fleet of basic types of equipment in agricultural organizations decreased by 1.97 times; the number of livestock decreased by 147.3 thousand heads, and the output per 1 employee increased by 1.86 times.

Table 1 Indicators characterizing agricultural production in the Chuvash Republic in 2010–2019

Indicator	Year									
	2010	2011	2012	2013	2014	2015	2016	2017	2018	2019
Agricultural production, million rubles	21,388.3	32,122.2	32,625.4	30,862.4	37,054.2	39,347.2	39,338.3	38,736.9	37,370.9	41,200.2
Including: – plant-breeding	6766.5	16,669.1	15,857.0	14,928.6	19,663.8	20,191.7	19,369.6	17,467.6	17,183.4	18,375.5
– animal husbandry	14,621.9	15,453.1	16,768.3	15,933.7	17,390.5	19,155.5	19,968.7	21,269.3	20,187.5	21,824.7
Gross harvest of major crops, thousand tons	1538.6	1574.3	1507.8	1209.0	1318.0	1167.6	1268.1	1249.2	1254.4	1165.5
Production of basic livestock products, thousand tons	1260.4	1304.2	1266.4	1204.3	1082.4	1074.1	1082.8	1070.4	1117.3	1124.8
Agricultural land, thousand hectares	1036.2	1035.9	1035.7	1035.4	1035.3	1035.2	1035.0	1034.5	1034.4	1034.4
Sown area of agricultural crops, thousand hectares	561.1	562.8	529.6	535.9	541.1	554.3	555.3	542.0	537.3	541.5
Application of mineral fertilizers for crops, thousand tons	15.3	15.0	14.7	12.8	12.8	12.1	7.4	8.5	8.6	8.6
Application of organic fertilizers for crops, thousand tons	444.3	462.9	425.7	500.0	578.4	556.2	591.4	604.0	688.4	646.2
The main park types of equipment in agricultural organizations, units	10,466	9625	9387	8506	7487	6898	6214	5912	5614	5315
Livestock, thousand heads	628.6	614.5	587.2	587.2	535.1	538.7	536.4	514.1	504.7	481.3
Production per 1 employee, thousand rubles/person	276.4	398.0	403.3	394.6	465.5	573.1	520.3	497.3	530.2	513.8

Note Table 1 is compiled on the basis of data from the statistical collection "Chuvashia in figures. 2019" [18]

For some indicators characterizing agricultural production in the Chuvash Republic in 2010–2019 there were no significant changes: agricultural land, sown areas of agricultural crops, the introduction of mineral and organic fertilizers for crops. In this regard, we consider it inappropriate to carry out their forecast for the next three years. Also, considering the production of industry products, we will not build separate forecasts for crop and livestock production due to the limited volume of the article.

3 Research Results

Thus, based on the dynamics of indicators for 2010–2019, characterizing agricultural production in the Chuvash Republic, and using the Excel program, in Table 2, the equations of three options for the development of the situation for 2020–2022 are derived (optimistic, probabilistic and pessimistic).

4 The Discussion of the Results

Figures 1, 2, 3, 4, 5 and 6 show forecast graphs (numerals on the x-axis mean years) for the six indicators in Table 2 with the highest approximation coefficient R^2. It is an indicator of the quality of forecasts: the closer its value is to one, the higher the probability of execution. So, for the production of agricultural products in the Chuvash Republic until 2022, R^2 has a maximum value with a pessimistic forecast—0.9052.

Consequently, it is more likely, about 0.91%, to be realized it. For the gross harvest of major agricultural crops and production of livestock products, the pessimistic forecast will also be fulfilled with a probability of 0.83% and 0.81%, respectively. For such indicators as: the park of the main types of equipment in agricultural organizations, the number of livestock and the output per one employee in the industry, a probabilistic forecast will be implemented (with values of 0.99; 0.98 and 0.89%).

The derivation of the eighteen graphs involved in the formation of Table 2 into this article is not possible, due to the limited volume of the article. The range of forecast values for all indicators characterizing the agricultural production of the Chuvash Republic until 2022 R^2 ranges from 0.7514 (with an optimistic forecast for the production of basic livestock products) to 0.9889 (with a probabilistic forecast for a fleet of basic equipment). Eleven out of eighteen constructed forecasts have values of $R^2 > 0.9$, which proves the accuracy of the calculations performed with 90% probability; six variants of forecasts have $R^2 > 0.8$ (80% probability).

Table 2 Forecast of changes in indicators characterizing agricultural production in the Chuvash Republic until 2022

Option forecast	The equation	Year				2022 by 2019%
		2019	2020	2021	2022	
Agricultural production, million rubles						
Optimistic	y = 9327.8 ln (x) + 21,825	41,200.2	47,410.2	49,057.6	47,567.1	115.5
Probabilistic	y = 8357.1 ln (x) + 22,801		44,115.5	45,772.0	44,946.6	109.1
Pessimistic	y = 7757.7 ln (x) + 23,412		42,723.5	43,556.8	42,983.8	104.3
Gross harvest of main crops, thousand tons						
Optimistic	y = 3.7129 × 2 – 84.736x + 1646.5	1165.5	1182.1	1175.4	1123.9	96.4
Probabilistic	y = 1.7496 × 2 – 66.089x + 1615.3		1096.5	1054.9	1013.3	86.9
Pessimistic	y = 1584e–0.035x		1053.3	1027.1	1001.8	85.9
Production of basic livestock products, thousand tons						
Optimistic	y = 3.4579 × 2 – 62.229x + 1368.1	1124.8	1133.1	1128.6	1105.2	98.2
Probabilistic	y = 1.3288 × 2 – 42.127x + 1334.6		1029.3	1005.8	982.3	85.5
Pessimistic	y = 1.141 × 2 – 40.433x + 1331.9		1018.2	991.2	973.5	84.5
The park of the main types of equipment in agricultural organizations, units						
Optimistic	y = 43.611 × 2 – 1066.7x + 11,748	5315	5322	5318	5314	99.9
Probabilistic	y = 8.1568 × 2 – 715.87x + 11,137		4233	3632	3029	24.5
Pessimistic	y = 6.3202 × 2 – 700.42x + 11,115		4101	3427	2977	21.5
Livestock, thousand heads						
Optimistic	y = 1.0128 × 2 – 26.167x + 658.84	481.3	501.2	495.8	488.4	101.5

(continued)

Table 2 (continued)

Option forecast	The equation	Year				2022 by 2019%
		2019	2020	2021	2022	
Probabilistic	$y = 648.62e{-}0.03x$		465.8	450.0	434.2	90.2
Pessimistic	$y = -0.0534 \times 2 - 16.212x + 642.46$		444.3	432.7	428.1	88.9
Production per employee, thousand rubles/person						
Optimistic	$y = 292.87 \times 0.2877$	513.8	602.5	623.2	649.8	126.5
Probabilistic	$y = 292.87 \times 0.2877$		591.1	615.5	639.8	124.5
Pessimistic	$y = 296.73 \times 0.275$		575.3	601.7	599.6	116.7

Note Calculated by the authors based on their own research

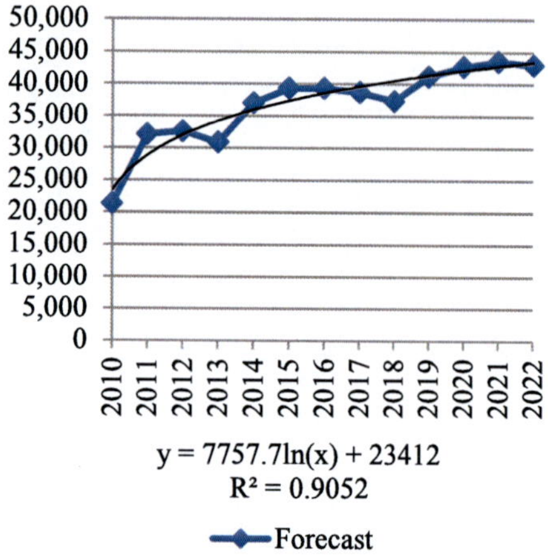

Fig. 1 Pessimistic forecast of agricultural production in the Chuvash Republic until 2022, million rubles

5 Findings

Based on the three forecast options (Table 2), it can be seen that the maximum growth is planned for such indicators as: output per worker and agricultural production in the Chuvash Republic. Even with a pessimistic forecast, the volume of production (in million rubles) will amount to 104.3% in 2022 in relation to 2019. However, for certain items, such as gross harvest of basic crops, production of basic livestock products, etc., it is expected decrease in values in 2022 (for four indicators out of six) even with an optimistic forecast. This situation should serve as a trigger for the development of corrective actions, since without this, sustainable functioning of

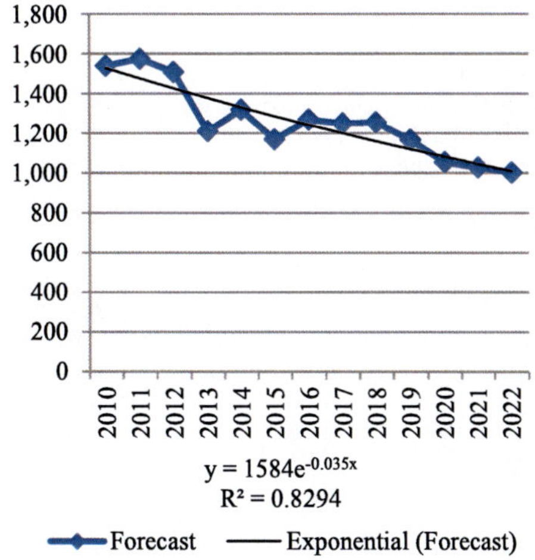

Fig. 2 Pessimistic forecast of the gross yield of major agricultural crops in the Chuvash Republic until 2022, thousand tons

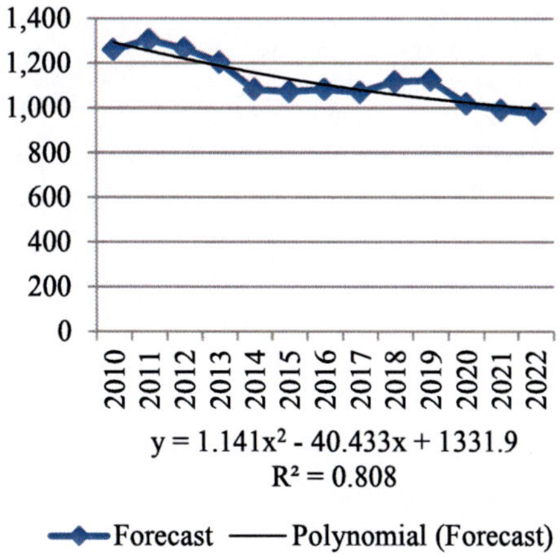

Fig. 3 Pessimistic forecast of production of basic livestock products in the Chuvash Republic until 2022, thousand tons

agriculture is impossible. If you do not take any action in a timely manner, then the process will only worsen. The task of state bodies is to react in time to the prevailing forecast trends in order to mitigate their influence.

However, even if investments flow into agriculture, then in this case it will be impossible to solve problems only by administrative methods; a new level of development of the village and its new dimension are associated with the formation of the peasant as a citizen. The village as a single territorial-production and socio-cultural complex can only develop with a combination of two mutually stimulating factors: positive changes in the outlook of rural residents should be supported by economic transformations.

Fig. 4 Probabilistic forecast of the availability of a fleet of basic types of equipment in agricultural organizations of the Chuvash Republic until 2022, units

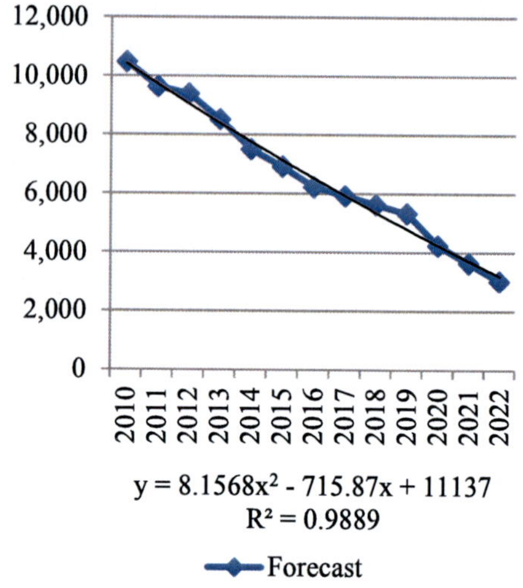

Fig. 5 Probabilistic forecast of livestock population in Chuvash Republic until 2022, thousand heads

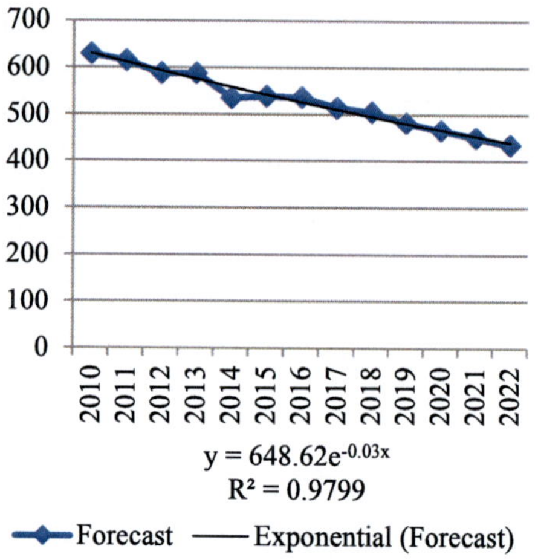

In order to effectively manage agriculture, a single "Chuvash Center for the Development of Rural Areas" should be created, which will be entrusted with a number of functions.

- Promote the creation of a new type of rural settlements with ensuring equal accessibility of social infrastructure and services for all groups of the rural population.
- To regulate the infrastructure of the rural labor market and to orient its facilities towards providing year-round employment.
- Monitor the implementation of direct obligations of the state and employers to rural residents and employees of agricultural organizations.

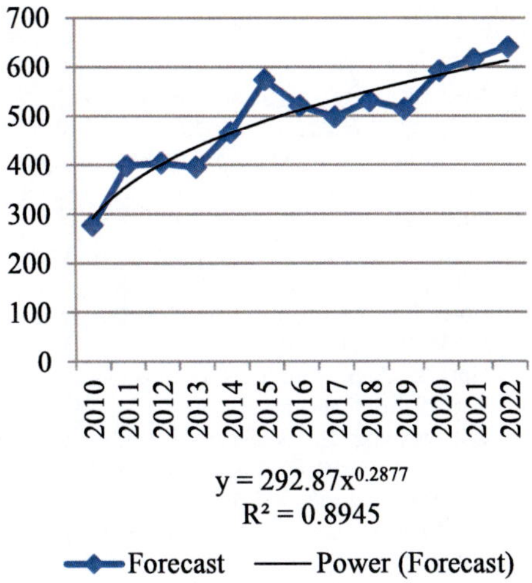

Fig. 6 Probabilistic forecast of production per 1 worker in agriculture of the Chuvash Republic until 2022, thousand rubles/person

- Contribute to ensuring a high educational, professional and living standard of rural residents.
- To develop a methodology for assessing the cost of living necessary for the reproduction of biophysical, labor, intellectual and social capital of various categories of rural residents and oblige agricultural organizations to use it when calculating wages.
- Contribute to the formation of a middle class and an entrepreneurial environment in the countryside.

Today in the republic these issues are dealt with by the Department of Investments and Social Development of the Village of the Ministry of Agriculture of the Chuvash Republic, consisting of 4 people, which is a negligible figure for the region where 36.6% live in the countryside. Moreover, the functions of the department are only in regulating the processes of development of rural areas, and the functional itself should be located "in the field"—in districts, rural settlements, etc. This recommendation is especially relevant in connection with the abolition of the Agriculture Departments and their transformation into departments within the regional administrations.

Thus, effective agricultural management largely depends on competent and timely government support, including through the creation of appropriate infrastructure. And the developed forecast is a mathematical substantiation of the decisions made, representing one of the grounds for legislative and executive authorities in the development of socio-economic projects and programs. In addition, the results of the study can be used in practice by agricultural enterprises, university professors, graduate students, as well as by those who are interested in management and organization of agricultural production.

References

1. Regions of Russia. Socio-economic indicators—2020 http://gks.ru/bgd/regl/b20_14p/Main.htm. Last accessed 2021/02/01
2. About the structure of the gross regional product of the Chuvash Republic http://minec.cap.ru/news/2019/03/13/o-strukture-valovogoregionaljnogo-produkta-chuvashs. Last accessed 2021/02/01
3. Bongiorno, E.G., Goia, A.: Describing the concentration of income populations by functional principal component analysis on Lorenz curves. J. Multivariate Anal. **170**:10–24 (2019)
4. Cole, S.A., Xiong, W.: Agricultural insurance and economic development. Ann. Rev. Econo. **9**, 235–262 (2017)
5. Hilty, L., Aebischer, B.: ICT Innovations for Sustainability, p. 470. Springer, Heidelberg, New York, Dordrecht, London (2015).
6. Kosiorowski, D., Rydlewski, J.P.: Centrality-oriented causality—a study of EU Agricultural subsidies and digital developement in Poland. arXiv:1908.11099v2 16 September (2019)
7. Lee, S.J., Lee, E.H., Oh, D.S.: Establishing the innovation platform for the sustainable regional development: tech-valley project in sejong city, Korea. World Technopolis Rev. **6**(1), 75–86 (2017)
8. Minina, N.N.: Formation of equity capital of agricultural organizations of the Republic of Belarus. Achiev. Sci. Technol. Agro-ind. Complex **32**(1), 50–56 (2018). https://doi.org/10.24411/0235-2451-2018-10111
9. Njos, R., Jakobsen, S.E.: Cluster policy and regional development: scale, scope and renewal. Reg. Stud. Reg. Sci. **3**(1), 146–169 (2016)
10. Stavtsev, A.N., Miroshnikov, G.A.: State support for updating the material and technical base of the agro-industrial complex of the Republic of Belarus. RJOAS **1**(49) (2016). https://doi.org/10.18551/rjoas.2016-01.02
11. Tacchella, A., Mazzilli, D., Pietronero, L.: A dynamical systems approach to gross domestic product forecasting. Nat. Phys. **14**(8), 861–865 (2018)
12. Tahamipour, M., Mahmoudi, M.: The role of agricultural sector productivity in economic growth: the case of iran's economic development plan. Macrothink Inst. Res. Appl. Econ. **10**(1) (2018). https://doi.org/10.5296/rae.v10i1.12809
13. Yoon, D.: The regional-innovation cluster policy for R&D efficiency and the creative economy: with focus on daedeok innopolis. J. Sci. Technol. Pol. Mana. **8**(2), 206–226 (2017)
14. Ivanova, T.V.: Methodology for assessing the personnel potential of agricultural workers in the Russian Federation and determining the rate of its reproduction. Sci. Works Inst. Econ. Nat. Acad. Sci. Azerbaijan **4**, 5–11 (2017)
15. Ivanova, T.V., Ivanova, A.A.: Formation of an effective system for managing the reproduction of personnel potential in agriculture in the region. IOP Conf. Ser. Earth Environ. Sci. **613**, 012049 (2020). IOP Publishing. https://doi.org/10.1088/1755-1315/613/1/012049
16. Orlov, V.N., Ivanova, T.V., Arkhipova, V.A., Ivanitskaya, I.P.: Assessment of the influence of social factors on reproduction of personnel potential in agriculture of Russia. In: PbWOSCE-2018: Business Technologies for Sustainable Urban Development. E3S Web Conferece, vol. 110 (2019). https://doi.org/10.1051/e3sconf/201911002143
17. Orlov, V.N., Ivanova, T.V., Brechagova, S.A., Rumbayeva, N.N.: Mathematical modeling of economic factors impact: reproduction of personnel potential in agriculture sector of Russia. IOP Conf. Ser. Earth Environ. Sci. **433**, 012012 (2020). IOP Publishing. https://doi.org/10.1088/1755-1315/433/1/012012
18. Chuvashia in numbers: 2019: Brief statistical compilation. Chuvashstat, Cheboksary, p. 210 (2019)

Advanced Solution for Improving Safety of Agricultural Machinery During Storage

Vyacheslav Terentyev, Konstantin Zabara, Alexandr Shemyakin, and Konstantin Andreev

Abstract Currently, agricultural machinery (AM) is usually stored in indoor areas, under a shelter and in open equipped areas. However, the indicated storage technologies of AM cannot provide optimal storage conditions and have a low degree of economic efficiency. The purpose of the research is theoretical substantiation and practical implementation of a new method of AM storage in an open area using an experimental design of an active thermal shelter (ATSh). The study presents a comparison of the three most common methods of AM storage, namely storage in an indoor area, under a shelter and in an open equipped area. The analysis showed that a new method of AM storage in an open area with the use of an experimental design of the ATSh requires less investment in arranging the storage place, does not require additional labor and material resources to prepare an AM object for storage, and the total costs for preparation and storage of an AM object are comparable to storing it indoors. The use of the experimental design of ATSh is a more rational way to ensure the reliable safety of the insulated AM object from the impact of the environment for the entire storage period.

1 Introduction

An important task in the process of operating agricultural machinery (AM) is to ensure its safety during the non-working period of long-term storage [1, 2]. Long-term storage of an AM object consists in keeping it in good working order and specially prepared condition, which will ensure its safety and readiness for use at the appointed time [3]. To ensure the readiness of an AM object, it becomes necessary to use effective conservation means at its storage sites, which provide reliable protection against harmful environmental influence for the entire storage period, which increases the labor intensity of work and requires additional material resources [4].

V. Terentyev · K. Zabara · A. Shemyakin · K. Andreev (✉)
Ryazan State Agrotechnological University Named After P.A. Kostychev, Ryazan, Russia

M. Shamtsyan et al. (eds.), *Robotics, Machinery and Engineering Technology for Precision Agriculture*, Smart Innovation, Systems and Technologies 247,
https://doi.org/10.1007/978-981-16-3844-2_37

Based on this, it is proposed to use an active thermal shelter (ATS) to increase the safety of equipment during long-term storage in an open area by using an experimental design of the active thermal shelter (ATSh).

2 Materials and Methods

The purpose of the constructive solution is to reduce the cost of maintaining the operability of the AM object, and, therefore, to ensure the reliable safety of the object from the environmental impact on it for the entire storage period.

This purpose is achieved by the fact that the experimental design of the ATSh contains an individual sealed air-supported structure, which is made of two-layer airtight shell with a heat-reflecting coating, separated by a dome-shaped air space. Air overpressure is created inside the structure, which ensures the necessary stability of the fabric shells without the use of rigid structures. There is a unit for pumping clean air with an electronic sensor for controlling air humidity of a range type and an overpressure valve. Moreover, to create a difference in air pressure inside the room and in the cavity between the layers of the shell of the structure, holes are provided on the inner screen of the shell along the line of the top of the vault, allowing to regulate the flow of air into the cavity. Inside the structure, there is an air gap between the inner screen of the shell and the AM object. Figure 1 shows the design of the ATSh.

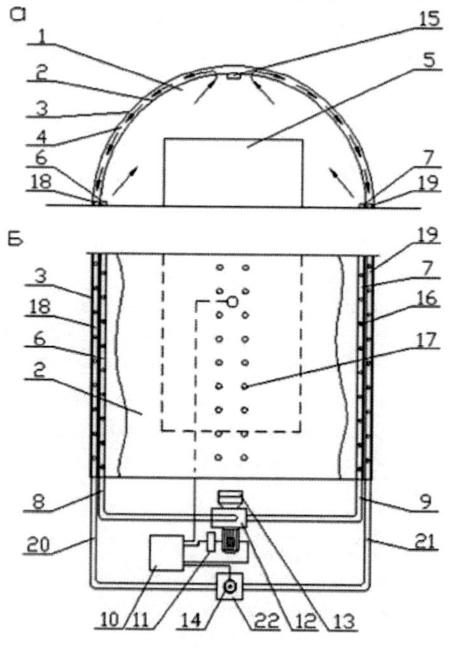

Fig. 1 Experimental design of the ATSh

The experimental design of the ATSh contains: an individual sealed air-supported structure 1, the shell of which is made of two layers of polyester fiber fabric with a silver polyvinyl chloride coating of high density (inner 2 and outer 3), separated by air space 4, with AM object 5 located inside the structure that is dome-shaped and has the structural integrity created by excess internal air pressure acting on the flexible shell. There is an air gap between the inner screen of the shell and AM object 5. The air pressure inside room 1 and in air space 4 is the same. Inside room 1, there are two air distribution pipelines 6 and 7 installed opposite one another and connected by air ducts 8 and 9 with a clean air injection unit, which includes control unit 10, starter 11, centrifugal electric fan 12, sealed air valve 13, electronic air pressure control sensor 14 and electronic air humidity control sensor 15, which is attached to the inner screen of the shell. Air distribution pipelines 6 and 7 are made with calibrated holes 16. Ventilation holes 17 are made on the inner screen of the shell along the line of the top of the vault. There are air intake pipelines 18 and 19 installed in air space 4 opposite one another and connected by air ducts 20 and 21 with an excess valve pressure 22. Air intake pipelines 18 and 19 are also made with calibrated holes 16.

A fabric made of polyester fiber coated with silver polyvinyl chloride of high density prevents the transfer of heat from one body to another in the form of thermal energy of the beam, and also prevents moisture penetration, while protecting the insulated object CXT 5 from the effects of precipitation. Air space 4 between the inner 2 and outer 3 layers of the shell of the structure provides the necessary thermal insulation properties and allows to maintain optimal climatic conditions inside room 1.

So that centrifugal electric fan 12 can operate in automatic mode, the clean air injection unit is equipped with electronic air pressure control sensor 14. When the upper limit of the excess air pressure previously set on electronic air pressure control sensor 14 is reached, an electrical signal will be sent to control unit 10, which will stop centrifugal electric fan 12. And, conversely, when the lower limit of the excess air pressure previously set on electronic air pressure control sensor 14 is reached, an electrical signal will be sent to control unit 10, which will start centrifugal electric fan 12.

Sealed air valve 13 is a device in which the blades of the damper itself, located on the axis, move in such a way that the air entering the damper will not enter the surrounding atmosphere. That is, air in an airtight valve move only in one direction, and if it changes abruptly, then the device closes, which prevents air from entering the surrounding atmosphere. Since air intake ducts 18 and 19 are located in the lower corners of air space 4, the air in it moves from top to bottom, uniformly ventilating the entire surface of the coating. The accepted large cross sections ensure very low-air velocity. This allows ventilation of larger structures at very low-air velocities, thereby providing optimal conditions for long-term storage of the AM object. In addition, poorly ventilated dead zones are eliminated. Overpressure valve 22 is used to stop automatically the removal of air from the structure after turning off centrifugal electric fan 12. Moreover, the return of the damper to the closed position when the pressure drops is adjusted on the valve directly when testing the system.

A positive result when using the experimental design of the ATSh is achieved in the following way. Injection of clean air into an individual sealed air-supported structure and reduction of air humidity in it is carried out in two stages. At the first stage, the unit for blowing clean air and reducing air humidity is switched on in manual mode. When pumping clean air with centrifugal electric fan 12, two streams are formed. The first flow maintains the pressure inside room 1, which is slightly above atmospheric, the second flow is directed into air space 4 between the inner and outer layers of the structure. At the same time, overpressure valve 22 is opened, and the upper and lower limits for regulating the relative humidity inside the structure are set on electronic air humidity control sensor 15. The installation is switched on by starter 11 connecting control unit 10 with centrifugal electric fan 12. At the same time, clean air is supplied from centrifugal electric fan 12 according to the air exchange scheme through air ducts 8 and 9, air distribution pipelines 6 and 7, the internal volume of structure 1 with AM object 5 located in it, as well as through existing ventilation holes 17 provided in the inner screen of the shell, along the line of the top of the vault. Humidified air enters air space 4, between the inner and outer layers of the structure shell, air intake pipes 18 and 19, air ducts 20 and 21 and through valve overpressure 22 is discharged into the surrounding atmosphere. When the lower limit of the set relative air humidity is reached, the unit is switched off manually by starter 11, after which overpressure valve 22 is closed and the unit is switched to automatic operation.

Further, in the process of storing AM objects in individual sealed air-supported structure 1, the relative humidity of the air increases and condensation can form both on the inner surface of outer layer 3 of the shell and on the inner screen of the shell due to sharp changes in ambient air temperature. When the upper limit of the relative humidity, previously set on the electronic humidity control sensor 15 is achieved it sends an electrical signal to control unit 10, which starts centrifugal electric fan 12.

Thus, the process of reducing air humidity will be repeated. The decrease in air humidity will be carried out until the relative humidity inside room 1 reaches the lower limit previously set on electronic humidity control sensor 15. In this case, the electrical signal from humidity sensor to control unit 10 will again be received, which will stop centrifugal electric fan 12. After reaching the required humidity level, the circuit is returned to its original position.

3 Results

To confirm the theoretical studies, preliminary tests of the experimental design of the ATSh were carried out. The objects of study in the process of long-term storage were grain seeders SZU-3.6 in the amount of four units. The storage process for four units of grain seeders SZU-3.6 was carried out as follows: 1 AM unit was in the open area, 1 unit was stored under a shelter, 1 unit was kept indoors and 1 unit was stored in an open area using the ATSh.

In the course of experimental studies, it was found that the loss of metal for the entire storage period of 1 grain seeder SZU-3.6 in an indoor area were about 22 g/m^2 in comparison with the case when storing 1 unit in an open area or under a shelter, which are approximately 148 g/m^2 and 129 g/m^2, respectively. Metal loss during storage of 1 grain seeder SZU-3.6 in an open area using ATSh amounted to approximately 27 g/m^2. To determine the values 200B200B of the corrosion resistance of the metal, the method of evaluating the weight loss of the metal was used. Indicators of the intensity of metal rusting during storage are shown in Fig. 2.

The degree of chalking of the paint coat for the entire storage period of 1 grain seeder SZU-3.6 indoors was about 12% in relation to the degree of chalking of the paint coat of 1 unit of the same AM object when stored in an open area, and the concentration of microcracks was 0.4 pcs/cm^2. The degree of chalking of the paint coat of 1 grain seeder SZU-3.6 when stored in an open area using ATSh was 18% in relation to the degree of chalking of the paint coat of 1 unit of the same AM object when storing it in an open area. The concentration of microcracks did not exceed 0.45 pcs/cm^2. Indicators of chalking and cracking of the paint coat are presented in Figs. 3 and 4.

The assessment of the decorative and protective properties of the paint coat was carried out by measuring the indicators of the appearance with the following devices: a photoelectric gloss meter BF5M-45/0/45 and a pulsed electric spark flaw detector.

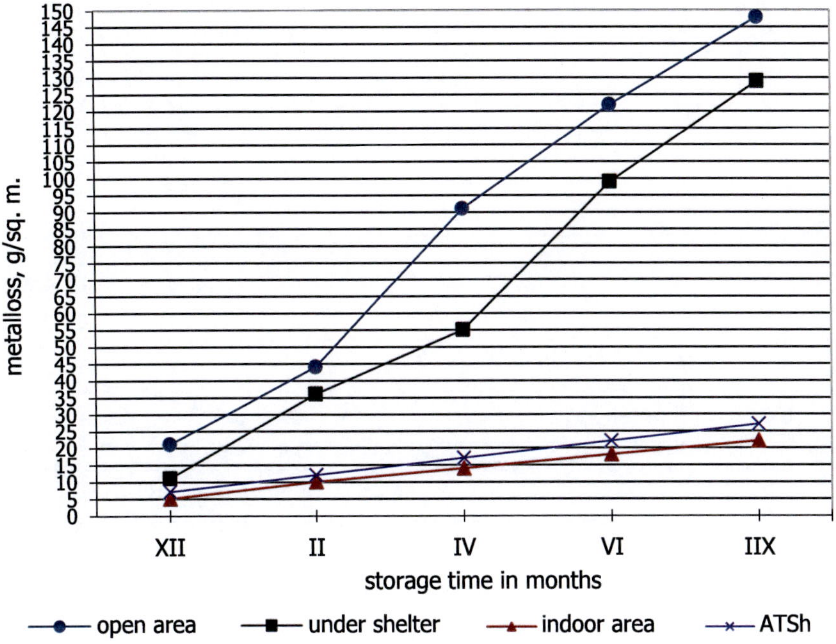

Fig. 2 Intensity of metal rusting during storage

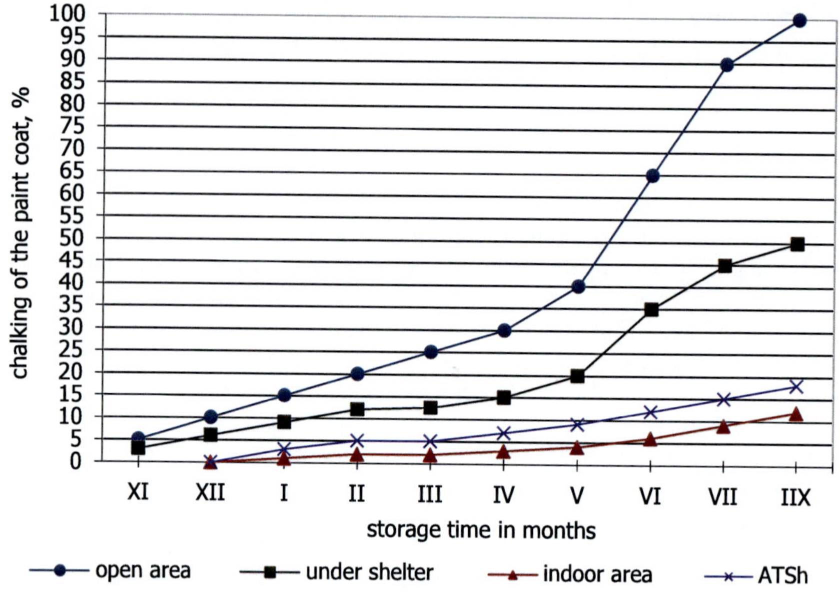

Fig. 3 Chalking of the paint coat

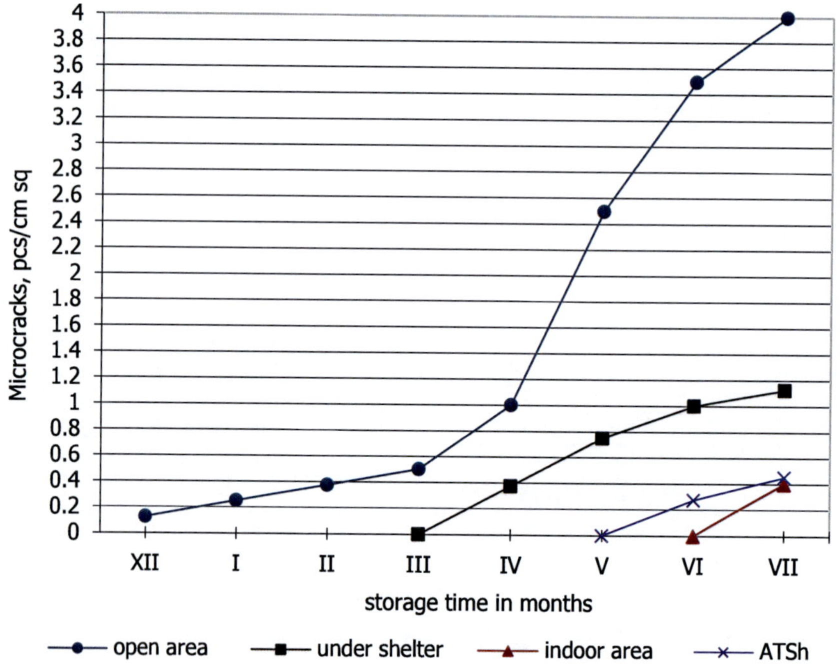

Fig. 4 Paint coat cracking

4 Discussion

A distinctive feature of the work of equipment in agriculture is its short-term periodic and intense use with subsequent long-term storage [5, 6]. After the completion of work related to the production of agricultural products, the agricultural facility must be cleaned of plant residues and technological pollution, washed. Its components and assemblies must be conserved and then put into long-term storage [7, 8]. The reliability of the operation of the AM object largely depends on the conditions of its long-term storage [9, 10]. It should be noted that during the storage period, the insulated AM object is under the influence of certain meteorological conditions prevailing for a given area, namely, air humidity, precipitation, temperature regime, solar radiation. The impact of meteorological conditions has a significant effect on the change in the physical—chemical properties of such structural materials as rubber, leather, wood, metal, textiles, plastic, etc. As a result of damage to agricultural machines from corrosion, the costs of maintaining their performance are increasing [11]. Such changes, as a rule, cause negative consequences and worsen the operational properties of the insulated AM object.

The surrounding atmosphere consists mainly of gases and various impurities (dust, water droplets, ice crystals, sea salts, combustion products) and at the same time is an endlessly acting carrier of aggressive corrosive reagents. The concentration of gases near the Earth's surface that make up the atmosphere is practically constant, with the exception of water (H_2O) and carbon dioxide (CO_2). Almost all components of the atmosphere have an effect on corrosive processes, and can also enhance the actions of each other [12]. Moisture and oxygen have the greatest influence on the course of corrosion processes. The atmospheric corrosion of metal structures represents one of the most severe forms of corrosion, resulting in huge economic and structural performance losses [13, 14].

At present, it is common to store AM in indoor areas, under a shelter and in open equipped areas [15, 16]. The first method is optimal, since it excludes any effect of solar radiation and precipitation on a specific object of AM. At the same time, this method insignificantly improves the quality of storage of the AM object. Its main disadvantage will be the high-relative humidity of the indoor air.

Storage in indoor areas requires significant investments in the design, construction and operation of facilities for AM storage. However, such investments pay off as a result of the increase in the service life of AM and the reduction in the cost of its repair. Storage in open equipped sites requires additional labor costs and material resources (conservation, sealing, dismantling of parts and assemblies which require warehouse storage). Therefore, there is a need to develop promising methods of AM storage using universal and inexpensive shelters that allow high-quality storage.

The results of statistical data show a constant increase in the number of AM objects with various kinds of failures and malfunctions. Analysis of the causes of AM failures and malfunctions when long-term storage showed that they are due to both the low quality of the work performed to prepare the machine for storage and the impact of climate and its components. The most dangerous climate factors affecting the reliability of AM parts and assemblies are solar radiation, precipitation and dew that is mainly observed in the evening, night and morning hours. If storage in a indoor area, under a shelter or a protective cover can fully or partially protect a specific AM object from exposure to solar radiation or precipitation, then such shelter methods will not be able to protect from dew falling on the surface of a stored AM object.

Dew-fall on the surface of the AM object is noted from the moment when the transfer of heat from the sun in the form of thermal energy of the beam ceases to flow into the near-earth space and the surface of the earth instantly begins to cool down. At the same time, the lower layers of air are cooled down rather quickly. At this moment, thermal energy is radiated by every heated body, depending on the density of each. In this case, bodies emitting thermal energy are cooled. The AM object, being a denser body, is cooled more intensively than the surrounding air. And if the dew point exceeds the surface temperature of the AM object, located, for example, in a indoor area, then, without any doubt, dew will fall on its surface. In this case, in order to protect the AM object from dew-fall on its surface, it will be necessary to create certain conditions. Conditions under which the cooling rate of the AM object would be equal to or less than the cooling rate of the surrounding air. It is possible to achieve such results by installing a barrier on the way of heat transfer from the AM object to the environment in the form of thermal energy of the beam, provided that of the object are completely protected from solar radiation and precipitation [17].

5 Conclusion

The obtained data allow to analyze the ratio of resource spending and time, as well as to determine the possible level of performance in certain conditions. So, for example, the indicators of the labor intensity of work when preparing grain seeder SZU-3.6 for storage and the total costs of the preparation and storage of a given AM object with various methods of its storage are presented in Table 1.

Table 1 shows that the labor intensity when preparing 1 grain seeder SZU-3.6 for storage in an open area is higher than with a method of storage in an indoor area. This is due to the fact that this method of storage requires additional labor costs and

Table 1 Indicators of labor intensity of work when preparing grain seeder SZU-3.6 for storage and the total cost of preparation and storage of this AM object with various storage methods

Storage method	Labor intensity of works when preparing an AM object for storage, mhrs	Total costs for the storage of an AM object, rub
Open area	2.065	2 646.07
Under the shelter	2.065	2 603.14
Indoor area	1.24	1 546.38
ATSh	1.25	1 478.07

material resources (conservation, sealing, dismantling and installation of parts and assemblies of an AM object that require warehouse storage). In addition, the costs of restoring the operability of parts and assemblies of the AM object after long-term storage also increase.

Labor intensity of work when preparing 1 grain seeder SZU-3.6 for storage under the shelter is practically the same as when storing in an open area. However, the safety of this AM object with this method of storage increases somewhat. Costs for preparation and storage indoors are lower than when stored in an open area or under the shelter.

The results of experimental studies allow to conclude that the use of the experimental design of the ATSh is a more effective way to ensure the reliable safety of the SZU-3.6 grain seeder from the impact of the environment on it for the entire storage period. Since, firstly, this method of storage requires less investment in arranging the storage place, and secondly, it does not require any additional labor and material resources to prepare an object for storage, and the costs of preparing and storing it are comparable to its storage in an indoor area. In this regard, this method of storage can be recommended to businesses that do not have sufficient funds to build capital structures for storing agricultural facilities.

References

1. Komarov, V.A., Nuyanzin, E.A., Kurashkin, M.I.: Investigation of the process of storing combine and self-propelled equipment in the regional agro-industrial complex. Rural Mach. Equipment **5**, 32–36 (2019)
2. Skuryatin, N., Novitsky, A., Zhiltsov, A., Soloviev, E.: Development of constructive-technological scheme of parking for agricultural machinery. Engineering for rural development 18. In: Proceedings of the 18th International Scientific Conference Engineering for Rural Development, vol. 369, pp. 239–246 (2019)
3. Mironov, E.B., Voronov, E.V., Shisharina, A.N.: Modern developments and technologies in the field of storage of agricultural machinery. Agric. Sci. J. **6**, 93–96 (2019)
4. Andreev, K.P., Terentyev, V.V., Shemyakin, A.V.: Preparation of agricultural machinery for storage. Repair. Recovery. Modernization **9**, 36–39 (2018)
5. Ushanev, A.I., Uspensky, I.A., Yukhin, I.A.: Pilot installation for applying protective coating on the surface of the agricultural equipment. In: IOP Conference Series: Earth and Environmental Science, vol. 488, no. 1, pp. 012049 (2020)
6. Ushanev, A.I., Uspenskiy, I.A., Yukhin, I.A., et al.: Influence of the droplet size on the uniformity of the distribution of protective material over the surface of agricultural machinery. In: IOP Conference Series: Earth and Environmental Science, vol. 488, no. 1, pp. 012048 (2020)
7. Polischuk, S.D., Churilov, D.G., Byshov, N.V., Borychev, et al.: Corrective maintenance and hardening of agricultural machinery parts. Key Eng. Mater. **836**, 158–167 (2020)
8. Shemyakin, A.V., Terentyev, V.V., Kuzin, E.G.: Cleaning of engines of agricultural machines before repair (experimental research). Bull. Ulyanovsk State Agricult. Acad. **1**, 171–175 (2017)
9. Byshov, N.V., Borychev, S.N., Uspensky, I.A., Yukhin, I.A. Golikov, A.A., Filyushin, O.V.: Improving the performance parameters of vehicles for intrafarm transport in the agro-industrial complex. In: IOP Conference Series: Earth and Environmental Science vol. 341, no. 1, pp. 012145 (2019)

10. Polischuk, S.D., Stekolnikov, Y.A., Churilov, D.G., Byshov, N.V., Borychev, S.N., Uspenskiy, I.A., Arapov, I.S.: Peculiarities of chromium coating in low-concentrated electrolytes when repairing details. Key Eng. Mater. **836**, 151–157 (2020)
11. Gaidar, S., Nizamov, R., Golubev, M., Golubev, I.: Protective efficiency of water-soluble corrosion inhibitors Mordovia University. Bulletin **28**(3), 429–444 (2018)
12. Ye, M., Ye, L., Krupin, A., Ye, T.: A device for making protective compounds at the agricultural equipment conservation. Mordovia Univ. Bull. **26**(4), 490–498 (2016)
13. Landolfo, R., Cascini, L., Portioli, F.: Sustainability **2**, 2163 (2010)
14. Andreev, K.P., Zabara, K.A., Terentyev, V.V., Shemyakin, A.V.: Improvement of storage technology for agricultural machinery. Repair. Recov. Modernization **7**, 32–38 (2020)
15. Komarov, V.A.: Research of technical service enterprises for promoting equipment reliability (Case Study of Agro-Industrial Complex of the Republic of Mordovia). Mordovia Univ. Bull. **28**(2), 222–238 (2018)
16. Byshov, N.V., Borychev, S.N., Uspensky, I.A., Fadeev, I.V.: Development of a new tool for protecting agricultural machines during storage. Mach. Equip. Village **6**, 38–42 (2019)
17. Zabara, K., Shpak, A., Shemyakin, A., Morozova, N., Melkumova, T., Podyablonskiy, A.: Prevention of corrosion fracture of agricultural equipment during storage. In: E3S Web of Conferences, vol. 164, p. 06002 (2020)

Regression Analysis of Immersion of Grains in Liquid

Aleksey Saitov, Vyacheslav Farafonov, and Viktor Saitov

Abstract In addition to grain, the grain heap of rye contains weeds and harmful impurities. Poisonous ergot sclerotia are harmful impurities. Modern grain cleaning machines do not provide the allocation of ergot sclerotia in one technological process by air flow and on sieves. Ergot sclerotia have a density lower than that of rye grain. To create a machine for the extraction of poisonous ergot from grain, practical experiments were carried out on throwing single grains of winter rye of the Falenskaya 4 variety into a liquid with a density of $\rho_{zh} = 1000$, 1090 and 1150 kg/m^3. Sodium chloride (NaCl) was used to prepare the solution. The array of the obtained experimental data is presented in the form of histograms by the distributions of the relative frequency of overcoming the surface tension and immersion of the grain in the liquid from the height h of the fall. Regression analysis of the experimental data set was carried out using the SigmaPlot 11.0 software. Regression models are obtained that adequately describe the real process of immersion of grains in a liquid. From the analysis of experiments, it follows that almost 100% of the immersion of grains in water occurs at a height h of at least 43×10^{-3} m and an aqueous solution of salt ($\rho_{zh} = 1150$ kg/m^3) – 58×10^{-3} m. Regression models can be used to develop a device for the isolation of poisonous ergot sclerotia by density in an aqueous solution of salt and seed dressing using the wet method.

1 Introduction

People have been using rye grain for food preparation for a long time. Rye grain is also the most valuable component for the preparation of feed for farm animals.

The grown rye crop is harvested by self-propelled combine harvesters [1–3]. The grain material in the harvesters is accumulated in the hopper. In addition to grain,

A. Saitov · V. Saitov (✉)
Federal Agrarian Scientific Center of the North-East, 166 A, Lenin Str., Kirov 610007, Russia
e-mail: vicsait-valita@e-kirov.ru

V. Farafonov · V. Saitov
Vyatka State Agrotechnological University, 133, October ave., Kirov 610017, Russia

M. Shamtsyan et al. (eds.), *Robotics, Machinery and Engineering Technology for Precision Agriculture*, Smart Innovation, Systems and Technologies 247,
https://doi.org/10.1007/978-981-16-3844-2_38

this grain material contains weeds and harmful impurities. Poisonous ergot sclerotia are among the harmful impurities [4, 5].

At present, the separation of various impurities from the grain heap of rye is carried out by machines of preliminary, primary and secondary cleaning, mainly by air flow and on sieves [6–8].

In one technological process, these machines do not provide the release of poisonous ergot sclerotia in one technological process due to the proximity of their physical and geometric properties (windage, color, width, thickness and length) [9–12].

Research on improving the design and increasing the efficiency of the technological process of these machines does not solve the problem noted [13–18].

Ergot sclerotia have a density lower than that of rye grain. Therefore, the release of poisonous ergot sclerotia is possible in an aqueous solution of inorganic salts. For this, you can use solutions of sodium chloride or potassium salt [19].

Accordingly, an urgent issue is the creation of a machine for the isolation of poisonous ergot sclerotia from rye grain by density in an aqueous solution of salt. When developing such a device for high-quality performance of the technological process for the isolation of ergot sclerotia from rye grain, it is necessary to determine the height of the location of the outlet of the loading hopper with grain relative to the surface of the aqueous salt solution [20].

As a result of practical experiments, an array of experimental data can be obtained on the effect of the height of the fall of a grain on the surface of a salt solution on its overcoming of surface tension and immersion in liquids of various densities. An adequate assessment of the obtained experimental data can be given as a result of the analysis using modern programs for statistical data processing, issued by a set of criteria for analyzing the resulting regression model.

The performed regression analysis of the experimental data set will make it possible to determine the minimum height of the location of the outlet of the loading hopper with grain relative to the surface of the liquid of various densities in the ergot separating device in order to ensure the overcoming of the surface tension forces by the weevil. The obtained regression models can be used in the development of a device for the isolation of poisonous ergot sclerotia by density in an aqueous solution of salt and a seed dressing agent by the wet method.

2 Materials and Methods

To obtain an array of experimental data in order to identify the rational height of the location of the outlet of the loading bunker with grain relative to the surface of an aqueous solution of salt, practical experiments were carried out on throwing single grains of winter rye variety Falenskaya 4 into a liquid with a density of $\rho_{zh} = 1000$, 1090 and 1150 kg/m^3. To change the density ρ_{zh} of the liquid, sodium chloride salt (NaCl) was used. A general view of a sample of winter rye grains of the Falenskaya 4 variety and sodium chloride (NaCl) in the placer is shown in Fig. 1.

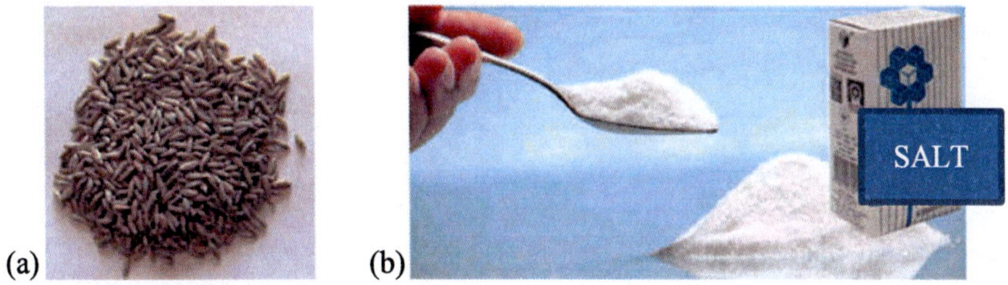

Fig. 1 General view in a placer of a sample of grains of winter rye variety Falenskaya 4 (**a**) and sodium chloride (NaCl) (**b**)

According to the geometric shape, the winter rye grains of the Falenskaya 4 variety have an elongated round shape. Therefore, the rye grain is considered as an elongated ellipsoid of revolution. This geometrical model of a grain has major and minor axes.

For the experiments, an experimental setup was made, which is shown in Fig. 2. The setup is a vessel in the form of a transparent glass bath. The bath has a length of 0.35 m, a width of 0.20 m and a height of 0.15 m. The internal cavity of the bath contained a volume of liquid in the amount of 9 L.

Above the bath, metal plates were installed, which had a thickness of 1.5×10^{-3} m, a length of 0.28 m and a width of 0.04 m. With the help of a set of these metal plates, the height h of the fall of rye grains into a liquid of various density ρ_{zh} was varied. The grains in the amount of 20 pieces were installed on the surface of the upper plate and dropped into the liquid along the main minor axis (flat). For the liquid to enter the surface along the main major axis, the grains were clamped with tweezers and dropped from the investigated height h in the direction of this axis. The experiments were carried out in triplicate, and the total number of discarded grains is 60 pieces.

The relative frequency of immersion in water ($\rho_{zh} = 1000$ kg/m^3) and an aqueous solution of sodium chloride (NaCl) with density $\rho_{zh} = 1090$ and 1150 kg/m^{3}of

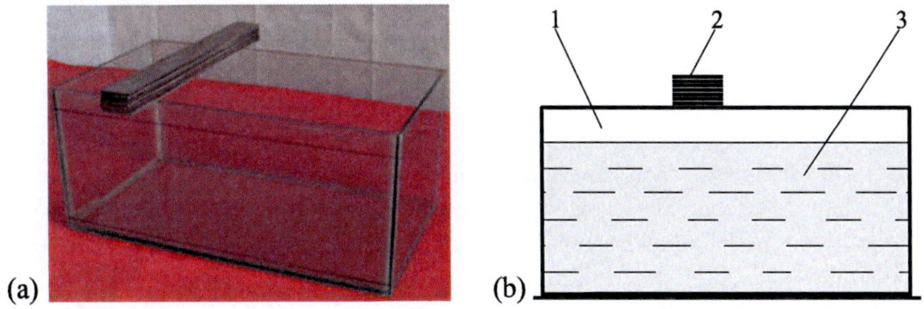

Fig. 2 General view (**a**) and scheme (**b**) of an experimental setup for conducting experiments on immersion in water ($\rho_{zh} = 1000$ kg/m^3) and an aqueous solution of sodium chloride (NaCl) with density $\rho_{zh} = 1090$ and 1150 kg/m^3 of individual grains of winter rye variety Falenskaya 4 when varying the height h of the fall: (1) bath; (2) metal plates; (3) aqueous salt solution

individual grains of winter rye variety Falenskaya 4 in varying the height h of the fall were determined by the expression (%):

$$p = \frac{n_2}{n_1} \cdot 100, \quad (1)$$

where n_1—the number of individual grains entering the liquid, pc;

n_2—the number of grains that are drowned in liquid, pc.

When fixing the obtained experimental data, the temperature of the liquid and the outside air in the laboratory was maintained at 20 °C.

The experimental data were processed using a SAMSUNG NP-R540H personal computer using the SigmaPlot 11.0 statistical data processing package. The reliability of the obtained regression models was assessed by confirming the significance of the obtained values of the coefficients of the regression models by comparing their calculated values of the t-statistics ($t_{\text{cal.}}$) with the tabular value of the student's t-test ($t_{\text{tab.}}$), comparing the actual values from the calculated ones, the coefficient of determination R^2 (Rsqr), which explains all the variability of the obtained function from the corresponding studied variables.

The accuracy of the obtained nonlinear regression model was assessed by the standard error of estimate.

The autocorrelation between the experimental data was assessed by the Durbin–Watson statistic (DW) criterion, which varies from 0 to 4. The indicator DW = 0 corresponds to positive autocorrelation, with negative autocorrelation DW = 4, and when autocorrelation is unambiguously absent, the statistics $1.5 < \text{DW} < 2.5$. When the DW index decreases or increases from these values, when the statistics are $0 < \text{DW} < 1.5$ and $2.5 < \text{DW} < 4$, then the autocorrelation between the experienced ones is questioned. In the case of autocorrelation between the experimental data, the obtained regression formula is usually considered unsatisfactory. Autocorrelation between the experimental data indicates an incorrect specification of the model, which must be corrected by choosing a different dependence formula or to include factors that were not previously taken into account.

Regression models for adequacy to the real process as a whole were also assessed by comparing their calculated values of F-statistics ($F_{\text{cal.}}$) with the tabular value of Fisher's F-test ($F_{\text{tab.}}$) for a significance level of 0.05 and the number of degrees of freedom (DF) k_1, k_2, determined as a result of analysis of variance. If the calculated value of the F-statistic ($F_{\text{cal.}}$) is greater than the tabular value of the Fisher's F-test ($F_{\text{tab.}}$), then the explained variance is significantly greater than the unexplained, and the regression model as a whole is significant in terms of quality and its parameters.

The adequacy of the description of the ongoing process by the regression model was also generally assessed by the P-level of significance (P-Value) in comparison with the accepted significance level of 0.05.

The resulting array of experimental data is presented in the form of histograms using the vector graphics editor CorelDRAW 12.

3 Results and Discussion

The array of experimental data when the grains of winter rye variety Falenskaya 4 are immersed in a liquid of different density ρ_{zh} along the main minor axis (flat) from the height h of the fall is processed and presented as a distribution of the relative frequency p, which is shown in Fig. 3.

The theoretical models of the distribution of the relative frequency p of grains of winter rye variety Falenskaya 4 when immersed in a liquid of different density ρ_{zh} along the main minor axis (flat) from the height h of fall have the form:

$$p = \mathrm{y0} + \mathrm{ah} + \mathrm{bh}^2 + \mathrm{ch}^3. \quad (2)$$

when dropping from the plate along the main minor axis (flat) of winter rye grains into the water ($\rho_{zh} = 1000$ kg/m^3), the relative frequency p_{1000} of immersion from the height h of the fall is described by the following equation (%):

$$p_{1000} = 51.043 - 8.948h + 0.453h^2 - 0.005h^3,\ R^2 = 0.997; \quad (3)$$

when rye grains fall into an aqueous solution of salt with a density of $\rho_{zh} = 1090$ kg/m^3, the dependence $p_{1090} = f(h)$ has the following form (%):

$$p_{1090} = 61.075 - 8.789h + 0.367h^2 - 0.004h^3, \quad R^2 = 0.999; \quad (4)$$

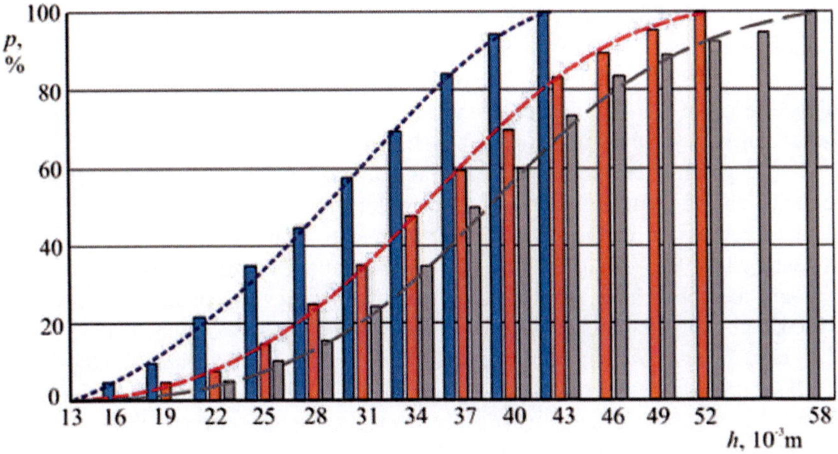

Fig. 3 Distributions of the relative frequency p of grains of winter rye variety Falenskaya 4 when immersed in liquid along the main minor axis (flat) from the height h of fall. [blue]—Water with density $\rho_{zh} = 1000$ kg/m^3; [orange]— sodium salt of hydrochloric acid aqueous solution with density $\rho_{zh} = 1090$ kg/m^3; [grey]—sodium salt of hydrochloric acid aqueous solution with density $\rho_{zh} = 1150$ kg/m^3

and when rye grains fall into an aqueous solution of salt with a density of $\rho_{zh} =$ 1150 kg/m^3, the dependence $p_{1150} = f(h)$ is described by the following expression (%):

$$p_{1150} = 73.596 - 9.725h + 0.369h^2 - 0.003h^3, \quad R^2 = 0.997. \tag{5}$$

For the convenience of analyzing the immersion of a grain in a liquid of various densities ρ_{zh}, in the obtained expressions, Eqs. (3–5), the height h of the fall of a grain on the surface of an aqueous salt solution is considered in millimeters. This is due to the fact that the value of the height h of the fall of the grain is important in these units. The regression model was chosen as a third degree polynomial because the regression curve must reach a plateau (constant), and a linear model cannot give such a curve behavior. The coefficient R^2 (Rsqr) of the reliability of the approximation of the experimental data by the second-order model is less than this parameter of the third-order model, and other statistical parameters show the insignificance of the second-order model.

To exclude the acceptance of the hypothesis about the random nature of the values of the coefficients of the obtained equations, their calculated values of the t-statistics (t_{cal}.) were compared with the tabular value of the student's t-test ($t_{tab.}$). The tabular value of the student's t-test ($t_{tab.}$) for the significance level $\alpha = 0.05$, and the number of degrees of freedom (DF) $n = 7$ for $p_{1000} = f(h)$ is 2.36. The calculated values of t-statistics ($t_{cal.}$) for all coefficients of the regression model $p_{1000} = f(h)$ are in the range 2.402…4.663, and the P-level of significance (P-Value) is 0.003…0.047. For p_{1090} = f(h) at $\alpha = 0.05$ and (DF) $n = 10$ $t_{tab.} = 2.23$, and $t_{cal.} = 7.564\ldots12.889$ (P-Value = 0.0001). For $p_{1150} = f(h)$ at $\alpha = 0.05$ and (DF) $n = 12$ $t_{tab.} = 2.18$ and $t_{cal.} = 7.549\ldots12.5$ (P-Value = 0.0001).

To visualize the results of the analysis when the grains of winter rye fall along the main minor axis (flat) on the surface of a liquid of different density ρ_{zh}, a graphical interpretation of the significance of the regression coefficients in comparison with the tabular value of the student's t-test is shown in Fig. 4.

On the graphs, the tabular value of the student's t-test ($t_{tab.}$) corresponds to a vertical line. The calculated values of the t-statistics for all the coefficients of the obtained regression models Eqs. (3–5) are greater than the tabular values of the student's t-test, ($t_{cal.} > t_{tab.}$). This confirms the statistical significance of the intercept and the coefficients for the variable h of the regression models. This is also evidenced by the fact that for each coefficient of the regression models, the P-level of significance is less than the accepted level of significance (P-Value < 0.05).

The reliability of the obtained regression models Eqs. (3–5) is also expressed by small deviations of the actual values from the calculated ones, the maximum values of which do not exceed 2.9…3.7%. This indicates an extremely high degree of agreement between the obtained trend models and experimental data.

At the same time, the coefficient R^2 (Rsqr) of the approximation reliability shows that 99.7% of the variations of the functions $p_{1000} = f(h)$ and $p_{1150} = f(h)$, as well as 99.9% of the variation of the function $p_{1090} = f(h)$, are explained by the variation

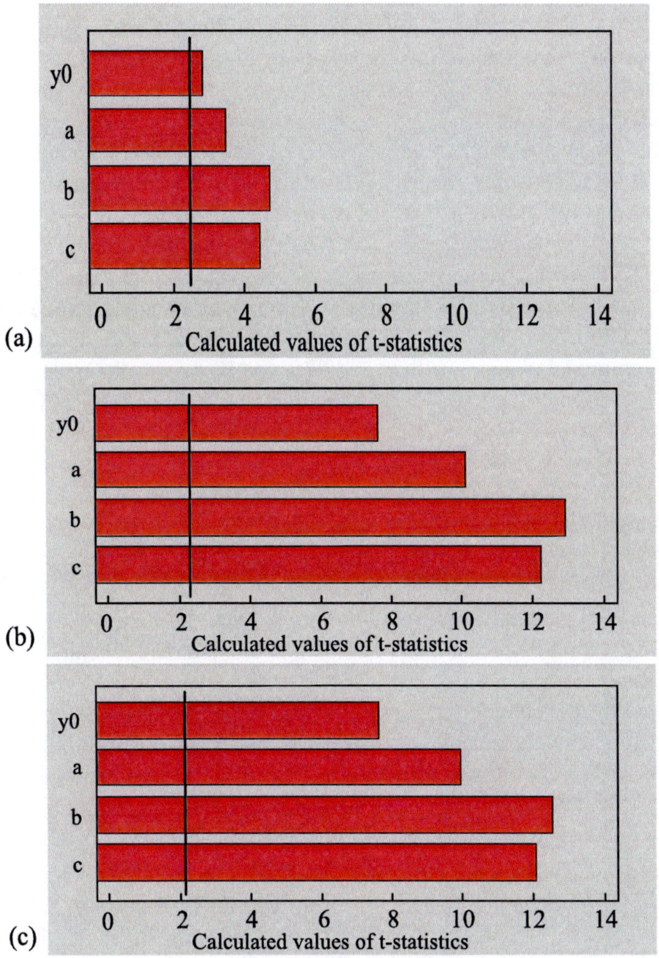

Fig. 4 Graphical interpretation of the significance of the regression coefficients for the immersion of winter rye grains variety Falenskaya 4 into the liquid along the main minor axis (flat). **a** for the regression model $p_{1000} = f(h)$. **b** for the regression model $p_{1090} = f(h)$. **c** for the regression model $p_{1150} = f(h)$

of their variable *h*, so the resulting models can be considered to be of fairly good quality.

The standard error of estimates of the regression models Eqs. (3–5), showing the standard deviation of the residuals, are 2.445, 1.468 and 2.238, respectively, which are significantly less than those for linear and second-order regression models. This speaks to the specific choice of the third-order regression models and their significance.

The Durbin–Watson statistic (DW) indices of the obtained regression models Eqs. (3–5), respectively, have the following values 2.323, 2.023 and 1.554. These values are in the range of $1.5 < DW < 2.5$, and therefore, there is definitely no autocorrelation between the experimental data.

The calculated value of the F-statistic ($F_{cal.}$), equal to 756.314 for the dependence $p_{1000} = f(h)$, is 173.9 times greater than the tabular value of Fisher's F-test ($F_{tab.}$), which for a significance level of 0.05 and the number of degrees of freedom (DF) $k_1 = 3$, $k_2 = 7$ is 4.35. For the function $p_{1090} = f(h)$, the calculated value of the F-statistic ($F_{cal.}$) is 2832.763, which is 763.6 times greater than the tabular value of the Fisher's F-test ($F_{tab.}$), which is 3.71 for a significance level of 0.05 and the number of degrees of freedom (DF) $k_1 = 3$, $k_2 = 10$. The calculated value of the F-statistic ($F_{cal.}$) for the regression model $p_{1150} = f(h)$ is 1522.976, and it exceeds the table value of the Fisher's F-criterion ($F_{tab.} = 3.49$) for a significance level of 0.05 and the number of degrees of freedom (DF) $k_1 = 3$, $k_2 = 12$ by 436.4 times. Accordingly, for the considered regression models, the calculated values of the F-statistic are greater than the table values of the F-Fisher's test ($F_{cal.} > F_{tab.}$). This means that the explained variance is significantly larger than the unexplained one, and the regression models as a whole are significant in terms of quality and its parameters.

The statistical significance of the research results described by regression models Eqs. (3–5) is also generally confirmed by the fact that their P-level of significance (P-Value), corresponding to a value of 0.0001, is less than the accepted significance level (P-Value < 0.05).

Consequently, the obtained regression models Eqs. (3–5) show an approximation that is almost absolutely reliable, and therefore, informationally significant characterizes the distribution of the relative frequency p of immersion in liquid of different density ρ_{zh} of winter rye grains variety Falenskaya 4 from the height h of fall when dropping from plates along the main minor axis (flat).

From the regression models for the fall of single grains on the surface of a liquid of different density ρ_{zh} in the direction of the main minor axis (flat), it follows that one hundred percent immersion in water with a density $\rho_{zh} = 1000$ kg/m^3 occurs at a height of $h \geq 43.0 \times 10^{-3}$ m, and an aqueous solution of sodium chloride (NaCl) with density $\rho_{zh} = 1090$ and 1150 kg/m^3 at a height of $h \geq 52.0 \times 10^{-3}$ and 58.0×10^{-3} m, respectively.

The array of experimental data upon immersion of grains of winter rye variety Falenskaya 4 in liquid of different density ρ_{zh} along the main major axis from the height h of the fall were processed and presented as a distribution of the relative frequency p, which is shown in Fig. 5.

The theoretical models of the distribution of the relative frequency p of grains of winter rye variety Falenskaya 4 when immersed in a liquid of different density ρ_{zh} along the main major axis from the height h of the fall have the following form:

$$p = y0 + \mathrm{ah} + \mathrm{bh}^2. \tag{6}$$

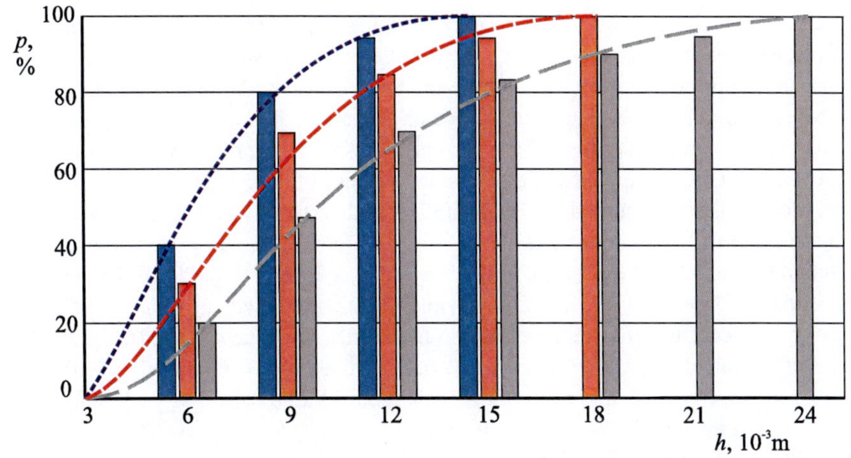

Fig. 5 Distributions of the relative frequency p of grains of winter rye variety Falenskaya 4 when immersed in liquid along the main major axis from the height h of fall —water with density $\rho_{zh} = 1000$ kg/m^3; —sodium salt of hydrochloric acid aqueous solution with density $\rho_{zh} = 1090$ kg/m^3; —sodium salt of hydrochloric acid aqueous solution with density $\rho_{zh} = 1150$ kg/m^3

when the grains of winter rye fall along the main major axis, the relative frequency p_{1000} of immersion in water ($\rho_{zh} = 1000$ kg/m^3) from the height h of the fall is described by the following equation (%):

$$p_{1000} = -61.0 + 22.071h - 0.754h^2, \quad (7)$$

when rye grains fall into an aqueous solution of salt with a density of $\rho_{zh} = 1090$ kg/m^3, the dependence $p_{1090} = f(h)$ has the following form (%):

$$p_{1090} = -48.0 + 16.970h - 0.486h^2, \quad R^2 = 0.992; \quad (8)$$

and when rye grains fall into an aqueous solution of salt with a density of $\rho_{zh} = 1150$ kg/m^3, the dependence $p_{1150} = f(h)$ is described by the following expression (%):

$$p_{1150} = -34.679 + 11.369h - 0.242h^2, \quad R^2 = 0.994. \quad (9)$$

For the convenience of analyzing the immersion of a grain in a liquid of different density ρ_{zh}, in the obtained expressions Eqs. (7–9), the height h of the fall of a grain on the surface of an aqueous salt solution is considered in millimeters. This is due to the fact that the value of the height h of the fall of the grain is important in these units. The regression model was chosen as a polynomial of the second degree because the regression curve must reach a plateau (constant), and a linear model cannot give such a curve behavior. The coefficient R^2 (Rsqr) of the reliability of approximating the experimental data by the third-order model and other statistical parameters characterizing the significance of the model are either no more than the values of the same parameters of the second-order model, or differ slightly.

To confirm the significance of the obtained values of the coefficients of the regression models, their calculated values of the t-statistics ($t_{\text{cal.}}$) were compared with the tabular value of the student's t-test ($t_{\text{tab.}}$). For the function $p_{1000} = f(h)$, the tabular value of the student's t-test ($t_{\text{tab.}}$), for the significance level $\alpha = 0.05$ and the number of degrees of freedom (DF) $n = 2$ is 4.3, and the calculated values of the t-statistics ($t_{\text{cal.}}$) for all coefficients of the regression model have a value of 6.264…9.995 (P-level of significance (P-Value) is 0.010…0.025). For $p_{1090} = f(h)$ at $\alpha = 0.05$ and (DF) $n = 3$ $t_{\text{tab.}} = 3.18$, and $t_{\text{cal.}} = 5.643\ldots9.183$ (P-Value = = 0.011…0.003). For $p_{1150} = f(h)$ at $\alpha = 0.05$ and (DF) $n = 5$ $t_{\text{tab.}} = 2.57$, and $t_{\text{cal.}} = 7.321\ldots14.122$ (P-Value = = 0.0001…0.0007).

To visualize the results of the analysis when the grains of winter rye fall along the main major axis on the surface of a liquid of different densities, a graphical interpretation of the significance of the regression coefficients in comparison with the tabular value of the student's t-test is shown in Fig. 6.

On the graphs, the tabular value of the student's t-test ($t_{\text{tab.}}$) corresponds to a vertical line. The calculated values of the t-statistics for all the coefficients of the obtained regression models Eqs. (7–9) are greater than the tabular values of the student's t-test, ($t_{\text{cal.}} > t_{\text{tab.}}$). This confirms the statistical significance of the intercept and the coefficients for the variable h of the regression models. This is also evidenced by the fact that for each coefficient of the regression models, the P-level of significance is less than the accepted level of significance (P-Value < 0.05).

The reliability of the obtained regression models is expressed by insignificant deviations of the actual values from the calculated ones, the maximum values of which for the function $p_{1000} = f(h)$ is equal to 4.3%, for the function $p_{1090} = f(h)$ does not exceed 5.8%, and for the function $p_{1150} = f(h)$ is 4.8%. This indicates an extremely high degree of agreement of trend models with experimental data.

At the same time, the coefficient R^2 (Rsqr) of the approximation reliability shows that 99.5% of the variation of the function $p_{1000} = f(h)$ and 99.4% of the variation of the functions $p_{1150} = f(h)$, as well as 99.2% of the variation of the functions $p_{1090} = f(h)$ is explained by the variation of their variable h. Therefore, the obtained models can be considered of sufficiently good quality.

The standard error of estimates for the regression models Eqs. (7–9), showing the standard deviation of the residuals, are 4.053, 4.737 and 3.395, respectively, which are significantly less than those for linear and third-order regression models. This speaks to the specific choice of the second-order regression models and their significance.

The Durbin–Watson statistics (DW) of the obtained regression models Eqs. (7 and 8) are 3.276 and 3.209, respectively. These values correspond to the limit of 2.5 < DW < 4, and then the autocorrelation between the experienced ones is questioned. The Durbin–Watson statistic (DW) indicator for the regression model Eq. (9) is

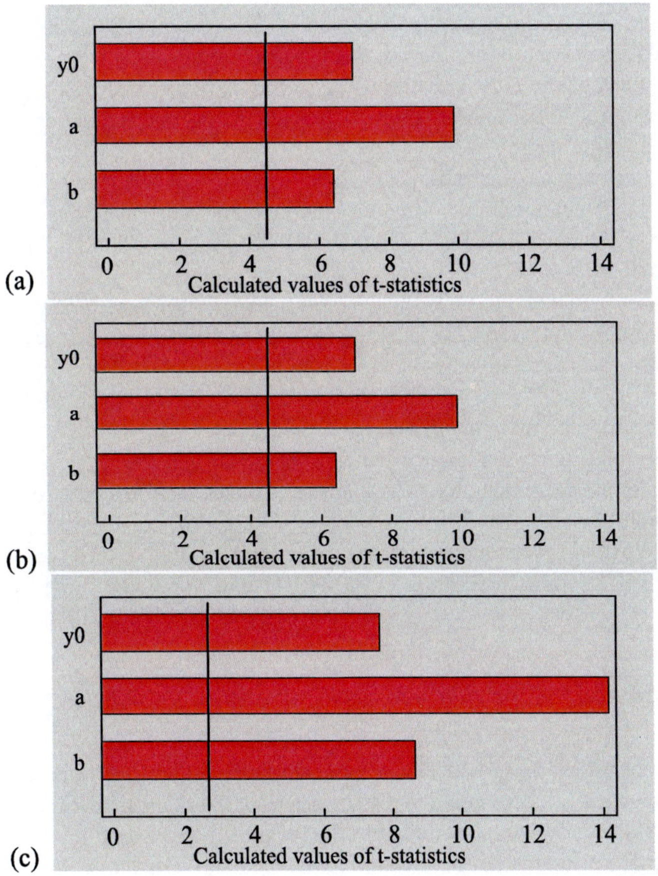

Fig. 6 Graphical interpretation of the significance of the regression coefficients of the immersion of winter rye grains variety Falenskaya 4 into the liquid along the main major axis. **a** for the regression model $p_{1000} = f(h)$. **b** for the regression model $p_{1090} = f(h)$. **c** for the regression model $p_{1150} = f(h)$

2.075, which is greater than 1.5 and less than 2.5, and therefore there is definitely no autocorrelation between the experimental data.

The calculated value of the F-statistic ($F_{\text{cal.}}$), equal to 217.522 for the dependence $p_{1000} = f(h)$, is 11.449 times greater than the tabular value of Fisher's F-test ($F_{\text{tab.}}$), which for a significance level of 0.05 and the number of degrees of freedom (DF) $k_1 = 2$, $k_2 = 2$ is 19.0. For the function $p_{1090} = f(h)$, the calculated value of the F-statistic ($F_{\text{cal.}}$) is 176.378, which is 18.469 times greater than the tabular value of Fisher's F-test ($F_{\text{tab.}}$), which is 9.55 for a significance level of 0.05 and the number of degrees of freedom (DF) $k_1 = 2, k_2 = 3$. The calculated value of the F-statistic ($F_{\text{cal.}}$) for the regression model $p_{1150} = f(h)$ is 417.565, and it exceeds the tabular value of Fisher's F-criterion ($F_{\text{tab.}} = 5.79$) for the level significance of 0.05 and the number of degrees of freedom (DF) $k_1 = 2$, $k_2 = 5$ in 72.118 times. Accordingly, for the considered regression models, the calculated values of the F-statistic are greater than the table values of the F-Fisher's test ($F_{\text{cal.}} > F_{\text{tab.}}$). This means that the explained variance is significantly larger than the unexplained one, and the regression models are generally significant.

The adequacy of the description of the ongoing processes by the regression models Eqs. (7–9) is also generally confirmed by the fact that their P-value is less than the accepted significance level ($P_{1000} = 0.0046 < 0.05$; $P_{1090} = 0.0008 < 0.05$; $P_{1150} = 0.0001 < 0.05$;).

Consequently, the obtained regression models Eqs. (7–9) characterize the approximation almost absolutely reliable and therefore informationally significantly describe the distribution of the relative frequency p of immersion in liquid of different density ρ_{zh} of winter rye grains variety Falenskaya 4 from the height h of fall when dropping from plates along the main major axis.

From the regression models for the fall on the surface of a liquid of different density ρ_{zh} of single grains along the main major axis, it follows that one hundred percent immersion in water with density $\rho_{zh} = 1000$ kg/m^3 occurs at a height of $h \geq 15.0 \times 10^{-3}$ m and in an aqueous solution of chloride sodium (NaCl) density $\rho_{zh} = 1090$ and 1150 kg/m^3 at a height $h \geq 18.0 \times 10^{-3}$ and 24.0×10^{-3} m, respectively.

4 Conclusion

Thus, as a result of the practical experiments carried out when single grains of different density ρ_{zh} are supplied to the surface of a liquid, an array of experimental data was obtained, which is presented in the form of the distribution of the relative frequency p of grains of winter rye varieties Falenskaya 4 when immersed in a liquid along the main minor axis (flat) and along the main major axis from the height h of the fall.

The performed regression analysis by confirming the significance of the obtained values of the coefficients of the regression models by comparing their calculated values of t-statistics ($t_{\text{cal.}}$) with the tabular value of the student's t-test ($t_{\text{tab.}}$), using the coefficient R^2 (Rsqr) of the accuracy of the approximation, the standard error of estimate, the Durbin–Watson statistic (DW), the calculated value of the F-statistic ($F_{\text{cal.}}$) in comparison with the tabular value of the Fisher's F-test ($F_{\text{tab.}}$) and the P-level of significance (P-Value) shows that the obtained regression models by immersion of individual grains in a liquid of different density ρ_{zh} correspond to a real process, informationally capable of characterizing the distributions of the relative frequency p of immersion of grains in a liquid.

It follows from the regression models that 100% immersion of grains in water occurs at a height h of at least 43×10^{-3} m and an aqueous salt solution – 58×10^{-3} m. Regression models can be used to develop a device for the isolation of poisonous ergot sclerotia according to density in an aqueous solution of salt and seed dressing in a wet way for setting the outlet of the loading hopper with grain at a certain height above the surface of the liquid of different density. The experiments performed are consistent with theoretical studies [21].

References

1. Orobinsky, V.I., Gievsky, A.M., Baskakov, I.V., Chernyshov, A.V.: Seed refinement in the harvesting and post-harvesting process. Adv. Eng. Res. 870–874 (2018)
2. Aldoshin, N., Didmanidze, O.: Harvesting Lupinus albus axial rotary combine harvesters. Res. Agric. Eng. **64**(4), 209–214 (2018)
3. Aldoshin, N., Didmanidze, O., Lylin, N., Mosyakov, M.: Work improvement of air-and-screen cleaner of combine harvester. In: Engineering for Rural Development: Proceedings of 18th International Scientific Conference (Latvia University of Life Sciences and Technologies, Faculty of Engineering), vol. 18, pp. 100–104 (2019)
4. Schekleina, L.M.: The influence of weather factors on certain periods of development of the fungus. Slavicep spurpurea (Fr.) Tul and the level of ergot erosion in the Kirov region. Agric. Sci. Euro-North-East **20**(2), 134–143 (2019)
5. Sheshegova, T.K., Schekleina, L.M., Zhelifonova, V.P., Antipova, T.V., Baskunov, B.P., Kozlovsky, A.G.: The resistance of rye varieties to ergot and the content of ergot alkaloids in sclerotia claviceps purpurea in the conditions of the Kirov region. Mycology Phytopathol. **53**(3), 177–182 (2019)
6. Volkhonov, M.S., Zimin, I.B., Ostrovsky, Yu.N.: Analysis of the state of preliminary cleaning of grain in farms of the north-western region of the Russian Federation and prospects for improvement. Bull. Kazan State Agrarian Univer. **2**(58), 82–86 (2020)
7. Burkov, A.I., Glushkov, A.L., Lazykin, V.A.: Comparative studies of the effectiveness of the functioning of the pneumatic separation channels of the fractional pneumatic separator of seeds. Bull. Voronezh State Agrarian Univer. **3**(62), 26–31 (2019)
8. Kharitonov, M.K., Gievsky, A.M., Orobinsky, V.I., Chernyshov, A.V., Baskakov, I.V.: Improving the efficiency of sieve cleaning of grain cleaning machines. Bull. Voronezh State Agrarian Univer. **1**(64), 19–27 (2020)
9. Orobinsky, V.I., Tarasenko, A.P., Gievsky, A.M., Chernyshov, A.V., Baskakov, I.V.: Improving the mechanization of high-quality seed production. Adv. Eng. Res. 849–852 (2018)
10. Badretdinov, I., Mudarisov, S., Tuktarov, M., Dick, E., Arslanbekova, S.: Mathematical modeling of the grain material separation in the pneumatic system of the grain-cleaning machine. J. Appl. Eng. Sci. **4**(17), 529–534 (2019)
11. Gievsky, A.M., Gulevsky, V.A., Orobinsky, V.I.: Ways to increase the performance of universal grain cleaning machines. In: Bulletin of the Federal State Educational Establishment of Higher Professional Education Moscow State Agroengineering University named after VP Goryachkin, vol. 3, no. 85, pp. 12–16 (2019)
12. Savinyh, P., Sychugov, Y., Kazakov, V., Ivanovs, S.: Development and theoretical studies of grain cleaning machine for fractional technology of flattening forage grain. In: Engineering for Rural Development: Proceedings of the 17th International Scientific Conference Engineering for Rural Development, pp. 124–130 (2018)
13. Gievsky, A.M., Orobinsky, V.I., Tarasenko, A.P., Chernyshov, A.V., Kurilov, D.O.: Substantiation of basic scheme of grain cleaning machine for preparation of agricultural crops seeds. In: IOP Conference Series: Materials Science and Engineering, vol. 327, p. 042035 (2018)
14. Orobinsky, V.I., Gievsky, A.M., Tarasenko, A.P., Chernyshov, A.V.: Study of strength and sowing qualities of seeds of winter wheat with fractional technology of post-harvest grain processing. Bull. Voronezh State Agrarian Univer. **62**(3), 13–18 (2019)
15. Andreev, V. L.: Calculation of the effective separation of light impurities in the inertial jalousie-countercurrent dust collector. In: IOP Conference Series: Materials Science and Engineering International Workshop Advanced Technologies in Material Science, Mechanical and Automation Engineering, p. 32097 (2019)

16. Ermoliev, Yu.I., Doroshenko, A.A., Belov, S.V.: Simulation of the process of separation of chopped straw in a pneumatic separator with three pneumatic channels. Bull. Don State Tech. Univer. **85**(2), 59–68 (2016)
17. Orobinsky, V.I., Gievsky, A.M., Tarasenko, A.P., Chernyshov, A.V., Baskakov, I.V.: A study of the effectiveness of cleaning heaps of spring wheat for seed purposes with an air-sieve separator. Bull. Voronezh State Agrarian Univer. **61**(2), 34–42 (2019)
18. Kharitonov, M.K., Gievsky, A.M., Orobinsky, V.I., Chernyshov, A.V., Baskakov, I.V.: Studying the design and operational parameters of the sieve module of the grain cleaning machine. In: IOP Conference Series: Earth and Environmental Science, p. 012021 (2020)
19. Saitov, V.E., Farafonov, V.G., Saitov, A.V.: Experimental substantiation of the effective height of a grain falling by a stream of liquid in an ergot release device. IOP Conf. Series: Earth Environ. Sci. **341**(012123), 1–6 (2019)
20. Saitov, A., Gataullin, R., Saitov, V.: A machine for separating ergot from rye seeds. Patent RF 2689470 (2019)
21. Sysuev, V.A., Saitov, V.E., Farafonov, V.G., Suvorov, A.N., Saitov, A.V.: Theoretical background of calculating of the parameters of the device for grain cleaning from Ergot Sclerotia. Russ. Agric. Sci. **3**, 273–276 (2017)

Simulation of Particle Motion in a Cylindrical Grain Separator

Vladimir Zaitsev, Vladimir Konovalov, and Victor Konovalov

Abstract The purpose of this work is to perform numerical simulation of particle motion in a cylindrical drum of a grain separator based on solving three-dimensional systems of force analysis and determining the influence of the main acting factors. The research methodology included the force analysis of particles motion along the inner cylindrical part of the rotating drum of the grain separator on the basis of three-dimensional force analysis system, determining the conditions of particles falling out of the cells and determining the trajectory of the flight. As a result of mathematical modelling, the influence of structural, kinematic and technological factors on the angle of rise of particles not included in the cell, the angle of beginning of particle exit from the cell and the trajectory of particle flight was established. The main factor influencing the violation of the nature of particle flow is the coefficient of friction, which is determined by both the type of plant (seed parameters) and its moisture content. An increase in moisture content leads to an increase in the friction coefficient and an additional rise of the granulate. This causes wheat grains to enter the short impurity fraction or oat seeds to enter the cleaned wheat grain fraction, which causes malfunctioning of the grain separator.

1 Introduction

Plant seeds are traditionally used as a source of human food [1], a raw material for industry [2] and a source of energy in modern animal husbandry [3]. Their role in human life is therefore important.

During harvesting, foreign objects inevitably get into the grain part of the crop. To remove these, measures are taken to clean the crop.

Not only foreign mechanical objects, but also seeds of poisonous plants [4] and weeds [5] are removed. Oats [6, 7], barley, rice and millet are also husked to obtain a quality product [8–10].

V. Zaitsev · V. Konovalov (✉) · V. Konovalov
Penza State Technological University, 1A/11, Pr. Baydukova/Gagarina St., Penza, Russia

M. Shamtsyan et al. (eds.), *Robotics, Machinery and Engineering Technology for Precision Agriculture*, Smart Innovation, Systems and Technologies 247,
https://doi.org/10.1007/978-981-16-3844-2_39

The main purpose of pre-cleaning is to provide better drying conditions for the grain. To do this, the first priority is to remove stones, branches and clumps of earth, which reduce the bulkiness of the grain and hinder its movement in grain dryers. Weed seeds are also disposed of if possible—in most cases, they are more wet than the main crop [8].

After pre-cleaning, primary cleaning of the grain (seed) is carried out to further cleanse it of impurities and to bring the grain to the basic conditions and the seed to the standards of the relevant class. Difficult to separate impurities are extracted in the following cleaning stages. At the first stage, the difference between the useful fraction and the rejects in terms of linear dimensions and aerodynamic properties is exploited. The difference in length is used for separation in grain separators [5] and in thickness, width and weight of the particles—in air-sieve grain cleaning machines [11]. Secondary clarification is carried out after the primary clarification in order to separate out seed grains as well as to separate out impurities that are difficult to separate. It is carried out in complex air-sieve grain cleaning machines [11]. Air-sieve grain cleaning machines use different velocities of light and heavy fractions in a uniform flow of air. Modern authors are mainly focused on this direction of technical development [11–13].

There are currently very few works devoted to the separation of cereal seeds from impurities in grain separators [6, 14].

The use of computer technology is aimed at modelling the processes carried out by technological machines. This allows solving specific tasks on specialization of machines for specific conditions of their operation and increase their qualitative and quantitative indicators, reducing the cost of the experimental part of the projects. The works on interaction of air flow with particles [15, 16], determination of product flow quality [17] are well known. Numerical modelling of grain movement in a grain separator has not been established.

The purpose of this work is to simulate numerically the movement of particles in a cylindrical drum of a grain separator on the basis of solving three-dimensional systems of force analysis and establishing the influence of the main acting factors.

2 Methods and Materials

The research methodology involved a force analysis of the movement of particles along the inner cylindrical part of the rotating drum of the grain separator on the basis of a three-dimensional force analysis system, the determination of particle fallout conditions from the cells and the determination of the flight path.

Grain separators are used to separate impurities that differ in length from the grains of the main crop. The impurities separated in grain separators include corncockles seeds, which are shorter than wheat grains, or oatgrass seeds, which are longer than wheat grains [5]. As the cylinder 1 (Fig. 1) rotates, a grain mixture consisting of short and long seeds enters it. Grains, the length of which is less than the diameter of the cells, are placed in them, and long—are located on the surface of the cylinder. As

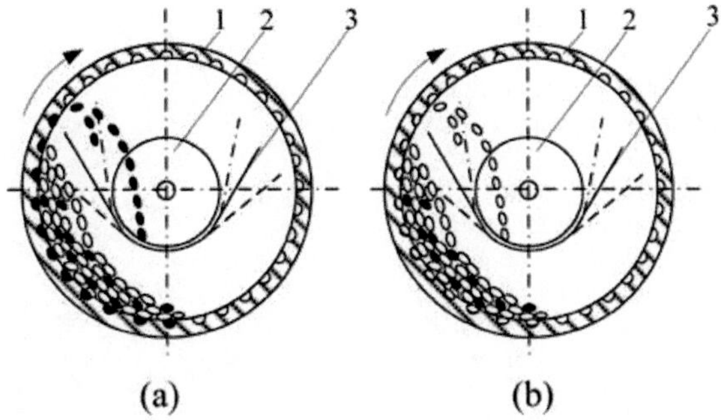

Fig. 1 Technological scheme of a grain separator drum: **a** corncockle separator; **b** wild oat separator; 1—rotating cylinder of the drum; 2—tray; 3—discharge auger; white particles—cultivated seeds; black particles—short **a** and long **b** weeds

it rotates, the cylinder lifts the seed mixture upwards. At a certain rotation relative to the axis of rotation, the grains under the action of their own mass fall out of the cells into the tray 3, located inside the cylinder. At the bottom of the tray there is a screw 2 which is used to remove the seeds from the machine. Long grains under the action of friction forces rise to a certain height, and then fall down, simultaneously moving on the surface of the cylinder along its axis downwards and come out from the opposite side of the cylinder. The grain separators for separating long impurities are called the wild oat remover, whilst those for separating short impurities are called the cockle remover.

Grain falls out of the cells in a certain area, the size of which is determined by the friction angles of the particles and the inclination angle of the drum. In practise, the inclination angle of the cylinder and the intake tray to the horizon is 3° ... 50°. The distance between the walls of the cylinder and the tray is 5 mm. The inclination angle of the drum cylinder varies between 5 and 20° for corncockles and 15–35° for wheat. For picking out corncockles and other fine impurities, the angle of inclination is 39° ... 50°, and for picking out oatgrass, the angle is 48° ... 62° from the horizon. The humidity of the processed grain is not higher than 16–19% [5].

3 The Results of the Studies

To determine the limit height of a particle lift in a rotating drum of the grain separator, we consider: (a) a rotating hollow cylinder with constant angular velocity ω, of the radius R, which axis of rotation forms with the horizontal angle α; (b) a spherical particle, of the diameter d, located in the initial position at the lower point inside the cylinder and removed from the bottom of the cylinder at some distance z. The calculation scheme of a grain separator mechanism with a particle is given in Fig. 2. Let us plot coordinate axes Oxyz and Aτn and consider conditions of motion of the particle in the cylinder cavity of the grain separator.

As the cylinder rotates, the particle moves together with the cylinder wall up to a certain limit position 3, which is determined by the limit angle of lift φ. At the value of the limiting angle of rise of the particle $\varphi < \pi/2$, the particle begins to slide along

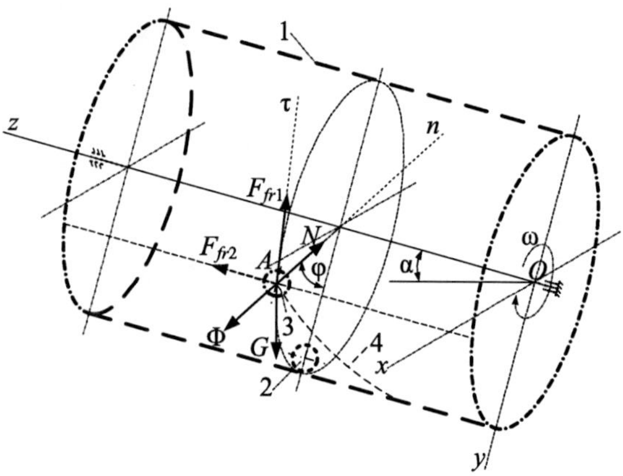

Fig. 2 The calculation scheme of particle motion in the cylinder cavity (drum) of the grain separator: 1—the hollow rotating cylinder (drum); 2—a particle in the initial position; 3—a particle in the limiting position; 4—the trajectory of particle motion in case of failure of the limiting position of equilibrium (at $\varphi < \pi/2$)

the side surface of the cylinder along a curvilinear trajectory 4. In case of violation of the limit position of equilibrium at a value of angle $\varphi > \pi/2$, the particle starts free fall in the cylinder cavity.

The following forces act on a particle at an arbitrary position: gravity force G, normal reaction of the cylinder wall N, frictional force F_{fr} (which projection on the Oxy plane is marked F_{fr1}, and on the Oz axis is marked F_{fr2}) and centrifugal force of inertia Φ (Fig. 3).

Consider a particle in two mutually perpendicular planes Oxy, Oyz (Fig. 3) and make equations of equilibrium in projections on coordinate axes $A\tau$ and An

$$\Sigma F_{k\tau} = 0,\ F_{fr1} - G \cdot \cos(\alpha) \cdot \sin(\varphi) = 0. \tag{1}$$

$$\Sigma F_{kn} = 0,\ N - G \cdot \cos(\alpha) \cdot \cos(\varphi) - \Phi = 0, \tag{2}$$

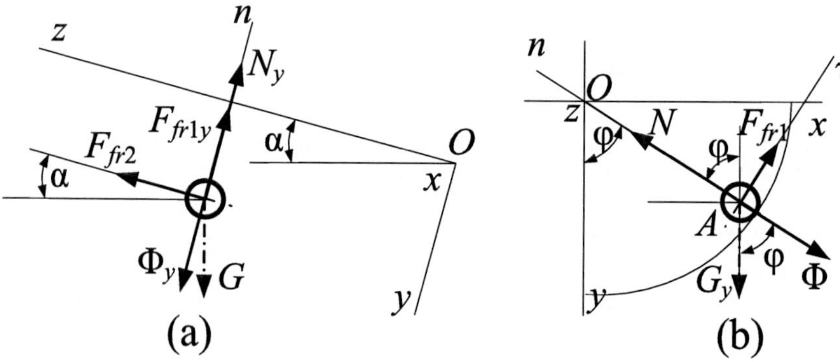

Fig. 3 The forces acting on a particle in the planes: **a**—plane Oyz; **b**—plane Oxy

where $G = m \cdot g$, m—the mass of a particle, g—the acceleration of gravity; $F_{fr1} = N \cdot f$, f—the coefficient of friction between the cylinder wall and the particle; $\Phi = m \cdot \omega^2 \cdot R$.

Equations (1) and (2) determine the value of the limiting angle of lift of the particle in the drum cavity of the grain separator.

The results of numerical simulation of the effect of different frequencies of rotation of the grain separator drum and friction coefficient on the limiting angle of rise of the particle (for the tilt angle of the drum to the horizon $\alpha = 3°$ and $\alpha = 50°$) are shown in Fig. 4. From the charts, it is seen that under unfavourable operating conditions (relative humidity of the product over 19%), the friction coefficient of the particles against the side surface of the drum exceeds the limiting value of 0.67 and may reach values greater than 0.8, which for the fast running grain separators leads to an increase in the limiting angle of approach up to $\pi/2$ and more. This contributes to poor separation of the initial product.

During the operation of the grain separator, the grain mass is divided into separable particles and the flow-through flow. Accordingly, the particles in the flowing stream must not reach the height of the tray, i.e. their lift must be kept to a minimum. For dry mass, the friction coefficient is lower than with increasing humidity due to the stickiness of the particles and the formation of grain lumps (granules), which worsen the rollability of the granulate. This leads to the adhesion of the grain mass on the inner surface of the cylinder and, consequently, to the rise of particles of the flow-through to the tray. This impairs the quality of the operation, i.e. is unacceptable.

On the other hand, the particles to be separated, corresponding in length to the dimensions of the cell cavities, are unable to roll down the inner surface of the grain separator. When gravity pulls the particle downwards and centrifugal forces cannot

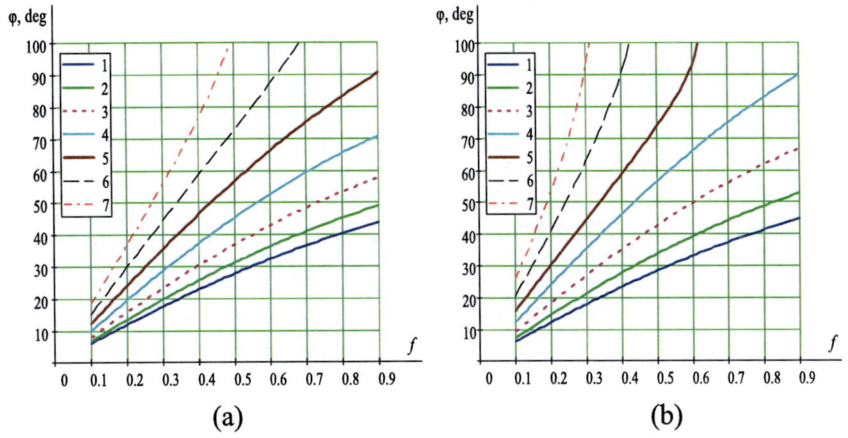

Fig. 4 The dependence of the lift angle on the friction coefficient: **a**—drum tilt angle 3°; **b**—drum tilt angle 50°; 1—drum revolutions—10 min^{-1}; 2—drum revolutions—20 min^{-1}; 3—drum revolutions—30 min^{-1}; 4—drum revolutions—40 min^{-1}; 5—drum revolutions—50 min^{-1}; 6—drum revolutions—60 min^{-1}; 7—drum revolutions—70 min^{-1}

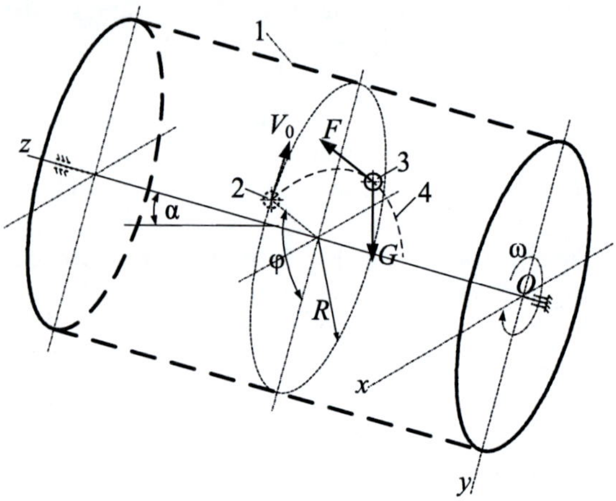

Fig. 5 The free fall of a particle in the drum cavity: 1—hollow cylinder; 2—particle in the limit position of equilibrium; 3—particle in any position; 4—the trajectory of a particle in case of violation of the limit position of equilibrium

hold the particle inside the cell, the particle leaves the cell and begins to fly with the available initial velocity. The beginning of the detachment of the particle from the cell is the absence of a positive value of the normal reaction force of the lateral inner wall of the cylinder N directed along the radius of the drum of the grain separator.

In case of violation of the limiting equilibrium position at a value of angle $\varphi > \pi/2$ the particle starts to fall freely in the cavity of the drum with the initial velocity $V_0 = \omega - R$ under the action of gravity force G and air resistance force F. The forces acting on the particle in the case of free fall are shown in Fig. 5.

The vector of the air resistance force F is numerically equal to [18, 19]

$$F = k \cdot k' \cdot \rho_2 \cdot A \cdot \frac{V^2}{2}, \tag{3}$$

where ρ_2—air density, $A = \frac{\pi \cdot d^2}{4}$—the cross-sectional area of a particle, m^2; d—the nominal diameter of a non-cylindrical particle, m; k—the air resistance coefficient; k'—particle shape coefficient, for a spherical particle $k' = 1$; V—the absolute velocity modulus of the particle, m/s.

The air resistance coefficient is determined according to the formula

$$k = \frac{24 \cdot \mu}{d \cdot \rho_2 \cdot V}, \tag{4}$$

where μ—the dynamic viscosity of air, Pa s.

Differential equations of motion of a particle in projections onto the Cartesian coordinate system Oxyz

$$\left.\begin{aligned}\frac{d^2x}{dt^2} &= -\frac{18\cdot\pi\cdot k'\cdot\mu\cdot A}{\rho_1\cdot\pi\cdot d^2}\cdot V_x,\\ \frac{d^2y}{dt^2} &= g\cdot\cos(\alpha) - \frac{18\cdot\pi\cdot k'\cdot\mu\cdot A}{\rho_1\cdot\pi\cdot d^2}\cdot V_y,\\ \frac{d^2z}{dt^2} &= -g\cdot\sin(\alpha) + \frac{18\cdot\pi\cdot k'\cdot\mu\cdot A}{\rho_1\cdot\pi\cdot d^2}\cdot V_z,\end{aligned}\right\} \tag{5}$$

where V_x, V_y, V_z—the projections of the absolute velocity vector of the particle on the coordinate axes Ox, Oy and Oz, respectively; ρ_1—particle density, kg/m^3.

The solution of the system of Eqs. (16) determines the trajectory of the particle in the drum cavity.

The condition for detachment of the particle from the side wall of the drum of the grain separator consists in equality to zero of the normal reaction of the drum wall ($N = 0$) in Eq. (3). As a result, the dependence of particle's angle of detachment is

$$\varphi = \pi - \arccos\left(\frac{R\cdot\omega^2}{g\cdot\cos(\alpha)}\right) \tag{6}$$

The dependence of the angle of detachment of the particle from the drum wall depending on the rotational speed and inclination angle of the drum is shown in Fig. 6.

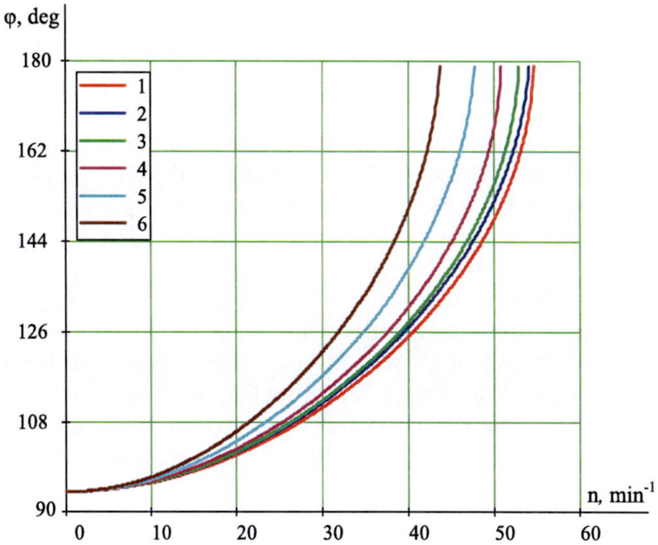

Fig. 6 The dependence of the separation angle of the particles on the number of revolutions and tilt angle of the drum: 1—drum tilt angle 0°; 2—drum tilt angle 10°; 3—drum tilt angle 20°; 4—drum tilt angle 30°; 5—drum tilt angle 40°; 6—drum tilt angle 50°

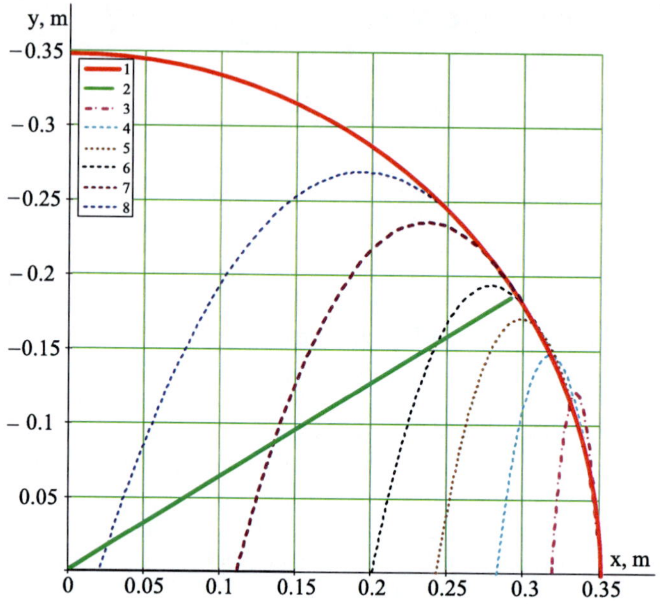

Fig. 7 Particle trajectories in the drum cavity: 1—drum; 2—receiving tray; 3—particle trajectory at the ascent angle of 94.5°; 4—particle trajectory at the ascent angle of 99°; 5—particle trajectory at the ascent angle of 103.5°; 6—particle trajectory at the ascent angle of 108°; 7—particle trajectory at the ascent angle of 117°; 8—particle trajectory at the ascent angle of 126°

The trajectory of particles in the cavity of drum with diameter of 0.7 m, for rotational speed of drum $n = 38\ \text{min}^{-1}$, depending on critical angle of rise of the particle is shown in Fig. 7.

The conducted numerical analysis has shown that the differences in particle size allow the fine seeds, lodged in the cell, to be lifted into the upper part of the drum cylinder of the grain separator for the remaining mass to be separated. Particles that do not fit into the cells are forced to slide along the smooth cylindrical surface of the grain separator drum, remaining at the bottom of the cylinder and not entering the tray. Existing grain separator designs are in accordance with these regulations. The disturbance of the technological process by the amount of moisture of the cleaned mass allowing the seed particles to stick together leads to an increase in the friction force via replacing rolling friction by sliding friction with the possibility of partial anchoring to the cells of the drum, and thus, provides the conditions of disturbance of the technological process of separating the particles along the length in the grain separator.

4 Conclusion

As a result of the dimensional force analysis, there were set the systems of expressions describing movement of particles on inner surface of inclined rotating cylindrical drum of the grain separator, the conditions of particles output from the drum cells and their flight path to justify the position of collector tray. As a result of the mathematical modelling, there was determined an influence of constructive (cylinder radius, drum inclination angle), kinematic (rotation frequency) and technological (friction coefficients, grain parametres) factors on the rise angle of particles not included in the cell, the angle of beginning of particles exit from the cell and the trajectory of particles flight. The main factor influencing the disturbance of the particle flow pattern is the friction coefficient, which is determined by both the plant species (seed parameters) and their moisture content. As a result of increasing moisture content, there is a danger of the seeds sticking together and sticking in the cells, which can lead to an increase in the friction coefficient and an additional rise in the particle granulation. This leads to the entry of wheat grains into the short impurity fraction (grain losses in the corncockles dryer) or the entry of wild oats seeds into the cleaned wheat grain fraction (contamination with cleaned grain waste in the wild oat separator).

References

1. Bucklin, R., Thompson, S., Montross, M., Abdel-Hadi, A.: Grain storage systems design. In: Handbook of Farm, Dairy and Food Machinery Engineering, pp. 175–223 (2019)
2. Dai, F., Zhao, W., Liu, G., Zhang, S., Shi, R., Wei, B.: Nongye Jixie Xuebao/Transa. Chin. Soc. Agric. Mach. **50**(8), 140–147 (2019)
3. Savinykh, P.A., Sychugov, Yu.V., Kazakov, V.A.: Development and research of the MZU-20D air-sieve grain cleaning machine. Agric. Mach. Technol. **14**(2), 59–66 (2020)
4. Saitov, V., Gataullin, R., Saitov, A.: Electrical control scheme of the machine for isolation of toxic ergot from rye grain. IOP Conf. Ser. Mater. Sci. Eng. **918**, 012135 (2020)
5. Tishaninov, N.P., Anashkin, A.V., Alshinayyin, H.J.J.: Study of the grain mixtures trier cleaning process from short impurities. Int. J. Agric. Stat. Sci. **16**(2), 713–718 (2020)
6. Li, X., Zhao, G., Ji, J., Wang, W.: Design and test of cylinder screen oat cleaning device with arc grid sieves and re-cleaning. Nongye Jixie Xuebao/Trans. Chin. Soc. Agric. Mach. **51**(5), 124–133 (2020)
7. Geng, L., Sun, C., Zuo, J., Jin, X., Kong, L., Ji, J.: Design on oat-grains purification device with rubbing roller. Nongye Gongcheng Xuebao/Trans. Chin. Soc. Agric. Eng. **35**(11), 38–47 (2019)
8. Krzysiak, Z., Samociuk, W., Zarajczyk, J., Beer-Lech, K., Bartnik, G., Kaliniewicz, Z., Dziki, D.: Effect of sieve unit inclination angle in a rotary cleaning device for barley grain. Trans. ASABE **63**(3), 609–618 (2020)
9. Masoumi, M., Mahmoudi, A., Alizadeh, M.R., Abdollahpour, S.: Effect of awn on aerodynamic properties of paddy grains (Oryza sativa l.). Agric. Eng. Int. CIGR J. **22**(2), 208–216 (2020)
10. Li, X., Meng, Y., Zhang, J., Geng, L., Ji, J.: Design and test of cleaning device for roller rubbing cylinder sieve of millet. Nongye Jixie Xuebao/Trans. Chin. Soc. Agric. Mach. **49**(10), 92–102 and 136 (2018)
11. Chernyakov, A.V., Koval, V.S., Yatsunov, A.N., Begunov, M.A., Shevchenko, A.P., Evchenko, A.V.: Research of a conical pneumatic grain separator with a device for forced light fraction removal. IOP Conf. Ser. Earth Environ. Sci. **659**, 012041 (2021)

12. Panasiewicz, M., Mazu, J., Zawi´slak, K., Kulig, R., Łysiak, G.: The process of separation of husked soybean in oblique airflow. Sustainability **12**, 7566 (2020)
13. Shatsky, V.P., Orobinsky, V.I., Axeonov, I.I., Kornev, A.S.: Analysis of the beats of separation sieve pans. IOP Conf. Ser. Earth Environ. Sci. **659**, 012106 (2021)
14. Myoung, H.K., Seung, J.P.: Analysis of broken rice separation efficiency of a laboratory indented cylinder separator. Journal of Biosystems Engineering **38**, 2 (2013)
15. Zhou, X., Wang, Z., Chen, N., Chen, D.: Numerical simulation and experiment of airflow field of cleaning room under action of conical fan. Nongye Jixie Xuebao/Trans. Chin. Soc. Agric. Mach. **50**(3), 91–100 (2019)
16. Li, H., Li, Y., Tang, Z., Xu, L.: Numerical simulation of material motion on vibrating screen of air-and-screen cleaning device based on CFD-DEM. Nongye Jixie Xuebao/Trans. Chin. Soc. Agric. Mach. **43**(2), 79–84 (2012)
17. Li, X.P., Zhang, J.L., Tao, J.J.: Analysis of airflow velocity field characteristics of an oat cleaner based on particle image velocimetry technology. Appl. Eng. Agric. **35**(2), 193–203 (2019)
18. Zaitsev, V., Kravtsov, A., Konovalovi, V.: Modelling particle movement in conical guide of seeder pneumatic pipe. E3S Web Conf. **175**, 12019 (2020).
19. Mudarisov, S., Khasanov, E., Rakhimov, Z., Gabitov, I., Badretdinov, I., Farchutdinov, I., Gallyamov, F., Davletshin, M., Aipov, R., Jarullin. R.: Specifying two-phase flow in modeling pneumatic systems performance of farm machines. J. Mech. Eng. Res. Dev. **40**(4), 706–715 (2017)

Mathematical Modeling of Physical Processes in Metals and Ordered Alloys

Alexander Semenov, Maria Semenova, Yuriy Bebikhov, and Ilya Yakushev

Abstract This study presents the results of mathematical modeling of physical processes in metals and ordered alloys using the Lennard-Jones potential and the MATLAB software package. The theoretical part describes the differential equations used for modeling, their initial and boundary conditions, as well as the difference approximation. The molecular dynamics principle has been selected as a modeling technique together with one of the pair potentials. The practical part describes a computational algorithm, as well as a step-by-step development model in the modeling environment. The following results were obtained: The chaotic motion (displacement) of atoms has been studied in two-dimensional and three-dimensional crystal lattices, as well as the distribution of atoms in the modeled cell and outward, the ratio of binding energies in real metals and the computational model has been considered, the interaction potential was determined, which turned out to be positive, and phase–frequency characteristics have been determined, which have passed the stability check.

1 Introduction

Mathematical modeling and computational experiment approaches are developing effectively and are currently considered as powerful tools for the theoretical study of nonlinear problems in physics, chemistry, biology, materials science, and other areas. Being an important link uniting theory and a natural experiment, the computational experiment approach requires for its formulation the development of appropriate models, determination of all properties of the studied substances, development of computational algorithms, and creation of program codes. The modern description of physical processes in metals and ordered alloys is carried out using two classes of models—continuous and continuous-atomistic [1]. Continuous models are based

A. Semenov (✉) · M. Semenova · Y. Bebikhov · I. Yakushev
Polytechnic Institute (Branch) in Mirny, Ammosov North-Eastern Federal University, 678170 Mirny, Russia
e-mail: as.semenov@s-vfu.ru

M. Shamtsyan et al. (eds.), *Robotics, Machinery and Engineering Technology for Precision Agriculture*, Smart Innovation, Systems and Technologies 247,
https://doi.org/10.1007/978-981-16-3844-2_40

on the equations of continuum mechanics, which are typically partial differential equations with the corresponding boundary conditions and equations of state [2]. The continuous-atomistic approach is based on molecular dynamics models and is used to describe the kinetics of various processes, including the kinetics of homogeneous phase transitions [3]. The molecular dynamics method directly finds the trajectories of atoms in accordance with the specified properties of intermolecular interaction [4–6].

The objective of this study is to describe the mathematical modeling of physical processes in metals and ordered alloys. To achieve this objective, the following tasks are solved—systems of differential equations are described, mathematical description of processes is provided using the molecular dynamics method, the interaction potential of particles in the material's crystal lattice is selected, a computational algorithm is developed, the molecular dynamic method is applied to study physical processes, and the developed algorithm is validated using the model of chaotic motion of *n* atoms in the crystal lattice using the MATLAB software package.

2 Theoretical Part

2.1 System of Equations

The mathematical model consists of a system of differential equations, their difference analog, the interatomic interaction potential, as well as specifically defined initial and boundary conditions [7].

The molecular dynamics (MD) method is based on a model concept of a polyatomic molecular system where all atoms are represented by material points, in which motion is described in the classical case by Newton's equations [8]. Then, the evolution of this model can be described by a system of *2 N* ordinary differential equations of motion:

$$\begin{cases} m_i \frac{d\vec{\upsilon}_i}{dt} = \vec{F}_i + \vec{F}_i^{\text{ext}} \\ \frac{d\vec{r}_i}{dt} = \vec{\upsilon}_i \end{cases}, \tag{1}$$

where m_i, $\overrightarrow{r_i}$, $\overrightarrow{\upsilon_i}$ is mass, radius vector, and velocity, respectively; $i = 1 \ldots N$; N is the number of point particles; $\overrightarrow{F_i}$ is the force of interaction with the rest of the particles; $\vec{F}_i^{\text{ext}}$ is the force of interaction with external fields.

The coordinates and velocities of all N particles at the initial moment $t = 0$: $\left(\overrightarrow{r_i}, \vec{\upsilon}_i\right)_{t=0}$ shall be known to integrate this system of equations.

2.2 *Initial Conditions*

Let us consider the condensed medium evolution modeling [9]. At the initial moment, the simulated medium can be a crystal, polycrystal, or liquid. In the case of a crystalline medium, the initial conditions are prepared in three stages [10]. At the first stage, the approximate value of the lattice constant a_T at temperature T is calculated for a sample, infinite in all directions, being at rest, and in equilibrium at zero pressure. At the second stage, particle coordinates are provided. At the third stage, particle velocities are provided in accordance with the Maxwell distribution for the temperature T or $2\ T$, depending on whether deviations from the equilibrium positions were provided at the previous stage of preparation. Further, the simulated object is relaxed for the more accurate derivation of the macroscopic parameters to obtain the initial values, as well as to ensure the system's stationary state [11–13].

2.3 *Difference Approximation*

Molecular dynamics modeling requires calculating the trajectories of a huge number of particles. For instance, the estimated number of particles is $N = 4L^3/a^3 \approx 6 \cdot 10^7$ for an aluminum cube with a face-centered lattice with constant $a = 0.405$ nm and edge $L = 0.1$ μm. In this regard, to ensure both computational efficiency and accuracy requirements, the MD uses the velocity Verlet algorithm [14], which has the following form. Assuming all quantities are known at time t_k, transition to the next moment $t_{k+1} = t_k + \Delta t$ is made by the formulas:

$$\begin{cases} \vec{\upsilon}_*^{\,k+\frac{1}{2}} = \vec{\upsilon}^k\left(1 - \frac{\Delta t}{2}\right) + \frac{\vec{a}^k \Delta t}{2} \\ \vec{r}^{\,k+1} = \vec{r}^{\,k} + \vec{\upsilon}_*^{\,k+\frac{1}{2}} \Delta t \\ \vec{a}^{\,k+1} = -\frac{\mathrm{grad}\left(U^{k+1}\right)}{m} \\ \vec{\upsilon}^{\,k+1} = \frac{\vec{\upsilon}_*^{\,k+\frac{1}{2}} + \frac{\vec{a}^k \Delta t}{2}}{1 + \frac{\Delta t}{2}} \end{cases}, \qquad (2)$$

where the superscript k is the time step and U is the system's potential interaction energy.

2.4 *Boundary Conditions*

System (1) is a system of ordinary differential equations. Only the initial conditions are required for its solution, and no boundary conditions are provided. However, additional requirements often arise when modeling objects formed by a particle system, which can be satisfied by imposing certain conditions on the object's boundaries [15].

If the X- and Y-coordinate axes are infinite for the modeled object and the computational domain is finite, then periodic boundary conditions with the periods L_x and L_y are used along the X- and Y-coordinate axes, respectively, to simulate the interaction with the part not included in the computational domain.

Periodic boundary conditions along x imply that particles with an x-coordinate within the limits $0 \leq x < L_x$ accurately reproduce particles for $kL_x \leq x < (k+1)L_x$ for any integer $k \neq 0$, that is:

$$\begin{cases} \vec{v}' = \vec{v} \\ x' = (x - L_x) \in [0, L_x) \text{ for } L_x \leq x < 2L_x \\ y' = y \\ z' = z \end{cases} \quad (3)$$

Similarly, the same applies to periodic boundary conditions along the Y-axis. The modeled object is finite along the Z-axis, while its surfaces are free from external influences, so no special boundary conditions are required in this direction.

3 Modeling Methods and Techniques

The key element of the entire molecular dynamics method is the choice of the particle interaction potential. Two main approaches currently exist to solve the problem of interatomic and intermolecular interactions—the so-called first principle approach [16–19] and the method of semi-empirical potentials [20–22]. The "first principle" approach appeared with the development of the quantum mechanics concepts. It is based on solving the Schrödinger equation for a condensed medium, which is considered as a set of interacting nuclei and electrons. Such a solution is only possible when simplifying approximations are applied, but the computational complexity associated with the simulated object size limits the applicability of such approximations in molecular dynamics.

Conversely, semi-empirical potentials have proven themselves well in molecular dynamics modeling, since they are constructed within certain limits of applicability. After estimating how well each of the potentials can describe the compounds with different atomic environments, let us choose several of them for mathematical modeling of physical processes in metals and ordered alloys, as well as provide their main parameters.

3.1 *Paired Potentials*

Models of interatomic interaction based on pair potentials address the interaction between two particles depending only on their mutual arrangement. This type of potential allows to correctly describe many physical phenomena and properties of substances at a qualitative level [23]. At the same time, pair potentials are known to provide stability only for sufficiently closely packed crystal lattices.

The system's energy approximation for paired potentials can be written as follows:

$$U(\vec{r_1} \ldots \vec{r_N}) = \frac{1}{2} \sum_{\substack{i,j=1 \\ i \neq j}}^{N} \varphi(r_{ij}), \tag{4}$$

where $\varphi(r_{ij})$—is the interaction potential of ith and jth atoms located at a distance r_{ij}.

3.2 *Lennard-Jones Potential*

To study the thermodynamic properties of inert gases, Lennard-Jones proposed a pair potential [24], which describes the van der Waals interaction of neutral atoms. This potential has the following common form:

$$\varphi(r) = \frac{\lambda_n}{r^n} - \frac{\lambda_m}{r^m}$$

The most typical formula for it has the following form:

$$\varphi(r) = 4\varepsilon\left[\left(\frac{\sigma}{r}\right)^{12} - \left(\frac{\sigma}{r}\right)^{6}\right], \tag{5}$$

where σ is the interatomic distance at which $\varphi(\sigma) = 0$; ε is the potential well depth located at a distance $\sigma\sqrt[6]{2}$[6]2. In this expression, the term $\sim r^{-6}$ dominates at large distances and corresponds to the dispersive dipole–dipole attraction. The term $\sim r^{-12}$ models a strong repulsion between a pair of atoms due to exchange interaction if they are very close to each other.

3.3 Morse Potential

Morse proposed an expression [25] that ensures the consistency of the calculated vibrational energy levels of diatomic molecules, in which he completely abandoned the power law dependence:

$$\varphi(r) = \varepsilon\{\exp[-2\alpha(r - r_{\min})] - 2\exp[-\alpha(r - r_{\min})]\}, \tag{6}$$

where ε is the potential well depth located at a distance $r_{\min}$ and α characterizes the potential well width. This potential is finite $\varphi(0) = \varepsilon\exp(\alpha r_{\min})\left[\exp(\alpha r_{\min}) - 2\right]$ at $r = 0$, which turned out to be very useful in numerical modeling.

The number of parameters in the Morse potential is increased to three, in contrast to the Lennard-Jones potential, making it more flexible when fitting the obtained data to experimental values. However, when studying metals with FCC and BCC lattices, as well as a graphene crystal lattice, the authors encountered delocalized nonlinear vibrational modes (DNVMs), which are determined exclusively by the point symmetry of the lattice and are exact solutions of the atom motion equations for any interatomic potential [26]. Consequently, we have chosen the Lennard-Jones potential as the interaction potential for modeling 2D and 3D crystal lattices.

4 Research Results

4.1 Computational Algorithm

Let us linearize the number of operations by the number of particles. This is necessary when performing a large number of calculations during the molecular dynamics experiment, as the requirements for the computational process arrangement in this case increase significantly. To do this, let us divide the integration step of the system of motion Eq. (1) into two stages—calculation of interaction potentials and total forces acting on each atom from the other atoms and calculation of new coordinates and velocities of all particles using Verlet algorithm formulas (2).

The first stage turns out to be the most demanding in terms of computational costs since the calculation of potentials requires performing arithmetic operations in the general case N^2, where N is the number of particles constituting the system. However, the Lennard-Jones potential we have chosen is short range. This means that there is a certain critical distance r_{cr} at which the potential for $r > r_{cr}$ can be neglected. Due to this, an algorithm was created for calculations with a linearly increasing number of particles, which recalculates the number of "neighbors" in the interacting environment once every M time steps. In this case, the neighbors are considered to be a pair of interacting atoms located at a distance not exceeding $l_{\max} = r_{cr} + \Delta l$.

Thus, the linearized algorithm and obviously velocity Verlet integration allowed us to simultaneously perform a number of calculations.

The calculation of basic thermodynamic quantities (pressure and temperature) in the molecular dynamics modeling of a condensed medium is based on the assumption of the rapid establishment of local thermodynamic equilibrium that can be reached in several picoseconds. The temperature with a translational degree of freedom $\alpha = x, y, z$ can be obtained by averaging the chaotic component of the kinetic energy over an ensemble of particles with a steady-state equilibrium distribution. The overall temperature is defined simply as the average for all the degrees of freedom:

$$T = \frac{T_x + T_y + T_z}{3}$$

The pressure tensor is calculated by averaging over the volume and additionally over time. The diagonal elements of the tensor represent stresses in three spatial directions. In the case of isotropic media, the pressure can be determined by simply averaging these components:

$$P = \frac{P_{xx} + P_{yy} + P_{zz}}{3}$$

Let us discuss the results of the development and validation of a mathematical model describing physical processes in the crystal lattices of metals and ordered alloys.

4.2 Model Validation in MATLAB Environment

MATLAB software package and the molecular dynamics method were used to model the behavior (chaotic movement) of N atoms in a crystal lattice. Previously, the authors used dedicated MD modeling software for this purpose, for example LAMMPS package [27–29]. Velocity Verlet scheme was chosen for calculations. The formulas have the following generic form:

$$\vec{r}(t + \Delta t) = \vec{r}(t) + \vec{v}\Delta t + \vec{a}\Delta t^2/2$$

$$\vec{v}(t + \Delta t/2) = \vec{v} + \vec{a}\Delta t/2$$

$$\vec{a}(t) = -\text{grad}(U(\vec{r}(t)))/2$$

$$\vec{v}(t + \Delta t) = \vec{v}(t + \Delta t/2) + \vec{a}\Delta t/2$$

Here, U is the Lennard-Jones potential (5), which was selected from a variety of other pair potentials as the optimal one for the conditions of our problem.

Two-dimensional (2D) and three-dimensional (3D) spaces were modeled; therefore, the gradient in the third formula was taken first in two and then in three directions. The potential $U(r(t))$ was transformed into $U(x(t), y(t))$ and $U(x(t), y(t), z(t))$ using the formulas:

$$\vec{r}_{2D} = x\vec{i} + y\vec{j},$$

$$\vec{r}_{3D} = x\vec{i} + y\vec{j} + z\vec{k}.$$

The initial data for 2D modeling were 100 atoms in a 10*10 crystal lattice (cell) and 1000 atoms in a 10*10*10 cell for 3D modeling. As an example of computational algorithm validation, we present the calculation of a mathematical model in two-dimensional space. Figure 1a shows the calculated two-dimensional crystal lattice with L*W dimensions. The main initial data also include r—the distance between particle centers, ε—the potential well depth, and σ—the distance of zero interaction energy. The parameters ε and σ characterize the atoms of the corresponding substance. A typical form of the Lennard-Jones potential is shown in Fig. 1b with its minimum being at point $r_{\min} = \sigma\sqrt[6]{2}$.

To speed up the calculations, let us limit the Lennard-Jones potential to a distance of $r_c = 2.5\Delta\sigma$. This is done because the interaction energy is only *≈0.0163* of the well depth ε at this distance. The parameters ε and σ can be found using the Joule-Thomson coefficient or by comparing the experimental value of the viscosity coefficient with the value obtained from the formula for the potential energy. Tables of values of parameters ε and σ for certain substances are provided in the book [30].

The program code was developed and saved in the *.m file of the MATLAB software package [31]. Figure 2 shows the program code fragment with initial data for 2D modeling.

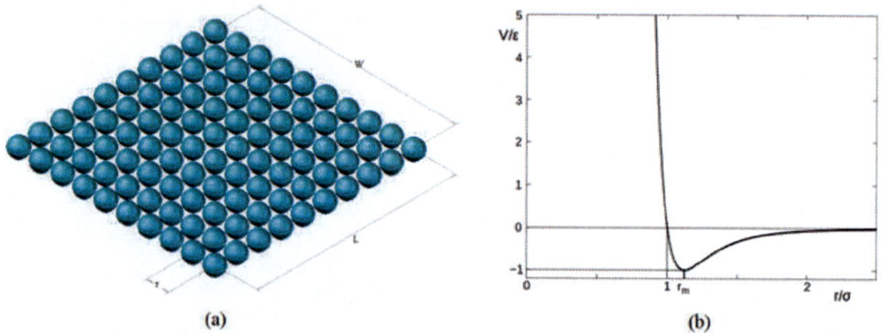

Fig. 1 Modeled cell structure **a** and typical form of the Lennard-Jones potential **b**

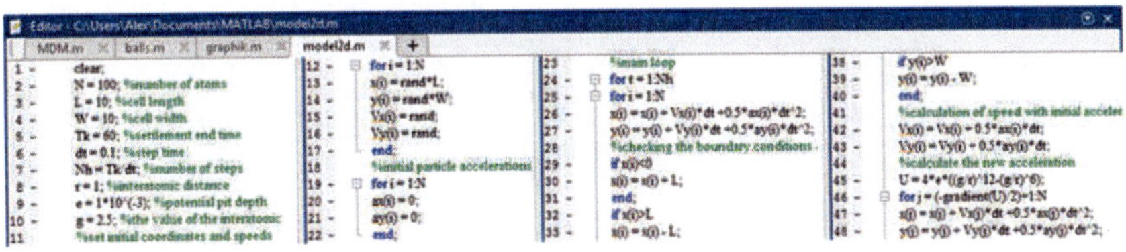

Fig. 2 Screenshot of MATLAB editor window with the program code fragment

Modeling results are presented on screenshots of plot plotter and analysis filters with computational operations in two-dimensional space, namely a selection of graphs of atom distribution during the chaotic motion of atoms inside the modeled cell and outward (Fig. 3); comparison of binding energies (Fig. 4a); amplitude–phase–frequency characteristics of the interaction potential (Fig. 4b).

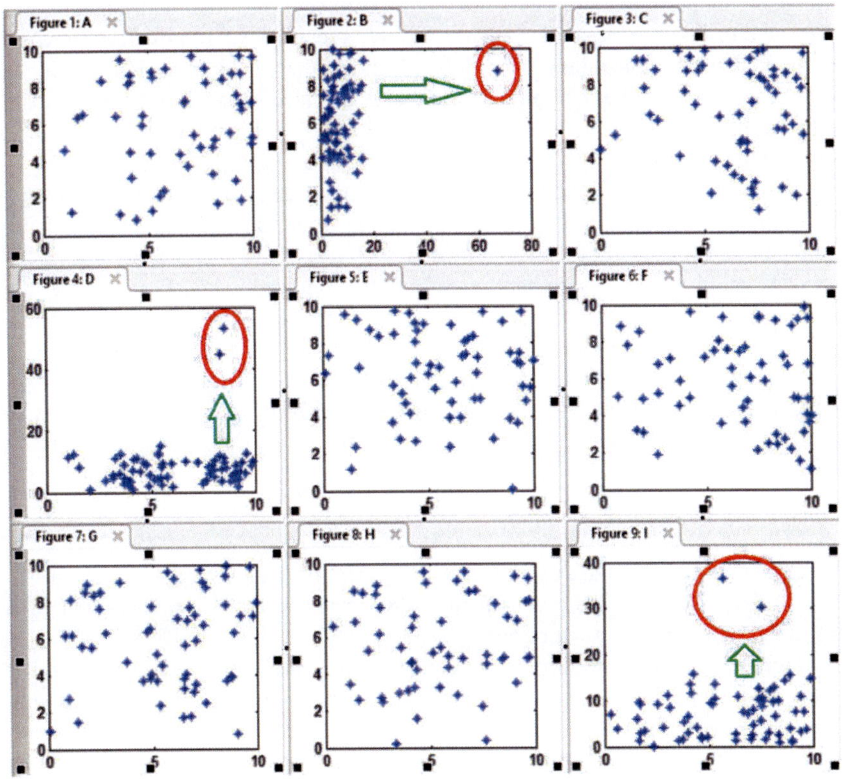

Fig. 3 Selection of atom distribution graphs of chaotic motion inside the modeled cell (**a**, **c**, **e**–**h**) and outward (**b**, **d**, **i**)

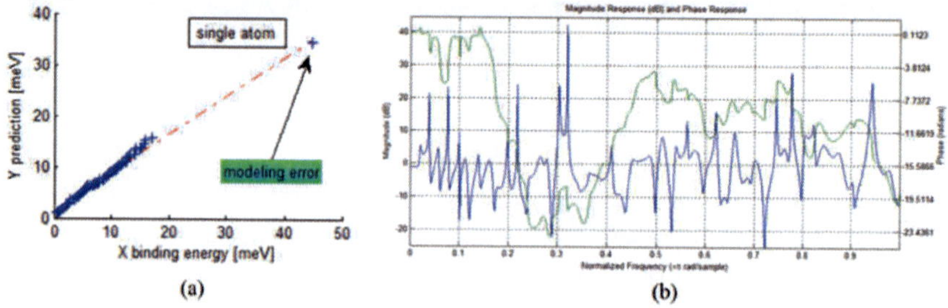

Fig. 4 Comparison of binding energies between the current model and standard data for an HCP metal cell **a** and amplitude–phase–frequency characteristics of the interaction potential **b**

5 Results and Discussion

A large number of simulations have been carried out, requiring a dedicated database to be created to evaluate the results. Figure 3 shows a selection of graphs characterizing the atom distribution resulting from a chaotic motion within the modeled cell. It can be seen that in a number of isolated cases, such as those shown in Figures b, d and i, the atoms leave the modeled cell and go far beyond its limits. Such cases are not possible for HCP metals, so we consider this fact as a modeling error.

When comparing the binding energies (Fig. 4a) obtained by modeling with standard data for HCP metals, large absolute and relative errors can be observed during the modeling of long-range interactions. In this graph, the abscissa represents the standard value, while the ordinate represents the modeled value. For instance, the modeling error was 23% for the HCP metal Be studied in [32] by the authors. Thus, we have confirmed that the Lennard-Jones potential is well applicable for pairwise short-range interactions. The modeling provided a positive value of the potential, in which the sample mean was determined by the formula:

$$\overline{U} = \frac{\sum_{i=1}^{n} U_i}{n} = 1.0978,$$

where n is the sample size and U_i is the ith element of the sample.

A positive value of the potential indicates strong forces of repulsion between atoms as a result of chaotic motion.

Figure 4b shows the amplitude–frequency and phase–frequency characteristics of the Lennard-Jones potential. The amplitude takes on a negative value when passing through the potential well, while the phase values become positive in this case. All criteria were met when checking the characteristics for stability, including the Nyquist hodograph describing the vector's behavior in the plane. Amplitude and phase extrema are 22.5 dB and 0.157 rad, respectively.

6 Conclusions

Mathematical modeling of physical processes in metals and ordered alloys included and provided the following:

1. Description of systems of equations, initial and boundary conditions, difference approximation—as a theoretical part of the computational algorithm development.
2. Selection of modeling methods and techniques, description of pair interaction potentials for molecular dynamics modeling, selection of the Lennard-Jones potential.
3. A computational algorithm for mathematical modeling is presented, which includes linearization of the number of operations by the number of particles and the calculation of basic thermodynamic quantities.
4. The results of modeling in the MATLAB environment are presented, the program code is provided, the numerical and graphical parameters and characteristics are presented, and the results are discussed.

The main results of modeling include the obtained graphical dependence of the binding energies for the current model and standard data for a HCP metal cell, the obtained stable amplitude–phase–frequency characteristics, the obtained positive value of the Lennard-Jones potential and its sample mean, and identification of isolated cases of atoms leaving the modeled cell due to chaotic motion.

Acknowledgements Semenov A.S. thanks the Russian Science Foundation for supporting the project no. 18-72-00006.

References

1. Volkov, I.A., Igumnov, L.A., Lutvinchuk, S.Y.: A continual model of damage for analyzing long-term strength of materials and structural elements. Adv. Struct. Mater. **136**, 147–174 (2020)
2. Makarov, E.S., Gvozdev, A.E., Zhuravlev, G.M., Minaev, I.V., Dobrovol'sky, N.N., Sapozhnikov, S.V., Kalinin, A.A.: Determination of material mathematical functions in the conditions of the duration of dilatating media from powder and ingot metal systems. Chebyshevskii Sbornik **20**, 538–561 (2019)
3. Woodliffe, J.L., Ferrari, R.S., Ahmed, I., Laybourn, A.: Evaluating the purification and activation of metal-organic frameworks from a technical and circular economy perspective. Coordination Chem. Rev. **428**, 213578 (2021)
4. Savin, A.V.: Hydroxygraphene: dynamics of hydrogen bond networks. Physica E **127**, 114561 (2021)
5. Yang, S., Zhang, L., Xie, H., Liu, W.: Interaction potential function for the deformation analysis of potassium dihydrogen phosphate using molecular dynamics simulation. Comput. Mater. Sci. **187**, 110122 (2021)
6. Krylova, K.A., Korznikova, E.A., Semenov, A.S., Bachurin, D.V., Dmitriev, S.V.: Linking tracks in mica crystals with phase transitions in a bistable lattice. Eur. Phys. J. B **93**, 23 (2020)

7. Gepreel, K.A.: Analytical methods for nonlinear evolution equations in mathematical physics. Mathematics **8**, 2211 (2020)
8. Zalizniak, V.E., Zolotov, O.A., Zolotova, O.P.: Mathematical model of conducting nanopore for molecular dynamics simulations. Siberian J. Sci. Technol. **19**, 677–682 (2018)
9. Kuzemsky, A.L.: Nonequilibrium statistical operator method and generalized kinetic equations. Theor. Math.Phys. (Russian Federation) **194**, 30–56 (2018)
10. Dmitriev, S.V., Korznikova, E.A., Baimova, Y.A., Velarde, M.G.: Discrete breathers in crystals. Phys. Usp. **59**, 446–461 (2016)
11. Mishra, L., Sundararajan, M., Bandyopadhyay, T.: MD simulation reveals differential binding of Cm(III) and Th(IV) with serum transferrin at acidic pH. Proteins Struct. Funct. Bioinform. **89**, 193–206 (2021)
12. Ngo, S.T.: Estimating the ligand-binding affinity via λ-dependent umbrella sampling simulations. J. Comput. Chem. **42**, 117–123 (2021)
13. Kelkar, A.S., Dallin, B.C., van Lehn, R.C.: Predicting hydrophobicity by learning spatiotemporal features of interfacial water structure: combining molecular dynamics simulations with convolutional neural networks. J. Phys. Chem. B **124**, 9103–9114 (2020)
14. Verlet, L.: Computer experiments on classical fluids. I. Thermodynamical properties of Lennard-Jones molecules. Phys. Rev. **159**, 98–103 (1967)
15. Iskhokov, S.A., Yakushev, I.A.: On solvability of variational Dirichlet problem for a class of degenerate elliptic operators. Chebyshevskii Sbornik **19**, 164–182 (2018)
16. Yin, S., Ding, J., Asta, M., Ritchie, R.O.: Ab initio modeling of the energy landscape for screw dislocations in body-centered cubic high-entropy alloys. NPJ Comput. Mater. **6**, 110 (2020)
17. Lobzenko, I.P., Chechin, G.M., Dmitriev, S.V.: Comparison of classical molecular dynamics and ab initio modeling of discrete breathers in grapheme. Solid State Phenom. **258**, 81–84 (2017)
18. Lobzenko, I.P., Chechin, G.M., Bezuglova, G.S., Baimova, Y.A., Korznikova, E.A., Dmitriev, S.V.: Ab initio simulation of gap discrete breathers in strained graphene. Phys. Solid State **58**, 633–639 (2016)
19. Chechin, G.M., Dmitriev, S.V., Lobzenko, I.P., Ryabov, D.S.: Properties of discrete breathers in graphane from ab initio simulations. Phys. Rev. B **90**, 045432 (2014)
20. Pei, Q.X., Jhon, M.H., Quek, S.S., Wu, Z. (2021) A systematic study of interatomic potentials for mechanical behaviours of Ti-Al alloys. Comput. Mater. Sci. **188**, 110239 (2014)
21. Jaimes, J.H.B., Alvarez, M.E.T., de Moraes, E.B., Macie, M.R.W., Filho, R.M.: Separation and semi-empiric modeling of ethanol—water solutions by pervaporation using PDMS membrane. Polymers **13**, 93 (2021)
22. Valimukhametova, A., Ryan, C., Paz, T., Grote, F., Naumov, A.V.: Experimental and theoretical inquiry into optical properties of graphene derivatives. Nanotechnology **32**, 015709 (2021)
23. Drautz, R., Fähnle, M.: On the pair-potential modelling of alloy surfaces. Surf. Sci. **585**, 108–112 (2005)
24. Jones, J.E.: On the determination of molecular fields. II. From the equation of state of a gas. Proc. Royal Soc. A **106**, 463–477 (1924)
25. Morse, P.M.: Diatomic molecules according to the wave mechanics II. Vibrational levels. Phys. Rev. **34**, 57–64 (1929)
26. Sunagatova, I.R., Subkhangulova, A.M., Semenova, M.N., Borisov, D.I., Semenov, A.S., Dmitriev, S.V.: Properties of one-dimensional nonlinear vibrational modes in triangular lattice with Lennard-Jones interactions. IOP Conf. Ser. Mater. Sci. Eng. **1008**, 012073 (2020)
27. Abdullina, D.U., Semenova, M.N., Semenov, A.S., Korznikova, E.A., Dmitriev, S.V.: Stability of delocalized nonlinear vibrational modes in graphene lattice. Eur. Phys. J. B **92**, 249 (2019)
28. Shepelev, I.A., Korznikova, E.A., Bachurin, D.V., Semenov, A.S., Chetverikov, A.P., Dmitriev, S.V.: Supersonic crowdion clusters in 2D Morse lattice. Phys. Lett. A **384**, 126032 (2020)
29. Krylova, K.A., Lobzenko, I.P., Semenov, A.S., Kudreyko, A.A., Dmitriev, S.V.: Spherically localized discrete breathers in bcc metals V and Nb. Comput. Mater. Sci. **180**, 109695 (2020)
30. Qian, X.S.: Physical Mechanics. Mir, Moscow (1965)

31. Bebikhov, Y.V., Semenov, A.S., Yakushev, I.A., Kugusheva, N.N., Pavlova, S.N., Glazun, M.A.: The application of mathematical simulation for solution of linear algebraic and ordinary differential equations in electrical engineering. IOP Conf. Ser. Mater. Sci. Eng. **643**, 012067 (2019)
32. Bachurina, O.V., Murzaev, R.T., Semenov, A.S., Korznikova, E.A., Dmitriev, S.V.: Properties of moving discrete breathers in Beryllium. Phys. Solid State **60**, 989–994 (2018)

Reliability of Robosystem Heterostructures

Mikhail Volnikov

Abstract The relevance is caused by insufficient knowledge of the processes occurring in heterostructures, leading to failures of robotics. The aim is to study the reliability of complex robotics systems, to identify and eliminate the causes of failures. The materials and methods were based on information on failures, statistical processing of data, as well as the results of computational experiments. Integral reliability estimates, multivariate neural network and information analyzes of information were used. The main reasons affecting robosystems reliability are considered. The theory of robosystems reliability in terms of system analysis of failures of complex systems is developed. A robosystem model has been created in the presence of degradation. Estimates of the influence of the main external factors on the occurrence of degradation of robosystems are given. Roboprophylaxis is proposed of failures based on systematization of analysis of causes causing failures. The reliability of the use of artificial intelligence in the control of robosystems was evaluated.

1 Introduction

Modern robotics are ultra-complex systems, including mechanical, electrical, electronic, microelectronic heterostructures, as well as artificial intelligence—software and biological intelligence.

In the twentieth century, technologies appeared that made it possible to create systems for controlling technical systems without human participation, the so-called robotic systems. Technologies for the production of robotic systems, their components and software are becoming increasingly widespread.

Recently, in the field of robotics, especially in the military sphere, there have been dramatic changes associated with the mass introduction of artificial intelligence. The ability to control systems without human intervention allows symbiosis of analytical and command programs based on neural networks.

M. Volnikov (✉)
Penza State Technological University, Baidukova Proezd/ Ul. Gagarina, 1a/11, 440039 Penza, Russia

M. Shamtsyan et al. (eds.), *Robotics, Machinery and Engineering Technology for Precision Agriculture*, Smart Innovation, Systems and Technologies 247,
https://doi.org/10.1007/978-981-16-3844-2_41

Progress in the field of intellectualization of decision-making processes during group control of robots has led to mass production and testing of robotic complexes in industry. This direction is especially fast developing in the space field and is accompanied by mass tests and applications in real conditions. For such systems, the question of the reliability of robosystems is acute.

Robosystems failures lead to dangerous situations for people lives, can lead to their death, and recently, with the massive introduction of artificial intelligence, to explosive military situations and major environmental disasters [1].

The quality and reliability of systems using artificial intelligence consists of a computer, software and a person managing the system [2]. Synthesis of information on reliability of the considered robotic systems allows us to conclude that there is insufficient reliability, primarily hybrid intelligence of these complex systems, which makes this direction promising and relevant [3–8].

2 Setting the Task

To solve the problem, using a systematic approach, it is necessary to find out the reasons for the decrease in reliability during the operation of robosystems, having examined the processes taking place in robosystems and the factors affecting the stable and reliable operation of these complex heterostructures. The problem is solved using general heterology in terms of system analysis of complex systems failures.

Consider the main reasons for reliability reduce.

Cause 1. The low reliability of robotic systems is associated with insufficient study and development of mathematical models of system connections and process patterns in robotic and technology heterostructures. The behavior of complex systems is ambiguous. As a result, some uncertainty is formed when making decisions, system behavior options arise. The choice of an adequate model of system behavior affects the reliability of robotic systems.

Cause 2. Untimely preventive maintenance and repair. While industrial conditions place emphasis on predicting failures, for timely preventive maintenance, such maintenance is usually not possible in combat conditions or during space flights, which means that the robotic system must continue to exist with the loss of some functions due to failures.

Cause 3. Extreme operating conditions of robotic systems, leading to a deterioration in the quality of information and the robotic system as a whole. In the use of unmanned aerial vehicles, ground robots, military equipment without human intervention, the idea of creating robotic systems operating in extreme conditions is increasingly used, when the probability of failure of some nodes increases in proportion to the complexity of the task being solved. Degradation of information occurs both over time in general and with an increase in the operating time of robotic systems. This is due both to the physical and moral obsolescence of the elements of the system, leading to the appearance of microdefects, and to the presence of external influences: temperature, vibrations, etc.

Degradation of information corresponds to conversion F_5 (with parameter φ_5) [1].

$$I\left(\vec{\lambda}, \vec{\xi}, \varphi_1, \varphi_2, \varphi_3, \varphi_4, \varphi_5\right) = F_5\left[I\left(\vec{\lambda}, \vec{\xi}, \varphi_1, \varphi_2, \varphi_3, \varphi_4\right)\right] \quad (1)$$

where F_5—degradation transformation, φ_5—degradation transformation parameter, $\vec{\lambda}$—information situation, $\vec{\xi}$—resource-creating parameters.

Studies of the temporal change in the parameter of information transformation under the influence of destabilizing factors (temperature, vibration) make it possible to monitor the process of information degradation and evaluate this process, as well as to develop recommendations for reducing information degradation in systems with heterogeneous structures.

Cause 4. Insufficient consideration of the heterogeneity of structures in the general methodology of design systemology, production and operation of technical devices leads to increased fault tolerance of systems.

The high heterogeneity of robosystems is one of the most important factors affecting their reliability. Modern product made according to hybrid technology, for example, 90% consists of connections of heterogeneous materials, while up to 95% of failures are due to failures of connections heterogeneous materials [9].

The most tragic example is the disaster of the Shuttle spacecraft with the death of a large group of astronauts, which gave rise to many hypotheses about the causes of disasters. In descending order of probability: destruction of thermal protection, control system failure, spontaneous opening of cargo compartment, abnormal operation of bypass valves, human factor [10].

NASA executives considered failures in the thermal protection system to be the main cause; in Roscosmos, it was considered that the main thing—the disadvantages of adhesive joints in the nasal coca and the front edges of the wings and steering wheel.

Neural network analysis of information on failures and failure rates of adhesive joints showed that the most failing in space conditions are RTV silicone adhesive joints.

Informological analysis (which included mathematical modeling, statistical modeling, experimental-instrumental analysis, system analysis and evolutionary helical-network algorithms of the ascending type) created for complex heterogeneous structures, showed that the following classification of the causes of disasters should be recognized as the most likely:

- 80% of disasters are due to insufficient knowledge of the processes taking place in heterostructures
- 18% with deficiencies in design, fabrication, assembly and operation methods
- 2% with performer errors [11].

Cause 5. Errors in the software and hardware of computer systems: typical system failures, errors in determining goals and tasks; programming errors in program texts; algorithmic errors; software-related errors; errors in memory elements.

Table 1 Failure rate in different heterostructures

No. P/N	Modules	Heterostructures	$\lambda_i^{min} \cdot 10^{-6}$, 1/h	$\lambda_i^{max} \cdot 10^{-6}$, 1/h
1	Mechanical	Alloy nanostructures	0.2	4.5
2	Electrotechnical	Material nanostructures	0.8	6.0
3	Electronic	Silicon nanostructures, etc	0.6	1.7
4	Microelectronic	Material nanostructures	0.04	0.3
5	Software (artificial intelligence)	System, application, electronics	1.0	6.0
6	Natural intelligence (human intelligence)	Circulation, nervous system, brain	3.0	15.0

Based on the above, judging by the information about the failures of systems produced in our country, as well as foreign industry, from 60 to 90% of system failures are in heterostructures [2].

The integrated information on failure rates of heterostructures is given in Table 1.

Cause 6. Artificial intelligence. Man controls the robot and interacts with it during military operations, and this is the main vulnerability of robosystems. The creation of fully autonomous robots for operational and tactical tasks is associated with the creation of artificial intelligence. The development of artificial intelligence is accompanied by many problems.

Table 1 shows that most of the malfunctions of robosystems are due to failures of computer heterostructures, including software and humans. The main failure hypothesis proposes insufficient fault tolerance of neuro-like heterostructures [12].

When creating artificial intelligence, scientists choose human intelligence for the prototype, analogs of neural networks of robosystems are created on its basis. In natural neural networks, the dendrites of neurons join the axons of other neurons, the connection of one neuron to another is a synapse. The number of synaptic compounds in different parts of the nervous system is very different, the body of the neuron is covered with synapses, on one neuron there are 1500. In peripheral nervous system neurons, axons contact axons of muscle tissue. Thinking and process are controlled by electrical impulses in the neural network of the central and peripheral nervous systems. Increasing the complexity and levels of impulse impacts when creating artificial intelligence of robosystems leads to an increase in the number of possible types of malfunctions in the functioning of systems and a decrease in the threshold levels of their occurrence.

3 Materials and Methods

Having found out the main reasons for reducing the reliability of robosystems, we will consider the directions of solving the problem of reliability and improving the reliability of heterogeneous robosystems.

Way 1. The robosystem model should adequately reflect the systemic relationships and patterns of processes in heterostructures. As a generalized physical model, a heterogeneous multicomponent heterogeneous structure with significantly different physical characteristics of structural units and subjected to various kinds of technological, physical influences, as well as the effects of operational factors should be considered [1].

In general, the robosystem model in the presence of disturbances and degradation of an object or information has the form [13].

$$\frac{d\widehat{X}}{dt} = F\left(\widehat{X}, t, \Delta\widehat{u}, v, \omega\right) \tag{2}$$

where $\widehat{X} = X_H + \Delta\widehat{X}$—is the n-dimensional vector of random parameters of the robosystem motion, $\hat{}$—random variable, X_H—normal or calculated value of a vector of parameters of the movement of a robosystem, $\Delta\widehat{X}$—a vector of accidental deviations of parameters of the movement of a robosystem from the nominal movement, t—the time of the movement of a system, $\Delta\widehat{u}$—a vector of the accidental uncontrollable indignations arising, for example, as a result of degradation of an object or information, v—a vector of control effects, ω—a vector of controlled disturbances.

Way 2. Intervals of preventive work can be modeled using models of systems evolution, which allow to investigate dynamics and evaluate the behavior of systems over time, as well as predict the behavior of systems over time in the process of aging [14]. Based on this model, you can define the time intervals at which preventive work should be carried out to restore the health of the management object. However, this requires the termination of systems, which is unacceptable in extreme operating conditions of robotic systems.

The following mathematical model can be used as an integral characteristic of the probability of failure-free operation of robosystems.

The system is loaded after a random time τ_i, and the damage from each load is a random value y_i.

The probability of failure of the system in a fixed time t is determined taking into account the complete group of events. There is a probability that exactly K loads will occur in time t

$$P_K = F^K(t) - F^{K+1}(t) \tag{3}$$

where $F^K(t) = p\left(\sum_{i=1}^{K} \tau_i \leq t\right)$—is the probability that at K loads the system works. Convolutions $F^K(t)$ are related by recurrence ratio

$$F^K(t) = \int_0^t F^{K+1}(t)(t - z)\mathrm{d}F(z) \tag{4}$$

Multiplying the obtained probability by the conditional probability of failure-free operation P_K at exactly K loads and summing over all $K = 1, 2, \ldots$, the desired full probability of failure-free operation for the time t is obtained:

$$P(t) = \sum_{K=1}^{\infty} p_K P_K(t) \tag{5}$$

A generalized model of the evolution of reliability of complex systems is presented in Fig. 1.

The value of the full probability of failure-free operation of the system during time t can be used in future to determine the intervals of preventive work of robosystems.

Way 3. Improving the reliability of robotic systems due to the use of a fail-safe element base, multiple redundancy of vulnerable parts of the system, a set of technical solutions that allow maintaining the operability of the robotic system and

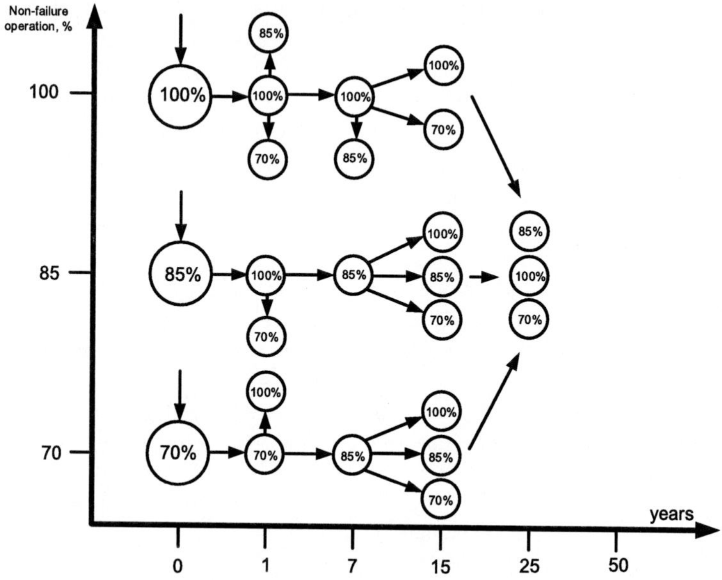

Fig. 1 Generalized model of evolution of reliability of complex systems

continuing to perform the main task along with the expected deterioration in technical characteristics, as well as using artificial intelligence algorithms.

An example is the transition in the orientation of the cruise missile from using satellite navigation systems to determining the location of the missile by comparing the readings of the radar display of the terrain with reference mapping, as well as the automatic transition to non-sensor algorithms for controlling systems in case of sensor failure.

Using artificial intelligence algorithms in the process of controlling robosystems allows you to include recovery processes in the system by obtaining updated information and maintain homeostasis in the system.

A systematic approach in combination with artificial intelligence allows you to predict the state of the robosystem, to generate optimal control actions to maintain the robosystem in a stable working position even if there is degradation. For this purpose, information accumulated during operation of the robotic system by artificial intelligence is used, which is a reference for forming control actions on the robotic system in order to restore its homeostasis.

Way 4. Consideration of heterostructures properties heterogeneity under external disturbing influences, among which vibration and temperature influences have the strongest influence on heterostructures.

Vibration effects are a strong factor that reduces the reliability of the robotic system. Typically, vibrations have a wide range of frequencies and acting on elements of the robosystem of a heterogeneous nature cause resonant phenomena at various levels of heterostructures, which in its turn disrupt the stability of both the mechanical and electronic, and therefore, the information components of the robosystem.

Heterogeneity of physical properties of heterostructures leads to destruction of multilayer boards, peeling of contacts, disruption of resonators, etc. Thus, differences in thermal expansion coefficients of heterostructures elements lead to increased stresses in heterostructures, the emergence of microcracks, depressurization of robosystems elements. Studies have found that when the temperature of air-cooled radio-electronic equipment heterostructures increases from +40° to +70 °C, the number of failures increases by 2–6 times.

The way out of this situation is application of vibration dampers, including using discrete media to eliminate harmful effects of vibration disturbances [15], as well as careful selection of elements of heterostructure with high-internal friction, some models of which are presented in [16].

One method of eliminating the temperature effect on the reliability of heterostructures is to put the operation of sensors into a non-stressed thermal mode, which is an effective way to reduce temperature error [1].

Other methods for preventing failures due to temperature effects are the use of heterostructures with close temperature expansion coefficients at the junction points, and in case of impossibility of using such materials, the use of heat protections that allow increasing the time of transient processes of temperature deformation of heterostructures.

It should be noted that the combination of vibration and temperature effects, each of which does not have a strong effect on the reliability of the technical object, can lead to a sharp increase in the failure rate.

Way 5. Address of technical and programmatic errors that occur at all levels of structure vulnerability.

Faults and reversible faults due to external influences result in software faults in the form of errors contained therein.

The reason for the errors on which attention should be focused on is the unconscious self of the programmer's intelligence, as well as the electromagnetic "noise": unconscious motives, the true meaning of which is not recognized due to contradictions with other motives; behavioral automatisms acting in the usual situation, the awareness of which is unnecessary; sub-threshold sensation that is not recognized due to low intensity; subconscious processes: intuition, inspiration, insight.

It is necessary to take into account the health indicators of the human operator of robosystems—the so-called contacts that determine the reliability of human performance of prescribed functions.

- Immune protection.
- Level and harmony of physical development.
- Functional state of the organism and its reserves.
- Development defects—diseases

Thus, in order to ensure the reliability of software-robotic complexes and technologies, especially defense applications, serious attention should be paid to protecting the immunity of personnel, personnel selection, etc.

Way 6. Improve the quality and reliability of systems with artificial intelligence.

In most developed countries, the improvement of systems with artificial intelligence is of great importance. So in Russia, in Decree No. 460 of 10.10.19 "On the development of artificial intelligence in the Russian Federation," V.V. Putin clearly formulated the tasks of the state in this area.

In the United States, the "American Initiative in the Field of Artificial Intelligence" has been put forward, which specifically states that artificial intelligence is the central element of the competitiveness and national defense of this country. To achieve superiority, a human learning strategy has been chosen, for example, the development of deep-learning neural networks, computer vision algorithms, etc.

To fulfill the tasks, it is necessary to build high-quality models of artificial intelligence of robosystems. In a structural approach, artificial intelligence is built by modeling the human brain.

As models of heterostructures of intelligence, statistical models based on big date and date mining software packages are used, as well as dynamic deterministic models of heterostructures based on fundamental dynamics models, in particular:

$$\frac{\mathrm{d}}{\mathrm{d}t}\left(\frac{\partial P}{\partial \dot{s}}\right) - \frac{\partial P}{\partial s_i} = F_i, \quad i = 1, 2, \ldots, n \tag{6}$$

where P—is the power of intelligence, s—is the intelligent solution, $\dot{s}$—is the speed of decision making, F_i—is the power of intelligence.

When processing information of small samples, the classical criteria of Sijel-Tuki and Mann-Whitney are used, as well as the modern approach of A.I. Ivanov, based on quantum representations and discrete spectra of artificial mathematical molecules, as well as the protection of artificial intelligence [7].

In order to gain competitiveness in the field of robotic systems, in most cases they develop IQ (mental intelligence) [6], which is based on logical, production and information presentation models.

Logical models implement objects and rules based on first-order predicates, are formalized models with deductive, monotonous logical outputs "from target to information." The advantage of the models is the high expressiveness of the language, well-defined semantics, the presence of various procedures for outputting and converting information, the ability to display in terms of calculating predicates all kinds of forms of information presentation. The disadvantages of logical models include the difficulties of expertise, since there is no clearly expressed representation scheme, a very thin structure of language constructs, difficulties in fragmentation and poor implementability of generalization and generic classifications.

Production models are based on the use of heuristic methods of presenting information, which provide tuning of the output mechanism to the features of the problem and taking into account uncertainty. The advantages of the models include the understanding of experts, the availability of information presentation, the implementability of fuzzy reasoning, explanations of conclusions, and the mirroring of expert logic. Disadvantages include weakness in the representation of concepts, the connectedness of complex concepts, difficulties in maintaining databases and their maintenance, difficulties in working with large volumes of information of a heterogeneous structure.

The model of presenting information in knowledge engineering in the form of semantic networks provides a very simple and understandable way of describing based on the relationship between nodes and arcs. Disadvantage as the size of the network increases, search time increases, there is a problem of ensuring the suitability of the output and inheritance of properties.

The frame model uses attached procedures to implement operating information, all object attributes are collected into one heterostructure of data—a frame. Frames provide some ways to connect procedural information to heterostructures described in different languages, most of which emphasize the representation of system components by functional elements oriented to data processing [8].

The imperfection of mathematical models and algorithms by which artificial intelligence works leads to the presence of failures and failures in the system. Problems are resolved by external intervention, such as a robosystem operator.

All the above directions ensure the reliability of the robotic system, and to one degree or another should be implemented during the development and operation of such systems.

4 Results and Conclusions

- Studies of models of a wide class of complex robosystems have been carried out. Integral reliability estimates of robotic heterostructures are obtained.
- A model of degradation of complex systems in the presence of perturbing factors is presented. The system analysis of causes of failures of complex human-controlled robotic systems is given.
- Reliability theory is developed in terms of system analysis of failures of complex robosystems. Roboprophylaxis of reliability of complexes based on analysis of causes of robosystems failures is proposed.
- The main directions of solving the problem of reliability and improving the safety of heterogeneous robosystems have been developed.
- At the present stage of the development of artificial intelligence, it is necessary to limit the scope of its action and nevertheless leave a decisive role in decision-making to a person.

References

1. Vasiliev, V.A., Volchikhina, N.I., Smogunov, V.V.: Dynamics of heterogeneous structures. Modeling of heterogeneous structures of information transducers of rocket and space systems. Monograph: in 12 t.T.5. Publishing House of Penz. State. University, Penza (2004)
2. Kuznetsova, O.V., Smogunov, V.V., Yakimov, A.M., et al.: Evolution of Intelligence Models. Publishing House of PenzGTU, Penza (2008)
3. Mikhailov, V.S., Yurkov, N.K.: Integral Estimates in Reliability Theory. Technosphere (2020)
4. Zikratov, I.A., Kozlova, E.V., Zikratova, T.V.: Vulnerability analysis of robotic complexes with swarm intelligence. In: LITMO Cyberleninka (2018)
5. Yuzvishin, I.I.: Fundamentals of Information Science (2000)
6. Volchikhin, V.I.: Information Technology Security. Publishing House of PSU, Penza (2019)
7. Ivanov, A.I.: Protection of Artificial Intelligence. Publishing house of PSU, Preprint (2020)
8. Becker, T.M., Kuznetsova, O.V., Boskebeev, K.D., et al.: Education of intelligence. Ilit, Bishkek (2009)
9. Smogunov, V.V., Yurkov, N.K., Kuznetsov, N.S.: Engineering of algorithms and nano-destruction models of heterostructures. Reliab. Qual. Complex Syst. **1**(21), 10–20 (2018)
10. Smogunov, V.V., Kliminov, I.P., Volchikhina, Z.K., et al.: Dynamics of heterogeneous mechatronics structures. In: System Analysis, Management and Processing of Information. PenzSU, Penza (2005)
11. Smogunov, V.V.: Dynamics of Heterogeneous Structures. T.1. Evolution of Rocket and Space Structures. PenzSU, Penza (2001)
12. Kuznetsov, N.S., Fionova, L.R., Yurkov, N.K., et al.: Hybrid intelligence of engineering of computer heterostructures. Reliab. Qual. Complex Syst. **3**, 85–97 (2018)
13. Volnikov, M.I., Smogunov, V.V.: Models of robosystems with degradation. Models Syst. Netw. Econ. Technol. Nat. Soc. **1**(37) (2021)
14. Pyukke, G.A.: On the issue of degradation models of multicomponent systems of various physical nature. Bull. Kamchatka State Tech. Univ. **44**, 21–29 (2018)
15. Volnikov, M.I., Smogunov, V.V., Vdovikina, O.A.: Modeling of vibration shock protection of heterogeneous structures. News of higher educational institutions. Volga region. **6**, 373–379 (2006)

16. Volnikov, M.I., Smogunov, V.V.: Mathematical models of the dynamics of heterostructures with friction. News of higher educational institutions. Volga region. Techncal Sciences. **3**(55), 98–108 (2020)

Operation of the Levelling Device

Vladimir Teterin, Mikhail Kostenko, Dmitry Blagov, Sergey Mitrofanov, and Olga Teterina

Abstract Balers of various designs are used to harvest compressed stalk forage, for example, compressed hay. Constant chamber roll balers are simpler and more reliable. However, the dried stalk mass coming from the rolls, due to peculiarities of the roll formation, rotates randomly in the bale chamber until it is full. Analysis of the density of hay inside the roll shows that the highest density of hay across the roll width (pickup width) is observed in the middle part. This is due to the shape of the swath of hay entering the roll baler. To determine the effect of the levelling device on the change in the density of stalk particles inside the roll, a series of experimental studies was carried out during which the density of rolls at various sites was determined when harvested using a roll baler with a constant volume chamber with and without the levelling device. The obtained research results show that when using a roll baler with a constant volume chamber and an installed levelling device, the density of hay at the edges of the bale increases to 84 kg/m^3, while the density remains the same (180–200 kg/m^3) in the middle of the roll.

1 Introduction

Currently, there are many technologies for making hay; but despite their diversity, they all share the same requirements for mowing and drying. It is known that during field drying of herbs, stems dry much more slowly than leaves, for example, in leguminous herbs with a moisture content in leaves of about 20%, the moisture content of stems is in the range of 35–40%. This factor increases the timing of harvesting hay from the field and, as a consequence, can lead to the loss of some of the leaves. In order to ensure uniform drying of all parts of the plant, special measures are applied, such as turning, squashing the stems, etc. After prewilting of

V. Teterin (✉) · M. Kostenko · D. Blagov · S. Mitrofanov
VIM Federal Scientific Agro-Engineering Center, Moscow, Russia

M. Kostenko · O. Teterina
Ryazan State Agrotechnological University Named After P.A. Kostychev, Ryazan, Russia

M. Shamtsyan et al. (eds.), *Robotics, Machinery and Engineering Technology for Precision Agriculture*, Smart Innovation, Systems and Technologies 247,
https://doi.org/10.1007/978-981-16-3844-2_42

hay in the field to the required moisture content in swaths or in rolls, it is harvested with subsequent laying in storage [1–3].

Roll balers of various designs are used to harvest compressed stalk forage, for example, compressed hay. The principle of their operation is based on compaction of the stalk mass to the specified values. For this, spring fingers pick up a roll of prewilted plant mass, which is fed into the bale chamber, where it is compressed to the required density, after which the resulting compressed volume of plant mass is fixed with twine and unloaded [1, 4].

Currently, piston and roll balers are widely used, and the latter can be produced both with a variable bale chamber and a constant volume bale chamber. Piston and round balers fundamentally differ in the organization of the baling process [5] (Fig. 1).

Fig. 1 General view of piston and roll balers: **a**—baler Metal-Fach Z2690; **b**—baler PELIKAN 1200

a)

b)

When the piston baler is moving in an aggregate with a tractor, spring fingers of the baler pick up plant mass from the swath and feed it to the wadboard packer, which pushes it into the bale chamber, which is formed from the bale press, the piston and previously compressed bales. During each working stroke, the piston picks up a portion of plant mass and pushes it through the bale chamber. An advantage of the balers of this type lies in the shape of the obtained bales, which facilitates their transportation and storage. The main drawbacks of these machines are associated with the decreased density of the first bales, which has a negative effect on their safety, and the increased energy consumption due to the operation of the cutting blades (a knife and a crossbar) to cut hay.

The use of endless elastic belts in the design of roll balers with a variable volume bale chamber ensure higher density in the middle of the cylinders, that prevents the penetration of moisture and air, thereby increasing the safety of the pressed forage. At the same time, uneven distribution of the swath over the pickup width causes deformation of belts in the midportion of the bale chamber, which leads to parametric failures.

Constant chamber roll balers are simpler and more reliable. However, the dried stalk mass coming from the rolls, due to peculiarities of the roll formation, rotates randomly in the bale chamber until it is full, after which, as a result of the arrival of the next portions, the roll is gradually compacted from the outside, and therefore, the transfer of pressure from the outer layers to the inner ones occurs with some delay (stress relaxation), while the arrangement of stems will have a significant effect on the uniformity of the density distribution inside the roll [6, 7].

Analysis of the density of hay inside the roll shows that the highest density of hay across the roll width (pickup width) is observed in the middle part. This is due to the shape of the swath of hay entering the roll baler, its cross-section being an inverted parabola. As a result of the roll compaction, it has a higher density in the middle of the cylinder. This factor leads to the formation of zones with a low density inside the roll, which contributes to the penetration of moisture, the development of microbiological processes and, as a consequence, a decrease in the safety of hay [5–8].

To reduce the influence of the environment, atmospheric air and weather conditions on the safety of the pressed material, balers can be additionally equipped with devices for packing bales (rolls) in plastic film or devices for introducing liquid preservatives [9].

Roll balers can be used either separately from the pickup baler, or in conjunction with it. The advantage of using these machines is that each roll of fodder is packed in plastic wrap and is a sealed mini storage [10]. However, this technology has some drawbacks; in particular, there is a problem of disposing of polyethylene film after unpacking and feeding hay to animals.

The most rational way of introducing liquid preservatives during the preparation of stalk fodder is the technology of direct introducing when pressing the material. For these purposes, various hydraulic systems are used, which are mounted either directly on the pickup baler or in front of the tractor. It is also worth noting that there are several approaches for applying preparations: continuous application, differentiated application depending on the volume of the supplied material and differentiated application depending on the degree of compaction of the material [9, 10]

However, it should be noted that in order to increase the safety of the pressed hay, it is also rational to use the operations of levelling the hay swath before it enters the pressing chamber of the pickup baler.

2 Materials and Methods

In order to study the effect of compressive stresses on the density of hay, a laboratory setup was used, consisting of a frame, a hydraulic jack, a thrust pipe, a sealing platform installed inside a guide cylinder with a diameter of 200 mm, placed on a base spring-mounted using calibrated springs.

The setup worked in the following way, a sample of hay with an ordered arrangement of stems either along the direction of deformation or across it was laid in the guide cylinder. The weight and moisture content of the sample were preliminarily determined. When the hay sample was compressed, the spring-loaded platform was displaced, and the compression force was determined by the average compression of the calibrated springs mounted between the platform and the base.

Five hay samples were prepared for insertion into the guide cylinder along and across the compression direction. The verification of the reliability of the experimental results was carried out using Fisher's ratio test.

To study the change in the density of hay inside the roll, a series of experiments was carried out, during which hay of natural herbs was harvested using a roll baler with a constant volume bale chamber with and without a levelling device installed on it. Rolls were randomly selected from each batch of rolls harvested by one roll baler. They were divided into zones, and at the boundaries of the zones, the force of insertion of the tip was determined using a Revyakin hardness tester equipped with special tips. At each border of the plot, the tests were carried out with three replications, and the obtained data were averaged according to the following formula:

$$h_{\text{aver}} = \frac{h_1 + h_2 + \cdots + h_n}{n} \tag{1}$$

where h_1, h_2,..., h_n are heights of diagrams, *m*.

The data obtained were statistically processed and entered into a table of values, on the basis of which a topographic picture of the roll density was created.

3 Results

When studying the influence of compressive stresses on the density of hay, regression equations were obtained for stalked particles located across the direction of deformation:

$$\sigma_c = -0.0586 + 1.8218 \cdot \gamma \tag{2}$$

where

σ_c is compression stress, kPa;
γ is density of the pressed hay, kg/m^3.

and stalked particles located longitudinally to the direction of deformation:

$$\sigma_c = -0.0376 + 2.0389 \cdot \gamma \tag{3}$$

Statistical analysis of Eqs. 2 and 3 showed high reliability, the coefficient of determination for Eq. 2 was 0.76, and for Eq. 3, it was 0.95, and the significance of all coefficients of the regression equations was high.

The analysis of the regression equations showed that the deformation stresses when pressing hay for longitudinally oriented stem particles were significantly higher than for transversely directed ones. In this connection, one can say that for the best process of compaction of stalked particles during the formation of a roll, their additional orientation in the transverse direction relative to the compressive forces is necessary.

To determine the effect of the levelling device on the change in the density of stalk particles inside the roll, a series of experimental studies was carried out during which the density of rolls was determined when harvested using a roll baler with a constant volume chamber with and without the levelling device. In the course of the research, the density values of the rolls were obtained at various sites, on the basis of which topographic pictures of the density distribution in the rolls were created. Using STATISTICA V10 program, the density models were built for the pressed hay in rolls prepared by a roll baler with a constant volume chamber when using the levelling device and without it. They are shown in Figs. 2 and 3.

To describe the presented surface characterizing the density distribution of hay inside the roll, a regression equation was drawn up:

$$\begin{aligned} V_3 = 112.523 + 144.617 \cdot V_1 + 197.146 \cdot V_2 - 86.218 \cdot V_1^2 \\ - 27.213 \cdot V_2^2 - 261.13 \cdot V_1 \cdot V_2 \end{aligned} \tag{4}$$

where

V_3 is the density of hay in a roll without levelling, kg/m3;

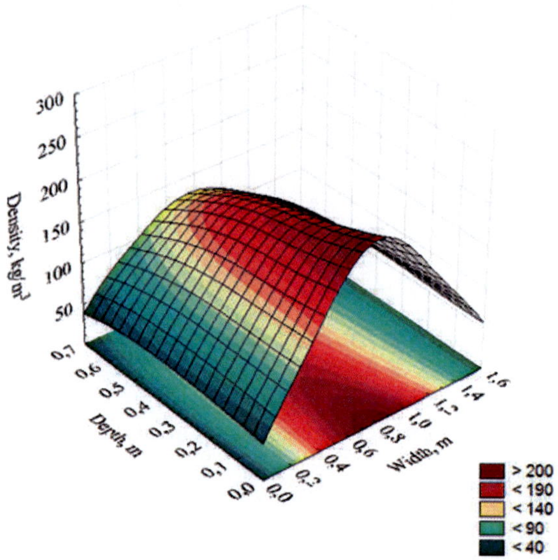

Fig. 2 Distribution of hay density inside a roll harvested by a roll baler with a constant volume chamber without the levelling device

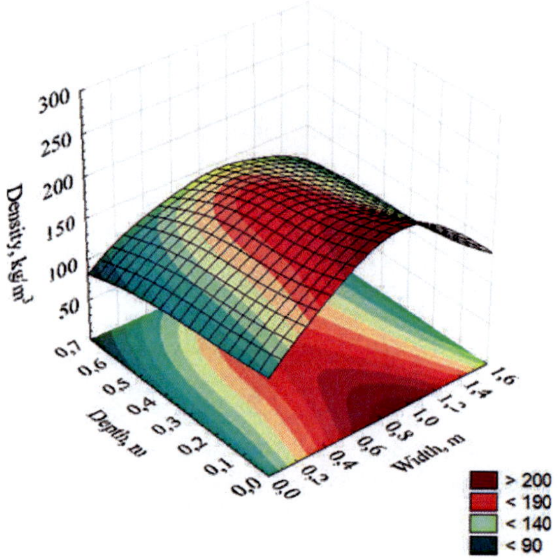

Fig. 3 Distribution of hay density inside a roll harvested by a roll baler with a constant volume chamber with the levelling device

V_2 is the tip penetration depth, m;
V_1 is the roll width, m.

The carried out statistical analysis of the obtained model shows that the correlation coefficient has a high degree of connection between the variables and is $R = 0.87$, and the coefficient of determination, showing the adequacy of the model, is 0.76.

The regression equation for a given surface is as follows:

$$V_4 = 78.531 + 160.275 \cdot V_1 + 39.44 \cdot V_2 - 121.259 \cdot V_1^2 \\ - 415.77 \cdot V_2^2 - 205.945 \cdot V_1 \cdot V_2 \quad (5)$$

where

V_4 is the density of hay in a roll with levelling, kg/m^3;
V_2 is the tip penetration depth, m;
V_1 is the roll width, m.

Statistical analysis of the model shows that the correlation coefficient is $R = 0.90$, and the determination coefficient showing the adequacy of the model is 0.81.

4 Discussion

Analysing the obtained graphs of the distribution surfaces of the density of hay inside the roll, it is found that when using a roll baler with a constant volume chamber and an installed levelling device, the density of hay at the edges of the roll increases from 40 to 84 kg/m^3; while in the middle of the roll, the density remained the same at the level of 180–200 kg/m^3. It should be noted that the compaction density at the edges decreases due to stress relaxation thanks to low values of the friction coefficient, especially that of grass stems. Experimental studies of the operation of a roll baler equipped with a developed levelling device show that the use of the levelling device allows to increase the roll weight by an average of 50 kg, thereby increasing the performance of the roll baler by 8–15%, when the levelling device has the following rational parameters and modes: the distance between the fingers is 0.15 m, the angular speed of the discs is 25 rad/s and the diameter of the discs is 0.74 m.

5 Conclusion

The considered theoretical and experimental studies have shown that in the process of roll formation, zones with different densities of hay stem particles appear. The emergence of zones with insufficient density of hay will facilitate the penetration of atmospheric moisture into the roll and, as a result, create favourable conditions for the development of colonies of mould fungi, which can lead to the formation of mycotoxins dangerous for the animal and human body. This factor will also contribute to a decrease in the quality indicators of hay, in particular its nutritional value, and lead to a deterioration in organoleptic properties.

Experimental and theoretical studies have shown that the use of a roll baler with a levelling device in the form of discs with fingers makes it possible to ensure uniform distribution of stem particles across the pickup width, ensuring the preliminary compaction of the hay swath before supply into the bale chamber, which in turn makes it possible to reduce or even eliminate harmful effects from the penetration of moisture and air, and also contributes to the production of rolls with a greater mass, thereby increasing the productivity of the roll baler by 8–15%.

It should be noted that in order to increase the safety of pressed hay in balers with a constant volume chamber, it will be rational to additionally use the technology of differentiated application of liquid preservatives. In this case, the technology of differentiated application of the preparation, depending on the degree of compaction of the material, will be more preferable. In this case, the dose of application of the preparation is higher in places with a lower density. For pickup balers with a chamber of variable volume, it is preferable to use the technology of differentiated application depending on the volume of the supplied material, where the application rate is adjusted according to the width of the pickup and the shape of the swath being picked up.

References

1. Johnston, D.J., Theodoridou, K., Ferris, C.P.: The impact of field bean inclusion level in dairy cow diets on cow performance and nutrient utilization. Livest. Sci. **220**, 166–172 (2019)
2. Fang, J., Zhang, Y., Yang, M., Wang, A., Wang, J., Liu, D., Gao, J., Li, H.: Stress relaxation behavior and modeling of alfalfa during rotary compression. Nongye Gongcheng Xuebao. Trans. Chin. Soc. Agric. Eng. **34**(16), 50–56 (2018)
3. Wei, X., Shangli, S., Juan, Q., Squires, V.R., et al.: Effect of different processing methods on the quality of Alfalfa Hay. J. Anim. Vet. Adv. **12**(6), 689–693 (2013)
4. Gao, D., Wang, D., Li, J., Liu, Y., Qiao, X.: Design and test of logarithmic spiral round baler chamber. Nongye Jixie Xuebao. Trans. Chin. Soc. Agric. Mach. **46**(7) (2015)
5. Terentyev, V., Andreev, K., Anikin, N., Novikov, N., Teterin, V., Blagov, D., Mitrofanov, S.: The study of forming a roll of pressed hay and the influence of its density on quality parameters. In: E3S Web of Conferences. Ecological and Biological Well-Being of Flora and Fauna (EBWFF-2020), p. 04013 (2020)
6. Uijttewaal, A., Chapuis, S., Crocq, G., Lépée, P.: Advances in methods for harvesting and storing forage legumes. Fourrages **227**, 157–166 (2016)
7. Byshov, N.V., Borychev, S.N., Teterin, V.S., Uspenskiy, I.A., Kostenko, M.Yu., Rembalovich, G.K.: Investigation of the pressed hay density distribution inside the roll. Herald Ryazan State Agrotechnol. Univ. Named after P.A. Kostychev **4**(28), 48–52 (2015)
8. Lei, J., Wang, D., Zhang, Q., Yang, X., Li, L., Li, C.: Experiment on stress relaxation characteristics of intact rice straw during rotary compression. Trans. Chin. Soc. Agric. Eng. **31**(8), 76–83 (2015)
9. Skorkin, A.V.: Efficiency of technologies for the preparation of stalked forages of high humidity. Bull. All-Russ. Res. Inst. Livestock Mech. **1**(9), 16–19 (2013)
10. Drinkcha, V.M.: Alternative ways to reduce the loss of hay. Fodder Prod. **7**, 44–48 (2012)

Automated Soil Microbiology Using Lensless and LDI MS Imaging with Buried Slides

Feodor Orekhov and Oleg Gradov

Abstract The complex of technologies for soil microbiological and agrochemical monitoring studies using an active chip is being developed based on the following principles: 1. CCD/CMOS chip monitoring and calibration systems, including polarization microphotometric system for the reconstruction of the brightness profiles (3D luminance surface) with the parallel display of the vectoroscopy results, a system for particle counting with oscilloscopic–photographic registration for video microscopy based on a lensless luminescent microscope, a stand for verification and calibration using a stroboscopic oscilloscope reflectometer with a discreteness up to the nanosecond range, and 2. telemetry system of the active chips with radio frequency identification and a receiver with the fine-tuning (in the earlier versions—manual tuning using agometric procedure. In fact, it is possible to use the data of a telemetric signal (in real time) after digitization without compression as a source of kinetic descriptors of the processes on chip-to-soil interface. There are several areas that could be developed within the framework of the paper, which seem to be rather interesting and promising for the soil science, geochemical microbiology and ecological studies: implementation of the microfluidic capillary soil planar chip technique in an alternative version with preservation of the capillary/microfluidic characteristics in Perfil'ev–Gabe method variations (from capillary analysis methods of soil microflora to microfluidic LDI- and SPIM-compatible pedoscopes, etc.), and combination of capillary methods of soil microbiology with capillary methods of mass spectrometry techniques (in chromatographic micro-columns and nano-columns on a chip; in ESI capillaries; data collection from MS-patch-clamp microcapillaries, etc.).

1 Introduction

The feasibility of using devices that combine the properties of buried slides and soil chambers with the properties of active CMOS/CCD detectors and targets (in passive conditions) for MALDI or other LDI is determined by the wide range of

F. Orekhov (✉) · O. Gradov
N.N. Semenov Institute of Chemical Physics, Kosygina 4, Moscow 119334, Russia

M. Shamtsyan et al. (eds.), *Robotics, Machinery and Engineering Technology for Precision Agriculture*, Smart Innovation, Systems and Technologies 247,
https://doi.org/10.1007/978-981-16-3844-2_43

applicability of the prototype samples, Rossi–Cholodny devices. The following are some bibliographic reference examples of the use of soil slides and chambers for the analysis of natural media:

- Soil microorganisms were studied, including those for technical or applied microbiology [1, 2].
- Soil fungi [3, 4] and fungal spores [5] were studied, as well as the effects of mycelium vegetation [6].
- Soil microalgae were studied as a community and the effect of a number of chemical agents, in particular, herbicides [7, 8].
- The effects of soil conditions on root system diseases and mycorrhizal effects were studied [9].
- The effects of colonization of artificial rock outcrops by microorganisms were studied [10].
- The freshwater microflora of drinking water sources was studied [11].
- Bacterial populations of seawater were studied, using a series of comparative complementary descriptors obtained from different sources and counting methods [12].

The nomenclature of the objects studied using soil chambers and buried slides is not inferior to the nomenclature of the similar items for MALDI biotyping, which includes, in addition to pathogenic bacteria, fungi, in particular, mold, yeast and pathogenic fungi [13–22]; lower plants (unicellular green algae, such as Chlamydomonas) [23, 24]; cyanobacteria [25], etc. At the same time, the species-specific chemical–taxonomic aspect of this problem and the implementation of the method allows to identify not only the samples belonging to a certain taxon (e.g., *Nocardia, Rhodococcus, Kocuria, Gordonia, Tsukamurella, Listeria* [26]) or its individual species (e.g., for *Staphylococcus* [27]), *Acinetobacter* [28], *Acanthamoeba* [29]), or record the presence of a special ("targeted") species in this sample, specific to a particular condition of its source (e.g., *Staphylococcus aureus* in biomedical and veterinary diagnoses samples [30] or a direct diagnostic multiplex test for multitarget contamination in blood sepsityping [31]), but also individual strains or biochemically determined races of organisms due to a direct molecular–chemical analysis provided by the use of mass spectrometry. The acceleration of identification in such cases [32] is usually achieved not only by iteratively narrowing the spectrum of identification according to taxa and identifiable groups of organisms (e.g., the first narrowing for bacteria may appear to be database-mediated division into gram-positive and gram-negative ones for clinical diagnosis), but also by the use of systems for the separation of forms during sample preparation and sample supply within the framework of advanced techniques of the so-called automated microbiology [33].

2 Limitations of Standard Buried Slide Techniques

There are a number of obvious issues that do not allow the use of buried slides and soil chambers of Rossi–Cholodny for their intended purpose—for in situ (real-time) analysis of soil microbiota.

First, the inability to monitor the state of the colony/biofilm in real time, due to the presence of the sample in the soil, makes it impossible to monitor a number of key growth indicators and reconstruction of the growth kinetics. Visualization of the so-called microbial landscapes (according to Tepper) requires sample preparation, which is time-consuming and not nondestructive with respect to one or another set of microorganisms and fragments of the bioinert medium.

Second, the need for staining (often selective staining) of microflora in order to ensure effective cell counting using microscopy (especially luminescent one) shifts the real biochemical characteristics of the sample components, so effective staining and population (species-specific) analysis of the sample become the antonyms of its compartmental biochemical analysis.

Third, the absence of kinetic correlates (due to the lack of telemetric broadcasting from the buried slide or soil chamber to the receiver) makes it impossible to reconstruct the stages of development of the colony/microbial population/biofilm/microbial mat, while the development stage determines the chemical composition of the sample and its surrounding media, as well as their adequacy to a certain standard model (inverse to the residual coefficient). It is impossible to identify the phase of development of the population (lag phase or another one), which leads to the fact that by observing the result of sampling at a separate time point, it is impossible to determine whether this point belongs to one or another kinetic interval and, as a result, the real objective characteristic of the sample state and the surrounding medium state in the biogeosystem/biogeochemical sense, that is, to characterize optimality by the criterion of the effect on the growth of a population in a given type of medium and phenological growth rhythms for a given type of medium, which influence the dynamics of production processes of biofilms, mats, etc., that are selective or specific in terms of biogeochemical criteria to it as a substrate or an active regulatory agent.

Fourth, all the efforts to perform selective removing of the single slides from the soil at different time intervals after multiple burring (up to several thousand samples) lead to a gradual change in the sample sizes, statistical parameters and heteroscedasticity of the results of different experiments. Suppose the final result of counting/cytometry is determined for 50–100 slides that underwent a full course of contamination (3–7 months), and during the episodic analysis for this period of time five to ten samples removing from the soil were carried out (the question "by what criteria were the slides selected" cannot be solved due to the subjectivity of visual assessments of microbial contamination in the field) under different (not controlled at the level of individual slides or in general in the experimental field) conditions, and the initial number of slides was from several hundreds up to a thousand or more. During hydrobiologically and phenologically determined changes in some microbial populations relative to the other ones (occurring due to natural competition, complete elimination of a number of species due to the non-optimal climate or other biogeochemical/biogeosystem conditions, physically different resistance of these species/populations to sample preparation processes at different stages and phases of sampling, individual reactivity and resistance of individual colonies/micropopulations, mapped by the areas of their localization in the soil, due to allelopathy), there was a multiple non-unidirectional shift in the medium balance, including the statistics-determining population–biochemical indicators, resulting in the statistical incompatibility of the early-stage and later-stage samples under equal conditions.

It is obvious that overcoming the factors of statistical non-representativeness and kinetic incompatibility for a biogeosystem microbiological complex of this experiment is feasible not within the framework of unification of the biogeochemical and allopathic conditions in the field (to the level of a pure laboratory experiment), but through taking into account the full range of factors and telemetry/time lapse kinetic monitoring for correlated analysis of different populations/colonies of a different taxonomic and geoecological nature.

This requires a hardware basis that includes, at least, a set of active sensors with no less than one element mapping the population density distribution; a receiver and transducer for data acquisition with a time resolution; and a hardware–software module/crate for multi-channel synchronous analysis (acquisition and processing) of the data and sending it to the PC memory for subsequent processing in statistical packages with the establishment of differences in localization and colocalization of the corresponding variables on the map registered by a position-sensitive sensor.

The implementation of morphometric operations in protocols of this kind is ensured under laboratory conditions by laser scanning techniques or in the field conditions using conventional microscopy. A good example of this was provided by one of the earliest initiative biogeochemical expeditions, in which in the field (and then in laboratory conditions to ensure biostatistical comparability of the measurement results), microscopy was used with MLD-2—a luminescent microscope, which after delivery of the soil samples was installed into the automatic system for panoramic counting with oscilloscopic registration during video microscopy (Fig. 1). The disadvantages of this approach, however, are obvious: The transfer of samples from one device to another provides incomplete reproducibility of patterns and the absence of the possibility of combining or comparing them, especially since this is not suitable for dynamic measurements in real time, because in these protocols with time resolution, either parameter *A* or parameter *B* is measured (either cytomorphometry or mass spectrometry, that is, either morphological or biochemical analysis is performed).

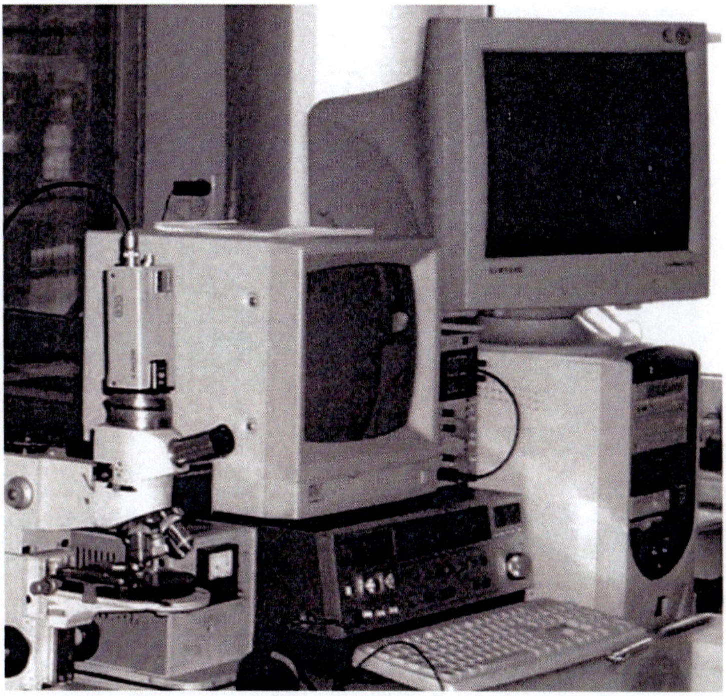

Fig. 1 Old-type setup for luminescence microscopy of microbiological species (GEOKHI RAS, before 2010th)

3 Our Earliest Development in Soil Slide-Based Techniques

We have proposed hardware technologies for the above tasks and analysis methods that satisfy the above requirements. In particular, it was proposed to use CCD and CMOS array sensors as position-sensitive (mapping) sensors.

The prototypes of a discrete matrix on the surface or in the screening of a CCD or CMOS soil-laying chip were: modification of chambers and buried slides of Rossi–Cholodny with electron microscopic grids for SEM known from 1970th [34]; technology of synchronous retention and analysis on a chip—a method for fixing plankton bacteria on a membrane filter replacing the chambers and buried slides of Rossi–Cholodny known from 1950th; and technology of introducing polymer coatings directly bonded to the polymer planar plate of the chip (especially when 3D-printed from PET), which was anticipated in a more simple implementation also using PET material by our Ukrainian colleagues [35]. Our SEM/TEM grid-based version of the buried slide-based chip geometry is provided in Fig. 2.

To identify the microorganism type (gram-positive/gram-negative) on the sensor surface, a magneto-optical converter was used ("flux detector" without passing through the Curie point for field conditions, or a MALDI-compatible laser variant

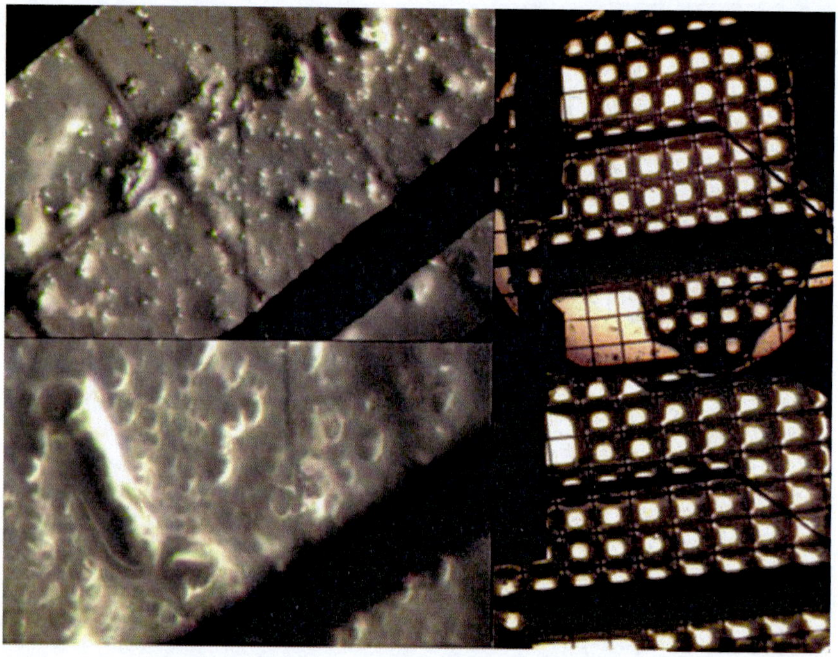

Fig. 2 Example of an immobilized soft matter on a chip with grids from the electrode elements of electronic and ionic devices for bioelectrophysical measurements and electrophysical monitoring of the soil during morphometry of microorganisms with the transduction of the analytical signal through the in situ telemetry channel in real time. **a**—a soft matter sample under high magnification on a grid on a chip; **b**—polymer deposition for intercalation of the transducer and application of a microsample on a grid on a chip

with passing through the Curie point for specified laboratory conditions). As a result, the possibility of combining the described complex with a magnetic variation of staining techniques according to Gram was also implemented [36].

It is obvious that the use of a planar or a surface-structured principle of interphase coupling between the soil and the interface, in which the interface is a surface with specified properties determined by the properties of the converters used, provides the possibility of conducting kinetic measurements in real time (i.e., in situ—in biological terms) without pretreatment with dyes and cytometric labels, allowing subsequent analysis by atmospheric mass spectrometry using laser desorption–ionization (not only MALDI). The main principles of compatibility of optical fluorescence measurements with MALDI were described by the authors in 2014–2016 [37]. However, the studies were not completed due to organizational reasons before the COVID pandemy.

4 Conclusion and Possible Perspectives

Thus, a complex of technologies for soil microbiological and agrochemical monitoring studies using an active chip is being developed based on the following principles: 1. CCD/CMOS chip monitoring and calibration systems, including polarization microphotometric system for the reconstruction of the brightness profiles (3D luminance surface) with the parallel display of the vectoroscopy results, a system for particle counting with oscilloscopic–photographic registration for video microscopy based on a lensless luminescent microscope, a stand for verification and calibration using a stroboscopic oscilloscope reflectometer with a discreteness up to the nanosecond range, and 2. telemetry system of the active chips with radio frequency identification and a receiver with the fine-tuning (in the earlier versions—manual tuning using agometric procedure; in the latest versions—with a tracking generator based on a scalar or, which is better from the standpoint of phase analysis, vector technology for data acquisition/processing). In fact, it is possible to use the data of a telemetric signal (in real time) after digitization without compression as a source of kinetic descriptors of the processes on chip-to-soil interface.

There are several areas that could be developed within the framework of the paper, which seem to be rather interesting and promising for the soil science, geochemical microbiology and ecological studies: implementation of the microfluidic capillary soil planar chip technique in an alternative version with preservation of the capillary/microfluidic characteristics in Perfil'ev–Gabe method variations; combination of the capillary methods of soil microbiology with capillary methods of mass spectrometry techniques (in chromatographic micro-columns and nano-columns on a chip; in ESI capillaries; data collection from MS-patch-clamp microcapillaries, etc.); implementation of a hanging drop method on a coverslip under soil conditions and a longitudinal experiment with a transition through the dew point and condensation on a chip of natural moisture with a psychrometric detection (including monitoring with a polymer capacitor built into the chip surface); using a flexible polymer "woven" microfluidics with an extremely developed surface and compartmentalized percolation plastics (by our group) simultaneously as a chip material and an adhesive layer; and using more effective luminescent methods and exotic signal converters and transducers.

References

1. Hirte, W.F.: Arbeitsmethoden in der naturwissenschaftlichen und technischen Mikrobiologie. I. Die anwendung der verdünnungsplattenmethode zur erfassung der bodenmikroflora. 1. Der quantitative nachweis. Zentralbl. Bakteriol. **119**, 429–447 (1965)
2. Prasad, M.: Microbial society and its activity in the soil. Zentralblatt für Bakteriologie, Parasitenkunde, Infektionskrankheiten und Hygiene. Zweite Naturwissenschaftliche Abteilung: Allgemeine, Landwirtschaftliche und Technische Mikrobiologie **132**(7), 659–665 (1977)
3. D'aeth, H.R.X.: A survey of interaction between fungi. Biol. Rev. **14**(2), 105–131 (1939)

4. Chesters, C.G., Thornton, R.H.: A comparison of techniques for isolating soil fungi. Trans. Br. Mycol. Soc. **39**(3), 301–313 (1956)
5. Nair, N.G.: Studies on the behaviour of mushroom spores in relation to the epidemiology of virus disease. Aus. J. Agric. Res. **30**(6), 1123–1132 (1979)
6. Brown, J.C.: Fungal mycelium in dune soils estimated by a modified impression slide technique. Trans. Br. Mycol. Soc. **41**(1), 81–88 (1958)
7. Forest, H.S.: Analysis of the soil algal community. Trans. Am. Microscop. Soc. **81**(2), 189–198 (1962)
8. Pipe, A.E., Cullimore, D.R.: An implanted slide technique for examining the effects of the herbicide diuron on soil algae. Bull. Environ. Contam. Toxicol. **24**(1), 306–312 (1980)
9. Sulochana, C.B.: Soil conditions and root diseases. Proceedings: Plant Sci. **36**(6), 234–242 (1952)
10. Cooke, W.B.: Colonization of artificial bare areas by microorganisms. Bot. Rev. **22**(9), 613–638 (1956)
11. Langmark, J., Ashbolt, N.J., Szewzyk, U., Stenström, T.A.: Adequacy of in situ glass slides and direct sand extractions to assess the microbiota within sand columns used for drinking water treatment. Can. J. Microbiol. **47**(7), 601–607 (2001)
12. Oppenheimer, C.H., ZoBell, C.E.: Bacterial populations in seawater as determined by different methods of enumeration. Limnol. Oceanogr. **4**, 128–139 (1952)
13. Schulthess, B., Bloemberg, G.V., Zbinden, R., Böttger, E.C., Hombach, M.: Evaluation of the Bruker MALDI Biotyper for identification of Gram-positive rods: development of a diagnostic algorithm for the clinical laboratory. J. Clin. Microbiol. **52**(4), 1089–1097 (2014)
14. Schmidt, A.M., Fagerer, S.R., Jefimovs, K., Buettner, F., Marro, C., Siringil, E.C., Boehlen, K.L., Pabst, M., Ibáñez, A.J.: Molecular phenotypic profiling of a Saccharomyces cerevisiae strain at the single-cell level. Analyst **139**(22), 5709–5717 (2014)
15. Yan, Y., He, Y., Maier, T., Quinn, C., Shi, G., Li, H., Stratton, C.W., Kostrzewa, M., Tang, Y.W.: Improved identification of yeast species directly from positive blood culture media by combining Sepsityper specimen processing and Microflex analysis with the matrix-assisted laser desorption ionization Biotyper system. J. Clin. Microbiol. **49**(7), 2528–2532 (2011)
16. Theel, E.S., Schmitt, B.H., Hall, L., Cunningham, S.A., Walchak, R.C., Patel, R., Wengenack, N.: Formic acid-based direct, on-plate testing of yeast and Corynebacterium species by Bruker Biotyper matrix-assisted laser desorption ionization-time of flight mass spectrometry. J. Clin. Microbiol. **50**(9), 3093–3095 (2012)
17. Lohmann, C., Sabou, M., Moussaoui, W., Prévost, G., Delarbre, J.M., Candolfi, E., Gravet, A., Letscher-Bru, V.: Comparison between the Biflex III-Biotyper and the Axima-SARAMIS systems for yeast identification by matrix-assisted laser desorption ionization time of flight mass spectrometry. J. Clin. Microbiol. **51**(4), 1231–1236 (2013)
18. Mancini, N., De Carolis, E., Infurnari, L., Vella, A., Clementi, N., Vaccaro, L., Ruggeri, A., Posteraro, B., Burioni, R., Clementi, M., Sanguinetti, M.: Comparative evaluation of the Bruker Biotyper and Vitek MS matrix-assisted laser desorption ionization-time of flight (MALDI-TOF) mass spectrometry systems for identification of yeasts of medical importance. J. Clin. Microbiol. **51**(7), 2453–2457 (2013)
19. Chen, Y.S., Liu, Y.H., Teng, S.H., Liao, C.H., Hung, C.C., Sheng, W.H., Teng, L.J., Hsueh, P.R.: Evaluation of the matrix-assisted laser desorption/ionization time-of-flight mass spectrometry BrukerBiotyper for identification of Penicillium marneffei, Paecilomyces species, Fusarium solani, Rhizopus species, and Pseudallescheria boydii. Front. Microbiol. **6**, 679 (2015)
20. Fraser, M., Brown, Z., Houldsworth, M., Borman, A.M., Johnson, E.M.: Rapid identification of 6328 isolates of pathogenic yeasts using MALDI-ToF MS and a simplified, rapid extraction procedure that is compatible with the Bruker Biotyper platform and database. Med. Mycol. **54**(1), 80–88 (2016)
21. Lee, H.S., Shin, J.H., Choi, M.J., Won, E.J., Kee, S.J., Kim, S.H., Shin, M.G., Suh, S.P.: Comparison of the Bruker Biotyper and VITEK MS matrix-assisted laser desorption/ionization time-of-flight mass spectrometry systems using a formic acid extraction method to identify common and uncommon yeast isolates. Ann. Lab. Med. **37**(3), 223–230 (2017)

22. Wilson, D.A., Young, S., Timm, K., Novak-Weekley, S., Marlowe, E.M., Madisen, N., Lillie, J.L., Ledeboer, N.A., Smith, R., Hyke, J., Griego-Fullbright, C., Jim, P., Granato, P.A., Faron, M.L., Cumpio, J., Buchan, B.W., Procop, G.W.: Multicenter evaluation of the Bruker MALDI Biotyper CA system for the identification of clinically important bacteria and yeasts. Am. J. Clin. Pathol. **147**(6), 623–631 (2017)
23. Krismer, J., Sobek, J., Steinhoff, R.F., Fagerer, S.R., Pabst, M., Zenobi, R.: Screening of Chlamydomonas reinhardtii populations with single-cell resolution by using a high-throughput microscale sample preparation for matrix-assisted laser desorption ionization mass spectrometry. Appl. Environ. Microbiol. **81**(16), 5546–5551 (2015)
24. Krismer, J., Steinhoff, R.F., Zenobi, R.: Single-cell MALDI tandem mass spectrometry: unambiguous assignment of small biomolecules from single Chlamydomonas reinhardtii cells. Chimia **70**(4), 236–239 (2016)
25. Imanishi, S.Y., Nakayama, T., Asukabe, H., Harada, K.I.: Application of MALDI Biotyper to cyanobacterial profiling. Rapid Commun. Mass Spectrom. **31**(4), 325–332 (2017)
26. Hsueh, P.R., Kuo, L.C., Chang, T.C., Lee, T.F., Teng, S.H., Chuang, Y.C., Teng, L.J., Sheng, W.H.: Evaluation of the Bruker Biotyper matrix-assisted laser desorption ionization-time of flight mass spectrometry system for identification of blood isolates of Acinetobacter species. J. Clin. Microbiol. **52**(8), 3095–3100 (2014)
27. Richter, C., Hollstein, S., Woloszyn, J., Kaase, M., Gatermann, S.G., Szabados, F.: Evaluation of species-specific score cut-off values for various Staphylococcus species using a MALDI Biotyper-based identification. J. Med. Microbiol. **61**(10), 1409–1416 (2012)
28. Alvarez-Buylla, A., Culebras, E., Picazo, J.J.: Identification of Acinetobacter species: Is Bruker biotyper MALDI-TOF mass spectrometry a good alternative to molecular techniques? Infect. Genet. Evol. **12**(2), 345–349 (2012)
29. Del Chierico, F., Di Cave, D., Accardi, C., Santoro, M., Masotti, A., D'Alfonso, R., Berrilli, F., Urbani, A., Putignani, L.: Identification and typing of free-living Acanthamoeba spp. by MALDI-TOF MS Biotyper. Exp. Parasitol. **170**, 82–89 (2016)
30. Szabados, F., Woloszyn, J., Richter, C., Kaase, M., Gatermann, S.: Identification of molecularly defined Staphylococcus Aureus strains using matrix-assisted laser desorption/ionization time of flight mass spectrometry and the Biotyper 2.0 database. J. Med. Microbiol. **59**(7), 787–790 (2010)
31. Buchan, B.W., Riebe, K.M., Ledeboer, N.A.: Comparison of the MALDI Biotyper system using Sepsityper specimen processing to routine microbiological methods for identification of bacteria from positive blood culture bottles. J. Clin. Microbiol. **50**(2), 346–352 (2012)
32. Sogawa, K., Watanabe, M., Sato, K., Segawa, S., Ishii, C., Miyabe, A., Murata, S., Saito, T., Nomura, F.: Use of the MALDI BioTyper system with MALDI-TOF mass spectrometry for rapid identification of microorganisms. Anal. Bioanal. Chem. **400**(7), 1905–1911 (2011)
33. Saffert, R.T., Cunningham, S.A., Ihde, S.M., Jobe, K.E., Mandrekar, J., Patel, R.: Comparison of Bruker Biotyper matrix-assisted laser desorption ionization-time of flight mass spectrometer to BD Phoenix automated microbiology system for identification of gram-negative bacilli. J. Clin. Microbiol. **49**(3), 887–892 (2011)
34. Hirsch, P., Pankratz, S.H.: Study of bacterial populations in natural environments by use of submerged electron microscope grids. J. Bas. Microbiol. **10**(8), 589–605 (1970)
35. Moshynets, O.V., Koza, A., Dello Sterpaio, P., Kordium, V.A., Spiers, A.J.: Up-dating the Cholodny method using PET films to sample microbial communities in soil. Biopolym. Cell **27**(3), 199–205 (2011)
36. Budin, G., Chung, H.J., Lee, H., Weissleder, R.: A magnetic Gram stain for bacterial detection. Angew. Chem. Int. Ed. **51**(31), 7752–7755 (2012)
37. Orekhov, F.K., Jablokov, A.G., Skrynnik, A.A.: Hybridization of laser-induced spectrofluorescence analysis (LIFS), matrix-assisted laser desorption/ionization mass spectrometry (MALDI), fluorescence recovery after photobleaching (FRAP) and fluorescence loss in photobleaching (FLIP) microtechnics. J. Biomed. Technol. **2**, 42–52 (2016)

An Alternative Approach to Managing the Nitrogen Content of Cereal Crops

Olga Mitrofanova, Vyacheslav Yakushev, Elena Zakharova, and Vitaly Terleev

Abstract Fertilizers are actively used in agriculture, and nitrogen is a main introduced agrochemical for cereal crops. It's necessary to study crops nitrogen status managing approaches by determining reasonable doses of fertilizer applied. The crop production process dynamic modeling method is difficult to apply in practice. In this connection, an alternative approach to control the crops nitrogen content based on the remote sensing processing is proposed. Cereal crops aerial photographs are used as the initial data of the problem under consideration. In addition, ground-based measurements can be used. Two approaches to solve this problem based on image processing are proposed. The first method is based on the analysis of plant color characteristics. In the case when in addition to the aerial photograph of crops there is a set of ground-based measurements of nitrogen content in plants, it is advisable to carry out spatial interpolation using the kriging method. It should be noted that in spite of the high accuracy of the plant nitrogen content prediction by both methods, the approach using test sites seems to be more effective.

1 Introduction

Providing our country with food has always been an actual problem, and this task is one of the key ones in the conditions of epidemiological situation. In today's world, fertilizers are actively used to grow agricultural products, for cereals, a significant proportion of agrochemicals is made up of nitrogen fertilization. However, despite the rapid development of technology, the rules and methods of using such fertilizers

O. Mitrofanova · V. Yakushev · E. Zakharova · V. Terleev (✉)
Agrophysical Research Institute, Grazhdansky Pr., 14, St. Petersburg 195220, Russian Federation

O. Mitrofanova
St.Petersburg State University, University embankment, 7/9, St. Petersburg 199034, Russian Federation

V. Terleev
Peter the Great St.Petersburg Polytechnic University, Polytechnicheskaya, 29, St. Petersburg 195251, Russian Federation

M. Shamtsyan et al. (eds.), *Robotics, Machinery and Engineering Technology for Precision Agriculture*, Smart Innovation, Systems and Technologies 247,
https://doi.org/10.1007/978-981-16-3844-2_44

are often violated, in most cases, they are applied in unreasonably high doses, which have a number of extremely negative consequences: impact on human health [1, 2], pollution of surface and groundwater [3, 4], as well as soil and air pollution [5, 6], etc. Thus, there is a need to explore approaches to the management of nitrogen crop regimen by identifying reasonable doses of fertilizers [7].

The method of dynamic modeling of the crop production process possesses high precision in solving such problems. The essence of the method is presented in general form [8], Eq. 1:

$$x(k+1) = f(x(k),\ a,\ u(k),\ w(k)), \qquad x(0) = x_0 \tag{1}$$

where x is the model state vector, a is the vector of parameters, u is the vector of controlled external influences (for example, the nitrogen fertilizers application), w is the vector of uncontrolled disturbances (for example, weather conditions), k is the model time step. In this case, dynamic modeling in crop production includes a description of a number of processes in the system «soil – vegetation – surface air layer»: radiation and turbulent regimes, photosynthesis, plants transpiration and soil water evaporation, etc.

However, such a method is difficult to apply in practice, as its implementation requires obtaining numerous initial parameters. In this connection, an alternative approach to control the crops nitrogen content based on the remote sensing processing is proposed. In recent years, imageprocessing methods and informatics have appeared to be perspective, inexpensive, and approachable techniques for controlling various processes in crop production [9–14]. Analysis of plants color characteristics based on remote sensing data has been the subject of many modern studies [15, 16].

So, for example, in [17], the possibilities of using various parameters to control winter wheat nitrogen content were investigated, namely, several vegetation indices obtained using RapidEye satellite images (NDVI—normalized difference vegetation index, NDRE—normalized difference red edge index), terrain curvature indicators, slope, and others. Prediction models based on a combination of the above parameters obtained using remote sensing data and proximal sensors have shown high efficiency in assessing the nitrogen content of plants. In the work [18], the authors created their own sensor system AS-Scanner as a payload for an unmanned aerial vehicle (UAV), which allows them to quickly and efficiently obtain an NDVI distribution map. The development was tested on the example of assessing the nitrogen status of crops in rice fields. The results were compared with mapping using MicaSense RedEdge camera, and a sufficiently high accuracy of constructions with greater efficiency was obtained. And in the work [19], the authors compared the possibilities of using the NDVI index for assessing the crops nitrogen status obtained with a multispectral camera, with the possibilities of using a cheaper DGCI (Dark Green Color Index) index obtained with RGB camera. As a result, a fairly high correlation of the DGCI index values with the nitrogen content and NDVI was obtained.

An increasingly promising and relevant direction in solving such problems is the use of satellite and hyperspectral images. So, for example, in the work [20], Sentinel-2 satellite images were used for a comparative analysis of eight spectral vegetation indices for the assessing the nitrogen content in corn crops. The research was carried out on the example of three farms located in different climatic zones. In the work [21], the Airborne Visible Infrared Imaging Spectrometer (AVIRIS) sensor was used to obtain hyperspectral images of wheat fields. On their basis, the nitrogen status of crops was assessed using the NDVI and VIG (Vegetation Index Green) indices. In the work [22], the authors proposed several indices that allow monitoring nutrients in rice fields (for example, NDVI hyper and NDVI broadbands). In the article [23], the authors presented an experiment conducted in corn fields. The test sites were used to create conditions for water and nitrogen stress, and then spectral bands were determined to detect stress. The study was carried out using a ground-based hyperspectral camera. In [24], a hyperspectral camera was used as a UAV payload to quantify the nitrogen status of winter wheat. The results obtained were compared with the values of the nitrogen nutrition index (NNI) calculated with ground-based measurements.

In the work [25], in addition to the multispectral camera on the UAV, Planet Lab's CubeSat images were also used. At the same time, the correlation between the results obtained by the two methods turned out to be very high (above 0.9).

It should be noted that most of the considered methods involve working with a big data. In this connection, it is also advisable to take into account the possibilities of using machine learning methods in the problems of assessing plant nitrogen stress [26].

The article [27] presents an approach to managing the wheat nitrogen regime based on an MK-SVR (multiple-kernel support vector regression) algorithm. At the same time, the authors took vegetation indices obtained with satellite images as the initial data. The proposed method also demonstrated a fairly high accuracy in determining the nitrogen content in plants.

In addition, geostatistical methods are increasingly used to solve such problems, which allow analyzing the spatial structure of an agricultural field and mapping various indicators, including nitrogen content in plants [28, 29]. So, for example, in the work [30], the authors mapped various agrochemical indicators based on ground measurements in order to determine the management zones in the agricultural fields of Ethiopia. In the work [31], to determine Nitrogen use efficiency (NUE) indicators, the authors used the methods of space–time statistics.

2 Materials and Methods

Precision agriculture (PA) allows the rational use of agrochemicals with differential application. PA is a new plant management strategy that enables technology to be introduced to growers primarily with regard to the use of chemicals. Compared to traditional technology, performance depends on the field variability.

Aerial photos of grain crops are used as the initial data for this task. In addition, ground-based measurements can be used (laboratory studies of the nitrogen content in crops or the readings of a manual N-tester). It is necessary to allocate homogeneous technological zones on the agricultural field and determine the most reasonable dose of nitrogen fertilizers for each elementary site. To solve this problem, we propose two approaches based on digital image processing: classification with training, and spatial interpolation using geostatistics methods.

The initial data were obtained on the experimental fields of the Agrophysical Research Institute (ARI), located in the Leningrad region in the Menkovo village. The Leningrad Region territory is characterized by a zone of excessive moisture, which in turn creates conditions for the soils emergence with varying degrees of hydromorphism. Agroecological monitoring of the state of agricultural fields by means of remote sensing and contact research, including an automated complex for georeferenced sampling, as well as a wide range of field experiments and experiments are carried out annually at the experimental fields.

To carry out experiments related to the approbation and implementation of resource-saving PA technologies developed at the ARI, modern agricultural machinery is used. For a wide range of experiments carried out at the biopolygon, unmanned aerial vehicles are used as a means of obtaining remote sensing data.

2.1 Classification with a Training Data

The method is based on the analysis of plant color characteristics. To create training standards on an agricultural field, so-called test sites are laid (Fig. 1)—small sections

Fig. 1 Wheat field with test sites (numbers indicate nitrogen doses in kg of active substance per 1 ha)

of the field where the nitrogen regime of crops is known, each section corresponds to a certain dose of fertilizer (0, 30, 50, 70, 90, 110 kg of active substance per 1 ha). A comparative analysis of the characteristics of the color of plants in each elementary section of the field with the corresponding parameters of the standards allows you to determine the necessary doses of agrochemicals for application.

For the most accurate determination of the nitrogen needs of plants in each section of the field, it is necessary to build a calibration curve based on the characteristics of the standards, reflecting the dependence of the fertilizer content on the color parameters.

2.2 Geostatistical Approach

In the case when in addition to the aerial photograph of crops there is a set of ground-based measurements $Z(x_i)$ of nitrogen content in plants, it is advisable to carry out spatial interpolation using the kriging method. The basic kriging formula is formed as a linear combination of source data [32], Eq. 2:

$$\hat{Z}(x_0) = \sum_{i=1}^{n} \lambda_i Z(x_i) \tag{2}$$

where $Z(x_i)$ is the observed value at the location x_i; λ_i is the unknown weight; $\hat{Z}(x_0)$ is the predicted value at the location x_0; n is the number of observations.

It is proposed to use the most studied ordinary kriging, in which it is assumed that the average value of the constant is unknown, which imposes an additional requirement on the weights λ_i, Eq. 3:

$$\sum_{i=1}^{n} \lambda_i = 1 \tag{3}$$

To use the kriging method, it is first necessary to conduct a variogram analysis based on an experimental variogram, Eq. 4:

$$\hat{\gamma}(h) = \frac{1}{2N(h)} \sum_{i=1}^{N(h)} [Z(x_i) - Z(x_i + h)]^2 \tag{4}$$

where $N(h)$ is the number of experimental points pairs separated by a distance h. The analysis of the resulting experimental variogram builds its theoretical model (spherical, exponential, and other models can be used).

As a result of the analysis of the initial data using one of these approaches, homogeneous zones of the agricultural field are determined, and for each elementary plot certain doses of fertilizer are prescribed. Based on the results obtained, a technological map of the field is constructed, which is downloaded to the on-board computer of agricultural machinery capable of scattering agrochemicals differentially.

3 Results and Discussions

Aerial photography of grain crops was carried out using the unmanned automated complex Geoscan-401, which allows us to get geo-referenced images in the visible and near-infrared spectra. Table 1 shows the main technical characteristics and equipment of the unmanned system.

The general scheme of actions for obtaining aerial photographs is as follows:

1. Choosing a territory for aerial photography (checking the necessary shooting conditions no obstacles, the object area, etc.).
2. Flight planning (except for the landing point).

Table 1 Technical characteristics and equipment of the unmanned automated complex Geoscan-401

Specifications	Value
Flight time	up to 1 h
Action radius	15 km
Min/Max speed	vertical speed up to 5 m/s, horizontal speed 0–60 km/h
Cruise speed	30 km/h
Flight altitude	up to 500 m
Operational altitude	30–200 m
Survey area per one flight (at a scale of 2 cm GSD)	up to 50 ha
Weight (including payload)	6,5 kg
Payload weight	up to 1,5 kg
Motors	electric
Preparation time	5 min
Start	platform 5 × 5 m
Landing	platform 5 × 5 m
Operation temperatures	−20… + 40 °C
Wind resistance	10 m/s
Basic configuration	
Sony Nex5 camera	
GPS navigation receiver	
Geoscan autopilot	
Additional	
Micasense RedEdge camera	

3. Loading of aerial photography equipment (battery charging, setting up cameras, etc.).
4. Departure of flight team to the place of aerial photography.
5. Loading of equipment for flight; landing and take-off points selection.
6. UAV starting.
7. Flight support by the ground control station operator.
8. UAV landing, saving aerophotos.
9. Equipment dismantling, flight team leaving.
10. Results processing.

Data pre-processing is carried out automatically in the programs GeoScan Planner and PhotoScan. The following are the results of applying the two proposed approaches to controlling the nitrogen regime of crops.

3.1 Construction of Calibration Curves

The average values of the *L*, *a*, *b* parameters were obtained for each test site as the initial characteristics of plant color (*L* is always positive and indicates lightness; $a > 0$ indicates red component; $a < 0$—green; $b > 0$—yellow; $b < 0$—blue). Therefore, it is necessary to turn the vector of color parameters into one value to construct a calibration curve that reflects the dependence of the test sites nitrogen dose on the plants color. Then, the so-called generalized color characteristic C_{Lab} is calculated by the formula [33], Eq. 5:

$$C_{Lab} = \alpha \cdot L + \beta_1 \cdot a + \beta_2 \cdot b \tag{5}$$

where $\alpha \geq 0$, $\beta_1 \geq 0$ and $\beta_2 \geq 0$ are empirically selected coefficients for each digital image, and, Eq. 6:

$$\alpha + \beta_1 + \beta_2 = 1 \tag{6}$$

Moreover, the main criterion for coefficients selecting is to ensure the most pronounced linear relationship between the nitrogen dose and the generalized color characteristic.

The general algorithm for constructing calibration curves is presented in Fig. 2. The algorithm is automated using the C# programming language and implemented to carry out experimental research work at Agrophysical Research Institute.

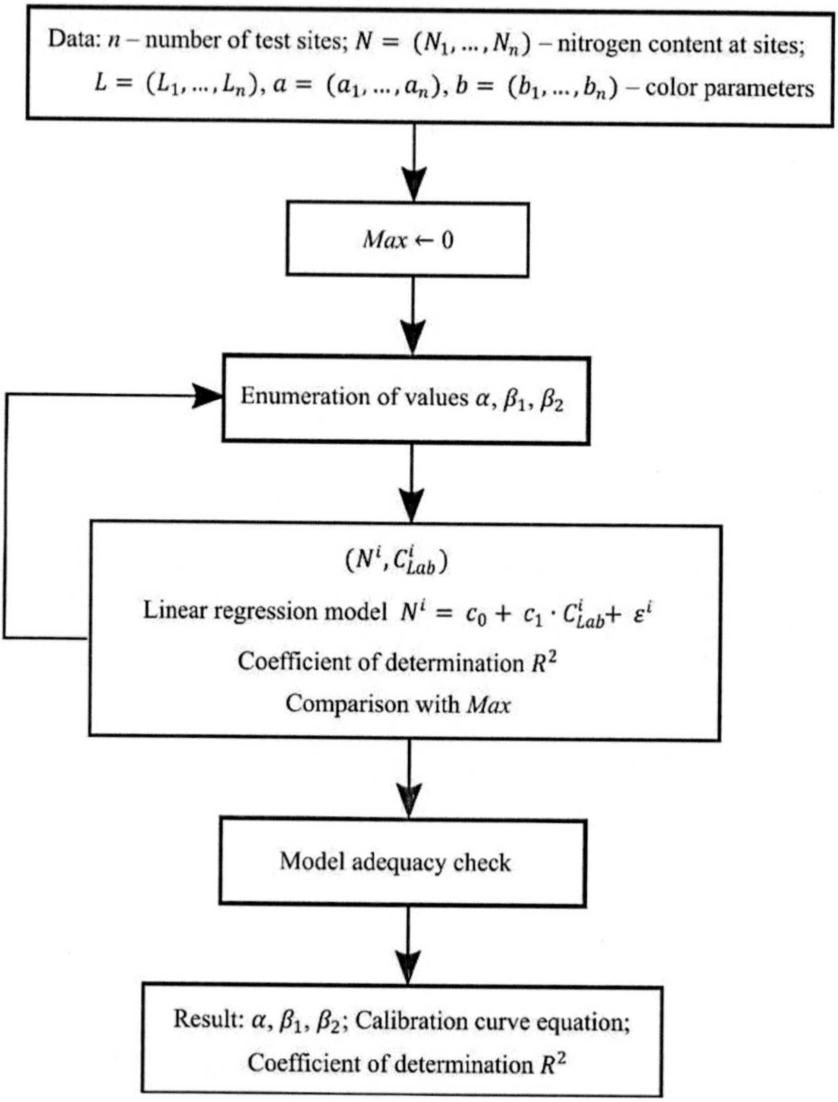

Fig. 2 General algorithm for constructing calibration curves

3.2 *Geostatistical Approach*

Let us consider the selected approach to identifying homogeneous zones for controlling nitrogen content in plants using the example based on a set of georeferenced ground-based measurements of soil electrical conductivity obtained in an experimental agricultural field using a mobile complex.

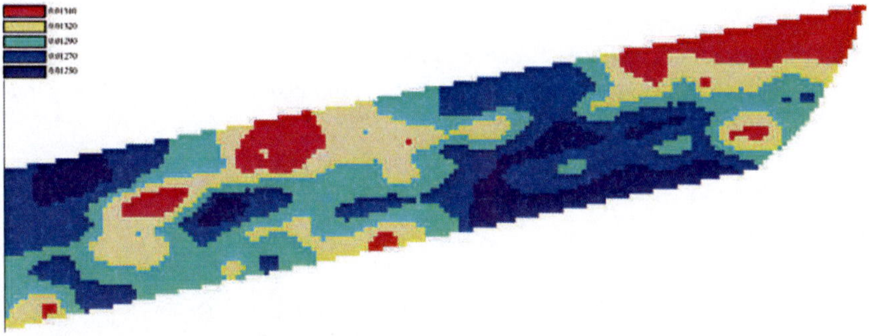

Fig. 3 The constructed map of the parameter distribution

Using the ordinary kriging method, the spatial distribution of soil conductivity parameters was predicted for a dataset of 702 points. All constructions were carried out in the SAGA GIS program. In the course of calculations, an experimental variogram was built with an approximating theoretical model, the spherical model turned out to be the closest. Then, the number of required clusters for the indicator gradation (5 zones) was entered and a map of the parameter distribution was built (Fig. 3). In the same way, the proposed approach for ground-based measurements of nitrogen content in plants is carried out for further decision-making on the volumes of applied doses for each identified zone of homogeneity.

4 Conclusion

As an alternative to the dynamic modeling of approaches to the management of the nitrogen regime of agricultural crops, methods for processing remote sensing data are promising and available. To use the classification method with training standards, it seems appropriate to construct a calibration curve for each aerial photograph, reflecting the dependence of the nitrogen dose on the generalized color characteristic. This approach allows you to determine the required dose of fertilizers for each elementary section of the agricultural field.

In addition, conventional kriging can also accurately estimate the required nitrogen dose at each location in the field based on a set of ground measurements without using reference locations.

It should be noted that, despite the high accuracy of predicting the nitrogen content in plants by both methods, the approach using test plots seems to be more effective. The issue of choosing the optimal number and distribution of ground measurement points for geostatistics is a serious problem. For each specific field, the problem with the selection of points is solved individually and in most cases, it is necessary to take samples at more than 100 points, which is more labor-intensive and more expensive than using test sites. A promising direction in the development of the method proposed here is the regulation of the moisture supply of agricultural crops on irrigated lands. The successful development of this method will be ensured by use of a physically substantiated model of soil hydrophysical properties, taking into account hysteresis [34–40].

References

1. Aktar, W., Sengupta, D., Chowdhury, A.: Interdiscip Toxicol. **2**, 1–12 (2009)
2. Brouwer, A., Longnecker, M.P., Birnbaum, L.S., Cogliano, J., Kostyniak, P., Moore, J., Schantz, S., Winneke, G.: Environ. Health Perspect. **107**, 639 (1999)
3. Kole, R.K., Banerjee, H., Bhattacharyya, A.: Bull Envirron. Contam. Toxicol. **67**, 554–559 (2001)
4. US EPA, Source water protection practices bulletin: Managing small-scale application of pesticides to prevent contamination of drinking water. Office of Water, Washington, DC (2001)
5. Bhardwaj, T., Sharma, J.P.: Int. J. Agric. Food Sci. Technol. **4**, 817–822 (2013)
6. Robertson, G.P., Vitousek, P.M.: Ann. Rev. Environ. Res. **34**, 97–125 (2009)
7. Bachmaier, M., Gandorfer, M.: Precis. Agric. **10**, 95–110 (2009)
8. Poluektov, R.A., Smolyar, E.I., Terleev, V.V., Topazh, A.G.: Models of the production process of agricultural crops. St. Petersburg University, SPb (2006)
9. Betzek, N.M., de Souza, E.G., Bazzi, C.L., Schenatto, K., Gavioli, A., Magalhaes, P.S.G.: Comput. Electron. Agric. **157**, 49–62 (2019)
10. Yakushev, V.P., Yakushev, V.V.: Her. Russ. Acad. Sci. **88**, 330–340 (2018)
11. Yakushev, V.P., Bure, V.M., Mitrofanova, O.A., Mitrofanov, E.P., Petrushin, A.F., Blokhina, SYu., Yakushev, V.V.: Sovremennye problemy distantsionnogo zondirovaniya Zemli iz kosmosa **17**(2), 114–122 (2020)
12. Fu, Y., Yang, G., Pu, R., Li, Z., Li, H., Xu, X., Song, X., Yang, X., Zhao, C.: Eur. J. Agron. **124**, 126241 (2021)
13. Poluektov, R.A., Oparina, I.V., Terleev, V.V.: Three methods for calculating soil water dynamics. Russ. Meteorol. Hydrol. **11**, 61–67 (2003)
14. Dunaieva, Ie., Mirschel, W., Popovych, V., Pashtetsky, V., Golovastova, E., Vecherkov, V., Melnichuk, A., Terleev, V., Nikonorov, A., Ginevsky, R., Lazarev, V., Topaj, A.: GIS services for agriculture monitoring and forecasting: development concept. Adv. Intell. Syst. Comput. **983**, 236–246 (2019)
15. Emadi, M., Baghernejad, M., Pakparvar, M., Kowsar, S.A.: Environ. Monit. Assess. **164**, 501–511 (2010)
16. Yakushev, V.P., Kanash, E.V., Rusakov, D.V., Blokhina, SYu.: Adv. Anim. Biosci. **8**, 229–232 (2017)
17. Wang, K., Huggins, D.R., Tao, H.: Int. J. Appl. Earth Obs. Geoinformation **82**, 101921 (2019)
18. Jiang, R., Sanchez-Azofeifa, A., Laakso, K., Wang, P., Xu, Y., Zhou, Z., Luo, X., Lan, Y., Zhao, G., Chen, X.: J. Cleaner Prod. **289**, 125705 (2021)
19. Caturegli, L., Gaetani, M., Volterrani, M., Magni, S., Minelli, A., Baldi, A., Brandani, G., Mancini, M., Lenzi, A., Orlandini, S., Lulli, F., de Bertoldi, C., Dubbini, M., Grossi, N.: Int. J. Remote Sens. **41**(2), 455–470 (2020)
20. Sharifi, A.: IEEE J. Sel. Top. Appl. Earth Observations Remote Sens. **13**, 2656–2662 (2020)
21. Nandibewoor, A., Kini, P., Konnur, A., Hegadi, R.: Int. J. Syst. Softw. Eng. **6**(1), 10–15 (2018)
22. Mahajan, G.R., Pandey, R.N., Sahoo, R.N., Gupta, V.K., Datta, S.C., Kumar, D.: Precision Agric. **18**, 736–761 (2017)
23. Balaji Naik, B., Naveen, H.R., Sreenivas, G., Karun Choudary, K., Devkumar, D., Adinarayana, J.: J. Indian Soc. Remote Sens. **48**(12), 1787–1795 (2020)
24. Liu, H., Zhu, H., Li, Z., Yang, G.: Int. J. Remote Sens. **41**(3), 858–881 (2020)

25. Cai, Y., Guan, K., Nafziger, E., Chowdhary, G., Peng, B., Jin, Z., Wang, S., Wang, S.: IEEE J. Sel. Top. Appl. Earth Observations Remote Sens. **12**(12), 5153–5166 (2019)
26. Chlingaryan, A., Sukkarieh, S., Whelan, B.: Comput. Electron. Agric. **151**, 61–69 (2018)
27. Wang, L., Zhou, X., Zhu, X., Guo, W.: Comput. Electron. Agric. **140**, 327–337 (2017)
28. Yakushev, V., Petrushin, A., Mitrofanova, O., Mitrofanov, E., Terleev, V., Nikonorov, A.: E3S Web of Conferences 164, 06030 (2020)
29. Yakushev, V.P., Bure, V.M., Mitrofanova, O.A., Mitrofanov, E.P.: Vestnik of Saint Petersburg University. Applied Mathematics. Computer Science. Control Processes **16**(1), 31–40 (2020)
30. Iticha, B., Takele, C.: Geoderma **351**, 85–91 (2019)
31. Liu, Y., Heuvelink, G.B.M., Bai, Z., He, P., Xu, X., Ma, J., Masiliunas, D.: Eur. J. Agron. **115**, 126032 (2020)
32. Dem'yanov, V., Savel'eva, E.: Geostatistics. Theory and practice (M.: Nauka) (2010)
33. Bure, V.M., Mitrofanova, O.A.: Constructive nonsmooth analysis and related topics (dedicated to the memory of V. F. Demyanov), 7973945 (2017)
34. Terleev, V., Nikonorov, A., Togo, I., Volkova, Yu., Garmanov, V., Shishov, D., Pavlova, V., Semenova, N., Mirschel, W.: Modelling the Hysteretic Water Retention Capacity of Soil for Reclamation Research as a Part of Underground Development. Procedia Eng. **165**, 1776–1783 (2016)
35. Terleev, V., Nikonorov, A., Togo, I., Volkova, Yu., Ginevsky, R., Lazarev, V., Khamzin, E., Garmanov, V., Mirschel, W., Akimov, L.: Hysteretic water-retention capacity of sandy soil. Mag. Civil Eng. **2**, 84–92 (2017)
36. Terleev, V., Ginevsky, R., Lazarev, V., Nikonorov, A., Togo, I., Topaj, A., Moiseev, K., Abakumov, E., Melnichuk, A., Dunaieva, Ie.: Predicting the scanning branches of hysteretic soil water-retention capacity with use of the method of mathematical modeling. IOP Conference Series: Earth and Environmental Science, vol. 90, pp. 012105 (2017)
37. Terleev, V., Nikonorov, A., Ginevsky, R., Lazarev, V., Togo, I., Topaj, A., Moiseev, K., Pavlova, V., Layshev, K., Arkhipov, M., Melnichuk, A., Dunaieva, Ie., Mirschel, W.: Hysteresis of the soil water-retention capacity: estimating the scanning branches. Mag. Civil Eng. **1**, 141–148 (2018)
38. Terleev, V., Mirschel, W., Nikonorov, A., Ginevsky, R., Lazarev, V., Topaj, A., Moiseev, K., Layshev, K., Arkhipov, M., Melnichuk, A., Dunaieva, Ie., Popovych, V.: Five models of hysteretic water-retention capacity and their comparison for sandy soil. MATEC Web of Conferences **193**, 02036 (2018)
39. Terleev, V., Mirschel, W., Nikonorov, A., Lazarev, V., Ginevsky, R., Topaj, A., Moiseev, K., Pashtetsky, V., Dunaieva, Ie., Popovych, V., Melnichuk, A., Arkhipov, M.: Models of hysteresis water retention capacity and their comparative analysis on the example of sandy soil. Adv. Intell. Syst. Comput. **983**, 462–471 (2019)
40. Terleev, V., Mirschel, W., Nikonorov, A., Ginevsky, R., Lazarev, V., Pavlova, V., Topaj, A., Pashtetsky, V., Dunaieva, Ie., Popovych, V., Melnichuk, A., Layshev, K.: Improved hydrophysical functions of the soil and their comparison with analogues by the Williams-Kloot test. Adv. Intell. Syst. Comput. **983**, 449–461 (2019).

Functional Representation of the Soil Hydrophysical Properties Using the Example of Loam

Vitaly Terleev, Roman Ginevsky, Viktor Lazarev, Alexander Chusov, Ielizaveta Dunaieva, Kirill Moiseev, Wilfried Mirschel, Luka Akimov, and Aleksandr Nikonorov

Abstract The study examines the physical foundations of modeling the water-retention capacity and hydraulic conductivity of the soil. A functional description of the hydrophysical properties of the soil is given using a pair of formulas that use the same physically and statistically interpreted parameters. The proposed functional description was verified on independent data from an authoritative literary source—the Mualem catalog. In addition, the proposed description was compared with the world's most famous analogue—the Van Genuchten functions. For this, the parameters of the hydrophysical functions were identified by the method of point approximation of the measured data on the water-retention capacity of the soil, and then, using these parameters, it was carried out the predictive calculating the values of the function of the relative hydraulic conductivity of the soil. The advantages of the hydrophysical functions of the soil proposed by the authors are shown. The

V. Terleev · R. Ginevsky · Viktor Lazarev · A. Chusov · L. Akimov
Peter the Great St.Petersburg Polytechnic University, Polytechnicheskaya, 29, St. Petersburg 195251, Russian Federation

V. Terleev
St.Petersburg State Agrarian University, Peterburgskoe shosse, 2, St. Petersburg-Pushkin 196601, Russian Federation

I. Dunaieva
Research Institute of Agriculture of Crimea, Kievskaya street, 150, Simferopol 295543, Russian Federation

K. Moiseev
Agrophysical Research Institute, Grazhdansky Pr., 14, St. Petersburg 195220, Russian Federation

W. Mirschel
Leibniz-Centre for Agricultural Landscape Research, Eberswalder Straße 84, 15374 Müncheberg, Germany

L. Akimov
Politecnico Di Milano, 32 Piazza Leonardo da Vinci, 20133 Milano, Italy

A. Nikonorov (✉)
Gazprom Pererabotka Blagoveshchensk, P. Gazoprovod, P. Sosenskoe, Moscow 108814, Russian Federation

M. Shamtsyan et al. (eds.), *Robotics, Machinery and Engineering Technology for Precision Agriculture*, Smart Innovation, Systems and Technologies 247,
https://doi.org/10.1007/978-981-16-3844-2_45

proposed description of the soil hydrophysical properties has the prospect of being applied in calculating the soil moisture dynamics for predicting the crop yield in precision farming.

1 Introduction

The hydrophysical properties of the soil include its water-retention capacity and hydraulic conductivity [1, 2]. For a quantitative description of these properties, the following indicators are used: for water-retention capacity, the relationship between the volumetric water content of soil θ [cm3·cm-3] and capillary pressure (capillary-sorption potential) of soil moisture ψ [cm H_2O]; for hydraulic conductivity (or the ability of soil as a porous body to conduct water)—the dependence of the moisture conductivity coefficient k [cm·day-1] from capillary pressure ψ or from volumetric water contentθ. To solve a number of problems of land reclamation and predicted crop yields, a functional representation of these dependencies is required. These tasks include the problem of calculating the dynamics of soil moisture. The accuracy of solving this problem is determined by the physical adequacy of the applied functional dependencies $\theta(\psi)$, $k(\psi)$ or $k(\theta)$, as well as by the accuracy of direct measurements, the results of which are used for parametric identification of the corresponding hydrophysical functions. As known, direct measurements of dependencies $\theta(\psi)$, $k(\psi)$ or $k(\theta)$ are very laborious. For these reasons, research in the direction of creating mathematical models of water-retention capacity and hydraulic conductivity of soil, which can reduce the amount of direct measurements of dependencies $\theta(\psi)$, $k(\psi)$ or $k(\theta)$ and accurately enough calculate the dynamics of soil moisture, is very relevant. The aim of the research carried out by the authors is to obtain a physically grounded representation of the hydrophysical properties of the soil in the form of functional dependencies.

Before proceeding to the mathematical modeling of the hydrophysical properties of the soil, it seems appropriate to dwell on the determination of the ψ value. This value characterizes the energy of the interaction of water with the solid phase of the soil that occurs due to the forces of a capillary-sorption nature. The quantitative indicator of this value is the specific (potential) energy of soil moisture. The concept "specific" is used if the value of energy refers to a certain amount of a substance, for example: to a unit of mass, weight or volume of water. If the amount of water to which its energy belongs is a unit of volume, then the specific energy is called pressure. In soil hydrophysics, pressure is usually measured in cm of water column [cm H_2O] by analogy with meteorology, which uses mm Hg for this purpose. The pressure of a deep vacuum (for example, pressure in open space) is taken as a starting point for the pressure calculation in classical physics. This pressure is called absolute pressure. All values of the absolute pressure of any medium with a presence of substance are positive. In soil hydrophysics, a starting point for the moisture pressure calculation in the soil (as a conditional zero) is taken to be the pressure under the flat surface of the water, which is equal to the normal atmospheric pressure. In this case, the deviation of the absolute pressure of soil moisture from normal atmospheric pressure can be either negative or positive: It quantitatively characterizes the deficit or excess of absolute pressure relative to the accepted reference point. Soil pores are predominantly capillaries, and the proportion of non-capillary pores in the total porosity is usually very small. If water wets the surface of the solid phase of the soil (water molecules are attracted to the surface of soil particles), then the absolute pressure of water decreases. In this case, the deviation of the absolute water pressure from the conditional reference point takes negative values.

Otherwise, the deviation will be positive. The deviation of the absolute pressure of water in the capillaries from the normal atmospheric pressure is called the additional or capillary water pressure. The Young–Laplace formula is used to quantify this pressure. In accordance with this formula, the capillary pressure of soil moisture is proportional to the average curvature of the surface of the interface between the liquid and gas phases of the soil (soil water and soil air). In this case, the coefficient of proportionality is equal to the doubled coefficient of the surface tension of soil moisture, and the average curvature is calculated as the half-sum of the reciprocal values of the two radii of curvature of the surface of the boundary between the liquid and gas phases of the soil. If the center of curvature of the interface and the observation point are on one side of this surface, then the corresponding radius of curvature is positive; otherwise, the radius of curvature is negative. Therefore, under the convex surfaces of the water menisci (with respect to the liquid phase of the soil), the capillary pressure of soil moisture takes positive values, and under the concave surfaces—negative values.

It should be noted that in the scientific literature of the beginning of the second half of the twentieth century, the value of the capillary pressure of water is defined as the difference between the pressure of the "nonwettable" phase (soil air) and the pressure of the "wetted" phase (soil moisture). This value turns out to be equal in absolute value to the capillary pressure in the classical formulation, but has the opposite sign. There is a certain contradiction here. At the same time, in hydrophysical calculations, relative values (having no physical dimension) are often used: In these cases, the indicated contradiction does not affect the final result of calculations. However, it should be noted that the determination of the capillary pressure of phase 1 (for example, soil moisture) bordering on phase 2 (for example, with soil air), as the difference in which the absolute pressure of phase 1 is the decreasing value and the subtracted value is the absolute pressure of phase 2, is methodologically more preferable. Such a definition of the capillary pressure of soil moisture ψ is consistent with the aforementioned Young–Laplace law, as well as with the concepts of classical physics in the vast majority of publications and, in fact, is a scientific paradigm. Note that there is no such contradiction in the modern scientific literature. However, it should be borne in mind that this contradiction took place in earlier publications.

2 Materials and Methods

The maximum volume of water per unit volume of soil is called the volumetric saturation moisture θ_s [$cm^3 \cdot cm^{-3}$]. The minimum volume of water per unit volume of soil, at which water has the properties of a liquid, is called residual volumetric moisture θ_r [$cm^3 \cdot cm^{-3}$]. To assess the saturation of the soil with water having the properties of a liquid, an indicator called the effective moisture saturation of the soil is used: $S_e = (\theta - \theta_r)/(\theta_s - \theta_r)$. If this indicator is zero, then hygroscopic moisture is adsorbed on the surface of soil particles, and the main part of the pore space is occupied by the gaseous phase (soil air, which includes water vapor). If this indicator is equal to one, then the pore space of the soil is completely occupied by the liquid phase (soil solution). Soil pores can vary significantly in shape and size (volume). Soil saturation with moisture begins with the smallest pores and ends with the largest pores. In unsaturated soil, the pore space occupied by the liquid phase is topologically closed. Therefore, on the one hand, the fraction of the pore space of the soil occupied by the liquid phase, i.e., menisci of soil moisture, is identical to the effective moisture saturation of the soil S_e On the other hand, this fraction can be used as the cumulative probability that moisture is in a liquid, and not in a gaseous state at a randomly selected observation point in the pore space. At the same time, any point on the interface of the space formed by the pores, sequentially (starting from the smallest) filled with the liquid phase, and the space formed by larger pores not filled with the liquid phase is characterized by the same value of the average curvature of the surface of the soil moisture menisci. According to the Young–Laplace law, this value uniquely corresponds to a certain value of capillary pressure ψ. As the pores are sequentially filled (starting with the smallest) with the liquid phase, the average curvature of the surface of the soil moisture menisci decreases, and the capillary pressure ψ increases. Therefore, it will be true that between the capillary pressure ψ and the accumulated probability of the liquid state of soil moisture, which is characterized by this capillary pressure ψ there is an unambiguous relationship. Therefore, we can assume that it is possible to describe this dependence in the form of an integral probability distribution function over the values of some random variable determined by capillary pressure ψ.

Let us move on to the question of the integral function of the probabilities distribution and the description of this random variable. The space of soil pores has a complex structure. This space is formed by cavities and gaps between soil particles and aggregates of very different sizes and shapes. The components of the solid phase of the soil have a highly differentiated granulometric and mineralogical composition. However, theoretically, it can be assumed that soil pores can be classified according to some criterion of similarity in shape. Further, it can be assumed that within each class, a description of the pore size distribution is possible. The distribution can be expected to be unique for each class. It is quite possible that none of these distributions will turn out to be normal. Nevertheless, as a consequence of Lyapunov central limit theorem, it can be argued that: the more similarity classes of pores with respect to their shape (with different distributions) are taken into account, the more the overall pore size distribution should correspond to the normal law. For this reason, it seems logical to suggest using just the normal law of probability distribution in the form of dependence $S_e(\psi)$ over the values of a random variable determined by capillary pressure.

As for the description of this random variable, it is proposed to consider one well-known phenomenon that occurs when water is displaced from the initially saturated soil in the Richards pneumatic press. With gradual increase in air pressure in the press, first, air dissolves in water, the volume of which does not increase significantly, but moisture remains in the soil. However, due to an increase in the competition between dissolved air molecules and water molecules for sorption centers on the surface of soil particles, there is a gradual weakening of the interaction of moisture with the solid phase of the soil, while the average curvature of the surface separating soil moisture and air gradually increases. At a certain value of the average curvature, the surface tension of water can no longer balance the increasing gas pressure, and air bubbles penetrate into the pore space filled with water (and these bubbles, in addition, can slightly enlarge due to the molecules of dissolved air). This phenomenon is called "air entry" ("bubbling") into the initially water-saturated soil. The entry of the first portion of air into the soil is accompanied by the displacement of the first portion of moisture from the soil. The capillary pressure of soil moisture, at which the entry of air into the soil and the displacement of water begins, will hereinafter be called "the pressure of the entry of air" and will be denoted by ψ_e (index "e" stands for "entry"). The discussed fact means that: $S_e = 1$ at $\psi \geq \psi_e$; $S_e < 1$ at $\psi < \psi_e$. Therefore, to formulate the dependence $S_e(\psi)$ within the interval $\psi < \psi_e$ it seems appropriate to use a random variable that contains the difference $(\psi - \psi_e)$. However, this difference within the interval $\psi < \psi_e$ does not change its sign; therefore, it cannot be accepted as a normally distributed random variable. Therefore, the authors propose to use a random variable $-\ln((\psi - \psi_e)/(\psi_0 - \psi_e))$ with zero general mean and variance σ^2. The parameter ψ_0 is the value of the capillary pressure at which the probability distribution density over the values of the proposed random variable reaches its maximum.

In the papers [3, 4], the following description is proposed to formulate the dependence $S_e(\psi)$:

$$S_e = \left[\begin{array}{l} \frac{1}{2}\mathrm{erfc}\left(\frac{n\sqrt{\pi}}{4}\ln(-\alpha(\psi - \psi_e))\right), \ \psi < \psi_e; \\ 1, \quad \psi \geq \psi_e, \end{array} \right. \tag{1}$$

where:

$\mathrm{erfc}(x) = 1 - \frac{2}{\sqrt{\pi}}\int_0^x \exp(-t^2)dt$ is the complementary error function;

$n = 4/\left(\sigma\sqrt{2\pi}\right)$, $\alpha = -1/(\psi_0 - \psi_e)$ [cm H_2O^{-1}], ψ_e [cm H_2O], ψ_0 [см H_2O] $< \psi_e$ are interpreted parameters (taking into account hysteresis: for drying branches $\psi_e \leq 0$; for wetting branches $\psi_e \geq 0$).

As is known, the phenomenon of hysteresis is characteristic of the water-retention capacity. A pair of values for ψ_e, ψ_0, σ parameters allows describing this phenomenon using relation (1) [5–9].

In a particular case, for $\psi_e = 0$ relationship (1) is reduced to the function of soil water-retention capacity proposed by Kosugi [10].

In the study [3], a continuous approximation of relation (1) in the class of elementary functions was proposed using the simplified Winitzki formula [11]:

$$S_e = \left[\begin{array}{l} \left(1 + (-\alpha(\psi - \psi_e))^n\right)^{-1}, \psi < \psi_e; \\ 1, \ \psi \geq \psi_e. \end{array} \right. \quad (2)$$

In a particular case, for $\psi_e = 0$ relationship (2) is reduced to the function of soil water-retention capacity proposed by Haverkamp et al. [12].

The function of soil water-retention capacity proposed by Van Genuchten [13] is among the world's most well-known analogs of ratios (1) and (2):

$$S_e = \left[\begin{array}{l} \left(1 + (-\alpha\psi)^n\right)^{-(1-1/n)}, \ \psi < 0; \\ 1, \psi \geq 0, \end{array} \right. \quad (3)$$

where α [cm H2O-1], n are empirical parameters (n > 1).

Mualem [14] proposed the formula for calculating the ratio of the hydraulic conductivity of the soil to the moisture filtration coefficient k_s[cm·day^{-1}] (relative hydraulic conductivity k/k_s).

Using this formula in the study [3], the function of the relative hydraulic conductivity of the soil was obtained, which we write here in the form of the following relation:

$$\frac{k}{k_s} = \left[\begin{array}{l} \frac{\sqrt{S_e}}{4}\left(\mathrm{erfc}\left(\mathrm{inverfc}(2S_e) + \frac{2}{n\sqrt{\pi}}\right)\right)^2, \theta < \theta_s; \\ 1, \theta = \theta_s \ , \end{array} \right. \quad (4)$$

where $\mathrm{inverfc}(\mathrm{erfc}(x)) = x$

In a particular case, for $\psi_e = 0$ relation (4) is reduced to the function proposed by Kosugi [15].

In the study [3], a continuous approximation of relation (4) in the class of elementary functions was proposed using the simplified Winitzki formula [11]:

$$\frac{k}{k_s} = \left[\begin{array}{l} \sqrt{S_e}\left(1 - \left(1 - S_e^{-1}\right)\exp\left(\frac{8}{n\pi}\right)\right)^{-2}, \ \theta < \theta_s; \\ 1, \theta = \theta_s \ . \end{array} \right. \quad (5)$$

An obvious advantage of functions (1), (2), (4) and (5) is that they have a common set of physically and statistically interpreted parameters.

The method for calculating the relative hydraulic conductivity of soil according to the Mualem formula using function (3) is widely known in the world as the Mualem–Van Genuchten method. The function of the relative hydraulic conductivity of the soil, calculated by this method, has the form:

$$\frac{k}{k_s} = \left[\begin{array}{l} \sqrt{S_e}\left(1 - \left(1 - S_e^{1/(1-1/n)}\right)^{(1-1/n)}\right)^2, \ \theta < \theta_s; \\ 1, \ \theta = \theta_s \ . \end{array} \right. \quad (6)$$

Functions (3) and (6) have a common set of empirical parameters of the position and shape of curves that graphically represent these functions, but these parameters have no physical meaning.

Let us introduce a list of notations. Hydrophysical functions (1, 2 and 3) will be denoted, respectively: WRC-KT, WRC-HT, WRC-VG (WRC stands for water retention capacity). Hydrophysical functions (4, 5 and 6) will be denoted, respectively: RHC-MKT, RHC-MT, RHC-MVG (RHC stands for relative hydraulic conductivity). The listed functions are used in pairs. Therefore, we will group these functions as follows: system 1 (WRC-VG and RHC-MVG); system 2 (WRC-KT and RHC-MKT), system 3 (WRC-HT and RHC-MT). The most famous and widely used system is system 1. The advantage of all three systems of functions is that a pair of functions that form a particular system has the same set of parameters. System 1 has significant drawbacks: first, the lack of physical meaning of the parameters; secondly, the presence of a restriction $(n > 1)$. These drawbacks are a consequence of the computational difficulties encountered by Van Genuchten when calculating the function of the relative hydraulic conductivity of the soil using the Mualem formula using the WRC-HT function (at $\psi_e = 0$). This is why Van Genuchten offers the WRC-VG function, which is the result of the conversion of the WRC-HT function (at $\psi_e = 0$) by raising the denominator of this function to the power of $(1 - 1/n)$ and imposing a constraint $(n > 1)$.

At large values of the parameter n, the Mualem–Van Genuchten method allows one to obtain a fairly low error in the estimation of the relative hydraulic conductivity of the soil using the RHC-MVG function with the parameters identified by the point approximation of the experimental data on the water-retention capacity of the soil using the WRC-VG function. However, there is a problem of small values of the parameter *n*, when the Mualem–Van Genuchten method has a higher error, as evidenced by the dubious result for the alluvial Beit Netofa clay [3]. The use of the hydrophysical functions of systems 2 and system 3 helps to avoid this problem. Here follows the results of a comparative analysis of three systems of soil hydrophysical functions.

3 Results and Discussions

Authors used the experimental data on the water-retention capacity of the soil “3403 Pachapa loam” from the Mualem catalog to identify the parameters of the three compared systems of hydrophysical functions [16]. The parameter values were calculated using the original “Soil-Hydrophysics” software package by means of a point approximation of the experimental dependence $\theta(\psi)$ using the WRC-VG, WRC-KT and WRC-HT functions. Table 1 shows the parameters values for each of the three compared hydrophysical functions systems.

Table 2 demonstrates the point approximation errors of the experimental data on the water-retention capacity of the soil, presented in the form of the dependence $\theta(\psi)$, as well as in predicting of the relative hydraulic conductivity of the soil $k/k_s(\theta)$ for each of the three compared systems of hydrophysical functions. Minimum errors are shown in bold.

From Table 2, it can be seen that the errors of systems 2 and system 3 in predicting the relative hydraulic conductivity are significantly lower than the error of system 1.

In Fig. 1, the experimental data on the water-retention capacity (round dots) are present on the left; the right-hand side of the figures shows the experimental data on the relative hydraulic conductivity of the soil (square dots). Solid curves in these figures show the results of identification by point approximation of the data on water-retention capacity and the results of predicting the relative hydraulic conductivity of

Table 1 Results of parameters’ identification of soil-hydrophysical functions of three systems by approximating experimental data on water-retention capacity of soil using the functions WRC-VG, WRC-KT and WRC-HT

Number of soil according to catalog [16] and soil title	System number	Function parameters				
		θ_s, $cm^3 \cdot cm^{-3}$	θ_r, $cm^3 \cdot cm^{-3}$	ψ_e, cm H_2O	α, cm H_2O^{-1}	n
“3403 Pachapa loam”	1	0.456	0.034	–	0.0063	1.889
	2	0.456	0.043	0	0.0033	1.445
	3	0.456	0.042	0	0.0033	1.494

Table 2 Comparison of the results of the measured water-retention capacity point approximation and the prediction of soils’ relative hydraulic conductivity with experimental data

Number of soil according to catalog [16] and soil title	Square root of squares’ arithmetic mean of the calculation results’ deviations from the experimental data (*RMSE—root mean square error*)					
	System 1		System 2		System 3	
	WRC-VG	RHC-MVG	WRC-KT	RHC-MKT	WRC-HT	RHC-MT
“3403 Pachapa loam”	**0.0156**	0.1463	0.0198	0.1261	0.0193	**0.0960**

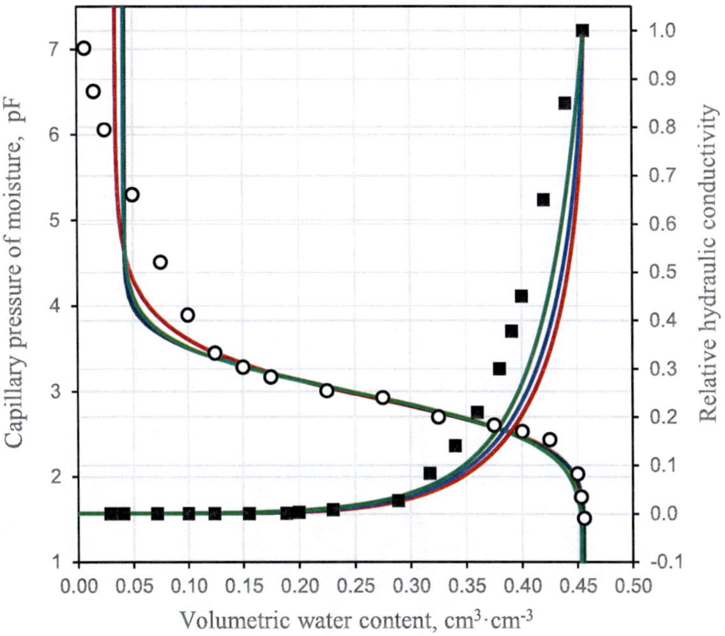

Fig. 1 Comparison of simulation results with experimental data for the soil "3403 Pachapa loam": $pF = \lg(-\psi)$

the studied soil. At the same time, different colors are used for different systems of functions: system 1—red, system 2—blue, system 3—green.

Table 3 shows the values of the correlation coefficients between the simulation results and experimental data. The highest values are highlighted in bold.

The results of reliability evaluation of differences of errors in predicting the relative hydraulic conductivity of the studied soils with the use of Williams-Kloot criterion [17] (with trust probabilities 0.95 and 0.975) are given in Table 4.

From Table 4, it can be seen that the system 1 has reliably the highest forecast error for the relative hydraulic conductivity. At the same time, in the error of the measured water-retention capacity's point approximation, no significant differences were found according to the Williams-Kloot criterion. Ratio system 3, being an approximation of

Table 3 The point approximation results' correlation of the measured water-retention capacity, as well as the results of predicting the relative hydraulic conductivity with experimental data

Number of soil according to catalog [16] and soil title	Correlation coefficient					
	System 1		System 2		System 3	
	WRC-VG	RHC-MVG	WRC-KT	RHC-MKT	WRC-HT	RHC-MT
"3403 Pachapa loam"	**0.9955**	0.9288	0.9927	0.9502	0.9931	**0.9729**

Table 4 Assessment of the reliability evaluation of differences of errors in predicting the relative hydraulic conductivity of the soil for three pairwise compared systems using the Williams-Kloot test

Number of soil according to catalog [16] and soil title	$y - (y_1 + y_2)/2 = \lambda(y_1 - y_2)$, where y is the experimental data								
	y_1 : RHC-MVG, y_2 : RHC-MKT			y_1 : RHC-MVG, y_2 : RHC-MT			y_1 : RHC-MKT, y_2 : RHC-MT		
	λ	$\lambda_{0.95}$	$\lambda_{0.97}$	λ	$\lambda_{0.95}$	$\lambda_{0.97}$	λ	$\lambda_{0.95}$	$\lambda_{0.97}$
"3403 Pachapa loam"	−6.109	0.824	0.999	−2.012	0.376	0.456	−2.912	0.651	0.789
	y_2 more accurate than y_1			y_2 more accurate than y_1			y_2 more accurate than y_1		

system 2 in the class of elementary functions, not only matches the original system in accuracy, but also reliably surpasses it when predicting the relative hydraulic conductivity of the soil "3403 Pachapa loam". In addition, according to Table 2, system 3 has the lowest forecast error $k/k_s(\theta)$ among all the systems presented. The aforementioned allows us to recommend system 3 for the use in modeling the hydrophysical properties of soils as a set of physically grounded, sufficiently accurate, briefly formulated, and convenient for calculating elementary functions.

4 Conclusion

For all compared models, sufficiently high values of the correlation coefficient between the experimental and calculated data were achieved with respect to the water-retention capacity of the studied soil. However, according to the Williams-Kloot test with a confidence level of 0.95, the compared models have certain differences. With regard to the point approximation of the data on the water-retention capacity (ψ), the errors of all models do not differ significantly. With regard to the estimates (prediction) of the data on the relative hydraulic conductivity $k/k_s(\theta)$, the error of the RHC-MVG model is significantly higher than the errors of the RHC-MKT and RHC-MT models; at the same time, the RHC-MT model is more accurate than the RHC-MKT model. The rather low error of the hydrophysical functions forming system 2 and system 3 indicates their physical adequacy. Thus, using the functions of system 2 and system 3, one should expect a significant increase in the accuracy of engineering calculations in the problems of hydraulic engineering and precision irrigation farming [18].

Acknowledgements The reported study was funded by RFBR according to the research projects N. 19-04-00939-a, N. 19-016-00148-a.

References

1. Poluektov, R.A., Oparina, I.V., Terleev, V.V.: Three methods for calculating soil water dynamics. Russ. Meteorol. Hydrol. **11**, 61–67 (2003)
2. Terleev, V., Nikonorov, A., Togo, I., Volkova, Yu., Garmanov, V., Shishov, D., Pavlova, V., Semenova, N., Mirschel, W.: Modelling the hysteretic water retention capacity of soil for reclamation research as a part of underground development. Procedia Eng. **165**, 1776–1783 (2016)
3. Terleev, V.V., Mirschel, W., Badenko, V.L., Guseva, IYu.: An Improved Mualem-Van Genuchten method and its verification using data on Beit Netofa clay. Eurasian Soil Sci. **50**(4), 445–455 (2017)
4. Terleev, V., Mirschel, W., Nikonorov, A., Ginevsky, R., Lazarev, V., Pavlova, V., Topaj, A., Pashtetsky, V., Dunaieva, Ie., Popovych, V., Melnichuk, A., Layshev, K. Improved hydrophysical functions of the soil and their comparison with analogues by the Williams-Kloot test. Adv. Intell. Syst. Comput. **983**, 449–461 (2019)

5. Terleev, V., Nikonorov, A., Togo, I., Volkova, Yu., Ginevsky, R., Lazarev, V., Khamzin, E., Garmanov, V., Mirschel, W., Akimov, L.: Hysteretic water-retention capacity of sandy soil. Magazine Civil Eng. **2**, 84–92 (2017)
6. Terleev, V., Mirschel, W., Nikonorov, A., Lazarev, V., Ginevsky, R., Topaj, A., Moiseev, K., Pashtetsky, V., Dunaieva, Ie., Popovych, V., Melnichuk, A., Arkhipov, M. Models of hysteresis water retention capacity and their comparative analysis on the example of sandy soil. Adv. Intell. Syst. Comput. **983**, 462–471 (2019)
7. Terleev, V., Ginevsky, R., Lazarev, V., Nikonorov, A., Togo, I., Topaj, A., Moiseev, K., Abakumov, E., Melnichuk, A., Dunaieva, Ie. Predicting the scanning branches of hysteretic soil water-retention capacity with use of the method of mathematical modeling. IOP Conference Series: Earth and Environmental Science vol. 90, pp. 012105 (2017)
8. Terleev, V., Nikonorov, A., Ginevsky, R., Lazarev, V., Togo, I., Topaj, A., Moiseev, K., Pavlova, V., Layshev, K., Arkhipov, M., Melnichuk, A., Dunaieva, Ie., Mirschel, W. Hysteresis of the soil water-retention capacity: estimating the scanning branches. Magazine Civil Eng. **1**, 141–148 (2018)
9. Terleev, V., Mirschel, W., Nikonorov, A., Ginevsky, R., Lazarev, V., Topaj, A., Moiseev, K., Layshev, K., Arkhipov, M., Melnichuk, A., Dunaieva, Ie., Popovych, V. Five models of hysteretic water-retention capacity and their comparison for sandy soil. MATEC Web of Conferences 193, 02036 (2018)
10. Kosugi, K.: Three-parameter lognormal distribution model for soil water retention. Water Resour. Res. **30**, 891–901 (1994)
11. Winitzki S. A handy approximation for the error function and its inverse (in https://sites.google.com/site/winitzki/sergei-winitzkis-files/erf approx.pdf?attredirects=0). Last approached 21 Feb 21 (2021)
12. Haverkamp, R., Vauclin, M., Touma, J., Wierenga, P.J., Vachaud, G.: A comparison of numerical simulation model for one-dimensional infiltration. Soil Sci. Soc. Am. J. **41**, 285–294 (1977)
13. Van Genuchten, MTh.: A closed form equation for predicting the hydraulic conductivity of unsaturated soils. Soil Sci. Soc. Am. J. **44**, 892–989 (1980)
14. Mualem, Y.: A new model for predicting hydraulic conductivity of unsaturated porous media. Water Resour. Res. **12**, 513–522 (1976)
15. Kosugi, K.: Lognormal distribution model for unsaturated soil hydraulic properties. Water Resour. Res. **32**, 2697–2703 (1996)
16. Mualem, Y.: A catalogue of the hydraulic properties of unsaturated soils. Research Project 442. Technion, Israel Institute of Technology, Haifa, Israel (1976)
17. Kobzar, A.I.: Prikladnaya matematicheskaya statistika. Dlya inzhenerov i nauchnyx rabotnikov. Izd.Fizmatlit, Moskva, (rus), (2006)
18. Dunaieva, Ie., Mirschel, W., Popovych, V., Pashtetsky, V., Golovastova, E., Vecherkov, V., Melnichuk, A., Terleev, V., Nikonorov, A., Ginevsky, R., Lazarev, V., Topaj, A.: GIS Services for Agriculture Monitoring and Forecasting: Development Concept. Adv. Intell. Syst. Comput. **983**, 236–246 (2019)

Using the Model of Hysteresis to Calculate the Precise Irrigation Rate for Silt Loam

Vitaly Terleev, Roman Ginevsky, Viktor Lazarev, Alexander Chusov, Kirill Moiseev, Ielizaveta Dunaieva, Aleksandr Nikonorov, Wilfried Mirschel, and Luka Akimov

Abstract The water-retention capacity of the soil is the most important hydrophysical property. It is formulated as the dependence between the volumetric water content of the soil and the capillary pressure of soil moisture. This study presents three models of the hysteretic water-retention capacity. The identification of the parameters was carried out by means of a point approximation of a measured data from the Mualem catalog on the main (boundary) drying and wetting branches of the silt loam using the original computer program «Soil-Hysteresis». A comparative analysis of hysteresis models is carried out on the basis of identifying significant differences between the errors of point approximation of measured data using the Williams–Kloot criterion. The advantages of the models improved by introducing an additional additive parameter are revealed. Applying the condition of equality of the exponential parameter values for drying and wetting branches eliminates the

V. Terleev · R. Ginevsky · V. Lazarev · A. Chusov · L. Akimov
Peter the Great St.Petersburg Polytechnic University, Polytechnicheskaya 29, St. Petersburg 195251, Russian Federation

V. Terleev
St.Petersburg State Agrarian University, Peterburgskoe shosse, 2, St. Petersburg-Pushkin 196601, Russian Federation

K. Moiseev
Agrophysical Research Institute, Grazhdansky Pr.14, St. Petersburg 195220, Russian Federation

I. Dunaieva
Research Institute of Agriculture of Crimea, Kievskaya street, 150, Simferopol 295543, Russian Federation

A. Nikonorov (✉)
Gazprom Pererabotka Blagoveshchensk, P. Gazoprovod, P. Sosenskoe, Moscow 108814, Russian Federation

W. Mirschel
Leibniz-Centre for Agricultural Landscape Research, Eberswalder Straße 84, 15374 Müncheberg, Germany

L. Akimov
Politecnico Di Milano, 32 Piazza Leonardo da Vinci, 20133 Milano, Italy

M. Shamtsyan et al. (eds.), *Robotics, Machinery and Engineering Technology for Precision Agriculture*, Smart Innovation, Systems and Technologies 247,
https://doi.org/10.1007/978-981-16-3844-2_46

undesirable (methodical) «pump effect». The use of physically based models of hysteretic water-retention capacity has prospects for development and application of resource-saving technologies in precision irrigation farming.

1 Introduction

Changes in climate conditions lead to a meridional shift of farming zones. At the same time, even in those zones that were previously characterized by excessive moisturizing, it becomes necessary to use irrigation to obtain sustainable yields. Along with climate change, the problem of providing agriculture with a sufficient amount of irrigation water becomes acute. For this reason, there is a growing need for the development of scientifically grounded approaches to the determination of such irrigation rates, at which: unproductive consumption of irrigation moisture is excluded; the loss of fertilizers, ameliorants and chemical plant protection products due to the leaching of agrochemicals outside the root layer of the soil is prevented; the risk of pollution of natural waters by means of agricultural chemicals is minimized. One of the most efficient approaches of calculating such irrigation rates is the method of mathematical modeling. The intellectual core of this method is a mathematical model that adequately describes the hydrophysical properties of the soil [1–3].

The most important hydrophysical property of the soil is its water-retention capacity. It is customary to formulate an indicator that quantitatively characterizes this property in the form of a functional dependence of the volumetric water content of soil θ [$cm^3 \cdot cm^{-3}$] from capillary pressure of soil moisture ψ [cm H_2O] [4]. Within the framework of the concepts of classical thermodynamics, the dependence $\theta(\psi)$ is an isotherm of the equilibrium states of soil moisture. The water-retention capacity of the soil is characterized by the phenomenon of hysteresis. This phenomenon manifests itself, for example, in the fact that the branch of dependence $\theta(\psi)$ formed by the sequence of desorption equilibriums differs from the branch formed by the sequence of sorption equilibriums of moisture in the soil [5–8]. This means that a set of values of θ corresponds to one given value of ψ. It is known that among this set, the maximum value corresponds to the so-called main branch of drying, and the minimum value corresponds to the main branch of wetting of the hysteresis of the water-retention capacity of the soil. Intermediate values correspond to the so-called scanning drying and wetting branches that fill the area between the main branches of the hysteresis loop [9–11].

The aim of this study is to assess the prospects of the proposed hysteresis model for calculating the precise rates in practical irrigation farming. This goal is achieved by solving the following tasks:

(1) Comparative analysis of models of soil hydrophysical properties taking into account hysteresis;

(2) Formulation of an original system of mathematical relationships describing the main and scanning branches of the hysteresis loop of the water-retention capacity of the soil;
(3) Development of a software package for modeling the hydrophysical properties of soil, taking into account hysteresis;
(4) Calculation of precise irrigation rates and assessment of its benefits in relation to the prevention of unproductive consumption of irrigation water.

2 Materials and Methods

In most modern models of the hysteresis of the water-retention capacity of the soil, the calculation of the scanning branches is carried out according to the algorithm proposed by Scott et al. [12]. In this model, various functions are used to formulate the dependence $\theta(\psi)$. For example, the original model of Scott et al. [12] uses the Haverkamp et al. function [13], while the Kool and Parker model [14] uses the Van Genuchten function [15]. The advantages of the function formulating the dependence $\theta(\psi)$ that is used in this study are proved in [16–18]. It uses three physically and statistically interpreted parameters: exponential, multiplicative and additive. The first two parameters take into account the cause of the hysteresis caused by the variability of the cross-sectional area of the soil pore in the direction of its longitudinal axis («beading effect»), and the third (additive) parameter takes into account the presence of dead-end pores trapping the soil air and determining the difference between the «air entrance pressure» and «water entrance pressure». This function is applied in the third hysteresis model $\theta(\psi)$, in which the calculation of scanning branches is also carried out according to the algorithm of Scott et al. [12].

For hysteresis models with enhanced functions of Kosugi and Haverkamp et al., the study uses the designations Hys-SKT and Hys-SHT, respectively. Hysteresis models with original functions of Kosugi [19] and Haverkamp et al. [13], meaning at $\psi_e = 0$, are designated as Hys-SKT$_0$ and Hys-SHT$_0$, respectively. For the Kool and Parker model with the function $\theta(\psi)$ of Van Genuchten, the designation Hys-KPVG is used (see Table 1). Our research compares three hysteresis models based on different functions $\theta(\psi)$. As a basis for the comparison, the error of the point approximation of the data on the main (boundary) drying and wetting branches of the hysteresis of the water-retention capacity of the soil «3305 Ida silt loam (>15 cm)» [20] is used.

Table 1 uses the following notations: $erfc(x) = 1 - \frac{2}{\sqrt{\pi}} \int_0^x \exp(-t^2)dt$—the complementary error function; $S_e = (\theta - \theta_r)/(\theta_s - \theta_r)$—the effective water saturation of the soil, θ_s[cm^3 cm^{-3}]—the saturated volumetric water content;θ_r[cm^3 cm^{-3}]—residual water content of the soil; in model #1 α [cm H_2O^{-1}], n and m—empirical parameters ($m = 1 - 1/n, n > 1$); in models #2 and #3 $n > 0$, $\alpha = -1/(\psi_0 - \psi_e)$ [cm H_2O^{-1}], ψ_e [cm H_2O], ψ_0 [cm H_2O] $<\psi_e$—interpreted parameters (for drying branches $\psi_e = \psi_{e,d} \leq 0$, for wetting branches $\psi_e = \psi_{e,w} \geq 0$);

Table 1 Functions $S_e(\psi)$ used in the hysteresis models

Hysteresis model	Model denotation	Model #	$S_e(\psi)$ description
Kool and Parker	Hys-KPVG	1	$S_e = \begin{cases} \left(1+(-\alpha\psi)^n\right)^{-m}, \psi < 0; \\ 1, \psi \ge 0. \end{cases}$
Scott et al with enhanced $\theta(\psi)$ $\theta(\psi)$ $S_e(\psi)$ by Kosugi	Hys-SKT	2	$S_e = \begin{cases} \frac{1}{2}\mathrm{erfc}\left(\frac{n\sqrt{\pi}}{4}\ln(-\alpha(\psi-\psi_e))\right), \psi < \psi_e; \\ 1, \psi \ge \psi_e. \end{cases}$
Scott et al with enhanced $\theta(\psi)\theta(\psi)$ $S_e(\psi)$ by Haverkamp et al	Hys-SHT	3	$S_e = \begin{cases} \left(1+(-\alpha(\psi-\psi_e))^n\right)^{-1}, \psi < \psi_e; \\ 1, \psi \ge \psi_e. \end{cases}$

under $\psi_e = 0$ models #2 and #3 accordingly reduced to the original model by Kosugi [19], as well as the original model by Haverkamp et al. [13].

Currently, the most commonly used method for calculating irrigation rates is the Kostyakov method [21]. In this method, the rate is calculated from the difference between FC (field capacity) and pre-irrigation soil moisture. As known, FC characterizes the maximum amount of capillary-suspended moisture that the soil can retain. FC can be measured in the field, for example, by the method of flooding areas. In addition, FC can be determined from the main drying branch $\theta(\psi)$ at value $\psi_{FC} = -330$ [cm H_2O]. The θ at $\psi_{FC} = -330$ [cm H_2O] is called «normative» FC. At $\psi > \psi_{FC}$, the soil is unable to retain moisture. Therefore, in this study, it is proposed to apply such an irrigation rate at which, after watering, the value ψ was lower than ψ_{FC}. The value $\psi_{FC} = -330$ [cm H_2O] is the average value for different soils. In each case, the value ψ_{FC} can differ from -330 [cm H_2O]. To calculate the FC of a certain soil, it is proposed to use the empirical dependence of Voronin [22], which intersects $\theta(\psi)$ in the FC point. However, during irrigation, the change in the state of soil moisture is characterized not by the main drying branch, on which the point of «normative» FC is located, but by scanning wetting branches of hysteresis $\theta(\psi)$. For this reason, to calculate the precise irrigation rate, data on the scanning wetting branch is required. The scanning wetting branch starts from the point determined by the pre-irrigation soil moisture and ends at the point of intersection of this scanning branch with the empirical Voronin dependence [22]. It is impossible to predict in advance what scanning wetting branches will be needed in the upcoming growing season. Therefore, the use of the mathematical model in this case is uncontested.

3 Results and Discussions

The above-mentioned three hysteresis models $\theta(\psi)$ are implemented as a software package «Soil-Hysteresis». Using this package, the parameters of the compared models (see Table 2) were identified by point approximation of the experimental data on the main (boundary) drying and wetting branches of the water-retention capacity of silt loam soil «3305 Ida silt loam (>15 cm)» [20].

The root mean square errors (RMSE) of the point approximation of the data on the main (boundary) branches of the hysteresis of the water-retention capacity of silt loam soil are given in Table 3. The results of revealing the differences between the errors of the compared hysteresis models in relation to the point approximation of the experimental data on the main branches of the hysteresis of the soil for the two options of the computational capacity of the experiment ($n_d \neq n_w$ and $n_d = n_w$) are provided in Table 4. To identify reliable differences, the Williams–Kloot criterion was used [23].

The results of two versions of the computational experiment with the Hys-SKT model based on the point approximation of the main hysteresis branches are shown in Figs. 1 and 2. Solid curves show the results of point approximation of the data on the main (boundary) branches of the hysteresis of the water-retention capacity of the studied soil. The dots represent the experimental data.

For $n_d \neq n_w$, the errors of the point approximation are lower due to the greater number of variable parameters. However, this condition is characterized by a «pump effect». To eliminate this effect, it is proposed to use the condition $n_d = n_w$. The

Table 2 Parameters of the models for «3305 Ida silt loam (>15 cm)»

Models	$n_d \neq n_w$									
	θ_r	θ_s	$\psi_{e,d}$	$\psi_{0,d}$	α_d	$\psi_{e,w}$	$\psi_{0,w}$	α_w	n_d	n_w
#1	0.0119	0.5300	–	−73.16	0.0137	–	−23.87	0.0419	1.369	1.283
#2	0.1605	0.5300	0	−202.15	0.0049	19.11	−125.72	0.0069	1.193	1.064
#3	0.1538	0.5300	0	−210.15	0.0048	15.16	−129.53	0.0069	1.214	1.039
#2 $\psi_e = 0$	0.1227	0.5300	–	−261.49	0.0038	–	−146.21	0.0068	0.9759	0.7687
#3 $\psi_e = 0$	0.1295	0.5300	–	−215.24	0.0046	–	−142.53	0.0070	0.961	0.8151
$n_d = n_w$										
#1	0	0.5300	–	−56.78	0.0176	–	−32.23	0.0310	1.306	1.306
#2	0.1657	0.5300	0	−195.53	0.0051	27.65	−125.18	0.0065	1.200	1.200
#3	0.1614	0.5300	0	−200.07	0.0050	26.26	−127.69	0.0065	1.226	1.226
#2 $\psi_e = 0$	0.1244	0.5300	–	−257.41	0.0039	–	−145.48	0.0069	0.868	0.868
#3 $\psi_e = 0$	0.1292	0.5300	–	−249.44	0.0040	–	−141.31	0.0071	0.925	0.925

Table 3 Errors of the point approximation of data on the main (boundary) branches of hysteresis

Hysteresis branches	$n_d = n_w$					$n_d \neq n_w$				
	Models									
	#1	#2	#3	#2 $\psi_e = 0$	#3 $\psi_e = 0$	#1	#2	#3	#2 $\psi_e = 0$	#3 $\psi_e = 0$
Main (boundary): identification by 61 dots	0.0174	0.0065	0.0069	0.0167	0.0155	0.0147	0.0061	0.0063	0.0118	0.0135

Table 4 Reliability of differences in errors of the point approximation of data on the main (boundary) branches of «3305 Ida silt loam (>15 cm)» hysteresis

$y - (y_1 + y_2)/2 = \lambda(y_1 - y_2)$, where y—exmerimental data

$n_d \neq n_w$

y_1 : Hys-KPVG, y_2 : Hys-SKT			y_1 : Hys-KPVG, y_2 : Hys-SHT			y_1 : Hys-SKT, y_2 : Hys-SHT		
λ	$\lambda_{0.95}$	$\lambda_{0.975}$	λ	$\lambda_{0.95}$	$\lambda_{0.975}$	λ	$\lambda_{0.95}$	$\lambda_{0.975}$
−0.480	0.098	0.118	−0.472	0.101	0.121	0.440	0.805	0.964
y_2 *more* accurate than y_1			y_2 more accurate than y_1			y_1 and y_2 are equal		
y_1 : Hys-KPVG, y_2 : Hys-SKT_0			y_1 : Hys-KPVG, y_2 : Hys-SHT_0			y_1 : Hys-SKT_0, y_2 : Hys-SHT_0		
λ	$\lambda_{0.95}$	$\lambda_{0.975}$	λ	$\lambda_{0.95}$	$\lambda_{0.975}$	λ	$\lambda_{0.95}$	$\lambda_{0.975}$
−1.832	0.481	0.576	−0.208	0.312	0.373	0.468	0.373	0.447
y_2 more accurate than y_1			y_1 and y_2 are equal			y_1 more accurate than y_2		

y_1 : Hys-SKT, y_2 : Hys-SKT_0			y_1 : Hys-SHT, y_2 : Hys-SHT_0			y_1 : Hys-SKT, y_2 : Hys-SHT_0			y_1 : Hys-SHT, y_2 : Hys-SKT_0		
λ	$\lambda_{0.95}$	$\lambda_{0.975}$	λ	$\lambda_{0.95}$	$\lambda_{0.975}$	λ	$\lambda_{0.95}$	$\lambda_{0.975}$	λ	$\lambda_{0.95}$	$\lambda_{0.975}$
0.438	0.126	0.151	0.473	0.113	0.136	0.454	0.106	0.127	0.434	0.131	0.157
y_1 more accurate than y_2			y_1 more accurate than y_2			y_1 more accurate than y_2			y_1 more accurate than y_2		

$n_d = n_w$

y_1 : Hys-KPVG, y_2 : Hys-SKT			y_1 : Hys-KPVG, y_2 : Hys-SHT			y_1 : Hys-SKT, y_2 : Hys-SHT		
λ	$\lambda_{0.95}$	$\lambda_{0.975}$	λ	$\lambda_{0.95}$	$\lambda_{0.975}$	λ	$\lambda_{0.95}$	$\lambda_{0.975}$
−0.465	0.086	0.103	−0.451	0.091	0.109	0.930	0.802	0.960
y_2 more accurate than y_1			y_2 more accurate than y_1			y_1 more accurate than y_2 (0.95)		

(continued)

Table 4 (continued)

$y - (y_1 + y_2)/2 = \lambda(y_1 - y_2)$, where y—exmerimental data

y_1 : Hys-KPVG, y_2 : Hys-SKT_0			y_1 : Hys-KPVG, y_2 : Hys-SHT_0			y_1 : Hys-SKT_0, y_2 : Hys-SHT_0		
λ	$\lambda_{0.95}$	$\lambda_{0.975}$	λ	$\lambda_{0.95}$	$\lambda_{0.975}$	λ	$\lambda_{0.95}$	$\lambda_{0.975}$
−0.960	1.410	1.688	−1.529	0.788	0.944	−2.454	1.120	1.341
y_1 and y_2 are equal			y_2 more accurate than y_1			y_2 more accurate than y_1		

y_1 : Hys-SKT, y_2 : Hys-SKT_0			y_1 : Hys-SHT, y_2 : Hys-SHT_0			y_1 : Hys-SKT, y_2 : Hys-SHT_0			y_1 : Hys-SHT, y_2 : Hys-SKT_0		
λ	$\lambda_{0.95}$	λ	$\lambda_{0.95}$	λ	$\lambda_{0.95}$	λ	$\lambda_{0.95}$	$\lambda_{0.975}$	λ	$\lambda_{0.95}$	$\lambda_{0.975}$
0.459	0.088	0.105	0.458	0.104	0.124	0.455	0.096	0.115	0.449	0.093	0.112
y_1 more accurate than y_2			y_1 more accurate than y_2			y_1 more accurate than y_2			y_1 more accurate than y_2		

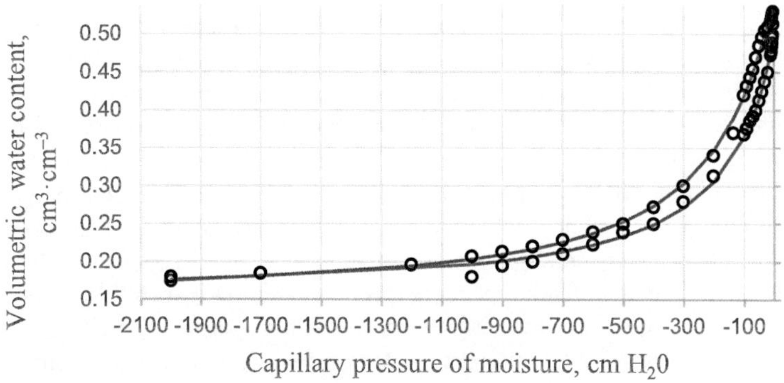

Fig. 1 **a** Point approximation of data on the main (boundary) using Hys-SKT ($n_d \neq n_w$)

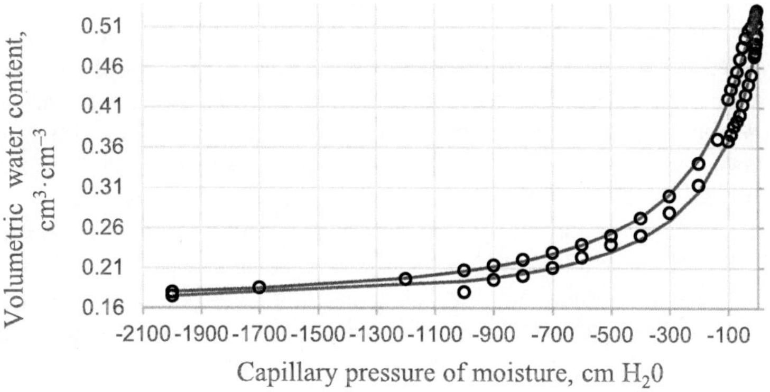

Fig. 2 **b** Point approximation of data on the main (boundary) branches using Hys-SKT ($n_d = n_w$)

use of the Hys-SKT and Hys-SHT models (in both the $n_d \neq n_w$ and $n_d = n_w$ options) for the point approximation of the data on the main (boundary) drying and wetting branches of the water-retention capacity of the soil is more preferable in comparison with the Hys-KPVG model. It is recommended to use the Hys-SHT model for calculating precision irrigation rates as it applies the practical preferred soil water-retention capacity function.

As an example, when calculating irrigation rates for pre-irrigation volumetric water content of soil θ_0, a certain value is set, corresponding to the point on the main drying branch, from which the scanning (primary) wetting branch begins. Such a situation in practical ameliorative agriculture is quite rare, since, due to atmospheric precipitation, the state of soil moisture is described by a sequence of scanning branches of wetting and drying. For this reason, the point on the hysteresis loop, where watering begins, usually falls between the main drying and wetting branches of the hysteresis of the soil water-retention capacity. Accounting the indicated sequence of scanning hysteresis branches is very problematic. Therefore, in this article, a simplified example is considered in which the starting point of irrigation is located directly on the main drying branch of the hysteresis (red solid curve in Fig. 3) at $\theta_0 = 0.189$ [$cm^3{\cdot}cm^{-3}$]. In the following publications, the authors intend to consider a more general case, in which the starting point of irrigation is located between the main branches of the hysteresis loop. It will be possible when the problems of calculating the scanning branches of moisture from the point, the position of which will be determined by the readings of two sensors: a soil moisture sensor and a moisture capillary pressure sensor, will be solved. In the considered simplified example, when constructing the scanning (primary) wetting branch (blue solid curve in Fig. 3) to the point of intersection with Voronin's «secant» (black solid curve in Fig. 3), the value of the volumetric water content $\theta_3 = 0.267$ [$cm^3{\cdot}cm^{-3}$] was obtained. In this case, an excess of gravitational moisture is not formed. The precise irrigation rate is equal to:$\theta_3 - \theta_0 = 0.078$ [$cm^3{\cdot}cm^{-3}$].

As a comparison with the obtained result, the «traditional» method was also applied to determine the irrigation rate, using the «normative» FC, which corresponds to the point of intersection of the main drying branch with the Voronin's «secant»: $\theta_1 = 0.290$ [$cm^3{\cdot}cm^{-3}$]. The irrigation rate (in units of θ), when calculating with the use of «traditional» method, is determined as the difference between the «normative» FC and θ_0; this difference is 0.101 [$cm^3{\cdot}cm^{-3}$]. It can be seen that the precise irrigation rate is lower than the rate calculated by the «traditional» method by the value $\delta = \theta_1 - \theta_3 = 0.023$ [$cm^3{\cdot}cm^{-3}$] (see Fig. 2). For an irrigated plot with an area of 1 ha with a 50-cm layer of soil moistened, the difference in irrigation rates would be: $0.023 \times 0.5 \cdot 10^4 = 115$ [$m^3{\cdot}ha^{-1}$].

When using the «traditional» rate after reaching the volumetric water content of soil $\theta_1 = 0.290$ [$cm^3{\cdot}cm^{-3}$], excessive moisture drains off under the influence

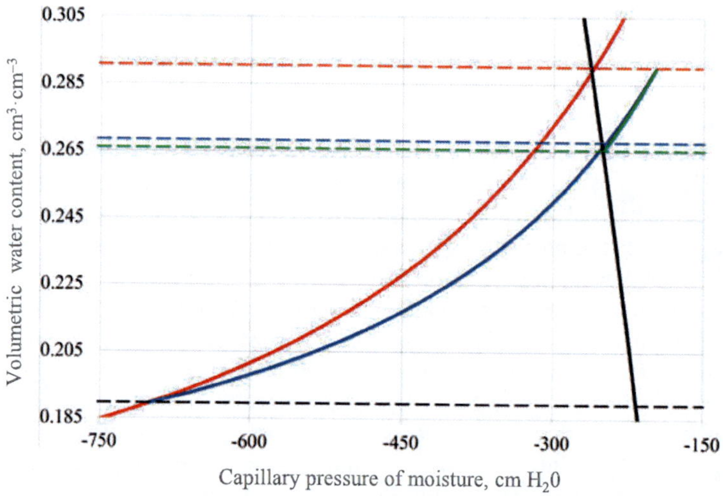

Fig. 3 Comparison of «traditional» and precision irrigation rates: Red solid curve is the main drying branch; blue solid curve is the primary wetting branch; black solid curve is Voronin's «secant»

of gravity. This runoff describes the scanning drying branch (green solid curve in Fig. 2); in this case, the θ reaches a value corresponding to the point of intersection of the scanning drying branch with Voronin's «secant»: $\theta_2 = 0.266$ [cm^3·cm^{-3}]. Hence, it follows that at the irrigation rate calculated by the «traditional» method, the unproductive consumption of irrigation water flowing out of the irrigated layer (in units of θ) is $\Delta = \theta_1 - \theta_2 = 0.290 - 0.266 = 0.024$ [cm^3·cm^{-3}]. For an irrigated plot with an area of 1 ha with a 50-cm layer of soil moistened, the total non-productive water consumption reaches $0.024 \times 0.5 \cdot 10^4 = 120$ [m^3·ha^{-1}].

It should be noted that reclamation practitioners know about these water losses. However, until now, it was not possible to more accurately calculate the irrigation rates, since the measurements of scanning branches (especially wetting branches) are very laborious.

4 Conclusion

The following conclusions can be drawn from the analysis of the study results:

1. The use of the Hys-SKT and Hys-SHT models (in both $n_d \neq n_w$ and $n_d = n_w$ options) for the point approximation of the data on the main (boundary) drying and wetting branches of the water-retention capacity of the soil is more preferable to the Hys-KPVG model.
2. Reliable differences using additional additive parameter ψ_e in both $n_d \neq n_w$ and $n_d = n_w$ options prove that the use of this parameter contributes to a significant reduction in errors. The Hys-SKT and Hys-SHT models are preferred over the Hys-SKT$_0$ and Hys-SHT$_0$ models, respectively.
3. The error of the point approximation of the data on the main (boundary) branches of the hysteresis of the water-retention capacity of silt loam soil is smaller for the models with $n_d \neq n_w$. However, in this case, a "pump effect" is possible when predicting the scanning hysteresis branches. The undesirable artificial (methodical) "pump effect" is eliminated when the condition of equality of the values of the exponential parameter n for the drying and wetting branches is fulfilled.
4. The use of the Hys-SKT and Hys-SHT models will be of great practical importance in the development of resource-saving technologies in precision ameliorative agriculture [24]. It is recommended to use the Hys-SHT model for calculating precise irrigation rates as it uses the practically preferred function $S_e(\psi)$. Precise rates are recommended to use for irrigation of crops (plantings), as well as industrial crops used for various kinds of lawns (decorative, sports, etc.).

Acknowledgements The reported study was funded by RFBR according to the research projects N. 19-04-00939-a and N. 19-016-00148-a.

References

1. Poluektov, R.A., Terleev, V.V.: Modeling of the water retention capacity and differential moisture capacity of soil. Russ. Meteorol. Hydrol. **11**, 70–75 (2002)
2. Poluektov, R.A., Oparina, I.V., Terleev, V.V.: Three methods for calculating soil water dynamics. Russ. Meteorol. Hydrol. **11**, 61–67 (2003)
3. Terleev, V., Petrovskaia, E., Nikonorov, A., Badenko, V., Volkova, Y., Pavlov, S., Semenova, N., Moiseev, K., Topaj, A., Mirschel, W.: Mathematical modeling the hydrological properties of soil for practical use in the land ecological management. MATEC Web Conf. **73**, 03001 (2016)
4. Poluektov, R.A., Terleev, V.V.: Modeling the moisture retention capacity of soil with agricultural and hydrological characteristics. Russ. Meteorol. Hydrol. **12**, 73–77 (2005)
5. Terleev, V.V., Topaj, A.G., Mirschel, W.: The improved estimation for the effective supply of productive moisture considering the hysteresis of soil water-retention capacity. Russ. Meteorol. Hydrol. **40**(4), 278–285 (2015)
6. Terleev, V., Nikonorov, A., Togo, I., Volkova, Yu., Garmanov, V., Shishov, D., Pavlova, V., Semenova, N., Mirschel, W.: Modelling the hysteretic water retention capacity of soil for reclamation research as a part of underground development. Procedia Eng. **165**, 1776–1783 (2016)
7. Terleev, V., Nikonorov, A., Togo, I., Volkova, Yu., Ginevsky, R., Lazarev, V., Khamzin, E., Garmanov, V., Mirschel, W., Akimov, L.: Hysteretic water-retention capacity of sandy soil. Mag. Civil Eng. **2**, 84–92 (2017)
8. Terleev, V., Ginevsky, R., Lazarev, V., Nikonorov, A., Togo, I., Topaj, A., Moiseev, K., Abakumov, E., Melnichuk, A., Dunaieva, Ie.: Predicting the scanning branches of hysteretic soil water-retention capacity with use of the method of mathematical modeling. IOP Conference Series: Earth and Environmental Science 90, 012105 (2017)
9. Terleev, V., Nikonorov, A., Ginevsky, R., Lazarev, V., Togo, I., Topaj, A., Moiseev, K., Pavlova, V., Layshev, K., Arkhipov, M., Melnichuk, A., Dunaieva, Ie., Mirschel, W.: Hysteresis of the soil water-retention capacity: estimating the scanning branches. Mag. Civil Eng. **1**, 141–148 (2018)
10. Terleev, V., Mirschel, W., Nikonorov, A., Ginevsky, R., Lazarev, V., Topaj, A., Moiseev, K., Layshev, K., Arkhipov, M., Melnichuk, A., Dunaieva, Ie., Popovych, V.: Five models of hysteretic water-retention capacity and their comparison for sandy soil. MATEC Web Conf. 193, 02036 (2018)
11. Terleev, V., Mirschel, W., Nikonorov, A., Lazarev, V., Ginevsky, R., Topaj, A., Moiseev, K., Pashtetsky, V., Dunaieva, Ie., Popovych, V., Melnichuk, A., Arkhipov, M.: Models of hysteresis water retention capacity and their comparative analysis on the example of sandy soil. Adv. Intell. Syst. Comput. **983**, 462–471 (2019)
12. Scott, P.S., Farquhar, G.J., Kouwen, N.: Hysteretic effects on net infiltration. In: Proceeding of National Conference on Advances in Infiltration, Publication 11–83, American Society of Agricultural Engineers, St. Joseph, Michigan, pp. 163–170 (1983)
13. Haverkamp, R., Vauclin, M., Touma, J., Wierenga, P.J., Vachaud, G.: A comparison of numerical simulation model for one-dimensional infiltration. Soil Sci. Soc. Am. J. **41**, 285–294 (1977)
14. Kool, J.B., Parker, J.C.: Development and evaluation of closed-form expressions for hysteretic soil hydraulic properties. Water Resour. Res. **23**(1), 105–114 (1987)
15. Van Genuchten, MTh.: A closed form equation for predicting the hydraulic conductivity of unsaturated soils. Soil Sci. Soc. Am. J. **44**, 892–989 (1980)
16. Terleev, V., Nikonorov, A., Badenko, V., Guseva, I., Volkova, Y., Skvortsova, O., Pavlov, S., Mirschel, W.: Modeling of hydrophysical properties of the soil as Capillary-Porous Media and improvement of Mualem-Van Genuchten method as a part of foundation arrangement research. Adv. Civil Eng. **2016**, 8176728 (2016)
17. Terleev, V.V., Mirschel, W., Badenko, V.L., Guseva, IYu.: An improved Mualem-Van Genuchten method and its verification using data on Beit Netofa clay. Eurasian Soil Sci. **50**(4), 445–455 (2017)

18. Terleev, V., Mirschel, W., Nikonorov, A., Ginevsky, R., Lazarev, V., Pavlova, V., Topaj, A., Pashtetsky, V., Dunaieva, Ie., Popovych, V., Melnichuk, A., Layshev, K.: Improved hydrophysical functions of the soil and their comparison with analogues by the Williams-Kloot Test. Adv. Intell. Syst. Comput. **983**, 449–461 (2019)
19. Kosugi, K.: Three-parameter lognormal distribution model for soil water retention. Water Resour. Res. **30**, 891–901 (1994)
20. Mualem Y.: A catalogue of the hydraulic properties of unsaturated soils. Research Project 442. Technion, Israel Institute of Technology, Haifa, Israel (1976)
21. Kostyakov, A.N.: Basics of land reclamation. M. (1960)
22. Voronin, A.D.: Fundamentals of soil physics. M.: MGU. (1986)
23. Kobzar, A.I.: Applied mathematical statistics for engineers and scientists. Izd. Fizmatlit, Moskva (rus) (2006)
24. Dunaieva, Ie., Mirschel, W., Popovych, V., Pashtetsky, V., Golovastova, E., Vecherkov, V., Melnichuk, A., Terleev, V., Nikonorov, A., Ginevsky, R., Lazarev, V., Topaj, A.: GIS Services for Agriculture Monitoring and Forecasting: Development Concept. Adv. Intell. Syst. Comput. **983**, 236–246 (2019)

Mathematical Simulation of Seismic Impact for Seismic Zones

Vladimir Lapin, Syrymgali Yerzhanov, Kamadiyar Kassenov, Nurakhmet Makish, Dauren Kassenov, and Zhassulan Omarov

Abstract For the seismically active region of the southern part of the Republic of Kazakhstan, a mathematical model of seismic action has been developed based on the representation of seismic action by a non-stationary random process, which is obtained from a stationary multiplication by a deterministic envelope function. For the territory of the city of Taraz, the seismic regime of its territory and soil conditions were studied. Accelerograms of strong earthquakes were selected from the world database, the source parameters of which correspond to the parameters of past earthquakes. It was found that the calculated sample of accelerograms for the region of Taraz should include instrumental records with acceleration maxima in the range of 124–275 cm/s^2, recorded on soils of the second category. For each accelerogram, the parameters of the correlation functions of the random process, the values of the effective duration and the frequency characteristics of the seismic impact are determined. This information allows to perform digital modeling of seismic impact, given as a random process. The results can be used to calculate buildings with various seismic isolation systems, high-rise and extended buildings.

1 Introduction

The city of Taraz is located in the northern part of the Talas flat-hilly plain between the Talas and Assa rivers. The seismicity of the urban area and adjacent areas is mainly determined by two seismically active zones: the Fergana Valley, the Chatkal ridge and the Central Tien Shan—in the south; Kyrgyz ridge—in the southeast. These zones are located relatively close to the city.

V. Lapin (✉) · S. Yerzhanov · N. Makish · D. Kassenov · Z. Omarov
Kazakh Research and Design Institute of Construction and Architecture, 21 Solodovnikova, 050043 Almaty, Kazakhstan
e-mail: lapin_1956@list.ru

K. Kassenov
Satbayev University, 22a Satbayev st, 050013 Almaty, Kazakhstan

M. Shamtsyan et al. (eds.), *Robotics, Machinery and Engineering Technology for Precision Agriculture*, Smart Innovation, Systems and Technologies 247,
https://doi.org/10.1007/978-981-16-3844-2_47

The most significant earthquakes were individual impulse accelerations of the soil no more than one to three times recurrence [1]. The abovementioned zones are characterized by a sharp release of energy in the source, which excites seismic shock waves in the environment. According to eyewitnesses, the strongest impacts are characterized as separate shocks or several shocks, often accompanied by a hum. In some cases, a sharp jolt at the beginning of an earthquake was replaced by more ordered vibrations of the earth's surface, which was a consequence of the excitation of natural vibrations of alluvial deposits, enhanced by resonance phenomena. In this case, the destructive actions of the earthquake were determined by the first sharp shocks, carrying much more energy than the subsequent vibrations.

Table 1 shows the characteristics of soil conditions in the zones of Taraz [1].

Since 1908, four 6–7-point earthquakes have been recorded in the city, two of which had a 7-point intensity at the epicenter. Until 1971, strong earthquakes with foci in the region did not occur. However, there were numerous transit earthquakes, the epicenters of which were located in adjacent seismically active zones. For example, a 9-point Chatkal earthquake on November 2, 1946, with an epicenter on the border of the Kyrgyz Republic and Uzbekistan in the city of Taraz manifested itself with an intensity of 5–6 points, and a 9-point Suusamyr earthquake on August 19, 1992, with an epicenter west of Lake Issyk-Kul—up to 4–5 points.

In 1971, on May 10, the Dzhambul earthquake with a magnitude of 5.5 occurred (the intensity at the epicenter was 8 points), the depth of the focus was 15–25 km, which manifested itself in the city as 6–7 points. The epicenter of the earthquake was 8–12 km southwest, which is the southern outskirts of modern Taraz.

The task is to develop a mathematical model of seismic impact for the region of the city of Taraz to calculate the values of reliability and seismic risk of buildings and structures, as well as other infrastructure facilities of the city. The actuality of question is very high. There are large industrial enterprises of the chemical industry in the city, the destruction of which will create great risks for the city's population.

2 Material and Methods

Experimental data on the values of the registered PGA accelerations are used [2–6]. Table 2 lists some of the known empirical formulas for determining the maximum acceleration values as a function of the magnitude M and the hypocentral distance *R*.

3 Results

Table 3 shows the maximum acceleration values for two values of magnitude and $R = 20$ km, calculated using the formulas from Table 2.

Table 1 Seismogeological conditions of the territory of Taraz city

Zones and subzones	Soil conditions	Intensity
A	The relief of zone A is gently wavy, with a low (up to 2 m) ledge of terrace II and a ledge of terrace III, reaching a height of 8–10 m, along the North Kirghiz fault. The upper layers of soils of this zone are represented by loam and sandy loam with a thickness of 0.6–2.3 m. They are underlain by pebble deposits with gravel–sand filler with the inclusion of boulders 25–50 m thick. The porosity coefficient of loams is 0.74. The bulk density of pebbles increases with depth and ranges from 2.1 to 2.4 t/m^3. At the base of the pebble, conglomerates, sandstones and clays of the Neogene age lie. The depth of the Paleozoic basement exceeds 200 m. Groundwater is observed at a depth of 3 m in the north and up to 6–10 m deeper in the south of the zone	7
B_1	Subzone B_1 has the upper layers of soils, represented by loam and sandy loam with sand lenses 0.5–2 m thick; the coefficient of porosity of loams and sandy loams ranges from 0.62 to 0.68. The lower layers of soils are represented by pebbles with gravel–sandy aggregate 25–50 m thick. Sandstones, conglomerates and Neogene clays lie at the base of pebble deposits. According to geophysical data, the Paleozoic basement in the southern part of the subzone is located at a depth of 300–350 m, in the northern–250 m. Ground waters–at a depth of 0–3 m, in some places – 3–6 m. In places where groundwater wedges out, waterlogging was noted (during the development of the territory)	8
B_2	Subzone B_2 occupies a part of the territory of the above–floodplain terrace overlain by lacustrine–proluvial sediments. In the north, the subzone is bounded by a clear scarp 8–10 m high. Lacustrine–proluvial deposits 2–6 m thick are represented by complexly interbedded loams, sandy loams and sands. Loams and sandy loams are not subsiding; their porosity coefficient is about 0.62, and their consistency is from solid to fluid. Sands are fine and silty, both loose and dense in structure; sands of medium size, as a rule, have an average density. Under these deposits are pebbles with gravel–sandy aggregate 9–23 m thick. Below are conglomerates, sandstones and Neogene clays. The Paleozoic basement lies at a depth of 300–350 m in the southern part of the subzone and 200 m in the northern. Ground waters were noted at a depth of 0–4 m, in some places – 5–5 m. On the western border of the subzone, groundwater came to the surface in spring and was kept at this level for 2–3 months, swamping the territory. Development of micro–district No. 1 and No. 2 changed the groundwater regime (mainly due to a decrease in irrigated agricultural land in the vicinity and led to a significant decrease in their level. Groundwater outflow to the surface has not been observed for several years	8

Table 2 Empirical formulas for maximum effort cm/s^2

Formula, cm/s^2	Source	Notes
$lg\ a_{max} = 0.28\text{M} - 0.81\text{gR} + 1.7$	[2]	$a_{max} > 160\text{cm}/s^2$
$a_{max} = 13.5\frac{\exp(0.896M_s)}{R}$	[3]	5 < M < 7 10 < R < 100
$a_{max} = 4800(R + 30)^{-2}\exp(0.8M)$	[4]	Soils of 2nd category
$a_{max} = 1080(R + 25)^{-1.32}\exp(0.5M)$	[5]	Medium

Table 3 Maximum acceleration values (cm/s^2)

Magnitude	Formula (1)	Formula (2)	Formula (3)	Formula (4)
5.5	199	124	156	130
6.0	275	194	233	166

Analysis of Table 3 shows that the calculated sample of accelerograms for the region of Taraz should include instrumental records with acceleration maxima in the range of 124–275 cm/s^2, recorded on soils of the second category.

Thus, during the formation of the computational sample of earthquakes, the selection is made for the accelerograms of earthquakes with $M = 5.5$–6.0 and $R = 15$–20 km and the maximum values of accelerations from the table. Table 4 shows a list of earthquakes with the specified parameters.

With the use of the PTC Mathcad Prime 3.1 computer mathematics system, calculations for processing accelerograms from Table 5 were performed.

At present, in problems of the theory of seismic resistance among domestic and foreign scientists, two types of approximations of correlation functions are most popular [6, 7]:

$$K(\tau) = \sigma^2\, e^{-\alpha|\tau|} \cos \omega\tau \tag{1}$$

corresponds to a non-differentiable random process, but

$$K(\tau) = \sigma^2 e^{-\alpha|\tau|}(\cos \omega\tau + (\alpha/\omega) sin\omega\tau) \tag{2}$$

Table 4 Engineering sampling of accelerograms

Earthquake	Date
WHITTIER NARROWS EARTHQUAKE	01.10.1987
COYOTE LAKE EARTHQUAKE	06.08.1979
MAMMOTH LAKES AFTERSHOCK	25.05.1980
WESTMORLAND EARTHQUAKE	26.04.1981
COALINGA AFTERSHOCK	22.07.1983
Average acceleration by 26 accelerograms, cm/s^2	175.85

Table 5 Accelerogram parameters

№	Accelerogram cipher	Maximum acceleration, cm / s^2	Root–mean–square acceleration, cm / s^2	Effective duration, s	α_2, 1/s	ω_2, 1/s	α_1, 1/s	ω_1, 1/s	β_{max}
1	USACA53–05–1	229.2	29.8	0.42	2.88	23.06	3.86	21.38	2.50
2	USBCA53–05–3	199.6	33.2	3.14	10.30	32.43	9.08	27.12	3.19
3	USBCA53–09–1	217.7	41.84	2.74	0.26	19.14	0.44	18.98	3.26
4	USBCA53–10–3	208.1	36.75	2.68	7.16	22.41	6.16	18.75	3.23
5	USBCA53–08–1	204.8	31.67	2.68	4.97	20.16	4.98	17.48	3.05
6	USBCA18.013	312.4	42.94	1.16	4.43	21.34	4.27	18.88	4.46
7	USBCA51.071	243.2	20.76	3.86	7.62	44.82	12.10	40.48	2.27
8	USBCA51.066	213.6	20.70	3.42	0.376	17.26	0.93	17.02	2.75
9	USACA51.064	176.4	18.20	3.02	–	–	–	–	4.2
10	USACA52.003	218.8	32.63	5.82	–	–	–	–	3.26
11	USACA52.004	174.5	33.03	5.50	13.16	27.97	10.11	21.85	2.78
12	USACA52.006	188.1	33.23	5.54	–	–	–	–	3.8
13	USACA52.015	168.9	11.42	3.04	10.13	35.17	9.65	29.84	3.0
14	USACA33.009–1	230.6	26.27	1.56	0.19	3.56	0.33	3.44	2.57
15	USACA33.022–1	250.8	53.34	3.60	1.77	7.01	1.65	6.05	3.13
16	USACA33.027–1	330.5	51.33	6.70	1.34	5.96	1.22	5.24	4.41
17	USACA33.028–1	201.9	29.46	21.42	–	–	–	–	2.57
18	USACA33.029–1	251.6	45.20	26.92	1.03	4.79	1.12	4.23	3.28
19	USACA33.037–1	215.6	29.96	7.16	–	–	–	–	2.93
20	USACA33.039–1	209.9	33.81	8.32	3.22	4.91	–	–	3.28

(continued)

Table 5 (continued)

№	Accelerogram cipher	Maximum acceleration, cm / s^2	Root–mean–square acceleration, cm / s^2	Effective duration, s	α_2, 1/s	ω_2, 1/s	α_1, 1/s	ω_1, 1/s	β_{max}
21	Average	222.31	32.71	5.94	4.60	18.78	4.71	17.26	3.11
22	Standard	40.13	10.13	6.45	3.91	11.86	3.93	10.18	0.59
23	Coefficient of variation	0.181	0.31	1.087	0.85	0.63	0.84	0.59	0.19

twice differentiated random process. Here, σ^2—variance of a process characterizing the intensity of an earthquake; α—correlation parameter characterizing the degree of attenuation of the curve; ω—frequency.

The values of the effective duration of the seismic impact d were determined, rms acceleration $\sigma_1, a_1, w_1, a_2, w_2$, respectively, the parameters of the correlation function of a stationary random process for cases of the form (1) and (2), β_{max}—spectral coefficient value.

Table 5 shows the results of calculations, and Figs. 1 and 2 graphs for one of the accelerograms—correlation functions, their approximations (1) and (2), spectral

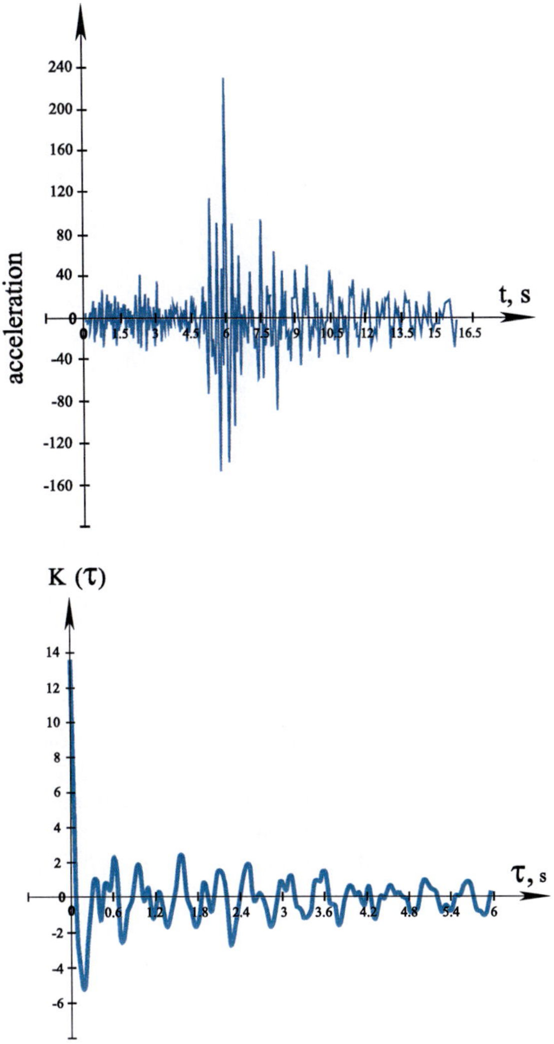

Fig. 1 Instrument recording USACA53-05-1 and its correlation function

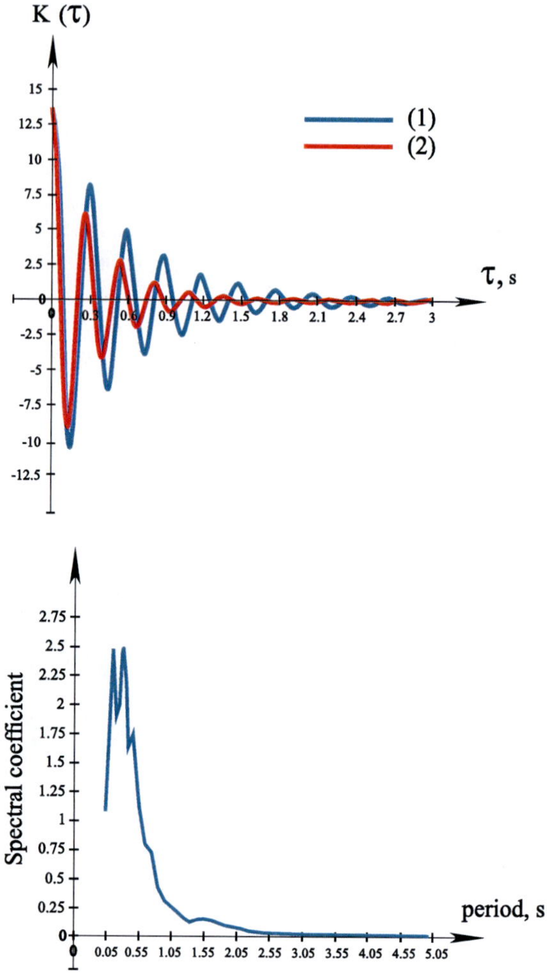

Fig. 2 Approximating correlation functions and spectral curve of instrumental recording USACA53-05-1

curves β. Curve 1 corresponds to expression (2), and curve 2 corresponds to the cosine-exponential approximation (1). Figures are from PTC Mathcad Prime 3.1 worksheets.

It should be noted that traditional methods of approximation of correlation functions (1) and (2) in 6 cases did not lead to any results in the field of real numbers. In this case, one should use other methods of approximation of correlation functions.

4 Discussion and Results

Table 5 in lines 21–23 shows the results of statistical processing of the results of calculating probabilistic characteristics, the sample mean, the root-mean-square value and the coefficient of variation. The spread for the values of accelerations and values of the dynamism coefficient β is the smallest, and for the parameters of correlation functions, it is quite significant.

The prevailing frequency of the impact here is in the range of 17.26–18.78 Hz, and the average effective duration of the seismic impact is about 6 s.

The results from Table 5 can be applied to generate artificial accelerograms with subsequent probabilistic calculation of buildings and structures for the seismic impact model for the city of Taraz. For this, methods from [8–13] can be used.

Accelerograms of the specified sample can also be directly used to perform calculations of buildings and structures in the city of Taraz for seismic effects. This method cannot be recommended for calculating critical buildings and structures. Preferred here are the methods of the probabilistic theory of seismic stability, which, however, are very laborious even with the use of modern multiprocessor computers.

Examples of determining the reliability of buildings and structures using the regional model for Taraz should be included in a separate article.

It should also be noted that this work is the first in the field of seismic impact modeling for the city of Taraz, where significant industrial capacities and human resources are concentrated—in 2014 the population is 352 thousand people. The accuracy of results, of course, can be improved.

5 Conclusion

1. For the first time, the problem of developing a seismic impact model for the region of the city of Taraz (southern Kazakhstan) has been solved. Estimates of parameters, both intensity and frequency composition of seismic impact, simulated by a random process, are obtained.
2. Accelerograms from Table 4 and the parameters of correlation functions from Table 5 can be used to perform calculations of seismic-resistant buildings and structures in Taraz for seismic effects in accordance with Eurocode 8.
3. Artificial accelerograms (implementation of a random process) can be used:
    - in calculations for seismic effects using a new regulatory framework based on Eurocodes;
    - when calculating systems of active seismic isolation of various types, modeled by nonlinear discrete systems [11–20];
    - in calculations for development Special Technical Conditions, drawn up during the design of buildings, the dimensions of which do not correspond to the current regulatory documents;

- for determination of the values of the building reliability (probability of failure-free operation) and the values of seismic risk.

The proposed algorithm for determining the parameters of a random process simulating seismic action can be used in the development of seismic action models for other regions of the Republic of Kazakhstan.

Acknowledgements The research was carried out using the grant AP 05130702 of the Ministry of Education and Science of the Republic of Kazakhstan.

References

1. Nurmagambetov, A., Sydykov, A.: Earthquakes in the Zhambyl region: past, present, future. KazNTU, Almaty (2007)
2. Mikhailova, N.N.: Seismic hazard in quantitative parameters of strong ground movements (i.e., Almaty): Diss for the degree of Doctor of Ph. and Math. Sciences. Moscow (1996)
3. Steinberg, V.V.: Ground vibrations during earthquakes. Nauka, Sources and impact of destructive seismic vibrations. Moscow (1990)
4. Duarte, R.T.: An investigation on the influence of large magnitude earthquakes on seismic risk. Proceedings of the Seventh world conference on earthquake engineering Sept 8–13. Istanbul: Turkey (1980)
5. Donovan, N.: Earthquake hazards for building practice for disaster mitigation. Build/Sci. **46**, 43 (1973)
6. Bolotin, V.V.: Statistical methods in building mechanics. Publishing house of literature on construction, Moscow (1965)
7. Bolotin, V.V.: Statistical modeling for earthquake resistance. Struct. Mech. calculation struct. **1**, 60–64 (1981)
8. Falamarz-Sheikhabadi, M.R., Zerva, A.: Two uncertainties in simulating spatially varying seismic ground motions: incoherency coefficient and apparent propagation velocity. Bull. Earthq. Eng. **16**(10), 4427–4441 (2018)
9. Guohuan, L., Yaqiang, L., Xiao, F., Dongsheng, W.: Simulation of spatially variable seismic underground motions in saturated double-phase media with overlying water excited by SV-wave and difference from P-wave incidence. Soil Dyn. Earthq. Eng. **0267–7261**(123), 144–161 (2019)
10. Kun, W., Guo-Kang, E., et al.: Nonlinear random vibrations of 3D cable-moored floating structures under seismic and wave excitations. J. Sound Vib. **452**, 58–81 (2019)
11. Lapin, V.A., Yerzhanov, S.Y., Aldakhov, Y.S.: Statistical modeling of a seismic isolation object under random seismic exposure. J. Phys.: Conf. Series **1425**, 1–9 (2020)
12. Wei, B., Wang, P., He, X., Jiang, L.: Seismic Isolation Characteristics of a Friction System. J. Test. Eval. **46**(4), 1–10 (2018)
13. Bulat, A.F., Dyrda, V.I., Lysytsya, M.I., Grebenyuk, S.M.: Numerical simulation of the stress-strain state of thin-layer rubber-metal vibration absorber elements under nonlinear deformation. Strength Mater. **50**(3), 387–395 (2018)
14. Dyrda, V., Kobets, A., Bulat, I., Lapin, V., et.al.: Vibroseismic protection of heavy mining machines, buildings and structures. E3S Web of Conferences **109**, 00022 (2019). https://doi.org/10.1051/e3sconf/201910900022
15. Bulat, A.F., Dyrda, V.I., Grebenyuk, S.N., Klimenko, M.I.: Determination of effective characteristics of the fibrous viscoelastic composite with transversal and isotropic components. Strength Mater. **51**(2), 183–192 (2019)

16. Giarlelis, C., Kofalis, D., Repapis, C.: Seismic Isolation: An Effective Technique for the Seismic Retrofitting of a Reinforced Concrete Building. Struct. Eng. **30**(1), 43–52 (2020)
17. Casablanca, O., Venture, G., Garesci, F., et.al.: Seismic Isolation of building using composite foundation based on metamaterials. J. Appl. Phys. **123**(17) (2018)
18. Ishii, K., Kikuchi, M.: Improved numerical analysis for ultimate behavior of elastometric seismic isolation bearings. Earthq. Eng. Struct. Dyn. **48**(1), 65–77 (2019)
19. Marsis, N.: Seismic isolation: early history. Earthq. Eng. Struct. Dyn. **48**(1), 269–283 (2019)
20. Tsang, H., Pitiliakis, K.: Mechanism of geotechnical seismic isolation system: Analytical modeling. Soil Dyn. Earthq. Eng. **122** (2019)

Stress–Strain State of Angular Areas of Structures

Lyudmila Frishter

Abstract In angle-shaped areas of combined structures, a complex stress–strain state (SSS) occurs stipulated not only by the geometry of the boundary cut-out, but also by rupture of the forced deformations coming to the vertex of the wedge-shape boundary notch. For the stress analysis (SA) of the wedge-shape boundary notches elements of the similarity theory and of the dimensional analysis. Problem solution of the elasticity theory for the complex area in one of the subareas of which forced deformation occurs is obtained experimentally on polymer models by photoelasticity method. The breach of the forced deformations comes to an irregular point of the boundary area, the vertex of the wedge-type notch. The objective of the paper is the study of the SA of structure's angular areas based on the theoretical and experimental approach using elements of the similarity theory and of the continual photoelasticity method. The combined usage of the elements of the similarity theory and of the experimental problem solution allows for unveiling the specifics of the tension distribution in the wedge-type boundary notch area. A conclusion is made on similarity and uniformity of the stress in the wedge-shape boundary notch area.

1 Introduction

In angle-shaped areas of combined structures, a complex stress–strain state (SSS) occurs stipulated not only by the geometry of the wedge-shape boundary notch, but also by the rupture of the forced deformations coming to the vertex of the wedge-shape boundary notch. Studies of the local SSS of a subarea of an essential structural non-uniformity are important during the calculation and design of structural elements at abrupt boundary shape change of a structure, e.g., at angular or stepped shape.

The subject of the study hereof is the complex area of the wedge-shape boundary notch, the vertex of which concentrate the rupture of the forced deformations along the contact lines of the subareas.

L. Frishter (✉)
National Research University Moscow State University of Civil Engineering, 26 Yaroslavskoye Shosse, Moscow 129337, Russian Federation

M. Shamtsyan et al. (eds.), *Robotics, Machinery and Engineering Technology for Precision Agriculture*, Smart Innovation, Systems and Technologies 247,
https://doi.org/10.1007/978-981-16-3844-2_48

Theoretical and experimental studies of the stress concentration stipulated by the boundary shape are presented in papers by R. Petersson, Neuber, N.G. Savin and V.I. Tul'chiy, B.N. Ushakov, I.P. Fomin, N.A. Makhutov, V.V. Vasilyev et al. [1–6].

The theoretical solution of the homogeneous boundary value problem of elliptic type [7–11] determines in the area of the wedge-shape boundary notch a power singularity of the solution. The study of the wedge-shape notch boundary areas is reduced to a consideration of the singular solution of the homogeneous boundary value problem.

The photoelasticity method [1, 2, 12–15] and the deformation "defrost" method as its sub-type allow for obtaining the SSS in an irregular point area of the boundary line in models of optically sensitive polymer material. The most effective and developed is the method in the solving of planar mechanical problems [1, 2, 12–16].

The objective of the paper is the study of the SA of structure's angular areas based on the theoretical and experimental approach using elements of the similarity theory and of the continual photoelasticity method.

For theoretical SA of the area, a resolving set of equations of the elasticity theory problem is analyzed in the wedge-shape boundary notch area and elements of the similarity theory [7, 13, 15, 17]; for experimental studies, the photoelasticity method is used [1, 2, 12–15]. The solution of the elasticity problem in the area of a wedge-shape boundary notch is obtained experimentally on complex polymer models with a wedge-shape boundary notch under forced thermal deformations in one of the areas. An analysis of the opportunities provided by the theoretical and the experimental approach to SS study in areas with an irregular boundary point allows for unveiling of the applicability areas of the linear elasticity problem solutions and of the stress distribution specifics in the wedge-shape boundary notch of the area.

2 Materials and Methods

2.1 SSS Analysis Using the Theory of Similarity and Dimensionality

The resolving set of equations of the elasticity theory in the small vicinity of the irregular point $O_\delta(0)$ in a special line of the elastic body border [2, 7, 13, 17] is reduced to dimensionless appearance by means of the following correlations:

$$\xi = \xi_0 \overline{\xi} \tag{1}$$

where ξ is the magnitude under analysis, ξ_0 is the characteristic value of the magnitude under analysis, $\overline{\xi}$ is the dimensionless value of such magnitude. The following correlations are analyzed:

$$\sigma = \sigma_0\overline{\sigma}; \quad \varepsilon = \varepsilon_0\overline{\varepsilon}; \quad U = U_0\overline{U}; \tag{2}$$

$$x = \frac{l_0}{t}\overline{x}; \quad y = \frac{l_0}{t}\overline{y}; \quad z = l_0\overline{z}\ , \tag{3}$$

where t is the dimensionless parameter of the similarity group, introduced for the analysis of the "degree of approximation to the irregular point". In order to obtain the homogeneous boundary value problem and to switch to the automatic model solution, it is required and sufficient that, in accordance with the resolving set of equations, criteria associated with the impacts are met, stipulated by the specified equations:

$$\frac{\sigma_{01}}{\sigma_0} = 0 \text{ or } \sigma_{01} \ll \sigma_0\ t \to \infty \tag{4}$$

$$\frac{\sigma_{01}}{\sigma_0} = 1 \text{ or } \sigma_{01} \sim \sigma_0\ t \to 1 \tag{5}$$

where σ_{01},ε_{01} are the characteristic values for stress, deformations in the subarea section $O_\delta(0)$ containing the irregular boundary point.

Criterion (4) means that the closer the irregular point of the boundary is approached from inside, the stress σ_0 stipulated by the singular problem solution is changing much faster than the stress σ_{01} stipulated by the impacts of the initial loads, that is, the "common stress field" in this vicinity is negligible.

The greater is the distance from the irregular point of the area boundary, the stress change orders σ_0, σ_{01} are equalized and, in certain cross-section location or parameter value t of the stress, σ_0 and σ_{01} become equivalent in accordance with criterion (5).

For automatic modeling of the boundary elasticity problem solution in the area of the irregular boundary point, the following criteria shall be met:

$$\left\{\frac{\sigma_0}{\varepsilon_0}\right\} = 1; \quad \left\{\frac{\varepsilon_{01}}{\varepsilon_0}\right\} = 0; \quad \left\{\frac{\alpha_0 T_0}{\varepsilon_0}\right\} = 0; \tag{6}$$

$$\text{or at } t \to \infty\ \sigma_0 \sim \varepsilon_0; \ \varepsilon_{01} \ll \varepsilon_0; \quad \alpha_0 T_0 \ll \varepsilon_0 \tag{7}$$

where ε_{01} is the characteristic value of the specified forced deformations ε_{ij}^0.

In accordance with criteria (6, 7), the specified deformations ε_{01}, $\alpha_0 T_0$ in the vicinity of $O_\delta(0)$ are changed much slower at approach to the irregular point or at the increasing of parameter t, than the deformations ε_0 stipulated by the irregularity of the boundary in point O. Thereby, at $t \to \infty$, that is, the closer to the irregular point O of the boundary, the more negligible are the specified deformations when matched to "singular" deformations.

The condition of the appearance:$\sigma_0 \sim \varepsilon_0$ at $t \to \infty$ means that the stress and the deformations have the same change order by the coordinates in the vicinity of the special point O, that is, in the vicinity $O_\delta(0)$, it is possible assume equivalence of the change order by coordinates for stress and deformations.

The farther from the irregular point O, that is, for $t \to 1$, the specified forced deformations $\varepsilon_{01}, \alpha_0 T_0$ have a change order equal to that of the deformations ε_0.

For displacements and deformations, the following criteria are met:

$$\left\{\frac{U_0 t}{\varepsilon_0 l_0}\right\} = 1 \ or \ U_0 t \sim \varepsilon_0 l_0 \ or \ U_0 \sim \frac{\varepsilon_0 l_0}{t} \tag{8}$$

In accordance with criteria (8) in the vicinity $O_\delta(0)$ of the irregular boundary point, the change order of the point coordinates' displacement function is by one unit bigger than the deformation change order, and then the change order of the stress, as per equivalence condition (7).

If the displacements in the vicinity $O_\delta(0)$ have the change order r^λ, where r is the distance from the irregular point $O(0, 0)$ to the specified one, then the deformations and, respectively, the stress have the order $r^{\lambda-1}$. At $\lambda \in (0, 1)$, the displacements are limited, whereas the stress and the deformations have order specifics $|\lambda - 1|$; thereby, the following evaluations are possible:

$$U_0 = O(r^\lambda); \qquad \sigma_0 = O((r - r_0)^{\lambda-1}) \tag{9}$$

where $r \neq r_0$, and the vicinity $O_\delta(0)$ is analyzed without the irregular point O proper and its certain small vicinity.

Due to the linearity of the homogeneous boundary value problem of elasticity, the change order of the displacements, dependent on the coordinates of the point $(x, \ y) \in O_\delta(0)$, has the following appearance:

$$U(tx) \sim C(tx)^\lambda \ or \ U(x_1) \sim Cx_1^\lambda; \quad V(x_1) \sim Cx_1^\lambda \tag{10}$$

where C is an arbitrary constant, $x_1 = tx, \quad y_1 = ty, \quad x_1, \ y_1 \in O_\delta(0)$.

At $\lambda \in (0, 1)$, the displacements $U(x_1, y_1), \quad V(x_1, y_1)$ in the vicinity $O_\delta(0)$ are continuous, limited, and, based on correlations (10), feature the property of uniformity.

In accordance with correlations (7, 9), the stress and the deformations have the change order of the point coordinates in the vicinity $O_\delta(0)$ less by one unit than that of the displacement:

$$\sigma_{ij} \sim C_1 \lambda x_1^{\lambda-1}; \quad \varepsilon_{ij} \sim C_1 \lambda x_1^{\lambda-1} \tag{11}$$

$x_1 = (x_1, \ y_1) \in O_\delta(0); \quad x_1 = tx; \quad y_1 = ty$. At $\lambda \in (0, 1)$, the stress and the deformations in the vicinity $O_\delta(0)$ have the specifics of the order $|\lambda - 1|$.

A solution of homogeneous planar problems in the vicinity of a point on a special line is advisable to be searched for in the polar system of coordinates, as follows:

$$U = r^{\lambda} f(\theta); \quad V = r^{\lambda} g(\theta); \quad W = r^{\lambda_1} p(\theta), \tag{12}$$

This matches the solutions shown in papers [2, 7, 17].

Due to the automatic modeling of the problem solution of the elasticity theory, the deformations, the displacements in a certain vicinity of the irregular point of the area boundary admit a similarity group and feature the property of the uniformity of the functions, distinguished by the property, that such functions can be represented as power complexes (10, 11 and 12).

The same properties (similarity, uniformity) must also be attributable to the experimental solution obtained as interference pattern on a model by photoelasticity method.

2.2 *Experimental Solution in the Wedge-Shape Boundary Notch*

Experimentally, the thermal elasticity problem is solved by the photoelasticity method with unfreezing of free thermal deformations [1, 2, 12–15]. The polymer beam model consists of two areas; in one of which free thermal deformations are generated $\alpha T \delta_{ij}$, whereas the other one does not accommodate any loads. After the "unfreezing", the model acquires the desired thermoelastic state. The leap of the forced deformation $\Delta\varepsilon_{ij} = \alpha T \delta_{ij}$ along the contact line of the areas constituting the model comes the irregular point O (0, 0) of the boundary, the vertex of the wedge-type notch of the beam face. One of the interference patterns of the model with opening 100° of the wedge-shape boundary notch is shown in Fig. 1. In the area of the beam end face with a wedge-shape boundary notch, stripe orders are plotted (isochromes) for a number of radial sections. The presence of areas was detected, which correspond to the "singular" and the non-singular solution of the problem, as well as the similarity of stripe order plots.

In accordance with the theoretical analysis [7, 18] of the SSS, there is such a vicinity, for which the singular solution of the homogeneous boundary value problem is true. The orders of the stripes in the area of the model's stress concentrator (singular solution areas) are illegible at any magnification of the irregular point vicinity.

In accordance with the theoretical SSS analysis, there is a vicinity where the non-singular boundary elastic problem is true with the same proper value min Re λ as in the singular problem. The stripe orders in the model corresponding to the non-singular solution area are legible except for some of them, as the case may be. The existence of such self-balanced radial SSS explains the growth of the stripe orders (isochromes) observed from inside the stress concentration area, but not in the very vertex of the wedge-shape notch of the area. The absence of the zero stripe is explained by the existence of another self-balanced SSS stipulated by the common stress field. For this area, criteria (7, 8) are true, that is, the similarity and the uniformity of the solution functions are observed.

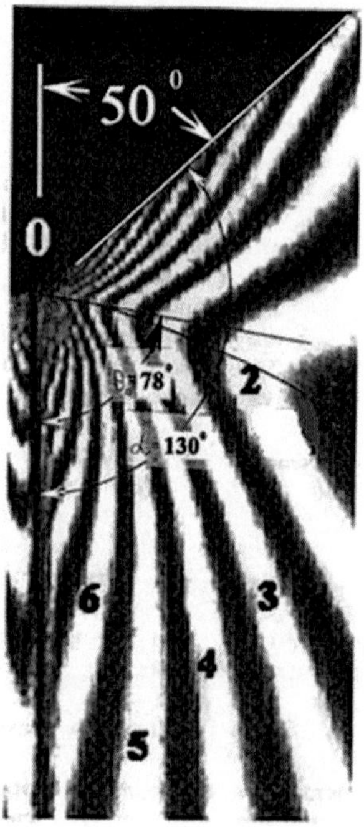

Fig. 1 Interference pattern in one of the model areas with the vertex angle 100° of the wedge-shape boundary notch

In accordance with the combined theoretical and experimental SSS analysis in the vicinity of an irregular boundary point of the flat area containing the leap of the forced deformations, the stripe similarity formula appears as follows:

$$m_{i+1} = \left(\frac{r_i}{r_{i+1}}\right)^{1-\lambda_0} m_i \ or \ m_{i+1}\ (r_{i+1})^{1-\lambda_0} = m_i\ (r_i)^{1-\lambda_0} \tag{13}$$

where m_i are the stripe orders, based on the experimental data, in r_i design section, in the vicinity of the non-singular solution of one boundary value problem, m_{i+1} are the stripe orders in the section of a smaller radius $r_{i+1} < r_i$ located in the area with illegible or "poorly" legible isochromes pattern of the model, $\lambda_0 = \min \mathrm{Re}\lambda$ is the minimum value of the effective part of the complex root of the characteristic equation of the homogeneous boundary value problem for the model wedge. Correlation (13) defines the power dependency of the isochromes order on the point coordinates in the area of the boundary notch: $m \sim C\lambda r^{\lambda-1}$.

The experimental solution obtained by the photoelasticity method on a model as an interference pattern confirms the presence of an area for which the properties of similarity and uniformity of the solution functions are true.

3 Results

Stress analysis of angular areas of structures was conducted, based on the theoretical and experimental approach, using elements of the similarity theory and the continual photoelasticity method. The combined usage of the elements of the similarity theory and of the experimental problem solution allows for unveiling the specifics of the tension distribution in the wedge-type boundary notch area. Formulas (11), (13), as the results of the study, confirm the presence of an area in the boundary notch vicinity, where the property of similarity and uniformity of the solution's functions is true. The stripe orders can be represented in the vicinity of an irregular boundary point of the model as power complexes: $m \sim C\lambda r^{\lambda-1}$, this is consistent with the results of the theoretical SSS analysis in this area.

4 Discussion

The vertex area of the wedge-shape boundary notch consists of several subareas: the subarea of plastic deformations where the final deformations must be considered, the subarea of elastic (linear and nonlinear) deformations, for which, in the framework of the linear elastic theory problem, the resolving set of equations of the boundary value problem is written with subsequent SSS analysis with the power singularity of (10), (11) or (12) appearance of the same order, as for SSS of the singular solution. The presence of the automatic modeling area of the elastic problem solution is experimentally proven by the fact that the stress, the deformations and the displacements in a certain vicinity of the irregular point of the area boundary admit a similarity group and feature the property of the uniformity of the functions, distinguished by the property, that such functions can be represented as power complexes.

5 Conclusions

The combined usage of the elements of the similarity theory and of the experimental problem solution by photoelasticity method allows for unveiling the specifics of the tension distribution in the wedge-type boundary notch area. The presence of the areas in the vicinity of the wedge-shape boundary notch where the power nature of the stress similarity is observed allows for detection of applicability zones of the solutions of the linear elasticity problem. The areas approximated to the irregular point of the boundary must be considered under recognition of plastic and finite deformations.

References

1. Koshelenko, A.S., Poznjak, G.G.: Theoretical foundations and practice of photomechanics in mechanical engineering. Granica, Moscow (2004)
2. Razumovskij, I.A.: Interference-optical methods of deformable solid mechanics. Publisher MGTU named after N. Je. Bauman, Moscow (2007)
3. Makhutov, N.A., Moskvichev, V.V., Morozov, E.V., Goldstein, R.V.: Unification of computation and experimental methods of testing for crack resistance: development of the fracture mechanics and new goals. Ind. lab. Diagn. mater. **83**(10), 55–64 (2017)
4. Matviyenko, Yu.: Fracture mechanics models and criteria. Fizmatlit, Moscow (2006)
5. Albaut, G.N., Kharinova, N.V., Sadovnichij, V.P., Semenova, J.I., Fedin, S.A.: Nonlinear problems of the mechanics of destruction. In: Proceedings of Lobachevsky Univ. of Nizhn. Novgorod, vol. 4, pp. 1344–1348 (2011)
6. Pestrenin, V.M., Pestrenina, I.V., Landik, L.V.: Proc. TGU. Math. Mech. **4**, 80–87 (2013)
7. Cherepanov, G.N.: Fracture mechanics. Izhevsk Computer Research Institute (2012)
8. Parton, V., Morozov, E.: Elastic-plastic fracture mechanics: Special problems of the fracture mechanics. Nauka, Moscow (2017)
9. Kuliev, V.D.: Singular boundary value problems. Nauka, Moscow (2005)
10. Feodosiev, V.I.: Strength of Materials. Publ. house N. E. Bauman, Moscow (2016)
11. Bakushev, S.V.: Differential equations and boundary value problems of deformable solid body mechanics. LENAND, Moscow (2020)
12. Frocht, M.M.: Photoelasticity. Edited by Prigorovskiy, N.I., vol. 2, pp. 432. GITTL, Leningrad, Moscow (1948)
13. Hesin, G.L., et al.: The photoelasticity method, vol. 3, p. 311. Stroyizdat, Moscow (1975)
14. Durelli, A., Riley, W.: Introduction to photoelasticity. Edited by Prigorovskiy, N.I. Mir, Moscow (1970)
15. Aleksandrov, A.Ja., Ahmetzjanov, M.H.: Polarization-optical methods for deformable solid mechanics. Nauka, Moscow (1974)
16. Frishter, L.Yu., Savostyanov, V.N.: Analytical opportunities of the experimental solution of the elastic problem in the stress concentration area. Proceedings of MGSU, Issue No.1, pp. 50–55 (2011)
17. Vardanyan, G.S.: Applied mechanics: use of methods of the similarity theory and analysis of dimensionality in approach to modeling problems of deformable solid body mechanics. INFRA-M, Moscow (2016)
18. Frishter, L.Yu.: Photoelasticity-based study of stress-strain state in the area of the plain domain boundary cut-out area vertex. In: Advances in Intelligent Systems and Computing, EMMFT 2017, Springer, vol. 692, Cham 836–844 (2017).

The Influence of Deformation and Thermal Effects on the Structure and Properties of the Metal of Welded Structure Elements of Lifting Cranes

Alexander Scherbakov, Elizaveta Lukashuk, Alexander Pushkarev, and Tamara Vinogradova

Abstract The paper studies the influence of deformation and thermal effects on the structure and properties of the metal of welded structure elements of lifting cranes. In the course of the study, a relationship between the structural, mechanical and magnetic parameters during cyclic elastic–plastic deformation of low-carbon and low-alloy steels was established. Based on the analysis, it was concluded that the dependence of the magnetic parameter H_p on the number of cycles during thermal cycling is influenced by the structure of the metal and its chemical composition.

1 Introduction

Depending on the supplied factory steel (hot-rolled, cold-rolled, heat-treated) for the manufacture of welded load-bearing metal structures of cranes from low-carbon and low-alloy steels, their initial structure significantly differs in grain size, which leads to a wide spread in the strength properties of the metal not only in different parts of the structure, but also in its individual elements [1].

In addition, welded joints are characterized by structural heterogeneity, which is determined by the initial structure of the materials being welded, the nature of the physical impact on it, and the degree of completeness of phase and structural transformations occurring during welding. In the welded joints of metal structures made of low-carbon and low-alloy steels, areas with different structures are formed during cooling: cast—in the weld zone, coarse-grained and fine-grained—in the heat-affected zone, which differently determine the mechanical properties, and, consequently, the reliability and performance of welded joints [2, 3].

It should be noted that during operation, structural changes occur in the elements of metal structures of lifting cranes, and plastic deformation can occur in hazardous areas of KL (tension, compression), leading to the destruction of the structure. Therefore, the presence of coarse-grained, fine-grained or deformed structures in

A. Scherbakov (✉) · E. Lukashuk · A. Pushkarev · T. Vinogradova
Saint Petersburg State University of Architecture and Civil Engineering, Vtoraya Krasnoarmeiskaya street, 4, Saint Petersburg 190005, Russia

M. Shamtsyan et al. (eds.), *Robotics, Machinery and Engineering Technology for Precision Agriculture*, Smart Innovation, Systems and Technologies 247,
https://doi.org/10.1007/978-981-16-3844-2_49

the elements of welded metal structures determines different mechanical properties in them and contributes to the formation of stress concentration zones [4, 5].

Numerous experimental studies show that the strength of annealed metals and alloys tested at sufficiently low temperatures, when the recovery processes are difficult, increases with grain refinement. So, in this paper, with a decrease in the grain size in the range from 10 to 1 μm, the yield stress, flow stress at various values of deformation, ultimate strength, hardness and fatigue strength of a significant number of metals and alloys increase. Studies have shown that one of the main practically important reasons due to the fine-grained structure of metals is an increase in their strength properties and impact toughness. For example, when the grain size in Jn-744 steel changes from 2 to 25 μm, the temperature of transition from ductile to brittle fracture shifts toward higher temperatures from −130 to −45 °C [6].

The analysis of the experimental data shows that at low temperatures, with a decrease in the grain size, the yield strength, strength, hardness, fatigue strength, and impact strength increase. In such materials, strength levels close to theoretical ones are potentially achievable. Consequently, obtaining a fine-grained structure is of independent importance for increasing the strength properties of metals.

In connection with the above, it is of scientific and practical interest to conduct research both to assess the effect of the initial structure, structural heterogeneity of welded joints and structural changes in the metal of structural elements on the operational reliability of metal structures of lifting cranes, and to control the process of forming a structure with varying degrees of dispersion in low-carbon and low-alloy steels, depending on the effect of various factors, with the search for simple and reliable ways to obtain a fine-grained structure.

2 Materials and Methods

A significant variety of welded supporting metal structures of lifting cranes, a growing number of steel grades, various initial structural states of metal of structural elements, the improvement and development of new technologies in order to obtain special physical and mechanical characteristics of metals do not allow, within the framework of the theory of ferromagnetism, giving such connections between magnetic and structural parameters of steels. Therefore, it was of interest to assess the capabilities of the passive flux-gate method for assessing the structural state of 08 ps low-carbon steel and 09G2S low-alloy steel in the process of formation of a structure with a given degree of dispersion during recrystallization annealing [7, 8].

The procedure for obtaining different initial microstructure in steels included preliminary cold plastic deformation (fractional rolling) of steels to a degree of deformation $\varepsilon = 50\%$ and subsequent recrystallization annealing at temperatures of 20 … 900 °C. The structure of 09G2S steel before and after rolling is shown in Fig. 1.

The beginning and end of recrystallization was determined by measuring the stray magnetic field strength Hp on the surface of the samples, using microstructural analysis and measuring the hardness of steels. The magnetic field strength Hs was

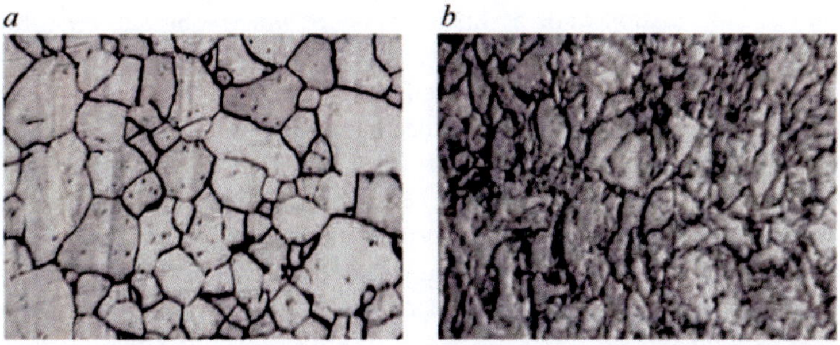

Fig. 1 Structure of 09G2S steel, × 650: *a*—as-delivered state, *b*—after rolling to a degree of 50%

measured at three points of the samples separated from each other at a distance of 50 mm after they were cooled from different temperatures.

The dependence of the magnetic field strength on the temperature of recrystallization annealing for specimens of steels 08 ps and 09G2S is shown in Fig. 2. It can be seen that in the studied steels in the temperature range from 150 to 300 °C, a significant decrease in the values of the magnetic field strength Hp is observed, which is probably associated with the occurrence of the first stage of recovery—rest, as a result of which the number of point defects (interstitial atoms and vacancies) is decreased by annihilation at edge dislocations, as well as a decrease in the distortion of the crystal lattice. In this case, no rearrangement of the dislocation structure occurs [9, 10].

An increase in the heating temperature above 300 °C leads to a further decrease in the values of the stray magnetic field strength Hp, however, with a lower intensity. In this case, due to the activation of diffusion processes, the redistribution of dislocations, their annihilation and the alignment of excess dislocations of the same

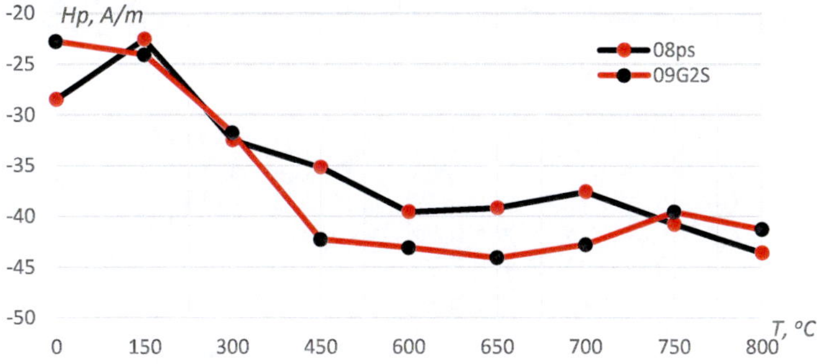

Fig. 2 Dependence of the stray magnetic field strength H_p on the recrystallization annealing temperature for 08 ps and 09G2S steels

sign in vertical dislocation walls occur, as a result of which subgrains are formed—the process of polygonization is underway. Due to the movement of dislocation boundaries, they become larger, which leads to a decrease in the density of defects.

At temperatures of 600…700 °C, the decrease in Hp values slows down, which corresponds to the passage of the primary recrystallization process. The resulting recrystallization centers are formed in those areas of deformed grains in which the dislocation density is increased, and the crystal lattice is most distorted. These centers grow as a result of the attachment (diffusion) of atoms from the deformed environment to them, while the boundaries of the recrystallization center migrate toward the deformed environment.

The end of the monotonic decrease in the stray magnetic field strength Hp from the annealing temperature indicates the end of primary recrystallization. Some rise in the Hp curve at 750 °C is probably associated with the partial phase recrystallization of steels. The sufficiently high sensitivity of the magnetic parameter Hp to temperature can be used to control structural changes in low-carbon and low-alloy steels during heat treatment.

3 Results and Discussion

According to the data of metallographic studies, the temperature of the beginning and end of recrystallization for steel 08 ps is 600 and 700 °C (Fig. 3), and for steel 09G2S

Fig. 3 Changes in the structure of 08 ps steel upon recrystallization annealing, × 650: **a** as-delivered, **b** after rolling at $\varepsilon = 50\%$, (*c-f*) after rolling at $\varepsilon = 50\%$ and annealing at 600, 700, 800, and 900 °C

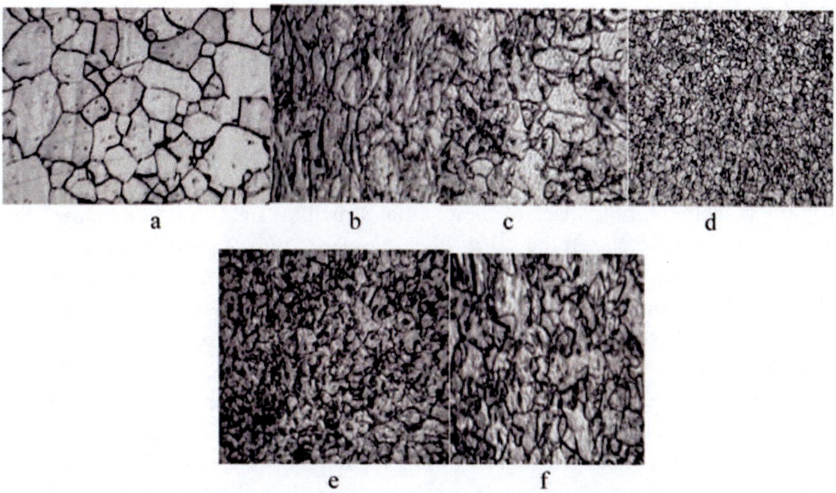

Fig. 4 Changes in the structure of 09G2S steel upon recrystallization annealing, × 650: **a** as-delivered, **b** after rolling at $\varepsilon = 50\%$, (*c-f*) after rolling at $\varepsilon = 50\%$ and annealing at 650, 725, 800, and 900 °C

is 650 and 725 °C (Fig. 4), respectively. Figures 3c and 4c show the microstructures of 08 ps and 09G2S steels after recrystallization annealing. It can be seen that at these temperatures, separate recrystallization nuclei appear, which are unevenly distributed in the deformed matrix. By the end of the primary recrystallization (Fig. 3d, 4d) at temperatures of 700 and 725°C, a rather fine-grained structure of 08 ps and 09G2S steels with grain sizes of 8 and 6 microns, respectively, is formed. A further increase in the heating temperature to 800 and 900 °C (Fig. 3e, f, 4e, f) leads to the passage of collective and secondary recrystallization, as a result of which a more equilibrium coarse-grained structure with increased grain size is formed. The smallest grain growth and grain size are characteristic of low-alloy steel 09G2S (compared to low-carbon 08 ps), which indicates a stabilizing effect on the structure of fine particles of the carbide phase.

Comparing the results of metallographic analysis with the change in the values of the magnetic field strength, it can be stated that the onset of recrystallization in steels 08 ps and 09G2S almost corresponds to the inflection points on the stress–annealing temperature curves (Fig. 2).

Thus, fixing a rather early stage of the completion of primary recrystallization, in steels 08 ps and 09G2S, it is possible to obtain a fine-grained equiaxed structure, which, with an increase in the degree of alloying of the steel, becomes finer-grained. A further increase in temperature leads to the formation of a coarser-grained structure, even in comparison with the initial one, which is important for obtaining coarse-grained structures when simulating various structural states of the steels under study.

It should be noted that during flux-gate control, not only the beginning and end of recrystallization is recorded, but also the return processes, which indicates a higher sensitivity of the magnetic method as compared to the metallographic method. Therefore, the passive flux-gate method can be recommended for monitoring structural changes in low-carbon and low-alloy steels during their heat treatment, as well as during the production of factory rolled products.

Typical structures for elements of welded load-bearing metal structures of lifting cranes and welded joints (possessing structural heterogeneity) with varying degrees of grain refinement, regardless of their initial structural state, can be obtained by thermal cycling of steels. At the same time, in order to reduce production costs and reduce the cost of products, it is desirable to carry out non-destructive testing not only after final processing, but also at its individual stages. Such possibilities are provided by the passive flux-gate control method, which was used in this work.

Studies were carried out using samples of 08 ps low-carbon steel and 09G2S low-alloy steel. The choice of these steels, as well as the production of a wide range of structural states on them due to various types of processing, allows assessing their advantages and disadvantages in determining the relationship between the structure, mechanical and magnetic properties, and recommending the obtained results for steels close to them in composition and properties. Steel samples were subjected to thermal cycling in the following conditions: as-delivered state; as-delivered + annealing at 900 °C; as-delivered + cold rolling to a degree of deformation $\varepsilon = 50\%$. In this case, the dimensions of the samples were 150 × 30x2 mm, and after rolling—280 × 30x1 mm. The TCT consisted of 5 cycles of steel heating to 770 °C, followed by cooling in a furnace to 690 °C, and at the end of the 5th cycle from 690 °C—in air. The heating and cooling rate was 3…5 degrees per minute.

The formation of a fine-grained structure in the process of TCT was controlled by the passive flux-gate method, as well as by microstructural analysis. The H_p values were measured with a two-channel flux-gate transducer installed perpendicular to the sample surface. On each sample, in its middle part along the axial line, the values of H_p were recorded in four control zones spaced 30 mm apart. The measurement of H_p values was carried out both before the first cycle of the TCT and at the end of each of the heating–cooling cycles. To plot the dependences of the H_p values on the number of cycles, the averaged H_p value of the indicated control zones was taken.

Figures 5 and 6 show the dependence of the stray magnetic field strength H_p on the number of cycles during TCT of 08 ps and 09G2S steels, depending on their initial state. It can be seen that the greatest changes in the values of the magnetic parameter H_p for the studied steels are observed during the first two cycles of TCT. In this case, the magnetic prehistory of the samples is erased, as evidenced by the change from the positive sign to negative and the convergence of the H_p values. With an increase in the number of cycles, the magnetic field strength values stabilize, which is noticeable from a decrease in the spread in H_p values.

As shown in Fig. 7, all samples in the as-delivered state have a significant change in H_p values after the first cycle. During subsequent cycles of heat treatment, negative values of H_p do not change their sign. Regardless of the steel grade, after the fifth heating–cooling cycle, the difference between the maximum and minimum values of H_p of the studied steels, depending on the initial state, is approximately the same and does not exceed 4 A/m, which indicates a relatively close structural

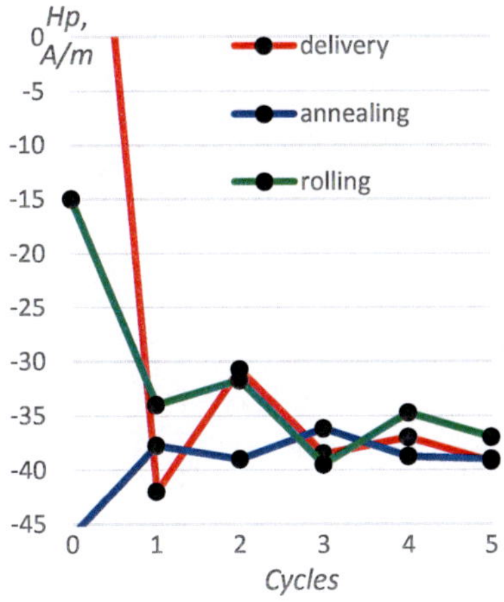

Fig. 5 Dependence of the magnetic field strength H_p on the number of cycles during thermal cycling of 08 ps steel in various initial states

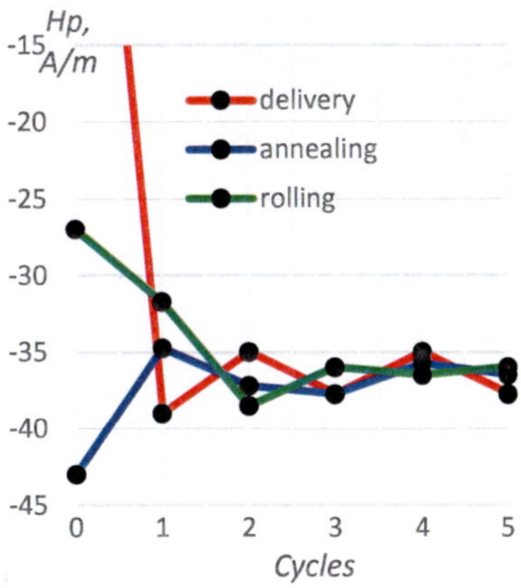

Fig. 6 Dependence of the magnetic field strength H_p on the number of cycles during thermal cycling of 09G2S steel in various initial states

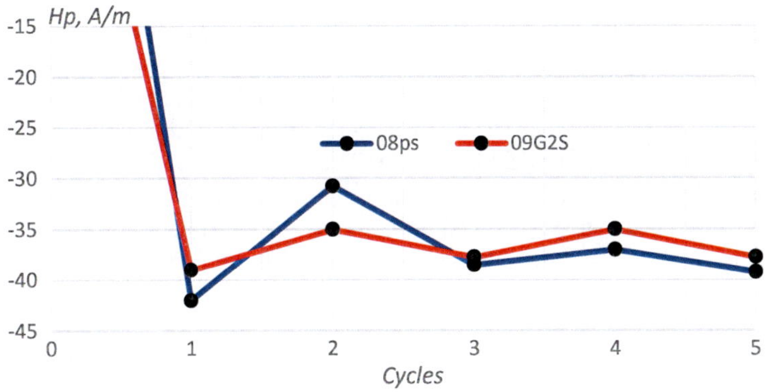

Fig. 7 Dependence of the magnetic field strength H_p on the number of cycles during thermal cycling of 08 ps and 09G2S steels in the as-delivered state

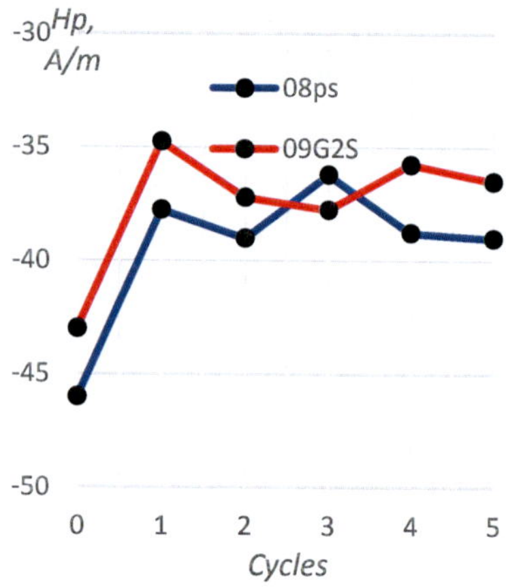

Fig. 8 Dependence of the magnetic field strength H_p on the number of cycles during thermal cycling of 08 ps and 09G2S steels after preliminary annealing at 900° C

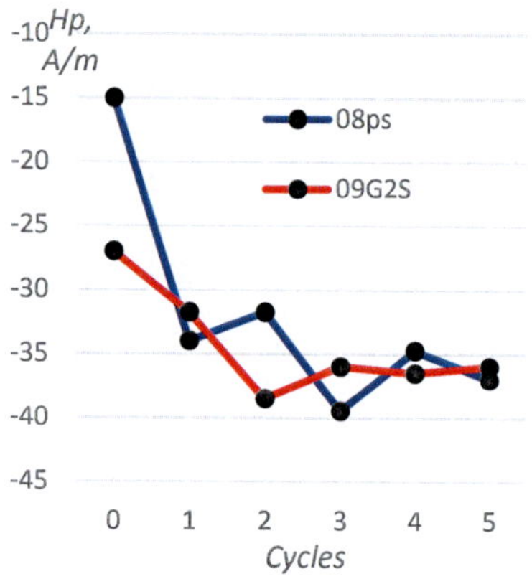

Fig. 9 Dependence of the magnetic field strength H_p on the number of cycles during thermal cycling of 08 ps and 09G2S steels after cold plastic deformation by ε = 50%

state of these steels (Fig. 8, 9) after thermal cycling. Significant changes in H_p are observed depending on the initial structure of the steels. Thus, steels with the coarse-grained and more equilibrium structure (as-delivered state + annealing at 900 °C), in comparison with finer-grained steels in the as-delivered state (Fig. 7, 8), have smaller changes in H_p, starting from the first cycle. Steels that have undergone preliminary cold plastic deformation are inherent in more significant fluctuations in the values of H_p, especially during the second and third cycles of TCT (Fig. 9).

A sharp change in the strength of the stray magnetic field Hp at the initial TCT cycles and a monotonic decrease in the magnetic amplitude during subsequent cycles can be used when carrying out thermal cycling in order to control the end of the formation of a structure with a given degree of dispersion in low-carbon and low-alloy steels.

Changes in the magnetic field strength Hp depending on the chemical composition of the steels, the initial microstructure, and the number of cycles are probably associated with structural changes in steels that occur during the TCT process. In contrast to the standard methods of heat treatment, during TCT, additional sources of influence on the structure appear, which are characteristic only of the process of continuous temperature change, the main of which are phase transformations, temperature gradients, thermal (volumetric) and interfacial stresses caused by the difference in thermophysical characteristics that make up the structure of phases.

Metallographic studies have shown that irrespective of the initial state of the samples, a fine-grained structure is formed in all the steels under study after the second cycle of TCT (Figs. 10, 11, 12, 13, 14 and 15). A further increase in the number of cycles less significantly increases the degree of its refinement, which is consistent with the data on changes in the magnetic field strength; in subsequent cycles, a decrease in the amplitude of fluctuations in the magnetic field strength is noted. Similar results were obtained in a work where it was shown using steels 30, 40, and 60 with different numbers of TCT cycles that 5–6 cycles were necessary in order to obtain the optimal structure and mechanical properties.

Figures 10, 11, 12 and 13 show that a microstructure with different grain sizes is formed after the first cycle. At the same time, in some areas there is a greater refinement of the grains than in other, neighboring areas. With an increase in the number of cycles, grain refinement is observed in those areas in which it was larger, and further refinement—in areas with fine grains obtained after the first heating cycle.

The chemical composition of steels has a significant effect on the formation of the structure during processing. So, if for steel 08 ps the grain size significantly decreases after the second cycle, then for steel 09G2S, in which ferrite is hardened

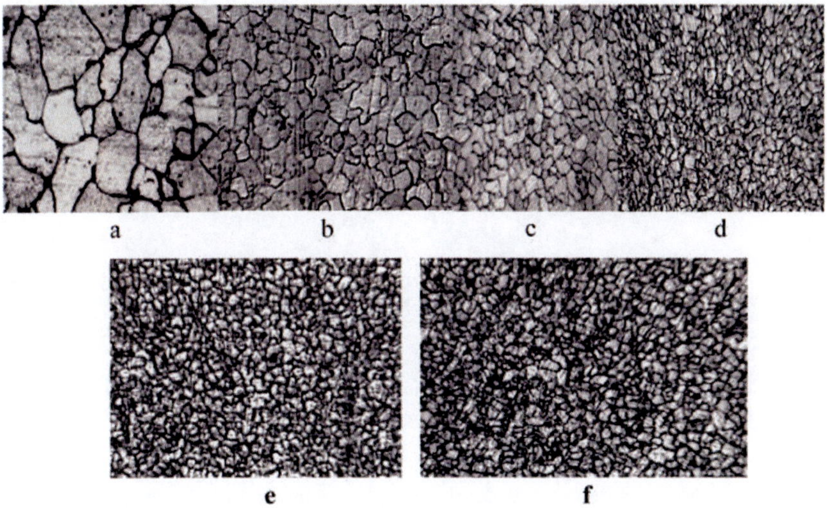

Fig. 10 Change in the structure of 08 ps steel during thermal cycling, × 650: *a*—as-delivered state, *b*–*f*—after 1–5 cycles, respectively

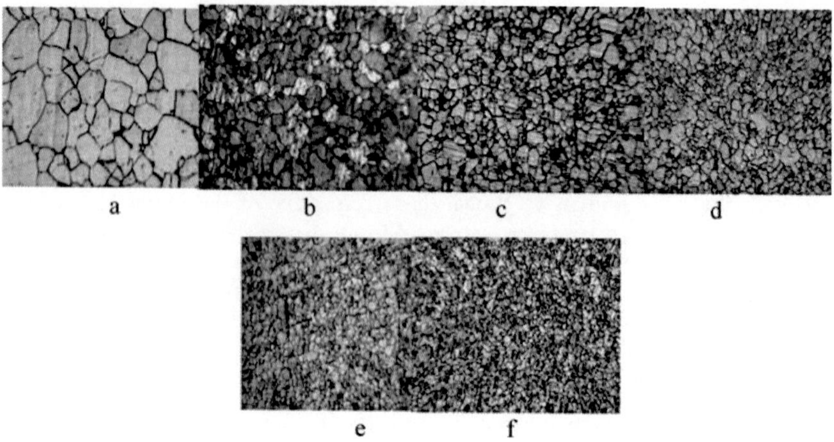

Fig. 11 Change in the structure of 09G2S steel during thermal cycling, × 650: *a*—as-delivered state, *b*–*f*—after 1–5 cycles, respectively

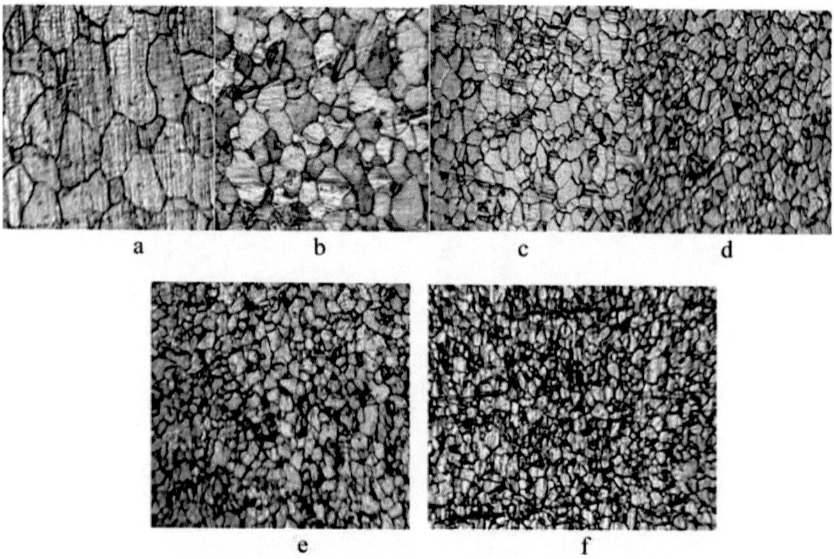

Fig. 12 Changes in the structure of 08 ps steel during thermal cycling, × 650: *a*—state (as-delivered + annealing at 900 °C), *b* – *f*—after 1–5 cycles, respectively

with manganese and silicon, after the third cycle, which is probably caused not only by the specifics of thermal cycling, but also by the influence of alloying elements.

Figure 10f shows that a fairly homogeneous structure with fine grain is formed in 08 ps low-carbon steel in the as-delivered state after five processing cycles, although in some areas, the grain size is slightly smaller than in others.

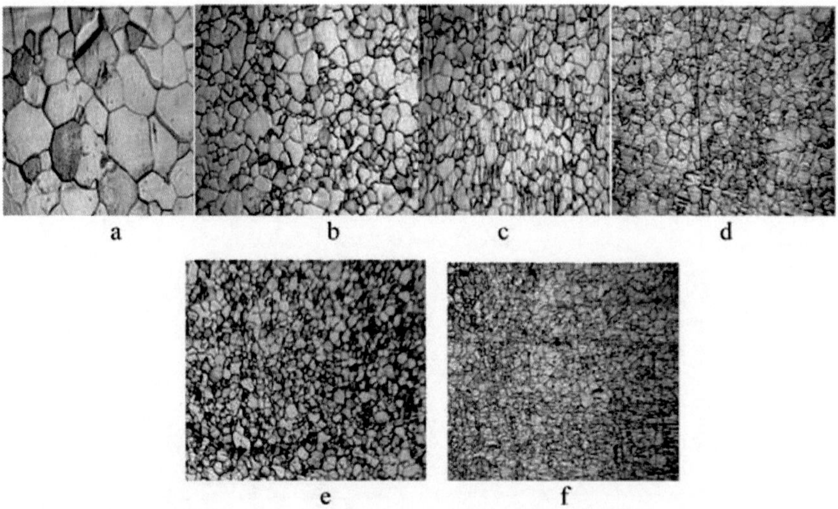

Fig. 13 Changes in the structure of 09G2S steel during thermal cycling, × 650: *a*—state (as-delivered + annealing at 900 °C), *b* – *f*—after 1–5 cycles, respectively

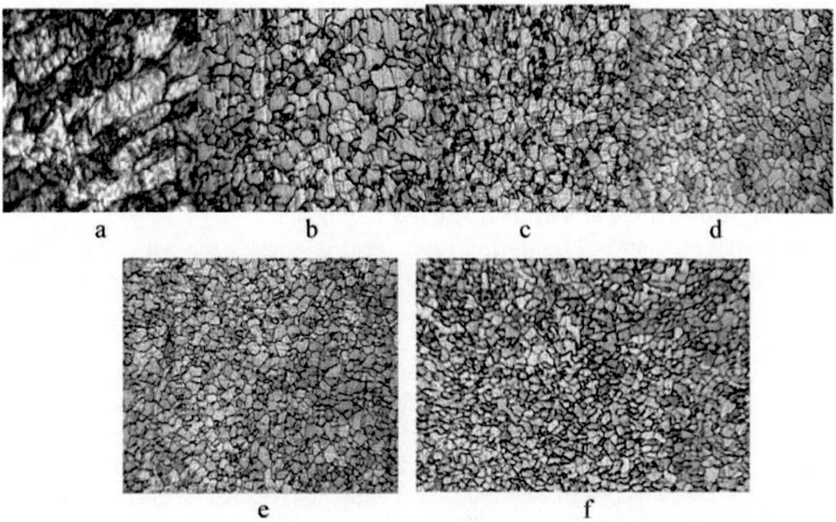

Fig. 14 Change in the structure of 08 ps steel during thermal cycling, × 650: *a*—state (as-delivered + rolling by $\varepsilon = 50\%$), *b* – *f*—after 1–5 cycles, respectively

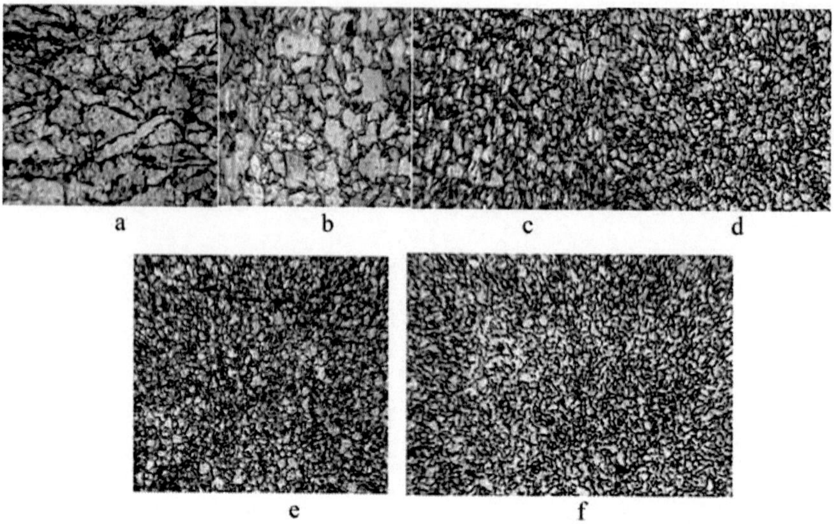

Fig. 15 Change in the structure of 09G2S steel during thermal cycling, × 650: *a*—state (as-delivered + rolling by $\varepsilon = 50\%$), $b - f$—after 1–5 cycles, respectively

The structure of 09G2S low-alloy steel, in which ferrite is hardened with manganese and silicon, somewhat differs from the structure of 08 ps steel. First, the original structure as-delivered has less non-homogeneity and grain size. However, as in the case of 08 ps steel, after the first treatment cycle, a refinement of the structure is observed with a noticeable difference in non-homogeneity in various areas. The final grain size after the fifth cycle is also less important.

Significant refinement of grains in some areas in comparison with others after the first cycle of TCT was also established by metallographic studies in the work. A further increase in the number of treatment cycles refined the microstructure in all areas.

Thus, it can be seen that in 09G2S low-alloy steel, after fivefold thermal cycling, in comparison with 08 ps low-carbon steel, a finer-grained structure is formed. Therefore, in alloyed steels, to obtain a fine-grained structure without noticeable grain variation, a greater number of TCT cycles is required than for carbon steel. Probably, the answer must be sought in the influence of alloying elements on the formation of the final steel structure during the TCT process.

Alloyed steels are distinguished by the fact that the thermodynamic activity of carbon in them is lower than in carbon steels. Therefore, a slowdown in the diffusion processes of cementite dissolution and its release from austenite leads to a shift of the C-shaped curve of isothermal decomposition of austenite to the right, making supercooled austenite more stable. It should be noted that the previously described methods of TCT of carbon steels can be applied to alloyed steels of the pearlite class without significant changes in technology, when the total content of alloying elements does not exceed 5%.

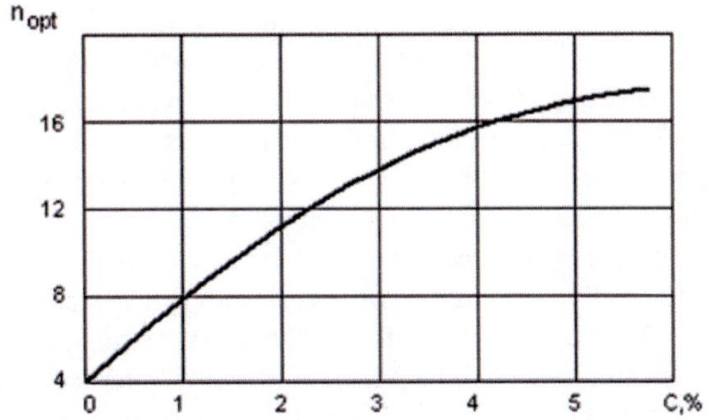

Fig. 16 Influence of the degree of alloying of steel with a carbon content of 0.4% on the number of cycles during medium-temperature thermal cycling in order to obtain a fine-grained structure

Structural changes during TCT of alloy steels occur more slowly. Therefore, the maximum temperature during heating should be increased from 30…50 to 50…70 °C above the Ac1 point. The influence of alloying elements on the number of cycles is such that with an increase in the content of alloying elements in steel, the required number of cycles also increases. Figure 16 shows the dependence of the optimal number of heating–cooling cycles of steel with a carbon content of 0.4% on the total content of alloying elements.

Thus, TCT methods for obtaining a fine-grained structure as applied to low-alloy steels should be corrected taking into account the influence of alloying elements.

In the light of the above, it becomes clear why in 09G2S low-alloy steel (compared to 08 ps low-carbon steel) the process of decreasing non-homogeneity is shifted toward a larger number of cycles. This is explained by the fact that in alloyed steels, the thermodynamic activity of carbon is lower than in carbon steels, and structural changes during TCT occur in them more slowly. Therefore, the influence of alloying elements on the number of cycles in TCT is such that with an increase in their percentage in steel, an increase in the number of cycles is required to obtain an equiaxed fine-grained structure.

The initial microstructure of the steels before TCT significantly affects the final grain size. The coarser-grained initial structure of steels (as-delivered + annealing at 900 °C), in comparison with the structure in the as-delivered state, also corresponds to the coarser-grained structure after heat treatment. Figure 12 shows the structure of 08 ps steel after fivefold heat treatment. Before TCT, the as-delivered samples were annealed for 30 min at 900 °C, as a result of which an equilibrium structure was formed in them. In this case, as in the case of TCT for samples in the as-delivered state, after the first cycle, a structure with different grain sizes is formed in different areas. However, the non-homogeneity in this case is somewhat less, and the grain size is larger. In subsequent cycles, further refinement of the grains occurs, and after the fifth cycle, a fairly homogeneous structure with fine grains is formed. For 09G2S steel, the same structural changes are characteristic as for 08 ps steel. However, after the fifth cycle, a finer-grained structure is formed (Fig. 13).

It should be noted that with an increase in the degree of alloying of steels (08 ps and 09G2S), a finer-grained structure is formed in them, both in the state (as-delivered + annealing at 900 °C) and in the as-delivered state.

The influence of the initial structure on the final grain size after TCT was noted earlier. So, for example, in the work carried out using 5% $Ni + Cr + Mo + V$ steel, the influence of the initial structure on the degree of grain refinement during thermal cycling was established. The preferred starting structure in this steel over others was a hot rolling structure in which carbides were present at the ferrite grain boundaries.

The presence of a deformed structure before TCT introduces significant adjustments for both 08 ps carbon steel and 09G2S low-alloy steel. A significant difference in TCT of pre-deformed samples is the obtaining of a finer-grained structure than that of the samples in the as-delivered state and in the state (as-delivered + annealing at 900 °C), which is probably caused by the activation of structural changes that occur during heating and cooling of steels.

The authors explain the significant refinement of the structure during TCT after cold plastic deformation by the fact that cold deformation redistributes and increases the density of imperfections in the crystal structure of dislocations, vacancies, stacking faults and, in addition, promotes the formation and development of low and high angle boundaries. Since crystal lattice defects strongly affect the formation of the structure of alloys during phase and structural transitions, plastic deformation before them, as well as during their passage, can be effectively used to create an optimal structure during TCT of steels. Thus, it has been shown in this work that TCT of preliminarily cold-deformed samples increases the number of centers of the formed γ-phase and, as a consequence, intensely refines the austenite grain. In addition, during deformation in the intercritical temperature range as a result of the dynamic $\gamma \rightarrow \alpha$ transformation, it is possible to significantly accelerate the recrystallization process, strongly rivet the constituent structures, and refine the grain. Thus, the use of preliminary cold plastic deformation can be effective in TCT of steels carried out in the region of pearlite transformations.

A more intense passage of structural changes during TCT of cold-shaping steels and the final formation of an equiaxed fine-grained structure after the third cycle is confirmed by a change in the H_p values during cyclic annealing. So, Fig. 9 shows that the values of H_p for all studied steels after the fourth and fifth cycles are almost equal, which indicates the end of the effective grinding of the grain size and obtaining a fine-grained structure. The high structural sensitivity of the magnetic method makes it possible to recommend it for monitoring the formation of a structure in the process of thermal cycling in industrial conditions as a more efficient method in comparison with the method of metallographic analysis.

The mechanical properties of the studied steels 08 ps and 09G2S largely depend on the number of TCT cycles. Therefore, the structure formed in the TCT process determines the strength and plastic properties.

Table 1 shows the results of experimental studies of the selected steels in the as-delivery state, depending on the number of TCT cycles. It can be seen that with an increase in the number of cycles, an increase in strength properties occurs, while the plasticity remains almost unchanged. An increase in the yield strength and strength

Table 1 Results of experimental studies of the selected steels in the state of delivery, depending on the number of TCT cycles

Material grade	Mechanical properties	Number of TCT cycles					
		0	1	2	3	4	5
08 ps	$\sigma_{0,2}$, MPa	166.0	172.0	181.5	188.0	190.5	193.0
	σ_B, MPa	281.5	295.5	301.0	309.5	315.0	317.5
	δ, %	24.0	23.5	23.0	23.0	23.5	23.0
09G2S	$\sigma_{0,2}$, MPa	357.0	371.0	385.0	392.0	404.0	405.0
	σ_B, MPa	446.0	468.0	482.0	491.0	499.0	502.0
	δ, %	21.0	20.5	21.0	20.5	20.5	20.5

of steels occurs mainly during the first three cycles of thermal cycling, and then these changes become less significant.

Mechanical tests have shown that an increase in the number of cycles increases both the yield strength and the ultimate strength with almost unchanged plasticity. Thus, the yield stress of the studied steels 08 ps and 09G2S with a coarse-grained structure increases by 12...15% after thermal cycling, and the tensile strength—by 13...14%, respectively, which must be taken into account during thermal reinforcement and subsequent diagnostics of elements of metal structures of lifting cranes in hazardous local zones of stress concentration. The studies carried out confirm the experimental data obtained by us earlier using the passive flux-gate control method and metallographic analysis.

Thus, when using the passive flux-gate method, it becomes possible to stage-by-stage control the formation of a structure with a given degree of dispersion in low-carbon and low-alloy steels already at the early stages of recrystallization annealing and thermal cycling, which is very important both when obtaining laboratory samples that correspond to the structural state of elements of real welded metal structures of lifting cranes and when carrying out the considered types of heat treatment of structural steels in industrial conditions, as well as recommend the use of the method for non-destructive testing of metal when assessing the technical condition of welded metal structures.

4 Conclusions

1. The main regularities of structural transformations and changes in the intensity of the stray magnetic field H_p in the process of heat treatment of pre-deformed structural steels have been established. The features of the formation of the structure of low-carbon and low-alloy steels in the process of heating the metal to high temperatures are revealed. It is shown that the introduction of alloying elements allows the formation of a finer-grained structure.

2. The modes of obtaining structures with a given degree of dispersion characteristic of elements of welded metal structures due to preliminary cold plastic deformation and subsequent high-temperature annealing with a step-by-step control of structural changes by changing the magnetic amplitude H_p have been developed.
3. The relationship between structural changes in steels and the strength of the stray magnetic field H_p during thermal cycling has been revealed. It was found that the value of the magnetic parameter H_p depends on the initial microstructure, the chemical composition of the steels and the number of TCT cycles. It is shown that the greatest changes in H_p are observed during the first three cycles of TCT, and their further increase decreases the magnetic amplitude, which indicates a decrease in the degree of refinement of the initial microstructure in subsequent cycles and is confirmed by metallographic analysis. This character of the change in the parameter H_p can be used to control the formation of a structure with a given grain size in structural steels under industrial conditions.
4. The features of structural changes in structural steels during thermal cycling are revealed. It is shown that an increase in the number of heating–cooling cycles to three leads to an intense refinement of the initial structure, a further increase in the number of cycles to five insignificantly decreases the grain size, but contributes to the elimination of grain size variation. The introduction of alloying elements and preliminary cold plastic deformation of steels makes it possible to form the finest-grained structure. The developed modes of obtaining a structure with a given degree of dispersion during TCT in structural steels with different initial structures with stage-by-stage passive flux-gate control make it possible to use them in industrial conditions.
5. A relationship has been established between the structure formed in the process of thermal cycling, the change in the mechanical properties of structural steels and the magnetic parameter H_p. It is shown that the greatest increase in strength properties occurs during the first three cycles (when an intense refinement of the structure occurs), and insignificantly during subsequent ones, which makes it possible to reinforce the metal in local zones of stress concentration in the elements of operating metal structures.

References

1. Latypova, E.Y., Tsumarev, Y.A., Tsumarev, Y.N.: Svar. Proizvod. **2**, 34–37 (2019)
2. Tabatchikova, T.I., Tereshchenko, N.A., Morozova, A.N., Gudnev, N.Z.: Fiz. Met. Metalloved. **6**, 661–667 (2019)
3. Scherbakov, A., Babanina, A., Graboviy, K.: Adv. Intell. Syst. Comput. **1259**, 298–311 (2021). https://doi.org/10.1007/978-3-030-57453-6_26
4. Lupachev, A.V., Pavlyuk, S.K.: Zavodskaya laboratoriya. Diagnostika materialov **8**, 43–46 (2019)

5. Scherbakov, A., Babanina, A., Kochetkov, I., Khoroshilov, P.: Technical condition of welded load-bearing metal structures of operated agricultural hoisting cranes. E3S Web of Conferences **175**, 11005 (2020). https://doi.org/10.1051/e3sconf/202017511005
6. Scherbakov, A., Babanina, A., Matusevich, A.: Adv. Intell. Syst. Comput. **1259**, 312–323 (2021). https://doi.org/10.1007/978-3-030-57453-6_27
7. Scherbakov, A.P.: Material and method selection for increasing the wear resistance of construction machines components. Russ. Automobile Highw. Ind. J. **17(4)**, 464–475 (2020). https://doi.org/10.26518/2071-7296-2020-17-4-464-475
8. Fireushin, A.M., Fatkullin, M.R., Ulyabaev, R.R., Chernyatyev, R.R., Kadoshnikova, A.V.: Neftegazovoe delo **16**(3), 76–83 (2018)
9. Scherbakov, A., Monastyreva, D., Smirnov, V.: Passive fluxgate control of structural transformations in structural steels during thermal cycling. E3S Web of Conferences **135**, 03022 (2019). https://doi.org/10.1051/e3sconf/201913503022
10. Scherbakov, A.P.: Experimental studies of heat treatment influence on the welded connections properties in working mechanisms of road building. Russ. Automobile Highw. Ind. J. **17(6)**, 664–675 (2020). https://doi.org/10.26518/2071-7296-2020-17-6-665-675

Destruction of Welded Metal Structures of Construction Machines Operated in Corrosive Environments

Alexander Scherbakov, Anastasiya Sklyarova, Alexandr Pushkarev, and Andrey Petrov

Abstract When calculating welded metal structures of construction machines, taking into account the influence of corrosive environments, it is required to develop a design mathematical model, i.e., a set of equations describing the processes of deformation and destruction of a welded metal structure, as well as the processes of interaction of structural metal elements with a corrosive environment. The paper presents three types of mathematical models for calculating metal structures: with unprotected surfaces, taking into account the influence of the stress–strain state, with local corrosion damage. To prove the statements made, the reliability of the selected mathematical models was assessed using the example of a large-scale welded metal truss. Graphical dependencies of values are obtained. Based on the results of the study, functional dependences of the structure, magnetic and mechanical parameters were determined, which allow a reliable assessment of the structural state of the metal and the acting stresses in welded joints and elements of welded metal structures.

1 Introduction

When calculating welded metal structures of construction machines (hereinafter referred to as MS), taking into account the influence of corrosive environments, it is required to develop a design mathematical model, i.e., a set of equations describing the processes of deformation and destruction of a welded MS, as well as the processes of interaction of structural metal elements with a corrosive environment. This approach presupposes: identification of these equations, assessment of the values of the equation coefficients and selection of suitable functions based on the results of experimental studies, solution of the resulting set of equations and analysis of the behavior of welded MS under various conditions [1].

The main feature of the diagnostic model used for individual forecasting should be to provide the ability to combine a priori information about the object, which is

A. Scherbakov (✉) · A. Sklyarova · A. Pushkarev · A. Petrov
Saint Petersburg State University of Architecture and Civil Engineering, Vtoraya Krasnoarmeiskaya street, 4, Saint Petersburg 190005, Russia

M. Shamtsyan et al. (eds.), *Robotics, Machinery and Engineering Technology for Precision Agriculture*, Smart Innovation, Systems and Technologies 247,
https://doi.org/10.1007/978-981-16-3844-2_50

available at the design stage, and a posteriori data obtained during operation. Most of the mathematical models were created to assess the survivability of machines and structures at the design stage [2], but diagnostic models that provide a forecast of the residual life of a welded structure based on operational information obtained by nondestructive methods are of the greatest interest. In this regard, the question arises about the choice of suitable models or ways to develop them.

When studying the effect of an aggressive corrosive environment on the performance of welded MS of construction machines, many authors paid most attention to the widespread type of general (massive) corrosion [3], in which uniform dissolution or destruction of the surface of metal structure elements in a corrosive environment is observed.

The effect of acting stresses on the corrosion rate was first taken into account by V.M. Dolinsky, who proposed a solution to the problem of strength calculation of a stretched thin-walled pipe undergoing continuous corrosion, the rate of which was considered a linear function of the stress intensity.

In work [4], mathematical models and methods are considered and the possibility of a multicriteria approach for optimizing thin-walled elements of MS is discussed. To solve the problem of determining the stress–strain state and durability of thin-walled structures with arbitrary geometry and boundary conditions interacting with a corrosive medium, the finite element method (FEM) is often used.

The FEM method was used in [4]. Mathematical models of the optimal design problems for thin-walled structures with arbitrary geometry and boundary conditions subject to corrosive wear were developed. The works of S.A. Timashev, devoted to the assessment of the reliability of plates and shells [5], works of V.D. Raizer with colleagues on the development of the theory of reliability of structures subjected to corrosive wear and works of V.V. Filippov should be noted [6].

A significant number of works have been devoted to the problem of interaction between MS and corrosive media. However, most of them analyze and systematize experimental data on the effect of corrosive media on the technical state, durability and resistance of construction MS. There are significantly fewer works in which theoretical approaches to predicting the behavior of welded MS in corrosive environments are developed.

2 Materials and Methods

Prediction of the residual life of welded MS of construction machines operating in corrosive environments for a long time requires the involvement of models describing the kinetics of corrosion destruction of welded structures. But changing operating conditions and other factors affecting the course of corrosion processes do not make it possible to create an accurate model. Therefore, mathematical modeling of metal corrosive wear is based on experimental data and then comes down to a mathematical description of the process of changing some basic parameters over time, which include weight loss, corrosion depth, change in cross-sectional area, etc.

The existing mathematical models of corrosion wear of structural steels can be divided into two types: models that describe the physicochemical process of corrosion at the causal level and phenomenological models that describe the kinetics of changes in state parameters in the form of an empirical dependence over time. Many existing models belong to this type.

In this case, the phenomenological model must meet the following a priori requirements:

- it is obligatory to describe the obtained experimental data within the measurement error;
- the function describing the corrosion process should be monotonically continuously increasing, smooth and equal to zero at t = 0, and its derivative (corrosion rate) should be everywhere limited and be decreasing as t → ∞;
- the model should be suitable for describing any corrosion process with various combinations of external factors affecting the corrosion process;
- determination of its parameters does not require the involvement of complex programs.

3 Results and Discussion

3.1 Mathematical Models for Calculating Metal Structures of Construction Machines with Unprotected Surfaces

Many physicochemical and mathematical models often use such a parameter as the depth of corrosion damage to a structural element δ. At the same time, physicochemical models (Table 1) establish the relationship between the parameters of the

Table 1 Physicochemical models

Model type	Source	Formula number
$\delta = \frac{\delta_k^f \cdot t_w \cdot m}{1000 \cdot P}$	[6]	(1)
$\delta = (0,035 + 0,70[SO_2])t$	[7]	(2)
$V_9 = \left[V_{20} + \frac{\Delta k}{\Delta t}\,(T_1 - 20)\right]\tau$	[6]	(3)
$V = \left\lvert V_{01} + \Delta\tau\frac{\Delta k}{\Delta t}\right\rvert n$	[8]	(4)
$M = t_1 \cdot V[W] + t_z \cdot V[W_{a.t}]$	[6]	(5)
$V_1 = b_0[W]\,[SO_2] + b_1[W] + b_2[SO_2] + b_3$	[9]	(6)
$M = (A_{ads} + a[C])\left(t_{ans} - t_{faz}\right) + \left(V_{faz} + b[C]t_{faz}\right)$	[8]	(7)
$\delta = a_1\left(A_{ads}^B\right)([SO_2] + \tau)$	[10]	(8)
$\delta = 9.013\left[\exp(0.016)[SO_2]\right]\left[\left(4.768 \cdot t^{0.7512-0.00582Q_k}\right)\right]$	[3]	(9)
$\delta = A \cdot \exp\left(\frac{\theta}{\rho T_{abs}}\right)t^{\frac{1}{n}}$, where $n = \left[A_1 \exp\left(-\frac{A_2}{\rho T_{abs}}\right)\right] t^{(a+bT_{abs})}$	[7]	(10)

environment (chemical composition of the environment, temperature, etc.) and the parameters of the corrosion process.

In the presented models, the dependences of corrosion on the aggressiveness of the atmosphere and meteorological factors are found. However, the parameters that determine the intensity of atmospheric corrosion (wind speed, intensity of solar radiation, etc.) are not taken into account. It should be noted that these formulas do not take into account the stress–strain state (hereinafter—SSS) of the structure, the scale factor and the position of the structural elements in space.

In this regard, in scientific research and engineering calculations of corrosion damage of a stressed MS, mathematical models are used that represent algebraic or transcendental expressions connecting the damage parameter with time, or differential, integro-differential equations describing the kinetics of corrosion changes taking into account various factors over time.

Table 2 presents the equations related to this group of mathematical models. In this case, the most rational way would be to use a bank of mathematical models of corrosion, which contains detailed information about not only the types of models and the values of coefficients for different conditions, but also their areas of application,

Table 2 Equations related to mathematical models

Model type	Source	Formula number
$\delta = k\left[1 - \exp(\alpha t)\right]$	[5]	(11)
$\delta = k \cdot \lg(a + t)$	[6]	(12)
$\delta = k \cdot t^n$	[7]	(13)
$\delta = \sqrt{a^2 + bt} - a$	[5]	(14)
$\frac{d\delta}{dt} = \frac{t}{at^2+bt+c}$	[8]	(15)
$\delta = \frac{\delta_0 t}{t+T}$	[8]	(16)
$\delta = \alpha \ln^3 t + \beta \ln^2 t + \gamma \ln t + \chi$	[9]	(17)
$t = \frac{\delta}{k} - \left(\frac{\alpha}{2k^2}\right)\left[1 - \exp(-\frac{2k\delta}{\alpha})\right]$	[5]	(18)
$G_1 G_2 \frac{d^2\delta}{dt^2} + (G_1 + G_2)\frac{d\delta}{dt} + \delta = k\chi$	[10]	(19)
$\delta(t) = \gamma\left[1 - \left(\frac{G_1}{G_1 - G_2}\right)\exp\left(-\frac{t}{G_1}\right) + \left(\frac{G_2}{G_1 - G_2}\right)\exp\left(-\frac{t}{G_2}\right)\right]$	[5]	(20)
$\frac{d\delta}{dt} = k\delta$	[8]	(21)
$\frac{d\delta}{dt} = \alpha \exp(-\beta\delta)$	[2]	(22)
$\frac{d\delta}{dt} = a\delta^n \varphi^k$	[3]	(23)
$\frac{d\delta}{dt} = \nu_0 \cdot \exp(-\alpha\, t)$	[4]	(24)
$\frac{d\delta}{dt} = k\delta(b - \delta),\ 0 < \delta < b,\ k > 0$	[5]	(25)
$\frac{d\delta}{dt} = k\delta(b - \delta) + \int_0^1 F(t - \tau)\delta(\tau)\, d\tau$	[5]	(26)
$\frac{d\delta}{dt} = \frac{k}{\rho} 10^{(C\varphi - d)}$	[6]	(27)

recommendations for working with different models and data on structures calculated using models.

Often, different mathematical models are chosen to describe the same process. At the same time, simple and clear criteria cannot be recommended for choosing the required model. In this case, it is necessary to use the bank of mathematical models of corrosion.

3.2 *Mathematical Models for Calculating the Metal Structures of Building Machines Taking into Account the Influence of the Stress–Strain State*

When stress–strain behavior affects the rate of corrosion destruction, mathematical models should be used that take this effect into account. Various models of corrosive wear, taking into account the effect of acting stresses and/or plastic deformations on the kinetics of corrosive wear, are presented in Table 3.

Corrosion wear of MS elements is greatly influenced by their position in space. In this case, the degree of influence of the position of the element in space is estimated by the coefficient Kβ, which represents the ratio of the wear of an arbitrarily located

Table 3 Various models of corrosive wear

Model type	Source	Formula number
$\frac{d\delta}{dt} = V_0 + m\sigma$	[3]	(28)
$\frac{d\delta}{dt} = \alpha \cdot \varphi(t)\ (1 + k \cdot \sigma)$	[7]	(29)
$\frac{d\delta}{dt} = V_0 \exp\left(\frac{A\sigma_0}{RT}\right)$	[8]	(30)
$\frac{d\delta}{dt} = V_0(1 + k\varepsilon)\exp\left(\frac{A\sigma_0}{RT}\right)$	[8]	(31)
$\delta = \alpha t^b \exp(\gamma\sigma)$	[9]	(32)
$\delta = \alpha t^{(b+kS)} \exp(\gamma\sigma)$	[3]	(33)
$\frac{d\delta}{dt} = k\ f(\sigma)\ (b - \delta)$	[3]	(34)
$\delta =$ $\sqrt{t}\left\{K_0 \exp\left(\gamma_0\,\sigma - \frac{u_0}{RT}\right) + K_A\sqrt{C_A}\exp\left(-\frac{u_A}{RT}\right) + K_B\sqrt{C_B}\exp\left(-\frac{u_B}{RT}\right)\right\}$	[8]	(35)
$\frac{d\delta}{dt} = \alpha\exp(-\beta t)\{b - (b-1)\exp(-\gamma\sigma)\}$	[5]	(36)
$\frac{d\delta}{dt} = \alpha\varepsilon(\sigma - \sigma_{por}) + V_0$	[4]	(37)
$\frac{d\delta}{dt} = \begin{cases} A, & \sigma < \sigma_{por} \\ A + k(\sigma - \sigma_{por}), & \sigma > \sigma_{por} \end{cases}$	[5]	(38)
$\frac{d\delta}{dt} = (1 + kE)V(t)$	[6]	(39)

Table 4 The values of the coefficient K_β for various types of steel section

Angle of inclination, β	Value of K_β			
	Pipe, closed section	Sheet, single profile	I-beam	Composite profile
0	1	1	1	1
45	0.6	0.7	0.9	0.8
90	0.4	0.5	0.8	0.6

element to the wear of a horizontally located element. The values of the coefficient K_β for various types of steel sections are given in Table 4.

In these formulas: δ—depth of corrosion; t—time; σ—stress; T—temperature; ε—deformation. The rest of the quantities are determinable coefficients or functions.

3.3 Mathematical Models for Calculating Metal Structures of Construction Machines with Local Corrosion Damage

During long-term operation of welded MS of construction machines, local corrosion damage occurs in their elements, which requires the development of mathematical models that take into account changes in design schemes due to the appearance of such damage. But there are not so many such works devoted to the issues of mathematical modeling of the formation and development of local corrosion damage in welded MS.

The authors of [3] were the first to consider taking into account local operational damage, where it was proposed to take into account local corrosion damage by introducing weakening in the form of elastic hinges of less rigidity than the rest of the structure.

But, unfortunately, this and subsequent works did not take into account the kinetics of damage development over time, which could lead to a redistribution of forces in the structural elements during its long-term operation.

Many mathematical models that take into account local corrosion damage were presented and then developed in works. In these works, local corrosion damage is modeled by the introduction of a certain discrete elastic bond, the kinetics of the change in stiffness of which reflects the effect of a corrosive environment. In this case, discrete elastic constraints can transmit longitudinal, transverse and moment forces. In case of local corrosion damage to a rod element operating in tension or compression, the stiffness of the elastic constraint is determined by the following expression:

$$C = \frac{\mathrm{EA}}{a} \tag{1}$$

where EA—stiffness of the rod at the defect site; a—extent (base) of the defect.

When a rod element operates in shear and receives local corrosion damage, the stiffness of the elastic constraint is determined by the expression.

$$C = \frac{\mathrm{GA}}{a} \tag{2}$$

where GA—shear stiffness of the rod at the place of the defect.

When a rod structural element works in bending and has received local corrosion damage, the stiffness of the elastic constraint is taken to be equal to:

$$C = \frac{\mathrm{EJ}}{a} \tag{3}$$

where EJ—bending stiffness of the rod at the place of the defect.

When changing the stiffness of discrete elastic constraints that simulate local corrosion damage to a structure, both functional and differential dependencies can be used, which include the following:

power

$$C = C_0 - at^b \tag{4}$$

exponential

$$C = C_0 \exp(-at) \tag{5}$$

fractionally rational

$$C = \frac{C_0}{\left(1 + at^b\right)} \tag{6}$$

asymptotic

$$C = \frac{a}{t^b} \tag{7}$$

where a and b—coefficients.

When describing the kinetics of a change in the stiffness of a discrete connection, the differential equation is also used:

$$\frac{dC}{dt} = -\alpha \exp(-bC),\ C(0) = C_0. \tag{8}$$

The solution to this equation at constant coefficients has the form:

$$C = \frac{1}{b}\ln(\exp(bC_0) - abt). \tag{9}$$

The solution of a differential equation of the form:

$$\frac{dC}{dt} = -kC^b \tag{10}$$

with the initial condition $C(0) = C_0$ is the following expression:

$$C = \left(C_0^{1-b} - kt(1-b)\right)^{\frac{1}{1-b}}. \tag{11}$$

In all of the above formulas *a*, *b*, *k* are coefficients that depend on temperature, humidity and other factors characterizing the corrosiveness of the external operating environment.

The stress–strain state of a metal in the zone of local corrosion damage has a significant effect on the kinetics of its development, but due to the lack of experimental data, it is still difficult to take this effect into account.

The presence of localized corrosion damage in statically definable rod structures can increase the magnitude of deformations and violation of geometric immutability, but these effects are determined using conventional methods of strength analysis [7]. At the same time, the appearance of corrosion damage in statically indeterminate systems leads to a redistribution of forces in the elements of welded MS, while dangerous cross sections can move to other zones. It follows from this that it is impossible to preset a design scheme that would take into account the occurrence of such damage, since the distribution of the acting stresses and forces depends on the appearance and further development of corrosion damage. Therefore, when local corrosion damage appears and develops, it is necessary to consider more and more new design schemes each time, passing from one scheme to another, while the calculation of the structure, taking into account the kinetics of local damage, must be carried out by a step method, to determine the laws of distribution of forces and stresses at the end of each step and check the conditions of corrosion resistance of suspicious sections. In this case, it is necessary to pay attention to whether local corrosion damage will appear in this place. The appeared defect is modeled by a discrete elastic constraint with a certain stiffness, and the subsequent calculation is carried out according to the design scheme taking into account the already introduced elastic constraint. In the absence of changes in the design scheme, it is necessary to take into account and reflect in it all local changes occurring over time.

3.4 Assessment of the Reliability of the Selected Models Using the Example of a Large-Scale Welded Metal Truss

Many welded MS operate in a flat stress state, and accordingly, the calculation is given by the formula (Table 3):

$$\frac{d\delta}{dt} = V = V_0 + m\sigma_i \tag{12}$$

where δ—wear depth; t—time; V_0—corrosion rate of unstressed metal; σ_i—stress intensity; and m—coefficient taking into account the effect of the stress state on the corrosion rate.

The assessment took the experimental data on the kinetics of corrosion wear shown in Fig. 1.

Equation (51) is reduced to the form.

$$\frac{V}{V_0} = 1 + \gamma \cdot \sigma_i \tag{13}$$

where $\gamma = m/V_0$, and the target function is taken as the following expression:

$$\Phi(\gamma) = \sum_{j=1}^{n} \left(\frac{V_j}{V_0} - 1 - \gamma\sigma_{ij} \right)^2. \tag{14}$$

Minimizing the functional with respect to γ and equating the resulting expression to zero, we obtain:

$$\gamma = \frac{\sum_{j=1}^{n} \left(\frac{V_1}{V_0} - 1 \right)\sigma_{ij}}{\sum_{j=1}^{n} \sigma_{ij}^2} \tag{15}$$

where n—the number of experimental points on the curve of the dependence of the corrosion rate on the level of stress state.

At the end of the calculations, the following values of the coefficient γ were obtained:

- in the case of a uniaxial stress state $\gamma = 0.0120$ (kgf/mm^2)$^{-1}$;
- in the case of a plane stress state $\gamma = 0.0235$ (kgf/mm^2)$^{-1}$;
- the corrosion rate of the samples in the unstressed state was 0.008 mm/h.

Figure 1 shows the results of calculation by formula (52) using the found values of the coefficients (straight lines), and the dots show the experimental data. As can be seen from the figure, the calculation results coincide with the experimental data.

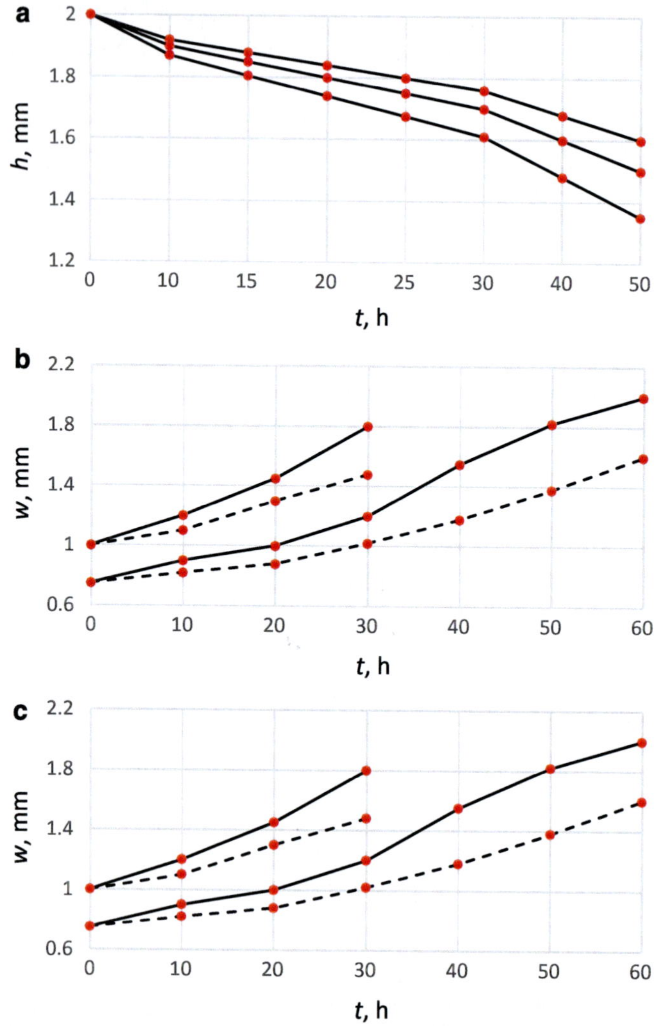

Fig. 1 Experimental and calculated values of **a** thicknesses, **b** deflections and **c** stresses in the center of a round plate made of St3 steel in a 30% HCl solution

When assessing the adequacy of the corrosive wear model according to formula (51), we used the problem of changing the stress state of a circular plate subjected to corrosive wear by the expression:

$$\frac{dh}{dt} = -\frac{d\delta}{dt} = -V_0(1 + \gamma\sigma_i). \quad (16)$$

Experimental data on plate bending under conditions of corrosive wear are taken from the work. In our experiment, pure bending of a circular plate with a bending moment unchanged in time was realized. Therefore, for the elastic case, the following relation will be valid:

$$\sigma_i(t) = \frac{\sigma_{i0} h_0^2}{h^2(t)} \tag{17}$$

where σ_{i0}—initial level of stress intensity in the plate and h_0—initial plate thickness.

Substituting (46) into (47), we obtain:

$$\frac{dh}{dt} = -\frac{\beta + V_0 h^2(t)}{h^2(t)} \tag{18}$$

where $\beta = \gamma\, V_0\, \sigma_{i0}\, {h_0}^2$.

Integrating (46) with the initial condition $h(0) = h_0$, we get the following expression:

$$t = \frac{1}{V_0}\left\{h_0 - h\sqrt{\frac{\beta}{V_0}\left[\operatorname{arctg} h_0\sqrt{\frac{V_0}{\beta}} - \operatorname{arctg} h\sqrt{\frac{V_0}{\beta}}\right]}\right\}. \tag{19}$$

Calculations using expression (58) were carried out for the values of the coefficients of the corrosive wear equation determined for the plane stress state. In Fig. 1, the dots show the experimental values taken from the experimental values, and the lines show the values of the thicknesses, deflections, and stresses calculated using (16) in the center of a round plate made of St3 steel in a 30% HCl solution at different times. The dashed lines in the same figure correspond to the calculation without taking into account the effect of stresses on the rate of corrosive wear.

A relatively good agreement between the calculation results and experimental data indicates the adequacy of the used model of corrosive wear to the described corrosion process, which makes it possible to recommend it for calculating elements of welded structures in a plane stress state [10].

In this regard, the following can be stated:

- increased SSS leads to a significant increase in the corrosion rate, and its rate is influenced not only by the level, but also by the type of SSS;
- when using different models of corrosive wear, their extrapolation capabilities should be checked not only outside the base time interval, but also under the influence of such accelerating factors as temperature, stress–strain rate, etc.;
- in a plane stress state, the stress intensity can be used as the equivalent stress in the models of corrosive wear, but this requires verification of the correctness of the use of the coefficients of the models determined under experimental conditions.

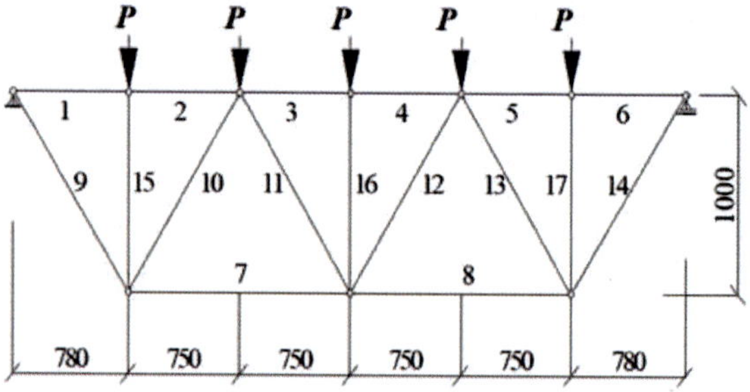

Fig. 2 Loading diagram of a large-scale model of a welded truss

Table 5 Large-scale flat truss bar section dimensions

Element number	Section	Forces (kN)
1, 2, 5, 6	$2L75 \times 6$	– 110.10
3, 4	$2L75 \times 6$	– 194.89
7, 8	$2L75 \times 6$	173.69
9, 14	$2L70 \times 50 \times 8$	179.58
10, 13	$2L70 \times 50 \times 8$	– 106.57
11, 12	$2L70 \times 50 \times 8$	35.33
15, 16	$2L40 \times 4$	– 58.29

An experimental study of the stress–strain state of a large-scale model of a flat truss operated for a long time in a corrosive environment was carried out (Fig. 2). The dimensions of the cross-sections of large-scale rods and the forces are given in the list of elements of a flat truss (Table 5).

The following models of corrosive wear were used in the work:

- exponential (85), where $\delta_0 = 0.04745$ mm, $R = 0.52$ year;
- fractional linear (86), where $\delta_0 = 0.4679275$ mm, $S = 0.2172$ year;
- logistic (87), where $\delta_0 = 0.57$ mm, $b = 0.014$ mm, $K = 117.1$ mm/month.

It should be noted that the calculations did not take into account the effect of SSS on the corrosion process, and its effect on the change in SSS was taken into account by reducing the cross-section of the rods. The cross-sectional area was calculated by the formula.

$$A_{\delta(t)} = k\,\delta^2_{(t)}\,L_0\delta_{(t)} + A_0 \tag{20}$$

where k—coefficient reflecting the type of profile; L_0—initial perimeter; A_0—initial area; and $\delta(t)$—thinning of the section.

The change in the calculated effective stresses over time during long-term operation under conditions of exposure to an aggressive corrosive environment is shown in Figs. 3, 4, 5, 6, 7, 8 and 9. In this case, the numbers of the curves shown correspond to the numbers of the truss rods in Fig. 2 and in Table 5.

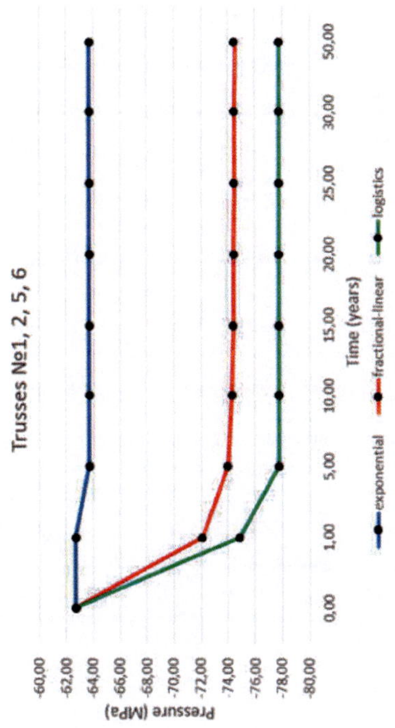

Fig. 3 Changes in the design internal stresses in the elements of large-scale welded truss Nos. 1, 2, 5 and 6 during operation under the influence of a corrosive environment

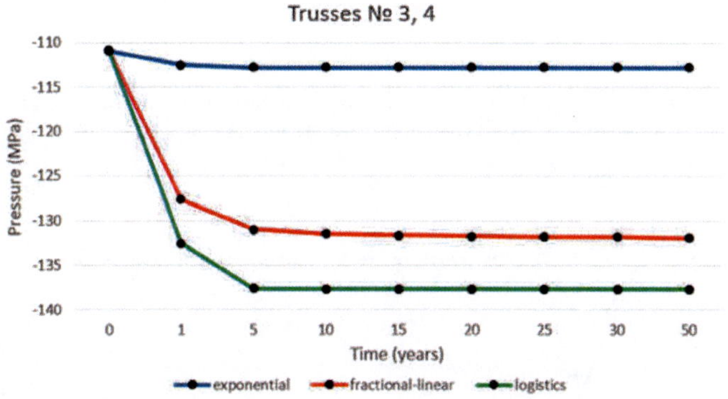

Fig. 4 Changes in the design internal stresses in the elements of large-scale welded truss Nos. 3 and 4 during operation under the influence of a corrosive environment

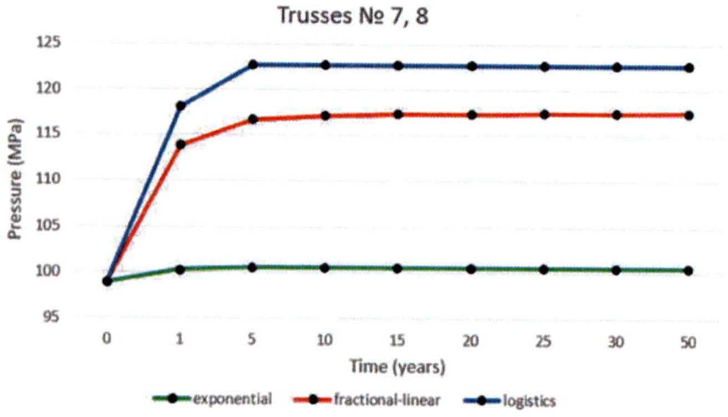

Fig. 5 Changes in the design internal stresses in the elements of large-scale welded truss Nos. 7 and 8 during operation under the influence of a corrosive environment

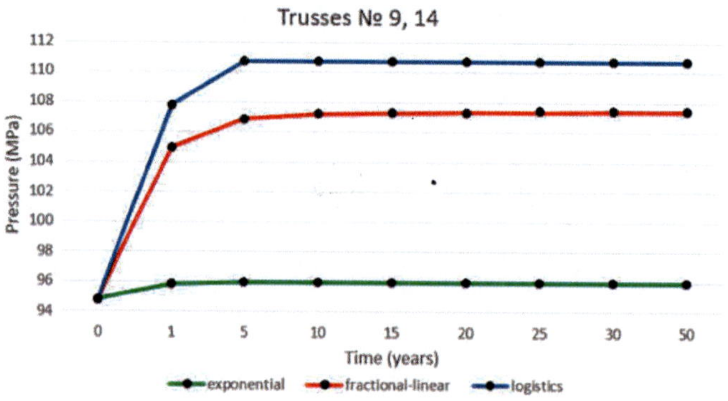

Fig. 6 Changes in the design internal stresses in the elements of large-scale welded truss Nos. 9 and 14 during operation under the influence of a corrosive environment

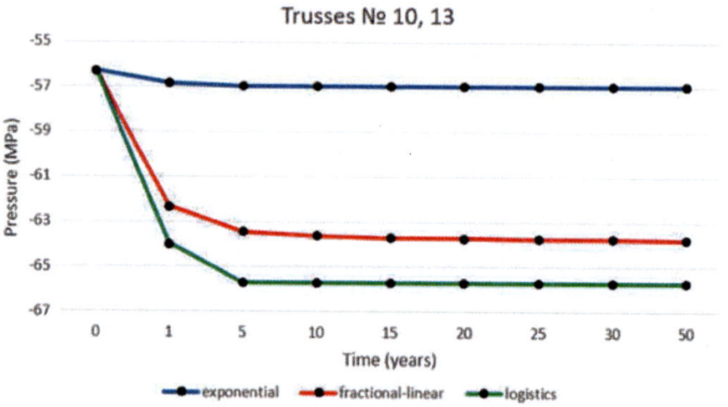

Fig. 7 Changes in the design internal stresses in the elements of large-scale welded truss Nos. 10 and 13 during operation under the influence of a corrosive environment

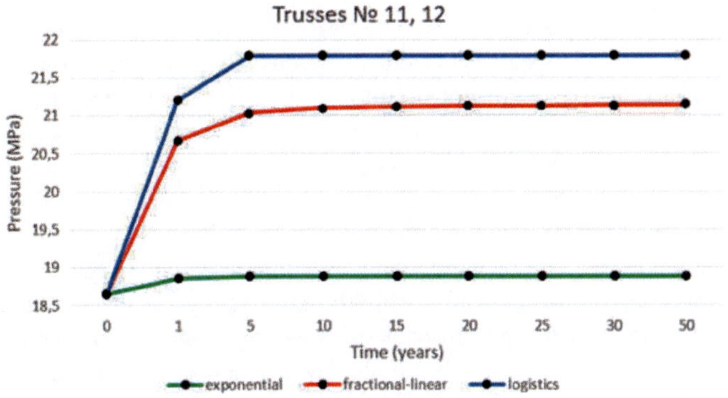

Fig. 8 Changes in the design internal stresses in the elements of large-scale welded truss Nos. 11 and 12 during operation under the influence of a corrosive environment

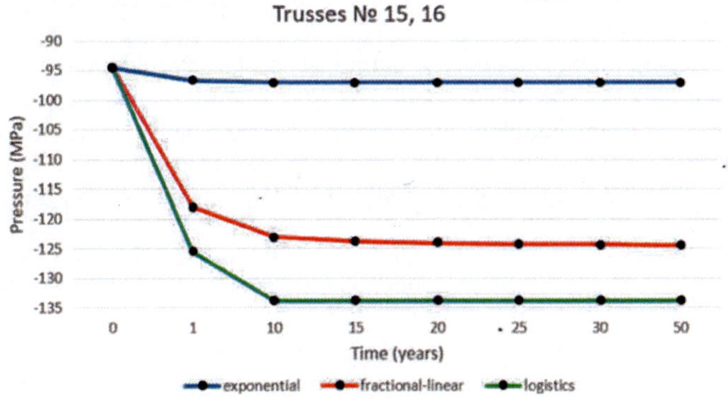

Fig. 9 Changes in the design internal stresses in the elements of large-scale welded truss Nos. 15 and 16 during operation under the influence of a corrosive environment

The calculation of the acting stresses in the elements of a large-scale welded truss, taking into account the influence of a corrosive environment, shows a fairly good coincidence of the calculated values of the linear fractional and logistic models, which makes it possible to recommend these models for calculating the elements of the truss structures in a plane stressed state (if the coefficients are determined based on the results of the flat stress tests).

4 Conclusions

In the process of long-term operation, all welded MS of construction machines, regardless of the purpose, are exposed to aggressive corrosive environments, which leads to a change in the mechanical characteristics of metals and, as a consequence, to a change in VAT and, accordingly, a decrease in MS reliability. Therefore, when calculating welded MS operating in corrosive environments for a long time, it is necessary to take into account not only uniform surface corrosion, but also local corrosion in individual sections, which can lead to a change in the design scheme and a significant redistribution of forces and the onset of a limiting state.

The performed analysis shows that many issues of strength and durability of welded structures operating in corrosive environments have been little studied. But at the same time, there is an increase in ongoing research on the interaction of welded MS with corrosive environment. In the existing regulatory documents, many issues of accounting for an aggressive corrosive environment are almost not reflected. Consequently, further studies devoted to the development of methods for calculating MS taking into account the effect of corrosive environments are relevant.

Calculation of welded MS operating in corrosive environments and subject to significant wear should include the following stages:

- analysis of experimental data and selection of the required model from the available data bank;
- determination of the values of the coefficients of the selected model of corrosive wear according to experimental data;
- development of a methodology and algorithm for calculating welded MS, creating a program;
- execution of calculations and analysis of the obtained calculation results.

Special difficulties in choosing a model lie in the fact that it is necessary to take into account both the obtained experimental data and the predictive capabilities of the chosen model.

The assessment of the suitability of the calculation results should be carried out using verification calculations consisting of solving a small number of specially selected problems that reproduce the qualitative results of the experiments quite accurately. In this case, the calculations performed must show the reliability of the results of the performed analysis.

However, sufficient consistency with the results of the verification calculation does not guarantee sufficient reliability, since the wrong choice of the corrosive wear model can be improved by a good choice of the metal model. At the same time, the insufficient correspondence of the calculation results can be explained by incompetent simplifications in the design scheme of the welded MS when constructing a metal model, although the model itself describes the occurring processes quite well.

Thus, when carrying out strength calculations and developing different methods, it is necessary to keep in mind that any type of corrosion (local or general) contributes to the appearance and development of hazardous stress concentration zones, i.e., potential places of destruction of welded joints and elements of welded MS, the identification of which is a rather difficult task. The use of the passive fluxgate method in assessing the technical state of welded MS in the process of continuous scanning of the test surface makes it possible to identify not only kinetic load zones and corrosion damage, but also the degree of their danger, as well as to indirectly determine the acting stresses, which contributes to an increase in the degree of reliability of the research results.

References

1. Scherbakov A., Babanina A., Kochetkov I., Khoroshilov P.: Technical condition of welded load-bearing metal structures of operated agricultural hoisting cranes. E3S Web of Conferences, vol. 175, 11005 (2020). https://doi.org/10.1051/e3sconf/202017511005
2. Scherbakov A., Monastyreva D., Smirnov V.: Passive fluxgate control of structural transformations in structural steels during thermal cycling. E3S Web of Conferences, vol. 135, 03022 (2019). https://doi.org/10.1051/e3sconf/201913503022
3. Scherbakov, A., Babanina, A., Graboviy, K.: Adv. Intell. Syst. Comput. **1259**, 298–311 (2021). https://doi.org/10.1007/978-3-030-57453-6_26
4. Lupachev, A.V., Pavlyuk, S.K.: Zavodskaya laboratoriya. Diagnostika materialov **8**, 43–46 (2019)
5. Vedyakov, I.I., Odessa, P.D., Gurov, S.V.: Stroitel'naya mekhanika i raschet sooruzhenij **2**(277), 68–75 (2018)
6. Scherbakov, A., Babanina, A., Matusevich, A.: Adv. Intell. Syst. Comput. **1259**, 312–323 (2021). https://doi.org/10.1007/978-3-030-57453-6_27
7. Scherbakov, A.P.: Material and method selection for increasing the wear resistance of construction machines components. Russ. Automobile Highw. Ind. J. **17**(4), 464–475 (2020). https://doi.org/10.26518/2071-7296-2020-17-4-464-475
8. Fireushin, A.M., Fatkullin, M.R., Ulyabaev, R.R., Chernyatyev, R.R., Kadoshnikova, A.V.: Neftegazovoe delo **16**(3), 76–83 (2018)
9. Plotnikov, D.G.: Nauchno-tekhnicheskie vedomosti SPbGPU **1**(214), 186–193 (2015)
10. Scherbakov, A.P.: Experimental studies of heat treatment influence on the welded connections properties in working mechanisms of road building. Russ. Automobile Highw. Ind. J. **17**(6), 664–675 (2020). https://doi.org/10.26518/2071-7296-2020-17-6-665-675

Design Process of Medical Devices for Robotic Rehabilitation: An Open Innovation-Inspired Approach

Raffaele Formicola, Federica Ragni, Alberto Borboni, and Cinzia Amici

Abstract The design process of innovative medical devices, and especially robotic systems for rehabilitation purposes, entails the collaboration among different stakeholders. Moreover, regulations strongly affect the products development, e.g. providing unexpected technical requirements and design constraints. In this paper, an innovative open innovation-inspired model for the design process of medical devices is presented. This model aims at optimizing the employed resources (such as time and cost) and allows overcoming the limits of the traditional linear design model, and specifies i) which stakeholders are involved in the different phases of the process, and ii) the role of the regulations. Special emphasis is placed on the contribution of users and operators (as physiotherapists or caregivers), in addition to physicians and engineers. The contribution of the regulatory aspect is analysed both from a general point of view, with international validity, and punctually, referring to MEDDEV 2.7/1 guideline as illustrative case study. Strength points and limits of the proposed model are highlighted, evaluating examples of incremental and disruptive innovation, like the adoption of new materials or the implementation of functional innovations.

1 Introduction

Availability and effective use of essential medical devices play an important role in the delivery of quality health services. Furthermore, the ageing of society and the necessity to satisfy the population needs have led to a more inclusive design of medical devices, considering the final user expectations.

In the biomedical field, a common and traditional design approach provides that the physician, as client, defines the functional requirements of the system, such as maximum force or acceleration needed, but potentially also expected dimensions (as portability) and desired final costs. The technician, e.g. the engineer, develops instead

R. Formicola · F. Ragni (✉) · A. Borboni · C. Amici
Department of Industrial and Mechanical Engineering, University of Brescia, via Branze 38, 25125 Brescia, Italy
e-mail: f.ragni@unibs.it

M. Shamtsyan et al. (eds.), *Robotics, Machinery and Engineering Technology for Precision Agriculture*, Smart Innovation, Systems and Technologies 247,
https://doi.org/10.1007/978-981-16-3844-2_51

more technical specifications of the device, such as actuation, kinematic structure, and constraints. After a first prototyping proposal for the client, the device is tested on the field, and at the end of this stage, a new approval of the system by the client is required in a final check.

The necessity to balance the users' needs with commercial considerations represents a possible limit in this linear design path. Moreover, since the medical, and in particular the rehabilitative, environment involves a complex interaction among design, user needs, and therapy procedures [1, 2], the limited access and the unfamiliarity with the environment [1] can increase the possible issues.

For these reasons, the principle of open innovation has frequently been applied in medical devices sector, allowing more open and collaborative approaches. This principle has accelerated innovation through the use of expertise coming from several knowledge fields [3, 4]. It also allows limiting costs, risk sharing, and time to market. Using the open innovation model in the development of medical devices enables to consider all the involved stakeholders of the project [5].

Actually, considering the possible design methods for medical devices, literature suggests that both physician and engineer participate with a strong interaction [6] in the development and dissemination phases of medical and rehabilitative devices since the earliest design phases [7]. Collaborations among engineers and physicians have increased innovative outcomes and patient satisfaction [8, 9], improved access and reduced healthcare costs [6]. Actually, this multi-disciplinary [6] and inter-disciplinary [10] team can actively cooperate to identify clinical issues and unmet needs and develop new technologies to address those needs and improve patient care [11]. By increasing clinical involvement from the earliest stages of the device design, a more practical approach to disruptive device development can be driven from basic research level to advance healthcare [10].

Furthermore, since the 1980s user's demands and needs have been increasingly considered in the design and development process [12] of the medical devices, and the scientific literature has enlightened and stated the increasing of innovation in medical equipment [13], if the final user can provide feedback during all the development stages [3, 14]. This is particularly true for robotic rehabilitation devices, such as robotic training for post-stroke patients [15–18], since for those applications engaging the user assumes a crucial importance for the efficiency of the rehabilitation training [19–21].

Besides, the cooperation between the involved stakeholders can fill knowledge gaps [10] and combine multiple perspectives to cause a significant paradigm shift in patient care innovations [8, 22].

The necessity of increasing (i) consistency within the system, (ii) the oversight on the highest risk medical devices, and (iii) transparency [10] led many countries to the introduction of regulatory instruments [23, 24]. Nevertheless, this can impact on device development processes and commercial competitiveness. Moreover, in the design and development process, the manufacturer must consider the requirements and obligations of the normative framework [3].

According to this scenario, this paper presents an integrated design process model for medical devices, inspired to the open innovation paradigm. The model grounds on two main pillars: (i) the involvement of all the potential stakeholders in the design process since the early stages of the new product development, and (ii) the evaluation of the normative framework and the subsequent identification of regulatory-related requirements and constraints since the concept conceiving.

2 Materials and Methods

2.1 The Normative Context

Medical devices vary greatly in field of application and in complexity. Each country's regulatory authorities define different classes of medical devices, depending on their use, complexity, characteristics, and other parameters. Moreover, for each class, each authority requires different modalities to place medical devices on the market [25].

Within this complex scenario, among the various references regulating medical devices, particular attention has been paid to the MEDDEV 2.7/1 guideline [26]. This document, produced by the European Commission and legally not binding, states the guidelines to correctly perform the clinical evaluation of a medical device, for enabling the product Conformité Européenne marking (CE Mark). According to MEDDEV 2.7/1, the result of the clinical evaluation is the Clinical Evaluation Report (CER), which contains two main sections: (i) clinical background and (ii) demonstration of equivalence.

In the first part, a systematic review concerning the field of application of the device under examination is required. For the second section, the demonstration of equivalence can be fulfilled through different possible methods: (a) comparing the new device with an equivalent device already on the market, (b) collecting clinical data of the device under examination to certify its safety, performance and clinical efficacy, or (c) thanks to a combination of the previous two options. For each of the proposed methods, the guideline requires the evaluation of clinical, technical, and biological factors.

2.2 The Soft Open Innovation Method Adopted by UniBS

Given the limitations of the traditional approach presented, in 2016 UniBS Applied Mechanics research team (AMrt) developed its soft open innovation model [7], reported in Fig. 1. This model is the result of more than 10 years of research and analysis of practical applications.

This model overcomes the intrinsic limits of the traditional linear approach through the cooperation of different expertise since the beginning of the design process. Nevertheless, this model does not consider possible constraints derived by the normative regulations.

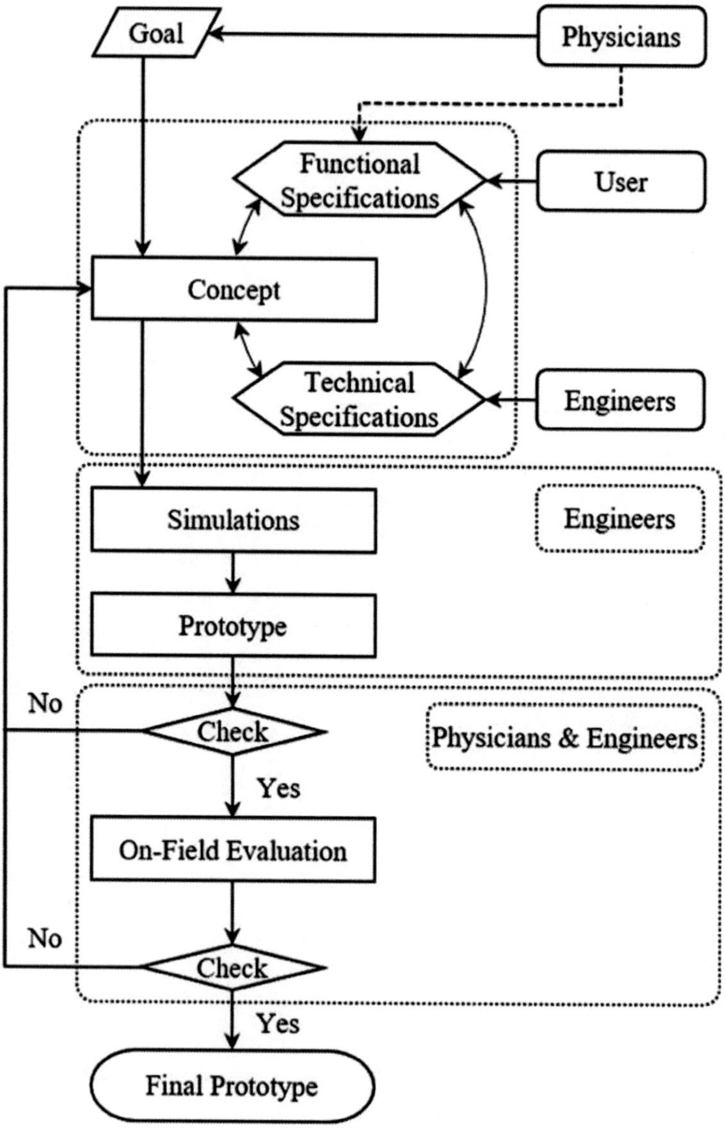

Fig. 1 Soft open innovation model

3 Results

Considering how the normative framework affects the design and development of medical devices, the open innovation model depicted in Fig. 2 has been conceived. In

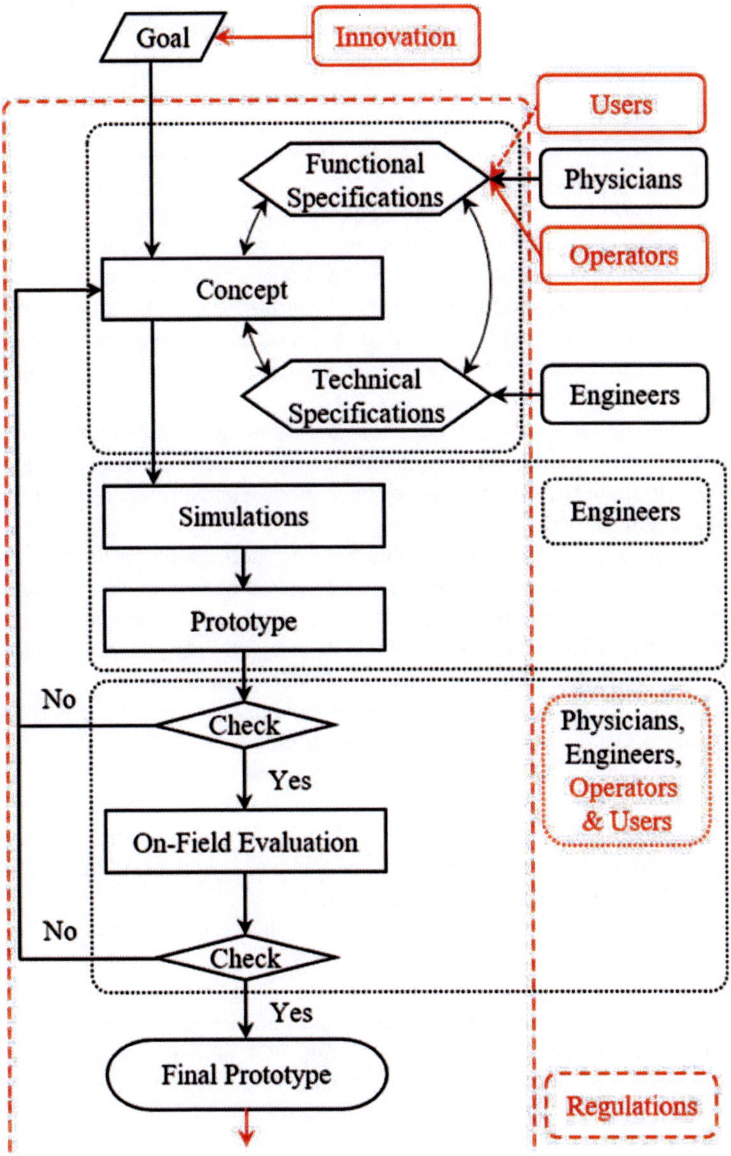

Fig. 2 Open innovation-inspired model for the design process of medical devices

particular, the model implements the regulatory constraints required to place medical devices on the market and specifies the stakeholders involved in the various phases.

According to this model, the medical device aims to fulfil a specific unmet need that requires innovation. The innovation that drives the project does not necessarily derive from a physician, but can arise from: (i) patients or operators, (ii) the market, or (iii) other stakeholders.

For the definition of the concept, the functional specifications have to be translated into the technical ones. Unlike the previous model of *soft* open innovation (Fig. 1), not only the physicians, but also operators (e.g. physiotherapists or caregivers) and users are involved in defining the functional specifications. Within the process, the user (associated with a dotted arrow in Fig. 2) plays a crucial role offering a fundamental contribution to the assessment of the project requirements, although often does not have decision-making power.

The detailed design and early production phases, including simulations and prototyping, remain tasks mainly oriented to engineering-related expertise. Although not explicitly stated in the model, even during these phases, update meetings with clinical actors are recommended.

Once the prototype has been produced and has successfully passed the relevant check, an "On-field evaluation" is required. In the soft open innovation model, only physicians and engineers were involved at this stage, whereas the new model includes operators and users. Their involvement at this stage is justified by the need of promptly verifying the fulfilment of requirements closely related to them: for instance, issues related to the setting of the machinery, which mainly interest operators, or ergonomic and usability measures, which mainly concern users.

After passing the "On-field evaluation" check, the result is the final prototype. Since the Beginning-of-Life (BOL) phase is over, the device is ready to follow the next stages of the life cycle [27]: the Middle-of-Life (MOL), which includes the product use and activities of support and maintenance, and the End-of-Life (EOL), which includes the product disposal or recycling.

In addition to specifying more precisely the involved stakeholders, the open innovation-inspired model explicitly considers the contribute of regulations. Within this context, the term regulations aims referring to the normative framework, i.e. standards, legislative regulations, directives, and guidelines. As depicted in Fig. 2, the regulatory aspect also affects the MOL and EOL phases, covering the entire product life cycle.

4 Discussion

The proposed open innovation-inspired model represents a tool to optimize the use of resources required to develop an innovative medical device. This model presents a general validity, but has been conceived and primarily optimized for the development of medical devices for both orthopedical and neurological [16, 28] robotic rehabilitation, as required in the training of post-stroke patients.

Setting the final purpose of the designed device allows a more precise definition of the involved stakeholders. Within the model, their classification into the proposed categories (e.g. user or operator) should be considered according to a functional rationale. In fact, in case of a portable device, which allows the subject an autonomous use, the patient would be both the device user, as the subject receiving the training, and the operator, as the subject who activates the system. Besides, a proper and complete identification of the potential stakeholders allows optimizing time and costs of the overall design process, reducing the efforts to achieve the final product. For instance, omitting users and operators in the "On-field evaluation" phase may lead to significant evaluation errors on the device feasibility or effectiveness. In fact, the development process could result in a technically adequate device that reproduces the exercises and movements required by physicians but does not pass the "On-field evaluation" check. This result could be due to a lack of ergonomic measures, which particularly affect users, or excessively long set-up times, which mainly affect operators.

The functional identification of the stakeholders represents a first classification attempt of roles and expertise, but each category could be further analysed, and sub-categories could be identified as well: for instance, design and process engineers could be described within the engineers class [29]. The accuracy level considered for the model, meant as the level of categories included in the analysis, reflects the aim of the study. For example, literature investigates the relation among expertise within the same field and states that cooperation within the engineering area is essential to achieve innovative solutions while optimizing available resources. To avoid the traditional "over the wall" approach, methods which promote the collaboration among engineers are proposed, like the concurrent engineering approach [30], also called simultaneous engineering or integrated product development (IPD).

Focusing on regulations, in the design process of innovative products, especially in the medical field, a fundamental role is played by the normative framework. As Fig. 2 depicts, the regulatory aspect affects the whole design process and is not limited to the definition of the final prototype. For instance, some regulations concern the post-market phases of the product life cycle [31], others may cover the entire device development process, such as standards related to the quality system [32] or the Design Control Guidance for Medical Device Manufacturers in the FDA world [33], and further regulations may intervene at specific stages. Among the latter we can mention the MEDDEV 2.7/1 [26], which intervenes in the "On-field evaluation" phase, demanding the comparison of the performance assured by the developing device and by the comparable products already on the market. This evaluation can be reasonably performed for products which represent incremental innovations, but could become a challenging task in case of disruptive innovations, such as brand new devices or systems presenting innovative characteristics, e.g. by functionality like innovative exercises or by construction like innovative materials.

For example, the MEDDEV 2.7/1 and other regulations require the biological compatibility [34] with the subject for the device parts coming into direct contact with the patient. Consequently, when innovative materials are implemented, it dedicated tests to demonstrate their biological compatibility need to be performed. In the same way, if an innovative rehabilitation exercise is proposed by the medical device, its clinical effectiveness must be demonstrated with experimental data. For both these examples, a considerable optimization of time, and therefore cost, can be achieved by considering the regulations since the earlier stages of the device project. In fact, the assessment of biological compatibility or the clinical relevance of a given innovative exercise can be proved while performing other activities, optimizing the overall time schedule, and anticipating possible criticalities.

Furthermore, regulations support, or in some cases require, the collaboration among experts from different knowledge fields as mean for the optimization of resources. For example, the MEDDEV 2.7/1 requires at least one evaluator with experience in the clinical field and one with experience in the technical field, when preparing the CER.

A potential limitation of the proposed open innovation-inspired model is that management figures, such as project managers, are currently not included in the schematization. Actually, since the project management tasks are performed during all the phases of the project, their contribute has been considered as a constant factor, running in background. A further potential limitation of this approach is that the proposed model focuses on the early stages of the product development, and only describes the process up to the final prototype phase (BOL). However, further investigations on the other stages (MOL and EOL), from industrialization to the end of product life, are currently ongoing, so that future improvements of the work could derive by the integration of these analyses.

5 Conclusions

This paper presents an open innovation-inspired model for the design of innovative medical devices for robotic rehabilitation, which aims at optimizing the use of resources, such as time and costs.

The design process of medical devices must involve the collaboration of different figures. Among them, in addition to physicians and engineers, operators and users are also crucial. Compared to the previous *soft* innovation framework, the proposed model specifies more in detail which stakeholders are involved in the different phases of the design process and makes explicit the contribution of the regulations. The method emphasizes the importance of the project team awareness about the regulations, especially in critical environments like the medical field, to detect possible hidden constraints or design requirements, and therefore anticipate potential criticalities of the product development.

References

1. Hagedorn, T.J., Grosse, I.R., Krishnamurty, S.: A concept ideation framework for medical device design. J. Biomed. Inform. **55**, 218–230 (2015)

2. Azman, N., Muhd Nor, N.H., Hassan, M.F., Jamian, S., Abd Rashid, M.H., Omar, B., Mohd Muji, S.Z., Mohamad, F., Ahmad, M.F., Turan, F.M., Bakar, E.A.: Review on design for medical device. MATEC Web Conf. **135**, 1–10 (2017)
3. Peter, L., Hajek, L., Maresova, P., Augustynek, M., Penhaker, M.: Medical devices: regulation, risk classification, and open innovation. J. Open Innov. Technol. Mark. Complex. **6**, 42 (2020)
4. Barrett, G., Dooley, L., Bogue, J.: Open innovation within high-tech SMEs: A study of the entrepreneurial founder's influence on open innovation practices. Technovation. **103**, 102232 (2021)
5. Davey, S.M., Brennan, M., Meenan, B.J., McAdam, R.: Innovation in the medical device sector: an open business model approach for high-tech small firms. Technol. Anal. Strateg. Manag. **23**, 807–824 (2011)
6. Ray, P.P., Amaral, J.F., Hinoul, P.: Innovation best practices in the medical device industry. Tech. Vasc. Interv. Radiol. **20**, 90–93 (2017)
7. Amici, C., Faglia, R., Taveggia, G., Mor, M.: Development of user-oriented mechatronic devices for post-stroke rehabilitation: the Experience of UniBS H & W. In: Proc. R&D Manag. Conf. 2016 From Sci. to Soc. Innov. Value Creat. 3–6 JULY 2016 I CHURCHILL Coll. I CAMBRIDGE Conf. Pap. 1–10 (2016)
8. Chatterji, A.K., Fabrizio, K.R.: Does the market for ideas influence the rate and direction of innovative activity? Evidence from the medical device industry. Strateg. Manag. J. **37**, 447–465 (2016)
9. Chatterji, A.K., Fabrizio, K.R.: Using users: when does external knowledge enhance corporate product innovation? Strateg. Manag. J. **35**, 1427–1445 (2014)
10. Miller, A.M., Behan, R., Smith, I., Griffin, M., Keane, F., Langan, J., O'Rourke, C., McAleenan, N., Pandit, A., Watson, M.: A multidisciplinary approach to online support for device research translation: regulatory change and clinical engagement. Heal. Policy Technol. **10**, 95–103 (2021)
11. Dharmawan, R., Ho, H., Ng, H.H.M., Iyer, N.G., Tan, H.K., Tan, N.C.: Implementing the biodesign process for medical device innovation in head and neck surgery. Surg. Innov. **27**, 653–658 (2020)
12. Qi, W., Zhou, L.: User-centered wearable product design for community elderly care. In: 2019 12th International Congress on Image and Signal Processing, Biomedical Engineering and Informatics (CISP-BMEI) 0–4 (2019)
13. Lettl, C., Rost, K., von Wartburg, I.: Why are some independent inventors 'heroes' and others 'hobbyists'? The moderating role of technological diversity and specialization. Res. Policy. **38**, 243–254 (2009)
14. Gherardini, F., Renzi, C., Leali, F.: A systematic user-centred framework for engineering product design in small- and medium-sized enterprises (SMEs). Int. J. Adv. Manuf. Technol. **91**, 1723–1746 (2017)
15. Amici, C., Ghidoni, M., Ceresoli, F., Gaffurini, P., Bissolotti, L., Mor, M., Fausti, D., Antonini, M., Ragni, F., Tiboni, M.: Preliminary validation of a device for the upper and lower limb robotic rehabilitation. In: 23rd International Conference on Mechatronics Technology (ICMT), Salerno (2019)
16. Borboni, A., Serpelloni, M., Borghetti, M., Amici, C., Aggogeri, F., Fausti, D., Antonini, M., Mor, M., Sardini, E., Faglia, R.: Hand robotic rehabilitation: from hospital to home. In: Advances in Service and Industrial Robotics. RAAD 2017. Mechanisms and Machine Science, pp. 877–884 (2018)
17. Serpelloni, M., Tiboni, M., Lancini, M., Pasinetti, S., Vertuan, A., Gobbo, M.: Preliminary study of a robotic rehabilitation system driven by EMG for hand mirroring. In: 2016 IEEE International Symposium Medical Measurement Application (MeMeA) (2016)
18. Tiboni, D., Legnani, M., Lancini, G., Serpelloni, M., Gobbo, M., Fausti, M.: ERRSE: Elbow robotic rehabilitation system with an EMG-based force control. In: Advances in Service and Industrial Robotics. RAAD 2017. Mechanical Machine Science, vol. 49, pp. 892–900 (2018)
19. Conti, A., Azzalini, E., Amici, C., Cappellini, V., Faglia, R., Delbon, P.: An ethical reflection on the application of cyber technologies in the field of healthcare. In: Advances in Service and

Industrial Robotics. RAAD 2017. Mechanisms and Machine Science, vol. 49, pp. 870–876 (2018)
20. Amici, C., Pellegrini, N., Tiboni, M.: The robot selection problem for mini-parallel kinematic machines: a task-driven approach to the selection attributes identification. Micromachines **11** (2020)
21. Amici, C., Ragni, F., Ghidoni, M., Fausti, D., Bissolotti, L., Tiboni, M.: Multi-sensor validation approach of an end-effector-based robot for the rehabilitation of the upper and lower limb. Electronics **9**(2020), 1751 (2020)
22. Glasgow, M.E.S., Colbert, A., Viator, J., Cavanagh, S.: The nurse-engineer: a new role to improve nurse technology interface and patient care device innovations. J. Nurs. Scholarsh. **50**, 601–611 (2018)
23. Henschke, C., Panteli, D., Perleth, M., Busse, R.: Taxonomy of medical devices in the logic of health technology assessment. Int. J. Technol. Assess. Health Care **31**, 324–330 (2016)
24. Formicola, R., Ragni, F., Mor, M., Bissolotti, L., Amici, C.: Design approach of medical devices for regulation compatibility: a robotic rehabilitation case study. In: Proceedings of the 7th International Conference on Information and Communication Technologies for Ageing Well and e-Health (ICT4AWE 2021), pp. 146–153 (2021)
25. Ramakrishna, S., Tian, L., Wang, C., Liao, S., Teo, W. E.: Medical Devices: Regulations, Standards and Practices. Elsevier (2015)
26. European Commission: MEDDEV 2.7/1 Rev.4: Clinical Evaluation: A Guide for Manufacturers and Notified Bodies Under Directives 93/42/EEC and 90/385/EEC. 1–65 (2016)
27. Stark, J.: Product Lifecycle Management. Springer International Publishing, Cham (2015)
28. Ceresoli, F., Aggogeri, F., Amici, C., Borboni, A., Faglia, R., Pellegrini, N., Tiboni, M., Antonini, M., Fausti, D., Mor, M., Petrogalli, G., Vertuan, A.: Differential system for gait rehabilitation. In: MESROB2018, pp. 1–8 (2019)
29. Stjepandic, J.W., Wognum, P.M.: Verhagen, Concurrent Engineering in the 21st Century. Springer International Publishing, Cham (2015)
30. Loureiro, B., Curran, A.: Complex Systems Concurrent Engineering. Springer London, London (2007)
31. International Organization for Standardization: Medical devices—Post-market surveillance for manufacturers (ISO standard No. ISO/TR 20416:2020) (2020)
32. International Organization for Standardization: Medical devices—Quality management systems—Requirements for regulatory purposes (ISO Standard No. 13485:2016) (2016)
33. Food and Drug Administration Center for Devices and Radiological Health: Design control guidance for medical device manufacturers. Des. Hist. File. **53** (1997)
34. International Organization for Standardization: Biological evaluation of medical devices—Part 18: Chemical characterization of medical device materials within a risk management process (ISO standard No. 10993–18:2020) (2020)

Development of Mathematical Models of Problems of Management the Production Division with a Discrete Unit Type Production

Ilyos Kalandarov

Abstract This article discusses the problem of managing production systems, the relationship of management of the organization of production, operational management with the efficiency of industrial production, formulates the tasks of managing the organization of production and operational management, solving the problem of increasing production efficiency, and also gives a brief overview of existing methods for solving the scheduling problem, how key task of managing the production unit. The developed mathematical models of the tasks of managing a production unit based on problem statements are considered. To solve the scheduling problem in the conditions of discrete unit production, it is necessary to develop a mathematical model that takes into account the specifics of this production. The developed information model of the control object allows you to fully describe the production process in the tool shop. The developed mathematical models of the problems of managing the organization of production and the tasks of scheduling take into account the specifics of unit production and an order-based planning system for the tool shop of the enterprise.

1 Introduction

A characteristic feature of improving the management of socialist production is that this activity covers the entire national economy and has a planned, directive and continuous character [1]. The management function is usually understood as the specialized part of the regular organizational activities of an information-management nature, characterized by the uniformity of goals, actions or objects of application of these actions.

To simplify the understanding of the problem, the management problem can be reduced to solving three interrelated tasks:

quick collection and analytical processing of information;

I. Kalandarov (✉)
Navoi State Mining Institute, Navoi city, Uzbekistan

M. Shamtsyan et al. (eds.), *Robotics, Machinery and Engineering Technology for Precision Agriculture*, Smart Innovation, Systems and Technologies 247,
https://doi.org/10.1007/978-981-16-3844-2_52

calculation of comparative options for the development of the enterprise, production planning or a particular type of product and on this basis the optimal solution; implementation of decisions made.

As you can see, these tasks correspond to the classical form of management: the plan, the identification of deviations and the development of regulatory impact.

In the organization of the management process can always be allocated work on the organization, planning, accounting and regulation of the production process.

One of the main stages of management is planning. The planning functions in management have a fundamental place for the following three reasons: the purpose, meaning and methods of enterprise management are set precisely by planning. Through this function, the enterprise is connected with the external environment: linking with the economic policy in the country, financing systems, incentives, pricing, etc.

Other management functions (such as accounting and regulation) are planning dependent. It is well known that accounting is carried out according to plan indicators, and regulation should constantly maintain the actual indicators of the object at the planned level. Thus, deficiencies in planning automatically turn into accounting and regulation. Thus, an incorrectly set or unclearly formulated goal can lead to negative consequences and reduce production efficiency.

The planning function is directive with respect to the control object. Therefore, all the advantages and disadvantages of planning are expressed in savings or loss of labor and material resources.

Planning will be effective in the case of proper organization of production.

When managing the organization of production, the following functions are performed:

- the organizational structure of the production unit is being formed;
- the specialization of equipment is determined;
- workers are assigned to equipment.

The regulatory functions of the control system are reduced to the elimination of current inconsistencies in production. Among specialists, these discrepancies are called production disturbances.

Most of the regulatory functions in the enterprise are carried out by the dispatching apparatus. Regulation carries out the following functions:

- Provides management tactics, all specific regulatory activities are subordinate to the goals formulated in the framework of planning;
- Creates specific actions necessary to eliminate unexpected disturbances;
- Not limited by a predetermined scope, regulation should comply with the only condition for intervention—its positive attitude to the implementation of the plan;
- In forced cases, the regulation adjusts the schedule of work approved by the planning, allocation of resources, etc., but the adjustment affects particulars and applies for a short time.

The link between planning and regulation is the accounting function of management. Accounting, as part of the management inherent in the following features:

- Accounting will generate information about the state of the managed facility used by the control system to generate control actions. Through accounting in production systems, the managed object is connected to the control system, called feedback;
- Accounting information is divided into two large groups that differ in different degrees of generalization: the first contains an assessment of the patterns of functioning and development trends of a managed facility, and the second contains a description of specific production disturbances and the circumstances of their occurrence;
- Information about patterns is used in planning when posed in front of an object aimed at the next period of time;
- Information on production disturbances is used in regulation during the development and implementation of specific measures to eliminate disturbances.

The organization of production processes in terms of planning and regulation is multivariate. The same plan, the same regulatory impact can have a large number of options. Among the many variations, there is the best, best option. But before you get it, you need to process a large amount of information. Computing technology, unlike a person, is capable of quickly sorting through a large number of managerial decision options, on the basis of which it seems possible for a person to select the best option justified by calculation [2].

Formation of the production program for the enterprise divisions involves the creation of conditions for its implementation, as well as its direct implementation. To create such conditions, it is necessary to carry out organizational intervention, especially in a single production, on the basis of a periodic analysis of the conformity of the production process to the organizational structure of production units, while the intervention should be consistent and flexible in achieving the main goals of the enterprise.

The efficiency of industrial production (in value form) is calculated as follows:

$$E = \frac{\mathrm{PR}}{\left(_{\mathrm{l.c.}} + _{\mathrm{p.c}}\right)}$$

where WP is the quantity of products or services produced per unit time, taking into account their quality;

$C_{\mathrm{l.c.}}$, $C_{\mathrm{p.c}}$—the costs of living and past labor, causing the production of these products or services.

It can be seen from the above relation that production efficiency can be improved in three different ways: either by increasing the quantity of output per unit of time and increasing its quality, or by reducing the cost of living and materialized labor per unit of output, or both simultaneously [3, 4].

In this paper, we consider improving production efficiency using the first method, implemented by managing the organization of production and the operational management of the production unit.

In order to determine the problems arising in the control process, we consider the standard description of a cybernetic system.

Any cybernetic system can be described using four sets of elements (indivisible)—workers P, jobs R, operations O and products P. The intersection according to the characteristics of the sets P and R will be called an aggregate and denoted by $A = \{P * R\}$. The composition of these sets is determined by the structure of the system.

Based on this description, we consider the statement of tasks for managing the organization of production and scheduling the tool shop of an industrial enterprise. To do this, we introduce the concept of production capacity. The literature discusses two main approaches to determining power. The first, which is called engineering or technical (technological), connects the value of capacity with the potential maximum of the product, and the second approach is the economic one, with the volume of production corresponding to the extreme of a certain cost or profit function [5]. In the future, we will use the first definition.

2 Submission of Manuscripts

Mathematical methods for solving the scheduling problem. In small-scale and single-unit production, with frequent changes in the nomenclature, with the participation in the production of parts for assembly of a large number of different sections and workshops, where extreme situations in turn arise, it is not so much impossible to set up a single criterion for evaluating the plan, but it is impractical. The planning and management system must be adaptive, with a changing production situation, and be able to adapt to the situation on the site and in production as a whole. Under these conditions, it is required to develop new approaches and methods for solving the problems of cladding and control [6].

The task of scheduling (scheduling) a production site is significantly different from the task of planning larger production units. When planning the work of an industry, enterprise, workshop, tasks are solved: how much, with what resources and for how long (how much, what, when). These tasks are well formulated in the language of operations research. The task of scheduling the work of the plots requires answers to the questions: what was processed, (which batch, part), on which unit of the resource, at what instant in time (what, where, when), i.e., it belongs to the class of combined tasks of complete ordering in time of various discrete processes, previously partially ordered according to technological routes. To solve these problems, methods and effective approaches developed in scheduling theory are used. The tasks of scheduling theory are close to those of queuing theory, which also studies the behavior of discrete processes. The difference between them is that, in the theory of schedules, mainly deterministic service systems are investigated, and in the theory of queuing, probabilistic ones. The problem can be reduced to integer programming problems by setting binary variables to express the constraints of the problem, but this does not lead to success due to the lack of effective algorithms for solving integer programming problems.

A lot of numerical research is devoted to the solution of the scheduling problem, and the following can be distinguished from existing methods for solving it:

- Analytical;
- Heuristic;
- Statistical.

The first group of methods, in turn, is divided into:

- exact methods, allowing for a finite number of steps to find the optimal plan;
- approximate methods to find a solution that is pretty close to optimal.

Exact solution methods include exhaustive search of options, the method of branches and boundaries, the method of sequential analysis of options, etc.

The method of exhaustive search of options for most problems is unacceptable, since the number of possible options usually increases exponentially with the dimension of the problem.

The method of sequential construction, analysis and screening of options allows you to reduce the enumeration of options, excluding from consideration several subsets of plans as unpromising. In some problems, this method gives satisfactory results, but in General, the number of operations required depends on the numerical data of the problem, and for some data, the method is not much more effective than a full search.

Statistical methods select possible plans depending on the values of some random variable, and the best of the selected plans is considered to be the solution to the problem. The simplest statistical method for finding the optimal solution is an omnidirectional random search or Monte Carlo method. The essence of the method lies in the fact that for a sequence $\pi_1,\ \pi_2\ \ldots,\ \pi_n$ of independent random processing order determined Xia the objective function $K(\pi i)$ constructed sequence

$$K_n(\pi) = \min\{K(\pi i)\}$$

To obtain a permutation $\pi i = [i_1, i_2, \ldots, i_n]$ [from the numbers 1, 2, …, n] uniformly distributed on the set of all permutations, a random variable ξ is generated$n$ times, uniformly distributed in the interval (0,1) and the kth implementation among all$n$ implementations in decreasing order assigns a number i_k. However, the Monte Carlo method, which does not take into account the specifics of the problem, gives insufficiently fast convergence to the optimal solution. Acceleration of convergence of a random sequence of plans to the optimal one can be achieved due to a more reasonable organization of a random search.

The methods associated with introducing the metric into the schedule space turned out to be very fruitful.

Most of these methods are local optimization models developed for continuous functions and modernized for a discrete argument. The essence of local optimization methods is that we introduce the concept of a neighborhood of the $U(\pi)$ plan π. To specify the $U(\pi)$ most often introduced the concept of distance $\rho(\pi_{1,}\pi_2)$ between the planes π_1 and π_2, wherein R—neighborhood $U_R(\pi)$ for any $\pi \in Q$ (Q—set consisting ofn all possible plans $\pi_i \in Q$) consists of a set of plans which satisfy the condition:

$$\rho(\pi, \pi_i) \le R$$

With this definition of a neighborhood, the local search algorithm can be formulated as follows:

1. Let the plan n be obtained at the nth step π_n. Select R_n—surrounding area of the plan π_n.
2. Simulate N_n a random plan time

$$\pi_n^{(s)} \in U_{R_n}(\pi_n), \text{где } s = 1, 2, \ldots, N_n$$

3. We find a plan such that

$$K(\pi_n^*) = \min\ K(\pi_n^3)$$

4. If $K(\pi_n^*) < K(\pi_n)$, then it is accepted as the initial plan π_n^* and the process continues in the same order, if $K(\pi_n^*) \ge K(\pi_n)$, then π_n it is taken as the local minimum.

The effectiveness of the described procedure is determined by the nature of the introduced metric. When constructing a particular metric, we proceed from purely intuitive considerations, so directed random search is essentially a combination of heuristic methods with the Conte-Carlo method.

For example, in [8], the distance between the permutations $\pi 1$ and $\pi 2$ is introduced as the number of violations of the pairwise arrangement of elements in one of the permutations relative to the other. If $\pi 1 = (1, 2, 3, 4, 5, 6, 7, 8)$, and $\pi 2 = (1, 2, 3, 4, 5, 6, 7, 8)$ then $\rho\ (\pi 1, \pi 2) = 1$.

In [7], a lexicographic metric is considered, based on the definition for each permutation π of some norm $N\ (\pi)$, and $\rho(\pi 1, \pi 2) = [N(\pi 1) = N(\pi 2)]$.

In [6], a directed search is organized using an inverse metric, in which the distance between $\pi 1$ and $\pi 2$ is defined as the number of inversions in one of them relative to the other.

As for heuristic methods, the latter can also be subdivided into a subclass of deterministic heuristic methods and a subclass of heuristic methods using statistical approaches. As an example of the first of the subclasses, we consider in more detail the algorithm for solving one Bellman-Johnson routing problem, in which the processing routes of all the parts coincide and the sequence of their execution on each machine coincides. This problem is the most studied, but despite this, currently there are no effective methods of solution, especially for problems of large dimension.

One route problem is often called a stream-type task, which indicates the advisability of studying cases where the number of parts is large and the parts themselves are relatively homogeneous. Solving the Bellman-Johnson problem in a similar situation allows you to load equipment more evenly and reduce the manufacturing time of all parts. Considering that with a large number of parts, the total processing time is relatively long; its reduction even by 2–3% is of great importance.

The algorithm described in [7] is most appropriate to apply in such a situation when the number of parts is large and the parts themselves (i.e., their complexity) are relatively homogeneous.

All the necessary information in a single routine task is given by the matrix $T = [tij]$ of dimension $n * m$, where*n* is the number of parts;*m* is the number of machines (parts); *t ij* is the processing time of the *i*th part on the *j*th machine (the execution time of the *j*th operation of the *i*th part).

Since the sequence of processing parts on the machines coincide, any possible solution to the problem is uniquely determined by the permutation $\pi = (i1, i2, \ldots, in)$, consisting of the first numbers of the natural series. Solving a problem means finding such a permutation

$$\pi * \in S(n)$$

where *S*(*n*) is the set of all permutations of numbers 1,2,...,*n* for which $K\pi\ n$ is the duration of the manufacturing cycle of parts, takes a minimum value.

The advantage of the algorithm is that since it is supposed to solve the problem approximately, it is advisable to be able to determine how much the solution found differs from the optimal one. For this, the concept of the lower boundary of solutions is used. Determining the exact lower boundary is also a very difficult task; however, for purposes it is sufficient to determine the lower boundary a:

$$K_{\min} = \max_{j=\overline{1,m}} A_j \tag{1}$$

where $A_j = \sum_{i=1}^{n} t_{ij}$—the sum of the elements *j*—th column of the matrix *T*.

Let the rows of the original matrix *T* be rearranged in an arbitrary order π. The resulting matrix is denoted by $[\tau ij]$. It can be shown that if π accept as solution of problem chi, then the absolute error of the solution has the estimate

$$K\_\pi - K\_\min \le (m-1)D\_\pi - (m-1)t\_\max \tag{2}$$

where

$$D_\pi = \max_{j=\overline{1,m}} \max_{l_1,l_2=\overline{1,n}} \sum_{i=l_1}^{l_2} \left(\tau_{ij} - \tau_{ij_0}\right) \tag{3}$$

Therefore, an approximate solution of one route problem can be found when solving the following problem: find a permutation $\pi*$ for which the quantity D_π calculated by Formula (3) takes the minimum value:

$$D\_(\pi^{\wedge *}) = \min \top(\pi \in S \wedge ((n))) f_0(D\_\pi) \tag{4}$$

For existing production systems (conveyor [3], cyclic and arbitrary), the writer describes many exact and approximate algorithms for solving scheduling problems, which are characterized by a large dimension and a significant time to solve the problem.

Recently, heuristic methods have been widely used to solve production management problems. The heuristic method is an ordered set of rules, which in some cases are introduced in a purely intuitive way or on the basis of common sense and which allow one to obtain an acceptable result that is considered to be knowingly acceptable, for example, to find a "completely satisfactory" solution (or several "completely satisfactory" solutions) varying optimization hydrochloric problem [8]. An important advantage of heuristic procedures is the convenience of their implementation on a computer even when solving relatively cumbersome tasks, as well as a good opportunity to simulate a production system, since the simulation process involves the sequential creation of conflict situations implemented through preference rules.

A fundamental feature of the system of economic and mathematical models and methods of organizing production is the ability to consistently refine and supplement them as a deeper study of the object of follow-up or obtain additional information about it. Since such practically oriented problems are often not solved in a general way, analytically, approximate optimal methods and simulation modeling are often used for calculations. Part of the variables in this case can be set tabular, statistically, and part of the dependencies is described by hypotheses, refined during the calculations. However, it is better to get an approximate solution of the problem on time than the exact one—with a delay or to get an answer not to the question posed by practice, but to the one for which there is already a mathematical theory [9].

Consider the economic description of the control object.

The instrumental production of the enterprise is single; a show system is used as a planning system. That is, the planning and accounting unit is an order for the production of a certain type of technological equipment in one or a small amount.

Operational planning is carried out ahead of schedule. Product launch dates are calculated in accordance with the release date and calendar lead time. Analytically, this can be represented in the form [10]: $Z = f(w, d)$—date of launch of the order; W is the release date of the order; d—calendar advance of launch relative to release.

Tool shop belongs to the auxiliary production and is designed to provide the basic production of the technologically equipment. The production structure of the workshop consists of the following production units:

- plot stamps;
- mold area;
- section of cutting and measuring tools;

- a plot of devices;
- procurement site;
- thermal section.

Production sites 1–4 are characterized by a constant range of technological equipment and individual (single) or small-scale type of production.

In the procurement section (Sect. 5), work pieces for Sects. 1, 2, 3 and 4 are processed.

Heat treatment of order details from Sects. 1, 2, 3 and 4 takes place in the thermal section.

All the main processing and assembly of technological equipment takes place inside the production sites 1–4, with the exception of heat treatment (intra-shop cooperation) and coatings and welding (inter-shop cooperation). As a result of this, Sects. 1–4 have a subject specialization.

The monthly plan for the workshop consists of 200–300 orders for the production of technological equipment.

A distinctive feature of the tool workshop is the development of route technological maps by the workshop technological bureau, according to the incoming design documentation for technological equipment. According to the technological cycle of manufacturing technological equipment in the tool shop, 17 technological operations are being processed, of which the main ones are turning, milling, coordinate-boring and grinding operations.

Equipment at production sites 1–4 is divided into the following groups of technological equipment (GTE):

1. turning;
2. milling;
3. coordinate boring;
4. coordinate grinding;
5. electro-erosive;
6. flat grinding;
7. circular grinding;
8. profile grinding;
9. threaded grinding;
10. internal grinding;
11. sharpening;
12. locksmith.

An order for the production of technological equipment reflects the need for the main production in this type of equipment by a specific date, so orders are opened in the tool shop taking into account design standards and the duration of the equipment manufacturing cycles, which ensures consistency in the work of the main and auxiliary production.

All orders carried out in the tool shop, according to the manufacturing cycle, are divided into three groups:

emergency;

forced;
ordinary.

Monthly plans for stamping and mold areas consist of 40–60 orders for the production of technological equipment, 20% of which are emergency and forced, the rest are processed according to the usual production cycle. For portions tonnage and cutting tools and fixtures monthly plans consist of 70–80 Coll basics of which are also 20% refer to emergency and forced.

The duration of the production cycle (TDPC) manufacturing equipment consists of the following elements:

$$T\ \mathrm{DPC} = T\ \mathrm{TECH} + T\ \mathrm{MO} + T\ \mathrm{PROM} + T\ \mathrm{CT},$$

where T TECH—the duration of the technological cycle;
T MO—time intraoperative bedding;
T PROM—time for operations in the intermediate workshops;
T CT—the time of the control and transport operations;

where in

$$T\ \mathrm{TECH} = T\ \mathrm{TECH}\ M + T\ \mathrm{TECH\ SB}, \tag{5}$$

where T TECH M—the duration of the technological cycle of machining;
T TECH SB—the duration of the assembly technological cycle.

The analysis of the complexity of orders shows that the average duration of the technological cycle (T cp those.) for each type of equipment is different and equal to:

- For stamps T f. tech = 250 n/h;
- For molds T f. tech = 450 n/h;
- For measuring and cutting tools T cf. tech = 20 n/h;
- For devices T cf. tech = 50 n/h;

The average duration of the production cycle (Tsr.dpc) for the manufacture of dies and molds, as the most time-consuming and complex types of technological equipment, is 10 work shifts for machining and 10 work shifts for the assembly of orders. The average number of parts included in an order for the manufacture of dies or molds is 12.

Operational planning in the tool shop is carried out on sites and WP (workplaces).

Based on Formula (1.5), the manufacture of technological equipment consists of two technological processing cycles, a machining cycle and an assembly cycle, while the work planning for each cycle is different. Planning the load of locksmith WP for the assembly cycle consists in the formation of monthly tasks for each WP (a monthly task consists of 2–8 orders). When planning WP loading for a machining cycle, a shift-daily task is formed for each WP.

Opening an order for an ordinary manufacturing cycle in tool production requires an average of one operation for each part of an order per shift.

Since the order for the manufacture of technological equipment is a set of parts with different durations of technological processing, the length of the production cycle for the order is equal to the duration of the production cycle of the most labor-intensive part (for dies and molds, such a part is a matrix as a rule), and Formula (1.4) taking into account Formula (1.5) can be represented in the form:

$$\mathrm{T_{PC}} = \mathrm{T_{TEX\,СБ}} + \mathrm{T^{KP}_{TEX\,M}} + \mathrm{T^{KP}_{MO}} + \mathrm{T^{KP}_{ПРОМ}} + \mathrm{T^{KP}_{KT}} \quad (6)$$

where $\mathrm{T^{KP}_{TEX\,M}}$, $\mathrm{T^{KP}_{MO}}$, $\mathrm{T^{KP}_{ПРОМ}}$, $\mathrm{T^{KP}_{KT}}$—parameter values for a part with a maximum duration of a work cycle (critical cycle).

wherein:

$\mathrm{T_{TEX\,СБ}} = \mathrm{const};;$

$\mathrm{T^{KP}_{TEX\,M}} = \mathrm{const};;$

$\mathrm{T^{KP}_{ПРОМ}} = \mathrm{const};,$

$\mathrm{T^{KP}_{KT}} = \mathrm{const};$

$\mathrm{T^{KP}_{MO}} = \mathrm{f}(\mathrm{T^{BH}_{MO}}, \mathrm{T^{M}_{MO}});$

where T^{BH}_{MO} is the time of intra-shift interoperation bedding along the critical cycle; T^{M}_{MO}—the time of inter-shift interoperation lying on the critical cycle;

$$T^{M}_{MO} \quad (7)$$

where $T^{BH}_{l\,MO}$ is the time of the lth inter-shift intraoperative bedding; $T^{M}_{n\,MO}$—time of the nth inter-shift interoperation bedding.

As a result, the duration of the production cycle of an order can be reduced in comparison with the ordinary production cycle only by reduction $\mathrm{T^{KP}_{MO}}$(up to 80% for some orders T_{PC}), it should be noted that $T^{M}_{MO} > T^{BH}_{MO}$. Based on Formula (1.7), the main way to reduce the duration of the production cycle of the order is to plan equipment loading in such a way as to obtain the minimum number of interpretational downtimes, both intra-shift and inter-shift.

3 Results

The construction of a mathematical model of a production system requires a preliminary analysis of the processes occurring in it. The input stream of the production system consists of works of various nomenclatures, each of which is characterized by a technological sequence of operations and the moment of receipt for processing. Any single operation of some work is characterized by a standard duration, a group of devices on which this operation can be performed, a place in the technological chain.

Interruptions to operations are not allowed. Each operation is performed on no more than one device. Each device at any given time is busy processing no more than one operation. All devices in the production system are divided into reconfigurable, i.e., those for which the transition from one operation to another is associated with the performance of some operations for reconfiguration, and not reconfigurable [11].

The unit is composed of a worker and equipment and is designed to perform a certain number of operations for processing a product.

Workers have a different level of professional skill, which is reflected in the assignment of a candy category to each worker.

Therefore, the correct arrangement of the workers of the workshop on the equipment of sites and the conditions of a single production plays an important role in the management of the organization of production.

Consider the mathematical model of the problem [4].

Much equipment available

$$V = \left\{ V_i^{kj} / i = \overline{1, \alpha},\ k = \overline{1, K},\ j = \overline{1, J} \right\},$$

where V_i^{kj} is the jth machine of the kth GTE of the ith section.

Many workers

$$W = \left\{ W_i^{kf} / i = \overline{1, \alpha},\ k = \overline{1, K},\ f = \overline{1, l} \right\},$$

where W_i^{kf} is the fth worker of the kth specialty of the ith sector;

wherein

$$W_i^{kf} = \left\{ F_i^{kf}, Z_i^{kf} \right\}$$

wherein F_i^{kf}—the fund working time for a month fth working kth function ith section, Z_i^{kf}—the discharge fth working kth function ith portion.

Many technological operations:

$$O = \left\{ O_n^{mk} / n = \overline{1, b},\ m = \overline{1, \alpha},\ k = \bar{1}, K \right\}$$

where O_n^{mk} is the kth technological operation of the mth part of the nth order, while.

$$O_n^{mk} = \left\{ T_n^{mk}, \overline{Z_n^{mk}} \right\},$$

where T_n^{mk} is the complexity of the kth technological operation of the mth part of the nth order, $\overline{Z_n^{mk}}$ is the discharge of the kth technological operation of the mth part of the nth order.

We introduce the concept of loss of skilled working time for the fth working zth category. Under the loss of qualified working time of the fth worker of the zth category, we will mean the performance of work by the worker that does not correspond to his qualifications $\left(Z_i^{kf} \neq \overline{Z_n^{mk}}\right)$.

It is required:

$$W \xrightarrow{A} V$$

where A must satisfy the following requirements:

$$P_{z\,i}^{kf} = \overline{F_{z\,i}^{kf}} - \overline{T_{z\,i}^{kf}} \rightarrow \min \tag{8}$$

where $P_{z\,i}^{kf}$ is the loss of qualified working time of the fth working kth specialty of the zth category of the ith section;

$\overline{F_{z\,i}^{kf}}$—fund of working time for the month of the fth working kth specialty of the zth category of the ith section; $\overline{T_{z\,i}^{kf}}$—the amount of work for the fth worker of the kth specialty of the zth category of the ith section with the restriction:

$$\overline{T_{z\,i}^{kf}} \tag{9}$$

where

$$F_{z\,i}^{\prime\,k} = \sum_{f=1}^{\bar{l}} \bar{F}_{z\,i}^{kf},$$

$$T_{z\,i}^{\prime\,k} = \sum_{f=1}^{\bar{l}} \bar{T}_{z\,i}^{kf},$$

The mathematical model of the task of organizing specialized WP.

The WP's specialization in the equipment used on equipment leads to an increase in the labor productivity of workers, a reduction in the loss of working time to perform auxiliary operations for the readjustment of equipment (i.e., a reduction in the preparatory-final time).

This problem is one of the main in managing the organization of production in the tool shop and is solved on the basis of the analysis of the equipment used in technological processing.

Consider the mathematical model of the problem [12].

There are many WPs:

$$R = \left\{R_i^{kj} / i = \overline{1, \alpha},\ k = \overline{1, K},\ j = \overline{1, J}\right\}$$

where R_i^{kj} is the jth WP of the kth of the ith section, while

$$R_i^{kj} = \left\{\phi_i^{kj}\right\}$$

where ϕ_i^{kj} is the fund of available time to change the jth WP of the kth ith section.

To efficiently load equipment, you must:

$$\phi_i^k = \overline{T_i^k}$$

where ϕ_i^k is the fund of available time to change the kth of the ith plot, $\overline{T_i^k}$ is the complexity of the technological operations performed on the kth of the ith plot, in turn

$$\overline{T_i^k} = t_i^{\prime k} + t_i^{\prime k}$$

wherein $t_i^{\prime k}$—the processing time parts of kth, ith lan ka, $t_i^{\prime k}$—changeover time by kth ith portion.

Many equipment used in the processing of order parts:

$$Q = \left\{Q_n^{k\tau} / n = \overline{1, b},\ k = \overline{1, K},\ \tau = \overline{1, e}\right\}$$

where $Q_n^{k\tau}$ is the τth rig of the kth technological operation of the nth order, while

$$Q_n^{k\tau} = \left\{t_n^{k\tau}\right\}$$

where $t_n^{k\tau}$ is the readjustment time of the τth rigging of the kth technological operation of the nth order.

It is required:

$$Q \overset{B}{\rightarrow} R$$

where B—must satisfy the following requirements:

$$\overline{t_i^{\prime kj}} = \phi_i^{kj} - \overline{t_i^{\prime kj}} \rightarrow \min \tag{10}$$

where $\overline{t_i^{\prime kj}}$ is the equipment changeover time at the jth WP of the kth of the ith section, $\overline{t_i^{\prime kj}}$—the processing time is detailed at the jth WP of the kth of the ith section with the restriction:

$$\overline{\phi_i^k} - t_i'^{kj} \geq 0 \tag{11}$$

where

$$\overline{\phi_i^k} = \sum_{j=1}^{J} \phi_i^{kj}$$

$$t_i'^{kj} = \sum_{j=1}^{J} \overline{t_i'^{kj}}$$

Now we will solve the mathematical model of the problem of managing the production structure of the workshop. The organizational structure of production sites is formed, as a rule, at the stage of development of technological plans. However, due to frequent changes in the nomenclature and heterogeneity of orders for the duration of their production (which is typical for a single production), there are capacity reserves and "bottlenecks" in production sites. This leads to a deterioration in the results of equipment loading and an increase in work-in-progress.

In this regard, to solve the issue of efficient use of production capacity, it is necessary to form the production structure of the shop based on the analysis of equipment utilization (i.e., to reorganize the structure of production sites).

Consider the mathematical model of the problem [13].

There are many:

$$M = \left\{M_i^k / i = \overline{1, \alpha},\ k = \overline{1, K}\right\}$$

where is the kth of the ith section, while

$$M_i^k = \left\{N_i^k, \overline{\phi_i^k}\right\}$$

where N_i^k is the number of WPs that are part of the kth TRP of the ith plot, $\overline{\phi_i^k}$ is the fund of available time for the month of the kth of the ith plot.

A lot of WP:

$$R = \left\{R_i^{kj} / \ i = \overline{1, \alpha},\ k = \overline{1, K},\ j = \overline{1, J}\right\}$$

where R_i^{kj} is the jth WP of the kth of the ith section, in turn

$$R_i^{kj} = \left\{\phi_i^{kj}, Z_i^{kj}\right\}$$

where ϕ_i^{kj} is the fund of available time for a month for the jth WP of the kth of the ith section, Z_i^{kj} is the category of a worker working on the jth WP, the kth of the ith section, and

$$\bar{\phi}_i^k = \sum_{j=1}^{J} \phi_i^{kj}$$

A lot of production capacities is required for the production program of the workshop in the planning period:

$$S = \left\{ S_i^k / \, i = \overline{1, \alpha}, \; k = \overline{1, K} \right\}$$

where S_i^k is the required production capacity of the kth of the ith plot.

It is required:

$$R \xrightarrow{C} M$$

where C—must satisfy the following requirements:

$$\theta_i^k = \bar{\phi}_i^k - S_i^k \rightarrow \min$$

wherein—the overload (<0) or reserve (>0) Capacity kth GTE ith section when the area under restriction:

$$\phi'^k - S'^k \geq 0 \tag{12}$$

4 The Mathematical Model of the Problem of Capacity Utilization

The basis for constructing a mathematical model is a system of quantitative relationships and dependencies between the quantities characterizing the real production process (equipment, labor, materials, work pieces, cards of technological processes for processing parts, assembly of products, etc.) and determining the movement of production (assembly, processing and moving parts, labor productivity of workers, use of equipment, etc.).

To develop a mathematical model of the problem in Table 1, the information model of the production area of the tool shop is given [14].

Table 1 Develop a mathematical model of problems

No	Name	Designation
1. Orders (products)		
1. Plan period $\bar{W}$ pl		
2. The production program of the site for the period $\bar{W}$ pl $\bar{A} = \{\bar{A}_n / n = \overline{1, b}\}$:		$\bar{A}_n$—order n—order number
3. Order Parameters: $\bar{A}_n = \{H_n, G_n, \bar{p}_n, \bar{D}_n\}$, $H_n = \sum_m \bar{H}_n^m$, $G_n = f(u_n, v_n, x_n, y_n)$, $G_n = \overline{0, 1}$, $G_n = 1$, if $u_n \le u$, $v_n = V_n$, $x_n = X_n$, $y_n = Y_n$ $G_n = 0$, if any condition is not met		H_n—the complexity of the order, G_n—order security function, $\bar{p}_n$—order priority, $\bar{D}_n$—date of launch of the order, u_n—the need for materials, $\bar{u}_n$—availability of materials, v_n—the need for blanks, $\bar{V}_n$—the availability of blanks, x_n—the need for a tool of the 2nd order, X_n—the presence of a tool of the 2nd order, y_n—the need for control programs, Y_n—availability of control programs
4. Details: $\bar{B} = \{\bar{B}_n^m / n = \overline{1, b},\ m = \overline{1, d}\}$, $\bar{B}_n^m = \left\{\left(O_n^{m1}, \ldots, O_n^{mf}\right), \bar{H}_n^m\right\}$, $\bar{H}_n^m = \sum_f T_n^{mf}$		$\bar{B}_n^m$—order detail, m—part number O_n^{mf}—technological operation, f—operation number, $\bar{H}_n^m$—the complexity of the part
2. Operations		
1. Processing Operations: $O = \left\{O_n^{mf} / n = \overline{1, b},\ m = \overline{1, d},\ f = \overline{1, l}\right\}$		
2. Operation Parameters: $O_n^{mf} = \left\{T_n^{mf}, k_n^{mf},\ \bar{Z}_n^{mf},\ q_n^{mf}, \tau_n^{mf}\right\}$		T_n^{mf}—type of processing for the operation, k_n^{mf}—the complexity of the operation $\bar{Z}_n^{mf}$—discharge of work on the operation, q_n^{mf}—priority of the operation, τ_n^{mf} }—the complexity of the part
3. Equipment and workers:		
1. $M = \{M_i^k / i = \overline{1, \alpha},\ k = \overline{1, K}\}$, $M_i^k = \{N_i^k, \bar{\phi}_i^k\}$		M_i^k—GTE, k—group number i—plot number, N_i^k—the number of WP, $\bar{\phi}_i^k$—fund of available time

(continued)

Table 1 (continued)

No	Name	Designation
2. WP: $R = \left\{R_i^{kj} / i = \overline{1, \alpha}, k = \overline{1, K},\ j = \overline{1, J}\right\}.$		R_i^{kj}—WP, j—WP number
3. PM parameters: (a) specialized $R_i^{kj} = \left\{\phi_i^{kj}, Z_i^{kj}, \bar{\tau}_i^{kj}\right\},$ (b) ordinary: $R_i^{kj} = \left\{\phi_i^{kj}, Z_i^{kj}\right\}$		ϕ_i^{kj}—fund of available time, Z_i^{kj}—qualification of the worker, $\bar{\tau}_i^{kj}$—snap
4. Equipment: $Q = \{Q_\tau^k / \tau = \overline{1, e},\ k = \bar{1}, K\},$ $Q_\tau^k = \{t_\tau^k\}$		Q_τ^k—equipment τ—snap number t_τ^k—changeover time

To load equipment, it is necessary to solve the problem of scheduling, i.e., create a schedule that allows you to solve the load issue, taking into account the following optimality criteria:

(1) Reducing the duration of the production cycle;
(2) Reduction of equipment downtime.

Based on the classification scheme of reserves for reducing the production cycle (Fig. 1) [9], the fulfillment of this criterion is achieved during the current manufacturing of parts by the time of operations and interruptions by specializing production capacities and improving production planning and organization.

At enterprises of a single, small-scale type with an irregular and rarely repeated launch of the same products, the uniformity of output of finished products means the fulfillment in each planning period on the basis of a schedule of equal or systematically increasing volumes of finished products in monetary or labor units.

Is required

$$O \xrightarrow{D} R$$

where D—must satisfy the following requirements:

$$E_n = H_n + \Pi_n \rightarrow \min \tag{13}$$

where E_n—the duration of the production cycle of the nth order;
H_n—The complexity of the nth order;
Π_n—Dwell time of the nth order under limitation:

$$\bar{D} \subset \bar{A};\ \bar{D} = \bigcup_{n=1}^{\bar{c}} \overline{A_n},\ 1 \leq \bar{c} \leq \bar{e} \tag{14}$$

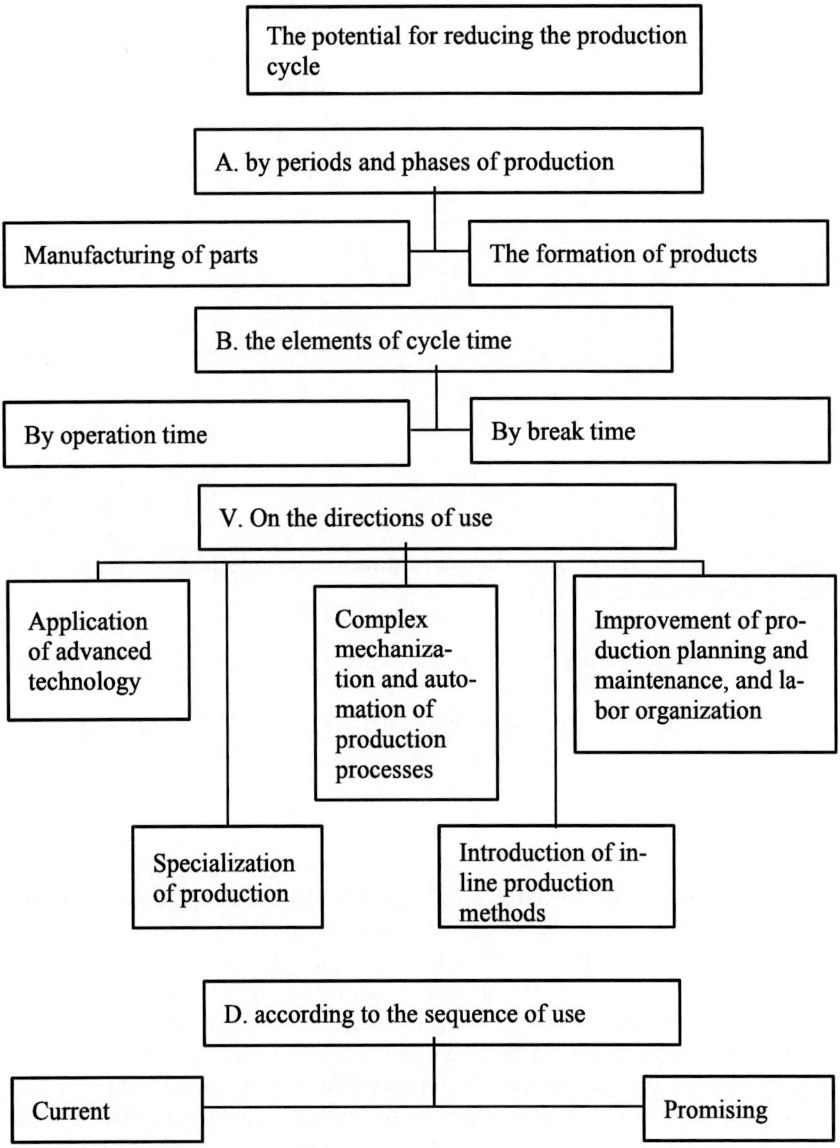

Fig. 1 Scheme for the classification of reserves to reduce the production cycle

$$\bar{D} = \left\{\overline{A_n} / \, n = \overline{1, \bar{c}}, \bar{\alpha} \le \overline{p_n} \le \bar{b}\right\}$$

where $[\bar{\alpha}, \overline{b}]$ are the boundaries in which the priorities of orders are changed, for which a reduction in the duration of the production cycle is required.

$$\bar{\Pi}_i^{kj} = \phi_i^{kj} - \bar{T}_i^{kj} \to \min \quad (15)$$

where $\bar{\Pi}_i^{kj}$—the duration of downtime but *j*th WP *k*th GTE *i*th section;
ϕ_i^{kj}—Fund of available time for the *j*th WP of the *k*th GTS of the *i*th section;
$\bar{T}_i^{kj}$—The complexity of technological operations performed on the *j*th WP of the *k*th GTE of the *i*th section.

Under limitation:

if

$$k_n^{mf} = M_i^k,\ \bar{Z}_n^{mf} = Z_i^{kj} \quad (16)$$

then

$$\phi_i^{kj} - \bar{T}_i^{kj} \geq 0 \quad (17)$$

where k_n^{mf} is the type of processing according to the *f*th technological operation of the *m*th part of the *n*th order;
M_i^k—*k*th TRP of the *i*th section;
$\bar{Z}_n^{mf}$—Discharge of the *f*th technological operation of the *m*th part of the *n*th order;
Z_i^{kj}—Qualification of a worker working on the *j*th *k*th TRP of the *i*th section.

5 Conclusion

From the foregoing, it follows that in order to effectively solve the scheduling problem in terms of the operational efficiency of the production process management, it is necessary to develop a heuristic method for solving this problem.

Therefore, the main task of this management problem is the task of scheduling. The solution to the problem of scheduling is associated with the state of a number of internal factors (organizational structures of the arrangement of workers, equipment, and specialization of equipment) of the organizational environment of the workshop. Therefore, to successfully solve the problems of managing production units, a comprehensive solution to the problems of managing the organization of production and operational management is required.

In the conditions of discrete unit production, management of the organization of production should consist in managing the internal factors of the organizational environment of the workshop: the organizational structure of the arrangement of workers; organizational structure of the arrangement of equipment; equipment specialization. The main issues of managing production systems are issues related to the achievement of the main goal of the functioning of the production unit, i.e., release of the maximum number of products of a given quality.

In conclusion to this article, it is important to answer the following. The most difficult task of algorithmization is the construction of mathematical models of production processes taking into account their specifications and features. In this article, a mathematical model of the production process was developed, taking into account the features of a single type of production of the control object, its discrete nature for solving scheduling problems. Mathematical models of production management and production division were also developed. This made it possible to describe the object under study in a formalized form and proceed to solving problems of managing such an object.

References

1. Kabulov, V.K.: Questions of the standard description of systems and aggregative systems NP Buslenko. Sat. Questions of Cybernetics **130**, 3–20 (1985)
2. Zemlin, A., Kholikov, I.V., Mamedova, I.: Lect. Notes in Civil Eng. **130**, 236–247 (2021)
3. Deterministic and Stochastic Scheduling.In.:Proc. NATO Adv. Study and Res. Inst. Theor. Approaches Scheduling Probl., Durham, 1981/M.A.H Dempster et al. Dordrecht (1982)
4. French, I.: Sequencing and Scheduling. An introduction to the mathematics of the job-shop. Chi Chester, Harwood (1982)
5. Horovite, E., Sahni, S.: Exact and approximate algorithms for scheduling non-identical processors. Technical report, computer science program, University of Southern California, 1974. J. ACM **23**(2), 317–327 (1976)
6. Kohler, W.H.: A preliminary evaluation of the critical path method for scheduling tasks on multiprocessor systems. Technical report NSF GK 37400–74–2, Department of Electrical and Computer Engineering, vol. 24(12), pp. 1235–1238. University of Massachusetts, 1974, IEEE. Transactions on Computer, (1975).
7. Vijayanandh, R., et al.: Comparative fatigue life estimations of Marine Propeller by using FSI. J Phys: Conf. Series **1473**(1), 012018 (2020)
8. Uskov, V.L., Howlett, R.J., Jain, L.C. (ed.).: Smart educ. and smart e-learning, vol. 41. Springer, (2015)
9. Fisun N.T.: Simulation modeling of production in the hull-processing shop of a shipbuilding enterprise. In: Automated Control Systems for Enterprises. IK AN UkrSSR, Kiev, 28–29 (1980)
10. Awang, M., Isa, M. H.: The Advances in Civil Engineering Materials
11. Kabulov, A.V., Kalandarov, I.I., Karimov, A.A.: Algorithmic and mathematical methods for solving the problem of calendar planning based on dynamic functioning tables. Int J Adv Sci Technol **29**(7), 9090–9097 (2020)
12. Kabulov, A., Kalandarov, I., Boltaev, S.: J Adv Res. Dyn. Control Syst.—JARDCS **12**(6), 778–792 (2020). ISSN: 1943–023X http://www.jardcs.org/abstract.php?id=5233 DOI: https://doi.org/10.5373/JARDCS/V12SP6/SP20201095
13. Kabulov, A.V., Normatov, I.H.: About problems of decoding and searching for the maximum upper zero of discrete monotone functions. J Phys **1260** (2019).
14. Kabulov, A., Normatov, I., Seytov, A., Kudaybergenov, A.: optimal management of water resources in large main canals with cascade pumping stations. In: 2020 IEEE International IOT, Electronics and Mechatronics Conference (IEMTRONICS), pp. 1–4, Vancouver, BC, Canada (2020). https://doi.org/10.1109/IEMTRONICS51293.2020.9216402

Multi-Agent Reinforcement Learning for Robot Collaboration

Yury Dubenko, Eugeny Dyshkant, and Dmitry Gura

Abstract The paper is devoted to the problems of implementation of intelligent multi-agent systems for monitoring the state of complex infrastructure objects by applying the paradigm of reinforcement learning. The aim of the work is to analyze intelligent multi-agent systems (using the example of monitoring complex infrastructural objects) and adapt their structural components, as well as the methodological apparatus for applying the reinforcement learning method. The following problems are being solved: analysis of intelligent multi-agent systems for monitoring the state of complex infrastructure facilities; analysis of the typical structure of individual intelligent agents; analysis of computer vision subsystems, training, implementation of agent's behavioral strategies, identification of the main problems of the implementation of these components, taking into account the characteristics of complex infrastructure objects. The following results were obtained: the structure of the block for the reconstruction of three-dimensional scenes was proposed, which is based on the use of convolutional neural networks of various types (Mask R-CNN, Siamese networks, generative adversarial networks), which makes it possible to construct three-dimensional models of complex infrastructure objects in an automatic mode with the ability to work with unstructured set of images, as well as performing occlusion processing; an algorithm for segmentation of macro-actions based on the method of distribution of labels has been developed, which makes it possible to take into account the weight coefficients of the edges of the transition graph; an algorithm for the functioning of a centralized multi-agent system has been developed within the framework of the reinforcement learning paradigm.

1 Introduction

According to [1], infrastructure objects are understood as: "the totality of all income-generating assets, equipment and working capital in the economy, which provide

Y. Dubenko · E. Dyshkant · D. Gura (✉)
Kuban State Technological University, 2, Moskovskaya, Krasnodar, Russian Federation
e-mail: authorgurada@kubstu.ru

M. Shamtsyan et al. (eds.), *Robotics, Machinery and Engineering Technology for Precision Agriculture*, Smart Innovation, Systems and Technologies 247,
https://doi.org/10.1007/978-981-16-3844-2_53

energy supply, transport services and telecommunications." Objects of this kind include industrial enterprises, transport infrastructure facilities (bridges, tunnels, etc.), residential and office buildings, utility facilities (collectors, etc.). The listed objects can be combined under the general term—complex infrastructure object (CIO). Damage and collapse of objects of this kind entails significant economic damage, as well as a threat to the health and life of the population. In this regard, the task of monitoring the state of such objects is urgent.

2 Literature Review

The characteristic features of the CPI that affect the choice of the structure, software and hardware of monitoring systems are: the complexity and variability of the structure, large area, inaccessibility of some components, the presence of massive reinforced concrete and metal structures, the stochastic nature of the external environment [1]. These characteristics influence the choice of the structure and methodological apparatus of the individual components of the agents that make up the multi-agent intelligent monitoring system of the CPI, namely, the subsystems of computer vision, training, and the implementation of behavioral strategies.

One of the activities associated with monitoring the state of the CPI, according to the Russian State Standard GOST 31937–2011, is a visual inspection in order to obtain a preliminary assessment of the state of the structure. For this, shooting or scanning of the CPI structural elements is carried out. The collected visual information can also be used to form three-dimensional models of the CPI, as well as its individual components. The works [2–7] are devoted to the implementation of algorithms for the functioning of computer vision subsystems, as well as their individual components, and the reconstruction of three-dimensional scenes—[2, 3]. The main problems of the implementation of the subsystems of this class are the problem of processing occlusions (when a part of an object is obscured by another object, i.e., it is hidden); the need to automate the process of reconstruction of three-dimensional scenes, as well as work with an unstructured set of images. In the considered sources, solutions to these problems are either not given or not fully covered.

The stochastic nature of the external environment, which is characteristic of the CPI, and the variability of the internal structure complicate the task of using supervised learning, because there is a need to record the conditions for the occurrence of all situations that may arise during the functioning of an intelligent agent (IA), which is almost impossible [8]. The application of an approach associated with the formation of a set of situations, which are most likely to occur, entails the risk of damage or loss of the IA in the event of situations that are not included in this set. However, unsupervised learning methods also require some training data set. In this case, the use of algorithms based on decision trees (for example, alpha–beta pruning) [9, 10] also does not solve the problem, since requires preliminary evaluations of actions. Rule-based systems also have a similar disadvantage [10]. The solution to this problem is the application of the reinforcement learning method, where the external environment is used as a teacher [2].

It is obvious that the implementation of the process of visual inspection of the CPI by using one AI is unproductive. This requires a group of agents. However, in this case, the application of training is associated with a number of problems [3]:

- Situations are possible when agents repeat the trajectories previously traversed by other agents, wasting the battery charge, or interfere with each other, for example, in collisions at the intersection of their trajectories.
- The complexity of the repeated application of experience and the long convergence time of the problem when solving typical problems (for example, to leave the room) at objects that are of the same type, but at the same time differ from each other in some parameters (for example, several rooms of different sizes with different locations of doorways).

The ways to solve these problems are to implement hierarchical and multi-agent reinforcement learning.

The works [11–14] are devoted to the study of the main practical and theoretical aspects of hierarchical reinforcement learning. It follows from them that the main problem of hierarchical reinforcement learning is the implementation of the process of segmentation of macro-actions. However, the existing methods of segmentation of macro-actions do not take into account the number of transitions performed for one or another edge, which does not allow separating the "popular" trajectories from the unclaimed ones.

As already noted, the task of visual inspection of the CPI requires the coordination of the actions of a group of agents. The following works [15–20] are devoted to the study of the main issues of multi-agent reinforcement learning, the results of the analysis of which make it possible to formulate the following problems:

- The number of possible combinations of joint actions increases exponentially with the growth of the number of agents.
- The need to calculate individual remuneration for each agent.
- Direct training of coordination agents, which may consist in the formation of an appropriate policy.
- Ability to reproduce the experience generated by various agents.

3 Problem Statement

The aim of the study is to analyze intelligent multi-agent systems (using the example of the problem of CPI monitoring) and adapt their structural components, as well as the methodological apparatus for applying the reinforcement learning method.

The main problems to be solved within the framework of this work:

- analysis of intelligent multi-agent systems using the example of the problem of CPI monitoring;

- analysis of the typical structure of individual intelligent agents;
- analysis of computer vision subsystems, training, implementation of agent's behavioral strategies, determination of the main problems of the implementation of these components, taking into account the characteristics of the CPI, as well as the possibility of using reinforcement learning.

4 Materials and Methods

4.1 Analysis of the Structure of the CPI State Monitoring System

The system for monitoring the CPI state includes the following components [4]: a decision-maker (DM), a data collection subsystem (DCS), a decision support system (DSS), a database (DB), a direct monitoring object—CPI, acting as an external environment.

Figure 1 shows a block diagram of the CPI state monitoring system.

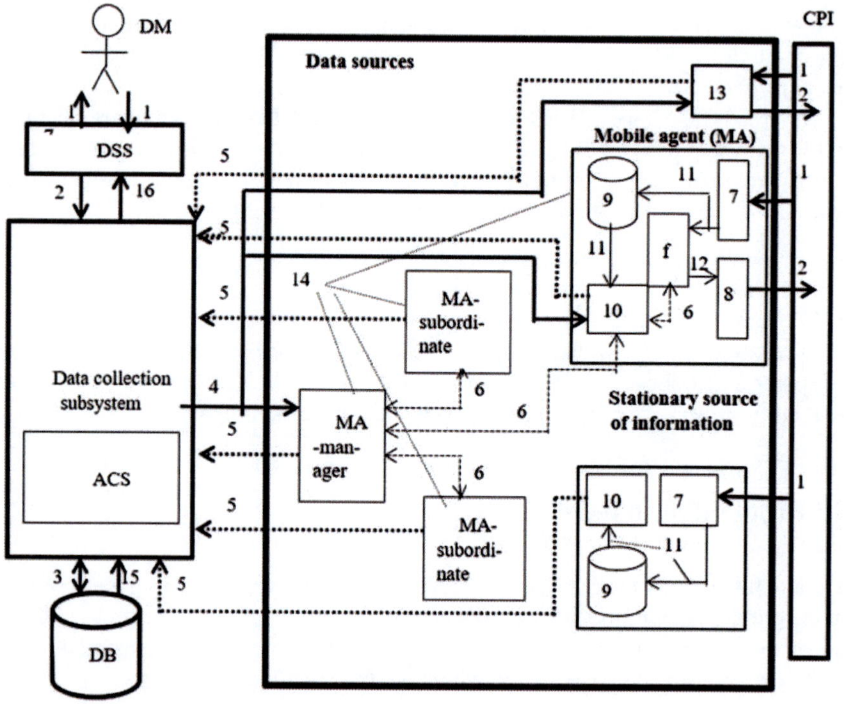

Fig. 1 Structure of the CPI state monitoring system

Figure 1 shows the following notations: *f*—agent function (is a projections of acts of perception (11) into actions (12)); ACS—an automated control system that distributes subtasks between agents and controls their implementation; MA—mobile agent; MA-manager—MA that manages a group of subordinate agents; MA-subordinate—a subordinate MA carrying out its activities in accordance with the coordinating influences of the MA-manager; 1—the set task; 2—request for data on the status of the CPI from the DCS; 3—request/receipt of data stored in the database (structure, technical characteristics of CPI elements, retrospective values of CPI parameters); 4—request for data on the status of the CPI from stationary sources of information or MA; 5—data on the state of the CPI received from stationary sources of information or MA; 6—transfer of coordinating influences from the MA-manager to subordinate agents / receipt by the MA-manager of data on the results of the functioning of subordinate agents; 7—sensor system of a stationary information source or MA; 8—MA actuators; 9—memory of a stationary information source or MA; 10—transceiver device; 11—perception (parameters of the external environment, recorded by the agent's sensory system); 12—actions (special manipulations performed by the agent by means of executive mechanisms in order to modify the state of the external environment, which are the agent's response to observation); 13—single MA; 14—group MA; 15—data on the state of the CPI received from stationary sources of information and MA; 16—data on the state of the CPI obtained from the database, as well as from stationary sources of information and MA; 17—recommendations for solving the problem.

The system for monitoring the state of the CPI belongs to the class of multi-agent systems, since it possesses their inherent characteristics, namely [21]: "is a set of system units, in which a subset of active units is distinguished" (MA); interacts with the environment (in this case—CPI) in which there are agents and objects; "contains many tasks (functions, roles) that are assigned to agents, as well as many relationships (interactions) between agents"; "contains many organizational structures formed by agents" (groups of agents managed by MA-managers); contains many actions of agents. The system under consideration also has a hierarchical management structure (DM, ACS, MA-manager) and can be classified as centralized multi-agent systems.

Thus, from the point of view of the classical theory of system analysis, the system for monitoring the state of the CPI can be represented as a multi-agent system in which the control object is MA, and the ACS is a distributed system that includes methods and algorithms for centralized control of agents, as well as methods and algorithms of collective interaction of agents with each other to achieve the goal set by the DM. This approach will expand the methodological apparatus of system analysis from the point of view of managing multi-agent systems.

4.2 Analysis of the Structure of an Intelligent Agent

Data sources are an important component of the CPI state monitoring system (Fig. 1). Complex infrastructural objects are characterized by a large area, and their full equipping with stationary sources of information is not possible due to high financial costs. The use of personnel for data mining is complicated both by the inaccessibility of some elements of the CPI structures and by the presence of risks to human life and health. For a number of CPI monitoring activities included in the list of works provided for by GOST 31937–2011, an alternative solution, free of the listed disadvantages, is the use of mobile robots (MA), both ground-based and aerial ones. According to [8], by the type of control, mobile robots can be divided into the following classes: remote-controlled, semi-autonomous, autonomous. CPI is characterized by the presence of massive reinforced concrete and metal structures that impede the radio signal, which makes it difficult to use remotely controlled robots. The variability of the structure and conditions for the CPI, the stochastic and partially observable external environment significantly limit the possibility of using semi-autonomous robots that function strictly in accordance with the set program. Thus, autonomous robots are optimal for use in CPI state monitoring systems.

To ensure the rational functioning of the agent under the indicated conditions, the mobile robot must have the ability to predict the state of the external environment, self-organization, and learning. Further in this work, the term intelligent agent (IA) [9] will be used to designate a mobile robot with the indicated functions. The general structure of the IA, built in accordance with [9], is shown in Fig. 2.

Figure 2 shows the following notations: DCS—data collection subsystem; sensors—technical means designed to perceive the parameters of the external environment for the purpose of their further processing; P_t—observation—parameters of the external environment, recorded by the sensors of the agent at time t; $f_{\text{sensor}}(P_t) = C_t$—a sensor function that formalizes observations P_t into a set of features $C_t = \left\{C_t^1, \ldots, C_t^{C_N}\right\}$; $f_{st}(C_t) = S_t$ —identification function S_t (agent state at time t); T—the task assigned to the agent; $f_{\text{cont}}(S_t, T) = U_t$—function of generating control actions U_t for the agent's actuators; actuators—technical means designed to modify the state of the external environment through the implementation of actions; $f_{\text{act}}(U_t) = A_t$ —function of actuators that transforms control actions U_t into actions A_t; A_t—special manipulations performed by the agent by means of actuators in order to modify the state of the external environment, which are the agent's response to the observation P_t; Nm—performance standard, which is a scale for assessing the rationality of an agent's actions; $f_{\text{perf}}(Nm, S_{t-1}, A_{t-1}, S_t) = r_{t-1}$—function for assessing the rationality of the agent's actions, where S_t—the state of the agent at time t (current state of the agent), S_{t-1}—the state of the agent at time $t-1$, A_{t-1}—the action performed at time $t-1$, r_{t-1}—assessment of agent rationality at time $t-1$; $f_{\text{learn}}(S_{t-1}, A_{t-1}, S_t, r_{t-1}) = \text{corr}_t$—a learning function that forms ⟦corr_t—a corrective action for the function f_{cont}. Taking this into account, the function of forming control actions can be rewritten in the following form: $f_{\text{cont}}(S_t, T, \text{corr}_t) = U_t$.

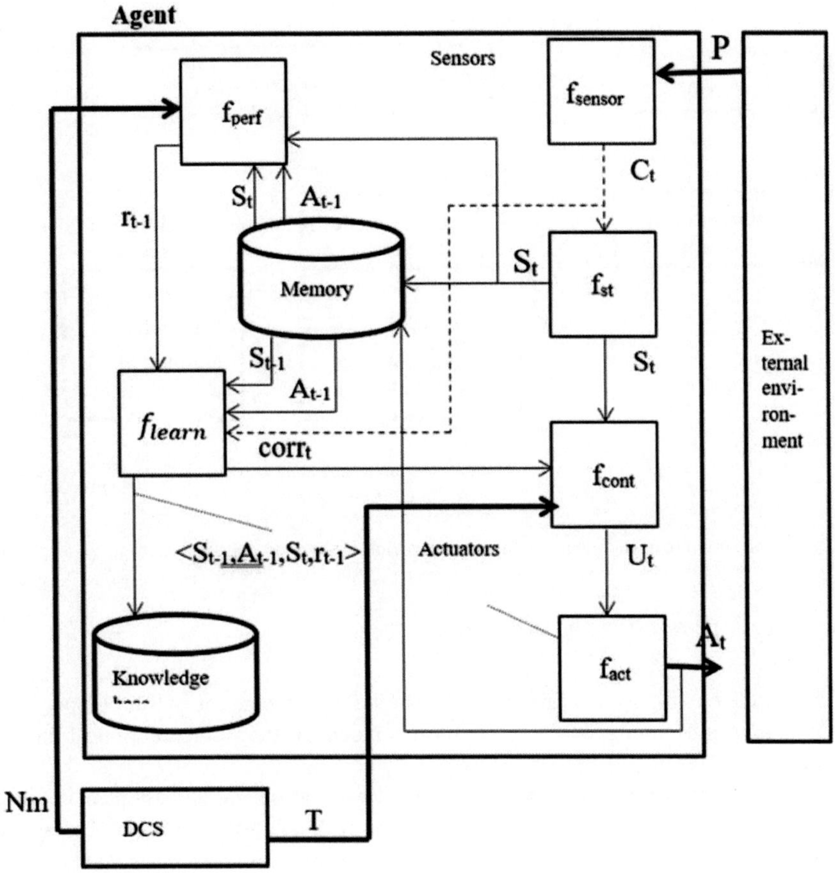

Fig. 2 The general structure of the IA

Thus, the following subsystems can be distinguished in the agent structure: computer vision subsystem (input: P_t; output: S_t; function describing the law of functioning: $f_{st}(f_{sensor}(P_t)))$; subsystem for the implementation of the agent's behavioral strategies (inputs: S_t, T, corr_t; output: A_t; function describing the law of functioning: $f_{act}(f_{cont}(S_t, T, \text{corr}_t)))$; learning subsystem (inputs: Nm, S_t, A_t; outputs: $corr_t$; function describing the law of functioning: $f_{learn}\left(S_{t-1}, A_{t-1}, S_t, f_{perf}(Nm, S_{t-1}, A_{t-1}, S_t)\right)$.

Further, the main problems of the implementation of these subsystems will be considered, and ways to solve them will be proposed.

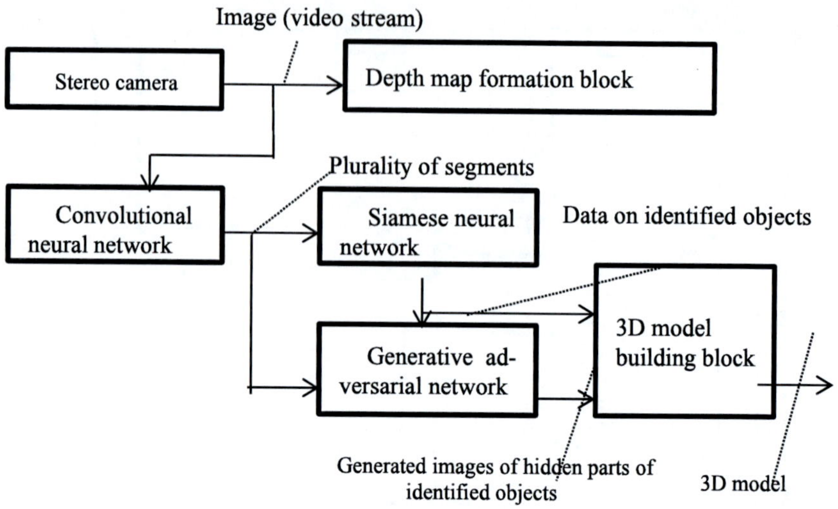

Fig. 3 The structure of the block for the reconstruction of three-dimensional scenes

4.3 Computer Vision Subsystem

To solve the previously identified problems of the implementation of this component, it is proposed to implement in its structure a block for the reconstruction of three-dimensional scenes (Fig. 3).

The block shown in Fig. 3 is based on the use of various modifications of convolutional neural networks, which allows working with an unstructured set of images. In this case, the convolutional neural network of the Mask R-CNN variety [4] implements image segmentation, and the Siamese neural network implements pattern recognition [5], forming tuples $\langle c, i, j, d \rangle$ for both stereopairs [22], where c is the class of the identified object, i, j—coordinates of the upper left point of the window, d—window width; generative adversarial network [23] constructs the hidden part of the identified object; the block for constructing a three-dimensional model carries out filtering of the point cloud by distance and using an octree [24], as well as reconstruction by applying the Delaunay triangulation [25]. The main features of the proposed block for the reconstruction of three-dimensional scenes are described in our following works: [26, 27].

4.4 Learning Subsystem

For a formal description of the concept of reinforcement learning, a Markov model of the following form is used [16]

$$\langle A, S, T, R \rangle,$$

where $A = \{a_1, a_2, \ldots, a_M\}$—a set of possible actions, $S = \{s_1, s_2, \ldots, s_N | \forall s_i \in S \exists A_{s_i} \subset A\}$—a set of possible states, where A_{s_i}—a set of activities available in the state s_i, $T : T(s, a, s' | s' \in S \wedge s \in S \wedge a \in A)$—state transformation function (transition function), $R : R = f(s, a, s')$—function of assessing the results of transformation of states $s \rightarrow s'$ by performing action a (reward function). In reinforcement learning, the term "strategy" is introduced $(\pi(s, a))$—the probability of choosing an action a when being in state s.

One of the types of reinforcement learning is Q-learning. Q-learning introduces the concept of the value of an action $Q(s, a)$, determined by the formula [16]:

$$Q(s, a) \leftarrow Q(s, a) + \alpha\left[r + \gamma Q(s', a') - Q(s, a)\right] \quad (1)$$

where α—learning rate parameter, $\gamma \in [0, 1]$—discount value, $r = R(s, a, s')$—agent's remuneration for making the transition $s \rightarrow s'$ by performing an action a.

This approach will also be used in the future to implement the behavioral strategies of agents.

4.5 *Subsystem for the Implementation of the Agent's Behavioral Strategies. Hierarchical Reinforcement Learning*

In order to reduce the convergence time of the problem when using the reinforcement learning method, to transfer experience, it is possible to combine several primitive actions that describe the solution of certain local subtasks in the framework of achieving a global goal. This approach is called hierarchical reinforcement learning [15].

The basic essence of hierarchical reinforcement learning can be summarized as follows. Let S be the set of states, A—the set of permissible primitive actions, while $\forall S_i \exists A_{s_i} = \{A_1, A_2, \ldots, A_N\}$, $A_{s_i} \subset A$, where A_{s_i} is the set of primitive actions, execution of which is allowed in the state S_i, $S_i : \exists M_k \in M_{s_i}$, where M_{s_i} is a set of macro-actions M_k that are allowed to be executed in the state S_i, where.

$$M_k = \{A_j | A_j \in A\} : \forall M \exists (S_\alpha \subseteq S) \wedge (S_\beta \subseteq S) \wedge \left(\pi_{M_k}(S_i, U_j | (S_i \in S) \wedge (A_j \in A_{s_i})) : S \times \cup A_{s_i} \rightarrow [0, 1]\right),$$

where S_α—the set of states in which a macro-action M_k can be initiated, S_β—the set of states in which a macro-action can be completed, π_{M_k}—"Policy" (stochastic or deterministic) [16], defined for a macro-action M_k.

Later in this paper, by a primitive action we mean an action satisfying the condition

$$A_j \neq \bigcup_{\gamma \in G} A_\gamma, \tag{2}$$

where G—some set of indices, A_γ—possible blocks or parts A_j. In this case, Condition (2) is not satisfied for macro-actions, therefore $M_k = \bigcup_{\gamma \in G} A_\gamma,\ A_\gamma : A_\gamma = \{A_1\} \vee A_\gamma = \{A_1, A_2\} \vee \ldots \vee A_\gamma = \{A_1, A_2, \ldots, A_W\}$.

A macro-action can be described by a tuple of the following form:

$$M_k = \langle I, \beta, \pi \rangle,$$

where $I \subset S$—a set of initialization of a macro-action (a set of states in which the execution of this option is allowed), π—a policy (defines a strategy for choosing primitive and macro-actions depending on the vertex, as well as assigning rewards and penalties [16]), $\pi\left(S_i, d | d \in D = \left\{A_j, M_k | \left(A_j \in A_{S_i}\right) \wedge \left(M_k \in M_{S_i}\right)\right\}\right)$—the probability of transition from state S_i to state S_i' when performing a primitive action A_j or macro-action M_k, admissible in this state, $\beta : S \to [0, 1]$—the condition for the completion of a macro-action, indicates the probability of the completion of a macro-action M_k in the state S_i [28].

One of the tasks of hierarchical reinforcement learning is to perform segmentation of macro-actions, within the framework of which there is a problem of accounting for the number of transitions performed for one or another edge. Let us consider the method of segmentation of macro-actions, based on the application of clustering by the method of label propagation, which allows solving the specified problem:

1. Generating a directed transition diagram $G = (V, E, w)$, where V—a set of vertices of a graph, $V = \left\{V_1, V_2, \ldots, V_{S_N}\right\} : \forall V_i \equiv S_i,\ S_i \in S$, where S—a set of states; E—a set of edges, $E = \{\}$, where T—a set of transitions, A—a set actions (primitive actions, other macro-actions); V_i, V_j, V_k—initial vertices of edges; V_i', V_j', V_k'—end vertices of edges; w—set of edge weights e_ϑ, $e_\vartheta \in E$, $w(e_z) = n_{V_i, V_i'}^{A_x} = \sum_{i=1}^{N} \delta\left(V_i, A_x, V_i'\right)$, where $n_{V_i, V_i'}^{A_x}$—a parameter indicating how often the action A_x is called in the state S_i (corresponds to the vertex V_i) led to a transition to the state S_i', $i = \overline{1, S_N}$, $j = \overline{1, S_N}$, $k = \overline{1, S_N}$, $x = \overline{1, A_N}$, $y = \overline{1, A_N}$, $z = \overline{1, A_N}$, $\vartheta = \overline{1, E_N}$, $E_N \in \mathbb{N}$, $T_N \in \mathbb{N}$.
2. A unique label L_i is assigned to each vertex, $V_i \in V$, i.e., $\text{Label}(V_i) = L_i$.
3. For each vertex $V_i \in V$, the following set is formed: $\text{Surr}^{V_i} = \left\{\text{Surr}_{e_k}^{V_i} = V_j, L_j, w(e_k) | V_j : V_i \in e_k \wedge V_j \in e_k\right\}$, where $L_j = \text{Label}\left(V_j\right)$, $w(e_k)$—weighting coefficient of the edge e, $k = \overline{1, E_N}$.

4. Label L_j is assigned to Vertex V_i: $L_j \in \mathrm{Surr}_{ek}^{V_i} \wedge w(e_k) \in \mathrm{Surr}_{ek}^{V_i} \wedge w(e_k) \to \max$. If there is the situation $w(e_x) = w(e_y)$, $w(e_x) \in \mathrm{Surr}_{e_x}^{V_i}$, $w(e_y) \in \mathrm{Surr}_{e_y}^{V_i}$, where $w(e_x) \in \mathrm{Surr}_{e_x}^{V_i}$, $w(e_y) \in \mathrm{Surr}_{e_y}^{V_i}$, then for the vertex V_i, the label is assigned randomly from the set $\left\{L_j, L_k | w(e_x) \in \mathrm{Surr}_{e_x}^{V_i} \wedge w(e_y) \in \mathrm{Surr}_{e_y}^{V_i}\right\}$, $j = \overline{1, S_N}$, $k = \overline{1, S_N}$, $x = \overline{1, E_N}$, $y = \overline{1, E_N}$.
5. As a result, clusters were formed $g : g \subset G \wedge (\forall V_i, V_j | V_i \in g \wedge V_j \in g : \mathrm{Label}(V_i) = \mathrm{Label}(V_j)) \wedge (\forall V_i \in g \exists V_j : V_j \in g \wedge V_j \in e \wedge V_i \in e) \wedge |g| > \varphi$, where φ—some threshold value of the cluster power, $i = \overline{1, S_N}$, $j = \overline{1, S_N}$.
6. Search for bottlenecks $e \in E_{g_a, g_b}$. Bottleneck [14]—edge that is border between two clusters g_a and g_b, while satisfying the condition $e = V_i, V_j \wedge V_i \in g_a \wedge V_i \in g_b \wedge g_a \cap g_b = \emptyset \to e \in E_{g_a, g_b}$.
7. Formation of macro-actions $M_{g_a, g_b} = I_{M_g}, \beta_{M_g}, R_{M_g}$ (transition from cluster g_a through cluster g_a into cluster g_b), where

$I_{M_g} = (g_a^* \cup g_b^*) \backslash E_{g_a, g_b}$—set of macro-action initialization M_{g_a, g_b}, where $g_i^* = \left\{V_i, V_i' | ((V_i \notin g_i) \vee (V_i \in g_i)) \wedge V_i' \in g_i\right\}$—a set of vertices connected with the vertices of cluster g_i, but not belonging to the cluster g_i—itself;

$((S_i \in I_{M_g} \wedge S_i \notin E_{g_a, g_b}) \to \beta_{M_g}(S_i) = 0) \wedge ((S_i \notin I_{M_g} \vee S_i \in E_{g_a, g_b}) \to \beta_{M_g}(S_i) = 1)$, where $\beta_{M_g}(S_i)$—probability of completion of a macro-action M_{g_a, g_b} in the state S_i;

$R_{M_g}(S_i, A_j, S_i')$—function for calculating rewards for a macro-action M_{g_a, g_b}, for which the following conditions hold: $(S_i' \notin I_{M_g} \to R_{M_g}(S_i, A_j, S_i') = -1) \wedge (S_i \in E_{g_a, g_b} \to R_{M_g}(S_i, A_j, S_i') = -1) \wedge (S_i \in I_{M_g} \to R_{M_g}(S_i, A_j, S_i') = 1 - \frac{1}{D})$, where D—distance from current state S_i to state $S_j \in E_{g_a, g_b}$.

4.6 Subsystem for the Implementation of the Agent's Behavioral Strategies. Multi-Agent Reinforcement Learning

The multi-agent reinforcement learning problem can be formalized as a tuple

$$\langle \alpha, S, U, R, P, \pi, \gamma \rangle,$$

where $\alpha = \{\alpha_1, \alpha_2, \ldots, \alpha_n\}$—set of agents of a multi-agent system, $S = \{S_1, S_2, \ldots, S_m\}$—state space, $A = \{A_1, A_2, \ldots, A\}$—action space. At each time interval t each agent α_i performs actions $A_{\alpha_i} \in A$, forming joint actions $A_{\mathrm{joint}} : A_{\alpha_i} \in A_{\mathrm{joint}} \wedge A_{\mathrm{joint}} \in A^N : |A^N| = N$. Thus, $R(S_t, A_{\mathrm{joint}}, \alpha_i) : S \times A^N \times \alpha \to R$—reward function, $P(S_{t+1} | S_t, A_{\mathrm{joint}}) : S \times S \times A^N \to [0, 1]$—transition function, $\pi(A_{\mathrm{joint}} | S_t) = \prod_{\alpha_i} \pi_{\alpha_i}(A_{\alpha_i} | S_t) : A^N \times S \to [0, 1]$—multi-agent system policy.

At each time interval t, agents $\alpha_i \in \alpha$, being in states $S_{\alpha_i} \in S$, carry out joint actions A_{joint}, receiving total reward $R_t^\alpha = \sum_{\alpha_i \in \alpha} \gamma_t^\alpha r_t^\alpha$, where $\gamma \in [0, 1]$—discount factor, while the value r_t^α can be calculated as follows [18]:

- agents receive the same reward:

$$r_t\left(S_t, A_{\text{joint}}, \alpha_i\right) = r_t\left(S_t, A_{\text{joint}}, \alpha_j\right), \forall \alpha_i, \alpha_j, \tag{3}$$

- the sum of rewards for all agents is equal to zero in all states:

$$\sum_{\alpha_i} r_t\left(S_t, A_{\text{joint}}, \alpha_i\right) = 0, \forall S_t \in S, A_{\text{joint}}. \tag{4}$$

However, the indicated options for calculating remuneration have a number of significant disadvantages:

- in (3), situations are not taken into account when agents can make different contributions to the solution of a common problem, and such an approach does not contribute to the "encouragement" of leaders and "punishment" of those who lag behind.
- Approach (4) is more flexible than (3), but it also does not allow taking into account the real contribution of agents to the solution of the general problem.

It would be more appropriate in this case to apply some of the principles of "feudal hierarchical reinforcement learning." In particular, for a group of agents, a manager agent can be assigned to distribute tasks among subordinate agents. Therefore, we have the problem of distributing tasks to agents $\alpha_i \in \alpha$ in such a way that

$$J_t^\alpha = M_{\mu_t \sim P(\mu_t)}\left(R_{\mu_t}\right) \to \max,$$

where $\mu_t = \left\{S_t^{\alpha_1}, A_t^{\alpha_1}, r_t^{\alpha_1 \in \alpha}, \ldots, S_t^{\alpha_n}, A_t^{\alpha_n}, r_t^{\alpha_n \in \alpha}\right\}$—scenario for performing actions by agents of a multi-agent system at a time t, $R_{\mu_t} = \sum_{\alpha_i \in \alpha} \gamma_t^{\alpha_i} r_t^{\alpha_i}$—the amount of rewards for the scenario μ_t, $M_{\mu_t \sim P(\mu_t)}$—the expected value of the sum of reinforcements when the scenario μ_t is executed, $P(\mu_t)$—probability distribution over scenarios μ_t. Moreover, the scenario μ_t is a chain of actions $\left\{A_t^{\alpha_1}, \ldots, A_t^{\alpha_n}\right\}$ performed by agents $\alpha_i \in \alpha$, if they are in the states $\left\{S_t^{\alpha_1}, \ldots, S_t^{\alpha_n}\right\}$ in the moment t and can be viewed as a macro-action that can be initiated in the state $S_\alpha = \left\{S_t^{\alpha_1}, \ldots, S_t^{\alpha_n}\right\}$, in this case, the terminal state $S_\alpha = \left\{s_{t+1}^{a_1}, \ldots, s_{t+1}^{a_n}\right\}$.

For $\forall \alpha_i \in \alpha$, sets $A_{\alpha_i} \subset A$ can be formed, denoting actions that can be performed by agent α_i. Thus, the set of possible joint actions $A_{\text{joint}} = A_{\alpha_1} \times A_{\alpha_2} \times \ldots \times A_{\alpha_n}$. Consequently, the set of possible combinations of joint actions increases exponentially with the growth of the cardinality of the set α. The solution to this problem is possible by forming groups $\alpha_g = \{\alpha_1, \alpha_2, \ldots, \alpha_l\} : l \leq n \wedge \alpha_g \subset \alpha$ based on some criteria $K = \{k_1, k_2, \ldots, k_h\}$, for example, the location of agents relative to scopes of each other:

$$\alpha_i \in O(\alpha_j) \wedge \alpha_j \in O(\alpha_i) \rightarrow \alpha_i \in \alpha_g \wedge \alpha_j \in \alpha_g,$$

where $O(\alpha_j)$—scope of agent α_j, $O(\alpha_i)$—scope of agent α_i. In this case, the term "scope" means a certain surrounding surr(α) around agent α, the state of which is identified by the sensory system of this agent.

For agent α_i, in accordance with its technical characteristics, the class $c \in C$ can be determined, according to which this agent has the technical ability to perform actions $A_c \equiv A_{\alpha_i}$. However, in addition to the standard primitive actions, such as "go forward," "turn right," "stop", for each agent $\alpha \in \alpha_g$, the set A_c must include the action A_{coord}, which assumes coordination of actions with other agents of the set α_g. At the same time, the format of the coordinating action A_{coord} must be defined, which can be implemented in a decentralized form:

- translation by agent α_i of its intentions at time step t by other agents of the group α_g, expressed as a tuple $\langle\langle S_t^{\alpha_i}, A_t^{\alpha_i}, S_{t+1}^{\alpha_i}\rangle$, where $S_t^{\alpha_i}$—the current state of the agent α_i, $A_t^{\alpha_i}$—the action that the agent plans to take to transfer to the state $S_{t+1}^{\alpha_i}$;
- getting tuples $\langle S_t^{\alpha_j}, A_t^{\alpha_j}, S_{t+1}^{\alpha_j}\rangle$ from other agents of the group $\alpha_j \in A_g$;
- tuple assessment $\langle S_t^{\alpha_i}, A_t^{\alpha_i}, S_{t+1}^{\alpha_i}\rangle, \langle S_t^{\alpha_j}, A_t^{\alpha_j}, S_{t+1}^{\alpha_j}\rangle : A_t^{\alpha_i} \in A \wedge A_t^{\alpha_j} \in A$, performing replacement $A_t^{\alpha_i} \rightarrow A_t^{\alpha\prime_j}$, if $Q_{\mu_t}^{A_t^{\alpha_i}} < Q_{\mu_t}^{A_t^{\alpha_i\prime}}$, where $Q_{\mu_t}^{A_t^{\alpha_i}}$ и $Q_{\mu_t}^{A_t^{\alpha\prime_j}}$—the projected values of joint actions if $A_t^{\alpha_i}$ or $Q_{\mu_t}^{A_t^{\alpha\prime_j}}$ is performed (see Formulas (1), (2)).

The disadvantage of this approach is, first of all, the complexity of its implementation and low reliability. In this case, in our opinion, the more preferable implementation of A_{coord} is in a centralized form:

- appointment of an agent-manager $\alpha_m : E_t^{\alpha_m} = \{e_1, e_2, \ldots, e_w\} \subset S_t^{\alpha_m} \wedge E_t^{\alpha_m} \rightarrow E_{t,\text{centroid}}^{\alpha_g}$, where $E_t^{\alpha_m}$—set of coordinates characterizing the position of the agent α_m, $E_{t,\text{centroid}}^{\alpha_g}$—some center of gravity of the group α_g;
- transmission of data on their current state $S_t^{\alpha_i}$ to α_m from each $\alpha_i \in \alpha_g$;
- forming on α_m tuples $S_t^{\alpha_i}, A_t^{\alpha_i}, S_{t+1}^{\alpha_i}$ and transferring them to $\alpha_i \in \alpha_g$;
- performing actions $A_t^{\alpha_i}$, transfer of tuples $S_t^{\alpha_i}, A_t^{\alpha_i}, S_{t+1}^{\alpha_i}, r_t^{\alpha_i}$ from agent α_m to each agent $\alpha_i \in \alpha_g$;
- receipt of reinforcements $r_t^{\alpha_m}$ by agent α_m.

An important aspect is also the formation of the π_{coord} policy that stimulates agents to coordinate their actions, which can be as follows: the value ρ is determined, which characterizes the probability of choosing A_{coord} instead of an uncoordinated action, the probability of choosing which is estimated at 1-ρ.

In the process of functiɔning, agents form "trajectories" representing chains of the following type $\tau = \{s_1, u_1, r_1, s_2, u_2, r_2, \ldots, s_{T-1}, u_{T-1}, r_{T-1}, s_T\}$.

In the memory of agent α_i the "trajectory" τ is represented in the form of an "experience buffer" [17]:

$$B_{\alpha_i} = \{s_1, u_1, r_1, s_2, s_2, u_2, r_2, s_3, \ldots, s_{T-1}, u_{T-1}, r_{T-1}, s_T\} \tag{5}$$

Accordingly, the problem of experience reproduction is to exchange tuples that make up the "experience buffer" (5) between agents α_i and α_j. Providing a process of exchange of experience between agents could significantly increase the speed of their learning. However, agent α_j can use the experience accumulated by agent α_i only if

$$\vartheta_{\alpha_i} = \vartheta_{\alpha_j} \wedge \varphi_{\text{surr}(\alpha_i)} = \varphi_{\text{surr}(\alpha_j)} : \vartheta_{\alpha_i} \in S_t^{\alpha_i} \wedge \varphi_{\text{surr}(\alpha_i)} \in S_t^{\alpha_i} \wedge \vartheta_{\alpha_j} \in S_t^{\alpha_j} \wedge \varphi_{\text{surr}(\alpha_j)} \in S_t^{\alpha_j} \quad (6)$$

where ϑ_{α_i} and ϑ_{α_j}—parameters of agents α_i and α_j, $\varphi_{O(\alpha_i)}$ and $\varphi_{O(\alpha_j)}$—parameters of the environment in the surroundings of agents α_i and α_j, respectively. But Condition (6) seems to be poorly fulfilled. Therefore, there is a need to develop methods for adapting the experience gained by one agent for use by another agent in similar conditions.

5 Results

When performing this work, the following results were obtained:

- the structure of the block for reconstruction of three-dimensional scenes is proposed, which performs the formation of three-dimensional models of the CPI in an automatic mode, which differs from analogs [2, 3, 6, 7] by the use of filtering the point cloud by distance and using an octree, as well as the ability to process occlusions. The main features of the proposed block for the reconstruction of three-dimensional scenes are described in our following works: [26, 27];
- an algorithm for segmentation of macro-actions is proposed, which differs from analogs [11–14] by the possibility of taking into account the weight coefficients of the transition graph edges;
- the problems of multi-agent learning with reinforcement are identified [29, 30] (problems of coordinating the actions of agents, calculating collective and individual reinforcements of agents, exchange of experience between agents).

6 Discussion

In addition to the considered monitoring of the CPI state, the following can be noted as examples of areas for the practical application of the obtained results:

- Survey, patrolling, protection of the territory by mobile robots. When solving this problem, agents can several times pass already surveyed areas, wasting the battery charge, which slows down the process of solving the problem. In the case of patrolling the territory, situations are likely when some areas will be unattended.

- Implementation of artificial intelligence in computer games (for example, sports simulators (football, hockey, etc.), strategies). Even if the characteristics of the agents that make up the team are, on average, rather high, the poor coherence of their actions can lead to the loss of a more "well-played" team with a poorer set of players.
- Exchange trade. In this case, the analysis of the behavior of other players will contribute to choosing the right strategy and making a profit.

7 Conclusions

As a result, the analysis of the problems of using intelligent multi-agent systems for monitoring the state of the CPI was carried out, and the structure of the block for the reconstruction of three-dimensional scenes was proposed, which allows the construction of three-dimensional models of the CPI in an automatic mode with the possibility of processing occlusions. To implement the learning process, as well as the behavioral strategies of intelligent agents in a multi-agent system, it is proposed to use the reinforcement learning method. However, the implementation of this approach is associated with the problems of coordinating the actions of a group of agents, transmission of experience, long convergence time of the problem in the case of using simple reinforcement learning. To solve these problems, it is proposed to use hierarchical reinforcement learning, as well as multi-agent reinforcement learning. An algorithm for segmentation of macro-actions has been developed, which makes it possible to take into account the weight coefficients of the edges of the transition diagram.

Further, it is planned to refine the block for the reconstruction of three-dimensional scenes in order to add the function of analyzing three-dimensional scenes in statics and dynamics, as well as conduct research in order to develop solutions aimed at eliminating the identified problems of multi-agent reinforcement learning.

References

1. Lantsov, A.: Infrastructure: concept, types and meaning. Stat. Econ. **3** (2013)
2. Shen, S.: Accurate multiple view 3d reconstruction using patch-based stereo for large-scale scenes. IEEE Trans. Image Process. (TIP) (2013)
3. Bailer, C., Finckh, M., Lensch, H.P.A.: Scale robust multi view stereo. In: European Conference on Computer Vision (ECCV) (2012)
4. Kaiming, H., Gkioxari, G., Dollár, P., Girshick, R.: Mask R-CNN. Computer Vision and Pattern Recognition (cs.CV). arXiv:1703.06870—Submittedon 20 Mar 2017 (v1) Last revised 24 Jan 2018 (this version, v3)
5. Yuechen, Y., Yilei, X., Weilin, H., Matthew, R.: Deformable siamese attention networks for visual object tracking. Computer Vision and Pattern Recognition (cs.CV). arXiv:2004.06711—Submitted on 14 Apr 2020
6. Saggese, A.: Detecting and indexing moving objects for behavior analysis by video and audio interpretation. Computer Science. Universit´e de Caen (2014)

7. Yazdi, M., Bouwmans, Th.: New trends on moving object detection in video images captured by a moving camera: a survey. Comput. Sci. Rev. **28**, 157–177 (2018)
8. Nakano, E.: Introduction to robotics (Translated from Japanese. M.: Mir) (1988)
9. Russell, S., Norvig, P.: Artificial intelligence: a modern approach, 2nd edn (Translated from English. M.). Publishing house "Williams" (2006)
10. Vagin, V., Golovina, E., Zagoryanskaya, A., Fomina, M.: Reliable and plausible conclusion in intelligent systems. In: Vagin, V.N., Pospelov, D.A., M. (eds.) Fizmatlit (2008)
11. Jonsson, A., Barto, A.: A causal approach to hierarchical decomposition of factored MDPs. In: Proceedings of the 22nd International Conference on Machine Learning, pp. 401–408 (2005)
12. Stolle, M., Precup, D.: Learning options in reinforcement learning/Abstraction, Reformulation, and Approximation, pp. 212–223 (2002)
13. Mannor, S., Menache, I., Hoze, A., Klein, U.: Dynamic abstraction in reinforcement learning via clustering. In: Proceedings of the Twenty-First. In: International Conference on Machine Learning, p. 71 (2004)
14. Fruit, R., Lazaric, A.: Exploration–Exploitation in MDPs with Options (2017). https://arxiv.org/abs/1703.08667v2 (2017)
15. Jeni, L., Istenes, Z., Hashimoto, H., Korondi P.: Hierarchical reinforcement learning for robot navigation using the intelligent space concept. In: 11th International Conference on Intelligent Engineering Systems (2007)
16. Sutton, R.S., Barto, A.G.: Reinforcement Learning: An Introduction 2nd edn. (Transl. from English A. A. Slinkina. M.). DMK Press (2020)
17. Leibo, J., et al.: Multi-agent Reinforcement Learning in Sequential Social Dilemmas. arXiv preprint arXiv: 1702, 03037 (2017)
18. Foerster, J.N.: Deep Multi-Agent Reinforcement Learning (PhD thesis). University of Oxford (2018) https://ora.ox.ac.uk/objects/uuid:a55621b3-53c0-4e1b-ad1c-92438b57ffa4
19. Ahilan, S., Dayan, P.: Feudal multi-agent hierarchies for cooperative reinforcement learning. In: Published in the Proceedings of the Workshop on «Structure & Priors in Reinforcement Learning» at ICLR 2019 (2019)
20. Tang, H., Hao, J., Lv, T., Chen, Y., et al.: Hierarchical Deep Multiagent Reinforcement Learning with Temporal Abstraction / https://arxiv.org/abs/1809.09332 (2018)
21. Tarasov, V.B.: From multi-agent systems to intelligent organizations: philosophy, psychology, informatics. M.: Editorial URSS (2002)
22. Abramenko, A.: Calibration of the relative position of a stereo camera and a three-dimensional scanning laser rangefinder. Comput. Opt. **2**, 220–230 (2019)
23. Airapetov, A., Kovalenko, A.: Types of generative adversarial networks. Achievements of science and education **4**(45). https://cyberleninka.ru/article/n/vidy-generativno-sostyazatelnyh-setey (2019). Date of access 18 July 2020
24. Liu, J.: RocNet: Recursive Octree Network for Efficient 3D Deep Representation. arXiv:2008.03875 [cs.CV] (2020)
25. Skvortsov, A.V.: Delaunay Triangulation and its Application. Publishing house of Tomsk University, Tomsk (2002)
26. Dubenko, Yu.V., Timchenko, N.N.: No. 2650857 System for determining the geometric parameters of three-dimensional objects from 17.04.2018 / Patent for invention
27. Dubenko, Yu.V., Dyshkant, E.E., Rudeshko, N.A., Timchenko, N.N.: Scientific and technical report on the results of the implementation of the stage under contract No. 3617GS1/60550 (code 0060550) dated 20.07.2020 "Development of a digital platform for ensuring the safety of road infrastructure using three-dimensional laser scanning", registration number ICRBS: AAAA-B20-220122990242-9, date of registration 29 Dec 2020, p. 64
28. Arulkumaran, K., Dilokthanakul, N., Shanahan, M., Bharath, A.: Classifying Options for Deep Reinforcement Learning. IJCAI 2016 Workshop on Deep Reinforcement Learning: Frontiers and Challenges (2016)
29. Dubenko, Y.V., Gura, D.A., Dyshkant, E.E.: Monitoring complex infrastructure facilities state using mobile robots: problem analysis. In: 2019 International Multi-Conference on Industrial Engineering and Modern Technologies, FarEastCon 8934179 (2019) https://doi.org/10.1109/FarEastCon.2019.8934179

30. Gura, D.A., Dubenko, Y.V., Shevchenko, G.G., Dyshkant, E.E., Khusht, N.I.: Three-dimensional laser scanning for safety of transport infrastructure with application of neural network algorithms and methods of artificial intelligence. Lect. Notes Civ. Eng. **50**, 185–190 (2020)

Mathematical Modeling of Heat and Moisture Regimes of Building for the Facade Thermal Insulation Composite System with Mineral Wool Insulation

Kirill Zubarev and Vladimir Gagarin

Abstract Nowadays, construction industry is running into a controversial problem of using unsteady-state calculation methods because the level of difficulty of physical processes calculation in modern building materials is increasingly higher than in classical materials which were used previously. It should be stated that it is crucial to develop new methods which allow engineers to use simple equations to calculate complicated tasks. Such method was obtained by Kozlov and it gives an opportunity not to calculate unsteady-state moisture behavior of enclosing structures and evaluate the moisture maximum plane position inside enclosures. The mathematical model of unsteady-state moisture regime was obtained to provide evidence of the Kozlov's theory. Two computer programs were created. The first program simulated unsteady-state moisture regime inside enclosing structures during a year. The second program calculated the moisture maximum plane position during the period with negative average monthly temperatures. The comparison between these calculation methods was made for an enclosing structure with aerated concrete base and mineral wool insulation. As a result, it was shown that according to the Kozlov's method the position of moisture maximum is between insulation and exterior stucco. The same result was obtained by unsteady-state moisture regime calculation, which made Kozlov's theory evidence-based.

K. Zubarev (✉) · V. Gagarin
Moscow State University of Civil Engineering, 26, Yaroslavskoye Shosse, 129337 Moscow, Russia

Research Institute of Building Physics of Russian Academy of Architecture and Construction Sciences, 127238 Moscow, Russia

M. Shamtsyan et al. (eds.), *Robotics, Machinery and Engineering Technology for Precision Agriculture*, Smart Innovation, Systems and Technologies 247,
https://doi.org/10.1007/978-981-16-3844-2_54

1 Introduction

1.1 Problems in the Moisture Transfer Calculations

Nowadays, evaluating moisture behavior inside enclosing structures is one of the crucial tasks in modern construction industry [1–4]. Moreover, moisture regime of enclosing structures has a vital impact on the heat losses through building enclosures [5–11]. It happens because the liquid and vapor water, which enclosing structures always contain, is a very good heat conductor in a winter season [12–15].

Although scientists know that walls in the building accumulate moisture during the period with negative temperatures of outdoor air and get rid of moisture during the period with positive temperatures of outdoor air, calculations of moisture content is an incredibly difficult problem [16]. There are two reasons which explain this complexity. Firstly, modern calculation methods usually take into consideration simple formulas which make it possible to calculate steady-state moisture regime, but moisture transfer inside enclosing structures is unsteady-state [17–20]. On the other hand, it is possible to solve an unsteady-state moisture transfer equation using numerical calculations, although this approach is extremely complicated and does not give an opportunity to include such methods in regulatory documents [21–23].

1.2 Developing Facade Thermal Insulation Composite Systems with Mineral Wool Insulation

It has to be stated that in recent years, it is becoming more and more popular to design walls which contain aerated concrete blocks as a base of the wall structure and mineral wool plates as an insulation layer. This type of enclosing structures gives an opportunity to reduce thickness of whole wall, because the insulation layer has an extremely low value of a heat conductivity coefficient and, at the same time, there are various technical advantages like quick installation speed of the enclosure, convenience of using, and low wall weight.

1.3 Developing Moisture Transfer Mathematical Models

Mathematical models are developed in different directions in science [24–26]. A new generation of scientists has to solve the issues of obtaining new approaches of moisture regime calculations for the facade thermal insulation composite systems. Kozlov proposed the new mathematical model which gives the possibility to determine the maximum wetting plane position in enclosing structures. The maximum wetting plane position is the place where the mathematical maximum of the moisture behavior function is located [27, 28]. This Kozlov's solution is so simple, but it was obtained for steady-state moisture behavior. It is obvious that we cannot use this method for enclosing structures which have an unsteady-state moisture distribution. However, the facade thermal insulation composite systems have not been studied for type of the moisture regime process. It is certain that in terms of providing the evidence of possibility of using or not using the Kozlov's mathematical model, it is so interesting to compare the results of the location of maximum wetting plane obtained according to the Kozlov's mathematical model and the unsteady-state mathematical model [29, 30], which is based on the same physics theory.

2 Problem

To compare the moisture maximum plane position for the facade thermal insulation composite system with mineral wool insulation by the Kozlov's calculation method of evaluating the moisture maximum plane position and unsteady-state mathematical model, which is based on the moisture potential *F*.

3 Materials and Methods

3.1 The Mathematical Model of Moisture Regime Based on the Moisture Potential F

The mathematical model is based on the moisture potential *F*, which was obtained by V.G. Gagarin and V.V. Kozlov. Current model has two great advantages. Firstly, it takes into account a vapor and liquid moisture transport simultaneously, which gives an opportunity to consider joint moisture transfer in sorption and ultrasorption zones. Secondly, the variable value (the moisture potential *F*) has the analytical expression, which simplifies the model.

The analytical expression of the moisture potential *F* is presented:

$$F(w,t) = E_t(t) \cdot \varphi(w) + \frac{1}{\mu} \int_0^w \beta(\zeta) d\zeta. \quad (1)$$

where μ—vapor permeability coefficient, kg/(m s ·Pa); F—moisture potential, Pa; β—moisture conductivity coefficient, kg/(m s kg/kg); w—material humidity, kg/kg (1 kg/kg = 100% by weight); E_t—saturated water vapor pressure, Pa;ζ—current material moisture value, kg/kg; φ—relative air humidity; t—temperature, °C.

The moisture transfer equation is based on the moisture potential *F*:

$$\frac{\partial F(w,t)}{\partial \tau} = \kappa(w,t) \cdot E_t(t) \frac{\partial^2 F(w,t)}{\partial x^2}. \quad (2)$$

where κ—material heat-humidity characteristic coefficient, m^2/(s Pa); x—coordinate, m; τ—time, s.

Third-order boundary conditions on enclosing structure surfaces and between enclosing structure layers for the moisture potential F distribution are presented:

$$-\mu_1 \frac{\partial F}{\partial x}\bigg|_{i=1} = \frac{1}{R_{\text{m.ext}}}(F_{\text{ext}} - F_1). \tag{3}$$

$$\mu_2 \frac{\partial F}{\partial x}\bigg|_{i=N} = \frac{1}{R_{\text{m.in}}}(F_{\text{in}} - F_N). \tag{4}$$

$$-\mu_1 \frac{\partial F}{\partial x}\bigg|_{i=v-0} = -\mu_2 \frac{\partial F}{\partial x}\bigg|_{i=v+0}. \tag{5}$$

where F_N—material moisture potential next to enclosing structure surface which contacts with inside air, Pa, F_1—moisture potential of enclosing structure section which contacts with outside air, Pa; v—section of multilayer enclosing structure, in which there is a material joint; μ_1—vapor permeability coefficient of enclosing structure layer which is the nearest to the building outside surface, kg/(m s Pa); $R_{\text{m.ext}}$—resistance to moisture exchange between outside air and the surface of the enclosing structure, (m^2 s Pa)/kg; μ_2—vapor permeability coefficient of enclosing structure layer which is the nearest to the enclosing structure inner surface, kg/(m s Pa) F_{in}—inside air moisture potential, Pa; F_{ext}—outside air moisture potential, Pa; $R_{\text{m.in}}$—resistance to moisture exchange between inside air and the surface of the enclosing structure, (m^2 s Pa)/kg.

3.2 A Discrete–continuous Solution of Unsteady-State Moisture Transfer Equation

It has to be stated that the efficiency solution of Eq. (3) was obtained by the discrete–continuous approach, which makes it possible to calculate unsteady-state moisture regime with final formula and avoid using numerical methods. This solution is based on formulating of the heat-mass transfer system in the matrix form. The discrete–continuous formula is presented [29, 30]:

$$+ (G + K \cdot E_t \cdot A)^{-1}\left(e^{(G+K\cdot E_t\cdot A)\cdot\tau} - E\right) \cdot \bar{B} + e^{(G+K\cdot E_t\cdot A)\cdot\tau} \cdot \bar{F}_0.$$

$$+ (G + K \cdot E_t \cdot A)^{-1}\left(e^{(G+K\cdot E_t\cdot A)\cdot\tau} - E\right) \cdot \bar{B} + e^{(G+K\cdot E_t\cdot A)\cdot\tau} \cdot \bar{F}_0. \tag{6}$$

where G—matrix of coefficients for materials joint; K—matrix, which takes into account the differences in the thermal and moisture properties of the materials of the calculating enclosing structure; A—matrix of coefficients before functions of the moisture potential F for a multilayer enclosing structure; $\bar{L}$—a column vector, the first element of which is equal to one, other elements are equal to 0 for a multilayer enclosing structure $\bar{B}$—a column vector, the first and last elements of which describe the boundary conditions on the outer and inner surfaces of the enclosing structure, other elements are equal to 0 for a multilayer enclosing structure; E_t—matrix of the saturated water vapor pressure;p—the coefficient of the external boundary condition for a multilayer building enclosing structure, Pa/s^2.

The computer program which based on the discrete–continuous approach was created to find the maximum wetting plane position in investigated enclosing structures.

3.3 Determination of the Moisture Maximum Plane Position

Determination of the moisture maximum plane position is obtained from steady-state moisture transfer equation. For this aim moisture transfer Formula (2) was reformulated in steady-state form:

$$\frac{\partial}{\partial x}\left(\mu\frac{\partial F(w,t)}{\partial x}\right) = 0. \tag{7}$$

To investigate Function (7) of a maximum, the partial derivative $F(w,t)$ was presented as a function of two independent variables: temperature and moisture:

$$\frac{\partial F(w,t)}{\partial x} = \frac{\partial F(w,t)}{\partial w}\cdot\frac{\partial w}{\partial x} + \frac{\partial F(w,t)}{\partial T}\cdot\frac{\partial T}{\partial x}. \tag{8}$$

where T—absolute temperature, K.

The moisture maximum position can be described by next expression:

$$\frac{\partial w}{\partial x} = 0. \tag{9}$$

If Eq. (8) is solved with a Condition (9), it will be possible to formulate the equation of the maximum wetting complex, which helps to calculate the moisture maximum plane position:

$$f_i(t_{m.m}) = 5330\frac{r_{t.v}\cdot\left(t_{\text{in}} - t_{\text{ext.neg}}\right)}{R_t\cdot\left(e_{\text{in}} - e_{\text{ext.neg}}\right)}\cdot\frac{\mu_i}{\lambda_i} \tag{10}$$

where $t_{\text{ext.neg}}$—outdoor air average temperature in the period of monthly average temperature below zero, °C; $e_{\text{ext.neg}}$—partial pressure of outdoor air water vapor in the period of monthly average temperature below zero, Pa; μ_i—vapor permeability coefficient of ith layer material, kg/(m s Pa); f_i—function that corresponds to the temperature of the layer i in the maximum moisture zone «maximum wetting complex», ((°C)^2/Pa); λ_i—thermal conductivity coefficient of ith layer material, W/(m^2 °C); $r_{t.v}$—vapor permeability total resistance of enclosing structure, (m^2 s Pa)/kg; t_{in}—inside air average temperature, °C; R_t—heat transfer total resistance of enclosing structure, (m^2 °C)/W; e_{in}—partial pressure of inside air water vapor, Pa; $t_{\text{m.m}}$—maximum wetting temperature, °C.

The moisture maximum plane position can be calculated using a special table according to the value of the obtained maximum wetting complex f_i [27, 28].

4 Results and Discussion

4.1 *The Facade Thermal Insulation Composite System for Calculation*

An enclosing structure was chosen to study the moisture maximum plane position according to steady-state and unsteady-state moisture transfer equations. In terms of the moisture maximum researching, the building wall with a mineral wool insulation layer and aerated concrete base is a curious case, which is extremely popular in modern construction industry.

The wall structure of the researched wall is presented in Table 1.

Data of a construction region and microclimate parameters are presented in Table 2.

Table 1 Characteristics of the researched enclosing structure

Structure of the wall		Thicknesses of wall layers, m	Thermal conductivity of wall layers, W/(m °C)	Vapor permeability of wall layers, kg/(m s Pa)
1	A plaster layer of an outer wall surface	0.007	0.93	$3.6 \cdot 10^{-11}$
2	Insulation plates of mineral wool	0.12	0.042	$1.4 \cdot 10^{-10}$
3	Base blocks of aerated concrete	0.3	0.15	$6.4 \cdot 10^{-11}$
4	A plaster layer of an inner wall surface	0.02	0.93	$2.5 \cdot 10^{-11}$

Table 2 Data of a construction region and microclimate parameters

Construction region	Average outdoor temperature during the period with negative monthly average temperatures, °C	Average outdoor partial pressure of water vapor during the period with negative monthly average temperatures, Pa	Indoor air temperature, °C	Indoor relative humidity, %
Russia Moscow	−4.58	364	20	55

4.2 *Investigation of the Moisture Maximum Plane Position According to the Steady-State Moisture Regime*

The enclosing structure was researched according to the steady-state moisture transfer equation (Fig. 1). In Fig. 1 on the left side—outer wall surfaces is presented, on the right side—inner wall surfaces is presented. In Fig. 1 there is temperature (line 1), maximum wetting temperature (line 2) and the position of the maximum moisture plane (line 3).

As a result, the moisture maximum plane position was between the insulation layer and the exterior stucco layer.

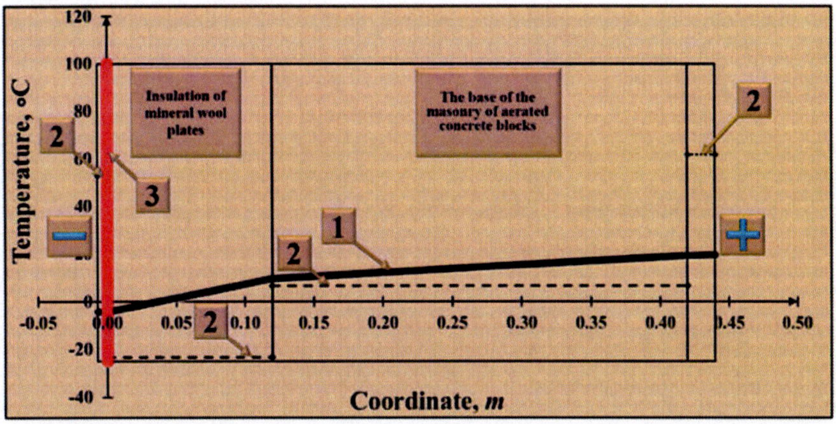

Fig. 1 Finding the position of the moisture maximum plane in the facade thermal insulation composite system with mineral wool insulation (1—temperature, 2—maximum wetting temperature, 3—the position of the maximum moisture plane)

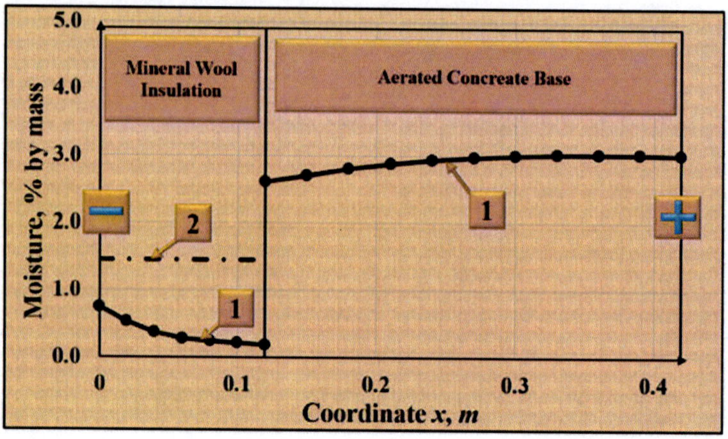

Fig. 2 Result of the moisture behavior calculation in the facade thermal insulation composite system with mineral wool insulation (1—a moisture behavior inside enclosing structure; 2—a maximum sorption zone)

4.3 Moisture Behavior According to Unsteady-State Moisture Regime

The facade thermal insulation composite system with mineral wool insulation was investigated by the steady-state moisture transfer Eq. (2) using discrete–continuous Formula (6). Moisture behavior calculation is presented in Fig. 2. In Fig. 2 there is a moisture distribution (line 1) and a maximum sorption zone (line 2).

As we can see, the moisture maximum is on the outer wall surface, which coincides with the result of the obtained moisture maximum by unsteady-state moisture transfer equation.

Thus, for the facade thermal insulation composite system with mineral wool insulation the position of the maximum wetting plane according to steady-state moisture regime is identical to the position of the maximum wetting plane according to unsteady-state moisture regime. It proves that a HVAC engineer is able to use steady-state equation and get the right result.

5 Conclusion

Comparison of the moisture maximum plane position for the facade thermal insulation composite system with mineral wool insulation was made by the Kozlov's calculation method of evaluating the moisture maximum plane position and the unsteady-state mathematical model, which is based on the moisture potential F. Both methods show that the maximum wetting plane position is located between the insulation layer and the exterior stucco layer. It means that if engineers use the Kozlov's calculation method, they will find the exact position of the maximum moisture plane. It gives a possibility to use steady-state moisture regime calculations for the facade thermal insulation composite system with mineral wool insulation in spite of unsteady-state moisture regime methods, which simplifies the work of practical engineers.

Current research expands knowledge of the moisture regime processes inside enclosing structures and proves the optionality of calculating unsteady-state moisture behavior inside the facade thermal insulation composite system.

References

1. Liuzzi, S., Rubino, C., Martellotta, F., Stefanizzi, P., Casavola, C., Pappalettera, G.: Characterization of biomass-based materials for building applications: the case of straw and olive tree waste. Indus. Crops Prod. **147**, 112229 (2020)
2. Chen, X., Qu, K., Calautit, J., Ekambaram, A., Lu, W., Fox, C., Gan, G., Riffat, S.: Multi-criteria assessment approach for a residential building retrofit in Norway. Energy Build. **215**, 109668 (2020)
3. Bai, H., Zhu, J., Chen, X., Chu, J., Cui, Y., Yan, Y.: Steady-state performance evaluation and energy assessment of a complete membrane-based liquid desiccant dehumidification system. Appl. Energy **258**, 114082 (2020)
4. Zubarev, K., Gagarin, V.: Determining the coefficient of mineral wool vapor permeability in vertical position. Adv. Intell. Syst. Comput. **1259**, 593–600 (2021)
5. Al-Tamimi, A.S., Al-Amoudi, O.S.B., Al-Osta, M.A., Ali, M.R., Ahmad, A.: Effect of insulation materials and cavity layout on heat transfer of concrete Masonry Hollow Blocks. Constr. Build. Mater, UNSP **254**, 119300 (2020)
6. Torres-Rodriguez, A., Morillon-Galvez, D., Aldama-Avalos, D., Hernandez-Gomez, V.H., Kerdan, I.G.: Thermal performance evaluation of a passive building wall with CO2-filled transparent thermal insulation and paraffin-based PCM. Sol. Energy **205**, 1–11 (2020)
7. Qu, M., Tian, S., Fan, L., Yu, Z., Ge, J.: An experimental investigation and fractal modeling on the effective thermal conductivity of novel Autoclaved Aerated Concrete (AAC)-based composites with Silica Aerogels (SA). Appl. Therm. Eng. **179**, 115770 (2020)
8. Kant, K., Anand, A., Shukla, A., Sharma, A.: Heat transfer study of Building Integrated Photovoltaic (BIPV) with nano-enhanced phase change materials. J. Energy Storage **30**, 101563 (2020)
9. Cabrera, V., Lopez-Vizcaino, R., Yustres, A., Ruiz, M.A., Torrero, E., Navarro, V.: A user-friendly tool to characterize the moisture transfer in porous building materials: FLoW1D. Appl. Sci.-Basel **10**, 5090 (2020)
10. Pooley, L.I., Abu-Bakar, A.S., Cran, M.J., Wadhwani, R., Moinuddin, K.A.M.: Measurements of specific heat capacity of common building materials at elevated temperatures: a comparison of DSC and HDA. J. Therm. Anal. Calorim. **141**(4), 1279–1289 (2020)
11. Berktas, I., Ghafar, A.N., Fontana, P., Caputcu, A., Menceloglu, Y., Okan, B.S.: Synergistic effect of expanded graphite-silane functionalized silica as a hybrid additive in improving the thermal conductivity of cementitious grouts with controllable water uptake. Energies **13**, 3561 (2020)
12. Zaborova, D., Vieira, G., Musorina, T., Butyrin, A.: Experimental Study of thermal stability of building materials. Adv. Intell. Syst. Comput. **692**, 482–489 (2017)
13. Gamayunova, O., Radaev, A., Petrichenko, M., Bogdanivics, R.: Predictive model of the dependence of the cost of insulation on thermal characteristics. In: E3S Web Conferences, vol. 140, p. 04018 (2019)

14. Gamayunova, O.S., Radaev, A.E., Petrichenko, M.R. (2019) the procedure for determination of the dependence of the cost of insulation materials on their thermophysical characteristics. In: IOP Conference Series: Materials Science and Engineering, vol. 660, p. 012018
15. Tarasova, D.S., Petritchenko, M.R.: Buildings Quasi-stationary thermal behavior. Mag. Civ. Eng. **4**(72), 28–35 (2017)
16. Castro, J.C.L., Zaborova, D.D., Musorina, T.A., Arkhipov, I.E.: Indoor environment of a building under the conditions of tropical climate. Mag Civ Eng. **8**(76), 50–57 (2017)
17. Lv, Y., Liang, J., Wang, B., Zhang, X., Yuan, W., Li, Y., Xie, J.: Effect of heat and moisture transfer on the growth of mould on the inner surface of walls: a case study in Dalian of China. In: Building Simulation (2020)
18. Chen, H., Li, S., Wei, P., Gong, Y., Nie, P., Chen, X., Wang, C.: Experimental study on characteristics of a nano-enhanced phase change material slurry for low temperature solar energy collection. Solar Energy Mater. Solar Cells **212**, 110513 (2020)
19. Fu, H., Ding, Y., Li, M., Cao, Y., Xie, W., Wang, Z.: Research and simulation analysis of thermal performance and hygrothermal behavior of timber-framed walls with different external thermal insulation layer: cork board and anticorrosive pine plate. J. Build. Phys. 1744259120936720 (2020)
20. Chung, D., Wen, J., Lo, L.J.: Development and verification of the open source platform, HAM-tools, for hygrothermal performance simulation of buildings using a stochastic approach. Build. Simul. **13**(3), 497–514 (2020)
21. Qu, K., Chen, X., Ekambaramd, A., Cui, Y., Gan, G., Økland, A., Riffata, S.: A novel Holistic EPC related retrofit approach for residential apartment building renovation in Norway. Sustain. Cities Soc **54**, 101975 (2020)
22. Fu, H., Ding, Y., Li, M., Li, H., Huang, X., Wang, Z.: research on thermal performance and hygrothermal behavior of timber-framed walls with different external insulation layer: insulation Cork Board and anti-corrosion pine plate. J. Build. Eng. **28**, 101069 (2020)
23. Lee, J., Wi, S., Chang, S.J., Choi, J., Kim, S.: Prediction evaluating of moisture problems in light-weight wood structure: perspectives on regional climates and building materials. Build. Environ. **168**, 106521 (2020)
24. Vorobyeva, I.V., Moshetova, L.K., Shcherbakova, E.V.: Therapy results in patients with primary open angle glaucoma and diabetic retinopathy. Oftalmologiya **14**(1), 40–46 (2017)
25. Vorobyeva, I.V.: Long-term results of treatment of patients with non-proliferative diabetic retinopathy angioprotectors. Oftalmologiya **14**(1), 67–77 (2017)
26. Vorobyeva, I.V.: Monitoring of separate pathogenetically significant biochemical markers in lacrimal fluid, ophthalmological parameters with combined pathology of diabetic retinopathy and age-related macular degeneration on the background angioprotective and antioxidant therapy. Oftalmologiya **15**(2), 189–199 (2018)
27. Gagarin, V.G., Akhmetov, V.K., Zubarev, K.P.: Position control of maximum wetting plane for building walls with foam polystyrene heat insulator. In: IOP Conference Series: Materials Science and Engineering, vol. 753, p. 022045
28. Gagarin, V.G., Akhmetov, V.K., Zubarev, K.P. (2020) Graphical method for determination of maximum wetting plane position in enclosing structures of buildings. In: IOP Conference Series: Materials Science and Engineering, vol. 753, p. 022046 (2020)
29. Gagarin, V.G., Akhmetov, V.K., Zubarev, K.P.: Unsteady-State moisture behavior calculation for multilayer enclosing structure made of capillary-porous materials. In: IOP Conference Series: Earth and Environmental Science, vol.177, p. 012021 (2018)
30. Gagarin, V.G., Akhmetov, V.K., Zubarev, K.P.: The Moisture Regime calculation of single-layer enclosing structures on the basis of the discrete-continuum method application. In: IOP Conference Series: Materials Science and Engineering, vol. 456, p. 012105 (2018)

Determination of the Current State and Forecast of the Remaining Life of the Catenary Supports of Electrified Railways

Valerij Li, Lyudmila Demina, and Sergey Vlasenko

Abstract The main task of reinforced concrete supports is to secure the contact and power lines for reliable power supply to consumers. The action of a constant load and the appearance of additional impacts lead to a gradual accumulation of defects and damage to the catenary supports, which gradually leads to breakage of the support and wires. In this regard, the determination of the residual bearing capacity of the catenary supports at any period of the life cycle is the main task of optimizing repair and reconstruction works aimed at extending the remaining life of the catenary system as a whole and becomes an urgent technical and economic problem. The fatigue curve of the reinforced concrete catenary support has been proposed, which takes into account additional factors affecting the defect formation. A simulation model of the current state of reinforced concrete catenary supports has been obtained, which allows evaluating the presence of damage and the degree of damage to concrete. A method has been developed for predicting the residual life of reinforced concrete catenary supports which allows determining the residual load-bearing capacity based on the service life and the size of defects for specific operating conditions.

1 Introduction

Material fatigue is the process of gradual accumulation of failures in an object under the influence of variable (often cyclic) stresses, which leads to changes in the mechanical and physical properties of the material, the formation of cracks, their development and destruction of the structure over a specified time, even if the induced mechanical stresses are much less than the permissible load limits [1]. Fatigue is the main cause of failure of many structures [2].

Linear and nonlinear studies of structures do not allow predicting fatigue failure [3]. They only calculate the response of a structure exposed to specified environmental conditions, constraints, and loads at a certain point in time. If the calculated

V. Li · L. Demina · S. Vlasenko (✉)
Far Eastern State Transport University, 47, Seryshev Street, Khabarovsk 680021, Russia

M. Shamtsyan et al. (eds.), *Robotics, Machinery and Engineering Technology for Precision Agriculture*, Smart Innovation, Systems and Technologies 247,
https://doi.org/10.1007/978-981-16-3844-2_56

stresses are within the permissible limits, they indicate that the structure is safe in this environment, regardless of how many times the load is applied [4].

The results of a static, nonlinear, or dynamic study of a modal time diagram can be used as a basis for determining the fatigue study [5]. The number of cycles required for fatigue failure at a location depends on the material and stress fluctuations.

2 Problem Statement

Fatigue strength is determined by applying different levels of cyclic stress to individual test samples and finding the number of cycles before failure [6]. The graphical representation of points on the fatigue curve (Fig. 1) is the ratio of the amplitude of cyclic (or alternating) stress (vertical axis) and the number of cycles before failure (horizontal axis). Fatigue strength is defined as the stress at which fatigue failure occurs at a given number of cycles [7].

Samples are tested at increasing load levels until failure occurs within the selected maximum number of cycles [8]. A section of the curve that is close to the horizontal section sets the fatigue limit of the tested material [9]. If the stress amplitude is below the fatigue limit of the material, the sample is considered to have an unlimited service life [10].

The fatigue curve for a material is plotted for a given stress coefficient R, which is defined as the ratio of the minimum cyclic stress to the maximum cyclic stress [11]. The coefficient for a fully reversing load is $R = -1$ for the load which is applied and removed (non-reversing) $R = 0$.

Accumulated damage theory assumes that a cycle with alternating stress is above the fatigue limit determined by the measured permanent damage [12]. It is also assumed that the total failure caused by multiple stress cycles is equal to the sum of failures caused by individual stress cycles.

Fig. 1 Typical fatigue curve of the material

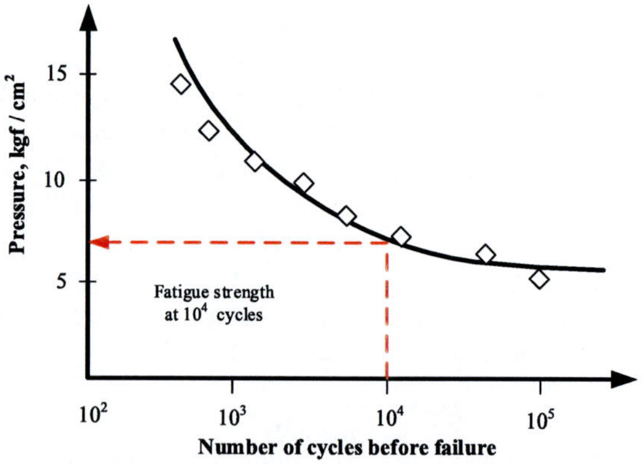

If we assume that the fatigue curve indicates that it takes a certain number of cycles of alternating stress to cause fatigue failure, then the theory states that each cycle causes a degree of damage that consumes an amount inversely proportional to this number of cycles called structural durability [13].

This rule is called the linear damage rule or the Miner rule. Fatigue failure occurs when the degree of damage is 1.0 [14].

3 Research Results

The fatigue curve shows the quantitative period at which destructions occur [15]. It is obvious that at the value of the alternating stress equal to infinity, the state of the support will be correlated as "new."

To estimate the remaining service life of a reinforced concrete support, as the initial data was taken the stress coefficient of a fully reversible load which compensated for the linear directions of wind and ice loads [16]. The values of the maximum and minimum stresses are taken from static analysis under the condition of icy conditions with wind. Loads are determined from zero to maximum load by the amplitude set by the fatigue curve [17]. The developed program for calculating fatigue phenomena proposes the amplitudes of cyclic stress and additionally sets the number of cycles under study [1]. Given that the full amplitude takes place once a season, we will consider 4 cycles = 1 year of actual time. The analysis has been carried out for the number of cycles ranging between 1 (3 months) and 300 (75 years). Every 40 cycles were breakdown with reference to the nominal 10 years (nominal indicators of wind and ice, the maximum repeatability in the 10-year period).

The resulting calculated fatigue curve (Fig. 2) is based on data obtained in static analysis, where the cyclic load is the total nonlinear load on the support, taking into

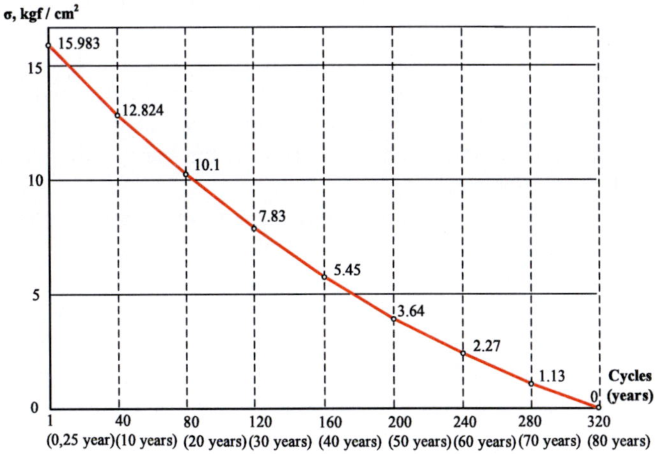

Fig. 2 Fatigue curve of a reinforced concrete catenary support

account a given time range, and each subsequent cycle summarizes the previously occurred damage [18]. The quantitative data of the life curve takes into account the dependences of the fatigue coefficient for each specified material [19].

According to our data, the first minor damage occurs after 25 years, while the fatigue limit is limited to 160 cycles (40 years). The analysis of practical experience in operating the catenary supports confirmed these results. Indeed, supports serve 60 and 70 years, almost their "second" term [20]. But, when designing a structure, in accordance with the reliability theory, a nominal period with a margin is always used [21].

It is also very important to note that electrified railway sections of direct and alternating current have distinctive features [22]. In particular, a phenomenon of electrical corrosion, which is more typical for direct current, rapidly destroys the support (due to the destruction of the reinforcement), thus contributing to a service life reduction [23]. An example is some sections of the Sverdlovsk DC railway, where almost no reinforcement was found on a number of reinforced concrete supports, according to the results of diagnostics of the underground section. For alternating current, this phenomenon is weak (hundreds of times less), so the destruction process (corrosion of reinforcement) is clearly slowed down [24]. Based on this symptom, it is quite difficult to judge the support integrity, because the key reason for reducing the load-bearing capacity of supports for alternating current is defect formation (destruction of the concrete part). Therefore, the support structures are designed, as a rule, with a margin factor equal to three.

The present work does not take into account the dynamic effects that are disturbing factors (wind gusts, forced vibrations of track devices, and the catenary), and their impact is unknown [25]. Based on the theory of dynamic vibrations, it can be assumed that the vibrations range is 3σ. We believe that this is a topic for further research on the phenomenon of defect formation depending on such additional factors.

We assumed that when modeling the initial data, consideration of dynamic vibrations can be taken into account by introducing dynamic coefficients, which, according to the theory of vibrations, usually reduce the overall load-bearing capacity by about 1.3–1.5 times.

Analyzing the fatigue phenomena resulted in the lifetime diagrams (Fig. 3) and damage diagrams (Fig. 4), presented in a color gradient (blue to red). Based on the load factor, the program developed in the course of this work determines the corrected alternating stress (from the stress values in the static study) for each node and a certain type of fatigue event.

The damage diagram clearly shows the damage to the support (color gradient) depending on the number of cycles. The areas that characterize the destruction are shown in red. According to the data obtained, we can see that damage occurs after 25 years (in accordance with the fatigue curve) and primarily in the areas of embedded parts, console heels, rod, and the conditional edge of a foundation. In the second place, approximately in 29–30 years, the areas of ventilation holes suffer. This fact is obvious, because they are close to the conditional edge of a foundation where the load is the biggest.

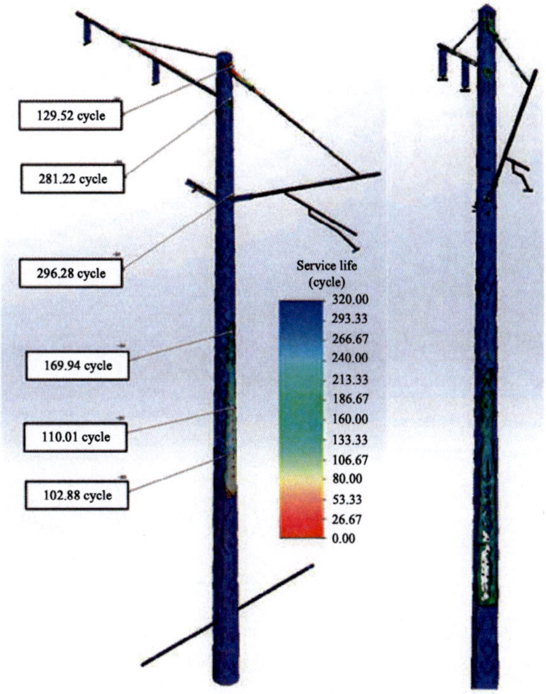

Fig. 3 Lifetime diagram of the reinforced concrete catenary support

The lifetime diagram shows the service life of the support, where the blue color shows areas that will not become unusable in 75 years, also marked typical zones (areas with red to green gradient), which will be first to break down, with the time when the destruction occurs.

In Fig. 3, the second lifetime diagram is shown rotated. This is done to make visible, when modeling, the areas (light spots) that characterize the complete destruction in these areas.

In the last five years, there has been an increase in the number of supports that have been in operation for over 60 years [26]. This could be explained by the fact that specified wind and ice deposit loading limits have never happened to been reached simultaneously. It should be remembered that supports have a safety margin, and it is currently not possible to take it into account in the calculations. The presence of cracks that exceed the permissible size indicates that such supports must be replaced because eventually the support will fall suddenly [27].

The resulting calculated fatigue curve (lifetime curve) takes into account many operational factors affecting the support, and many issues of defects are included in it. If the reliability factor (the service life factor) is neglected, the curve will only be higher or lower relative to the resulting curve, but it will follow the same law of decreasing exponent [28].

Fig. 4 Damage diagram of the reinforced concrete catenary support

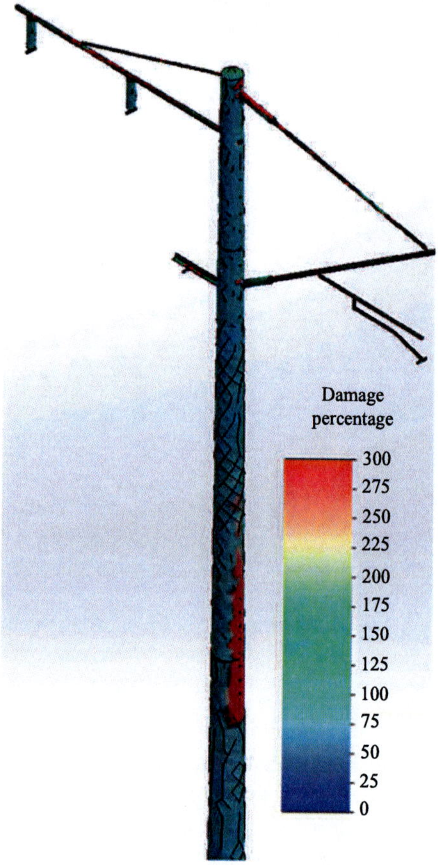

Solving a scientific problem, we note that the model is, of course, imperfect, but having taken almost the maximum load values with an account of various circumstances like additional influences, dynamics factors in the form of equations, which were set statically without defining dependency, we obtained a method which allows evaluating forecast values of residual life of concrete pillars [29].

The support is a columnar conical structure. From the perspective of the construction mechanics theory, the method for calculating the strength of columns is not quite suitable for catenary supports, although the proposed method is based on the basic rules for calculating columns. Therefore, if we take into account the factors to the maximum degree, we can say how long the support should be used under such conditions.

According to the instructions [1], the operation is possible at certain values of the crack width, length, number, and location. When defects are large, the support is no longer considered load-bearing. Using the normative requirements for the calculation of reinforced concrete supports, the work solved the problem of determining additional factors of crack formation and the degree of their effect on the determination of the support resource to assess the remaining service life.

Since, according to the instructions [1], three longitudinal cracks (the most common type of defect) with a width of up to 1.5 mm (with prestressed wire reinforcement) are allowed, it is permissible to determine the total crack width when determining the defect of the support. At this point, we assume that the total crack width of 4.5 mm will correspond to the exhaustion of the support resource. On the other hand, the nominal lifetime of 40 years determined by modeling is equal to the total crack width of 4.5 mm.

Using the above assumptions, a nomogram is proposed for determining the remaining service life of supports, taking into account the climatic conditions of the city of Khabarovsk, depending on the crack width and the life curve, assuming a straight-line dependence of the bending moment on the crack width (Fig. 5). This assumption will only reduce the residual load-bearing capacity.

Rules for using the nomogram (according to Fig. 5):

1. The total crack width is determined (3.5 mm in the figure).
2. The perpendicular line is drawn to the intersection with the defect curve to get point 1.

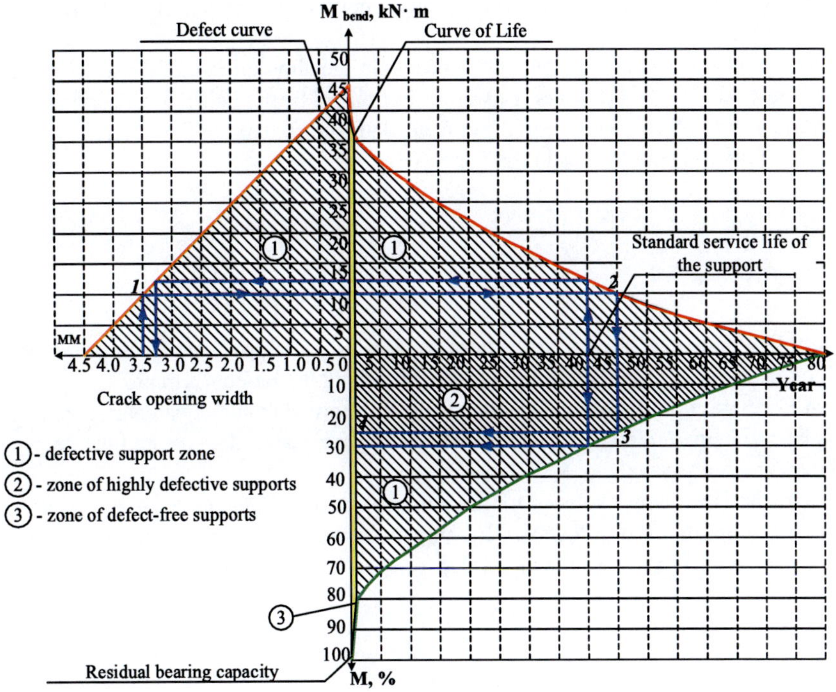

Fig. 5 Nomogram for determining the remaining life of supports

3. From point 1, a horizontal line is drawn to the intersection with the life curve to get point 2.
4. A vertical line is drawn to the intersection with the residual load-bearing capacity curve to get point 3.
5. A horizontal line is drawn from point 3 to the intersection with the coordinate axis to get point 4, corresponding to the value of the remaining resource 26% (according to figure), located in the zone of defective supports.

If there are no longitudinal cracks, the axis of the support service life must be used. In our example, 40 years of operation corresponds to a crack width of 3.3 mm, while the residual load-bearing capacity is 30%, which is located at the boundary between the two zones.

For climatic conditions of other power supply distances, the lifetime curve should be adjusted depending on the types and brands of supports used. The shape of the curves will not change.

If using the data of a survey of concrete part of the supports with ultrasonic instrument (UK-1401 [30]), the defectiveness of support on the nomogram can be assessed by indicator P2. The method for determining the residual load capacity remains the same.

To date, a number of forecasting methods are known and developed [31]. They are based on various analytical approaches based on statistics. The most widely used methods are the trends method, artificial neural networks, correlation and regression analysis, etc.

The proposed calculation method is based on the simulation modeling, which is implemented in the SolidWorks software environment by the SolidWorks Simulation module. It will allow forecasting the current state and remaining life of supports for specific operating conditions in the Far East.

4 Conclusions

The fatigue curve of the reinforced concrete catenary support has been proposed, which takes into account additional factors affecting the defect formation.

A simulation model of the current state of reinforced concrete catenary supports has been obtained, which allows evaluating the presence of damage and the degree of damage to concrete.

A method has been developed for predicting the residual life of reinforced concrete catenary supports which allows determining the residual load-bearing capacity based on the service life and the size of defects for specific operating conditions.

References

1. Li, V., Demina, L., Vlasenko, S.: Assessment of the concrete part of the contact system supports in the field. In: E3S Web Conference, vol. 164, p. 03028 (2020). https://doi.org/10.1051/e3sconf/202016403028
2. Schlicke, D., Hofer, K., Nguyen, V.T.: Adjustable restraining frames for systematic investigation of cracking risk and crack formation in reinforced concrete under restrained conditions. In: Advanced Techniques for Testing of Cement-Based Materials (2020)
3. Zhao, X., Gong, P., Qiao, G., Lu, J., Lv, X., Ou, J.: Brillouin corrosion expansion sensors for steel reinforced concrete structures using a fiber optic coil winding method. Sensors **11**, 10798–10819 (2011). https://doi.org/10.3390/s111110798
4. Tsunemoto, M., Shimizu, M., Kondo, Y., Kudo, T., Ueda, H., Ijima, T.: Replacement criteria for concrete catenary poles. Q. Rep. RTRI **58**, 270–276 (2017). https://doi.org/10.2219/rtriqr.58.4_270
5. Sannikov, D.V., Kolevatov, A.S., Vavilov, V.P., Kuimova, M.V.: Evaluating the quality of reinforced concrete electric railway poles by thermal nondestructive testing. Appl. Sci. **8**, 222 (2018). https://doi.org/10.3390/app8020222
6. Li, V.N., Demina, L.S., Vlasenko, S.A., Tryapkin, E.Y.: Assessment of the impact of the electromagnetic field of the catenary system on crack formation in reinforced concrete supports. In: IOP Conference Series: Materials Science Engineering, vol. 918, p. 012118 (2020). https://doi.org/10.1088/1757-899X/918/1/012118
7. Frankowskib, P.K., Sikora, R., Chadya, T.: Identification of rebars in a reinforced mesh using eddy current method. In: AIP Conference Proceedings, vol. 1706, p. 090008 (2016). https://doi.org/10.1063/1.4940545
8. Romanenko, E.Y., Trubitsin, M.A.: Combined way of diagnostics of reinforced concrete conical structures. Mater. Sci. Forum **931**, 346–351 (2018). https://doi.org/10.4028/www.scientific.net/msf
9. Ignatenko, I.V., Vlasenko, S.A.: Determination of criteria to find the quality of an electrical connection. Russian Electrical Eng. **87**(2), 65–67 (2016). https://doi.org/10.3103/s1068371216020085
10. Prakoso, P.B., Lechner, B.: Finite element analysis of CRCP slab track system designed with active crack control. In: 6th International Conference on Electric Vehicular Technology, pp. 359–367 (2019). https://doi.org/10.1109/ICEVT48285.2019.8993995
11. Li, W., Fan, S., Ho, S.C., Wu, J., Song, G.: Interfacial debonding detection in fiber-reinforced polymer rebar–reinforced concrete using electro-mechanical impedance technique. Struct. Health Monit. **17**, 461–471 (2018). https://doi.org/10.1177/1475921717703053
12. Ignatenko, I., Vlasenko, S.: Diagnostics of electrical connections of electric traction network. Adv. Intell. Syst. Comput. **1115**, 69–78 (2020). https://doi.org/10.1007/978-3-030-37916-2_8
13. Rozli, M., Safiuddin, C., Harun, M., Ahmad, J., Amin, N., Dora, A.: The behaviour of Prestressed Concrete Sleeper (pcs) sitting on railway track. In: IOP Conference Series Material Science Engineering, vol. 615, p. 012123 (2019). https://doi.org/10.1088/1757-899X/615/1/012123
14. Szymanik, B., Karol-Frankowski, P., Czady, T., Chelliah, C.: Detection and inspection of steel bars in reinforced concrete structures using active infrared thermography with microwave excitation and eddy current sensors. Sensors **16**, 234 (2016). https://doi.org/10.3390/s16020234
15. Ignatenko, I.V., Vlasenko, S.A.: Search for additional criteria to estimate the electrical connection state under the impact of currents of different value. In: IOP Conference Series: Material Science Engineering, vol. 760, p. 012024 (2020). https://doi.org/10.1088/1757-899X/760/1/012024
16. Maierhofer, C., Reinhardt, H., Dobmann, G.: Non-Destructive Evaluation of Reinforced Concrete Structures: Non-Destructive Testing Methods, vol. 2. Woodhead Publishing, UK (2010). https://doi.org/10.1533/9781845699536

17. Apostolopoulos, C.A., Koulouris, K.F., Apostolopoulos, A.C.: Correlation of surface cracks of concrete due to corrosion and bond strength (between steel bar and concrete). Adv. Civ. Eng. **2019**, 3438743 (2019). https://doi.org/10.1155/2019/3438743
18. Ignatenko, I., Tryapkin, E., Vlasenko, S., Onischenko, A., Kovalev, V.: Impact of return traction current harmonics on the value of the potential of the rail ground for the ac power supply system. Adv. Intell. Syst. Comput. **1115**, 117–127 (2020). https://doi.org/10.1007/978-3-030-37916-2_13
19. Zharkov, Y.I., Semenov, Y.G., Kolosov, D.V., Otýpka, J.: The principals of operation of the automated current collection diagnostic system for electrified railways. In: 2015 16th International Scientific Conference on Electric Power Engineering (EPE), pp. 760–763 Kouty nad Desnou (2015). https://doi.org/10.1109/epe.2015.7161202
20. Bao, Y., Valipour, M., Meng, W., Khayat, K., Chen, G.: Distributed fiber optic sensor-enhanced detection and prediction of shrinkage-induced delamination of ultra-high-performance concrete overlay. Smart Mater. Struct. **26**, 085009 (2017). https://doi.org/10.1088/1361-665X/aa71f4
21. Zhao, S.B., Yang, S., Feng, X.Z., Lu, M.J.: Study on thermal conductivity of reinforced concrete plate. Appl. Mech. Mater. **438–439**, 321–328 (2013). https://doi.org/10.4028/www.scientific.net/amm
22. Li, G., Liu, Q., Zhao, S., Qiao, W.: Ren X automatic crack recognition for concrete bridges by fully convolutional neural network and Naive Bayes data fusion based on visual detection system. Meas. Sci. Technol. (2020). https://doi.org/10.1088/1361-6501/ab79c8
23. Ignatenko, I.V., Vlasenko, S.A.: Health assessment of the electrical contact-line connections in view of the operational traction load pattern of the electric rolling stock. In: IOP Conference Series: Material Science Engineering, vol. 918, p. 012154 (2020). https://doi.org/10.1088/1757-899X/918/1/012154
24. Kobayashi, K., Banthia, N.: Corrosion detection in reinforced concrete by using induction heating and infrared thermography. J. Civ. Struct. Health Monit. **1**, 25–35 (2011). https://doi.org/10.1007/s13349-010-0002-4
25. Galkin, A., Kovalev, A., Okunev, A.: Assessing remained service time of contact-line support under the constant load. In: Popovic, Z., Manakov, A., Breskich, V. (eds.) VIII International Scientific Siberian Transport Forum. TransSiberia 2019. Advances in Intelligent Systems and Computing, vol. 1115, pp. 693–702. Springer, Cham (2020). https://doi.org/10.1007/978-3-030-37916-2_68
26. Romashko, O.V., Romashko, V.M.: Model of multilevel formation of normal cracks in reinforced concrete elements and structures. In: IOP Conference Series: Material Science Engineering, vol. 708, p. 012069 (2019). https://doi.org/10.1088/1757-899X/708/1/012069
27. Kats, V., Volkov, A.: Features extraction from non-destructive testing data in cyber-physical monitoring system of construction facilities. In *Journal of Physics: Conference Series*, vol. 1425, p. 012149 (2019). https://doi.org/10.1088/1742-6596/1425/1/012149
28. Uteuliyev, B.: Residual life assessment of overhead transmission lines 110 kV and above and determination of their reconstruction terms. In: E3S Web of Conferences, vol. 28, p. 04005 (2017). https://doi.org/10.1051/e3sconf/20172504005
29. Raziff, M., Sam, R., Masrie, M., Janin, Z.: Design and simulation of pick and place system using solidworks simulation. In: 2018 IEEE 5th International Conference on Smart Instrumentation, Measurement and Application (ICSIMA), pp. 1–6. Songkla, Thailand (2018). https://doi.org/10.1109/ICSIMA.2018.8688800
30. Mociran, B., Oglejan, R.: Methods of interconnecting designing programs simulation and optimization in engineering. In: 2018 International Conference and Exposition on Electrical And Power Engineering (EPE), pp. 0475–0478. Iasi (2018). https://doi.org/10.1109/ICEPE.2018.8559613
31. Tryapkin, E., Shurova, N.: The use of technology 'Big data' and 'Predictive analytics' in the power supply system of railways. Adv. Intell. Syst. Comput. **1115**, 60–68 (2020). https://doi.org/10.1007/978-3-030-37916-2_7

Effectiveness of Robotic Business Processes in the Enterprise

Alexander Sukhorukov, Nataliya Ulitskaya, Tatyana Fomicheva, Julia Sulyagina, and Sergey Eroshkin

Abstract Nowadays, most business processes of modern enterprises with different degree of automation are realized by means of information systems (Online Transaction Processing, OLTP). To increase the share of automatic operations, in recent years, a new information technology of robotization of business processes (Robotics Process Automation, RPA) on top of existing systems has gained great popularity. However, the cost-effectiveness of such digital transformation is not obvious and requires justification. In the article, methods of an estimation of efficiency of introduction of the software robots replacing the processes carried out by people in the automated control systems of the enterprises are considered. The analysis of two approaches to efficiency estimation is carried out: classical one, from the point of view of efficiency indicators of IT-projects of enterprise development and cost one, from the point of view of the complex analysis of services (the basic business processes). Practical recommendations on efficiency estimation of robotization of business processes are given.

1 Introduction

The enterprise management system implies qualitative execution of business processes, which can be classified into three types [1]:

- the basic business processes which provide housing and communal services and are directly directed on receiving profit by the enterprise;

A. Sukhorukov
Plekhanov Russian University of Economics, Stremyannyy Pereulok, 36, Moscow 115093, Russia

T. Fomicheva · J. Sulyagina · S. Eroshkin (✉)
Russian State Social University (RSSU), Ulitsa Vil'gel'ma Pika, 4, Moscow 129226, Russia
e-mail: eroshkin.sy@rea.ru

N. Ulitskaya
Moscow Automobile and Road Construction State Technical University (MADI), Leningradsky Ave, 64, Moscow 125319, Russia

M. Shamtsyan et al. (eds.), *Robotics, Machinery and Engineering Technology for Precision Agriculture*, Smart Innovation, Systems and Technologies 247,
https://doi.org/10.1007/978-981-16-3844-2_57

- providing business processes that support the execution of basic processes and create the structure of the enterprise;
- business processes of management, which are aimed at the management of the enterprise.

In some methodologies of business processes organization, there is a separate type of development processes, aimed at getting profit in the future, at improving the activity of the enterprise.

As a rule, performers and participants of business processes are people. If people do not participate in the process, such processes are called technological, or specifications. Despite the fact that most of the processes of modern enterprises are automated with the help of information systems, in recent years, a new information technology for robotizing business processes (Robotics Process Automation, RPA) has gained great popularity. The essence of this technology consists in replacement by software robots of business processes which are carried out by the person-operator in transaction systems of enterprise management (Online Transaction Processing, OLTP) [2, 3]:

- ERP (Enterprise Resource Planning)—enterprise resource planning;
- CRM (Customer Relationships Management)—customer relationship management;
- SRM (Supplier Relationships Management)—vendor relationship management;
- SCM (Supply Chain Management)—supply chain management;
- Specialized transaction systems that automate the management of individual types of resources, such as:
  - WMS (Warehouse Management System)—warehouse management systems;
  - CMMS (Computerized Maintenance Management System)—computerized maintenance (repair) management systems;
  - EAM (Enterprise Asset Management)—enterprise asset management. Represents transformation of CMMS-systems, realizing strategy EAM (reduction of expenses for maintenance service, repair and material and technical maintenance without decrease in level of reliability, or increase in production parameters of the equipment without increase in expenses);
  - MES (Manufacturing Execution System)—systems of operational (shop floor) management of production processes;
  - SCADA (Supervisory Control and Data Acquisition)—dispatching control and data acquisition, programs for automated control of technological processes, analog of automated process control systems (PCS);
  - ECM (Enterprise Content Management)—management of enterprise information resources. The analog of the domestic concept of "electronic document management systems" (EDMS).

In Fig. 1, the scheme of digital transformation of the listed automated systems of execution of business processes of the enterprises is shown. It is visible over what automated systems software robots (RPA) are written. For example, an employee of the enterprise (operator) manually enters data into the resource planning system

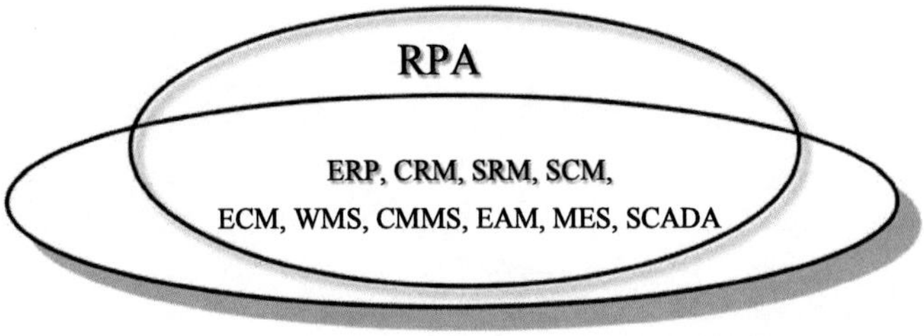

Fig. 1 Digital transformation of automated systems for the execution of business processes of enterprises using RPA

of the ERP enterprise, or sorts the customer letters received in the ECM electronic document management system. By means of RPA, such business processes can be robotized without changing the working information management systems (ERP, ECM). In spite of the fact that manufacturers of enterprise management information systems constantly improve their systems and robotize many business processes on the level of the systems themselves, it is nevertheless economically expedient in most cases to supplement them with external software robots [4].

If you look at the forecast of the research company Gartner in 2019 [5], the RPA technology in the next two years will be released on the plateau productivity curve maturity (Hype cycle). This confirms the relevance of RPA development in terms of its practical significance. Such rapid development of RPA in the next two years is justified by the replacement of simple routine business processes. However, the future of RPA is directly related to the development of artificial intelligence (Artificial Intelligence, AI). The use of machine learning (ML) on large data (Big Data, BD) will allow for the robotization and complex intelligent business processes of enterprises. On the maturity curve, you can see that most AI technologies are still at the point of technology launch or at the peak of over-expectations like ML. Plateau performance of such technologies will be achieved only in 5–10 years, which means that RPA in the future will enter a new round of its development.

At the same time, the intensive development of RPA raises an acute problem, which is to relieve some management personnel of their functions. Software robots are in many ways superior to humans in the performance of business processes. They perform many processes faster, make fewer operational mistakes, do not need salaries, rest, medical, pension, etc.

Introduction of project RPA in the automated control system of the enterprise influences its business model as a whole (Fig. 2) [6]. All kinds of business processes, organizational structure, supply chains of materials, interaction with clients, cash flows, and document circulation are changing. Thus, influence RPA on business processes of the enterprises should be innovative, that is economically effective. The decision of this problem is not obvious at first sight and demands a choice of such method (methods) of an estimation of efficiency of the offered digital transformation which at the least expenses will show the best result. In modern realities, it is necessary

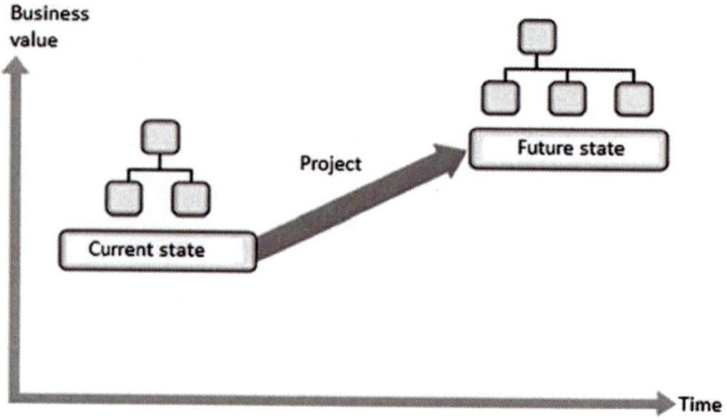

Fig. 2 Transformation of the enterprise through the development project (Guide to the Set of knowledge on project management. (PMBOK® guide)—Sixth edition. PMI website. https://www.pmi.org/)

to take into account the possibility of implementation of this method, or methods in modern digital modeling and analytical systems.

Traditional methods of estimation of efficiency of introduction of IT-solutions in information systems of management of the enterprise 1.

For some last decades from experience of introduction of information control systems, the whole system of methods of an estimation of economic efficiency of such projects has developed (Table 1). If to consider introduction RPA, IT-project of development of the enterprise use of the offered methods is quite proved.

The traditional analysis of efficiency of IT-project consists in forecasting of incomes and losses in the form of the report on income (Income Statement, IS),

Table 1 System of methods for assessing the economic efficiency of IT-solutions implementation in information systems of enterprise management

Financial	Qualitative	Probabilistic
NPV, net present value	BSC, balanced scorecard	ROV, real options valuation
IRR, internal rate of return	IT scorecard	AIE, applied information economics
ROI, return on investment	IE, information economy	
payback period	PM, portfolio management	
EVA, economic value added	TEI, total economic impact	
TCO, total cost Of ownership	REJ, rapid economic justification	

in forecasting of cash flow (Cash Flow, CF), in forecasting of balance of assets and enterprise liabilities (Balans Sheet, BS).

Each of the methods presented in Table 1 has been studied in sufficient detail and is described in the literature. Within the limits of the given researches, it is necessary to stop only on features of their applicability at an estimation of efficiency of robotization of business processes of the enterprise.

The method of return on investment (ROI) and the method of total cost of ownership of information systems (Total Cost of Ownership, TCO) became the most popular among financial methods in evaluating the effectiveness of RPA implementation.

ROI method shows the ratio of net profit increase to the volume of investments.

ROI it is also necessary to take into account the growth of customer satisfaction with housing and communal services, the growth of qualification of employees of enterprises, obtaining competitive advantages from the introduction of RPA.

The base price based on resource consumption by the TCO method is calculated by a formula:

$$C_{\mathrm{RPA}} = (1 + R) \sum_{\forall j} \mathrm{Tj}\,\mathrm{gj} \tag{1}$$

where R is the calculated standard of profitability of housing and communal services at their transfer to the client (R is formed under the influence of frequency, urgency and other factors);

Tj—the estimated objective costs of computing, labor, and material resources;

j—current number of used resources;

$\forall$—symbol, which shows that the summation occurs for each j;

gj—tariff per calculated unit of services at the consumption of jth resource, or the sum of resources (for example, computing systems) at the standard of profitability equal to zero.

In case of mutual calculations between subdivisions within the information services company, Formula (2) can be used in the case if $R = 0$. It means that in internal calculations, the profit of one subdivision at the expense of other subdivisions will not be formed. Weighting factors for different services and resources can be entered into Formula (2).

The TCO method allows to calculate the total direct and indirect costs on which the benefits of RPA implementation are estimated.

The total economic impact (TEI) method is popular among qualitative methods. In fact, software robots can be so different in functionality, complexity, and price that one has to choose from a number of solutions:

- Develop the software robot yourself;
- to buy a ready-made boxed solution;
- order an individual version of the software robot.

TEI method, as a rule, is used for making such decisions. It is used for balanced assessment of RPA benefits, risks, often resulting from flexibility, and cost estimation. This method, like the financial TEI method, uses the total cost of the project.

Since RPA directly affects the organizational structure of an enterprise, it is useful transformation. This indicator is calculated by Formula (3). The analysis of the indicator in dynamics will show how the FTE decreases with the implementation of RPA.

$$\mathrm{FTE} = (T_{\mathrm{df}} D_{\mathrm{f}} N_{\mathrm{f}} W + T_{\mathrm{di}} D_{\mathrm{i}} N_{\mathrm{i}} W)/T_{\mathrm{p}} \tag{2}$$

where T_{df} is full-time working hours on a full-time employee's day;
T_{di}—working hours per day of a part-time employee;
D_{f}—number of working days per week of a full-time employee;
D_{i}—number of working days per week of a part-time employee;
N_{f}—number of full-time employees;
N_{i}—number of part-time employees;
W—number of weeks for the reporting period;
T_{p}—working hours during the reporting period.

The method of calculation according to Formula (3) is to perform sequential actions:

- the period (month, quarter, half-year, year) for which FTE is calculated and expressed in weeks (for example, half-year equals 26 weeks);
- the number of full-time employees is determined;
- the number of hours worked by these employees per week is calculated;
- the number of part-time employees is determined;
- the number of hours worked by these staff members is calculated;
- the number of hours worked per week is calculated; the hours worked by these employees are calculated; the hours worked by these employees are summed up;
- this amount is divided by the number of working hours in the reporting period.

FTE is most often used in the calculation of such indicators as:

- employee productivity
  PTS = sales amount/FTE;
- human value added
  DS HCVA = (Income—(Expenses—Salary—Premium))/FTE;
- sales profit per staff member
  RPE = Annual Revenue Amount/FTE.

2 Materials and Methods of Functional Cost Analysis in Business Process Simulation

The traditional methods listed above do not allow to estimate all parties of digital transformation of business model of the enterprise at introduction of software robots in the automated control system. Taking into account that the number of RPA-transformation objects can be different parameters: operators who receive calls and fill in digital databases about clients, the number of clients served, the results of RPA-transformation can be absolutely different: from the dismissal of individual employees and the liquidation of entire departments to the introduction of new employees with digital competences. It is difficult to understand which of the options for RPA-transformation of enterprises is the most profitable. At the same time, it is too expensive to conduct full-scale experiments at a particular enterprise. In this case, creation of imitation models of business processes of enterprises with RPA-transformation can allow to predict results of functioning of real management system at lower costs.

Simulation modeling of performance of business processes is widely applied in projects on reengineering of activity of the enterprises (Business Process Reengineering, BPR) and also at the consecutive evolutionary changes realizing the approach of continuous (Kaizen) and general improvement of quality (Total Quality Management, TQM). The results of the simulation are also widely used for management decision making in such approaches as Just-In-Time (JIT) and KANBAN.

First of all, when simulating business processes, the following indicators are of importance:

- process costs;
- process duration;
- quantity of served clients on the set observation interval, if the basic process represents a service.

These indicators define efficiency of business processes. If business process has high cost, it increases expenses of the enterprise. Prolonged performance of business process may also lead to increased costs, or may lead to the loss of its relevance. Insufficient servicing of the number of clients or insufficient number of services produced reduces profits and may lead to complete failure of the business model of the enterprise.

To evaluate such high-tech digital transformation of business processes together with simulation modeling, it is reasonable to apply the method of functional cost analysis (FCA) [7], which has recently received a "second breath" due to its use in digital business process management systems (Business Process Management, BPM) and business intelligence systems (Business Intelligence, BI).

FCA differs from traditional financial methods of analysis of enterprise activity by the fact that the activity is evaluated not just by functional operations that spend the budget, but also by measuring the cost of the service (the main business process), taking into account its usefulness.

In some cases, it is enough to conduct the FSA of a separate software robot. Then not business process of top level describing service entirely is exposed to the analysis, and functions of the robot itself are systematically investigated for the purpose of search of balance between its cost price and utility. Such approach is considered classical and is described in works of the Soviet researcher Yu. M. Sobolev as elementary economic analysis (PEA) [8], the American engineer L.D. Miles as cost analysis / design of cost (Value Analysis/Value Engineering, VA/VE) [9]. At a stage of introduction of the program robot, the classical method of an estimation of efficiency—return on investment ROI can be applied also.

If to consider digital transformation of RPA of the enterprise as a whole, the generalized calculation of cost price by kinds of activity (Activity-Based Costing, ABC) [10] is carried out. Such method of the description of expenses identifies already business processes of the enterprise and assigns expenses to each process according to the present cost of each separately taken subprocess. In this method, overhead costs are converted into direct costs. As a matter of fact, calculation of the cost price of service (the basic business process) which differs from traditional that focuses on calculation of cost of the business processes of the bottom level necessary for service granting, instead of division of expenses into direct and indirect and grouping of these expenses on calculation articles is made.

The essence of the ABC method is:

- Identification of the main business processes of the enterprise;
- Identification of the cost carrier for each business process of the enterprise;
- distribution of overhead costs on "cost carriers," created for each business process;
- determining the rate of distribution of overhead costs for each "cost carrier" based on the relevant cost factor.

Cost carrier (Cost Driver) is an indicator of measuring the activity of a business process, reflecting its essence and is related to the calculated object.

Calculation of an expense carrier rate on all carriers of expenses of business processes is carried out by the formula:

$$\mathrm{Rd} = P/D \tag{3}$$

where Rd is the cost carrier's rate;
P is the cost of the process;
D—cost driver (number of operations) [10].

Examples of some cost drivers are presented in Table 2.

Table 2 Examples of cost drivers for business processes of the enterprise

Business process	Cost bearer
Ordering processing	Number of orders, pcs
Materials delivery	Number of lots received, pcs
Entering meter indicators into information system	Number of entries, pcs
Verification of water meters	Number of verifications, pcs
Pumping equipment operation	Machine/hour
Operation of freight transport	Ton/kilometers
Loading, unloading of materials	Weight of cargo in tons

3 Digital Tools to Assess the Effectiveness of Robotization of Business Processes of the Enterprise

Figure 3 shows digital tools, which can be used to assess the effectiveness of robotization of business processes of the enterprise. In the middle, the RPA object that is information systems automating business processes is shown. From above, information-analytical systems in which methods of an estimation of efficiency of introduction RPA are realized are shown.

Most of the traditional methods of assessing the effectiveness of RPA implementation in information systems of enterprise management, which are listed in Table 1, can be automated by means of specialized software such as "PROJECT EXPERT," "INVESTOR," "ALT-Invest," "COMFAR 3.0," "PROPSPIN," etc. Usually such software products are used for development of financial models and strategic development plans of enterprises, for development of financial section of business plans of investment projects. If we consider RPA implementation as an IT-project, then

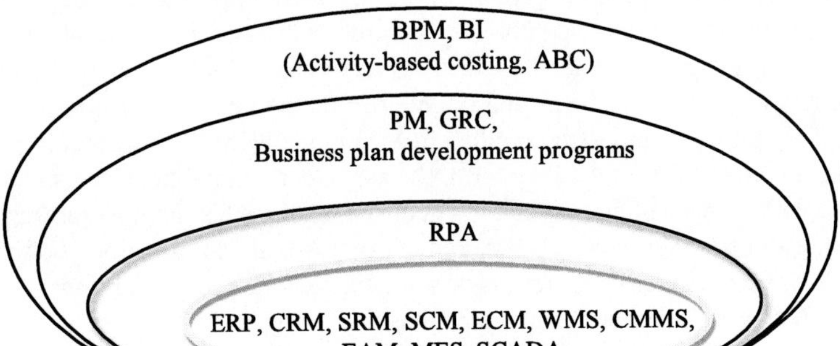

Fig. 3 Use of modern digital systems in assessing the efficiency of robotization of business processes of enterprises

according to PMBoK efficiency assessment is carried out at the stage of project organization and preparation and is included in the group of planning processes. These processes are automated by means of the listed software which is based on the indicators listed in Table 1. Simplified financial model can be implemented in calendar planning and budgeting programs in project management systems (Project Management, RM), such as Primavera Project Planner P3, Microsoft Project, OpenPlan, Spider, etc.

In some cases, evaluation of efficiency of robotization of several business processes at once can be embedded in integrated risk management and support systems of GRC internal control system (Governance, Risk, Compliance).

Last years with development (Building Information Model, BIM) and (Product Lifecycle Management, PLM) approaches, there were possibilities of financial modeling and in the modern CAD-systems using elements of information modeling. To them system Renga of company ASCON, Revit of company Autodesk, Bentley Systems, etc., concern. Methods of FCA are realized in the majority of BPM-systems which allow to carry out modeling of business processes [11]. Such systems include ARIS Software AG, Business studio, ELMA, and Business Engineer. Possibilities of joint use of FSA methods at simulation modeling are at some BI-systems.

4 Conclusion

Currently, there is no universal method that would accurately estimate the expected economic effect of RPA implementation. In order to assess the efficiency of RPA implementation, it is necessary to have a sufficiently long time interval for the operation of software robots, which may not exist, as the rapid development of information technologies quickly changes the market indicators. [15] In most cases, FSA methods are the most promising, as they are organically built into the simulation model of the enterprise. And functioning imitation model of enterprise activity together with digital analytical tool is a weighty step toward creation of digital double of enterprise. [14] The digital double allows to estimate in real time any changes including digital transformation. If the digital double of the enterprise to consider on all life cycle of a building or industrial product in such model BIM, PLM technologies will organically enter [12]. In the long term, the analysis of digital doubles can be made self-taught by means of an artificial intellect and machine learning. Algorithms FSA very well approach for intellectualization of processes of an estimation of efficiency. [13]. At the same time, FSA methods do not exclude the use of traditional methods of efficiency assessment at the planning stages.

References

1. August-Wilhelm, S.: ARIS-Toolset: Von Forschungs-Prototypen zum Produkt. Informatik-Spektrum Springer-Verlag **19**, 71–78 (1996)
2. Eroshkin, S.Y., Kameneva, N.A., Sukhorukov, A.I., Kovkov, D.V.: Conceptual system in the modern information management. Procedia Comput. Sci. IEEE, USA, pp. 609–612 (2017). https://doi.org/10.1016/j.procs.2017.01.079
3. Sukhorukov, A.I., Guo Shuhong, Koryagin, N.D., Eroshkin, S.Y.: Tendencies of information management development in the conditions of the origin of a new ecosystem of the digital economy. In: 2018 Eleventh International Conference "Management of Large-Scale System Development", MLSD. pp. 1–4. Moscow, Russia (2018). https://doi.org/10.1109/MLSD.2018.8551859
4. Sukhorukov, A., Eroshkin S., Vanyurikhin, P., Karabahciev, S., Bogdanova, E.: Robotization of business processes of enterprises of housing and communal services. In: E3S Web of Conferences, vol. 110, p. 02082 (2019). doi.org/https://doi.org/10.1051/e3sconf/201911002082
5. Gartner. Official website in Russia. https://www.gartner.com/en. Last accessed 21 Feb 2020
6. Guide to the Set of knowledge on project management. (PMBOK® guide)—Sixth edition. PMI website. https://www.pmi.org/. Last accessed 21 Feb 2020
7. Altshuller, G.S., Zlotin, B.L., Filatov, V.I.: Profession-search for the new (Functional and cost analysis and theory of solving inventive tasks as a system for identifying savings reserves). Kartya Moldovenyaske, Kishinev, p. 196 (1985)
8. Sobolev, Yu. M.: The designer chooses a solution. Perm: Perm book. ed., p. 110 (1979)
9. Miles, L.: Technigues of value analysis and engineering, N. Y. Mc-Craw—Hill, p. 203 (2014)
10. Gramp, E.A.: Organization of Functional and Cost Analysis Services in Industrial Firms in the USA. Informelectro, Moscow (1971)
11. Sukhorukov, A.I.: Use of functional and cost analysis for innovative transformation of business processes of passenger service on board. Innov. Civ. Aviat. **4**(4), 79–90 (2019)
12. Sukhorukov, A.I., Eroshkin, SYu., Kallaur, GYu., Papikyan, L.M.: Technologies of information modeling of all stages of the life cycle of a technical object. Bull. Mech. Eng. **4**, 84–86 (2018)
13. Sukhorukov, A., Koryagin, N., Sulyagina, J., Ulitskaya, N., Eroshkin, S.: Digital transformation of airline management as the basis of innovative development. Adv. Intell. Syst. Comput. **1115**, 845–854 (2020). https://doi.org/10.1007/978-3-030-37916-2_83
14. Fomicheva, T.V., Kataeva, V.I., Sulyagina, J.O., Evstratova, T.A., Chardimsky, M.G.: Digitization of the population in Russia: technologies and levels of interaction. Int. J. Rec. Technol. Eng. **8**(2), 4728–4731 (2019)
15. Kataeva, V.I., Fomicheva, T.V., Sulyagina, Ju., Kuvshinova, Ju., Evstratova T.A., Moskvitin, G.I.: Recommendation for the use of regional tourist resources by the example of the resort city of Sochi, Russia. J. Environ. Manag. Tourism, **10**(1(33)), 244–250 (2019). http://dx.doi.org/https://doi.org/10.14505/jemt

Increasing the Information Capacity of Materials Non-destructive Electrical Measurements

Nikolay Pshchelko, Olga Tsareva, and Vera Breskich

Abstract Various methods of material control are considered. It is shown that electrical methods of non-destructive testing of dielectrics, connecting the electrical and mechanical characteristics of materials, have not been sufficiently developed—the process of accumulation and relaxation of electric charges in dielectric materials is long-term. At room temperatures, migration polarization in ionic dielectrics can develop over the years. Therefore, the use of standard measuring bridges of alternating current for measuring capacitance and dielectric constant, tangent of dielectric loss angle, and leakage currents for these purposes is unacceptable. At the same time, measuring instruments for constant electric fields have not been developed enough due to the objective difficulties of direct measurement of these fields. This paper develops an approach to solving this problem, based on the use of low-frequency and static electrical parameters and characteristics as informative for non-destructive quality control. The theoretical foundations for increasing the sensitivity of electrical methods for non-destructive testing of high-resistance materials are presented. In this sense, the frequency and magnitude of the testing signal are analyzed. It was concluded that the most informative is low-frequency measurements on a weak signal. In this case, measurements of the permittivity of ceramic products in the form of a disk were carried out both in the microwave range on an 8-mm interferometer and in a constant electric field. Measurements in a constant electric field demonstrated the best information content and correlation with the ultimate strength in bending of the studied samples.

N. Pshchelko
Academy of Telecommunications Named After S.M. Budyonny, St. Petersburg, Russian Federation

O. Tsareva · V. Breskich (✉)
Peter the Great St.Petersburg Polytechnic University, Polytechnicheskaya, 29, St. Petersburg 195251, Russia

O. Tsareva
e-mail: tsareva_os@spbstu.ru

M. Shamtsyan et al. (eds.), *Robotics, Machinery and Engineering Technology for Precision Agriculture*, Smart Innovation, Systems and Technologies 247,
https://doi.org/10.1007/978-981-16-3844-2_58

1 Introduction

During the operation of buildings and structures, especially buildings of cultural heritage monuments, cracks and other deformations (torsion, heel, uneven settlement, tension, and compression) occur. To ensure the safety of the operation of such buildings and structures, it is necessary to monitor and control these deformations. There are various geodetic methods for controlling deformations (geometric and trigonometric leveling, laser scanning, GPS observations, etc.) [1], but their field of application is greatly reduced in tight working conditions and when measurements are required in hard-to-reach places. To control the mechanical properties, there are also electrophysical methods of non-destructive testing (X-ray, ultrasonic, eddy current, tensometric, methods based on the use of photoelasticity, scanning contact potentiometry, etc.), which make it possible to detect cracks with sizes from a few tenths of a millimeter at a depth of a few millimeters.

When detecting pores, cracks, and other defects by electrophysical methods, the authors of these methods proceed from the fact that a pore or crack, from an electrical point of view, is a macro electric dipole or capacitor, the charges of the "plates" of which are caused by the accumulation of charges of different signs on opposite sides of the crack. In fact, the principle of generating an informative signal is the same as when receiving an electrical signal from an electret microphone [2]. It seems that this approach to the detection of charges of structural defects is insufficiently effective, since the charges themselves are not investigated directly, but their manifestation is analyzed. Below we will consider our proposed method, in which the accumulation and relaxation of the charge on the "plates" of the defect are determined and simultaneously measured directly by using a constant electric field.

It should be emphasized that the process of accumulation and relaxation of electric charges in dielectric materials is long-term. So, at room temperatures, migration polarization in ionic dielectrics can develop over the years [3]. Therefore, the use of standard measuring bridges of alternating current for measuring capacitance and dielectric constant, tangent of dielectric loss angle, and leakage currents for these purposes is unacceptable. At the same time, measuring instruments for constant electric fields (EFs) have not been developed enough due to the objective difficulties of direct measurement of these fields.

The disadvantages of the above indirect methods [1] is that most of them allow you to obtain only the strength characteristics of the material, while its deformative properties are assumed to be unchanged [4]. Based on the experimental data, which will be given below, it can be assumed that the proposed method of using a constant electric field may be sensitive to this characteristic. The purpose of this work is to identify the most informative methods for non-destructive testing of the mechanical properties of high-resistance materials using electrical measurements.

2 Selecting the Test Signal Frequency

Electrical measurements are traditionally used to control the quality of metal-dielectric structures (MDS) for technical purposes. So, the quality of the working structures of capacitors is usually assessed by the values of the capacitance, the tangent of the loss angle, and the insulation resistance (or leakage current). The former are measured, as a rule, at frequencies not lower than industrial, and the latter are measured after a short, not more than one minute, time after the application of a constant voltage. When the specified parameters go beyond the limits regulated by the normative and technical documentation for the product, samples are rejected that do not meet the required functional characteristics.

At the same time, as the accumulated experience of testing and using technical devices based on MDS shows, the compliance of the specified parameters-criteria for the suitability of products with the agreed standards does not guarantee their reliability. So, the resource of capacitors with normal the values of capacitance, tangent of the loss angle, and leakage current may turn out to be significantly less than the guaranteed one [3, 5]. This means that the standard parameters-criteria for the suitability of technical devices based on MDS are not informative in terms of detecting potentially dangerous defects that limit the reliability of the product and makes it necessary to look for new approaches and develop corresponding electrophysical methods of non-destructive testing.

This paper develops an approach to solving this problem, based on the use of low-frequency and static electrical parameters and characteristics as informative for non-destructive quality control. The prerequisites for this approach are characteristic features of the structure and kinetic properties of MDS materials. The structural features of technical dielectrics are manifested in their kinetic properties. Thus, phase inhomogeneity creates the prerequisites for migratory polarization—the slowest of all polarization mechanisms. The most significant influence on the kinetic properties is exerted by the structural disorder and inhomogeneity of dielectrics. The dominant mechanism of charge transfer in such materials is hopping electrical conductivity, which is carried out through thermally activated hopping of charge carriers between localized states. The main charge transfer processes in this aspect—polarization and electrical conductivity in high-resistance disordered materials in a wide frequency range—are represented as follows. At high frequencies, the decisive factor is the two-site hopping electrical conductivity, when during the half-period of the external alternating voltage the charge carrier manages to make only one jump between the localization centers. With a decrease in frequency, the number of hopping of charge carriers in a voltage half-period increases and there is a transition to multiplet hopping electrical conductivity on alternating current along an extended chain of localized states. At the lowest frequencies (long times), the dimensions of this chain become so large that it closes the electrodes (in terms of the theory of hopping electrical conductivity, an infinite cluster of localized states participating in charge transfer is formed in the system), and static conductivity is established.

The considered picture of charge transfer in dielectric materials was used as the basis for a search for an approach to nondestructive quality control of MDS materials studied in this work. The following requirements were used as criteria:

- the sensitivity of the parameters and characteristics used to the impact of EF;
- information content of the used electrophysical parameters and characteristics in terms of identifying potentially dangerous defects.

Both requirements are met by low-frequency and static electrical parameters and characteristics of the studied MDS, measured, as will be shown below, in weak EF.

To determine the frequency range in which it is advisable to carry out non-destructive quality control of the MDS, the following model experiment was carried out. In a dielectric, artificially, by means of optical excitation inhomogeneous in thickness, an inhomogeneity in conductivity of the two-layer type was created and with time the interface between the layers shifted into the interior of the sample under the action of an external EF. The control characteristic was the frequency dependence of the dielectric loss tangent, which, as is known, has an extreme character for inhomogeneous dielectrics. The most informative was the range of infrasonic frequencies, in which this maximum is observed [3, 5].

This allows us to conclude that static and measured in the infrasonic frequency range electrical parameters and characteristics are informative for non-destructive quality control of MDS of technical devices. This conclusion is also confirmed by the obvious consideration that slowly proceeding processes of charge accumulation cannot be recorded when using a high-frequency voltage as a test signal.

3 Selection of Test Signal Magnitude

Let us now consider the question of the magnitude of the EF, and in practice—the value of the electric voltage, which should be used in measurements. For this, we will take into account that the volt-ampere characteristic of a dielectric, as a material with a small amount of free charge carriers, has a sublinear character: with an increase in the electric field strength, the extraction of charge carriers from the bulk of the dielectric onto the electrodes leads to a decrease in the number of these free charge carriers. Therefore, at large values of the EF, the current in such materials is saturated and is determined not by the EF value, but by the rate of generation of new charge carriers in the bulk of the material. A striking example of such a behavior of dielectrics is the volt-ampere characteristic of air: Ohm's law is fulfilled for it only in very weak EFs (up to 1 V/m), and practically, in the entire range of used EFs (from 1 V/m to 10^6 V / m), the current density is 10^{-15} A/m^2 and does not depend on the value of the EF strength [6].

The presence of a defect in a high-resistance dielectric layer is characterized by an increased through-conductivity local to the area. This, as a rule, is due to the presence of weakly bound ions (or electrons) in the defect region, which are capable of leaving their attachment points much more often at a given temperature than the

Fig. 1 Dependences of the current density in the dielectric layer on the electric voltage across it. 1—in the area of a defect with increased conductivity, 2—through a dielectric layer that does not contain defects

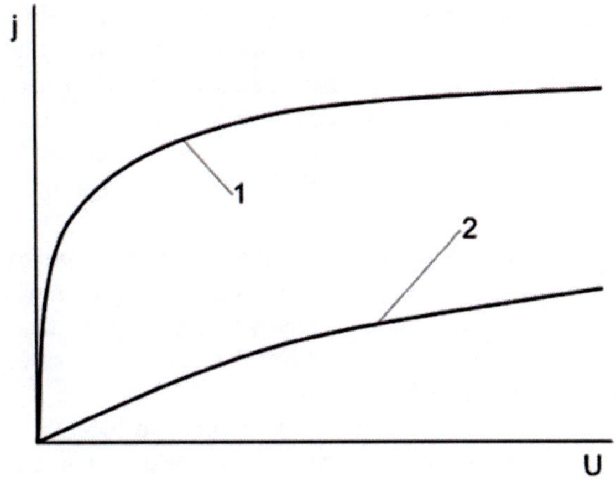

bulk of the ions (or electrons) of a dielectric. The low value of the activation energy of these weakly bound ions (or electrons) leads to the fact that the EF significantly more effectively affects the process of their "release" and involvement in electrical conductivity.

The considered representations are illustrated by curves in Fig. 1.

The curves analytically shown in Fig. 1 can be approximated by an expression of the form (1):

$$j = AU^n \tag{1}$$

where A and n are constants, and as can be seen from the figure, the exponent n_1 for curve 1 corresponding to the area of the defect is less than the value n_2 for curve 2 corresponding to the layer without defects. Let us analyze, taking into account the remark made, the contribution of the current through the defect I_d in the total current I_0 through the dielectric layer (2):

$$\frac{I_d}{I_0} = \frac{A_1 U^{n_1} S_d}{A_1 U^{n_1} S_d + A_2 U^{n_2}(S_0 - S_d)}, \tag{2}$$

where A_1 and A_2 are constants, S_d is the area (or volume) occupied by the defect, and S_0 is the total area (or volume) of the dielectric.

Expression (2) is reduced to the form (3):

$$\frac{I_d}{I_0} = \frac{1}{1 + BU^m}, \tag{3}$$

where B is a constant and the exponent is $m = n_2 - n_1 > 0$.

From expression (3), it follows that in order to identify defects by current measurements, these measurements should be carried out at low voltages, since as, $U \to 0$ we get $\frac{I_d}{I_0} \to 1$, i.e., the measured current will correspond to the current due to the presence of defects. At high voltages used for measurements, the contribution of the current caused by defects to the total current may not be noticeable, and in this case, it will be impossible to distinguish between the potentially unreliable and suitable samples.

The arguments made are confirmed by a simple experiment we have done: we measured the leakage resistances of identical alumina capacitors with an operating voltage of 100 V. At a nominal voltage, the resistance spread did not exceed 20%, and at a voltage of 5 V, these resistances differed several times and even more than order of magnitude.

Thus, we can conclude that in order to identify potentially dangerous defects that tend to grow long-term, current measurements should be carried out, firstly, at a constant voltage, and secondly, at low voltages, i.e., in weak EFs. In practice, the situation with the available measuring devices, as a rule, is directly opposite: to measure resistances and currents through dielectrics, large EFs are used; since at low voltages, current measurements are difficult due to their smallness. In addition, for measurements of dielectric characteristics, AC bridges are usually used because of the advantages of bridge circuits and their current principle of operation—current measurements are simpler than potential ones. Despite the relatively small measurement voltages used in this case, the current through the dielectric has a relatively large value (the current in this case has a reactive nature due to the use of alternating voltage) and is easily measured.

4 Results

In this work, the objects of testing were small-size ceramic products in the form of a disk—crystal holders of microwave devices. The technique for measuring the strength and elastic characteristics of these products has been sufficiently developed. Therefore, let us dwell in more detail on the determination of the relative dielectric constant ε of products and our comparison of the value of ε with the mechanical (strength) properties of ceramics.

In this work, measurements of the dielectric constant were carried out both in the microwave range with an 8-mm interferometer and in a constant electric field. In both cases, the measurement results were compared with the strength characteristics of the material, measured by the destructive method.

Consider first measurements in the microwave range, Fig. 2.

Most oxide-based ceramic materials are radiotransparent and allow the use of radio wave methods to determine ε. The length of the electromagnetic wave should be less than the diameter of the investigated item to avoid diffraction phenomena. The method for determining the high-frequency value of the dielectric constant ε_{HF}

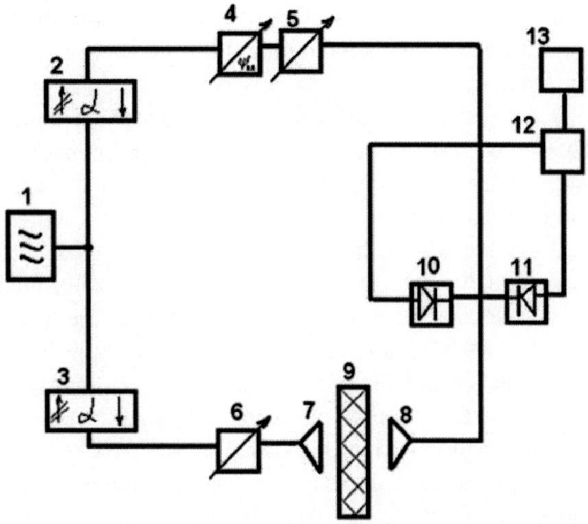

Fig. 2 Schematic diagram of an interferometer (wavelength—8 mm) for measuring the relative permittivity: 1—microwave generator; 2, 3—valves; 4—graduated phase shifter; 5, 6—attenuators; 7, 8—matching horn; 9—product; 10, 11—detector sections; 12—discriminator; 13—microvoltmeter

is as follows: when the test sample is placed between the matching horns of an 8-mm interferometer, the diagram of which is shown in Fig. 2, an imbalance of the phases of the electromagnetic wave occurs in the arms of the interferometer, which is eliminated using a graduated phase shifter. According to the indications of the phase shifter, the value of the additional phase incursion φ, due to the introduction of the sample, is taken. Further, according to the value of φ, the high-frequency value of the dielectric constant $\varepsilon_{HF} = n^2$ is calculated, where n is the refractive index. This technique and the corresponding calculations are considered in detail in [3].

Let us now consider measurements of the low-frequency value of the relative permittivity ε_{LF} in a constant EF. We recall that measurements in a constant EFs can be very informative, especially when studying slowly proceeding processes.

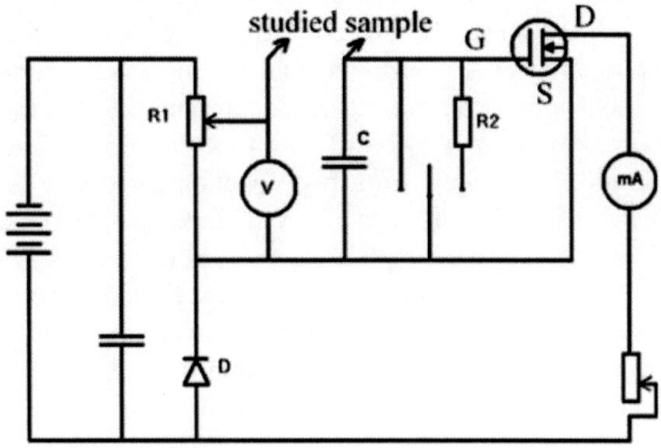

Fig. 3 Electrical circuit of the capacitance and resistance meter for high-resistance materials

Measurements of ε_{LF} were carried out according to the scheme shown in Fig. 3 [3]. Thermally deposited aluminum films were used as electrodes for the samples under study.

The physical foundations and features of the scheme are considered in detail in [3]. The device is based on the use of a transient process in series-connected capacitors, one of which is the sample under study, and the other is a capacitor with known parameters connected between the gate and the source of the field-effect transistor, Fig. 3. At the initial moment of time, when voltage is applied to these series-connected capacitors, the voltage across them will be established in accordance with the values of their capacitances and after some time of the transient process—in accordance with the values of their resistances. To measure the voltage across the sample, we used field-effect transistors with an extremely low leakage current (less than 0.001 nA) at the input (gate—source), which at operating voltages of the order of tens of Volts corresponds to an input resistance of 10^{12}–10^{14} Ω. Such a large resistance allows, by measuring the voltage at the input of the transistor, to determine the capacitance and resistance of the sample under study, connected in series with the input electrode (gate) of the transistor. In this case, the voltage at the input of the transistor can be determined by the current in the source-drain circuit of the transistor.

The transient process for high-resistance materials can be quite long. In this case, the change in time of the voltage measured at the input of the transistor is described by the expression [3]:

$$U_{\text{tr}} = \left(U - \left[\left(\frac{C}{C + C_s} - \frac{R_s}{R_{\text{tr}} + R_s}\right) \cdot \exp\left(-\frac{(R_{\text{tr}} + R_s)}{(C + C_s)R_{\text{tr}}R_s}t\right) + \frac{R_s}{R_{\text{tr}} + R_s}\right]U\right) \tag{4}$$

where U is the supply voltage of the circuit, U_{tr} is the voltage across the transistor, R_{s} is the electrical resistance of the sample, R_{tr} is the electrical input resistance of the transistor, and C is the capacitance of a capacitor with low leakage currents together with the input capacitance of the transistor (the latter is usually negligible).

The above expression for the dependence of the voltage at the input of the transistor on time makes it possible to find the resistance of the sample and its capacitance without waiting for the end of the transient process: for this, it is enough to measure the values of U_{tr} at two different times t_1 and t_2. Then, solving the corresponding system of equations from two equations, you can determine two unknowns—C_s and R_s. Knowing the capacity of the sample, based on the formula of a flat capacitor $C_s = \frac{\varepsilon_0 \varepsilon S}{d}$ (S is the sample area, d is its thickness), it is possible to calculate the value of the dielectric constant. To perform the considered operations, we used the software developed by us [3]. The difference between the moments of time t_1 and t_2 was about 1 h—during this time, on the one hand, the voltage U_{tr} changed significantly, and on the other, it did not reach steady-state values.

Despite the presence of a transient process, the EF strength in which the sample is located can be considered practically constant, since the value of the input voltage across the transistor during measurements is usually tenths of a volt, and the voltage of the regulated voltage source is tens of volts, i.e., $U_s \approx U$.

Table 1 Correlation between ultimate strength in bending σ with values of dielectric permittivity at high-frequency ε_{HF} and static value ε_{LF}

ε_{HF}	5.2	5.4	5.6	5.8	6.0	6.2
ε_{LF}	7.4	7.8	9.2	10.4	11.2	11.7
σ, MPa	286	281	274	268	230	205

5 Discussion

Table 1 presents data on the measurement of the static value of the dielectric constant ε_{LF} for beryllium ceramic samples in comparison with their mechanical strength (ultimate bending strength σ), and high-frequency value of the dielectric constant ε_{HF}, measured using the 8 mm interferometer, is considered above.

It should be noted that Table 1 shows the values with an accuracy corresponding to the measurement error. Thus, despite the relatively small changes in the dielectric constant depending on the sintering temperature and the ultimate bending strength, the indicated dependences are reliable.

The experiments carried out showed that the measurement of the dielectric constant in a constant EF is more informative for controlling the properties of a material than with the standard use of an alternating EF for such measurements—in the studied range in a constant EF, the dielectric constant varied in the range from 7.4 to 11.7, i.e., 1.58 times, while in the microwave range—from 5.2 to 6.2, i.e., 1.19 times. It seems that such a difference is associated with the possibility of fixing slowly proceeding processes (for example, migration polarization), fixing electric charges at grain boundaries, etc., when using a constant voltage.

Currently, various grades of concrete are widely used as building materials. Therefore, intensive research is underway aimed at the development of non-destructive methods of its control [7]. It seems that the above results may be useful for the development of these methods. At the same time, it should be noted that we have not completed the relevant studies in relation to concrete to date.

6 Conclusions

The paper compared the developed high-frequency (radio wave) method for measuring the dielectric constant and the method for measuring the same characteristic at a constant voltage. From the data obtained, it follows that the low-frequency value of the dielectric constant is more sensitive to the controlled mechanical parameter and, therefore, more reliably allows it to be estimated in a non-destructive way.

The obtained result seems promising for the study of a wide range of materials, including construction materials, since the considered non-destructive low-frequency method has demonstrated the sensitivity of the electrophysical characteristics of the material to its strength characteristics.

References

1. Ata, E., Hoşbaş, R.G., Pirti, A.: Monitoring the structural deformation of davutpasa barrack by using geodetic methods. Teh. Vjesn. **25**, 944–947 (2018)
2. Saveliev, A., Pshchelko, N., Krestovnikov, K.: Method of sensitivity calculation for electrete diaphragm capacitive sensors. In: Proceedings—International Conference on Developments in eSystems Engineering, DeSE. pp. 721–725. Institute of Electrical and Electronics Engineers Inc. (2019)
3. Pshchelko, N.S.: Electrophysical Methods of Non-destructive Testing and the Formation of Metal-dielectric Structures. St. Petersburg State Mining University, Diss (2011)
4. Kolokhov, V., Sopilniak, A., Gasii, G., Kolokhov, A.: Structure material physic-mechanical characteristics accuracy determination while changing the level of stresses in the structure. Int. J. Eng. Technol. **7**(4.8), 74–78 (2018)
5. Pshchelko, N.S., Khanin, S.D.: Electrical methods of non-destructive quality control of metal-dielectric structures of oxide capacitors. Notes Min. Inst. **189**, 268–271 (2011)
6. Koritskiy, Yu.V., Pasynkov, V.V., Tareev, B.M.: Handbook of Electrical Materials. Energoatomizdat, Moscow (1988)
7. Korotchenko, I.A., Ivanov, E.N., Manovitsky, S.S., Borisova, V.A., Semenov, K.V., Barabanshchikov, Y.G.: Deformation of concrete creep in the thermal stress state calculation of massive concrete and reinforced concrete structures. Mag. Civ. Eng. **69**, 56–63 (2017)

4D and 5D Design Processes Automation Using Databases, Classification and Applied Programming

Artur Kuzminykh, Anna Kukina, and Galina Bardina

Abstract The article presents the role of classification at 4D and 5D BIM stages and the automation entering of classification into the model element. Method of operation with a Revit model is presented by Dynamo scripts linking model elements with standard unit prices. Interaction is based on Excel format. The stepwise working method is proposed for comparing model elements and local regulatory codes. The method based on classifiers allows to calculate the work duration and cost, to link the construction schedule and model elements. Autodesk Navisworks Manage is used as the main visualization tool. Classification is the basis of achieving significant results at 4D and 5D stages. The uniqueness and readability of the classification system allow to redefine the information output format and assign attributes to the elements in the future. It forms the interaction approach with the model as the database.

1 Introduction

BIM as the part of investment and construction project implementation is not only a design optimization instrument but is also an excellent tool for financial planning and construction control [1]. It largely saves the money spent on the project and makes it «transparent» for the customer. Not long time ago information model (IM) of construction project's life cycle was designated by five levels: the 4th stage introduced time parameters and the 5th level added costs [2, 3]. Nowadays, BIM is framed by nD levels, where 8D is associated with safety [4, 5].

Mostly, the information modelling process is carried out within 3D BIM stage in design organizations [6]. One of the reasons hindering the industry development is the lack of shared interests among the project participants, which leads to the life cycle (BLC) stages' disunity. Since the BIM model should be used by all project participants, it is necessary to introduce the data that can be used by third-party specialists later.

A. Kuzminykh · A. Kukina (✉) · G. Bardina
Peter the Great St.Petersburg Polytechnic University, Saint-Petersburg, Russia
e-mail: kukina_aa@spbstu.ru

M. Shamtsyan et al. (eds.), *Robotics, Machinery and Engineering Technology for Precision Agriculture*, Smart Innovation, Systems and Technologies 247,
https://doi.org/10.1007/978-981-16-3844-2_59

Each transition to the next level, from 3D to 4D, for example, requires an analysis of a model that came from previous steps. For this purpose, automation tools can be used to reduce the time for the check and the number of errors caused by human factor.

In this paper, the role of IM elements' classification at 4D and 5D stages is considered as it's an integral part of quantities retrieving and linking the model with its real cost. Automation of 4D and 5D stages is proposed, with the use of database, applied programming and Autodesk products Revit, Navisworks and Dynamo.

Navisworks from Autodesk is one of the most widely used software at 4D modelling and Revit leads at the 3D stage (that includes geometry and elements' attributive information) [6, 7]. Modern construction and investment projects' design and management systems—BIM and PLM/BLM (product/building lifecycle management)—are tending to merge. The systematic usage of information flows and administration systems' integration into a single database are observed at modern information management software. Data exchange formats, such as industry foundation classes (IFC), were created for high-quality information exchange, which is key to information modelling policy.

The element's basic properties in a modelling environment are sometimes not enough for its definite description, especially for complex technological equipment. In this regard, additional attributes are created, such as classification codes (classifiers), which allow to describe the element's special properties uniquely, its position and interaction with other model elements. We can conclude that classification is one of the key works at the stage of creating a 3D model for a high-quality transition to 4D and 5D steps. Unequivocal identification allows to determine cost, required resources, and necessary work operations for the particular element with less error.

A useful tool that allows classifiers conversion in case we have at least one code (which is based on the systems OmniClass, Uniformat or MasterFormat) is an add-on for Revit—Autodesk Classification Manager. It also offers a Dynamo script to add new attributes automatically. The essence of the program is to have a database, which can be extended by the user in xlsx tables format, where each classifier is associated with a classifier of another system. The add-on assigns a mapped code as a new type property to the model element [8].

However, one of the classification difficulties is the necessity to identify at least three components of the classifier. It can be assigned to materials (e.g. concrete B25, F100, W2), to elements (e.g. column placed on the 1st floor at the intersection of axes A and 0) or to the types of model elements (concrete column, B25, F100, W2, 400 × 400, with height equal to 3 m). These components can be also defined as:

- model element's classifier;
- construction works and spends classifier;
- resources' classifier.

Speaking of the input method (input costing) in assessing the project costs, regarding BIM model, we can define three types of resources:

- based on real model components, that can be seen directly in a model (e.g. windows, break walls);
- based on description information or dependent on LOD (e.g. small details, such as reveals);
- unmodeled resources (spent for temporary and auxiliary works, e.g. scaffolding, temporary supports).

Thus, at 3D BIM stage, the elements' attributes and the rules for their compilation and coding play an important role for successful transition to the following model dimensions.

The authors of [9] concluded that combining information-analytical systems into a single digital environment contributes to a more accurate project's cost estimation and openness, as well as to the IM's adaptability. Unified rules defining the process of model creation provide ease of use and increase the BIM efficiency as a part of the investment and construction project implementation.

An example of assigning a compound classifier to elements when dividing a construction project into working zones is shown in [10]. At the 4D stage, it allowed to reduce the 4D modelling time to visualize the construction process faster. The developed classifier was based on elements material and corresponding working zone, which led to work grouping and quantities calculation simplification. This approach, however, when extracting specifications by material, if it is not separate type for each size, will require additional filtering and analysis by element properties. These properties can be thickness, area, volume, perimeter height, etc. if the cost estimation is based on the national regulatory framework.

Scheduling (4D) with cost-related information (5D) is also shown at [11] with add-ons demonstration and BIM scenarios descriptions. The authors proposed to calculate the cost by selecting model elements and matching them with collections of estimated prices linked with databases in the cloud service.

Making a conclusion on the information and papers review, we can say that methods for automating the work on to 4D and 5D BIM stages are being continuously improved. To develop a qualitative approach for simplification of certain stages using automation, first, it is important for all project participants to define a classification system. It becomes crucial if the project is realized at the expense of the state budget.

2 Materials and Methods

The main purpose of this article was to show the chain of interactions between BIM software and its automation through databases and applied programming. Integral value of element's classifier is taken according to the scheme depicted in Fig. 1. This strategy was chosen based on companies' practice and literature analysis. At the same time, its value is obtained in an automated way using scripts and coding.

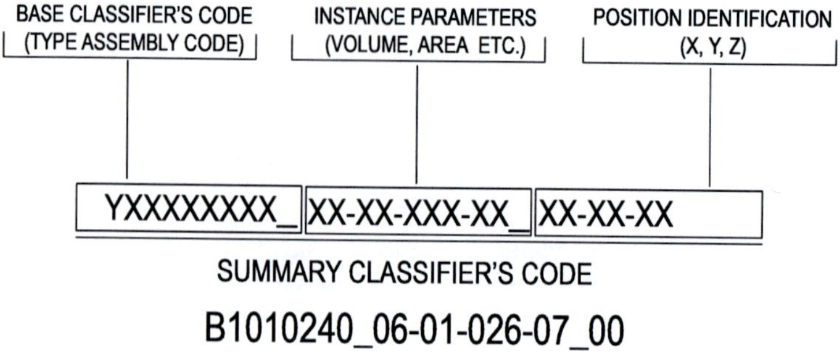

Fig. 1 Classifier's components

By defining the classification structure and the necessary parameters for analysis and recording into the BIM model elements' attributes, program codes were written in the visual programming environment—Dynamo. These codes transfer the work with a model into work with databases in tables and automate the transition steps towards 4D and 5D stages. The automation process can be described in eight stages using pouring works as an example.

First stage. For the IM's elements representing a building's bearing structure, the classifier's base value is assigned (assembly code). As a result of the classification hierarchy's analysis, the conclusion made in relation to foreign classification tables in paper [12] was confirmed.

Second stage. Using Dynamo script, elements' data is automatically extracted with filtering by level to the xlsx table with prepared format (Figs. 2 and 3). Based on the assembly code, in the Excel file, the work's code is filtered from the connected database of Federal Unit Rates. After, based on the obtained elements' properties (thickness, height, perimeter, volume, area, reference level), user selects a specific code, receiving the time standard and price (Fig. 4).

Third stage. Based on the data obtained from the IM (viz. scope of work) and the database of time and price standards, the work duration and costs are calculated. User assigns the number of workers, machines and labour hours.

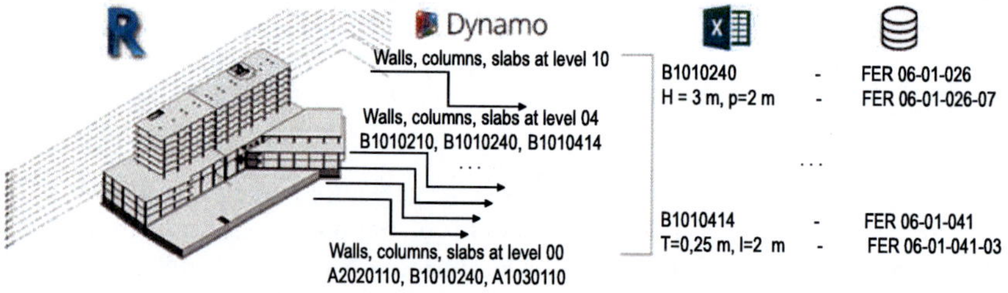

Fig. 2 Second stage. Data retrieving

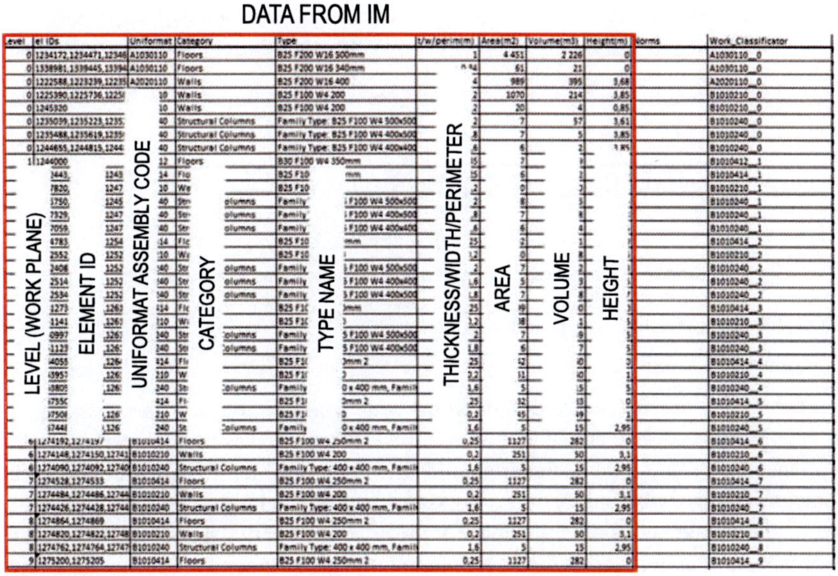

Fig. 3 Formatted table

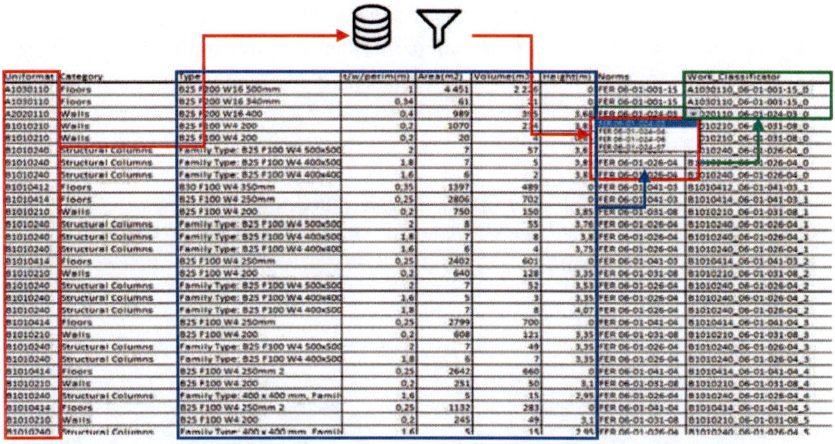

Fig. 4 Database filtration with local code setting based on elements' properties. Integral value of new classifier as taken as a summary of assembly code, standard's code and level

Fourth stage. After processing the information in a tabular format, a second code is used that automates the attributes assignment to the model elements. Each element (according to the unique ID obtained in the first step) receives a work classifier (instance parameter), generated automatically. Thus, for classification by type of work, the scheme depicted in Fig. 1, will have the form shown in Fig. 5.

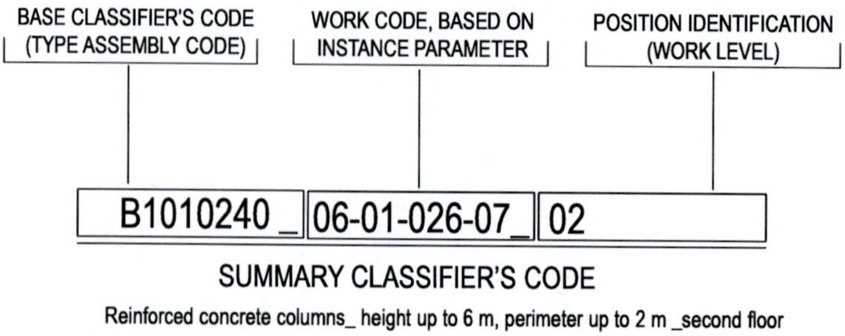

Fig. 5 Element's classifier by type of work

Fifth stage. For detailed 4D modelling with working zones, it is necessary to "divide" the structural element. For this purpose, the method of creating parts is used. In manual execution, the designer should divide the profile of each wall and each floor, indicating the boundary of the working zone. With a large number of elements, the operation takes considerable time. For the process of partial automation, the third code was written. It requires the user only to indicate the boundaries of the working zones with lines without constantly editing all the model elements (Fig. 6). In this case, the elements will be divided into parts along the selected lines.

Sixth stage. By selecting parts within one working zone's boundaries, its number is assigned (in Fig. 7—parameter "AK_Workzone"), then a detailed work code ("AK_WorkIntegral") is obtained for each part. Thus, after this stage, all the elements will be divided into parts according to the accepted working zones. Each part has a classifier that designates the basic element (wall, slab, column), work code and position in space, obtained by the level and working zone number (Fig. 8).

Seventh stage. Based on the work durations, their cost and the number of working zones, construction schedule is formed. In the schedule, summary classifiers for elements' parts are assigned to each corresponding specific work. MS Project or CSV tables environment can be used for this purpose.

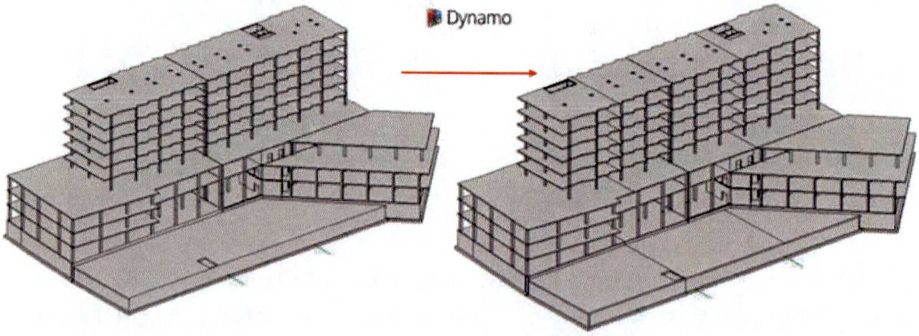

Fig. 6 The original model with the indicated boundaries of the working zones and the model after applying the code with parts created

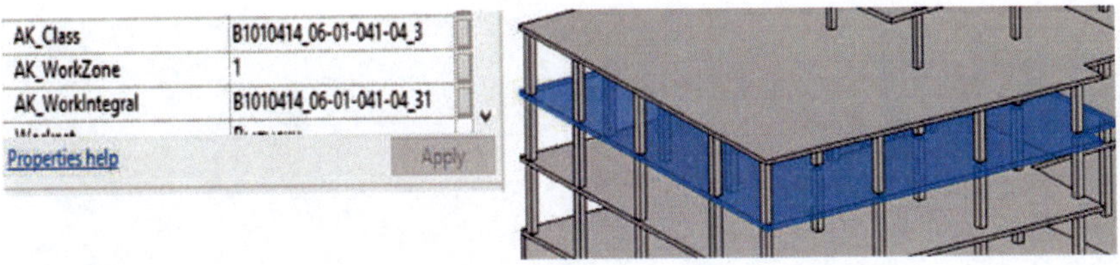

Fig. 7 Element's classifier by type of work

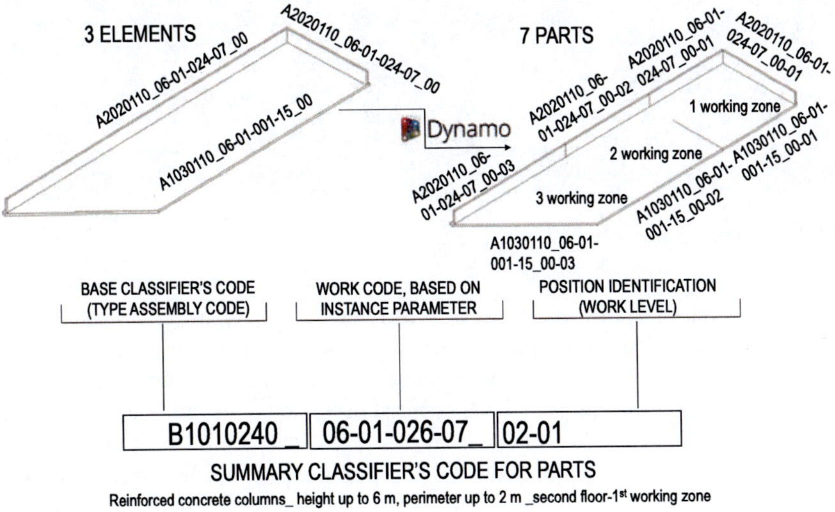

Fig. 8 Parts' classifier by type of work

Eighth stage. In order to visualize the construction process in Navisworks, the import of schedule with costs calculated in the previous steps is needed. In Timeliner module, we get an erection visualization with necessary expenses by mapping work codes assigned to model elements with schedule work codes. Attachment of time and cost values to model elements is automatized by applying program auto-rules (parts' summary classifiers are compared with classifiers from the schedule).

3 Results and Discussions

In this paper, the structure of the IM elements' classifier by the type of work was determined. Part of the code contains information about normative basis that was used for time and cost estimation. Using application programming, steps for transition to 4D and 5D BIM stages are automated. Also, it allowed to transfer the process of data analysis from BIM environment to table format editors that can be linked to the

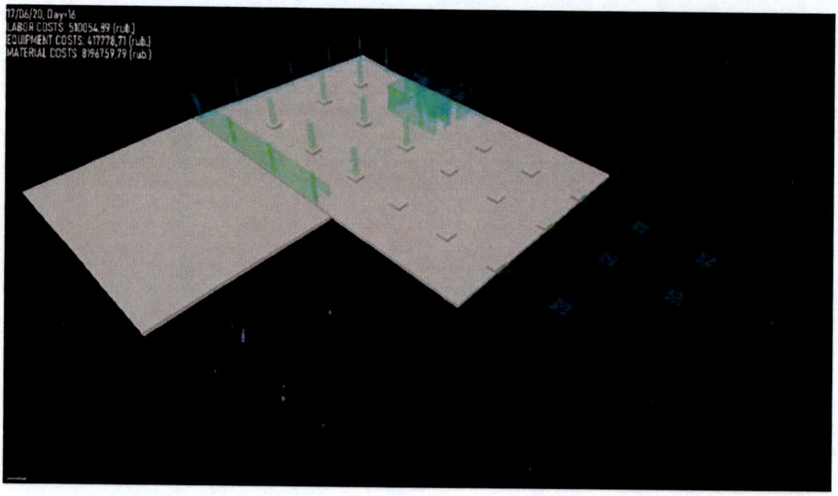

Fig. 9. 4D model with costs—5D BIM

databases. After finishing a work schedule forming, visualization of the construction progress is automatically obtained with the display of cost and time indicators (Fig. 9).

With the described technique, the amount of time spent on design is reduced by automating certain steps, which mostly consist of monotonous identical operations. It also directly reduces the number of errors that occur due to the human factor. However, such methods require more time at the beginning, like in comparison between BIM and CAD design processes. Additional time is required to debug scripts of codes, create the layout of the table in which the user will interact with the model, configure the database and compare its elements with the classification system.

A deeper study of the automation scripts is possible. For example, the step of setting working zone's numbers is excluded by adding a code: working zones' consecutive numbers will be assigned automatically in accordance with a workflow direction. The drawback of it is that the workflow directions may vary when developing a 4D model, in which the entire construction project and installation works are present. At this point, it will require additional filtering or elements separation, etc. In turn, the more detailed the codes, the more difficult the process of adapting and debugging them for new projects can become.

4 Conclusions

The demonstrated methodology for working with the IM and data analysis allows project managers to be part of the process even at early stages since it will require them to work more with databases instead of BIM model.

Classification is one of the key works at the 3D BIM stage for accelerated transition to 4D and 5D steps since it largely determines the volume and quality content of attributive information.

If the construction and investment project is international, the classification systems readability for all participants is important. Therefore, it makes sense to map it (compare) with existing ones. In the future, this will make it easy to redefine the format of the output information.

References

1. Pärn, E.A., Edwards, D.J., Sing, M.C.P.: The building information modelling trajectory in facilities management: a review. Autom. Constr. **75**, 45–55 (2017)
2. Mesaros, P., Smetankova, J., Mandicak, T.: The fifth dimension of BIM—implementation survey. In: IOP Conference Series: Earth and Environmental Science, vol. 222, p. 012003 (2019)
3. Heaton, J., Parlikad, A.K., Schooling, J.: Design and development of BIM models to support operations and maintenance. Comput. Ind. **111**, 172–186 (2019)
4. Koutamanis, A.: Dimensionality in BIM: Why BIM cannot have more than four dimensions? Autom. Constr. **114**, 103153 (2020)
5. Kamardeen, I.: 8D BIM modelling tool for accident prevention through design. In: ARCOM 2010: Proceedings of the 26th Annual Association of Researchers in Construction Management Conference, pp 281–289. Leeds (2010)
6. Piselia, M.: 4D BIM implementation to improve EPC project performance from contractor's perspective: a case study. In: IOP Conference Series: Materials Science and Engineering, vol. 930, no. 1, p. 012011 (2020)
7. Grishina, O.S., Savchenko, A.V., Marichev, A.P., Zalata, E.S., Petrochenko, M.V.: Monitoring of the construction site using an information model. Constr. Unique Build. Struct. **11**(62), 7–14 (2017)
8. Cerezo-Narváez, A., Pastor-Fernández, A., Otero-Mateo, M., Ballesteros-Pérez, P.: Integration of cost and work breakdown structures in the management of construction projects. Appl. Sci. **10**(4), 1386 (2020)
9. Churbanov, A.E., Shamara, Y.A.: Influence of information modeling technology on the investment and construction process development Vestnik MGSU **13**(7), 824–835 (2018)
10. Romanovich, M.A.: Features of construction project management using 4D and 5D. In: BIM Proceedings of 5th Internationaler BBB-Kongress (Graz) pp. 341–364 (2019)
11. Krivoy, S.A., Semin, A.I., Popov, A.V., Bebyakin, B.O.: Interrelation of BIM-scenarios in the investment and development project. Constr. Unique Build. Struct. **2**(65), 20–39 (2018)
12. Karakozova, I.V., Malyha, G.G., Pavlov, A.S., Panin, A.S., Tesler, N.D.: Research of preparatory work for BIM-technologies using on the example of medical organizations design Vestnik MGSU **15**(1), 100–111 (2020)

Study of Fine-Grained Cement Concrete with Ground Glass

Yuliya Schepochkina, Sergey Voinash, Viktoria Sokolova, Dmitry Koloshein, and Alexander Scherbakov

Abstract The large surface area of the aggregate in small-grained concrete causes an increased consumption of the binder (cement) and water, increased porosity, low bending, tensile and shock loads. To a large extent, these disadvantages can be reduced by enhancing the adhesive interaction of the cement stone (binder paste) with the surface of the aggregate. However, the adhesion strength of a cement stone to the surface of sand grains is less than the strength of the cement stone. Therefore, the latter is not completely used in cement–sand mortar. The substance from which the aggregate grains are composed may have an amorphous or crystalline structure. The amorphous structure of a material determines its isotropy. In this case, glass may be a suitable quartz-containing amorphous material. It was revealed that when using materials characterized by a vitreous structure as a filler, the adhesive interaction of their particles with the cement stone (binder paste) increases due to an increase in the quality of the mortar layer between them and the cement stone (binder paste), which is due to the fact that glass particles have significantly fewer defects in comparison with particles of materials characterized by a crystalline structure.

Y. Schepochkina
Ivanovo State Polytechnical University, 153000 Ivanovo, Russia

S. Voinash · V. Sokolova
Saint Petersburg State Forest Technical University Named After S.M. Kirov, 194021 Saint-Petersburg, Russia

D. Koloshein
Ryazan State Agrotechnological University Named After P.A. Kostychev, 390044 Ryazan, Russia

A. Scherbakov (✉)
Saint Petersburg State University of Architecture and Civil Engineering, 190005 Saint Petersburg, Russia

M. Shamtsyan et al. (eds.), *Robotics, Machinery and Engineering Technology for Precision Agriculture*, Smart Innovation, Systems and Technologies 247,
https://doi.org/10.1007/978-981-16-3844-2_60

1 Introduction

In recent years, multicomponent small-grained concrete has been actively introduced into construction [1–5]. The large surface area of the aggregate in small-grained concrete causes increased consumption of the binder (cement) and water, increased porosity, low bending, tensile and shock loads. To a large extent, these disadvantages can be reduced by enhancing the adhesive interaction of the cement stone (binder paste) with the aggregate surface.

The adhesive interaction depends on the dispersion of the aggregate particles and the binder, the structure, the chemical and mineral composition of the contacting phases, the surface roughness of the aggregate particles, the thickness of the water or mortar layer between the cement stone (binder paste) and the surface of the aggregate grains and a number of other factors. The cement stone is of great importance for the properties of concrete. It mainly consists of calcium hydrated silicates, forming a certain structure in space, including the unreacted part of cement grains covered with a membrane of hydrated new growths, and also intergranular space partially filled with hydrated new growths [3, 6]. Calcium hydrated silicates are different in structure. There are crystalline, semi-crystalline or amorphous structures of hydrated silicates. Hydrated silicates formed during heat and autoclave treatment, as well as during crystallization of new growths in pores and in intergranular space, have mainly crystalline structure. Differences in the structure and properties of these zones affect the quality of the material as a whole. The structure of the material obtained depends on the distribution of the liquid and air phases in the initially formed concrete structure by the time of setting.

2 Materials and Methods

In comparison with ordinary concrete with coarse aggregate, small-grained concrete has greater tensile strength. With a decrease in the binder content, the strength of concrete decreases sharply. With a water–cement ratio of 0.4 or more, the largest strength of small-grained concrete is ensured at the maximum density of the concrete mixture, which is possible with a rational ratio between the binder and the aggregate. Thus, by adjusting the ratio of the binder and the aggregate, as well as the water–cement ratio, it is possible to obtain small-grained concrete with specified strength characteristics.

It is explained in studies [6], that the product of the number of contacts per unit surface and the average strength of each contact determine the strength of porous materials. When compared with ordinary coarse concrete, small-grained concrete has a greater number of such contacts, due to a decrease in particle size and a denser packing. In this case, the total adhesion force of the aggregate and cement stone grains, which plays a decisive role in the material's ability to resist tensile stresses, increases. This, in accordance with the structural theory of strength, should positively affect the performance of concrete.

The consistency and structure of binder paste depend on the amount of cement and the water–cement ratio, which, accordingly, affects the quality of the formed cement stone. Binder paste fills intergranular voids in concrete and is used to coat aggregate grains with some extension. In the case of little binder paste, it is enough only for coating the aggregate grains and intergranular voids remain unfilled. An excess of binder paste does not only fill all the intergranular voids but also extends aggregate grains to an excessive distance. With the remaining parameters unchanged, the ratio of the aggregate consumption in the concrete mixture to cement consumption determines the structure of the future small-grained concrete. Two main types of conglomerate structure of small-grained concrete are distinguished. The first type of the structure is found in dense small-grained concrete with a minimum amount of air involved (up to 4–5%). The volume of the cement stone in relation to the volume of intergranular voids is increased in this concrete. It ensures complete filling of intergranular voids together with coating the surface of the aggregate. A similar structure of small-grained concrete is characterized by a large conventional thickness of the cement stone layer per unit surface area of the aggregate, which is about 25–30 microns [3]. At the same time, the strength of concrete with a constant water–cement ratio does not practically depend on the concentration of the cement stone.

The structure of the second type is characterized by a higher content of entrained air (more than 5%). A smaller amount of binder paste in the concrete mixture can only be provided with a coating of aggregate grains, and the volume of intergranular voids will remain unfilled. Then the thickness of the conditional layer of the cement stone per unit surface of the aggregate will be equal to 15–20 microns [3]. When considering the second type of the structure of small-grained concrete, one can note a clearer dependence of the compressive strength and bending on the cement–sand ratio. Factors such as the volume of the cement stone and the amount of entrained air in the material have a significant impact on the strength of concrete. The tighter the aggregate packing, the less the thickness of the cement stone layer between the grains. The concrete structure is significantly strengthened due to interactions in the contact layers of the aggregate grains [3, 6]. As a result of this, the structure-forming function of sand and the relative proportion of stresses acting on it increase in resistance to load. Tight packing of aggregate particles reduces binder paste and shrinkage. In addition, physical–chemical interactions with new growths of the liquid phase occur at different stages of concrete hardening in the contact zone of the binder paste with the aggregate. It leads to the formation of diffuse structured shells around aggregate grains in the cement stone. The volume of entrained air in the material during the transition to the structure of the second type increases, and the nature of the porosity of concrete changes, which can be determined by the coefficient of pore saturation. So it is 0.7–0.75 in the structure of the first type, but it is 0.6–0.65 in the structure of the second type, which may indicate the presence of conditionally closed pores in the system [3]. An increase in the relative volume of entrained air in the material with a conditionally closed nature of pores up to 5–6% provides concrete with increased frost resistance and durability. In this case, it is believed that the size of air pores should be within certain limits (from 20 to 500 microns) that are most effective. Air trapping is most effective when the pores are evenly distributed in the cement stone near the aggregate grains and have a small volume [3, 5].

Hydration of cement and its hardening have a great influence on the structure of concrete. Hydration occurs at the boundaries of the binder grains with the formation of a cement gel. Cement gel at the same time increases in volume both outward and inward. Some of the hydrated binder components diffuse in the direction of the cement grain shell, while the water moves in the opposite direction inward. Hydrated binder components attach to already formed gel crystals and form new ones. As a result, about 55% of new growths appear outside the original border of the clinker nodule and 45% remain inside. By the time of setting, the ratio of the dense and porous components of the cement stone depends on the concentration of the binder. The structure, consisting of primary new growths resulting from cement hydration, represents the "initial matrix" by which one can judge the future structure of concrete [3].

The processes of concrete structure formation are conditionally divided into three stages. During the first stage, the primary concrete structure is formed, during the second stage, the formed structure slowly grows and compacts, and during the third or the final stage, there are practically no changes in the concrete structure [3, 5].

The strength of a cement stone depends on forces of the condensation origin, which are based on a covalent bond of type Si–O–Si and bonds having a donor–acceptor origin, based on the strength of interaction of water molecules with surface calcium ions and other positive ions. Complete interactions of Ca2 + cations with oxygen atoms of water molecules provide the union of tobermorite-like C–S–H crystals into a single polycrystalline new growth. Water molecules can enter the coordinate sphere of calcium cations, as a result of which, during interactions inside C–S–H crystals, O–Ca–O bond is formed, and Ca–O bond is formed on the crystal surface [3].

Individual layers of crystals merge in the first variant and the second variant, due to the unsaturation of bonds, which leads to the unification of crystals at the points of their contact and the appearance of large polycrystalline intergrowths. If there is no direct contact between individual C–S–H crystals, then, when interacting with sorption moisture, O–Ca–H2O–Ca–O bonds are formed. Such connections, in turn, also affect the physical–mechanical characteristics of the polycrystalline aggregate since they can perceive some mechanical stresses. This type of bond is formed only if the distance between the surfaces of neighboring crystals is not more than 1.3 nm [1, 3, 5]. In this case, interlayer bonds increase and the crystals become wider due to an increase in the coordination number of calcium ions. When the number of water molecules in the chain grows further, the distance between crystals increases, and this ultimately leads to the formation of microcracks and a decrease in the strength of concrete.

The studies [7–16] concern mainly the interaction of the cement stone (binder paste) with the coarse aggregate. The difference in the structure of cement concretes with coarse and fine aggregates does not allow one to completely transfer the known features of the interaction of the cement stone (binder paste) with the coarse aggregate to interact with fine aggregate particles, especially if one takes into account that coarse aggregate particles do not interact with the cement stone (binder paste) but with cement–sand mortar.

Two technological stages are distinguished in creating structural density and strength of small-grained cement concrete [3]. The first stage includes the spontaneous merge of the dispersed system into structural blocks by reducing the excess energy of the cement–aggregate–water system. The second stage includes external mechemical action on the system to create a dense structure by changing the distance between solid particles of the aggregate.

The mechanical effect on the above system leads to the approach of particles with the destruction of their initial structure. This model involves the use of silica sand as an aggregate.

The purpose of this study is to increase the strength of small-grained concrete by replacing quartz sand or part of it by ground scrap glass.

Based on the stated theoretical provisions, there is a reason to believe that the use of ground scrap glass as an aggregate (of the same dispersion as quartz sand) causes the increase in the adhesive interaction of glass particles with the cement stone (binder paste) due to an increase in the quality of the mortar layer between them and the cement stone (binder paste). In our opinion, this is due to the fact that glass particles have a significantly smaller number of structural defects in comparison with quartz sand particles. The amorphous structure of ground scrap glass determines the isotropy of the aggregate particles, which is a positive factor.

Most of the compositions of small-grained concrete include quartz sand as an aggregate. At the same time, it is possible to use another aggregate, if its introduction into the composition of the concrete mix will increase the properties of small-grained concrete, primarily strength.

It is known that the strength of a cement-sand mortar with quartz sand is lower than the strength of a cement stone. In particular, the cement stone is obtained on ordinary cements, the strength of which exceeds by 2 times the grade (activity) of cement, which is determined by the standard method for testing samples prepared from a cement-sand mortar 1:3. If during the experiment with cement [14], similar to its standard test, natural quartz sand with rougher grains is used instead of normal (Volsky) sand, then the strength of the tested samples will increase by 15–25%, but will be lower than the strength of the cement stone. If crushed sand from rocks is used instead of natural sand, a slight increase in concrete strength can be achieved, although some grains in the crushed sand have smooth edges and are separate mineral crystals. The surface of the aggregate grains has a significant effect on adhesion, which is shown by an experiment using fine aggregate (fraction up to 0.8 mm) based on nitrile rubbers in concrete.

Some crystalline minerals being crashed are destroyed by breaking interatomic bonds. It was found [11] that, due to ionization of the grain surface, freshly crushed quartz aggregates acquire physical–chemical activity for a short time, which manifests itself in an increase in concrete strength due to better adhesion. However, the adhesion strength of a cement stone to the surface of sand grains is less than the strength of the cement stone. Therefore, the latter in cement–sand mortar is not used completely. The substance from which the aggregate grains are composed may have an amorphous or crystalline structure. The amorphous structure of a material determines its isotropy. Crystals are anisotropic, which is manifested in the unevenness of thermal deformations and in other undesirable effects [12]. Glass may be a quartz-containing amorphous material, partially devoid of these drawbacks.

Below the possibility of using ground scrap glass as an aggregate for small-grained concrete is considered. The use (utilization) of ground scrap glass is one of the priority tasks since about 1.5–2.4 billion tons of unused scrap glass is accumulated annually in the world. Modern principles of scrap glass disposal are based on the use of those glass properties that are close to the properties of the product being replaced, that is, quartz sand. Thus, ground scrap glass can become a substitute for quartz sand or part thereof when preparing mixtures for the production of small-grained concrete.

It should be noted that improving the operational properties of small-grained concrete is possible not only by using ground scrap glass from industry, but also, for example, by using natural glassy materials. However, the disposal of accumulated scrap glass produced by the industry and characterized by well-known properties is cheaper to use as an aggregate than the use of natural glass and can partially solve the existing environmental problems.

3 Results and Discussion

The practical part of the work involved the preparation of four series of mixtures to obtain standard (40 × 40 × 160 mm) beam samples of small-grained concrete with the inclusion of various proportions of scrap glass and the study of their properties.

White Portland cement CEM II/A-I 32.5 N was used as a binder in all compositions. Sand for construction work (GOST 8736-93) sifted through a 2.5 sieve and the most common window scrap glass (GOST 111-2014) were used as aggregates. The scrap glass was ground in an experimental ball mill (Fig. 1). The resulting material was sifted through a 2.5 sieve.

When choosing glass, the alkaline environment of the cement stone was taken into account, in which the degradation of unstable silicates proceeds with greater or lesser intensity. The alkaline medium slowly destroys the glass, the silica is released on the surface of its grains in the form of flakes, which to some extent can lead to softening of the concrete structure. In this regard, it becomes necessary either to use alkali-binding additives (which is not economically feasible) or to use an alkali-resistant glass aggregate. The industrially produced window glass is quite alkaline and quite resistant to the alkaline environment of the cement stone. Moreover, a possible very slight degradation of such glass particles under the influence of an alkaline medium theoretically can affect the change in the structure and properties of concrete, but such an effect will be secondary and due to the effect on the hydration of cement. With the properties of concrete stabilized over time, the processes taking place with a glass aggregate will have practically no effect on them.

The compositions of the mixtures to prepare small-grained concrete are presented in Table 1.

Fig. 1 Experimental ball mill for grinding scrap glass

Table 1 Compositions of mixtures for the production of small-grained concrete

Component	The content of components pts. wt. in series, No			
	1	2	3	4
White Portland cement CEM II/A-I 32.5 N	1	1	1	1
Quartz sand sifted through a 2.5 sieve	3	2	1	–
The window ground scrap glass sifted through a 2.5 sieve	–	1	2	3
Water cement ratio	0.5			

The results of the obtained test beams of small-grained concrete are presented in Table 2.

Based on the data obtained, there is a reason to believe that when using ground glass waste as an aggregate (of the same dispersion as quartz sand), the adhesive interaction of glass particles with the cement stone (binder paste) increases due to an increase in the quality of the mortar layer between them and the cement stone (binder paste). In our opinion, this is due to the fact that glass particles have a significantly

Table 2 Small-grained concrete properties

Properties	Test beams series, No			
	1	2	3	4
Compressive strength at the age of 28 days (MPa)	14.1	14.4	15.6	17.4
Bending strength at the age of 28 days (MPa)	1.05	1.08	1.13	1.18
Water absorption (%)	5.1	4.3	3.4	2.3

smaller number of structural defects in comparison with quartz sand particles. The amorphous structure of the scrap glass determines the isotropy of aggregate particles, which is a positive factor since in the concrete technology it is usually impossible to control the spatial orientation of aggregate particles and they are randomly located [16].

It can be seen from Table 2 that the use of ground scrap glass as an aggregate has a positive effect on small-grained concrete. Moreover, compositions 3 and 4 are most suitable for use (Table 1), which presupposes replacing 2/3 of the mass of quartz sand or all of it with ground scrap glass.

4 Conclusion

1. When using materials with a glassy structure as an aggregate, the adhesive interaction of their particles with the cement stone (binder paste) increases due to an increase in the quality of the mortar layer between them and the cement stone (binder paste), due to the fact that glass particles have a significantly smaller number of defects in comparison with particles of materials characterized by a crystalline structure.
2. When choosing glass for subsequent use as an aggregate in small-grained concrete, it is necessary to take into account the influence of an alkaline environment of the cement stone on it, in which the degradation of unstable silicates proceeds with greater or lesser intensity. It is advisable to use ground window scrap glass as an aggregate.
3. The replacement of quartz sand or part of it with ground scrap window glass in compositions for producing small-grained concrete leads to an increase in its strength and a decrease in water absorption. In this case, it is most appropriate to replace 2/3 of the mass of quartz sand or all of it with ground scrap glass.

References

1. Bazhenov, Yu.M.: Multicomponent small-grained concrete. In: Construction Materials, Equipment, Technologies of the 21st century, vol. 10, p. 24 (2001)
2. Gusev, B.V., In Yenlyan, S., Kuznetsova, T.V.: Cement and Concrete: Development Trends. Scientific World, Moscow (2012)
3. Krasnov, A.M.: The formation of the structure, composition and properties of high-strength small-grained concrete for precast road surface. Dis. Doct. Tech. Science, Ivanovo (2010)
4. Kharchenko, I.Y., Bazhenov, D.A.: Effective self-compacting small-grained concrete with compensated contraction. Build. Mater. **5**, 48–52 (2018)
5. Reshetnyak, I.A.: Investigation of road small-grained cement concrete. Dis. Cand. Tech. Sci. (1968)
6. Venyua, M.: Cement and concrete in construction (Transl. from Fr. Moscow). Stroyizdat (1980)
7. Leonovich, S.N., Litvinovsky, D.A., Chernyakevich, O.Yu., Stepanova, A.V.: Strength, crack resistance and durability of structural concrete when temperature and corrosion influences. In: Leonovich, S.N. (ed.). BNTU, Minsk (2016)
8. Lermit, R.: Concrete technology problems. (Transl. from Fr. Ed. by Desov A.E.). Publishing House of LCI, Moscow (2007)
9. Šavija, B., Zhang, H., Schlagen, E.: Assessing hydrated cement paste properties using experimentally informed discrete models. Mater. Civ. Eng. **31**(9), 04019169-1–04019169-13 (2019)
10. Shafaghat, Ja., Allahverdi, A.: Enhancing concrete properties by using silica fume as reactive powder and Portland cement-clinker as reactive aggregate. Mater. Civ. Eng. **31**(11), 04019278-1–04019278-9 (2019)
11. Ledzion, K.: Safety innovations in materials testing for tunnel construction. Constr. Mater. **168**, 215–218 (2015)
12. Dvorkin, L.I., Dvorkin, O.L.: Special Concretes. Infa-Engineering, Moscow (2012)
13. Itskovich, S.M., Chumakov, L.D., Bazhenov, Yu.M.: Concrete Aggregate Technology. Higher school, Moscow (1991)
14. Pat. 2353603 the Russian Federation. Raw mix for the manufacture of light concrete. Prepared by Fedosov, S.V., Bobrova, A.A., Sapronova, I.A., Schepochkina, Yu. A. No. 2007129744/03. Declared 08/02/2007, Publ. 04/27/2009, Bull. No. 12
15. Dvorkin, O.L.: Building materials from industrial waste. Rostov-on-Don, Phoenix (2007)
16. Fedosov, S.V., Schepochkina, Yu.A., Rumyantseva, V.E., Konovalova, V.S.: Secondary material resources for the construction industry, p. 188. IVGPU, Ivanovo (2017)

Author Index

M. Shamtsyan et al. (eds.), *Robotics, Machinery and Engineering Technology for Precision Agriculture*, Smart Innovation, Systems and Technologies 247,
https://doi.org/10.1007/978-981-16-3844-2